The Riemann Legacy

Mathematics and Its Applications

Volume 417

The Riemann Legacy

Riemannian Ideas in Mathematics and Physics

by

Krzysztof Maurin

Division of Mathematical Methods in Physics,
University of Warsaw,
Warsaw, Poland

KLUWER ACADEMIC PUBLISHERS
DORDRECHT / BOSTON / LONDON

A C.I.P. Catalogue record for this book is available from the Library of Congress.

ISBN 0-7923-4636-X

Published by Kluwer Academic Publishers,
P.O. Box, 3300 AA Dordrecht, The Netherlands.

Sold and distributed in the U.S.A. and Canada
by Kluwer Academic Publishers,
101 Philip Drive, Norwell, MA 02061, U.S.A.

In all other countries, sold and distributed
by Kluwer Academic Publishers,
P.O. Box 322, 3300 AH Dordrecht, The Netherlands.

Printed on acid-free paper

Part I is the translation of K. Maurin's Appendix to *Geometry of Riemann Manifolds*, M. Skwarczyński, PWN, ©1993. Chapters 1–9 and 14–26 of Part II are translated from *Analysis* (part II), K. Maurin, PWN ©1991 (second edition). The remaining chapters of Part II are the translation of Maurin's contribution to *Mathematical Lexicon*, Wiedza Powszechna ©1993. Translated from the original Polish by Jerzy Kowalski-Glikman.

Printed in the Netherlands.

Contents

Foreword: Riemann's Geometric Ideas and their Role in Mathematics and Physics

The study of the ways in which great mathematical ideas are born, develop and die out (i.e. their so called 'history') is undoubtedly one of the most fascinating branches of history. However, it requires an extensive and profound knowledge of contemporary mathematics. Being involved with the life of great mathematical ideas is fruitful not only for mathematics and physics but also for the person involved. It enables him (or her) to come into contact with and participate in the life of the world of ideas (the *cosmos noethos* of the Platonists.) For nowhere can we see more concretely, one is tempted to say almost palpably, the enormous spiritual energy which, although acting in people, still lacks clear contours and desires 'to be mounded' and developed by people – people called mathematicians.

Plato, being strongly influenced by the Pythagoreans, was aware of this. So was Eudoxos, one of the greatest mathematicians of antiquity active in Plato's Academy. Hence the famous inscription on the entrance to the Academy:

> Let no one enter this place, who does not know geometry.

It was Riemann, who probably more than anyone else, enriched mathematics with new ideas. These ideas display an unusual degree of vitality and impulse the whole of mathematics as well as many branches of physics.

The world of ideas is 'one' – i.e., it is a cohesive living organism in which all 'parts' interact and where even slight stimulations propagate producing echoes in the (seemingly) distant organs which may be called theories or 'branches of mathematics'. Similarly as in Weierstrass-Riemann principle of 'analytic continuation', a change of a (meromorphic) function even within a very small domain (environment) affects through analytic continuation the whole of Riemann surface, or analytic manifold .

Riemann was a master at applying this principle and also the first who noticed and emphasized that a meromorphic function is determined by its 'singularities'. Therefore he is rightly regarded as the father of the huge 'theory of singularities' which is developing so quickly and whose importance (also for physics) can hardly be overestimated.

Amazing and mysterious for our cognition is the role of Euclidean space. Even today many philosophers believe (following Kant) that 'real space' is Euclidean and other spaces being 'abstract constructs of mathematicians, should not be called spaces'. The thesis is no longer tenable – the whole of physics testifies to that. Nevertheless, there is a grain of truth in the 'prejudice': E^3 (three-dimensional Euclidean space) is special in a particular way pleasantly familiar to us – in it we (also we mathematicians!) feel particularly 'confident' and move with a sense of greater 'safety' than in non-Euclidean spaces. For this reason perhaps, Riemann space M stands out among the multitude of 'interesting geometries'. For it is:

1. Locally Euclidean, i.e., M is a differentiable manifold whose tangent spaces T_xM are equipped with Euclidean metric g_x;

2. Every submanifold M of Euclidean space E is equipped with Riemann natural metric (inherited from the metric of E) and it is well known how often such submanifolds are used in mechanics (e.g., the spherical pendulum).

And finally:

3. Every Riemann space M can be isometrically embedded in a certain E^N.

In other words, every Riemann space is a submanifold of a certain (maybe higher-dimensional) Euclidean space! This alone would be a convincing argument that Riemann space geometry is worth our attention.

Bernhard Riemann himself (1826-1866) created and concerned himself with his geometry for reasons that were cognitive-philosophical and physical-scientific, which is amply demonstrated by his famous habilitation lecture: *Über die Hypothesen, welche der Geometrie zu Grunde liegen.* That short classic text is unrivaled in magnitude of its influence upon mathematics and physics. The work no doubt contains the sources of Einstein's concept of gravity. Riemann already 'knew that the distribution of masses influences the (physical) geometry of space. The conception was later developed by great Clifford, the English translator of Riemann. Einstein great conception that space-time is a four-dimensional pseudo-Riemannian manifold' could not have arisen without the brilliant work of Minkowski, who as the first introduced (pseudo-Euclidean) four-dimensional space-time as the proper substrate for the special theory of relativity ('electrodynamics of moving bodies'.)

While thinking or talking about a mathematical idea, one must not, as we saw, truncate it, i.e., isolate it from other related ideas. It is also necessary to watch its birth, its first germs out of which (as from that evangelical mustard seed) develops and grows a large tree in whose crown many creatures find shelter, in whose shade many a weary traveler can rest and regain his strength!

Hermann Weyl (1885 - 1955), the greatest mathematician of this century, was the first who in Riemann geometry of M distinguishes (or gives prominence to) three structures: 1. a topological one, 2. a differentiable one, 3. a metric one:

1. M is locally homeomorphic with $\mathbb{R}^N$;
2. M possesses a differentiable atlas;
3. Every space tangent T_xM possesses an Euclidean metric g_x.

The author cannot help remarking that the three structures quoted above (after Weyl) although implicitly contained already in Riemann's work, were explicitly introduced by Weyl himself, who

1. as the first introduced (in his famous monograph*Idee der Riemannschen Fläche* (1913) axioms of (Hausdorff) topological space;
2. introduced the concept of the differentiable atlas;
3. as the first presented the axioms of Euclidean space by means of the scalar product g.

As to the global treatment of Riemann manifold M as space with distance $d(x, y)$, $x, y \in M$ we owe it the classic work by Heinz Hopf and Willy Rinow (1932).

As we can see, Riemann manifolds are distinguished by close relationship with Euclidean spaces, and what is more, the depth of Riemann ideas is characteristic of great conceptions:

a) the scope of hitherto existing theories is considerably broadened, but

b) the generalization is just right, i.e., it does not lose contact with reality, being comprehensive it possesses such a rich structure that makes it possible to create beautiful, rich constructions; leads to important new problems, and is open for many applications in other branches of mathematics, in physics, astronomy, technology.

The idea of Riemann surface

Another great deed of Riemann was the "liberation", or rather the placing of the function in its proper environment - the biotope. Plants, animals, or people living in unsuitable environments become stunted, or even die. They often look ugly, disfigured, or 'freakish'. Similarly, functions before Riemann were unable to reveal their true nature. Actually, every function f, or its germ f_x (at point x) grows on its Riemann surface. Or in other words: the germ f_x analytically extends to its full Riemann surface. The fundamental phenomenon here is the *Monodromy Theorem* which says that an extension along homotopic paths $c_1 \sim c_2$ that is, the ones capable of continuous transformation, leads to the same germ. This theorem is no doubt the source of Poincaré fundamental group $\pi_1(X)$ of the space X and his theory of covering spaces.

Leibnitz concept of isomorphism of two objects X_1 and X_2 and of a class of isomorphic objects plays a key role in mathematics as well as in philosophy. In Riemann surface theory isomorphism is called biholomorphism. The famous *Riemann mapping theorem* asserts that there exists (up to an isomorphism) only three simply connected Riemann surfaces $\mathbb{C}$, $\mathbb{P}^1(\mathbb{C})$ (the Riemann sphere) and $\mathfrak{H}$, the upper half plane $\{z \in \mathbb{C} : \operatorname{Re} z > 0\}$.

By far the most important is the third one $\mathfrak{H} \simeq \mathbb{D}$ (unit disc.) The upper half plane is equipped with the Poincaré–Beltrami metric and is a model of non-Euclidean geometry. Its shortest lines (geodesics) are semicircles (and half lines) orthogonal to the real axis.

Biholomorphic maps of $\mathfrak{H}$ are isometries of $(\mathfrak{H}, ds^2)$, they are Möbius transformations $\mathrm{GL}(2,\mathbb{R})$. Of great importance are discrete subgroups of $\mathrm{GL}(2,\mathbb{Z})$ (the modular group) denoted by Γ.

Another fundamental theorem of Riemann theory is the *Uniformization theorem* (Koebe–Poincaré) which tells that every compact Riemann surface X (of genus $p > 1$) is isomorphic to the quotient $\mathfrak{H}/\Gamma$ (the orbit space of Γ), where Γ is some subgroup of $\mathrm{PSL}(2,\mathbb{Z})$. This was the reason for the famous exclamation of young Herman Weyl: *Wir betreten damit den Tembel in welchem die Gottheit ... aus der irdischen Half ihrer Einzelverwirklichungen sich selbar zurückgegeben wird: in dem Symbol des zweidimensionalen Nicht-Euklidischen Kristals wird das Urbild der Riemannschen Flächen selbst ... rein und befreit von allen Verdunkelungen und Züfälligkeiten erschaubar.* Here the 'non-Euclidean crystal' is clearly $\mathfrak{H}/\Gamma$: the non-Euclidean plane $\mathfrak{H}$ divided by the lattice Γ.

It seems to me that the divine idea – 'Gottheit' – in Weyl's thought not only reveals itself, i.e., allows us to know it, but is something 'more' – it is a great energy, a creative power acting upon and transforming reality. But to do so, it needs the cooperation of people - in this case, mathematicians. An idea acts, i.e., it represents itself in other spaces, in different worlds, it incarnates itself over and over again in different forms, in different bodies.

It has to be emphasized here that even in Riemann thought the idea of his surface did not reveal itself fully. It was only Felix Klein and eventually H. Weyl, who gave it its present form ('definition') as a surface provided with conformal structure, i.e., as one-dimensional complex manifold. However, a mathematical idea manifests itself in great theorems. Besides the above mentioned ones, the famous Riemann–Roch Theorem plays a leading, though still unfathomable, role.

Riemann–Roch theorem

Thanks to Hirzenbruch and Atiyah–Singer we perceive this theorem as the *index theorem* for the Riemann–Cauchy operator d'', where $d'' : A^{0,0}(X) \to A^{0,1}(X)$, $\operatorname{ind} d'' = 1 - p_X$ for a compact Riemann Surface X. Here the left hand side has an analytic sense $\operatorname{ind} d'' = \ker d'' - \ker d''^*$, and the right hand side is a topological invariant: p_X, the genus of X is the number of handles attached to the sphere. Gelfand suggested that similar theorem should exist for any elliptic operator and compact Riemann space (of arbitrary dimension.) Atiyah–Singer index theorem unifies several very important theorems: Riemann–Roch, Chern–Gauss–Bonnet, Hirzenbruch.

But let us return to the Riemann idea of providing functions with adequate biotopes. The most important functions of mathematical physics and mathematics are solutions of differential equations. The most famous ordinary differential equation is

The hypergeometric equation. This is a second order ordinary differential equation with three singularities. It was Riemann, who, in his lectures and publications, placed the hypergeometric equation in its proper biotope: this is the sphere $X := \mathbb{P}(\mathbb{C}) - \{a, b, c\}$ with three points deleted. If we analytically continue solution u of the hypergeometric equation along closed path in X, it will be transported by a linear (invertible) transformation of $\mathbb{C}^2$. By virtue of the Monodramy theorem, we obtain therefore a linear representation (i.e., a homomorphism)

$$\rho : \pi_1(\mathbb{P}(\mathbb{C}) - \{a, b, c\}) \to \mathrm{GL}(2, \mathbb{C}).$$

The value of ρ ($\operatorname{im}\rho$) is called the monodromy of the hypergeometric equation. Riemann posed the following problem (called the monodromy problem, and also the Riemann–Hilbert problem, (21st)): Given a group $\Gamma \subset GL(2,\mathbb{C})$ construct a linear differential equation with 'regular singularities' such that its monodromy is the given Γ. Solutions of this problem in special circumstances were found by Poincaré, Hilbert, Birkhof, and others; but the full solution, due to H. Röhrl appeared only quite recently. Röhrl's solution was possible as a result of modern developments in complex analysis.

But the (Riemannian) idea has been developing steadily: it has led to the theory of Manin–Gauss connections of E. Brieskorn and the mighty theory of singularities of Arnold and others[1].

Riemann's approach to the theory of elliptic and, more generally, abelian integrals forced him to built the theory of *abelian varieties* being complex tori $\mathbb{C}^p/\Gamma$, where p is the genus of the Riemann surface of the abelian integral. The notion of a *period* of such integrals got a lucid geometric interpretation. An important question arises under which conditions the torus $\mathbb{C}^p/\Gamma$ is an algebraic (projective) variety. The answer is given in the famous Riemann bilinear relations. The crowning of these endeavors is the famous *Kodaira theorem* giving the intristic (cohomological) characterization of algebraic varieties: A compact Kähler manifold (X,ω) is algebraic if and only if the Kähler differential $(1,1)$ form ω is integral valued (Fields medal, 1954). Clearly, a compact Riemann surface is Kähler: $d\omega = 0$ and Riemann bilinear relations give Kodaira condition.

Thus the theory of Riemann surfaces necessarily leads to complex, multidimensional manifolds: with any Riemann surface there is associated a p dimensional torus – its jacobian $\operatorname{Jac}(X)$. A natural conjecture arises:

TORELLI, 1913. *Every compact Riemann surface X is determined (up to isomorphism) by its Jacobian* $\operatorname{Jac}(X)$ *together with* polarization. *The latter can be given in several ways, for example, by period matrix.*

Torelli theorem was proved many years later (by A. Weil, Andreotti, and many others.)

Investigation a Riemann surface, for example, and elliptic curve by means of its Jacobian is a powerful method, indispensable in number theory.

And here we have again an extremely interesting general epistemological

[1]To my infinite regret a large part on these problems had to be eliminated due to the shortage of space!

principle: To get to know a thing (not necessarily a geometric construct, or an object), one has to 'lift' it to its true nature (essence) and there, in that higher world, to get to know it thoroughly and then with knowledge so enriched descend to the original level of departure (to Plato's Cave). It seems that the symbol of Plato's parable from the VIII Book of the Republic is truly profound: ascending to the world of ideas and then descending again (mostly difficult and dangerous) represents the rhythm of the acquisition of truly profound knowledge. This principle has been illustrated by the famous Theorem concerning isometric smooth embedding of every Riemann space into a certain Euclidean space.

A similar but more difficult theorems were formulated by H. Grauert: a natural, open n dimensional complex manifold, in many respects similar to open Riemann surfaces, are Stein manifolds. They are the most important complex manifolds and can be intristically characterized. For these manifolds Weierstrass and Mittag–Leffler theorems hold.

Grauert and Remmert proved their famous theorems characterizing Stein manifolds:

1. X is Stein $\Longleftrightarrow X$ can be regarded as holomorphic submanifolds of some $\mathbb{C}^N$;

2. X is Stein $\Longleftrightarrow$on X lives a strictly convex (that is, plurisubharmonic) exhausting function p;

3. A real analytic (paracompact) manifold can be regarded as a real analytic submanifold of some $\mathbb{R}^K$.

We see once again that important spaces in mathematics are characterized by functions living on them. The first theorem of this type was the classic Riemann theorem:

(X, Y are isomorphic compact Riemann surfaces) $\Longleftrightarrow$(The function fields $\mathcal{M}(X) \simeq \mathcal{M}(X)$, where $\mathcal{M}(X)$ denotes the field of meromorphic functions on X)

Dirichlet principle was a very powerful method mastered by Riemann. One could call it the *energy principle.* The energy $\mathcal{E}$ (or the Dirichlet integral) of a function on X or a map $f : X \to Y$ is a functional on a family of maps. Its extremal points are called *harmonic* maps (or functionals.) With the help of his Dirichlet principle Riemann proved a number of his famous theorems. Some of the proofs have severe gaps, but the idea of Dirichlet principle was so suggestive and powerful that nobody doubted Riemann theorems. These apparent contradiction has led to two approaches: either to prove Riemann theorems by another means, or to 'rescue' the Dirichlet principle (Hilbert, Weyl). This second approach was proved most successful, it has led to the direct methods of calculus of variations.

Conjectures and Hypotheses. There is a general philosophic principle: good questions are more important than answers. History of philosophy shows that there are only very few questions, the famous Kant questions. A similar situation takes place in mathematics and physics (natural sciences.) Riemann put forward a very important conjectures:

A. **The moduli problem.** The space $\mathcal{M}_p$ of isomorphy classes of Riemann surfaces of genus $p > 1$ is $3(p-1)$ dimensional.

This problem generated a huge theory: the Teichmüller theory. Teichmüller constructed a $3(p-1)$ dimensional unified covering $\mathcal{T}_p$ of the moduli space $\mathcal{M}_p$, $\mathcal{M}_p = \mathcal{T}_p/\Gamma_p$, with Γ_p being the Teichmüller modular group. The manifold $\mathcal{T}_p$ has beautiful geometry: it is Stein, can be equipped with Kodaira metric of negative curvature, etc. There are many methods of proving these marvelous theorems (Ahlfors, Bers, Wolpert), but the most natural one is the 'purely Riemannian method' by Tromba (and later Wolf and Jost): one starts with construction of an energy integral on $\mathcal{T}_p$, and shows that it is strictly convex and exhausting, whence $\mathcal{T}_p$ is Stein. Its Hessian gives a Kähler metric, the so called Petersson–Weil metric.

The moduli problem was the impulse for the huge theory of *deformations of complex structures* (Teichmüller, Kodaira, Spencer, Kuranishi, Grauert, and others) This was the first moduli problem. Nowadays we have moduli problems for complex surfaces, Kähler submanifolds, stable vector bundles, Yang–Mills connections. N. Hitchin prophetess that we are at the beginning of the moduli problems era.

B. **The monodromy problem**, we talked about above.

C. **ζ functions and L series.** *These famous complex functions know everything about number fields, we have only to ask them humbly to tell us* (variation on G. Harder).

The ζ function was introduced for *real* arguments by Euler and generalized by Dirichlet to his famous L series $L(s,\chi)$. Riemann considered them for complex s in the half plane $\operatorname{Re} s > 3/2$, where $\zeta(s) = \sum_{n=1}^{\infty} n^{-s}$

is clearly absolutely convergent and he extends it meromorphically onto the whole complex plane $\mathbb{C}$. He also proves the famous functional equation $\xi(s) = \xi(1-s)$. Deep and important phenomena (for example, from the point of view of number theory) take place outside the original half plane $\mathrm{Re}\, s > 3/2$: *the important (nontrivial) zeros of* $\zeta(s)$ *lie on the line* $\mathrm{Re}\, s = \frac{1}{2}$. This is one of the most famous conjectures in mathematics – the 'Riemann hypothesis'. The kingdom of ζ and L functions increases: the Dedekind $\zeta_K(s)$ for a number field K, the Hecke L series $L(s, f)$ for modular (cusp) forms f, $L(s, E)$ for elliptic curve E over $\mathbb{Q}$, ζ_A function for an elliptic self adjoint operator on a compact manifold X, Selberg ζ function, etc. Erich Hecke has spent all his life meditating ζ functions and their functional equations. His great followers Eichler, Shimura, Taniyama, Serre, Barry Mazur, Ribet, Frey, and Andrew Wiles succeeded in obtaining marvelous insight into arithmetic. And in 1994 Andrew Wiles proved the Fermat Last Theorem by proving the Taniyama conjecture in the restricted form given by Shimura: *semistable elliptic curves over* $\mathbb{Q}$ *are modular.*

This last event was a great sensation: after 350 years of gigantic labor to prove the non-existence of triples $a, b, c \in \mathbb{N}$ such that $a^n + b^n + c^n = 0$, $n > 2$ was finally established. A prominent mathematician said 'I feel as if I have lost an old friend.'

But for the philosophy of mathematics the battles fought over Fermat equation are amazing: the consequences of proving the Fermat Last Theorem have been and still are unknown and yet people of genius (Fermat, Euler, Kummer and many, many others) sparing no effort have constructed great edifices of theories enriching the whole of mathematics.

Very strange indeed are these 'glory be to Mankind' oriented efforts, as a great mathematician put it. As we have seen, the most fascinating phenomena in mathematics are those which link seemingly disparate branches of the discipline: analysis and geometry, analysis and arithmetic, geometry and arithmetic, local and global. The last pair is probably the 'hermetic' relation between micro- and macrocosm.

Mathematics and physics make up one organism - man's task is to actualize this unity of the world of ideas. Riemann was deeply aware of this: he thought of himself as mathematician and physicist. Constantly repeated, puzzled questions about 'the mysterious and incomprehensible congruity between mathematics and physics' have as their source the unconscious and stubborn inclination to dissect this one (and the same) organism of mathematics–physics. Thus two different, artificially created, entities come into being which are in fact organs of one great reality.

Riemann outward life was brief, punctuated by the deaths of those he loved. It did not abound in any great worldly adventures. But his true life was devoted to the enrichment of the world of mathematical ideas, which was only natural as he was a profound philosopher, a disciple of G.T. Fechner and his 'Zend Avesta'.

Creator lives in his creations! Therefore this work could be regarded as a biograph of great Riemann!

Part I

Riemannian Ideas in Mathematics and Physics

CHAPTER 1

Gauss Inner Curvature of Surfaces

Gauss discovers a number of properties of two-dimensional submanifolds of $\mathbb{R}^3$, which can be understood as M^2, that is, as two-dimensional differential manifolds 'on their own' i.e., without any reference to the question if they are isomorphically imbedded in $\mathbb{R}^3$ or not. Many (including Gauss himself) found fascinating the great discovery that the so-called 'Gauss curvature' $K(x) = r_1(x)^{-1}r_2(x)^{-1}$, where $r_1(x)$, $r_2(x)$ are the main curvature radii of the surface M^2 at the point $x \in M$, is an invariant of folding, i.e., is an internal property of M^2. Why this fact called by Gauss *'Theorema egregium'* (wonderful theorem) is so surprising? The main curvatures $k_i(x) = \frac{1}{r_i(x)}$ are defined as external objects. A surface M^2 imbedded in $\mathbb{R}^3$ is cut by the normal plane N to obtain the plane curve contained in N. This curve has at the point $x \in N \cap M^2$ the curvature $r(x)$. By rotating N around the normal $n(x)$ at the point x, we obtain a family of curves and a family of curvatures (Fig. 1). These curvatures (as it was observed by Euler) have two extremal values, these are just the main curvatures $r_1(x)^{-1}$, $r_2(x)^{-1}$. There are therefore external objects: the normal $n(x)$ is external! However, their product $K(x) = r_1(x)^{-1}r_2(x)^{-1}$ is itself an internal property of the manifold, it can be described in terms of the coefficients of the 'second fundamental form'.

Another great discovery of Gauss is the fact that the total curvature *curvatura integra* derived by him and defined as an integral of the Gauss curvature

$$\int_{M^2} K(x)\, dx$$

(for compact surface M^2) is a topological invariant (not only an isometric invariant!), and, moreover, it is equal to the Euler characteristics (Euler

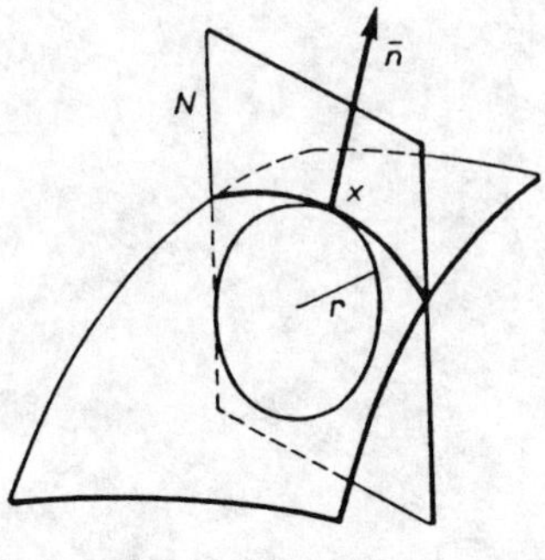

Fig. 1

number) χ_{M^2}

$$(G-B-D) \qquad \frac{1}{2\pi}\int_{M^2} K(x)\,dx = \chi_{M^2} = 2(1-g),$$

where p is the genus of M^2 equal to the number of holes. Another, also external, definition of Gauss curvature is obtained with the help of 'Gauss mapping' which is roughly defined as follows. A unit normal vector $n(x)$, $x \in \sigma \subset M$ taken at the point $x \in M^2$ defines a point on a unit sphere, which we denote by $Gn(x) \in S^2$. Thus a part (region) σ of the surface M^2 is mapped by G on the part $G\sigma$ of the sphere S^2. On S^2 and M^2 we have a natural measure $|\cdot|$, therefore, we can construct $\frac{|G\sigma|}{|\sigma|}$. When σ is contracted to the point $\{x\}$, then the limit $\lim_{|\sigma|\to 0} \frac{|G\sigma|}{|\sigma|} = K(x)$ (Fig. 2).

The famous G-B-D theorem called the *Gauss–Bonnet theorem*, was fully proved for compact manifolds M^2 only by von Dyck, a pupil of Klein. It is the first theorem of 'global differential geometry' and it is a close analogue of the most famous theorem of 'global complex analysis', the Riemann-Roch theorem, about whose we will talk later.

The next breakthrough was achieved by Riemann in his habilitation lecture *Über die Hypothesen, welche der Geometrie zu Grunde liegen.*

Let us note at this point that in his idea of geometry, Riemann allowed for completely arbitrary metrics in T_xM, tangent spaces to M at $x \in M$. Instead of a quadratic form g_x (in 'geometrical' language: instead of a quadric, a hypersurface of second kind, $g_x \leqslant 1$), Riemann allows for a homogeneous, convex, and symmetric function F_x (geometrically: symmetric, convex field $W_x \subset T_x$.) In this way, any space T_xM becomes a Minkowski space with the gauging field 'Eichköper' $W(x)$. This one sentence long remark by Rie-

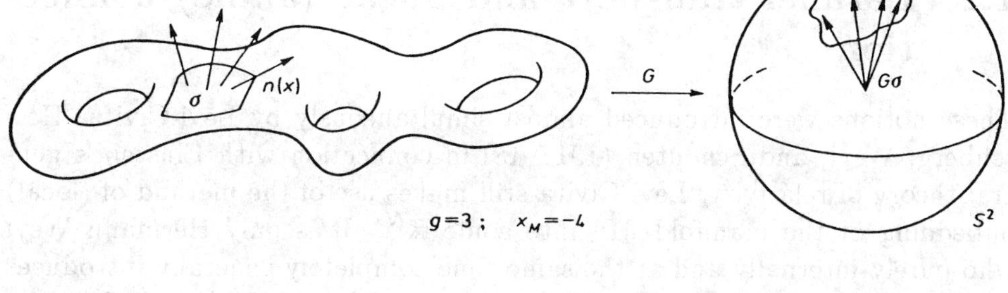

Fig. 2

mann was many years later reincarnated by Finsler (a pupil of a celebrated expert in calculus of variations, Carathéodory) in his Ph.D. thesis of 1918; this sentence have led him to the so-called 'Finsler geometry'. Another, more important consequence resulting from the Riemann habilitation was, mentioned already, the concept of Minkowski of his convex bodies leading to famous *Geometrie der Zahlen* and four-dimensional Minkowski space $\mathbb{R}^4$ with metric gauged by the hyperboloid $x_0^2 - x_1^2 - x_2^2 - x_3^2 = 1$. Namely, a Finsler manifold F^n is a manifold which is locally Minkowski, that is, T_xF^n is the Minkowski space (for all $x \in F^n$). This space geometrizes geometric optics in natural way: light rays are geodesics (that is, locally shortest curves) in this geometry. It can be expected that the boundary of W_x is in close relation with 'elementary waves' of the Huyghens principle (which is indeed the case, as it was shown by the author in his first publication.[1]) The Finsler geometry, even though there are hundreds of papers devoted to it, has never reached importance and beauty of Riemannian geometry. Here once again the genius of Riemann is clearly seen: restriction to the right metric (locally Euclidean). One can recall the famous saying of Goethe 'In der Beschränkung ist der Meister'; the right restriction characterizes a master.

We devote special section to geodesics in Riemann space. How did the Riemannian geometry develop? Christoffel (1829 - 1900) and Ricci (1853 - 1925) developed the theory of covariant differentiation, called today 'tensor analysis on Riemannian manifolds.' This is a natural generalization of the theory of directional derivatives and gradients. The geometrization of this

[1] Cf. Part 'Riemann and Calculus of Variations'

notion is

1.1 Parallel transport and linear (affine) connection

These notions were introduced almost simultaneously by Levi-Civita, Hessenberg, Weyl, and Schouten (1917–18) in connection with Einstein's general theory of relativity. Levi-Civita still makes use of the method of (local) imbedding of the manifold M^n into some $\mathbb{R}^N$. It is only Hermann Weyl who purely internally and at the same time completely generally introduces a notion of parallel transport (linear connection) on arbitrary differential manifold, that is, with no reference to the Riemann metric.

The brilliant idea of Weyl is 'very natural', great concepts, once understood, seems to be natural and even obvious, but it is for a genius to open our eyes, which looked but did not see.

Parallel transport (linear connection) in arbitrary differential manifold $M = M^n$, more precisely, in the tangent bundle $T_M = \coprod_{x \in M} T_x M$ (disjoint sum of the spaces $T_x M$) along smooth curve $c : [0,1] \to M^n$ is nothing but the isomorphism (of vector spaces)

$$P_c : T_{c(0)}M \xrightarrow{\sim} T_{c(1)}M$$

which does not only depend on the ends $c(0)$, $c(1)$.

The tangent spaces $T_{x_1}M$, $T_{x_2}M$ for $x_1 \neq x_2$ are n- dimensional vector spaces (and thus isomorphic with $\mathbb{R}^n$) with no relation to each other. It turns out however (and this is the great idea) that for any curve $c : [0,1] \to M$ one can define an isomorphism $P_c : T_{c(0)}M \xrightarrow{\sim} T_{c(1)}M$. This isomorphism is defined by second order differential equations.

The isomorphism P_c is called parallel transport (or parallelism) *along curve c*. We say that the vector $e \in T_{c(0)}M$ is parallel to the vector $P_c(t)e \in T_{c(t)}M$. A family of parallelisms $\{P_c\}$ where c runs through all smooth curves on the manifold M is called *linear connection* and, in analogy to the gradient operator, is often denoted by ∇.

Clearly, the construction above can be extended to arbitrary (smooth) vector bundles $\pi : E \to M$, where π denotes a smooth mapping (surjection) on M called the *projection* and any fiber $E_x := \pi^{-1}M$ (over $x \in M$) is isomorphic to given vector space $\mathbf{E}$ of dimension r ($\cong \mathbb{R}^r$); r is called *rank* of the bundle and $\mathbf{E}$ a typical fiber.

REMARK. Along with tangent bundles $TM \to M$, there are another bundles of rank 1 which are of great importance, the so-called line bundles. Obviously, the bundles of tangent tensors $\tau_l^k := \bigotimes^k TM \bigotimes^l T^*M$ k-times contravariant and l-times covariant are important as well.

A section σ of the bundle $E \xrightarrow{\pi} M$ is the mapping $\sigma : M \to E$ such that $\sigma(x) \in E_x$, for every $x \in M$.

Tensor fields of the type $\substack{k \\ l}$ are sections of the bundle $\tau_l^k \to M$. For Riemannian geometry, the metric tensor $g \in \tau_2^0$ is of utmost importance; instead of $g(x)$ one often uses the notation g_x

$$g_x(X,Y) := (X_x \,|\, Y_x)_x, \quad \text{where } X_x, Y_x \in T_xM.$$

Given a connection in the vector bundle $E \to M$, one can immediately define the covariant derivative ∇ as follows: For any curve $c \subset M$, one defines derivative $\nabla_c\sigma$ of section σ as a limit of the difference quotient

$$\lim_{t\to 0} \frac{\sigma(c(t)) - P_{c(t)}\sigma(c(0))}{t}.$$

We can immediately check that such derivative ∇_c satisfies the classical Leibnitz rule:

$$\nabla_c(f\sigma) = \nabla_c f \cdot \sigma + f\nabla_c\sigma, \quad f \in C^\infty(M), \sigma \in C^\infty(E)$$

(smooth section of the bundle $E \to M$)

where $\nabla_c f$ is the gradient of the function f.

Denoting a vector field (that is, section of the tangent bundle $TM \to M$) by X, we see that connection defines a bilinear mapping, denoted also by $\nabla : C^\infty(TM) \times C^\infty(E) \to C^\infty(E)$ satisfying the following two conditions (we denote $\nabla(X, \sigma)$ by $\nabla_X(\sigma)$):

(1) $$\nabla_X(f\sigma) = X(f)\sigma + f\nabla_X\sigma,$$

(2) $$\nabla_{fX}(\sigma) = f\nabla_X\sigma.$$

Here $X(f)$ is the directional derivative $\nabla_X f$ of the function f in direction X.

The connection ∇ is often defined as 'covariant differential' with the help of axioms (1), (2). Then one shows that ∇ exists (in local system of coordinates, this reduces to the theorem of existence of solutions of second order differential equation.)

REMARK. Of course, historically the development proceeded in reverse order: In 1869 Elwin Christoffel defined locally 'Christoffel derivative' and showed that it has invariant character. It was this paper where the famous 'Christoffel symbols' appeared for the first time; we will talk about them in a moment. Christoffel shows that the famous Riemann curvature tensor R^i_{jkl} can be defined in terms of partial derivatives of Christoffel symbols. It is therefore justified to call this tensor the 'Riemann-Christoffel tensor'. It is mainly the Italian school which develops 'tensor analysis' on Riemann manifolds, earlier called the 'Ricci calculus'. The new great impulse was provided by the general theory of relativity. The above mentioned works by Levi-Civita, Hessenberg, Weyl (1916–17), Schouten (1918), and Cartan appear. The general theory of parallel transport and connection is being formulated (Weyl and Cartan.) The theory of representation of Lie groups makes it possible to develop the 'theory of connections in principal bundles' (Steenrod and others), that is, bundles $\pi : E \to M$ such that the fibers $\pi^{-1}(x) := E_x$ are isomorphic (diffeomorphically or only topologically) to a given topological space $\mathbf{E}$ called *typical fiber*. This notion, this idea works without alternations in arbitrary 'vector bundle $\pi : E \xrightarrow{\pi} M$ over M' (E and M are (connected) differential manifolds, π is a smooth surjection called also *projection* on M), where all fibers $\pi^{-1}(x)$, $x \in M$ are vector spaces isomorphic to the given vector space $\mathbf{E}$, the typical fiber. Since the notion of vector bundle is of fundamental importance in geometry, let us discuss it now.

1.2 Vector bundles and operations on them

From the notion of tensor bundle $\tau^k_l \to M$, there arise a general notion of fiber bundle $\pi : F \to M$ over differential manifold M with typical fiber $\mathbf{F}$: this is a family $\{\varphi_x\}$, $x \in M$ of spaces F_x isomorphic to given topological space $\mathbf{F}$, called the typical fiber. If, along with the topological structure, $\mathbf{F}$ possesses some richer structure, e.g., $\mathbf{F}$ is a differential manifold (analytic, holomorphic etc.), then the projection $\pi : F \to M$ is a differential (analytic, holomorphic etc.) mapping. In the following we will mainly consider vector bundles, that is, the bundles whose typical fibers $\mathbf{F}$ are vector spaces. Usually the fiber $\mathbf{F}$ will possess additional structures, for example, it can be a vector space over $\mathbb{R}(\mathbb{C})$; then we say that the bundle $\pi : F \to M$ is real (complex) and its dimension $r := \dim \mathbf{F}$ is called the rank of the bundle F. When $\mathbf{F}$ has a structure of Euclidean (Hilbert) space with a scalar product $(\cdot|\cdot)$, then any fiber $F_x := \pi^{-1}(x)$ over $x \in M$ has the same structure and the bundle $F \to M$ is called *Riemann (Riemann-Hilbert) bundle.*

In the typical fiber $\mathbf{F}$ the group of automorphisms $G = \mathrm{Aut}(\mathbf{F})$ of this space acts, in the case $\mathbf{F} = \mathbb{R}^n$, this is the group $\mathrm{GL}(\mathbb{R}^n) \equiv \mathrm{GL}(r, \mathbb{R})$ (in the case $\mathbf{F} = \mathbb{C}^n$, $\mathrm{GL}(r, \mathbb{R})$); $G = \mathrm{Aut}(F)$ is called the *structural group of the bundle* $F \to M$.) But in order to define a bundle, we need one more structure, namely the local triviality.

Every point $a \in M$ possesses a neighborhood $U \ni x$ such that there exists the isomorphism $h_u : \pi^{-1}(U) \tilde{\to} U \times \mathbf{F}$ which maps the fiber $E_x := \pi^{-1}(x)$ on $\{x\} \times \mathbf{F}$ isomorphically. h_u is called *trivialization* of (the bundle) F over U; indeed $F_U := \pi^{-1}(U) \equiv F|U$ is isomorphic to the trivial bundle $U \times \mathbf{F}$ over $U \subset M$. The mapping h_U is called also the (local) *chart of the bundle.* Observe that for any pair of trivializations h_i and h_k over U_i and U_k, with $U_i \cap U_k \neq \emptyset$, there exists a mappings $a_{jk} : U_i \cap U_k \to \mathrm{Aut}(\{\mathbf{F}\})$ given by $a_{ik}(x) := (h_i \circ h_k^{-1})|_{\{x\} \times \mathbf{F}}$ and called the transition mapping or transition function (from chart to chart.) Therefore, over $U_i \cap U_k \cap U_l$ the cocycle relation $a_{ik} \circ a_{kl} = a_{il}$ is satisfied.

The family (a_{ik}) is called the cocycle corresponding to the atlas (h_i). Given a manifold M, a (standard) fiber $\mathbf{F}$, and a cocycle (a_{ik}) corresponding to the covering (U_i) of the manifold M, one can reconstruct the bundle $\pi : F \to M$ with the fiber $\mathbf{F}$.

If the bundle $\pi : F \to M$ is a Riemann–Hilbert bundle, it is meaningful to define in any fiber $(F_x, (\cdot|\cdot)_x)$ a unit sphere $S_x = \{f \in F_x : \|f\|_x = 1\}$ and a unit ball $B_x = \{f \in F_x : \|f\|_x \leqslant 1\}$ and we can talk about a bundle of spheres $SF \to M$ (or a bundle of balls $BF \to M$) over M.

Given two vector bundles $F, E \to M$ over manifold M, it is completely clear what is meant by

(1) direct sum $E \oplus F \to M$,

(2) dual bundle $E^* \to M$,

(3) tensor product $E \otimes F \to M$,

(4) external product $\Lambda^k E \to M$,

(5) $\mathrm{Hom}(E, F) \to M$, etc.

Namely, if E is defined by (or possesses) a cocycle (a_{ij}) with values in $\mathrm{GL}(\mathbb{C}^r)$, F is defined by (or possesses) a cocycle (b_{ij}) with values in $\mathrm{GL}(\mathbb{C}^k)$, then $E \otimes F$ is defined by the cocycle $a_{ij}(x) \otimes b_{ij}(y) \in GL(\mathbb{C}^r \otimes \mathbb{C}^k)$. The bundles of rank one, the so called line bundles $\mathbf{F} = \mathbb{C}$ (or $\mathbb{R}$) are particularly important; for example, the determinant bundles $\det E$ and canonical bundles $K_M := \Lambda^n T^*(M)$, $n = \dim M$ belong to this class:

(6) $\det E \to M$ with the cocycle $\det a_{ij}(x) \in GL(\mathbb{C}) = \mathbb{C}^* =: \mathbb{C} - \{0\}$,

(7) $K_M \to M$ with the cocycle x_{ij} being the Jacobian of transition mapping from one chart to another.

A section E over $U \subset M$ is a mapping $s : U \to E$ such that $s(x) \in E_x$, that is, $\pi \circ s = \mathrm{id}_U$. The space of smooth sections is denoted by $C^\infty(E)$. A section of tangent bundle $TM \to M$ is a vector field, a section of the bundle $\tau_l^k \to M$ is a tensor field of the type τ_l^k, sections of the bundles $\Lambda^k(M) \to M$ are differential k forms (we compactly write $\Lambda^k(M)$ instead of $\Lambda^k(T^*M)$), more precisely $C^\infty(\Lambda^k(T^*M))$. Sections of the bundle $E \otimes \Lambda^k(M)$ are called 'differential k forms with values in the bundle $E \to M$'.

When the bundle E is given by the cocycle (a_{ij}), then the section $f \in C^\infty(E)$ over U, which over $U_i \cap U$ has the representation $f_i : U_i \cap U \to \mathbb{C}^r$ ($\mathbf{F}$, in general), satisfies the relation

$$f_i(x) = a_{ij}(x) f_j(x) \quad \text{for } x \in U_i \cap U_j \cap U$$

(there is no summation over repeating indices!!)

A linear connection (parallel transport along curves) $\pi : E \to M$ is a family of isomorphisms $P_c : E_{c(0)} \xrightarrow{\sim} E_{c(1)}$ parametrized by smooth curves $c : [0,1] \to M$ in M. The connection P_c can be understood as a lift of curves c lying in M to the bundle space E.

REMARK. The notion of connection reminds the fundamental idea of Riemann-Weierstrass of analytic continuation (of a germ of analytic function) along curve c on Riemann surface M^2: In that case, one lifts the curve c to the space of sheaf of germs $\pi : F \to M^2$ being a covering of M^2. There are, however, two major differences:

1. the coverings of a fiber $\pi^{-1} =: F_x$ (called *stalks*) are discrete spaces, and moreover,

2. the germ of algebraic function f_x cannot be continued along arbitrary curves: the singular points x_j appear (their number is finite if M^2 is compact) and at these points continuation is impossible: over x_j the germ crawls hither and thither.

On the set $M' := S - \{x_1, \dots, x\}$, the germ f_x can be analytically continued along any curve in M'. In this way, we obtain the connected component R' of the germ f_x (in the space of sheaf of germs F.) Appending R' with the 'ramified germs' $r_1, \dots, r_p$, we obtain the Riemann surface $R = R' \cup r_1 \cup \dots \cup r_p$, being the Riemann surface of (germ of) algebraic function f.

1.3 Riemann surfaces

The famous Riemann theorem says that R is a compact two-dimensional manifold. And vice versa, any meromorphic function f on compact Riemann surface R (that is, holomorphic mapping $f : R \to S^2 := \mathbb{C} \cup \{\infty\}$) is an algebraic function, that is, it satisfies the equation $W(z, f) = a_n(z) f^n + \dots + a_1(z) f + a_0(z) = 0$, where a_j are polynomials of z and the n-th order polynomial W (of variable f) is irreducible.

I have dwelt on the subject of 'Riemann surfaces' – finally codified by young Hermann Weyl in his classic *Idee der Riemannischen Fläche* (1913) – to stress that Riemann is also a father of 'singularity theory', the major and important part of modern mathematics which develops very rapidly.

The idea of Riemann surface is probably the most powerful momentum of development of new mathematics. Algebraic and general topology, theory of functional fields, algebraic geometry, theory of discrete groups, theory of groups representation, theory of (ordinary) differential equations with singularities, problems of calculus of variations, potential theory, theory of many complex variables etc., etc., owe their creation and flourishing to the idea of Riemann surface.

It is known that in the theory of functions of one (real) variable, there is a gap between functions that are differentiable and analytical (expandable in convergent power series.) In exactly the same way, the n-dimensional differential manifold is a notion which is much more general than the notion of analytical (complex) manifold. At this point, the reader may rise the question:

When an (orientable) two-dimensional (Riemannian) differential manifold M^2 is a Riemann surface, that is, when one can introduce analytical coordinates (holomorphic atlas) on M^2?

The answer is provided by the famous

THEOREM. *Let g be a Riemannian metric on a two-dimensional ori-*

entable manifold M of the (smoothness) class $k \geqslant 3$. Then

(I) *in a neighborhood of any point $x \in M$ there exist isothermal coordinates (that is, $g = ds^2 = e^{2\beta}((du_r^1)^2 + (du_z^1)^2)$ with β differentiable), and therefore any two-dimensional Riemann surface is locally conformally equivalent to Euclidean plane $\mathbb{R}^2$;*

(II) *if a family $\{U_\alpha, u_\alpha\}$ is a set of all isothermal (for the metric g) local oriented coordinate systems and if $z_j := u_j^1 + iu_j^2$, then the family $\{U_\alpha, u_\alpha\}$ defines a complex structure on M, and therefore a structure of Riemann surface;*

(III) *two Riemannian metrics g and g' define the same complex structure on M if and only if they are conformally equivalent (that is, if $g' = \lambda g$ for some positive, nonzero function λ).*

REMARKS.

1. In point I, the assumption of orientability is (obviously) not needed. This observation is due to Korn and Lichtenstein. The problem reduces to the theorem on (local) solution of (nonlinear) elliptic equation of second order with smooth coefficients.

The proof of point I was substantially simplified (in 1955) by the celebrated Chinese geometer S.S. Chern; below we will hear about him a lot!

2. Clearly, the kernel of the theorem and the major difficulty of the proof reside in point I.

3. In general, the differential mapping $f : M \to M'$ of Riemann manifolds (M, g), (M', g') is conformal if

$$g'_{f(x)}(df \cdot X, df \cdot Y) = \lambda(x) g_x(X, Y), \quad \text{for } x \in M, X, Y \in TM_x, \ \lambda(x) > 0.$$

The conformal property of f means that the mapping f preserves angles between tangent vectors (thus between intersecting curves.)

4. A holomorphic function on compact domain in $\mathbb{C}$ (or locally holomorphic mapping of Riemann surface) is, obviously, conformal. This observation was a cornerstone of the 'geometric theory of functions', that is, the theory of Riemann surfaces. At this point we should recall the central theorem of the theory of Riemann surfaces, which is

UNIFORMIZATION THEOREM (Koebe-Poincaré (1907).)

I. *Any simply connected Riemann surface M is conformally (thus biholomorphically) equivalent to one of three following surfaces,*

a) $\mathfrak{H}_1$ *(upper half plane)* $= \{z \in \mathbb{C} : \operatorname{Im} z > 0\}$, *b)* $\mathbb{C}$ $(= \mathbb{R}^2) \simeq \mathbb{D} =$ *disc:* $= \{z \in \mathbb{C} : |z| < 0\}$,

c) $S^2 =$ *the Riemann sphere* $\simeq \mathbb{P}^1(\mathbb{C})$.

II. *Any Riemann surface (two-dimensional Riemann manifold) M is a space of orbits $\tilde{M}/\Gamma$ of the group of conformal diffeomorphisms Γ, and, moreover, the covering $\tilde{M} \to M$ is locally conformal.*

REMARKS.

1. Case a) is 'general' that is, most often encountered: indeed any compact Riemann surface of genus $p > 1$ (that is, a sphere with p handles) is of the form $\mathfrak{H}_1/\Gamma$.

2. Following Poincaré, on the upper half plane $\mathfrak{H}_1$ one introduces the Riemannian metric of the form

$$ds^2 = g = \frac{dzd\bar{z}}{y^2}, \quad z = x + iy.$$

It can be shown that all geodesics, that is, the analogues of straight lines on the Euclidean plane $\mathbb{R}^2$, are half-lines and half-circles perpendicular to the real axis $y = 0$ (Fig. 3.) It is seen from this that $\mathfrak{H}_1$ ($\simeq \mathbb{D}$) is a model of a non-Euclidean 'plane': through the point z_0 not belonging to the line l one can draw an infinite number of 'straight lines' (that is, geodesics, the simplest and shortest curves) which do not intersect l (Fig. 4.)

3. As we will soon see, the Gauss curvature of the manifold $\mathfrak{H}_1$ is constant and negative.

4. The famous Hilbert theorem shows that the surface $\mathfrak{H}_1$ (and $\mathbb{D}$ as well) *cannot* be isometrically (and analytically) embedded in the Euclidean space $\mathbb{R}^3$. This theorem shows that geometric intuition is often deceptive: not every surface M^2 can be regarded as a submanifold of 'our' Euclidean space $\mathbb{R}^3$! In this connection the 'Weyl problem' is very instructive, we will talk about it in the next section.

Let us return to generic differential manifolds M^n and vector bundles. The linear connection is an additional structure on the space M^n (more precisely, on the tangent bundle TM^n.) The question arises as to whether in any vector bundle $E \xrightarrow{\pi} M^n$ the linear connection can be introduced?

The answer is in affirmative: the proof of existence of parallel transport (connection) along any curve $c \subseteq M^n$ reduces to the classical theorem from the theory of (systems of) linear ordinary differential equations. What is important, is that we transport along curves, the one-dimensional objects

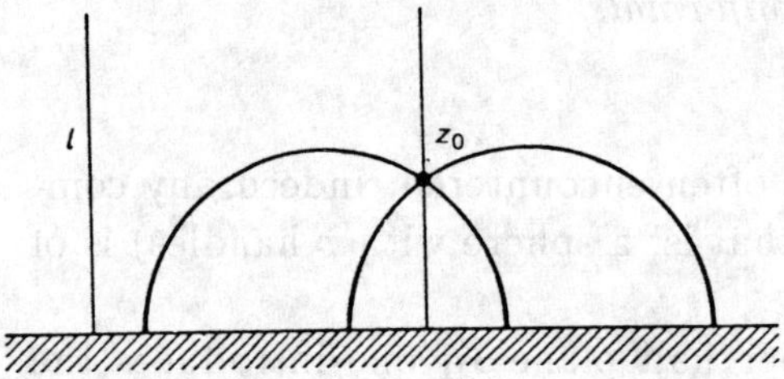

Fig. 3

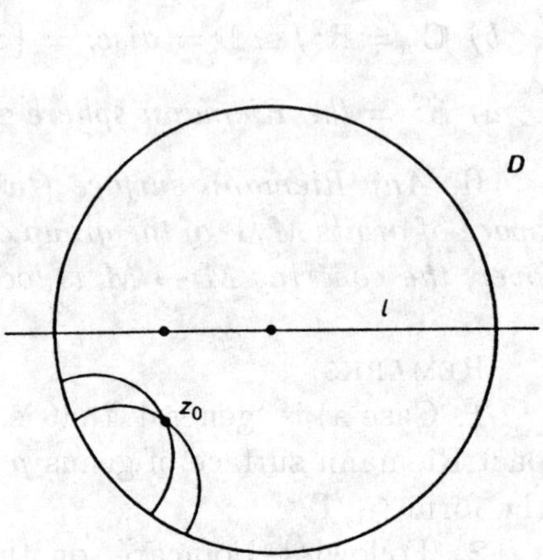

Fig. 4.

(thus we have ordinary differential equations.) But, in general, the parallel transport of a vector along different curves with the same ends does not give the same result. In other words, the transport along closed curve $\gamma \subset M^n$ does not result in the initial vector, and therefore the lifting $\tilde{\gamma} \subset E$ of the closed curve $\gamma \subset M^n$ may be a curve which is not closed. This phenomenon is called the *curvature of connection (curvature tensor of connection)* (Fig. 5.)

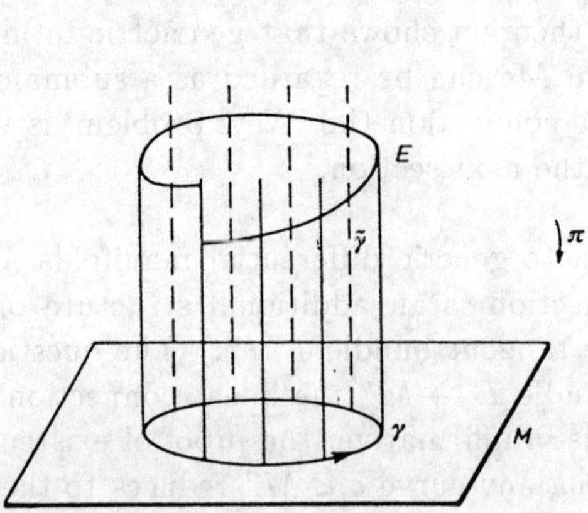

Fig. 5

It can be shown that the Gauss curvature $K(x)$ can be defined in terms of the parallel transport in tangent planes T_xM^2 preserving the metric g_x. In the multidimensional case, and generally, given a connection ∇ in a vector bundle $E \to M$, we have the following natural procedure.

Let Δ be a rectangle and isosceles triangle in $\mathbb{R}^2$ with both legs equal 1 (Fig. 6) and let $\sigma : \Delta \to M^n$ be such smooth mapping that $\sigma(0) = x$, $d\sigma_0(1,0) = h$, $d\sigma_0(0,1) = k$ (Fig. 7.) Let $\sigma_\tau(u) := \sigma(\tau u)$, where $0 \leqslant \sigma \leqslant 1$.

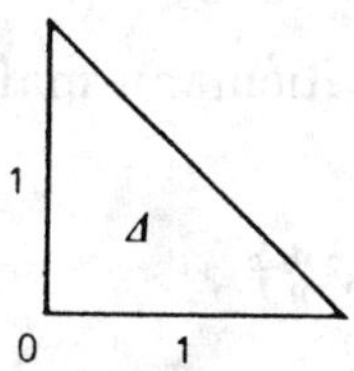

Fig. 6

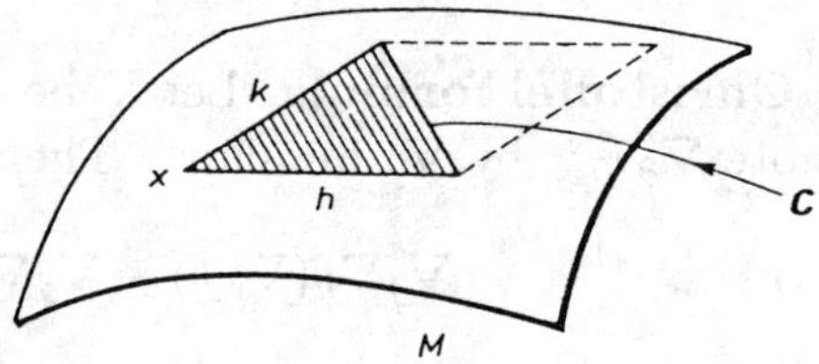

Fig. 7

Let us denote by Φ_τ the automorphism of the fiber E_x obtained as a result of parallel transport along $\partial\sigma_\tau$, the curve being the boundary of σ_τ. Let

$$K(x;h,k) := \lim_{\tau\to 0} \tau^{-1}\left(\Phi_\tau - \mathrm{id}_{E_x}\right).$$

DEFINITION. The bilinear mapping $K = K^\nabla$

$$(*) \qquad T_x(M) \times T_x(M) \ni (h,k) \to K(x;h,k) \in \mathrm{Aut}(E_x)$$

is called the *curvature of connection* ∇ at the point $x \in M$. Because bilinear mapping are also called tensors, the curvature K^∇ of connection ∇ is a tensor field on M (for short, a tensor on M) valued in Aut E. Given a coordinate system on M $(u^1, \dots, u^n)$ in a neighborhood of x and a basis in E_x, e_α, $\alpha = 1, \dots, r = \dim E_x$, we can write the curvature tensor in coordinates

$$K^\alpha_{ij\beta}(x), \quad i,j = 1,\dots,n, \ \alpha,\beta = 1,\dots,r.$$

This tensor is antisymmetric in the indices i, j. If we have a covariant derivative in the vector bundle $E \to M$ defined 'axiomatically', then for the curvature $K = K^\nabla$ of connection ∇ we can take $K : C^\infty(TM) \times C^\infty(TM) \to \mathrm{Hom}_{\mathbb{R}}(C^\infty(E) \times C^\infty(E))$

$$(K) \qquad K(X,Y) := \nabla_X\nabla_Y - \nabla_Y\nabla_X - \nabla_{[X,Y]},$$

where $[X,Y]$ is a commutator of the vector fields X, Y defined to be $XY(f) - YX(f)$ for $f \in C^\infty(M)$.

From this definition we have as an immediate consequence that

$$K(X,Y) = -K(Y,X.)$$

If in definition (K) we substitute for X, Y the components of the gradient ∇f of function f on M, we obtain an important formula first derived by Christoffel (1869.)

Christoffel formula. Let ∇ be a connection on Riemann manifold; we denote $\nabla_{\partial/\partial x^j}$ by ∇_j for short. Then

$$(Chr) \qquad \nabla_j \nabla_k (\nabla_r f) - \nabla_j \nabla_r (\nabla_k f) = K^m_{ikr} \nabla_m f.$$

Thus, the curvature tensor measures asymmetry of third covariant derivative of a scalar field f. This was the way which led Christoffel to his curvature tensor.

This formula is quite useful: we will make use of it while discussing the theory of Jacobi (geodesic) fields.

So what is a relation between Gauss curvature and the curvature tensor? The answer is given by, again due to Riemann

THEOREM. *Let R_{ijkl} be a R-Ch tensor (of Levi-Civita connection) on M^2. Then*

$$R_{1212} = -K(x) \det(g_{ij}(x)),$$

where $K(x)$ is the Gauss curvature at the point $x \in M^2$.

1.4 Riemannian connection. Levi-Civita connection

If in a vector bundle $\pi : E \to M$ we have a Riemannian structure, that is, if in any fiber $E_x = \pi^{-1}(x)$ we have defined inner product $(\cdot|\cdot)_x$ which varies smoothly on M, then $(E \to M, (\cdot|\cdot)_x)$ is called a *Riemann bundle.* With this richer structure in hand, it is natural to demand that the parallel transport $P_c : E_{c(0)} \tilde{\to} E_{c(1)}$ preserves the inner products (is an isometry.) If this holds, the connection P is called a *Riemannian connection.* It is not hard to show that any vector bundle can be equipped with Riemannian connection.

The most important example of such a situation is, of course, Riemannian manifold, that is, the tangent bundle $TM \to M$ with the metric $g_x = (\cdot|\cdot)_x$ in every tangent space T_xM. Such Riemannian connection was first constructed by Tulio Levi-Civita (1916) (for a manifold embedded into $\mathbb{R}^N$) and, almost simultaneously by Hessenberg; the general definition (without embedding Riemann manifold M^n into $\mathbb{R}^N$) was given by H. Weyl (1917.) The Levi-Civita connection is a Riemannian connection and is symmetric (that is, without torsion.) The following important theorem holds.

THEOREM (RICCI LEMMA). *There is single Riemannian connection on the tangent bundle of Riemann manifold which does not have torsion (that is, it is symmetric.) This connection is the Levi-Civita connection.*

This theorem was called 'the fundamental theorem of Riemannian geometry', the proof of this theorem can be found in any monograph on Riemannian geometry. It has a philosophical meaning as well: among many linear connection, people first found (constructed, discovered) this best one, the single one, the one 'given by God', one would be willing to say.

Historically speaking, the road leading from covariant derivative, through parallel transport of Levi-Civita to the connection of Weyl, who presented a full theory of linear connections on vector bundles.

How did Weyl reach this fundamental notion? The answer is surprising: by meditating the Einstein's theory of gravity and by looking for the 'unified theory of fields'.

Historical remark. It is very significant and fascinating that the general notion of connection arose in connection with physics, in the famous works of Hermann Weyl on general theory of relativity, and, more precisely, in his, first in the history of physics, attempt to unify the Einstein's theory of gravity with the Maxwell-Hertz electromagnetism. In his famous work of 1918 *Gravitation und Elektrizität* (the title of the 'purely mathematical' work written at almost the same time is *Reine Infinitesimalgeometrie*) in connection with the purely infinitesimal geometry, Weyl says that one should have a courage to reject the preservation of inner product in parallel transport (along curves), and thus he *implicitly* talks about general parallel transport. However, employing 'clever limitations' (Goethe), he then uses a conformal connection, that is, as a result of parallel transport, the metric form g is getting multiplied by a (positive) function λ. Weyl uses philosophical reasoning: a field theory should be purely infinitesimal, the Einstein's theory of gravity still contains an element of distance interactions (the Riemann-Levi-Civita connection.) Weyl shows that introduction of the factor $\lambda(r)$, $P_{\gamma(t)}g = \lambda(\gamma(t))g_{\gamma(t)}$ leads to 'gauging' of the theory: the additive one form $\varphi_j dx^j$ appears which is enthusiastically interpreted by Weyl as a vector potential of electromagnetic field. Initially, this theory aroused an enthusiasm not only of Einstein but also of celebrated British astrophysicist Eddington. However, the later criticism of Einstein forced also the creator of the first in history unified field theory to reject it (the quantum theory did not exist that time, it appeared only seven years later.) Paraphrasing biblical words of Jahwe, with the great sense of humor Weyl eventually said: 'Let no man put together what God put asunder'. However, the conformal theory of Weyl influenced

1. differential geometry a lot,

2. and, maybe even more the new physics, reincarnating in Yang-Mills theories (gauge theories) which, in turn, provide an important momentum for development of mathematics (important works of Donaldson on four-dimensional manifolds and differentiable structures.)

In my opinion, it was not an accident that Weyl discovered his unified conformal theory; as we know, already young Weyl was a celebrated expert in the theory of Riemann surfaces, and there the notion of conformal transformations is of fundamental importance.

But let us return to the parallel transport. This notion arose as a result of the need to 'geometrize' the old notion of *covariant derivative* introduced half a century before.

The directional derivative of a function in $\mathbb{R}^n$ is a limit of the differential quotient

$$(*) \qquad \frac{v(x+ht)-v(x)}{t} \xrightarrow{t} \nabla_h v.$$

In the case of differential manifold, the tangent spaces $T_{c(0)}M$, $T_{c(t)}M$ do not have anything in common, only the parallel transport along the curve c makes it possible to transport a vector $v \in T_{c(0)}M$ to $T_{c(t)}M$. Now, we can construct differential quotient (*), as a result of which, we obtain the covariant derivative $\nabla_c v$.

Since derivative is a local notion, we can make use of a local coordinate system, a map around the point $x = c(0)$, and we obtain the formula (for arbitrary vector bundle $E \to M$ of rank r.)

Let $c : t \to x(t) \in M, t \in [0,1]$ be a smooth curve and let σ be a smooth section of the bundle $E \to M$ along the curve, that is, $\sigma(t) = \sigma(x(t))$, where $\sigma \in C^\infty(E)$ is a section of the bundle E, $\sigma(m) \in E_m$, $m \in M$. Then

$$(\nabla_c \sigma)(t) = (\nabla s)\left(x(t)\frac{\mathrm{d}x}{\mathrm{d}t}\right).$$

We have $\nabla_c(f \cdot \sigma) = f' \cdot \sigma + f \cdot \nabla_c \sigma$, $f \in C^\infty(M)$ and $\nabla_c \sigma$ is called the *covariant derivative* of section σ along the curve c. In local coordinates $(x^1, \ldots, x^n)$, σ^α, $\alpha = 1, \ldots, r$ we have the fundamental formula

$$(\nabla_c \sigma)^\alpha = \frac{\mathrm{d}\sigma^\alpha}{\mathrm{d}t} + \Gamma^\alpha_{\beta j}\sigma^\beta \frac{\mathrm{d}x^j}{\mathrm{d}t} \quad \text{summation convention!}$$

The functions $\Gamma^\alpha_{\beta j}$ are called *the coefficients of connection* or *Christoffel symbols* in commemoration of this celebrated mathematician and physicists who first introduced formally the covariant derivative of tensor field on Riemannian manifold.

For components of the curvature tensor K^∇ of connection ∇ defined by the Christoffel symbols $\Gamma^\alpha_{\beta j}$ one derives the formula (known to Christoffel and Ricci-Curbastro)

$$R^\alpha_{\beta ij} = \frac{\partial \Gamma^\alpha_{\beta j}}{\partial x^i} - \frac{\partial \Gamma^\alpha_{\beta i}}{\partial x^j} + \Gamma^\alpha_{\kappa i}\Gamma^\kappa_{\beta j} - \Gamma^\alpha_{\kappa j}\Gamma^\kappa_{\beta i}$$

(summation over Greek indices from 1 to r, over Latin indices from 1 to n.) In the case of tangent bundle $TM \to M$, $r = n$ and one consider symmetric connections:

$$\Gamma^\alpha_{\beta j} = \Gamma^\alpha_{j\beta} \quad \text{(symmetry with respect to lower indices.)}$$

It is not difficult to show that coefficients of connection define connection. It only remains to present the formula for coefficients of Levi-Civita connection. Let in the coordinate system (u, x^i), g_{ij} be components of a metric tensor g in tangent bundle TM. Then

$$\Gamma^i_{jk} = \frac{1}{2} g^{il}\left(\frac{\partial g_{jl}}{\partial x^k} + \frac{\partial g_{kl}}{\partial x^j} - \frac{\partial g_{jk}}{\partial x^l}\right).$$

We see that this connection is indeed symmetric

$$\Gamma^i_{jk} = \Gamma^i_{kj}, \quad k, i, j = 1, \ldots, n, \quad \text{because } g_{jk} = g_{kj}.$$

Covariant derivative makes it possible to define a notion of *geodesic lines* (on manifold M with connection ∇.) A *geodesic* of connection ∇ is a curve c whose tangent vectors (velocity field $\dot{x}$) are parallel:

$$(G) \qquad \nabla_c \frac{\mathrm{d}x}{\mathrm{d}t} = 0,$$

and therefore the geodesic line c satisfies the following differential equation

$$(G') \qquad \frac{\mathrm{d}^2 x^i}{\mathrm{d}t^2} + \Gamma^i_{jk} \frac{\mathrm{d}x^j}{\mathrm{d}t} \frac{\mathrm{d}x^k}{\mathrm{d}t} = 0.$$

Therefore, the geodesic is the 'straightest' line on manifold (M, ∇) equipped with linear connection ∇.

We see that on such manifolds one can consider Newtonian mechanics, indeed a notion of acceleration makes sense.

The first law of Galileo-Newton mechanics can be therefore phrased as follows:

A body (material point) on which no force acts moves along geodesic of the manifold (M, ∇)

How could Newton and his followers investigate mechanics of material points without having any idea about linear connection and covariant derivative? In fact, they used the space $M = \mathbb{R}^n$ with classical Euclidean metric and natural linear connection, and they identified all tangent spaces $T_x\mathbb{R}^n \equiv \mathbb{R}^n$.

1.5 Geodesics in Riemann space (manifold) (M, g) as lines of extremal length. Euler-Lagrange equation

The Euclidean straight lines are not only 'straightest' (that is 'autoparallel'), they are also the shortest: one can construct a straight line l through any two points $x, y \in \mathbb{R}^3$ such that the interval bounded by these points has length $d(x, y) = \left(\sum_{i=1}^3 (x^i - y^i)^2\right)^{1/2}$.

The simplest example of geometry which is not planar is the sphere S^2. As it is well-known from geometry (geography) the shortest lines on S^2 are great circles, to be more precise, shorter arcs of these circles bounded by $x, y \in S^2$. As this example shows, to find differential equation describing geodesics in the case of arbitrary Riemann manifolds, one should look for curves for whose the functional of 'length of arc' takes extremal values (the curves are locally shortest.) The problem of geodesics in $M^2 \hookrightarrow \mathbb{R}^3$ is very old; it arose in early days of calculus of variations (Euler, Bernoulli, Leibnitz, Newton, Gauss.)

As we know, in the Riemann space, the length of a smooth curve $c : [0,1] \to M$ is given by the formula

$$(1) \qquad s = \int_0^1 W\,dt, \quad W = (g_{ij}dx^i dx^j)^{1/2},$$

and the velocity vector is $\frac{\mathrm{d}x^i}{\mathrm{d}t} =: \dot{x}^i$.

Let us compute the first (variational) differential of the functional (1) under the 'translation'

$$\bar{x}^i := x^i + \epsilon X^i + \epsilon^2(\ldots), \quad \text{that is, } X^i = \frac{\partial \bar{x}^i}{\partial \epsilon}(t,\epsilon)|_{\epsilon=0}.$$

Fixing ends of curves, and taking derivative with respect to ϵ, we have

$$\begin{aligned}\delta s &= \frac{\mathrm{d}s}{\mathrm{d}\epsilon}\Big|_{\epsilon=0} = \int_0^1 \left(\frac{\partial W}{\partial x^i}X^i + \frac{\partial W}{\partial \dot{x}^i}\dot{X}^i\right) dt \;= \text{(integration by parts)}\\ &= \int_0^1 \left(\frac{\partial W}{\partial x^i} + \frac{\mathrm{d}}{\mathrm{d}t}\frac{\partial W}{\partial \dot{x}^i}\right) X^i\,dt.\end{aligned}$$

Since X^i are arbitrary, the expression in the bracket vanishes, thus

$$(2) \qquad V_i := \frac{\partial W}{\partial x^i} + \frac{\mathrm{d}}{\mathrm{d}t}\frac{\partial W}{\partial \dot{x}^i} = 0.$$

These are the famous Euler–Lagrange equations. Taking

$$(3) \qquad L := \frac{1}{2}g_{ij}dx^i dx^j = \frac{1}{2}W^2,$$

we have

$$V_i = \frac{\mathrm{d}}{\mathrm{d}t}\left(\frac{1}{W}\frac{\partial L}{\partial \dot{x}^i}\right) - \frac{1}{W}\frac{\partial L}{\partial x^i}$$

or, if we take as a parameter of the curve the length of the arc $t = s$, then $W = 1$ and we have

$$V_i = \frac{\mathrm{d}}{\mathrm{d}s}\frac{\partial L}{\partial \dot{x}^i} - \frac{\partial L}{\partial x^i}.$$

Substituting into (3), we find

$$\frac{\partial L}{\partial x^r} = \frac{1}{2}\frac{\partial g_{ij}}{\partial x^r}\dot{x}^i\dot{x}^j, \quad \frac{\partial L}{\partial \dot{x}^r} = g_{ri}\dot{x}^i,$$

$$\frac{\mathrm{d}}{\mathrm{d}s}\frac{\partial L}{\partial \dot{x}^r} = g_{ri}\ddot{x}^i + \frac{\partial g_{rj}}{\partial x^i}\dot{x}^i\dot{x}^j,$$

$$\begin{aligned} V_r &= g_{ri}\ddot{x}^i + \frac{\partial g_{rj}}{\partial x^i}\dot{x}^i\dot{x}^j - \frac{1}{2}\frac{\partial g_{ij}}{\partial x^r}\dot{x}^i\dot{x}^j = \\ &= g_{ri}\ddot{x}^i + \frac{1}{2}\left(\frac{\partial g_{ri}}{\partial x^j} + \frac{\partial g_{ij}}{\partial x^r} - \frac{\partial g_{rj}}{\partial x^i}\right)\dot{x}^i\dot{x}^j = g_{ri}\ddot{x}^i + \Gamma_{ij,r}\dot{x}^i\dot{x}^j, \end{aligned}$$

where $\Gamma_{ij,r} := \frac{1}{2}\left(\frac{\partial g_{ri}}{\partial x^j} + \frac{\partial g_{ij}}{\partial x^r} - \frac{\partial g_{rj}}{\partial x^i}\right)$ are 'Christoffel symbols of first kind'. Raising the indices, we get $V^l := G^{lr}V_r = \ddot{x}^l + \Gamma^l_{ij}\dot{x}^i\dot{x}^j$. Since $\delta s = 0$, we have $V_l = 0 = V^l$, and finally we obtain the expected geodesic equation

$$(G') \qquad \ddot{x}^l + \Gamma^l_{ij}\dot{x}^i\dot{x}^j = 0$$

in the Riemann space (M, g). For Levi-Civita parallel transport we have again equivalent formulas

$$(L-C) \qquad \frac{\mathrm{d}}{\mathrm{d}t}X^l = -\Gamma^l_{ij}X^i\frac{\mathrm{d}x^j}{\mathrm{d}t}, \quad \frac{\mathrm{d}}{\mathrm{d}t}X_l = +\Gamma^i_{lj}X_i\frac{\mathrm{d}x^j}{\mathrm{d}t}.$$

REMARK. In connection with the theory of gravity, Weyl introduced a notion of *geodesic coordinate system*: this is a map (x^j) in the neighborhood of the point $x_0 \in M$, such that in this coordinate system g_{ij} has vanishing first partial derivatives $\frac{\partial g_{ij}}{\partial x^r} = 0$ in x_0. In the geodesic coordinate system, Christoffel coefficients vanish at x_0, and therefore equations (L-C) take the form

$$(W) \qquad \frac{\mathrm{d}}{\mathrm{d}t}X^l = 0 = \frac{\mathrm{d}}{\mathrm{d}t}X_l \quad i = 1, \ldots, n$$

Therefore, following Weyl, one can first introduce the Levi-Civita parallel transport demanding that (W) is satisfied and then show that the definition does not depend on the coordinate system (cf. H. Weyl, *Raum-Zeit-Materie*.)

1.6 Jacobi fields (curvature and geodesics)

In his famous lectures on mechanics Carl Gustav Jacobi considered, among others, the problem of geodesics on ellipsoids, introducing the *geodesic field*. The important tool are *Jacobi fields*. These are vector fields $Y(\cdot)$ along geodesics $c_{x,y}(s) = c(s)$ (joining the points x, y of the Riemann space (M, g), $0 \leqslant s \leqslant d(x, y)$; the distance between points x and y is defined naturally as a lower bound of the length of (broken) curves joining x and y) being velocity fields

$$Y(s) := \left.\frac{\partial c(s,\epsilon)}{\partial \epsilon}\right|_{\epsilon=0}$$

of some geodesic variation $c(s,\epsilon)$, $0 \leqslant \epsilon \leqslant 1$, $c(\cdot,0) = c(\cdot)$. The field $Y(\cdot)$ satisfies, *ex definitione*, the Jacobi equation

$$(J') \qquad \ldots \ddot{Y} + K(t)Y = 0,$$

where $K(t) := K(c(t))$ is the Gauss curvature of the surface M^2.

In the multi-dimensional case (M^n, g), $Y(\cdot)$ satisfies the equation

$$(J) \qquad \ldots \frac{\nabla}{ds}\frac{\nabla}{ds}Y(s) + R(c(s); Y(s), \dot{c}(s))\dot{c}(s) = 0,$$

where $R(\cdot; Y(s), \dot{c}(s))$ is a curvature tensor (form) of the manifold (M, g).

Particularly important are fields orthogonal to the geodesic c: $(Y|\dot{c}) = 0$. The manifold of such Jacobi fields has dimension $2n - 2$. It is particularly convenient to investigate these fields in *Fermi coordinates*; in these coordinates, the curvature tensor takes the simple form

$$R(c(s); Y(s), \dot{c}(s))\dot{c}(s)^j = R_i^j Y^i, \quad \text{where } R_i^j := R_{lil}^j = -\frac{1}{2}\frac{\partial g_{ll}}{\partial x^i \partial x^j}.$$

Taking $R = \{R_i^j : i, j \leqslant n-1\}$, we can rewrite (J) in the form

$$(J'') \qquad \ddot{Y} + RY = 0,$$

which, for $M = M^2$, is identical with the classical Jacobi equation (J'). We will return to Jacobi fields many times. But, as for now, this completes the description of 'analytical apparatus' of Riemannian geometry.

In mathematics, the objects of fundamental importance are concrete objects; in geometry, concrete Riemann spaces: investigating them, mathematicians gain experience and intuition. For this reason, in the next chapter, we will discuss the spaces of constant curvature.

CHAPTER 2

Sectional Curvature. Spaces of Constant Curvature. Weyl Hypothesis

As we have seen, by introducing the notion of curvature tensor K^∇, with any pair of tangent vectors $Y, Z \in T_xM$ we associated a linear transformation $K_x(Y, Z)$ of the tangent space T_xM. Let F_x be an oriented plane spanned by Y, Z (that is, Y, Z are basis vectors in F_x.) Let S_F be a surface (in M) generated by geodesics tangent to F, more precisely, if U is a sufficiently small neighborhood in F_x, then the exponential mapping $U \to \exp_x(U) =: S_F$ is a diffeomorphism and the inner product in T_xM induces a non-degenerate inner product in any tangent plane T_xS_F. Thus, the so obtained manifold S_F is Riemannian and one can talk about its Gauss curvature $K_{S_F} =: K(x, F) \equiv K_x(F)$ called the *sectional curvature in bi-direction F*. Let us denote by

$$|Y \wedge Z|^2 := g(Y, Y)g(Z, Z) - g(X, Y)^2$$

the square of the area of the parallelogram with sides Y, Z spanning F. We have

THEOREM (RIEMANN). *Let M be a (pseudo) Riemannian manifold with the metric tensor g and curvature tensor R. If Y, Z is a basis in F, then the sectional curvature*

$$K(x, F) = \frac{g(R(Y, Z)Y, Z)}{|Y \wedge Z|^2}.$$

REMARK. Riemann knew that $K_x(F)$ defines R (where F runs through all F above.)

DEFINITION. We say that Riemann manifold M has *constant curvature* $K(x)$ *in point* x if for any bi-direction $F \subset T_xM$, $K(x,F) = K(x)$; thus the sectional curvature does only depend on x; it is independent of a bi-direction. If, moreover, $K(x) = \text{const} = K$, we say that the manifold (M,g) has *constant curvature* $= K$.

We have the following surprising

THEOREM (F. SCHUR). *Let M^n be a connected Riemann manifold of dimension $n \geqslant 3$ having at any point x constant curvature $K(x)$. Then M^n has constant curvature.*

REMARK. This, obviously, does not hold for $n = 2$.

There is a rich theory of (Riemann) spaces of constant curvature and these spaces are of the utmost importance as an experimental field of differential geometry. These spaces are particularly homogeneous; for a long time it was conjectured that Euclidean spaces are characterized by a very rich group of (rigid) motions; it turned out, however that this is a property of all spaces of constant curvature ('Riemann-Helmholtz problem'.)

It is again Riemann who we owe an important step on the way to the classification and characterization of spaces of constant curvature; also from his habilitation thesis we have the following beautiful

THEOREM (RIEMANN). *Let (M^n, g) be a (pseudo) Riemannian manifold and K some real number. Then the following conditions are equivalent:*

(a) *M^n has constant curvature K;*

(b) *in some neighborhood of arbitrary $x \in M^n$ there exist the map (u^i) in which the metric g has the form*

$$ds^2 = \frac{e_1(du^1)^2 + \cdots + e_n(du^n)^2}{1 + \frac{1}{4}K\sum_{i=1}^n e_i(u^i)^2} \qquad e_i = \pm 1,\ i = 1, \ldots, n. \tag{1}$$

The following fundamental investigation of spaces of constant curvatures are due to W. Killing, H. Poincaré, and H. Hopf.

COROLLARY (Poincaré) *Let in* (1) *all* $e_i = 1$, $K < 0$ *(hyperbolic geometry.) Let us take* $K = -r^{-2}$, $r > 0$. *Then the manifold* M^n *denoted* $\mathcal{H}^n$ *(with curvature* $-1/r^2$*) is isometric to the disc* $\mathbf{D^n}$ *of radius* $2r$ $\{u \in \mathbb{R}^n : \|u\| < 2r\}$ *with the metric*

$$(2) \qquad ds^2 = \frac{(du^1)^2 + \cdots + (du^n)^2}{\left\{1 - \left(\frac{1}{2r}\right)^2 \sum_{i=1}^n e_i (u^i)^2\right\}^2}.$$

The spaces $\mathcal{H}^n$ *are also isometric to the half-space* $u^1 > 0$ *in* $\mathbb{R}^n$ *with the metric*

$$ds^2 = r^2 \frac{(du^1)^2 + \cdots + (du^n)^2}{(u^1)^2}.$$

The isometry of the disc $\|u\|^2 < 4r^2$ on the half-space $u^1 > 0$ in $\mathbb{R}^n$ is provided by famous

Generalized Cayley transformation $c = \xi_0 \rho_0 \xi^1$, where ξ denotes the stereographic projection (in $\mathbb{R}^{n+1}$) of the sphere $S^n \subset \mathbb{R}^{n+1}$ with radius r, the center in the point $(r, 0, \ldots, 0)$, and the north pole $p = (2r, 0, \ldots, 0)$, on hypersurface $\mathbb{R}^n$

$$\xi : S^n - \{p\} \to \mathbb{R}^n,$$

and ρ denotes rotation of the sphere S^n by the angle $\frac{1}{2}\pi$ around the $(n-1)$-plane to whose the center of the sphere belongs and given by equation $u^0 = r$, $u^1 = 0$. Thus, for $a \in \mathbb{R}^{n+1}$

$$\rho(a^0, a^1, a^2, \ldots, a^n) := (a^0 + r, r - a^1, a^2, \ldots, a^n).$$

In the case $n = 2$, the Cayley transformation has a particularly simple complex form

$$c(z) := -i\frac{z+i}{z-i}, \quad i = \sqrt{-1}.$$

If $r = 1/2$, the metric on $\mathbb{D}^2$ has the form

$$ds^2 = \frac{dz d\bar{z}}{(1 - |z|^2)^2},$$

and on ξ_1,

$$ds^2 = \frac{dz d\bar{z}}{4(\operatorname{Im} z)^2}.$$

HISTORICAL REMARK. At this point, it would be inappropriate not to say few words about one of the most celebrated mathematicians of XIX century Arthur Cayley (1821–1895.) He was a creator of the theory of algebraic geometry and invariants, which 'like Athena in full armor from the head of Zeus, leaped from the head of Cayley' (H. Weyl): *Memoire sur les Hyperdeterminants* (1846) and famous nine *Memoirs on Quantics* (1854–59.) The live and work of this great and quiet man were unbelievable: He was born in Richmond and raised in Petersburg as a son of an English merchant. After great mathematical success (the highest award) at the University of Cambridge, he became a solicitor in London and for twenty years he pursued very successful solicitor career. It was during his 'solicitor years' when the bulk of his fundamental mathematical works was written (13 large volumes.) Cayley developed, among others, projective geometry (descriptive) and introduced the famous Cayley metric in projective spaces; this is the reason behind his famous statement 'the metric geometry is a descriptive geometry'. In Cambridge he assembled a narrow circle of magnificent pupils (Sylvester, Salmon, and others.) He died on January 22, 1895.

The case $\mathcal{H}^2$ is particularly important. The disc $\mathbb{D}^2$ and the half plane $\mathfrak{H}_1$ in $\mathbb{C}$ play a role of universal (that is, simply connected) Riemann surfaces of hyperbolic type. These are the first models of non-Euclidean geometry.

As we mentioned above, the geodesics are half-circles (and half-lines) orthogonal to the boundary. Similarly, we have important

THEOREM (KILLING–HOPF). *Any complete and connected Riemann space M^n, $n \geqslant 2$ of constant curvature K has as a universal cover S^n, $\mathbb{R}^n$, or $\mathbf{H^n}$. More precisely, M^n is isomorphic to one of the following quotient spaces*

S^n/Γ, *where* $\Gamma \subset O(n+1)$ *for* $K > 0$;

$\mathbb{R}^n/\Gamma$, *where* $\Gamma \subset E(n)$, *the group of Euclidean motions in* $\mathbb{R}^n$ *for* $K = 0$;

$\mathbf{H}^n/\Gamma$, *where* $\Gamma \subset O^1(n+1)$ *for* $K < 0$,

moreover, Γ *is a group of isometries acting freely (that is, without fixed points and properly discontinuous.)*

Let us recall that Γ acts on a locally compact space S properly discontinuously if any $s \in S$ has a neighborhood U such that the set $\{\gamma \in \Gamma : \gamma(U) \cap U\}$ is finite.

REMARK. The theorem above is very much the same as the 'uniformization theorem' of the theory of Riemann surfaces; in the latter, we also have three kinds of covering spaces, the sphere S^2, the plane $\mathbb{C}$, and the upper half plane $\mathfrak{H}_1$ ($\simeq \mathcal{H}^2$.) Therefore, one could call the Killing–Hopf theorem the 'uniformization theorem for Riemann spaces of constant curvature'. H.

Hopf was certainly guided by this fundamental theorem!

If we had known (*a priori*) that on any Riemann surface X one can introduce a Riemannian metric such that its curvature was constant, then the uniformization theorem would have been a direct consequence of the Killing–Hopf theorem. But we do not know that! On the contrary, the uniformization theorem makes it possible to introduce a metric of constant curvature on Riemann surface.

Let us return to general Riemannian manifolds. For complete (M^n, g) for whose $K(F)$ has constant sign, two classical theorems hold (known to Bonnet and Hadamard for $n = 2$.)

THEOREM (MYERS, BONNET). *(M, g) is complete if $(k(F) \geqslant r^2 =$ const $> 0) \Longrightarrow$ (M has a diameter less than or equal π/r, that is, M is compact.)*

THEOREM (CARTAN, HADAMARD) *Let M^n be complete and $K(F) \leqslant 0$. Then the universal cover $\tilde{M}^n$ is diffeomorphic to $\mathbb{R}^n$. In particular, complete and simply connected M^n is diffeomorphic with $\mathbb{R}^n$ and the exponential mapping $\exp_n : \mathbb{R}^n \to M^n$ is a diffeomorphism (for any $x \in M^n$).)*

PROOF follows from the fact that for $K_x(F) < 0$ x does not have any focal points.

At this point I cannot resist mentioning deep theorems concerning two-dimensional Riemannian manifolds M^2 of positive curvature called *ovals* (or *convex surfaces.*)

2.1 Ovals

Following Blaschke, a surface $M \subset \mathbb{R}^3$ which 1. is compact, 2. has curvature $K > 0$ will be called an *oval.*

The examples of ovals are ellipsoids (and thus spheres.) Already in 1897, Hadamard proved the following

HADAMARD THEOREM. *Let $M \subset \mathbb{R}^3$ be an oval. Then*

(i) *M is orientable;*

(ii) *for fixed orientation the (normal) Gauss mapping* $G : M \to S^2$ *is a diffeomorphism;*

(iii) M *is strictly convex, that is, for any* $x \in M$*, the oval* M *lies on one side of the tangent plane* T_xM*, that is,* T_xM *is a supporting plane* (Stützebene *of Minkowski).*

Soon afterwards (1897) H. Liebmann showed a 'new property', namely

$$(M \text{ compact, convex surface in } \mathbb{R}^3 \text{ with constant } K) \Longrightarrow$$
$$\Longrightarrow (K = r^2 > 0 \text{ and } M \text{ is a sphere of radius } r^{-1}.)$$

Cohn-Vossen (1927) and Herglotz (1943) showed 'rigidity of ovals':

$$(M, M' \text{ are isometric ovals}) \Longrightarrow$$
$$\Longrightarrow (\text{there exists a motion in the space } \mathbb{R}^3 \text{ transforming } M \text{ into } M'.)$$

Already in 1915 – 16 Weyl posed and partially solved the so-called

Weyl problem. Let (M^2, g) be a two-dimensional Riemannian manifold of positive curvature K which is homeomorphic to the sphere S^2. Then there exists an isometry (sufficiently smooth) $f : M^2 \to M \subset \mathbb{R}^3$ on the oval M which is unique up to Euclidean motion.

As Weyl recalled himself (forty years later), he was able to solve this problem only partially: being already a professor at E.T.H. (Polytechnic University of Zurich), he has been drafted (as a German citizen) to the army: it was the period of the First World War. More than half year long service in barracks, before E.T.H. was able to reclaim him back, resulted in his leaving the problem. He only published a remarkable paper (1916) with partial solution. In 1938, Hans Levy, refugee from Götingen, generalized the Weyl problem in the analytic case (analytic metric g and analytic embedding f.) The case $C^{k,\alpha}$ (class C^k with Hölder coefficient α) was, with the help of Weyl method, solved only in 1953 by L. Nirenberger in his work *The Weyl and Minkowski Problems in Differential Geometry in the Large* (Comm. Pure Appl. Math. (1953), 337–394.)

Theorem of Weyl, Levy, and Nirenberger has a deep *philosophical* sense: the manifold (M^2, g) is the idea of a surface in $\mathbb{R}^3$ and the (W-L-N) theorem says that this idea is accomplished, in the case $K > 0$, in one and only one way (as an oval.) As Weyl says in the foreword of this classical paper of 1916 *Über die Bestimmung einer geschlossenen Konvexen Fläche durch ihr Linienelement*: The relation between 'idea' and 'reality' is here most perfect.

As the history of mathematics clearly shows, descending of an idea from the world of pleroma (cf. famous parable of Plato about the cave in VII book of his *Politea*) to the empirical world is very difficult and dangerous. It is usually much more difficult than to climb from this 'cave of empiria' to pleroma of *logoi spermatikoi* – the germs of ideas. The development of germs of ideas into full idea and its crystallization – materialization, realization – is an assignment for a man; in the case of mathematics this required efforts of, sometimes, entire generations of great mathematicians.

In order to at least draw a relation between geometry of surfaces and analysis (theory of elliptic partial differential equations), let me present three major steps required to solve the Weyl problem.

A set of Riemannian metrics g of class $C^{l,\alpha}$ on the sphere S^2 is a Banach space and a linear topological space, if equipped with weak topology. To prove the W-L-N theorem, it is sufficient to show that the set $C^{l,\alpha}$ of metrics which can be isomorphically imbedded (with isomorphism of the class $C^{l,\alpha}$), is in this topological Banach space: (a) open, (b) closed, (c) connected. Points (a) and (c) were, in principle, proved by Weyl himself. Point (b) turned out to be most difficult, it requires 'a priori estimates' for non-linear elliptic equations of the form

$$\frac{\partial z}{\partial u \partial u} \frac{\partial z}{\partial v \partial v} = \left(1 + \left(\frac{\partial z}{\partial u}\right)^2 + \left(\frac{\partial z}{\partial v}\right)^2\right) K(u, v.)$$

This was precisely this theory which developed due to efforts of distinguished scientists, among others, S. Bernstein, E. Hopf and L. Nirenberger, E. Heinz (1962.) Point (c) is proved with the help of the Koebe uniformization theorem. To prove point (a), one needs a strong inverse function theorem (so-called implicit function theorem in Banach space), in this case the theorem on solvability of Darboux equation (a particular case of Monge-Ampére equation) which plays a more and more important role in differential geometry, complex analysis and general relativity (cf. works of Yau which I will mention later.)

2.2 Riemannian manifolds as metric spaces (Hopf-Rinow). Geodesic completeness

As we know, among Hausdorff spaces, metric spaces play a particularly important role. Such are Euclidean, Hilbert, Banach spaces. Let us recall that metric space $(X.d)$ is complete if any Cauchy sequence (also called a converging sequence) has a limit in X. The space $(\mathbb{Q}, |\,|)$ of rational numbers with standard distance (modulus of difference) is not complete. The Meray-Cantor procedure of construction of the space of real numbers $(\mathbb{R}, |\,|)$ can be extended to any metric space (X, d), and such constructed complete metric space $(\tilde{X}, \tilde{d})$ is called a completion of (X, d). The natural embedding $i : X \to \tilde{X}$ assigns to any $x \in M$ a class of stationary sequences $x, x, x, \ldots$ and makes it possible to regard X as a dense subset in $\tilde{X}$. It has been known already for a long time that geodesics on Riemannian manifold (M, g) are locally the shortest curves. A construction of a distance in (M, g) is self-suggesting.

Let us denote by $\Omega(p, g)$ the set of piecewise smooth curves connecting p and q. For any curve $c \in \Omega(p, q)$ we define its length by the formula

$$l(c) := \sum_{i=1}^{k-1} \int_{t_i}^{t_{i+1}} g(\dot{c}(t), \dot{c}(t))^{1/2} dt,$$

where $[t_i, t_{i+1}]$ is the smooth segment of the curve. Let us denote

$$(*) \qquad d(p, q) := \inf(l(c) : c \in \Omega(p, q)).$$

It is easy to see that the function d is a distance on the set M. The fundamental theorem of celebrated geometer H. Hopf (one of the founders of modern topology) and his pupil W. Rinow formulated for two-dimensional manifold (M^2, g) holds, without alternations, for arbitrary (M^n, g), $n \geqslant 0$. This theorem reads

THEOREM (HOPF–RINOW).

A. *Topology defined by the distance (*) is identical with the initial topology of the differential manifold M.*

B. *The following four conditions are equivalent:*

1. *The metric space (M, d) is complete.*

2. *Any geodesics c on the manifold* (M, g) *can be arbitrarily extended, therefore, the set of arguments t can be extended to* $\mathbb{R}$.
3. *For any* $x \in M$, *the exponential mapping* $\exp_x : T_xM \to M$ *is a surjection.*
4. *Any subset* $K \subset M$ *which is d-bounded (that is, belongs to some d ball) possesses compact closure.*

C. *If at least one of the conditions 1–4 is satisfied, then arbitrary two points* $p, q \in M$ *can be joined by geodesics c such that* $l(c) = d(p, q)$.

The property 2. is called a *geodesic completeness* of the space (M, g); the equivalence $1 \equiv 2$ is very surprising, both these notions of completeness are defined very differently! Since completeness of metric space has important consequences, it is not surprising at all that this property is being assumed in many deep theorems on Riemannian manifolds. There is an immediate

COROLLARY. *Any compact Riemannian manifold is complete.*

REMARK 1. The notion of geodesic completeness requires only the notion of geodesics defined to be a curve satisfying differential equation $\nabla_{\dot{c}}\dot{c} = 0$. It can be therefore defined without alternations on any manifold (M, ∇) equipped with linear connection ∇.

REMARK 2. For pseudo-Riemann manifolds, for example, those appearing in general theory of relativity, points 1. and 2. are not equivalent; also the corollary above does not hold. This gives rise to interesting phenomena absent in Riemann spaces.

However, we have important

THEOREM (CARTAN–AMBROSE–HICKS). *Let the manifold* (M, ∇) *with linear connection* ∇ *and curvature tensor* $R \equiv R^{\nabla}$ *be geodesically complete and simply connected.*

Then the following conditions are equivalent.

(i) $\nabla R = 0$ *and the torsion tensor* $T = 0$ *(symmetry of* ∇*);*

(ii) *for any* $x, y \in M$, *the linear isomorphism* $f : T_xM \to T_yM$ *transforming* R_x *into* R_y *can be extended to a diffeomorphism* φ *of the space M which preserves* ∇*;*

(iii) *for any* $x \in M$, *the geodesic symmetry* $s_x : \exp_x(Z) \to \exp_x(-Z)$ *is a restriction to* $\exp_x(T_xM)$ *of some diffeomorphism of* M *preserving connection* ∇.

We will immediately present an important corollary due to Cartan, but before doing so, we must get acquainted with the most important class of Riemannian spaces, namely

2.3 Symmetric spaces

E. Cartan introduced this very important class of homogeneous Riemannian spaces and presented the basics of its theory.

DEFINITION. A Riemannian manifold (M, g), in general a manifold (M, ∇) with symmetric connection ∇, is *locally symmetric* if for any $x \in M$, the geodesic symmetry $s_x : \exp_x(Z) \to \exp_x(-Z)$ is an isometry of some normal neighborhood of the point x. M is (globally) symmetric if the local symmetry s_x can be extended to global isometry of M, more generally, to diffeomorphism preserving connection ∇.

The reader may formulate the corresponding three equivalent conditions for local symmetry, e.g., (i') $\nabla R = 0$; (ii') replacing φ in (ii) by local diffeomorphism.

In what follows, we will discuss only Riemannian spaces. The reader certainly guessed that the C-A-H theorem is due to Cartan in the case (M, g) and was generalized by Ambrose and Hicks to the case of manifold (M, ∇) with symmetric connection ∇.

As an example, we present the theorem which is fundamental for the theory of symmetric spaces.

THEOREM (CARTAN). 1. *((M, g) is locally symmetric)* $\Longleftrightarrow$ *($\nabla R = 0$).*

2. *Symmetric* (M, g) *is geodesically complete.*

3. *Simply connected, complete, and locally symmetric* (M, g) *is symmetric.*

4. *Symmetric* (M, g) *is homogeneous, that is,* $M = G/H$, *where* G *is an isometry group of* (M, g) *and* H *is an isotropy group of some point* $x \in H$ $=$ *{a set of isometries preserving* x*}.*

COROLLARY. *Connected spaces of constant curvature are locally symmetric.*

Using the theory of Lie algebras, Cartan gave also some other characteristics of symmetric spaces.

Point 4. of the theorem above suggest the question under which conditions the homogeneous space $M := G/H$ is symmetric? There is a quite general answer given by Cartan,

THEOREM (CARTAN). *If G is a semi-simple non-compact Lie group and H is some maximal compact subgroup of G, then the homogeneous space $M := G/H$ is symmetric.*

Here some clarifications are in order.

1. Let $\mathfrak{g} = LG$ be the Lie algebra of the group G, that is, the tangent space T_eG at the neutral element e of the group G; LG can be therefore identified with the space of left-invariant vector fields on G. Now we have a natural linear mapping $X \to [\ ,\ ]$ of the algebra LG denoted by ad.

Wilhelm Killing, the celebrated German geometer and the founder of the theory of semi-simple Lie algebras and precursor of Cartan and Weyl introduces on LG a bilinear symmetric form B, $B(X,Y) := \mathrm{tr}(\mathrm{ad}\, X \circ \mathrm{ad}\, Y)$ called a *Killing form.*

If the form B is non-degenerate, the group G and its algebra LG are called *semi-simple.* It can be shown that then LG is a direct sum of its simple ideals (that is, ideals not possessing nontrivial subideals), and vice versa, this latter condition is sufficient for semi-simplicity of LG. This clarifies the term 'semi-simple'.

2. When G is a semi-simple non-compact group, then the algebra $\mathfrak{g} = LG$ decomposes into direct sum $\mathfrak{g} = \mathfrak{p} \oplus \mathfrak{h}$, where $B|\mathfrak{p}$ is positive definite, and the restriction $B|\mathfrak{h}$ is negative definite. The subalgebra $\mathfrak{h}$ is the Lie algebra of the maximal compact subgroup H. The following important inclusions

$$[\mathfrak{h},\mathfrak{h}] \subset \mathfrak{h}, \quad [\mathfrak{h},\mathfrak{p}] \subset \mathfrak{p}, \quad [\mathfrak{p},\mathfrak{p}] \subset \mathfrak{h}$$

hold.

3. Therefore, $B|\mathfrak{p}$ defines a Riemannian metric g on $G/H = M$ in a natural way, and Cartan theorem says that (M, g) is a symmetric space.

DEFINITION. The spaces $M = G/H$ in theorem above are called *symmetric spaces of non compact type.*

The important examples of these spaces are our old friends, the disc $\mathbb{D} \subset \mathbb{C}$ and the upper half-plane $\mathfrak{H}_1 = G/H$, where $G = \mathrm{Sp}(2,\mathbb{R})$, $H = G_{(i)} = \mathrm{SO}(2)$.

The disc $\mathbb{D}$ is an example of a bounded region in $\mathbb{C}$ on which the group of holomorphisms acts transitively. E. Cartan undertook deep investigations of the regions $\mathbb{D} \subset \mathbb{C}^n$ which are symmetric spaces with hermitian structure. The important ingredient here is a good hermitian metric, that is, the metric invariant with respect to holomorphisms. The beautiful construction of such metric is due to Stefan Bergman. This metric is, moreover, Kählerian, so we can get full satisfaction! Let us present necessary notions.

Hermitian symmetric spaces are beautiful indeed; these are the complex manifolds (M,h), whose tangent spaces possess hermitian scalar product h_x. Therefore (T_xM, h_x) is a finite dimensional Hilbert space over $\mathbb{C}$. Then the decomposition of h into real and imaginary part $h_x = \operatorname{Re} h_x + i \operatorname{Im} h_x$ gives the Riemannian metric $g_x := \operatorname{Re} h_x$. Moreover, $\omega := \operatorname{Im} h$ defines a symplectic structure $\omega(X,Y) = -\omega(Y,X)$ which is skew-symmetric; the 2 form $\omega = \operatorname{Im} h$ is non-degenerate. Thus the hermitian metric on M has a very restrictive structure. Even more special is

DEFINITION. *Kähler manifold* is a hermitian manifold such that the 2 form $\omega := \operatorname{Im} h$ is closed, that is, $d\omega = 0$.

Of course, Riemann surfaces are Kähler manifolds. Martin Kähler introduced his spaces which soon became indispensable objects of complex and algebraic geometry in his short note *Über eine bemerkenswerte hermitesche Metrik* in 1932. We will return to these spaces later.

The examples below show the importance of Kähler manifolds, however one can ask why are they so distinguished among (complex) Riemannian manifolds? As we know, the main reason behind the beauty and importance of Riemann spaces is that they approximate (to first order) the 'given by God' Euclidean spaces. Kähler manifolds do even better, they approximate Euclidean spaces to the second order.

Another equivalent characteristics of Kähler metric will be presented later. Let us now present the most important class of Kähler spaces, these are

2.4 Bounded regions in complex plane. Bergman metric (for the first time)

Let D be a bounded region in $\mathbb{C}^n$, μ a Lebesgue measure in $\mathbb{R}^{2n}$ ($\simeq \mathbb{C}^n$.) Then the space $L^2(D)$ of complex μ-square integrable functions is a Hilbert space with hermitian scalar product

$$(f|g) := \int_D f(z)\overline{g(z)}d\mu(z), \quad \|f\|^2 := (f|f.)$$

Let $A^2(D)$ be a subset of holomorphic functions on D belonging to $L^2(D)$. It turns out that on $A^2(D)$ the convergence with respect to the norm $\|\cdot\|$ imply uniform convergence on any compact subset $K \subset D$

$$|f|_K := \sup_{z\in K} |f(z)| \leqslant c(K)\|f\|.$$

REMARK. It can be even shown that $A^2(D)$ equipped with seminorms $|\cdot|_K$ with K running through compact sets is a nuclear space and

$$A^2(D) \subset L^2(D) \subset A^2(D)'$$

is a Gelfand triple (cf. K. Maurin, *General Eigenfunction Expansions*) which is a decisive property for the general theory of eigenfunctions expansion in the space $L^2(D)$.

We have the following important

THEOREM (S. BERGMAN). *Let $(e_j)_1^\infty$ be an arbitrary orthonormal basis in the Hilbert space $(A(D), \|\,\|)$. Then the series*

$$\sum_{j=1}^{\infty} e_j(z)\overline{e_j(\zeta)} =: k(z,\bar{\zeta}) = k(\zeta,\bar{z}) \quad \text{hermitian symmetry}$$

is uniformly convergent on compact subscts.

The function $k(\cdot,\cdot)$ is independent of the choice of the orthonormal basis and has reproduction property: for any $z \in D$ and any $f \in A(D)$,

$$f(z) = \int_D f(\zeta)k(z,\bar{\zeta})d\mu(\zeta), \quad f \in A^2(D).$$

The function $z \to k(z,\zeta)$ belongs to $A^2(D)$.

DEFINITION. The function k is called the *Bergman kernel.* It plays a very important role in complex analysis. With the help of Bergman kernel, one defines on D a hermitian metric

$$h := \sum_{j,k=1}^{n} \frac{\partial \log k(z,\bar{z})}{\partial z_j \partial \bar{z}_k} dz^j d\bar{z}^k, \quad g := \operatorname{Re} h$$

called the *Bergman metric.*

PROPOSITION (KÄHLER–BERGMAN). *The Bergman metric h on D is Kählerian, that is, the 2 form $\omega := \operatorname{Im} h$ is closed, $d\omega = 0$.*

The Bergman metric has another beautiful property, namely it well behaves under holomorphic change of coordinates:

THEOREM (BERGMAN). *Let D and D' be bounded domains in $\mathbb{C}^n$ and let g and g' be metrics on D and D', respectively. Then every biholomorphism $\varphi : D \to D'$ is an isometry of Riemann spaces $(D,g) \xrightarrow{\varphi} (D',g')$.*

In particular, the group of holomorphisms of domain D $\operatorname{Hol}(D)$ is a subgroup of isometries of the space (D,g). When the group $\operatorname{Hol}(D)$ acts transitively on D, then the bounded domain D is called *homogeneous.*

It is known that the Ricci curvature *Ric* plays an important role in Riemannian geometry. For bounded domains we have surprising facts:

PROPOSITION. *Let (D,g) be a bounded and homogeneous domain. Let k be a Bergman kernel and g a metric. Then*

$$k(z,\bar{z}) = c\det(g_{ij})^{1/2}, \quad Ric = 2g.$$

We will return to these problems later in chapter devoted to Kähler spaces.

Already in 1935 E. Cartan in his work *Sur les domaines bornes homogenes de l'espace de n variables complexes*, Abh. Math. Sm. Unv. Hamburg 11 1935), 116–162, posed a fundamental question, namely, if homogeneous and bounded domains in $\mathbb{C}^n$ are symmetric spaces? The answer is in the affirmative for $n = 1,2,3$. In 1959, Piatecki-Szapiro presented counterexamples for $n = 4,5$.

Bounded symmetric domains in $\mathbb{C}^n$ seem to be a very special hermitian symmetric spaces of non-compact type. However, it turns out that these are all such spaces. This surprising fact was partially proved by Cartan in the paper in 1935. The full proof of this deep theorem is due to Harish-Chandra and result from his gigantic investigation on representations of semi-simple groups (1956):

THEOREM (HARISH-CHANDRA). *(a) Every bounded symmetric domain* $(D, g) \subset \mathbb{C}^n$ *is a hermitian symmetric space of non-compact type (that is,* $D = G/H$*, where G is a non-compact Lie group.) In particular, D is simply connected.*

(b) Let M be a hermitian symmetric space of non-compact type. Then there exists a bounded symmetric domain $D \subset \mathbb{C}^n$ *and holomorphic diffeomorphism of M onto D.*

Let us finish this chapter presenting the most important example of symmetric bounded domain $D \subset \mathbb{C}^n$, namely

2.5 Siegel half-space and Siegel disc

We saw that for the Riemann surface M^2 of genus $p > 1$, the universal covering space is an upper half-plane $\mathfrak{H}_1$ which isometric (biholomorphically) with the disc $\mathbb{D} = \mathbb{D}_1 \subset \mathbb{C}$. Both these two-dimensional manifolds are equipped with hermitian metrics discovered by Poincaré. It is not hard to show that these metrics are Bergman metrics. One of the greatest mathematicians of this century Carl-Ludwig Siegel in his fundamental papers on symplectic geometry, complex analysis of several variables, and theory of ϑ and automorphic functions introduced a natural generalization of the space $\mathfrak{H}_1 \simeq \mathbb{D}_1$ in n dimensions $\mathfrak{H}_n \simeq \mathbb{D}_n$. This spaces appear in the theory of abelian varieties, complex n-dimensional tori. It was already Riemann who started investigations of these tori. With any compact Riemann surface X of genus p one can associate a p-dimensional torus $Jac(X)$ called a *Jacobian* of the surface X. The construction of $Jac(X)$ proceeds as follows (we present it here because it makes it easier to understand the Hodge-Kodaira-de Rham theory of the next section.)

Let p be a genus of the surface X. Let α be a closed curve (cycle) on X which is noncontractible (Fig. 8), and let $\omega_1, \ldots, \omega_p$ be a basis of

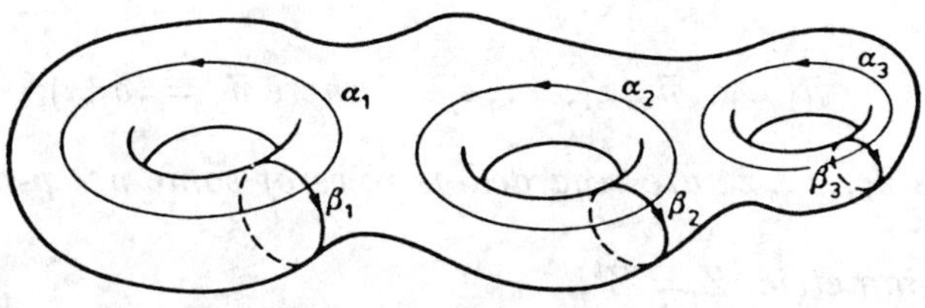

Fig. 8.

holomorphic one forms (of first kind): $\omega_j \in A^{1,0}(X)$, $j = 1, \ldots, n$. The vectors $\vec{z} = (z_1, \ldots, z_p)$, where $z_j := \int_\alpha \omega_j$, $1 \leqslant j \leqslant p$ (z_j is called *a period of the form* ω_j on the cycle α) form a subgroup $L \in \mathbb{C}^p$. In the theory of Riemann surfaces one proves that the group (of periods) L form a lattice (in $\mathbb{C}^p$), that is,

a) L is discrete, that is, there exists a neighborhood U of zero in $\mathbb{C}^p$ such that $L \cap U = \{0\}$;

b) L is not contained in any (proper) subspace in $\mathbb{C}^p$.
REMARK. Both these conditions defining lattice L in $\mathbb{C}^p$ are equivalent.

c) There exist $2p$ vectors $l_1, \ldots, l_{2p} \in \mathbb{C}^p$ such that $L = Zl_1 + \cdots + Zl_{2p}$.

DEFINITION. A *Jacobian* Jac(X) of a Riemann surface X is a torus Jac$(X) := \mathbb{C}^p/L$, where L is a group of periods of the surface X related to the basis $\omega_1, \ldots, \omega_p$.

Any Jacobi torus Jac(X) is an algebraic variety, that is, it can be holomorphically imbedded in some complex projective space $\mathbf{P}^N(\mathbb{C})$. The tori which are algebraic varieties are called *Abel varieties*; they have the form $\mathbb{C}^p/\Gamma$, where the lattice Γ satisfies Riemann conditions a) and b.) There is an important characteristics of Abel varieties given by

THEOREM (SIEGEL). *(Complex, compact manifold M of complex dimension n is abelian) $\Leftrightarrow$ (on M there exist n algebraically independent (and non-constant) meromorphic functions).*

THEOREM (RIEMANN). *For the basis of 1 forms of first kind $\omega_1, \ldots, \omega_p$ appropriately chosen, one can find a canonical Z-basis of the group L in the form*

$$\vec{n}_1, \ldots, \vec{n}_p, \vec{z}_1, \ldots, \vec{z}_p, \quad \text{where } \vec{n}_j = (\delta_{jk}.)$$

If the vectors $\vec{z}_1, \ldots, \vec{z}_p$ are regarded as rows of some $p \times p$-matrix Z, then

a) *Z is symmetric, $Z = Z^t$;*

b) *imaginary part of Z is positive definite ($\operatorname{Im} Z > 0$.)*

The matrix Z depends on choice of the basis, for different basis, one obtains a modular transformation

$$Z \to \frac{AZ + B}{CZ + D},$$

where A, B, C, D are $p \times p$-matrices with integer entries, and the $2p \times 2p$-matrix $\gamma := \begin{bmatrix} A & B \\ C & D \end{bmatrix} \in \mathrm{Sp}\,(2n, \mathbb{Z})$, that is, matrices γ form a discrete subgroup Γ_p of the symplectic group $\mathrm{Sp}\,(2n, \mathbb{R})$. The group Γ_p is called the *Siegel modular group* and, similarly to the symplectic group $\mathrm{Sp}\,(2n, \mathbb{R})$, this group acts on the space $\mathfrak{H}_n$ being (*ex definitione*) a set of complex $p \times p$-matrices satisfying a) and b), that is, symmetric and having positive definite imaginary part. This is the famous *Siegel half-space*. For $p = 1$, we have an upper half-plane $\mathfrak{H}_1$, and $\Gamma_1 = \mathrm{Sp}\,(2, \mathbb{Z})$ is the famous *modular group* playing an important role in the theory of elliptic functions.

With any compact Riemann surface X (of genus p) and, therefore, with any algebraic function, one can associate a point in the quotient space $A_p := \mathfrak{H}_p/\Gamma_p$.

Two biholomorphically equivalent surfaces correspond to the same point of the space A_p. Let us denote by $\mathcal{M}_p$ a set of equivalence classes of Riemann surfaces of genus p. We have therefore the mapping $\mathcal{M}_p \to A_p$. The famous *Torelli theorem* says that $\mathcal{M}_p \to A_p$ is an injection.

THEOREM (TORELLI). *If the matrices of periods Z and $\tilde{Z}$ of two compact Riemann surfaces are related by a modular transformation $\gamma : Z \to \tilde{Z}$, $\gamma \in \Gamma_p$, then the surfaces are equivalent. In other words, the mapping $\mathcal{M}_p \to A_p$ is an imbedding (injection).*

The famous *Riemann–Schotsky problem* is to characterize those abelian varieties $\mathbb{C}^p/\Gamma_p$, that are Jacobians of compact Riemann surfaces. This problem, almost a hundred years after being posed, is almost solved today. It is clear that problems like that cannot have a single solution: not all mathematicians are satisfied with the 'solution' presented by Tokahiro Shiota in the paper *Characterization of Jacobian varieties in terms of soliton equations,* Invent. Math. **83** (1986), 333–382. On the first sight, it may appear to be strange, that soliton waves (solitons) described by non-linear differential equations of Kadomcew-Petashwilli have anything to do with abelian tori. But it was a brilliant mathematician from Moscow, Novikow who put forward such a hypothesis, which, in turn, was proved by Shiota in a very ingenious way. After this historical 'justification' of the relevance of Siegel spaces, we present one more introductory theorem of the Siegel theory.

There is an exact analogue of the one-dimensional theorem

THEOREM (SIEGEL).
1. *The symplectic group $Sp(2p,\mathbb{R})$ acts transitively on $\mathfrak{H}_p$, where the action is defined by*

$$\mathfrak{H}_p \ni Z \to g\cdot Z := \frac{AZ+B}{CZ+D} \in \mathfrak{H}_p, \quad \textit{where } \gamma = \begin{bmatrix} A & B \\ C & D \end{bmatrix} \in Sp(2p,\mathbb{R}),$$

A, B, C, D are $p\times p$-matrices.
2. *The stabilizer K of the point $\sqrt{-1}\cdot 1_p$ is a compact subgroup $K = Sp(2p,\mathbb{R})\cap SO(2p)$,*
3. *$G_p \simeq Sp(2p,\mathbb{R})/K$.*

Of course, 3. follows from 1. and 2.

Like an upper half-plane $\mathfrak{H}_1$ which is biholomorphically (and isometrically) equivalent to a disc $\mathbb{D}\subset\mathbb{C}$, $\mathfrak{H}_p$ is biholomorphically equivalent to a disc $\mathbb{D}_n\subset\mathbb{C}^n$, $n=\frac{1}{2}p(p+1)$ defined to be a set of (all) complex symmetric $p\times p$-matrices Z $(=Z^t)$ such that $1_p-\bar{Z}Z$ is strictly positive definite (that is, $\bar{Z}Z<1_p$.) Therefore, $\mathbb{D}_p$ is indeed a bounded domain in $\mathbb{C}^{\frac{1}{2}p(p+1)}$. The biholomorphism $\mathfrak{H}_p\to\mathbb{D}_p$ is given by the generalized Cayley transformation $Z\to(1_p+iZ)/(1_p-iZ)$. Since $\mathbb{D}_p$ is a bounded symmetric domain, it possesses a Bergman metric. This metric can be obtained in yet another way: since the group $\mathrm{Hol}(\mathbb{D}_p)$ of holomorphisms of the domain $\mathbb{D}_p$ $(=\mathrm{Sp}\,(2p,\mathbb{R}))$ is semi-simple, the Killing form B of the Lie algebra $\mathrm{LHol}(\mathbb{D}_p)$ $(=\mathrm{LSp}\,(2p,\mathbb{R}))$

defines an invariant metric on $\mathbb{D}_p \simeq \mathrm{Hol}(\mathbf{D}_p)/K$ which is identical with the Bergman metric. Siegel found this metric explicitly. We see that $\mathfrak{H}_p$ and $\mathbb{D}_p$ are hermitian symmetric spaces of non-compact type.

In 1951, Chow showed that any holomorphic submanifold of complex projective space $\mathbb{P}^n(\mathbb{C})$ is an algebraic variety[1]. In 1954, Kunihiko Kodaira presented an intristic definition of algebraic varieties: these are such compact Kähler manifolds X that the Kähler 2 form $\omega \in H^2(X, \mathbb{Z})$ (see below.) The theory of Bergman kernels makes it possible to show that any compact complex manifold of the form D/Γ, where D is a bounded domain in $\mathbb{C}^n$ and Γ is a subgroup of $\mathrm{Hol}(D)$ acting properly discontinuously, is algebraic.

We will return to these problems later, one should constantly return to great ideas.

2.6 Jacobi fields once again. Focal points

As we know, geodesics on Riemann manifold (M, g) are extremals of the functional $c \to l(c)$ of the length (of the curve.) The differential equations describing geodesics cannot tell if the curve is indeed minimal. As it is in the case of a function of one variable, the second differential gives only the necessary condition for a minimum. Already an example of the sphere $S^2 \subset \mathbb{R}^3$ clearly shows this situation. It was several dozen years from the birth of geodesics theory, before in 1838 Jacobi in his short but fundamental paper made a substantial step on the way towards sufficient conditions: this step consists of considering a family of geodesics close to the given geodesics c. It also become a beginning of the theory of geodesic fields of Weierstrass, Hilbert, Carathéodory.

It has been thought for a long time that the curvature of a surface M^2 is related to the fields of geodesics. The famous

Jacobi equations show this relation explicitly. Let $c : [0, 1] \to M$ be the geodesics of interest. Let us encircle it with the family of geodesics $t \to a(t, u)$ where u is the parameter of the family such that $a(t, 0) = c(t)$, $t \in [0, 1]$. Such family is called the *variation of c by geodesics* and the vector field $t \to W(t) := \frac{\partial a}{\partial u}(0, t)$ is called the *field of variations* along c.

Jacobi noticed that field W satisfies a second order differential equation. Similar equations are satisfied in general Riemann spaces.

[1]The proof will be presented in Part V

THEOREM. *Let R be curvature tensor of the Riemann space M.*

1. *Then every field of variations along geodesics c satisfies Jacobi equation*

$$(J) \qquad \frac{\nabla^2 Y}{t^2} + R\left(\frac{\mathrm{d}c}{\mathrm{d}t}, Y\right) = 0.$$

2. *And vice versa, every solution of equation (J), called by physicists geodesic deviation, is a field of variations along the given geodesics c.*

PROOF. 1. If a is a variation of c by geodesics, then $\frac{\nabla^2 \partial a/\partial t}{t} \equiv 0$, because $\partial a/\partial t$ is a tangent vector of some geodesics. Thus, using fundamental properties of the tensor R, we have

$$0 = \frac{\partial}{\partial u}\frac{\nabla}{t}\frac{\partial a}{\partial t} = \frac{\nabla}{t}\frac{\partial}{\partial u}\frac{\partial a}{\partial t} + R\left(\frac{\partial a}{\partial t}, \frac{\partial a}{\partial u}\right)\frac{\partial a}{\partial t} =$$

$$= \frac{\nabla^2}{t^2}\frac{\partial a}{\partial u} + R\left(\frac{\partial a}{\partial t}, \frac{\partial a}{\partial u}\right)\frac{\partial a}{\partial t}.$$

□

DEFINITION. A vector field Y along geodesic c satisfying (J) is called the *Jacobi field.* Points p and q on geodesic c are called *focal* if there exists a Jacobi field Y along c which vanishes at p and q. A *multiplicity of focal points* p and q is dimension of the vector space of all such Jacobi fields. Since (J) is a second order differential equation on n-dimensional differential manifold M, the (linear) space $J(c)$ of Jacobi fields on c is $2n$-dimensional.

Now we can sketch the proof of point 2. of the theorem. As we know, geodesics are locally the shortest lines and two sufficiently close points can be joined by a unique minimal geodesic. This geodesic depends on its endpoints in a differentiable way. It is not hard now to construct a Jacobi field W (as a field of variations along $c|[0,\delta]$, where δ is small) with given values in $t = 0$ and $t = \delta$. We build the variation $a(\cdot, u)$ of the geodesics c as a minimal geodesic joining $p(u)$ and $q(u)$ such as to have $p(0) = c(0)$ and $\partial p/\partial u(0)$ to be a given vector in $TM_{c(0)}$; analogously, $p(0) = c(\delta)$ with $\partial q/\partial u(0)$ being arbitrary. In this way, the formula $W \to (W(0), W(\delta))$ defines the linear mapping

$$A : J(c) \to TM_{c(0)} \times TM_{c(\delta)}.$$

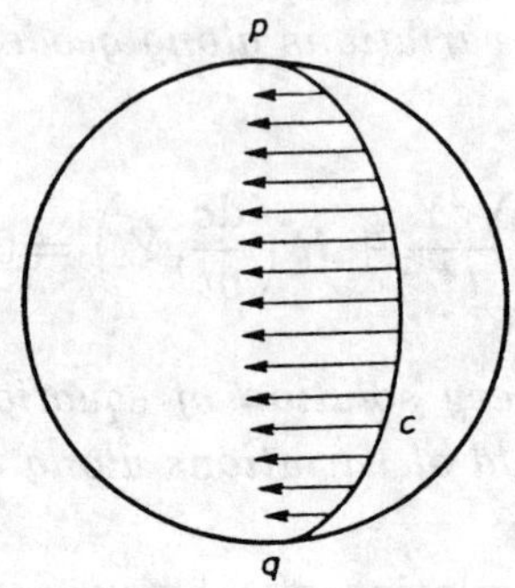

Fig. 9.

We have just showed that A is a surjection. Since both vector spaces have the same dimension $2n$, A is an isomorphism. Thus point 2. is proved. □

EXAMPLE. $M = S^n$. Let p, q be two poles (antipodal points) of the unit sphere S^n (Fig. 9). The geodesic c joints points p, q. Turning S^n around with fixed p, q, we obtain a field of variations vanishing at p, q. Turning S^n in this way in $n-1$ directions (with fixed p, q), we obtain $n-1$ linearly independent Jacobi fields. Therefore p and q are focal points along c with multiplicity $n-1$. Therefore, focal points may be defined as points where differential of the exponential mapping

$$(d\exp_p)_{\tilde{c}(t_1)} : T_pM \to T_{c(t_1)}M$$

with $p = c(0)$, $\tilde{c}(t) := t\dot{c}(0)$, $t \geqslant 0$ is not a bijection. The following important theorem 'justifies' introduction of focal points:

THEOREM. *Let $c : [0,1] \to M$ be a geodesic in M, $\|\dot{c}(t)\| = 1$. Then*

(i) *If c does not have any focal points inside, then any other curve b sufficiently close to c and joining ends of c has the length $l(b) \geqslant l(c)$.*

(ii) *But also vice versa: if the geodesic c contains an inner focal point $c(t_1)$, $t_1 < 1$, then in any neighborhood of c there exist curves b joining $c(0)$ and $c(1)$ which are shorter than c, that is, $l(b) < l(c)$.*

REMARK. The proof of (i) is simple, but that of (ii) requires investigations of the second differential (so-called second variation) of the functional

$l(\cdot)$.

Now we can imagine the proof of the brilliant

THEOREM (MYERS, 1941). *Let (M^n, g) be a complete Riemann space such that its Ricci tensor Ric (the trace of curvature tensor R) satisfies $Ric(v,v) \geqslant (n-1)/r^2$, $\|v\| = 1$, where r is a positive constant. Then*

(i) *every geodesics on M^n of the length $> \pi r$ contains focal points and, therefore, is not minimal.*

(ii) *M^n is compact with diameter $\leqslant \pi r$.*

REMARK. (ii) follows immediately from (i.) Point (i) makes use of the Morse index theorem. We can only mention here this famous theorem which forms a basis of the so-called global calculus of variations and was the starting point of nonlinear functional analysis and the so-called Morse theory, to which many monographs are devoted. Here one should recommend a delightful and extremely clearly written monograph by celebrated John Milnor *Morse Theory*, Princeton University Press, 1963, which contains, among other things, the beautiful proof of famous Bott's *periodicity theorem* which also has major consequences.

Like in the case of functions of k variables, where a decisive role is played by second differential (being a quadratic form) and its index, in the theory of geodesics, the object of major importance is the second differential

$$(*) \qquad E_{**} : T_c\Omega \times T_c\Omega \to \mathbb{R}$$

of the functional of energy E of a curve, taken at its critical point c, that is, for the geodesic c. (As we know, geodesics are critical points (zeros of the first differential) of both the length functional and the energy E of curves on Riemann space M.) These functionals can be regarded as an infinite dimensional manifolds Ω of smooth curves on M, and one can equip Ω with the structure of Riemann-Hilbert manifold, that is, the tangent spaces of the manifold have the structure of (real) Hilbert space. Such an approach to Morse theory was independently developed by Palais and Smale. The bilinear functional E_{**} above is called the *Hessian* (of energy $\mathcal{E}$.)

THEOREM (MORSE). *The index λ of Hessian E_{**} (that is, a maximal dimension of the space $T_c\Omega$ on whose E_{**} is negative definite) is equal to the number (counted with multiplicities) of points $c(t)$, $0 < t < 1$ focal to the* *starting point $c(0)$ of the geodesics $c : [0,1] \to M$.*

COROLLARY. *An arc of geodesic $c : [0,1] \to M$ may contain only finite number of points focal to $c(0)$.*

Let us turn to further

EXAMPLES OF JACOBI FIELDS. On the surface M^2 of constant curvature $K = K_0$, Jacobi equation (J) takes the form $\ddot{y}(t) + Ky(t) = 0$. We are interested in solution with initial condition $y(0) = 0$, $\dot{y}(0) = 1$. We immediately have

PROPOSITION. *If $K_0 > 0$, then $y(t) = \sin(t\sqrt{K_0})$. If $K_0 = 0$, then $y(t) = t$. If $K_0 < 0$, then $y(t) = \sinh(t\sqrt{K_0})$.*

Thus, focal points exist only for $k_0 > 0$. For $K_0 \leqslant 0$, geodesics (from $c(0)$) disperse.

The reader could visualize this phenomenon taking as examples the upper half-plane $\mathfrak{H}_1$ $(= \mathcal{H}^2)$ or the disc $\mathbb{D}$ with Poincaré metrics.

When we consider geodesic flow g_t on a surface M of negative curvature and eject from a (small) region $U \subset M$ geodesics $U_t = \{g_t x : x \in U\}$, for $t \to \infty$, the region will 'disperse'. This phenomenon is a starting point of the beautiful *ergodic theory on surfaces of negative curvature* founded by Eberhardt Hopf (cf. his beautiful monograph *Ergodentheorie*) and Hedlund. The reader can find modern approach in the textbooks by outstanding Moscow mathematicians Sinai and others.

While talking about complete surfaces of negative curvature, one should mention a famous philosophical fact which is

THEOREM (HILBERT). *A complete manifold (M^2, g) of negative curvature cannot be imbedded into the space $\mathbb{R}^3$.*

Thus, the pictures we make to illustrate $\mathfrak{H}_1$, or Riemann surfaces of the genus > 1 (having curvature $K = -1$ and inherited from $\mathfrak{H}_1$: $M^2 \simeq \mathfrak{H}_1/\Gamma$) cannot be isometrically imbedded in $\mathbb{R}^3$. Our pictures and models have therefore different, not natural metric!

The situation is completely different when we discuss compact manifolds (M^2, g) of positive curvature $K > 0$, or, as Weyl says, convex compact surfaces. Then the famous Weyl-Levy-Nirenberg theorem says that isometric embedding in $\mathbb{R}^3$ does exist and is a unique (up to Euclidean motion) oval $W^r \subset \mathbb{R}^3$.

The charm of Hilbert theorem is just the impossibility of imbedding into $\mathbb{R}^3$. No-go theorems are usually particularly interesting; recall, for example, laws of thermodynamics.

An analogous phenomenon takes place in a quite general situation, namely, for symmetric spaces, for whose (as we know) the curvature tensor is parallel, that is $\nabla R = 0$, and the class of whose contains spaces of constant curvature. In analogy with the proposition above, we have

THEOREM. *Let (M, g) be locally symmetric, that is, $\nabla R = 0$. Let $T, E_1, \ldots, E_n$ be an orthonormal basis for linear mapping*

$$(*) \qquad X \to R(T, X)T \quad for\ t = 0.$$

Then the Jacobi equation

$$(J) \qquad \nabla_T \nabla_T Y = R(T, Y)T$$

has constant coefficients and its solutions $Y(\cdot)$ vanishing at $t = 0$ (that is, $Y(0) = 0$), have the form

$$\sin(t\sqrt{\lambda})E(t), \;\; tE(t) \;\; \sinh(t\sqrt{-\lambda})E(t),$$

for $\lambda > 0$, $\lambda = 0$, $\lambda < 0$, where $E(t)$ is a parallel eigenvector of the mapping (*) corresponding to the eigenvalue λ.

The proof can be found in the book by J. Milnor.

CHAPTER 3

Cohomology of Riemann spaces. Theorems of de Rham, Hodge, Kodaira

We owe Riemann the first steps in algebraic topology of manifolds. The modest germs of Riemann ideas had grown, as a result of works of Klein, Poincaré, Brouwer, Lefschetz, Hopf, de Rham, Hodge, Kodaira, Leray, Serre, to mention only the names of the greatest, into big tree of the theory of (co)homology theory of Riemann manifolds. In this chapter, I will try to present the facts which are most spectacular and, at the same time, most clearly related to geometrical and analytical ideas of Riemann. These results are forever associated with the names of de Rham and Hodge. We owe Riemann a fundamental notion of the harmonic tensor field ω on M^2 and its period on a cycle c, that is, a close curve c on M^2: this is the integral $\int_c \omega$. We also owe Riemann the theorem on existence of a harmonic field of given periods on M^2.

The Hodge theory culminates in the corresponding theorem for oriented, compact Riemann spaces of arbitrary dimension. Clearly, we cannot describe fascinating and sometimes dramatic history of the birth of the theory of homology and cohomology, which shows once again the deep truth that great mathematical theories are delivered in, sometimes, long lasting pains.

3.1 Homology. Cohomology. De Rham cohomology

In mathematics, one often encounters the following situation. Let a sequence of abelian groups (moduli, etc.) be given

$$\{C^n, n \in \mathbb{Z}\}$$

along with homomorphisms $d_n : C^n \to C^{n+1}$ (called *differentials* or *homomorphisms of coboundaries*) for whose

$$d_{n+1} d_n = 0 \quad \text{(zero element of the group) for all n}$$

Then $K = \{C^n, d_n, n \in \mathbb{Z}\}$ is called a *(co)chain complex.*

Since $\operatorname{im} d_{n-1} \subset \ker d_n$, we can construct the quotient group

$$H^n(K) := \ker d_n / \operatorname{im} d_{n-1}$$

called the *n-th cohomology group of complex K.*

The dual notion is a complex of chains of abelian groups C_n

$$K_0 = \{C_n, \partial_n, n \in \mathbb{Z}\};$$

it is assumed that $C_n = 0$ for $n < 0$.

Let us consider homomorphisms $\partial_n : C_n \to C_{n-1}$ satisfying the relation

$$\partial_n \partial_{n-1} = 0 \quad \text{for all n.}$$

Since $\operatorname{im} \partial_{n+1} := \partial_{n+1}(C_n) \subset \ker \partial_n$, we can construct the quotient group

$$H_n(K_0) := \ker \partial_n / \operatorname{im} \partial_{n+1}$$

called the *n-th homology group of complex* K_0.

At this point one should say few words about

(Co)homology group valued in abelian group A. If A is an abelian group, e.g., a sheaf of germs of smooth or holomorphic functions, $K = \{C_n, \partial_n\}$ is a chain complex (of arbitrary polyhedron X), then one can construct the chain complex $K \otimes_{\mathbb{Z}} A = \{C_n \otimes_{\mathbb{Z}} A, \partial_n\}$ and the cochain complex $\operatorname{Hom}(K, A) := \{\operatorname{Hom}(C_n, A), d_n\}$. The group $H_n(K \otimes A)$ is denoted by $H_n(K, A)$; similarly, $H^n(\operatorname{Hom}(K, A))$ is denoted by $H^n(K, A)$. These are called groups of homology and cohomology with coefficients in group A. The most important example of cohomology is

De Rham cohomology. Let X be an oriented differential manifold (X does not need to be Riemannian), and let $C^r := \Lambda^r(X)$ be the space of differential r forms on X which, as we know, in local map are written as follows

$$\varphi = \sum f_{i_1,\dots,i_r} dx^{i_1} \wedge \cdots \wedge dx^{i_r}.$$

The exterior multiplication $\wedge$ and other notions of exterior algebra, also called Grassman algebra (e.g., elegant theory of determinants) were introduced by H. Grassman. E. Cartan observed that this notions can be extended into any cotangent bundle over differential manifold X. In this way, one obtains the so-called Cartan algebra of differential forms $\Lambda(X)$; this is an algebra with gradation, exterior differential r forms are denoted by $\Lambda^r(X)$. It was a great discovery of Cartan to introduce the exterior derivative $d : \Lambda^r(X) \to \Lambda^{r+1}(X)$ which satisfies the following axioms

1. when φ is a 0 form, that is, a function on X, then $d\varphi$ is its differential;
2. $d : \Lambda^r(X) \to \Lambda^{r+1}(X)$ is R-linear;
3. $d(d\varphi) = 0$, it is sufficient to assume that for 0 forms only;
4. if $\omega \in \Lambda^r(X)$, $\varphi \in \Lambda^p(X)$, then
$$d(\omega \wedge \varphi) = d\omega \wedge \varphi + (-1)^r \omega \wedge d\varphi.$$

It is easy to show that the operation d is 'given by God', that is, it does exist and, moreover, is a unique one satisfying properties 1–4. Even more, if $f : X \to Y$ is a differentiable mapping, $d_X(f^*\omega) = f^*(d_Y\omega)$ for any exterior form on Y. In coordinates, $d\varphi$ has the form

$$d\varphi = \sum df_{i_1,\dots,i_r} \wedge dx^{i_1} \wedge \cdots \wedge dx^{i_r}.$$

Both conditions of coboundaries are satisfied. Cohomology groups obtained in this way are called de Rham cohomology groups and denoted by $H^r_{dR}(X)$. The operation of external multiplication of differential forms turns a direct product of de Rham groups into a ring with gradation (and a superalgebra)

$$H^*_{dR} = \bigoplus_r H^r_{dR}(X.)$$

Now we present the fundamental 'example' of *homology.*

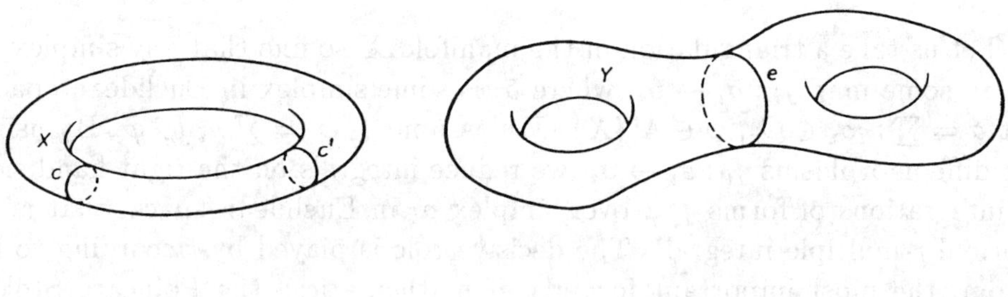

Fig. 10.

A *triangulation of manifold* X is an (appropriate) decomposition of X into simplices. A *complex* is a set of simplices with appropriately adjoining faces, and every point $x \in X$ belongs to finite number of simplices. As a topological space, the complex is defined by the set of its vertices, and by indicating which of them form simplices. A topological space which is homomorphic with a complex is called a *polyhedron* and the homomorphism is called its *triangulation.* With any complex X we can associate the chain complex $K = \{C_n, \partial_n; n \in \mathbb{Z}\}$. Here $C_n = \oplus \mathbb{Z}\sigma_i$ is a free $\mathbb{Z}$-module, whose generators are n-dimensional simplices σ_i. Each of these simplices is being oriented by the choice of its vertices. $\sigma_i = \overset{n}{\sigma}_i = \{x_0, \dots, x_n\}$. Its boundary $\partial \overset{n}{\sigma}_i$ defined by $\partial_n \sigma_i = \sum_{k=0}^{n} (-1)^k \epsilon_k \sigma_i^k$, where $\sigma_i^k := \{x_0, \dots, x_{k-1}, x_{k+1}, \dots, x_n\}$ and $\epsilon_k = \pm 1$ depending on whether the permutation leading from $x_0, \dots, x_{k-1}, x_{k+1}, \dots, x_n$ to the series of vertices of σ_i^k which defines the orientation is even or not. One extends the homomorphism ∂_n to the whole group C_n by additivity.

Elements of the group $\ker \partial_n$ are called *cycles.*

Elements of the group $\operatorname{im} \partial_{\nu+1}$ are called *boundaries.*

One can check that $\partial_n \partial_{n+1} = 0$, and thus, every boundary is a cycle. The group $H_n(X)$ is called the n-th group of homology of the complex X. A geometric meaning of an element $\xi \in H_n(X)$ is that it is a closed part of the space X, and two such parts are being identified (that is, belong to the same homology class) if they both bound a $(n+1)$-dimensional peace. In Fig. 10, cycles c, c' define the same element of the group $H_1(X)$, and the close curve e defines the zero element of the group $H_1(Y)$.

A relation between both these 'dual examples' is given by integration of differential forms on chains.

Let us take a triangulation of the manifold X so fine that any simplex σ_i lies in some map $f_i : \bar{\sigma}_i \to \sigma_i$, where $\bar{\sigma}_i$ is some simplex in Euclidean space. Let $\sigma = \sum n_i \sigma_i \in C_r$, $\varphi \in \Lambda^r(X)$. We assume $\int_\sigma \varphi := \sum n_i \int_{\sigma_i} \varphi$. By using the diffeomorphisms $f_i : \bar{\sigma}_i \to \sigma_i$, we reduce integrals on the right-hand side to integrations of forms $f_i^* \varphi$ over simplex σ_i in Euclidean space, that is to 'normal r-multiple integral'. The decisive role is played by, according to R. Thom, the most important formula of mathematics: the Poincare–Stokes formula

$$(P - S) \qquad < \partial c_r, \overset{r-1}{\varphi} >:= \int_{\partial \sigma_r} \overset{r-1}{\varphi} = \int_{\sigma_r} d \overset{r-1}{\varphi} =< c_r, d \overset{r-1}{\varphi} > .$$

The product $< c, \varphi >$ can be extended to $c \in C_r \times \mathbb{R}$ and then (P-S) expresses duality (adjointness) of the boundary operator in the space $C_r \times \mathbb{R}$ and the operator of differential $d\omega \in \Lambda^r(X)$. It follows immediately from the $(P - S)$ theorem that the product $< c, \varphi >= 0$ if $\partial c =$ or $d\varphi = 0$; if $c = \partial c'$ or $\varphi = d\varphi$. This means that the product $< c, \varphi >$ extends into cohomology and homology classes, that is, the spaces $H_r(X, \mathbb{R})$ and $H^r_{dR}(X)$. We have the famous

THEOREM (DE RHAM).
1. $< c, \varphi >$ *defines a duality between spaces* $H_r(X, \mathbb{R})$ *and* $H^r_{dR}(X)$, *thus the spaces* $H_r(X, \mathbb{R})$ *and* $H^r_{dR}(X)$ *are isomorphic;*
2. *for compact* X *these spaces are finite dimensional.*

REMARK. When $\partial c_r = 0$, that is, when c_r is a cycle, and $d\varphi = 0$, that is if φ is a closed form, the integral

$$< c_r, \varphi >:= \int_{c_r} \varphi$$

is called the *period of form* φ *on cycle* c_r and the mapping $\varphi \to \int_{c_r} \varphi$ is called the *period mapping*.

The theorem above contains

FIRST DE RHAM THEOREM. *When an orientable differential manifold* X *is compact and we have a basis of r-cycles* $\overset{1}{c}, \ldots, \overset{n}{c}$, $n = \dim H^r(X, \mathbb{R})$, *then there exists a closed form* φ, *which takes on these cycles given values.*

REMARK. De Rham theorem does not assume the existence of a Riemann structure on X.

As we will see in a moment in his theory of harmonic forms Hodge made the first de Rham theorem more precise, showing that on a compact Riemann space X there always exists a harmonic r form which on given cycles has definite periods.

This wonderful Hodge theorem comprises therefore the full generalization of the famous Riemann theorem on existence of harmonic 1 form with given periods on given closed curves on a compact Riemann surface X^2. In order to formulate Hodge theorem and to sketch the proof, we need to introduce some fundamental notions in which a Riemann metric tensor g_{ik} plays an important role.

3.2 Hodge theory of harmonic forms

As we have seen, the spaces of de Rham cohomology are 'natural' and for an expert in analysis they comprise the easiest route to topology of manifolds. As we know, on arbitrary differential manifold M one can introduce (in many ways) a Riemann structure, that is a symmetric, positive 2 form g. As Hodge, inspired by Riemann, observed in 1930th, on such an enriched manifold $M^n \equiv M$ one can introduce a notion of harmonic forms. This is done in two stages.

I. Duality operator $*$: $\Lambda^r(M^n) \to \Lambda^{n-r}(M^n), r \leqslant n$. If $e_1, \ldots, e_n$ is an orthonormal basis in Euclidean space, $*(e_{i_1} \wedge \cdots \wedge e_{i_r}) := \pm e_{j_1} \wedge \cdots \wedge e_{j_{n-r}}$, where $\{j_1, \ldots, j_{n-r}\}$ is a complement of $\{i_1, \ldots, i_r\}$ in $\{1, \ldots, n\}$; we have $(+)$ if $\{i_1, \ldots, i_r, j_1, \ldots, j_{n-r}\}$ is an even permutation of $\{1, \ldots, n\}$, and $(-)$ otherwise. For generic vectors, we define $*$ by linearity. Performing this procedure in all tangent spaces $T_x M^n$, we obtain the Hodge star $*$.

Since for any r forms $\omega, \varphi \in \Lambda^r(M^n)$, $\omega \wedge *\varphi \in \Lambda^n(M^n)$, we can integrate this expression, taking $(\overset{r}{\omega} \mid \overset{r}{\varphi}) = \int_M \overset{r}{\omega} \wedge * \overset{r}{\varphi}$ (assuming that M is compact and oriented) which has all properties of a scalar product on the space $\Lambda^r(M^n)$. We can define the operator δ adjoint to d:

$$(d \overset{r-1}{\omega} \mid \overset{r}{\varphi}) =: (\overset{r-1}{\omega} \mid \delta \overset{r}{\varphi});$$

therefore, $\delta : \Lambda^r(M^n) \to \Lambda^{r-1}(M^n)$.

One can easily check the following formulas collected in

THEOREM.
(i) $(*\omega| * \varphi) = (\omega|\varphi)$, *isometry of* $*$.
(ii) $\delta = (-1)^{r+r(n-r)} * d* = (-1)^{nr+n+1} * d*$.
(iii) $d(\omega \wedge *\varphi) = d\omega \wedge (*\varphi) - \omega \wedge (*\delta\varphi)$.

REMARK. The operation d is a natural generalization of rotation (rot), and δ of divergence (div) known from vector analysis of electrodynamics and hydrodynamics.

II. Harmonic forms. Hodge rediscovers the generalization of the Laplace-Beltrami operator, introduced earlier by Weitzenböck in 1923. This operator is now called the Hodge-Laplace operator and reads

$$(W) \qquad \Delta := -(d\delta + \delta d) = -(dd^* + d^*d)$$

Clearly, this operator maps $\Lambda^r(M^n) \to \Lambda^r(M^n)$, so, strictly speaking, we should write $\Delta = \Delta_r$, which we actually will do sometimes to avoid misunderstandings.

DEFINITION. An r form φ on compact oriented Riemann manifold M^n is *harmonic* if it satisfies the equation
(a) $\Delta\varphi = 0$, which is equivalent to the conditions
(b) $0 = d\varphi = \delta\varphi$.

Indeed, if $\Delta\varphi = 0$, then

$$0 = (-\Delta\varphi|\varphi) = (d\delta\varphi|\varphi) + (\delta d\varphi|\varphi) = (\delta\varphi|\delta\varphi) + (d\varphi|d\varphi),$$

and thus, $d\varphi = \delta\varphi = 0$. □

Therefore, the harmonic form φ is at the same time closed $d\varphi = 0$ and coclosed $\delta\varphi = 0$.

REMARK. For non-compact manifolds M, and also for manifolds with boundaries $\delta M \neq \emptyset$, conditions (a) and (b) are not equivalent; one should impose some boundary conditions or conditions at 'infinity'. These problems, which are important for applications in physics, for example, in hydrodynamics, are difficult and not fully understood.

HISTORICAL REMARK. The operator (W) is often incorrectly, that is, not in conformity with the history, called the Hodge-Laplace operator. In fact, this operator in all its generality (that is, for arbitrary Riemann space) has been introduced already in 1923 by R. Weitzenböck in his famous (but soon forgotten) monograph *Invariantentheorie*, Groningen, 1923. The operator δ was introduced there as well. Moreover, Weitzenböck computes the operator Δ in terms of covariant derivatives, this is the *Weitzenböck formula* (pp. 393–397 of the monograph.) This rather complicated formula involving, of course, components of curvature tensor will be presented below. Formulae of Weitzenböck were popularized by de Rham in his monograph *Varietes differntiables*, Paris 1955 comprising a systematic and comprehensive exposition of the theory which owes de Rham so much (theory of de Rham currents, theory of de Rham cohomology, harmonic forms etc.) The Weitzenböck formulas became an indispensable tool of differential geometry and complex analysis.

The following useful formulas for operators d, δ, Δ are easy to check

$$*\Delta = \Delta *, \quad d\Delta = \Delta d = d\delta d, \quad \delta\Delta = \Delta\delta = \delta d\delta.$$

3.3 Hodge decomposition

The famous Hodge decomposition

$$(H) \qquad \Lambda^r(M) = d\Lambda^{r-1}(M) \oplus d^*\Lambda^{r-1}(M) \oplus H^r(M),$$

where $H^r(M)$ denotes the (finite dimensional) space of harmonic n forms, corresponds to the classical decomposition of Euclidean space in which a linear operator $L : E \to E$ acts such that $L \circ L = 0$:

$$(H') \qquad E = LE \oplus L^*E \oplus H,$$

where $H = \{e \in E : Le = 0 = L^*e\}$.

(In the infinite dimensional case (where E is a Hilbert space) one should take completions $\overline{LE}$, $\overline{L^*E}$ because the operator L is not bounded; it is defined only on a linear dense set in E and (H′) should be replaced by

$$(H'') \qquad E = \overline{LE} \oplus \overline{L^*E} \oplus H.$$

We immediately note that if the first two terms are orthogonal, then

$$(Lx|L^*y) = (L^2x|y) = 0, \quad x, y \in E.$$

Let C be an orthogonal completion of $\overline{LE} \oplus \overline{L^*E}$. $H \subset C$, but for $x \in C$, we have

$$((Ly|x) = 0 \text{ for any } y) \Rightarrow (L^*x = 0)$$

Similarly $Lx = 0$, therefore $C \subset H$, and thus $C = H$.

In this way we showed the classical formula (H′) and 'almost' (H″.) What is important in the Hodge theorem, is the thesis that the space $H^r(M)$ of harmonic forms is finite dimensional. The proof of this deep fact requires 'elliptic engineering' (A. Weil), that is, not easy theorems from the theory of linear elliptic equations: equation $\Delta u = f$ belongs to this class! We will discuss them in a moment!

In linear algebra we have important

THEOREM. *Let V, W be finite dimensional Euclidean spaces and let $L : V \to W$ be linear. Then*

$$(Lx = y \ \textit{has solution}) \Leftrightarrow (y \perp \ker L^*),$$

because we have the orthogonal decomposition

$$W = LV \oplus \ker L^*, \quad \textit{thatis}, W = \operatorname{im} L \oplus KerL^*.$$

PROOF. $(Z^{\perp}LV) \Leftrightarrow$ (for any x, $0 = (Lx|z) = (x|L^*z) \Leftrightarrow (L^*z = 0)$. In other words, coker $L=$ ker L^*. □

At this point it is convenient to introduce a notion of *index* of mapping L, to wit,

$$\operatorname{ind} L := \dim \ker L - \dim \ker L^*.$$

The reader can show that ind $L = \dim V - \dim W$ (independently of L.) It turns out that in the case of elliptic operators a number of finite dimensional facts survives; this was a momentum of development of linear functional analysis.

The above observations suggest introducing

The Green operator. Let H be a perpendicular projection on H^r and let $(H^r)^{\perp}$ be the orthogonal completion of the subspace H^r in $W_s^2 = W_s^2(\Lambda^r(M))$. Then the linear mapping $G : W_s^2 \to (H^r)^{\perp}$, where $G\varphi$ is a solution of the equation $\Delta\omega = \varphi - H\varphi$, that is,

$$\Delta G\varphi = \varphi - H\varphi \quad (\Delta G = 1 - H)$$

is called the *Green operator* (in analogy with the Green function in potential theory.) Thus, the proof of Hodge theorem reduces to the proof (construction) of the Green operator G. Indeed, given G, we immediately obtain (recalling that $-\Delta = d\delta + \delta d$)

$$\varphi = d(\delta G\varphi) + \delta(dG\varphi) + H\varphi.$$

A very ingenious construction of the Green operator and the projection operator H is due to Milgram and Rosenbloom, who used

3.4 The method of heat transport (diffusion equation)

It seems that already Riemann himself dealt with this equation (on his surfaces.) Since in the last decades this method turned out to be very powerful, let me present its keynotes. The temperature $u(x,t)$, $x \in M$, $t > 0$ at the point x of the body M and at the moment of time t satisfies the following (parabolic) differential equation

$$(D) \qquad \frac{\partial u}{\partial t} - \Delta u = 0, \text{ if there are no sources of heat in } M.$$

If at the initial moment $t = 0$, $u(x,0) = f(x)$, $x \in M$, then the 'heat wave' travels under action of the semigroup $P(t) = e^{t\Delta}$, $t \geqslant 0$ of bounded operators in the Hilbert space $L^2(M)$. By 'semigroup', we mean the relation

$$P(t_1 + t_2) = P(t_1)P(t_2), \quad t_1, t_2 \geqslant 0, \; P(0) = 1.$$

The operator Δ is the *generator* (generating operator) of the semigroup $P(t)$, $t \geqslant 0$. Thus, the solution u of equation (D) is of the form $u(\cdot,t) = P(t)f(\cdot)$.

On transport phenomena and diffusion equation (heat transfer equation). If some physical system is away from thermodynamical equilibrium, then – as it is described by statistical physics (e.g., kinetic theory of gases) – as a result of motions and collisions of particles, we have a transport of mass, momentum, energy, etc. Macroscopically – as it is sometimes said – phenomenologically, these transport phenomena exhibit themselves in the form of

a) heat transfer,
b) viscosity,

c) diffusion.

Surprisingly enough, the kinetic theory of gases (Maxwell, Boltzmann, Gibbs) is capable of computing coefficients of the phenomena listed above. This was deservedly regarded as a triumph of the atomic theory of nature. At this point, it is worth commenting on the philosophy prevailing since Democritos till modern days saying that in order to understand a physical or mathematical phenomenon, one should 'decompose' it into elementary elements, a-thomos, and these indecomposable elements should be described, classified, and so on. In the case of mathematics these are prime numbers, simple groups and algebras, irreducible representations, irreducible symmetric spaces, etc. It is hoped that later, from these simple, harmonic vibrations – from these irreducible representations – by synthesis, integration, etc., one can reconstruct phenomena under investigation. This philosophy, as we know, was and often is productive, but, unfortunately, is also encumbered with a number of defects, and sometimes leads to fatal results if not supplemented by the complementary philosophy, namely, 'universalism', *Gestaltheorie*, organicism, etc. In these philosophies, the whole is primary and exhibits itself in its 'organs', 'members', specimen, and so on.

The transport phenomena a) – c) are irreversible, which on first sight seems to be not comparable e.g., with the kinetic theory of gases: the Newton equations are invariant with respect to the 'time reversal' $t \to -t$. The solution of this puzzle is that along with the Newtonian mechanics, one uses the rules of probability theory which results in this asymmetry.

Equations of heat transfer are derived from the law of heat conservation:

$$u = \int_\Omega \rho \vartheta c_w,$$

$$-\frac{\partial u}{\partial t} = -k \int_{\partial\Omega} \operatorname{grad} \vartheta = -k \int_\Omega \operatorname{div} \operatorname{grad} \vartheta = -k \int_\Omega \Delta\vartheta.$$

Thus

$$-\frac{\partial u}{\partial t} = -\int_\Omega \rho c_w \frac{\partial \vartheta}{\partial t} = -k \int_\Omega \Delta\vartheta$$

for any domain $\Omega \subset \mathbb{R}^3$. Therefore, the expressions under integrals must be equal

$$\frac{\partial \vartheta}{\partial t} = c\Delta\vartheta.$$

In the formulas above $\vartheta = \vartheta(x,t)$ denotes the temperature in point x at time t, $c = k/\rho c_w$, ρ is the density, and c_w is the specific heat.

It turns out that $P(t)$ is an integral operator with smooth kernel $K(x,y;t)$: $P(t)f(x) = \int_M K(x,y;t)f(y)dy$. The physical intuition says that when $t \to \infty$, then we obtain the stationary final distribution of temperature $P(t)f \to Hf$, $t \to \infty$ which is the harmonic distribution Hf. This happens indeed even in general case: Let M be a compact Riemann space and $\Delta = \Delta_W$ the Laplace-Weitzenböck operator. The elliptic theory says that the operator Δ, more precisely, its closure (we will denote both by the same symbol), is a positive, self adjoint operator whose spectrum is purely discrete $0 = \lambda_0 \leqslant \lambda_1 \leqslant \lambda_2 \leqslant \ldots$ tending to infinity. The eigenspaces are (therefore) finite dimensional $\Delta\varphi_k = \lambda_k\varphi_k$, $k = 0, 1, 2, \ldots$. If we denote $\varphi \otimes \psi(x,y) := \varphi(x)\psi(y)$, the operator $P(t)$ has the form

$$P(t) = \sum_{k=1}^{\infty} e^{-t\lambda_k}\varphi_{\lambda_k} \otimes \psi_{\lambda_k} \stackrel{t\to\infty}{\longrightarrow} \sum_j \varphi_{0,j} \otimes \psi_{0,j} =: H$$

and thus H is indeed the perpendicular projection on the eigenspace of the operator Δ corresponding to the eigenvalue $\lambda = 0$, and therefore, the perpendicular projection on H^r, we are looking for.

From definition of the operator H as the limit $P(\infty) = \lim_{t\to\infty} P(t)$, we see that $P(s)H = H$ for any $s \geqslant 0$, because

$$P(s)H = \lim_{t\to\infty} P(s)P(t) = \lim_{t\to\infty} P(s+t) = \lim_{t\to\infty} P(t) = H.$$

We have therefore

$$0 = \frac{\mathrm{d}}{\mathrm{d}t}H\varphi = \frac{\mathrm{d}}{\mathrm{d}t}(P(t)H\varphi) = -\Delta H\varphi, \quad \Delta H = H\Delta = 0,$$

and we see once again that $H\varphi$ is harmonic.

Since $P(t) = \exp(-t\Delta)$ tends exponentially to H for $t \to \infty$, we can construct the integral (Bochner) defining the Green operator

$$G\varphi := \int_0^\infty (P(t) - H)\varphi\, dt,$$

thus, since $\Delta H = 0$,

$$\Delta G\varphi = \int_0^\infty \Delta P(t)\varphi\, dt = \int_0^\infty \frac{\mathrm{d}}{\mathrm{d}t}(P(t)\varphi) = \lim_{t\to\infty} P(t)\varphi - \lim_{t\to 0} P(t)\varphi = H\varphi - \varphi.$$

□

Let us return to physics: Let φ be a 0 form, for example, a function of temperature of the 'world' $M \times \mathbb{R}_+$, $\mathbb{R}_+$ is the time half-line, M is a compact Riemann space. As we have shown, $P(t)\varphi \to H\varphi$ for $t \to \infty$, thus the distribution of temperature tends to harmonic distribution. But, as we now, the only harmonic functions on compact spaces are constant functions, and therefore distribution of temperature tends to constant distribution, that is 'thermal death'. The spirit of thermal death terrified many people on the turn of XIX century to the extend that some of them even committed suicide. But, of course, many unjustified assumptions have been made, for example, that the constant time sections M are compact ('closed universe'), that the laws of physics are invariable in space and time and hold in the whole universe. And, what is most important, some completely unjustified form of reductionism prevailed: This was a physicalism, the belief that all reality reduces to physical empiria. It is not hard to argue today that this – as any – reductionism is a very poor and impoverishing philosophy.

In this way we proved the Hodge-Kodaira theorem on orthogonal decomposition. But the method presented above gives even more: If φ is closed and o_i is its period on the cycle c_i, that is $o_i =< c_i, \varphi >$, then the form $\varphi(t) : P(t)\varphi$ has on any c_i the same periods as the from $\varphi = \varphi(0)$. Tending to the limit $t \to \infty$, we see that

$$o_i =< c_i, \varphi >=< c_i, P(t)\varphi > \overset{t\to\infty}{\longrightarrow} < c_i, H\varphi > .$$

This means that the harmonic form $H\varphi$, being the projection of the form $\varphi(t)$ has the same periods as the initial form φ. But, as de Rham has shown, there always exists an r form having given periods. We proved therefore

HODGE THEOREM. *There exists a harmonic form taking on given cycles c_i given (a priori) periods o_i, $i = 1, \ldots, \dim H^r$.*

REMARK. From the theorem on orthogonal decomposition, we have immediately

COROLLARY (Hodge.) *In any de Rham class $H^r(M)$ there exists exactly one harmonic form h. Indeed, if there were two such forms h_1 and h_2, then the difference $h_1 - h_2$ would have to be orthogonal to itself, thus $h_1 - h_2 = 0$. Therefore, $H^r(M) \simeq \mathcal{H}^r(M)$.*

The topology (cohomology) of compact Riemann space M is therefore defined by harmonic forms growing on it.

Now we can present the oldest topological invariant.

3.5 The Euler-Poincaré characteristic (Euler number)

This characteristic of a compact space M^n is

$$(E-P) \qquad \chi(M^n) = \sum_{r=0}^{n} (-1)^r \dim H^r(M^n.)$$

In the two dimensional case, for Riemann surface,we know since von Dyck that $\chi(M^2)$ is the only topological invariant; it is not surprising at all that it appears in two earliest and most important theorems of global analysis and differential geometry, the Gauss-Bonnet and Riemann-Roch theorems. It turns out that by virtue of $(E-P)$ formula

Euler-Poincaré characteristic is an index of elliptic operator. To see that, let us take the algebra $\Lambda = \Lambda(M)$ of all differential forms on the space M $\Lambda := \bigoplus_r \Lambda^r$ and the differential operator

$$D := d + d^* : \Lambda \to \Lambda.$$

Then $D^2 = DD^* = dd^* + d^*d = -\Delta$ and therefore D is elliptic as a 'square root' of elliptic operator. We decompose Λ into even and odd parts Λ^+ and Λ^-, respectively, as follows

$$\Lambda^+ = \bigoplus_{r \text{ even}} \Lambda^r, \quad \Lambda^- = \bigoplus_{r \text{ odd}} \Lambda^r.$$

Analogously, we define $D^\pm$ as the restriction of D to $\Lambda^\pm$: $D^+ : \Lambda^+ \to \Lambda^-$, $D^- : \Lambda^- \to \Lambda^+$ and observe that $D^- = (D^+)^* D^+$ is elliptic because D is elliptic. We check that $\ker D = \ker D^2 = \ker \Delta$. Indeed,

$$(D^2 u = 0) \Longrightarrow (0 = (u|D^2 u) = \|Du\|^2) \Longrightarrow (Du = 0.)$$

Therefore

$$\ker D^+ = \bigoplus_{r \text{ even}} H^r, \quad \ker D^- = \bigoplus_{r \text{ odd}} H^r.$$

Finally, we have

PROPOSITION. $\operatorname{ind}(D^+) = \ker D^+ - \ker D^- = \sum(-1)^r \dim H^r = \chi(M)$.

REMARK. Till now, we needed ellipticity of the operator A only to be able to talk about its index, that is in order that the spaces $\ker A$ and $\ker A^*$ ($=\operatorname{coker} A$) be finite dimensional.

3.6 Index theorem (for the first time)

HISTORICAL REMARK. Fritz Hirzenbruch was probably the first who observed that expressions similar to that of Euler number are indices of some differential operators. In his classical habilitation thesis of 1954, *Topologische Methoden ...* we encounter also the most important, very general case of the 'index theorem'. Hirzenbruch proves not only his version of the Riemann-Roch theorem for algebraic varieties, the theorem that was missing for many years, but also presents in the masterly way all notions and tools of both topological and analytical nature needed for this brilliant Hirzenbruch theorem. Around 1958 Israel M. Gelfand poses (was that under influence of the Hirzenbruch monograph?) the courageous conjecture, or research program that the index of an arbitrary linear elliptic operator should be expressible in topological terms. Finally, in 1963, Michael Atiyah and Isadore Singer prove the 'index theorem' – the result was announced in Bull. Amer. Math. Soc. **69** (1963), pp. 422–433; and the full proof is presented in the gigantic series of papers in Annals of Mathematics in 1968.

Today, the preferable and fastest way leading to index theorems, as the Hodge, Kodaira, de Rham, and general index theorems, is again

The method of equation of heat transfer and index theorems. From the theory of elliptic operators, we present (of course, without proof)

THEOREM. *Let M be a compact Riemann space, $E \to M$ vector bundles over M and let $Q : C^\infty(E) \to C^\infty(E)$ be an elliptic self adjoint (pseudo) differential operator with positive leading symbol (for example, with $Q = -\Delta_W$.) Then*

1. *P has purely discrete spectrum* $\operatorname{spec}P \subset [C, \infty[$, *and thus the eigenspaces $E(\lambda) = \{\varphi \in L^2(E) : Q\varphi = \lambda\varphi\}$ are finite dimensional;*

2. *Eigenvectors (that is, sections) of Q are smooth;*

3. *The kernel $K(x, y; t)$ of the operator e^{-tQ} is a smooth function for $t > 0$;*

4. *The trace of* $e^{-tQ} = \sum_k e^{-t\lambda_k} = \int_m tr_{E_x} K(x,x;t)dx$, *here* tr_{E_x} *is the trace of the* $(\dim E_x \times E_x)$*-matrix* $K(x,x;t)$, *with* E_x *being the fiber over* x *of the bundle* $E \to M$.

From the theorem above, we have important

COROLLARY. *Let* $L : C^\infty(E) \to C^\infty(F)$ *be an elliptic operator, where* $E \to M$, $F \to M$ *are vector bundles over compact Riemann space* M. *Then the operators* e^{-L^*L}, e^{-LL^*} *(which for* $t > 0$ *are semigroups of the equations*

$$\frac{du}{dt} = -L^*L, \quad \frac{du}{dt} = -LL^*$$

(of heat transfer) have smooth kernels and

$$index(L) = tracee^{-L^*L} - tracee^{-LL^*} = \sum_j \left(e^{-\lambda_j t} - e^{-\mu_j t}\right)$$

for any $t > 0$.

PROOF. Let λ_j and μ_j be eigenvalues of the operators L^*L and LL^*, respectively. We observe that if $\lambda_j \neq 0$ is an eigenvalue of L^*L, it is an eigenvalue of LL^* as well. Indeed, $(L^*L\varphi = \lambda\varphi) \Longrightarrow (LL^*(L\varphi) = \lambda(L\varphi))$.

When $E_0(\lambda)$, $E_1(\lambda)$ are eigenspaces of the operators L^*L and LL^* corresponding to the eigenvalue $\lambda \neq 0$, then the operator $L^*L : E_0(\lambda) \to E_1(\lambda) \to E_0(\lambda)$ is an isomorphism and therefore $\dim E_0(\lambda) = \dim E_1(\lambda)$. We have

$$\begin{aligned}
\operatorname{trace} e^{-tL^*L} - \operatorname{trace} e^{-tLL^*} &= \sum e^{-t\lambda}(\dim E_0(\lambda) - \dim E_1(\lambda)) \\
&= e^{to}(\dim E_0(0) - \dim E_1(0)) \\
&= \dim\ker L^*L - \dim\ker LL^* \\
&= \dim\ker L - \dim\ker L^* = \operatorname{index} L,
\end{aligned}$$

because $\ker L^*L = \ker L$, $\ker LL^* = \ker L^*$. □

REMARK. The kernels of the semigroup operators e^{-tQ}, being integral operators, are called *Green functions.* These functions appeared in physics and classical analysis much earlier (in Green's times) than functional analysis in Hilbert and Banach spaces. We will return to index theorems in the last part of this chapter. Now we must pause for a while to tell about some fundamental theorems from the theory of linear elliptic operators which we *implicitly* made use above (cf. *Analysis* part II and my *Methods of Hilbert Spaces.*)

3.7 Sobolev spaces. Theorems of Rellich, Sobolev, and Gårding

The development of great mathematical ideas has some necessity in it. One cannot help the feeling that such ideas tend to be realized and the great mathematicians are merely the means making possible the growth and development of these ideas from modest germ (*logos spermaticos* of Stoics and Plato) into the great tree, or even the forest, or, as some might prefer, a beautiful crystal shining in many colors.

Namely, the creation – construction – of the field of real numbers $\mathbb{R}$ as a completion (in the norm $\|\ \|$) of the field of rational numbers $\mathbb{Q}$ was necessary because convergent (Cauchy) sequences of rational numbers usually converge to irrational numbers; there is no such 'mess' in the field of real numbers $\mathbb{R}$.

The Dirichlet problem and direct methods of variational calculus 'forced' mathematicians to construct complete functional spaces $W_p^2(M)$ with Hilbert norms

$$\|f\|_p = \|\ \|^1_{W^2_p} := \left(\int_m \sum_{|\alpha| \leqslant p} |D^\alpha f|^2 dx \right)^{1/2},$$

$$D^\alpha = \frac{\partial^{|\alpha|}}{\partial x_1^{\alpha_1} \cdots \partial x_n^{\alpha_n}}, \quad |\alpha| = \alpha_1 + \cdots + \alpha_n,$$

where M is a domain in $\mathbb{R}^n$ with smooth boundary, or (compact) Riemann space, moreover, the derivatives $D^\alpha f \in L^2(M)$, that is, they are square integrable in Lebesgue sense. Other definition of these, so called Sobolev spaces, is obtained by completing the space of smooth functions $C^\infty(M)$ in the norm $\|\ \|_p$. What are these spaces good for? The norm $\|\ \|_1$ appeared first in connection with the classical Dirichlet integral: under integral sign we have there the first derivative of f, $|\text{grad}\, f|^2$. The *minimal* sequence $(f_j)_1^\infty$ is the sequence of (smooth) functions tending to the element which is the lower bound of Dirichlet integral. But in which sense should this sequence converge? The answer is: in the norm $\|\ \|_1$ – but the limit will be, in general, not smooth: it will belong to the space W_1^2. We are still not fully satisfied: we demand the limit f of the sequence (f_j) to satisfy given equation $\Delta f = 0$, but on the first sight we know only that this equation is only satisfied in the *weak* sense $(f|\Delta\varphi) = 0$ identically for $\varphi \in C_0^\infty(M)$. Here, the young Weyl enters the stage

Weyl lemma: $0 = (f|\Delta\varphi)$ for $\varphi \in C_0^\infty(M) \Longrightarrow (f \in C^\infty(M)$ and $\Delta f = 0)$, *thus a weak solution of Laplace equation (more generally, elliptic equation with smooth coefficients) is smooth and is a solution in the classical sense.*

This Weyl lemma is a forerunner of the theory of distribution of Laurent Schwartz. In this initial stage of distribution theory, an important role was played by Polish mathematicians S. Zaremba and O. Nikodym. We see that while solving completely classical problems (Dirichlet problem, problem of existence of geodesics, etc.), it is useful to introduce a full range or ladder of Sobolev spaces W_r^2, $r = 1, 2, \ldots$. These spaces are similarly defined for sections of vector spaces $E \to M$; as a results one obtains the Sobolev spaces $W_r^2(M)$, in particular, the spaces of differential forms

$$W_r^2(\Lambda^p(M)) \quad \text{for simplicity denoted by } W_r^2(\Lambda^p.)$$

As Sobolev showed (first for scalar functions)

Theorem (Sobolev). 1. $\bigcap_{r=0}^\infty W_r^2(E) = C^\infty(E)$.

2. *For $r > n/2$, the embedding $W_r^2(E) \hookrightarrow C^0(E)$ is continuous if one endow the space $C^0(E)$ with the norm* $\sup|f(M^n)|$.

Let us consider another fact due to Rellich

Rellich lemma. *The embedding $i : W_{r+1}^2(E) \hookrightarrow W_r^2(E)$ is compact, that is, the image of the ball $\|u\|_{r+1} \leqslant 1$ has a closure in $W_r^2(E)$ which is a compact set.*

For elliptic operators, the following theorem proved in 1953 by Lars Gårding is of fundamental importance

Gårding theorem. *Let $L : C^\infty(E) \to C^\infty(E)$ be an elliptic operator of order r. Then there exist constants c_1, c_2 such that for every $u \in W_{r+l}^2(E)$*

$$\|u\|_{r+l} = c_1\|Lu\|_l + c_2\|u\|_0. \tag{G}$$

When u is orthogonal, in the sense of $(\cdot|\cdot)_0$, to $\ker L$, then in the Gårding inequality one can take $c_2 = 0$.

Let us note the useful criterion of finite dimension of Hilbert (in general, Banach) spaces.

LEMMA. ($\dim H < \infty$) $\Longleftrightarrow$(*a ball closed in H is compact.*)

PROOF is immediate. $\Longrightarrow$is an elementary fact.

$\Longleftarrow$. Let $\dim H = \infty$. We take an orthonormal sequence e_j, $j = 1, 2, \ldots$. Obviously, from the elements this sequence one cannot form a convergent subsequence, because for any $e_j \neq e_i$, $\|e_j - e_i\| = 2$. □.

Above, for example in Hodge theory, we made use of the fact that $\ker L$ is finite dimensional if L is elliptic. This follows from Rellich lemma and Gårding theorem, as the simple observation of Jaak Peetre shows:

LEMMA (PEETRE.) *Let X, Y, Z be Hilbert spaces (it is sufficient to assume that they are reflexive Banach spaces), such that $L : X \to Z$ is continuous and linear, and the injection $X \hookrightarrow Y$ is compact. Then the following conditions are equivalent*

(a) *there exist constants c_1, c_2 such that for every $x \in X$*

$$(G') \qquad \|X\|_X = c_1\|Lx\|_Z + c_2\|x\|_Y;$$

(b) *the set LX (*$\operatorname{im} L$*) is closed and* $\dim \ker L < \infty$.

PROOF. (a)$\Longrightarrow$(b.) Since $X \hookrightarrow Y$ is compact, the unit ball in $\ker L$ has finite dimension.

Let us decompose $X = X_1 \oplus \ker L$. Since $L : X_1 \to Z$ is an injection and $X \hookrightarrow Y$ is compact, for any $x_1 \in X_1$ we have

$$(*) \qquad \|x_1\|_{X_1} \leqslant c\|Lx_1\|_Z.$$

Let for some sequence $x_j \in X_1$, $Lx_j \to z$. To show that $\operatorname{im} L$ is closed, we must find such $x \in X_1$ that $z = Lx$. But it follows from (*) that (x_j) is a Cauchy sequence in X, thus $x_j \to x \in X_1$. But from continuity of L we have $z = \lim Lx_j = Lx$.

(b)$\Longrightarrow$(a.) We again write $X = X_1 \oplus \ker L$, thus $L|_{X_1} \to Z$ is an injection. From the theorem on closed graph we obtain (G′.) □

COROLLARY. *Let L be an elliptic operator of order r. Taking in Peetre lemma $X = W^2_{r+l}(E)$, $Z = W^2_l(F)$, $Y = W^2_r(E)$, we see that* $\ker L$ *has finite dimension and* $\operatorname{im} L$ *is closed. Thus one can talk about an index of the operator L.*

In our considerations above, we needed Gårding inequality for (elliptic) Weitzenbröck–Hodge operator $\Delta = \Delta_W$ on $E = \Lambda^p(M^n)$, the bundle of differential p forms on compact manifold M^n. It is useful to describe this operator $-\Delta = d\delta + \delta d$ in terms of covariant derivatives. This was indeed done by Weitzenbröck in 1923. We cannot postpone presentation of these famous formulas any longer.

The elegant proof of Gårding inequality with the help of Weitzenbröck formulas can be found in the monograph by Griffith and Harris.

3.8 Weitzenbröck formulas

Let $M = M^n$ be a compact Riemann space and ∇ the covariant derivative of Levi-Civita. Then we have

THEOREM (WEITZENBRÖCK). *For any p form $\varphi \in \Lambda^p(M)$ the following Weitzenbröck formulas hold*

$$(W1) \qquad (d\varphi)_{k_1\ldots k_{p+1}} = \sum (-1)^{i-1} \nabla_{k_i} \varphi_{k_1\ldots\hat{k}_i\ldots k_{p+1}}$$

where $\hat{}$ means that the index under caret is omitted,

$$(W2) \qquad (\delta\varphi)_{k_1\ldots k_{p-1}} = -\nabla^i \varphi_{ik_1\ldots k_{p-1}}, \quad \text{where } \nabla^i := g^{ik}\nabla_k,$$

and the most important Weitzenbröck formula

$$-(\Delta\varphi)_{k_1\ldots k_p} = (d\delta + \delta d)\varphi_{k_1\ldots k_p} = \nabla^i\nabla_i\varphi_{k_1\ldots k_p} -$$

$$(W3) \qquad -\sum_{\nu=1}^{p}(-1)^{\nu} R^{hi}_{ik_\nu}\varphi_{hk_1\ldots\hat{k}_\nu\ldots k_p} - 2\sum_{\mu<\nu}^{p}(-1)^{\mu+\nu} R^{hi}_{k_\nu k_\mu}\varphi_{ihk_1\ldots\hat{k}_\nu\ldots\hat{k}_\mu\ldots k_p},$$

where R^{ij}_{kl} are coefficients of the curvature tensor (Riemann–Christoffel.)

REMARK. We see that in the first sum in formula ($W3$) the Ricci tensor appears

$$(\mathrm{Ric})^h_k := R^h_k := R^{hi}_{ik} \quad \text{(summation over } i.)$$

This tensor is also called *Ricci curvature*. We see once again how important a role of curvature tensor is. In the case of 0 form f (scalar field), we have

$$\Delta f = \nabla^i\nabla_i f.$$

In the case of 1 form, the last term in $(W3)$ drops out and we have

$$(W') \qquad -(\Delta\varphi)_k = \nabla^i\nabla_i\varphi_k - R^h_k\varphi_h.$$

As a consequence of this formula, we have interesting

VANISHING THEOREM (Bochner.) *If $Ric > 0$, then $H^1(M) = 0$, that is, the first Betti number $\beta_1 := \dim H^1 = 0$. Thus, if $\beta_1 \neq 0$, on compact Riemann space there does not exist a Riemann metric with positive Ricci curvature.*

PROOF. Integrating by parts, we obtain from (W′)

$$0 = (\varphi| - \Delta\varphi) = \int_M (|\nabla\varphi|^2 + R^{hl}\varphi_h\varphi_l)dx, \quad \text{thus } \varphi = 0.$$

Already this Bochner theorem shows how important is the Riemann tensor in geometry. We must stop here, but it would be unforgivable if we did not mention a fundamental role of *Ric* in general theory of relativity.

Einstein-Hilbert (E-H) equations of general theory of relativity. As we have already mentioned at the beginning of this chapter, the Einstein theory of gravity was – and still is – a source of inspiration for many mathematicians: we mentioned Hermann Weyl who introduced a general notion of linear connection and theory of gauge fields (*Eichtehorien.*) We know that Hilbert (a teacher and master of Weyl) was enthusiastic about the great concept of Einstein even when Einstein was still not able to find differential equations of motion of his theory of gravity. Without these equations, the theory was only a dream. Hilbert was an expert in calculus of variations (theory of geodesics, saving Dirichlet principle in the theory of Riemann surfaces ...) and the theory of invariants. After two weeks long Einstein's visit in Göttingen, (he was there on Hilbert's invitation) (Hilbert: 'every junk in Göttingen knows more about mathematics than Einstein, but Einstein is a genius!'), Hilbert, under influence of field theory of Gustav Mie, formulated 'axioms' which Einstein's theory should satisfy. He constructed the 'simplest' Lagrange function for the theory and obtained in this way – few days before Einstein did – the famous, desperately sought by Einstein, equations of the theory of gravity, which are, deservedly, called today

Einstein-Hilbert equations. Let (M^4, g) be a pseudo Riemannian space-time, (R^k_h) the Ricci tensor, R the scalar curvature of the metric g, (T^k_h) the energy-momentum tensor. Then the following equations hold

$$(E-H)\qquad R^k_h - \frac{1}{2}\delta^k_h = \lambda T^k_h.$$

One can find derivation of these equations and the discussion in the unsurpassed Weyl monograph *Raum-Zeit-Materie.*

DEFINITION. The Riemann space whose metric tensor satisfies $R_{ij} = g_{ij}$ is called *Einstein space.*

After this necessary digression, let us return to index theorem. We have seen that the Euler-Poincaré characteristics $\chi(M^n)$ of compact Riemann space is an index of the operator $d+\delta$ (the square root of the Weitzenbröch-Hodge operator $-\Delta_W$.) We have seen also that the index of elliptic operator can be expressed in terms of a trace of the kernel of heat transfer operators.

3.9 Euler form. Hopf theorem on index of vector field

The oldest topological invariant $\chi(M^n)$ of compact M^n, and the only one in the case of M^2, fascinated and still fascinates mathematicians. It appeared in earliest and probably still fundamental theorems of global analysis and geometry, Gauss-Bonnet (G-B) and Riemann-Roch (R-R) theorems. In the works of Poincaré, the Euler characteristics $\chi(M)$ was an alternating sum $\chi(M) := \beta_0(M) - \beta_1(M) + \cdots + (-1)^n\beta_n(M)$, where $\beta_k(M)$ is the kth Betti number of the (complex) M, that is, the number of k-dimensional cells of the manifold M. As we know from de Rham theorem, $\beta_k(M) = \dim H^{n-k}_{dR}(M)$, and from the Hodge theorem $\beta_k(M) = \dim \mathcal{H}^{n-k}(M)$, the dimension of the space of harmonic k forms, thus

$$\chi(M^n) = \sum_{k=0}^{n}(-1)^n \dim \mathcal{H}^k(M^n.)$$

But, as we showed, $\chi(M)$ is an index of the Hodge-Hirzenbruch operator $\chi(M) = \text{index}\,(d+\delta)$.

The third, perhaps most surprising equality, providing relation between the whole (topology) and the part is

Hopf theorem (on index of vector field) which in the two dimensional case was proved by Poincaré, and which is, for that reason, sometimes called Hopf-Poincaré theorem. This theorem was proved by Heinz Hopf in 1927 in his Ph.D. thesis, and is formulated as follows.

Let M^n be compact oriented Riemann manifold and let v be a vector field on M^n. Then

$$(H-P) \qquad \chi(M) = \sum_{v(x)=0} \operatorname{ind} v(x)$$

where $\operatorname{ind} v(x)$ is the index of the vector field v in his (non-degenerate) critical point x, being, *ex definitione*, a zero of this field, $v(x) = 0$.

Since M is compact, the set of critical points is finite for $v \not\equiv 0$. The $(H-P)$ equality is surprising indeed: the left hand side $\chi(M)$ does not depend on the vector field v, and therefore the right hand side, the sum of indices, is identical for any field. We will present a number of interesting consequences of Hopf theorem below, now we must describe $\operatorname{ind} v$. Let $v(x) = 0$ and let $x^1, \ldots, x^n$ be coordinates around x, therefore

$$(*) \qquad v(x) = \sum a_{ij} x^i \frac{\partial}{\partial x^j} + \text{higher order terms}$$

where the matrix $A = (a_{ij})$ is non-degenerate.

Integrating the vector field v, we obtain a flow (that is, a one-parameter group of diffeomorphisms $f_t : M \to M$.) For small t, the fixed points of f_t are just zeros of v, therefore, the Jacobian $J(f_t)(x)$ in vicinity of x has the form $J(f_t)(x) + e^{tA} +$ higher order terms. Thus

$$J(f_t)(x) - 1_n = t\left(A + \frac{t^1}{2}A^2 + \frac{t^2}{3!}A^3 + \ldots\right),$$

and for $t > 0$ sufficiently small, we have

$$(1) \qquad \operatorname{ind} v(x) := \operatorname{sgn}\det A = \operatorname{sgn}\det(J(f_t)(x) - 1_n) =: \operatorname{ind}(f_t(x).)$$

As the next important application of the Hodge theory, we present fundamental theorems of homology theory.

3.10 Poincaré duality. Künneth theorem

Poincaré duality. Let $M = M^n$ be an oriented compact manifold. Then

(i) the bilinear form $(\cdot,\cdot)\colon H^p(M)\times H^{m-p}(M)\to\mathbb{R}$ given by

$$(\alpha,\beta) := \int_M \alpha\wedge\beta$$

is non-degenerate, that is $H^p(M)^* \simeq H^{m-p}(M)$;

(ii) if (M^n, g) is a Riemann space, the isomorphism above is given by the Hodge star

$$*H^p(M) = H^{m-p}(M) \quad \text{for } p = 0,\dots,m.$$

The proof is immediate, because $(\alpha, *\alpha) = (\alpha|\alpha) > 0$ for $\alpha \neq 0$. Another simple observation: $H^*(M^m) := \bigoplus_{p=0}^m H^p(M)$ is an algebra with gradation (where multiplication is defined as the external product of differential forms.) If $N = N^n$ is also a compact, oriented Riemann manifold, we immediately obtain vector spaces $H^j(M\times N)$, $H^p(M)\otimes H^q(N)$. Identifying these spaces with the corresponding spaces of harmonic forms and calculating dimensions of these spaces (taking orthonormal bases), we almost immediately obtain famous

Künneth theorem.

$$\bigotimes_{p+q=j} H^p(M)\otimes H^q(N) \simeq H^j(M\times N)$$

$j = 0,\dots,m+n$, *which can be compactly written as*

$$\kappa : H^*(M)\otimes H^*(N) \overset{\approx}{\to} H^*(M\times N.)$$

The isomorphism κ is given by the formula

$$\kappa(\alpha\otimes\beta) := (p_1\alpha)\wedge(p_2\beta),$$

where $p_1 : M\times N \to M$, $p_2 : M\times N\to N$ are projections of the left and right term called Künneth isomorphisms.

Historical remark. In his classical paper of 1924, Künneth talks only about Betti numbers for *homologies* of manifold $M\times N$. The cohomology theory was created much later. As one could have expected, the analogous 'Künneth theorem' holds for vector bundles and sheaves over differential manifolds (cf. the monograph of Hirzenbruch.)

Poincaré form η_Δ of the diagonal $\Delta \hookrightarrow M \times M$ and Euler form. Let (α_i) be a basis in the vector space $\mathcal{H}^*(M) \equiv H^*(M)$ and let (β_j) be its dual basis, that is $(\alpha_i, \beta_j) = \delta_{ij}$, where

$$\int_M \alpha_i \wedge \beta_j =: (\alpha_i, \beta_j)$$

and

$$\begin{array}{ccc} & M \times M & \\ {}^{p_1}\swarrow & & \searrow^{p_2} \\ M & & M \end{array}$$

We know from Künneth theorem that $(p_1^*\alpha_i \wedge p_2^*\beta_j)$ is an additive basis of the space $H^*(M \times M)$. We have

PROPOSITION (LEFSCHETZ). *Let η_Δ be a Poincaré form of diagonal $\Delta \hookrightarrow M \times M$ and let $e := D^*\eta_\Delta$ be a differential m form on M, called the Euler form of manifold M (more precisely, of the tangent bundle $TM \to M$.) The following formulas hold*

$$\eta_\Delta = \sum (-1)^{\deg \alpha_i} p_1^*\alpha_i \wedge p_2^*\beta_i, \tag{1}$$

and therefore

$$e \equiv e_M := D^*\eta_\Delta = \sum (-1)^{\deg \alpha_i} \alpha_i \wedge \beta_i, \tag{2}$$

for dual bases (α_i), (β_j) of the spaces $H^q(M)$, $H^{m-q}(M)$. Moreover,

$$\int_M e = \int_\Delta \eta_\Delta = \chi(M.) \tag{E}$$

Therefore, the integral of Euler form $e = e(TM)$ of manifold M is equal to the Euler-Poincaré characteristics of this manifold, and this is the reason for the name of the form $e_M = e := D^*\eta_\Delta$.

PROOF. (1.) Certainly, η_Δ is a linear combination of the form

$$\sum a_{ij} p_1^*\alpha_i \wedge p_2^*\beta_i.$$

Let us compute the integral $\int_\Delta p_1^*\beta_k \wedge p_2^*\alpha_l$ in two ways: first, since $D^*(\kappa(\alpha \otimes \beta)) = \alpha \wedge \beta$, we have

$$\int_\Delta p_1^*\beta_k \wedge p_2^* a_l = \int_M D^* p_1^*\beta_k \wedge D^* p_2^* a_l = \int_M \beta_k \wedge \alpha_l = (-1)^{\deg \beta_k \deg \alpha_l} \delta_{kl}.$$

On the other hand, from the definition of Poincaré form of submanifold, we have

$$\begin{aligned}
\int_\Delta p_1^*\beta_k \wedge p_2^*\alpha_l &= \int_{M\times M} p_1^*\beta_k \wedge p_2^*\alpha_l \wedge \eta_\Delta \\
= & \sum_{ij} a_{ij} \int_{M\times M} p_1^*\beta_k \wedge p_2^*\alpha_l \wedge p_1^*\alpha_i \wedge p_2^*\beta_j \\
= & \sum_{ij} a_{ij} (-1)^{(\deg\beta_k+\deg\alpha_l)\deg\alpha_i} \int_{M\times M} p_1^*(\alpha_i\wedge\beta_k) p_2^*(\alpha_l\wedge\beta_j) \\
= & (-1)^{(\deg\beta_k+\deg\alpha_l)\deg\alpha_k} a_{kl}.
\end{aligned}$$

Thus $a_{kl}=0$ for $k\neq l$, $a_{kl}=(-1)^{\deg\alpha_k}$ for $k=l$, and therefore we have (1.)

(2.) Formula (2) immediately follows from (1.)

(E.) Exactly as in the proof of formula (1), we have

$$\int_\Delta \eta_\Delta = \int_M D^*\eta_\Delta = \sum(-1)^{\deg\alpha_i}\int_M \alpha_i\wedge\beta_i$$

$$=\sum(-1)^{\deg\alpha_i} = \sum_{p=0}^{m}(-1)^p \dim H^p(M)\chi(M.)$$

□

3.11 Intersection number (Kronecker index) of two cycles

We also owe Riemann a very important notion of the number of intersection of two oriented (closed) curves on a compact Riemann surface; this notion is intuitively very simple. With any point of (transversal) intersection we associate the number $+1$ if the tangent vectors of the curves are consistent with orientation on M^2 and -1 otherwise. This notion plays a decisive role in the theory of periods, which can be characterized by the following slogans: Jacobi theorem, Abel theorem, matrix of periods, abelian torus $\mathrm{Jac}(M^2)$. Leopold Kronecker generalized this notion to multi dimensional objects (1861): in m-dimensional oriented Euclidean space E^m, two oriented subspaces of dimensions p and $(m-p)$ intersect in one point and with this point one can associate the number $+1(-1)$ depending on the fact if orientations of these subspaces determine orientation consistent (opposite) with the orientation of the space E^m. This notion was in turn generalized to differential manifolds by Poincaré in 1881; he defined Kronecker index of intersecting cycles of complementary dimensions z^p, z^{m-p}. *This was the beginning of algebraic topology*: indeed, the Poincaré duality theorem. The following major steps were the classical papers by Brouwer, Lefschetz, and Hopf, and the crowning achievement, the theory of currents of de Rham–Hodge–Kodaira:

With any compact, oriented $(m-p)$-dimensional submanifold C^{m-p} one can associate harmonic p forms $\eta_{C^{m-p}} = HC^{m-p}$ (where H is the orthogonal 'projection' on $H(M^m)$.) The Kronecker index or the intersection number of the cycles C^p, C^{m-p} is

$$\sharp(C^{m-p} \cdot C^p) = I((C^{m-p}, C^p)) := \int_{C^p} \eta_{C^{m-p}}.$$

REMARK. C^{m-p} is understood here as a de Rham current, see Part V, Chapter 5.

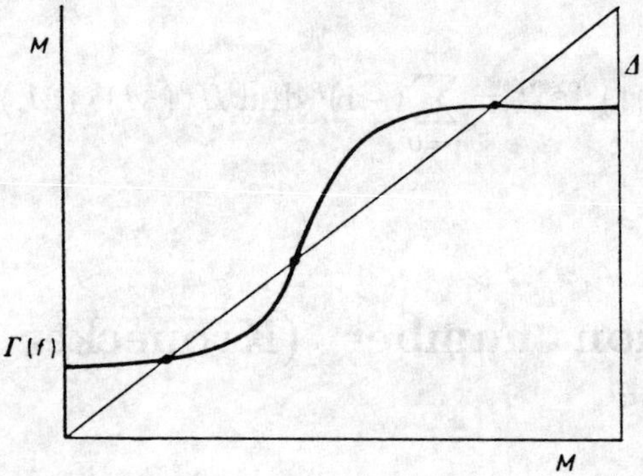

Fig. 11

In view of the Lefschetz proposition, we see that the E-P characteristic (Euler number) $\chi(M^m)$ of the a manifold $M = M^m$ is the Kronecker index of the intersection of the diagonal $\Delta \subset M \times M$ with itself

$$\chi(M) = \int_\Delta \eta_\Delta.$$

Lefschetz theorem generalizes this fact by giving the intersection number of the graph $\Gamma(f) \subset M \times M$ of the mapping $f : M \to M$ with the diagonal

$\Delta \subset M \times M$ (see Fig. 11.)

$$(L) \qquad {}^{\sharp}(\Gamma(f)\cdot\Delta) = \sum_{f(x)=x} \operatorname{ind} f(x.)$$

We already computed the left hand side of Lefschetz formula (L). Let us compute now the right hand side, and this will indeed comprise the proof of Lefschetz theorem (L). We will proceed as in the proof of Lefschetz proposition, but first we state a different definition of non-degeneracy condition of isolated fixed point $x = f(x)$ of the mapping $f : M \to M$, which, in local coordinates can be phrased as follows: *Jacobi matrix* $J(f)(x) : T_xM \to T_xM$ of the differential $df(x)$ satisfies the condition

$$(T) \qquad \det(J(f)(x) - 1_m)) \neq 0.$$

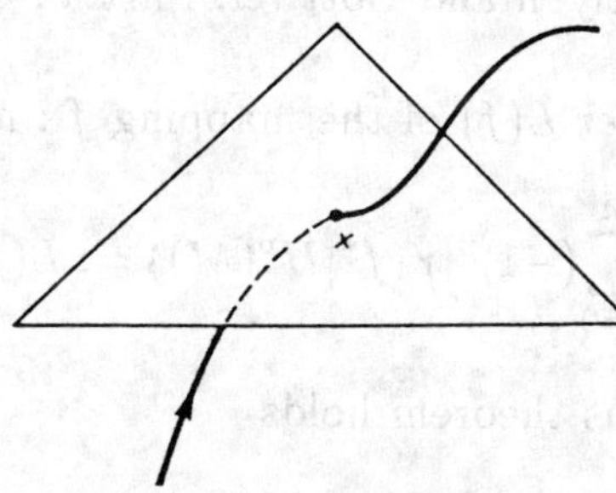

Fig. 12 .

We will show that (T) has geometrical meaning: this is the condition that the diagonal $\Delta \subset M \times M$ intersects with the graph $\Gamma(f)$ transversally, that is, the subspaces tangent to Δ and $\Gamma(f)$ at (x, x), $T_{(x,x)}\Delta$ and $T_{(x,x)}\Gamma(f)$, respectively, span the whole of $T_{(x,x)}(M \times M)$, the tangent space of $M \times M$. Indeed, in an appropriate coordinate system, this condition takes the form

$$0 \neq \det \begin{bmatrix} 1_n & 1_n \\ 1_n & J(f)(x) \end{bmatrix} = \det(J(f)(x) - 1_n),$$

thus condition (T) follows.

Let us compute the intersection number

$$\sharp(\Delta\cdot\Gamma(f)) = \int_{\Gamma(f)} \eta_\Delta = \sum(-1)^{\deg\alpha_i}\int_{\Gamma(f)} p_1^*\alpha_i\wedge p_2^*\beta_i$$

(because $(1\otimes f)p_1^* = 1^*$, $(1\otimes f)p_1^* = f^*$ (where $1 = \mathrm{id}_M$))

$$= \sum(-1)^{\deg\alpha_i}\int_M \alpha_i\wedge f^*\beta_i = \sum_{p=0}^{m}(-1)^p \operatorname{tr}(f^*|H^p(M)) =: L(f);$$

which is (*ex definitione*) the Lefschetz number of mapping f. □

Therefore we proved Lefschetz theorem and thus the proof of Hopf-Poincaré theorem is completed as well!

REMARK. The Lefschetz number $L(f)$ is an important invariant. This notion, along with Lefschetz theorem was generalized to the case of arbitrary elliptic complexes by Atiyah and Bott (cf. Part V, Chapter 3.)

The Lefschetz number $L(f)$ of the mapping $f: M\to M$ is

$$\sum_{p=0}^{m}(-1)^p \operatorname{tr}(f^*|H^p(M)) =: L(f) \tag{1}$$

The following famous theorem holds

LEFSCHETZ FIXED POINT THEOREM.

$$L(f) = \sum_{f(x)} \operatorname{ind} f(x.) \tag{L}$$

From the equalities above, we have also the important formula

$$L(f_t) = \chi(M) \quad \text{for any } t \geqslant 0 \tag{2}$$

and Hopf index theorem

$$\chi(M) = \sum_{v(x)=0} \operatorname{ind} v(x.) \tag{H – P}$$

We see therefore that $\chi(M)$ is the number of zeros of a vector field on M, appropriately counted (Hopf), and also that $L(f)$ is a number of fixed points of a mapping $f : M \to M$, again appropriately counted. In the course of construction of his characteristic classes (Chern classes), by employing Hopf theorem, Chern (in 1943) showed that there exists a far reaching generalization of Gauss-Bonnet theorem (G-B) for arbitrary dimension n, called

Chern-Gauss-Bonnet theorem, or Chern theorem, for short:

$$(C-G-B) \qquad \chi(M^n) = e_n(M^n) = \int_{M^n} e(TM^n),$$

where $e \in \Lambda^n(M^n)$ is called the *Euler form* of manifold M^n, or, rather, the bundle TM^n and can be expressed in terms of the curvature tensor R^∇ of the Levi-Civita connection ∇.

We devote a full chapter below to Chern-Gauss-Bonnet theorem.

3.12 Index of vector a field and degree of mapping. Kronecker integral

If we have two oriented m-dimensional manifolds M^m, N^m, then with any mapping $f : M \to N$ we can associate an integer $\deg(f)$, the degree of mapping f, by the following formula

$$(D) \qquad \int_M f^*\omega =: \deg(f) \int_N \omega,$$

where $\omega = \overset{n}{\omega}$ is the volume form of N.

THEOREM. 1. *If $x \in N$ is a regular value and $f^{-1}(x) = \{y_1, \ldots, y_l, y_{l+1}, \ldots, y_{l+k}\}$, where $T_{y_1}(f), \ldots, T_{y_l}(f)$ preserve the orientation and $T_{y_{l+1}}(f), \ldots, T_{y_{l+k}}(f)$ change the orientation, then* $\deg(f) = l - k \in \mathbb{R}$.

2. *If* $\deg(f) \neq 0$, *then $f(M) = N$ (f is a surjection).*

3. *If $f, g : M \to N$ are homotopic, then* $\deg(f) = \deg(g)$

4. $\deg(f \circ h) = \deg(f) \cdot \deg(h)$.

5. *If $f : M \to N$ is an ρ-fold covering and if orientations of M and N are chosen such that f is a diffeomorphism locally consistent with these orientations, then* $\deg(f) = r$.

PROOF.

3. is immediate because for homotopic f, g, the left hand side of (D) is

$$\int_M f^*\omega = \int_M g^*\omega.$$

4. follows also from the definition of deg, because $(f \circ h)^* = h^* \circ f^*$.
5. is an obvious corollary following from 1.

The only nontrivial point is therefore 1.

Sard theorem says that the set of regular values of a differential mapping $f : M \to N$ is dense and open in N.

We turn to the proof of point 1. Let $x \in N$ be a regular value of f. Since M is compact, the set $\{f^{-1}(x)\}$ is finite (because it is discrete and closed in M.) Let $B \ni x$ be a neighborhood of x and $B_i \ni y_i$ be neighborhood of y_i such that $B_i \cap B_j = \emptyset$ and let $f_i := f|B_i : B_i \to B$ be a diffeomorphism, $i = 1, \ldots, l + k$, and the orientation f_i be equal to the orientation $T_{y_i} f$. Let $\beta := f^*(\omega)$, $\int_M \omega = 1$, $\beta \sum \beta_i$, support $\beta_i \subset B_i$

$$\deg(f) = \int_M f^*(\omega) = \sum_i \int_{B_i} \beta_i,$$

but $\int_{B_i} \beta_i = \pm \int_B \omega = \pm \int_N \omega = \pm 1$ depending on the fact if f_i is consistent with the orientation of T_{y_i} or not. □

Corollary I. Fundamental theorem of algebra. *A polynomial $z \to P(z)$ of order $n \geqslant 0$ defines the mapping $f : S^2(:= \mathbb{C} \cup \{\infty\}) \to S^2$ of Riemann sphere whose critical points correspond to zeros of the derivative $P'(z)$ and to the point $\{\infty\}$. It is easy to see that* $\deg(f) > 0$, *thus, it follows from* 2. *that f is a surjection, and therefore f has a root.* □

Corollary II. *Let $g : S^m \to S^m$, $g(x) := -x$. Then* $\deg(g) = (-1)^{m+1}$.

Corollary III. *On even-dimensional sphere S^{2m} there are no vector fields without singular points (theorem on combing the sphere S^{2m}).*

The degree of mapping (*Abbildungsgrad*) was introduced by the celebrated Dutch Brouwer about 1909 and it is him whom we owe the notion of index of vector field (cf. historical remarks below), which is defined as follows.

Step I. A vector field on domain O of Euclidean space $\mathbb{R}^m$ can be regarded as a smooth mapping $v : O \to \mathbb{R}^m$; let us assume that $0 \in O$ and that 0 is an isolated singular point of the field v, that is, $v(0) = 0$. For any $r > 0$ we have the mapping $f_r : S^{m-1} \to S^{m-1}$, $f_r(x) := v(rx)/\|v(rx)\|$, $\|x\| = 1$. It is easily seen that for $r_1 > 0$, f_{r_1} is homotopic with f_r, and thus $\deg f_r$ does not depend on r.

DEFINITION. $\deg f_r$ is called the *index of field v* in 0 and is denoted $\operatorname{ind} v(0)$.

The *Kronecker integral* is the index of mapping $g = p \circ h$, where $h : M^{m-1} \to \mathbb{R}^m - \{0\}$ and $p : \mathbb{R}^m - \{0\} \to S^{m-1}$, where $p(x) := x\|x\|^{-1}$. We have therefore the diagram

$$\begin{array}{ccc} M^{n-1} & \xrightarrow{\quad h \quad} & R^m - \{0\} \\ & \searrow^{\,g = \rho \circ h} & \downarrow \rho \\ & & S^m \end{array}$$

In order to compute $\deg(g)$, we recall the form of the volume element, that is, the volume form of the unit sphere S^{m-1} which is a $(m-1)$ form in $\mathbb{R}^m$. Let $xdx := \sum x_j dx_j$, similarly, the $(m-1)$ form (dual to it)

$$\sigma := *(xdx) = \sum_1^m (-1)^j x_j dx_1 \wedge \cdots \wedge d\hat{x}_j \wedge \cdots \wedge dx_m$$

is $O(m)$-invariant. It is easy to see that $\sigma' := \sigma | S^{m-1}$ is the required volume form of the sphere S^{m-1}. It is not hard to check by explicit computation that

$$p^* \sigma' = \frac{\sigma}{\|x\|^m}.$$

Thus the degree of g can be expressed in terms of the Kronecker integral

$$(K) \qquad \frac{1}{|S^{m-1}|} \int_{M^{m-1}} g^* \sigma' = \frac{1}{|S^{m-1}|} \int_{M^{m-1}} h^* \tau,$$

where $\tau := p^* \sigma'$, because $g^*(\sigma') = f^* \circ p^*(\sigma') = f^*(p^* \sigma')$. In the coordinates $y_j = x_j / \|x\|$,

$$\sigma' = \sum_1^m (-1)^{i-1} y_i dy_1 \wedge \cdots \wedge d\hat{y}_i \wedge \cdots \wedge dy_m,$$

$$\frac{1}{|S^{m-1}|} = m \frac{\pi^{1/2}}{\Gamma(\frac{1}{2} m + 1)} \qquad \text{the volume of unit sphere } S^{m-1}.$$

Thus the Kronecker integral (K) measures the number of times the surface $h(M^{m-1})$ winds around $0 \in \mathbb{R}^m$. Writing $h(M^{m-1})$ in local coordinates $x = x(u_1, \ldots, u_{m-1})$, the Kronecker integral can be rewritten as

$$\frac{1}{|S^{m-1}|} \int_{Z^{m-1}} \frac{1}{\|x(u)\|^m} \det \left(x, \frac{\partial x}{\partial u_1}, \ldots, \frac{\partial x}{\partial um - 1} \right) du_1 \ldots du_{m-1}.$$

This is exactly the form which appeared in the classical Kronecker paper of 1861. We devoted so much time to this analytical expression because it is not only the germ of the theories of degree of mappings and index of vector fields, but it also comprises the basis of the proof of Hopf theorem, and also of the Chern theorem, the famous multi dimensional analogue of the Gauss–Bonnet–von Dyck theorem, to which we devote the following section.

COROLLARY. ind $v(0)$ *is an integer and is invariant with respect to small perturbations of* v *(because is a continuous function and* v *is integer valued.)*

VERY IMPORTANT EXAMPLE. Let v be a vector field in a neighborhood of $0 \in \mathbb{R}^2$ and let φ denote the angle between v and some given direction in $\mathbb{R}^2$. Then

$$\operatorname{ind} v(0) = \frac{1}{2\pi} \int_{S^2} d\varphi$$

SUB-EXAMPLE. If v is the field whose integral curves (field lines) are represented in Figure 14 has in 0 the isolated singular point with index -1. The field represented in Figure 13 has in 0 the isolated singular point with index +1. The additional properties of index in $\mathbb{R}^m$ which are required for general definition are contained in

PROPOSITION. *Let the vector field in an neighborhood* $0 \in O \subset \mathbb{R}^m$ *possess in 0 an isolated singular point.*

1. *If* $f : O \to \mathbb{R}^m$ *is a diffeomorphism such that* $f(0) = 0$, *then* $\operatorname{ind} f_* v(0) = \operatorname{ind} v(0)$.

2. *If 0 is a non-degenerate singular point of the field* v, *that is if* $dv(0)$ *is an invertible mapping,* $\det(dv(0)) \neq 0$, *then* $\operatorname{ind} v(0) = sgn \det(dv(0))$.

PROOF. 1. Recall that $f_* v$ is a mapping of the field v defined by the formula

$$f_* v(x) := v(f(x).)$$

Let us assume first that f is an orthogonal mapping in $\mathbb{R}^m$, thus $df = f$. From $f_*(v(x)) = f(v(f^{-1}(x))$ and multiplicativity of deg under superposi-

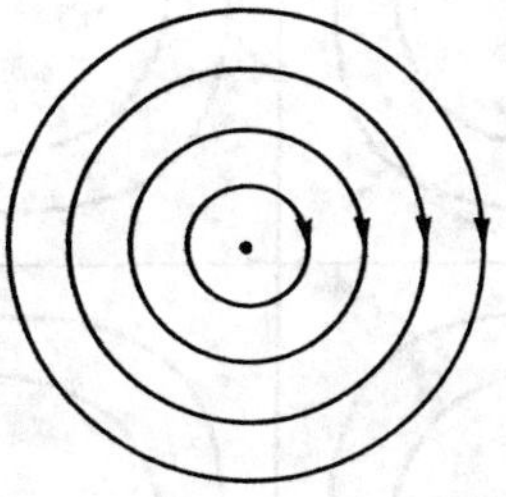

Fig. 13.

tion, we have therefore

$$\text{ind } f_*(v) = \deg(f|S^{m-1}) \text{ ind } v(0) \text{ ind}(f|S^{m-1})^{-1} = \text{ind } v(0.)$$

In the case of arbitrary diffeomorphism f, we can connect the matrix $df(0)$ in the Lie group $\mathrm{Gl}(m,\mathbb{R})$ with some orthogonal matrix by means of mapping $g(t)$ as follows: $t \in [0,1]$, $g(0) \in O(m)$, $g(1) = df(0)$. We have therefore $f(x) = df(0) + o(x)$, $h_t(x) = g(t) + to(x)$, $t \in [0,1]$; $h_1 = f$, $h_0 = g(0)$. Since deg is homotopy invariant, we have finally

$$\text{ind } f_* v(0) = \text{ind } h_{1*} v(0) = \text{ind } h_{0*} v(0) = \text{ind } v(0.)$$

2. If $\det(dv(0)) > 0$, we can, as in 1., connect (in $\mathrm{Gl}(m,\mathbb{R})$) $dv(0)$ with 1_m. Thus $\text{ind}(dv(0)) = +1 = \text{sgn} \det(dv(0))$. When $\det(dv(0)) = -1$, to the curve in $\mathrm{Gl}(m,\mathbb{R})$ we add a reflection σ, that is, a symmetry with respect to some hyperspace $\mathbb{R}^m$. But $\deg(\sigma|S^{m-1}) = -1$. □

REMARK. The point 2. shows that the definition of index of non-degenerate (regular) singular point of the field v as a sign of differential $dv(0)$, preceding the proof of Hopf and Lefschetz theorems, was justified.

Point 1. shows that the following definition is map–independent.

DEFINITION. Let M^m be a smooth manifold and $x \in M^m$ be an isolated critical point of a vector field defined in an neighborhood of x and let $(0, \varphi)$ be a map around x such that $\varphi(x) = 0$. Then we define

$$\text{ind } v(x) := \text{ind } v(\varphi_* v)(0.)$$

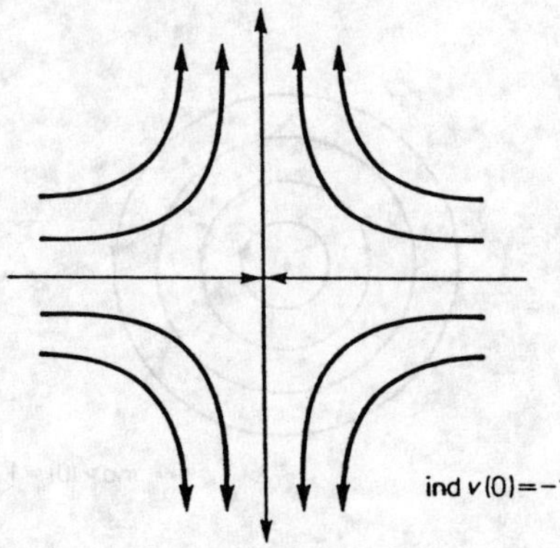

Fig. 14.

To finish this chapter we present

3.13 Relation between Morse index and index of a vector field

Let $x \in M^m$ be a critical point of smooth function $f : M^m\mathbb{R}$, that is, $df(x) = 0$. Then the second derivative d^2f defines a bilinear symmetric form $\operatorname{Hess} f(x) : T_xM \times T_xM \to \mathbb{R}$ called the Hessian or Hesse form of function f. The critical point x is non-degenerate if $\operatorname{Hess} f(x)$ is a non-degenerate form and its index, that is, the dimension of the maximal subspace T_xM on which this form is negative definite, is called the *Morse index* of function f in point x.

If M^m is equipped with Riemann metric, then with the differential $df(x)$ we can associate the vector field ∇f (often called gradient of f): $(f\nabla t) := df(t)$, $t \in T_xM^m$. We have almost immediately

PROPOSITION. *When x is a non-degenerate critical point of $f \in C^\infty(M^m)$, then x is a non-degenerate singular point of the vector field ∇f. If k is the Morse index of f in x, then*

$$\operatorname{ind} \nabla f(x) = (-1)^k.$$

PROOF is obvious. By choosing coordinates appropriately, we can make the Hessian $\operatorname{Hess} f(x)$ diagonal with non-vanishing diagonal elements, from whose exactly k have negative sign. The assertion follows from point 2. of the previous theorem.

Merging this proposition with Hopf index theorem, we obtain

Morse equality. Morse theorem. Let f be a smooth function on compact oriented manifold M. Let f possess only non-degenerate critical points and let the number of these points be finite. If A_k is a number of points where f has Morse index equal k, then the famous Morse equality holds:

$$\sum_{k=0}^{m}(-1)^k A_k = \chi(M.)$$

HISTORICAL REMARK. Systematic studies of vector fields on manifolds was first undertaken by H. Poincaré in his classical *Memoires* on curves defined by differential equations. These works grown from a completely new understanding of problems of dynamics ('sky mechanics', three-body problem, ...): Poincaré is certainly the greatest and most brilliant of all great French mathematicians and physicists: versatility of his interests, his speed of work, intuition, ingeniousness are indeed fantastic. He originated not only new theories, but completely new branches of mathematics. His ideas are so great, his way of thinking of and looking at mathematical reality has been so widely accepted, that to us, his descendants, it seems strange that people may have thought differently, e.g., that dynamical systems should be considered on differentiable manifolds and not only on $\mathbb{R}^n$ – because, indeed, a spherical pendulum is a motion on S^2. But even such a genius as Poincaré has its predecessors: Riemann, Kronecker, Gauss, and many others! But the bravery of some ideas of Poincaré was so great that some concepts, for example, the calculus of differential forms (we owe Poincaré possibly the most important theorem of analysis, Poincaré-Stokes theorem) must have waited for several dozen of years before, thanks to the works of E. Cartan, they attain acceptance of the society of mathematicians. The origin of the theory of intersection of cycles can be found in works of Gauss (was it because of his works on electricity?), Riemann, and Kronecker, to whose works from Monaschefe (1861) Poincaré *explicitly* refers to. But even in works of Poincaré we have many gaps and even errors: even his intuition sometimes failed (contrary to Riemann.)

CHAPTER 4

Chern–Gauss–Bonnet theorem

The idea of Chern is brilliantly simple: it is best to present the words of the master himself:

Let M^m be a compact oriented Riemann manifold of even dimension $(m = 2n.)$ We define on M^m a global m form. The C-G-B formula says that

$$(C) \qquad \int_{M^m} e(\Omega) = \chi(M^m.)$$

From M^m we turn to the bundle of spheres $SM^m := M^{2m-1} = \{t \in T(M^m) : \|t\| = 1\}$. We show that on M^{2m-1} the form $e(\Omega) = d\Pi$. Taking a field v of unit vector (that is, sections of the bundle $SM^m \to M^m$) with finite number of singular points (e.g., only one), we obtain as its image $v(M^m) =: V^m \to SM^m$ and the integral on the left hand side of (C) is equal to the integral

$$\int_{V^m} e(\Omega) = \int_{V^m} d\Pi = \int_{\partial V^m} \Pi \sum_{v(x)=0} \operatorname{ind} v(x) = \chi(M^m.)\text{"}$$

The first equality follows from Stokes theorem; the next two from Hopf index theorem. The main problem now is to construct the $(m-1)$ form Π. Here we can only present the result, referring to the beautiful original calculations or to its intelligible (and faithful) presentation in the monograph by R. Sulanke and P. Wintgen. We present the form of both fundamental forms $e(\Omega)$ and Π

$$(*) \qquad e(\Omega) = (-1)\frac{1}{2^{2n}\pi^n n!} \sum_{\sigma \in S_{2n}} \epsilon_\sigma \Omega_{\sigma(1)\sigma(2)} \cdot \Omega_{\sigma(3)\sigma(4)} \cdots \Omega_{\sigma(2n-1)\sigma(2n)},$$

here $\Omega_{ij} = -\Omega_{ji}$ are $m \times m$ matrices with elements being 2 forms. (Ω_{ij}) is called *curvature form* and is related to the curvature tensor R^∇ of the

Levi-Civita connection ∇ as follows $\Omega_{ij} = \sum R^{hk}_{ji} dx_h \wedge dx_k$. Expression $(*)$ is a form well-known from the theory of invariants: this is the *Pfaffian* of the skew-symmetric matrix (Ω_{ij}) (with values in commutative ring; multiplication of even order forms is commutative.) Thus

$$e(\Omega) = \frac{1}{(2\pi)^n} \mathrm{Pfaff}(R^{\nabla}.)$$

The form $e(\Omega)$ is called the *Euler form*, because we have

$$(C-G-B) \qquad \chi(M^m) = \frac{1}{(2\pi)^n} \int_{M^m} \mathrm{Pfaff}(R^{\nabla}) = \int_{M^m} e(\Omega.)$$

We will say few more words about Pfaffian of bilinear symplectic form $B(x, y)$ on vector space W of even dimension in the following section; here we observe only that if $A = (a_{ij})$, where $a_{ij} := B(e_i, e_j)$, where e_i is a basis in W, then the determinant $\det(A) = (\mathrm{Pfaff}(A))^2$. The Pfaffian is therefore a polynomial of elements of skew-symmetric matrix A, the form of whose was known already by Jacobi and Cayley

$$\mathrm{Pfaff}(A) = \frac{1}{2^n} \frac{1}{n!} \sum_{\sigma \in S_{2n}} \epsilon_\sigma a_{\sigma(1)\sigma(2)} \cdot a_{\sigma(3)\sigma(4)} \cdots a_{\sigma(2n-1)\sigma(2n)},$$

where S_{2n} is, clearly, the group of permutation of $2n$ elements, and $\epsilon(\sigma)$ is the sign of permutation σ.

The expression for the form Π ($d\Pi = e(\Omega)$) is less clear. Let us recall the form of the linear connection matrix ∇.

Let $\omega^1, \ldots, \omega^m$ be a (local) basis of the cotangent bundle $T^*(M)$, and let $\Gamma^i_{jk}(\cdot)$ be Christoffel symbols of connection ∇. Then the matrix (ω_{ij}) of 1 forms

$$\omega_{ij} := \sum_{k=1}^{m} \Gamma^i_{jk} \omega^k$$

is called the *connection matrix* ∇. The curvature matrix (Ω_{ij}) is obtained from the connection matrix as follows

$$\Omega_{ij} = d\omega_{ij} - \sum_k \omega_{ik} \wedge \omega_{kj}.$$

(In what follows we will omit the symbol of exterior multiplication $\wedge$, as we have already done in definition of the form $e(\Omega)$.) Following Chern, we take

$$(1) \qquad \Phi_k := \sum \epsilon_{\alpha_1 \ldots \alpha_{n-1}} \Omega_{\alpha_1 \alpha_2} \cdots \Omega_{\alpha_{2k-1} \alpha_{2k}} \omega_{\alpha_{2k+1}, m} \cdots \omega_{\alpha_{m-1}, m},$$

$$\text{(2)} \qquad \Pi := \frac{1}{\pi^n}\sum_{k=0}^{n-1}(-1)^k \frac{1}{1\cdot 3\cdots(2n-2k-1)2^{n+k}k!}\Phi_k.$$

One can check that indeed $-d\Pi = e(\Omega)$.

Since M^m is a Riemann space, in vicinity of any point $0 \in M$, we can introduce normal (geodesic) coordinates: if for the point $P \in M^m$ the (geodesic) distance from 0 is s, then the coordinates x^i of this point are given by $x^i = s\lambda^i$, where λ^i are directional cosines of the tangent vector of the geodesics $0P$. We transport parallelly a (unit) vector v from 0 to P along the geodesics $0P$. Let in the neighborhood $\{s \leqslant r\}$ (geodesic ball of radius r) a field of unit vector v be given and let coordinates of these vectors be smooth functions of normal coordinates (x^i), with possible exception of the point 0. Since the form Φ_k, $k \geqslant 1$ is at least of order 2 in x^i, there exists a constant N such that

$$\left|\int_S \Phi_k\right| < Ns, \quad k \geqslant 1,$$

where $S = S(x)$ is a geodesic sphere with center at 0 and radius s. If J is the index of v at the singular point 0, from Kronecker integral we have

$$J = \frac{1}{|S^{m-1}|}\int_S \omega_{1m}\cdots\omega_{m-1,m}.$$

There exists constant N_1 such that

$$\text{(3)} \qquad \left|J - (-1)^m \int_{S(x)} \Pi\right| < N_1 s,$$

thus

$$J = (-1)^m \lim_{s\to 0}\int_{S(x)} \Pi.$$

From formulas (1), (3), Stokes-Poincaré theorem, and Hopf theorem, we find therefore

$$\int_{M^m} e(\Omega) = (-1)^m J = (-1)^m \chi(M^m),$$

which completes the proof of Chern-Gauss-Bonnet theorem.

REMARKS. 1. In our sketch of the proof we omitted partially the construction of the field v with required singularity.

2. Some slightly different version of the proof is presented in the first paper by Chern, above we presented its modification presented in his second

paper. The first version is based on the following observation

PROPOSITION. *Let v be a vector field on M^m having singular points x_i, $i = 1, \ldots, k$. Thus the field $V : M^m - \{x_1, \ldots, x_k\} \to SM^m$ defines a submanifold $V^m \hookrightarrow SM^m$, whose boundary*

$$\partial V^m = \sum_{i=1}^{k} \operatorname{ind} v(x_i)\pi^{-1}(x_i), \quad \Pi : SM^m \to M^m,$$

where $\pi^{-1}(x_i) \hookrightarrow SM^m$ is a unit sphere in the space $T_{x_i}M^m$ over x_i. In particular, if the field v has only one singular point $0 = x_1$, then $\partial V^m = \chi(M^m) \cdot S^m$, where S^m is a unit sphere over 0.

Turning to the proof of C-G-B theorem, we immediately have

$$\int_{M^m} e(\Omega) = \int_{V^m} e(\Omega) = \int_{V^m} d\Pi = \int_{\partial V^m} \Pi =$$

$$\int_{\chi(M^m)S^m} \Pi = \chi(M^m) \int_{S^m} \Pi = \chi(M^m.)$$

□

4.1 Allendorfer-Weil formula

Chern proves this formula using the same kind of ideas. Let X^n be an oriented Riemann manifold (not necessarily compact) and let $P^n \hookrightarrow X^n$ be a compact submanifold with oriented boundary ∂P^n. Let again v be a field of unit vectors with finite number of singular points x_i, $i = 1, \ldots, k$ inside P^n; v does not have singular points on ∂P^n. Then

$$(A - W) \qquad \int_{P^n} e(\Omega) = -\int_{\partial P^n} \Pi + (-1)^n \chi'(P^n),$$

where $\chi'(P^n)$ is an internal Euler-Poincaré characteristics of P^n. The proof follows again from Poincaré-Stokes theorem and from proposition from Remark 2:

$$\int_{P^n} e(\Omega) = -\int_{\partial P^n} \Pi + (-1)^n \sum_i \operatorname{ind} v(x_i),$$

but

$$\sum_i \operatorname{ind} v(x_i) = \chi'(P^n.)$$

□

BROUWER FIXED POINT THEOREM follows immediately from (A-W). P^n is (now) an n-dimensional ball in $\mathbb{R}^n$ and let $f : P^n \to P^n$ be a continuous mapping of the ball into itself. Then f has at least one fixed point.

PROOF. Let f do not have any fixed point. We construct the vector field $v(x) := -(xf(x))^{\rightarrow}$(the vector from x to $f(x)$.) From assumption, v does not have singular points. But $\chi(P^n) = 1 \neq 0 = \sum \mathrm{ind}\, v$, the contradiction. □

Clearly, this proof is like breaking a butterfly on the wheel, but the original proof of Brouwer followed exactly this line.

Let us conclude this chapter with some historical remarks.

HISTORICAL REMARKS. As always, while contemplating great inventions, constructions, or notions – ideas, the question arises as to from where Chern took or as to how he invented the famous $2m$ form $e(\Omega)$ ($=$ Pfaff(R^∇)), being the natural generalization of the Gauss *curvatura integra* $K dx_1 \wedge dx_2$. One should look for the answer in the history.

Gauss-Bonnet theorem always fascinated geometers and topologists, for example, Heinz Hopf devoted a number of papers of his early period to the problem of *curvatura integra* in m-dimensional geometry ($m > 2$) of compact Riemann spaces. Led by brilliant instinct, he realized that the key is the notion of index of vector field in M^m: his predecessors were Kronecker, Poincaré, and Brouwer; the result was *Hopf index theorem.* In the case of a hypersurface M^m imbedded in $\mathbb{R}^{m+1}$, one could talk immediately about *curvatura integra* as a product $r_1 \cdots r_m$ of main curvatures, or define it with the help of Gauss mapping (cf. Chapter 1.) But in the case $M^m \subset \mathbb{R}^{m+k}$ one cannot proceed in this way. Here the major step was made by Carl B. Allendorfer in 1939 in his short paper *The Euler number of a Riemann manifold.* In the case $M^m \hookrightarrow \mathbb{R}^{m+k}$ Allendorfer finds the correct form $e(\Omega)$ which was later used by Chern. How he had found it? Here Allendorfer refers to thc, brand new at that time, work of Hermann Weyl *On volume of tubes* (1939); the tube is the Cartesian product $M^m \times S^q$, and to the Kronecker integral. This beautiful idea of Allendorfer has been revived by Allendorfer and André Weil in the celebrated work of 1944, we have mentioned already above.

There is no recipe for genius and inspiration! Even though Chern had in his disposal the Allendorfer form $e(\Omega)$, Kronecker integral, Hopf index theorem, and the work of Whitney on construction of a vector field with given singularities, but the brilliant construction of the forms $e(\Omega)$ and Π on the bundle of spheres $SM^m \to M^m$ and understanding of the manifold V^m as a manifold with boundary $\hookrightarrow SM^m$ defined by the vector field v on M^m is 'his own', and its simplicity and beauty take the breath away.

With the works of Chern of 1944 the theory of 'Chern characteristic classes' was born, and we turn to it now. Obviously, we cannot even mention here the concept of Stiefel characteristic classes which, few years earlier (1936), appeared in his Ph.D. thesis written in Zürich under supervision of Hopf, we cannot also discuss the works of Whitney (1937, 1941), Pontriagin (1942, 1944), Steenrod (1942–44), Ehresmann (1942, 1943) in which the topology of fiber bundles and their characteristic classes matures. I will only mention the splendid work by Chern of 1945 *Characteristic classes of hermitian manifolds* Ann. Math. 47 (1946), because this work has arisen directly from the works of Chern discussed in this chapter and presents, among other facts, the direct construction of Chern and Pontriagin (characteristic) classes.

CHAPTER 5

Curvature and Topology or Characteristic Forms of Chern, Pontriagin, and Euler

5.1 Chern forms

Let $\pi : E \to M$ be a smooth complex vector bundle of rank k (that is $\pi^{-1}(x) =: E_x \simeq \mathbb{C}^k$) over smooth manifold M equipped with connection ∇. Let R^∇ be the curvature of ∇. As we know, the curvature tensor $R = R^\nabla$ is related to the connection form $\omega = (\omega_{ij})$ and curvature form $\Omega = (\Omega_{ij})$ as follows $\Omega_{ij} = d\omega_{ij} - \omega_{ik} \wedge \omega_{kj}$, for short, $\Omega = d\omega - \omega \wedge \omega$, $\Omega_{ij} = R_{ij}^{hk} dx_h \wedge dx_k$. In what follows we will write indices below and we will sum, as usual, over repeating indices.

Because in the following considerations we will have to do with polynomials constructed out of elements of the curvature matrix Ω which are to have tensor character, that is, we will define on tensor fields M, we must introduce

Invariant polynomials. Let A_{ij} be elements of the matrix A, A is a $k \times k$ matrix with complex elements $A \in L(\mathbb{C}^k) \equiv M_k(\mathbb{C})$. Let $P : L(\mathbb{C}^k) \to \mathbb{C}$ be a complex polynomial of elements of $k \times k$ matrix satisfying the condition

$$(N) \quad P(AB) = P(BA), \quad \text{that is, } P(hAh^{-1}) = P(A) \text{ for } h \in GL(\mathbb{C}^k) \ .$$

Then P is called the *invariant polynomial.*

As an example we can mention the trace $\mathrm{tr} : (A_{ik}) \to \sum A_{ii}$ or the determinant $\det : (A_{ik}) \to \det(A_{ik})$. When P is an invariant polynomial, we can define the differential form $P(\Omega)$ of even order, because the curvature (that is, curvature form Ω) is a tensor, that is it transforms 'correctly' under change of coordinates. Sometimes, we will write $P(\Omega) \equiv P(\nabla)$ or $P(R^{\nabla})$ to stress that we have in mind the form resulting from given linear connection ∇ in the bundle $E \to M$. We know from elementary algebra that any symmetric polynomial of k variables $\lambda_1, \ldots, \lambda_k$ is a polynomial of elementary symmetric functions,

$$c_1 = \lambda_1 + \cdots + \lambda_k,$$
$$c_2 = \lambda_1\lambda_2 + \cdots + \lambda_{k-1}\lambda_k,$$
$$\cdots = \cdots\cdots$$
$$\chi_k = \lambda_1 \cdots \lambda_k.$$

As we know $1 + c_1 + \cdots + c_k = \prod_{j=1}^{k}(1 + \lambda_j)$.

DEFINITION. $c_j(E) = c_j(-i \cdot 2\pi)^{-1}\Omega \in H^{2j}(M)$ is called the *jth Chern form*;

$$c(E) := \det\left(1 + \frac{\sqrt{-1}}{2\pi}\Omega\right) = 1 + c_1(\Omega) + \cdots + c_k(\Omega) = 1 + c_1(E) + \cdots + c_k(E)$$

is called the *total Chern form*. Thus, $c_j(E)$ is the homogeneous summand of order j of the total form $c(E)$. Similarly one can construct analogous forms $P(\Omega)$, where now $P(\Omega)$ is a formally invariant power series, because the high powers $\Lambda^j\Omega$ vanish (for $2j > \dim M$.) Thus we have the very important form $\mathrm{ch}(E)$ called the Chern characteristics of bundle E

$$\mathrm{ch}(E) := \mathrm{tr}\, e^{(i/2\pi)}\Omega = \sum_j \frac{1}{j!}\,\mathrm{tr}\left(\frac{\sqrt{-1}}{2\pi}\Omega\right)^j.$$

REMARK. Clearly, above, all powers mean the powers of exterior multiplication, and there is no problem with multiplication ordering because for even order forms, this multiplication is commuting.

The forms $P(\Omega)$ play a fundamental role, because they define de Rham cohomology: they are closed. This is a fascinating relation between topology and curvature for differential manifolds. Moreover the such defined cohomology classes do not depend on choice of connection. This fact is expressed in important

THEOREM (CHERN–WEIL). *Let P be an invariant polynomial. Then*
(a) $dP(\Omega) = 0$ *(that is, $dP(\Omega)$ is a closed form);*
(b) *If we have two connections ∇_0, ∇_1 in the bundle $E \to M$, then the difference $P(\nabla_0) - P(\nabla_1)$ is a closed form: there exists the form $TP(\nabla_0, \nabla_1)$ such that $P(\nabla_0) - P(\nabla_1) = TP(\nabla_0, \nabla_1)$.*

PROOF is not difficult and can be found e.g., in the Gilkey's monograph (pp. 91, 92.)

From the theorem on symmetric functions, we have immediately

PROPOSITION. *If $P(\cdot)$ is an invariant polynomial, then there exists a unique polynomial Q such that $P(E) = Q(c_1(E), \ldots, c_k(E))$.*

This fact shows the great importance of Chern classes. Cohomology classes defined by Chern forms are called the *characteristic Chern classes.* Some important properties of these forms (classes) are collected in

THEOREM. *Let $E_1, E_2 \to M$ be vector bundles (equipped with connection.) Then*
(a) *Chern character is a homomorphism:*

$$ch(E_1 \oplus E_2) = ch(E_1) + ch(E_2), \quad ch(E_1 \otimes E_2) = ch(E_1) \wedge ch(E_2);$$

(b) $c(E_1 \oplus E_2) = c(E_1) \wedge c(E_2)$;
(c) $c(E^*) = 1 - c_1(E) + c_2(E) - \cdots + (-1)^k c_k(E)$;
(d) *if $f : Y \to M$ is a smooth mapping, f^*E is a pullback of the bundle E, then $c_j(f^*E) = f^*(c_j(E))$ – which makes Chern forms natural.*

PROOF. We present only the proof of point (c). Recall that if E^* is the bundle dual to $E = (E, \nabla)$, then E^* has connection ∇^* for which $< \nabla s, s_1^* > + < s, \nabla^* s_1^* >= d < s, s_1^* >$, where $s \in C^\infty(E)$, $s_1^* \in C^\infty(E^*)$.

For dual bases in the bundles E, E^*, the matrices of connection and curvature satisfy $\omega = -{}^t\omega_1$, $\Omega = -{}^t\Omega_1$. Thus

$$c(E^*) = \det\left(1 - \frac{i}{2\pi}{}^t\Omega\right) = \det\left(1 - \frac{i}{2\pi}\Omega\right)$$

from which the assertion follows.

REMARK concerning proposition and its proof. Here, we can make use of the theorem on symmetric functions, indeed, if P is an invariant polynomial and A is a diagonal matrix $A = \operatorname{diag}(\lambda_1, \ldots, \lambda_k)$, then $P(A)$ is a symmetric function of the variables $\lambda_1, \ldots, \lambda_k$. Proposition holds therefore for $P(A)$ with A diagonal. But diagonal matrices form a dense set (in the set of matrices) and P is a continuous function, therefore, $P(A) = Q(c_1, \ldots, c_k)$ for any A.

Todd forms. In his habilitation thesis, Fritz Hirzenbruch showed the great importance of forms introduced earlier (1937-39) by British mathematician I. A. Todd. These forms are given by formal series

$$\operatorname{td}(E) = \prod_{j=1}^{k} \frac{\lambda_j}{1 - e^{-\lambda_j}} = 1 + c_1(E) + \frac{1}{12} c_1^2(E) + c_2(E) + \frac{1}{24} c_1(E) c_2(E) + \ldots$$

Again the homogeneous expression of order r is denoted by td_r. We have

PROPOSITION. $\operatorname{td}(E_1 \oplus E_2) = \operatorname{td}(E_1)\operatorname{td}(E_2)$.

REMARK. If we have to do with the 'mother' of all bundles over M, the tangent bundle $E : TM \to M$, instead of $c(TM)$, $\operatorname{ch}(TM)$, $\operatorname{td}(TM)$, etc., we simply write $c(M)$, $\operatorname{ch}(M)$, $\operatorname{td}(M)$, ..., similarly as when we talk about metric tensor or connection on M, thinking about the tensor or connection in TM.

Now we present important

EXAMPLE. Let M be a Riemann surface of genus p. As we know, $\chi(M) = 2 - 2p$. Let ω be the volume 2 form on M, that is, a generator of the space $H^2(M)$. Then we have

$$c_1(TM) = (2 - 2p)\omega = \chi(M)\omega,$$
$$\operatorname{ch}(TM) = 1 + \chi(M)\omega,$$
$$\operatorname{td}(TM) = 1 + (1 - p)\omega.$$

This example is the basis of the famous Riemann-Roch theorem (R-R), which Hirzenbruch brilliantly generalized to arbitrary (complex) manifolds of arbitrary dimension. This theorem, which we will formulate in a moment, was, along with Chern-Gauss-Bonnet theorem, the main step on the way towards Atiyah-Singer index theorem.

HIRZENBRUCH-RIEMANN-ROCH THEOREM. *Let M be a complex compact manifold of dimension n, and let $V \to M$ be a holomorphic vector bundle* *over M. Hirzenbruch generalizes naturally the Euler-Poincaré characteristic by taking*

$$\chi(M,V)=\sum_{j=0}^{n}(-1)^j \dim H^j(M,V),$$

where $H^2(M,V)$ can be defined as a space of harmonic j forms on M valued in the bundle V. Then

$$(H-R-R) \qquad \chi(M,V)=\int_M \mathrm{ch}(V)\wedge \mathrm{td}(TM.)$$

Since, as we will see in a moment, $\chi(M,V)$, similarly to $\chi(M)$ is an index of some elliptic operator (d''_V), Hirzenbruch-Riemann-Roch theorem is an index theorem

$$\mathrm{index} d''_V=\int_M \mathrm{ch}(V)\wedge \mathrm{td}(TM.)$$

The fact that H-R-R makes indeed a gigantic evolution of the classical Riemann-Roch theorem can be most easily observed by recalling that the divisor D on a Riemann surface M is a formal integer combination of points $D=k_1x_1+\cdots+k_rx_r$, $k_i\in\mathbb{Z}, x_i\in M$ and can be represented by the holomorphic line bundle $E=E_D$ over M, whose holomorphic sections $M^0(M,E_D)$ are meromorphic functions f with multiplicity D, that is $(f)\geqslant -D$.

Let us pause here to recall some notions necessary to formulate the (R-R) theorem. As we know, Riemann discovered a fundamental role of singularities. The *divisor* D on compact Riemann space M is the mapping $D: M\to\mathbb{Z}$ such that there exists only finite number of points $x\in M$, for whose $D(x)\neq 0$. Divisors form the abelian group $\mathrm{Div}(M)$. A divisor is often written as a formal (linear) combination of points $D=n_{x_1}x_1+\cdots+n_{x_p}x_p$. The degree of D is denoted by $\deg D:=\sum_{x\in M}D(x)=\sum_i n_{x_i}$. With meromorphic function f one associates the divisor (f) as follows

$$\mathrm{ord}_a(f)=\begin{cases}0, & \text{if } f \text{ is holomorphic in } a \text{ and } f(a)\neq 0,\\ k, & \text{if } f \text{ has in } a \text{ a zero of order } k,\\ -k, & \text{if } f \text{ has in } a \text{ a pole of order } k,\\ \infty, & \text{if } f \text{ vanishes identically in some neighborhood of } a.\end{cases}$$

Thus the mapping $x\to\mathrm{ord}_x(f)$ is a divisor on M and is denoted by (f).

For a meromorphic 1 form $\omega=f\,dz$, we take $\mathrm{ord}_a(\omega):=\mathrm{ord}_a(f)$ to get the divisor denoted by (ω) and called the divisor of the form ω. In the group of divisors $\mathrm{Div}(M)$ we have partial ordering: if $D,D'\in\mathrm{Div}(M)$, we write

$D \leqslant D'$ if $D(x) \leqslant D'(x)$ for all $x \in M$. With a divisor D we can associate a holomorphic line bundle $E_D \to M$ such that its holomorphic sections are represented locally by meromorphic functions f such that $(f) \geqslant -D$.

Let $\psi_i \in M(U_i)$ (where $\{U - i\}$ is a covering of M) and let it be such that $(\psi_j) = D$ on U_i. Then $a_{ij} := \psi_i/\psi_j \in A(U_i \cap U_j)$, because over $U_i \cap U_j$ the functions ψ_i and ψ_j have the same poles. Let E_D be the line bundle (with fiber $\approx \mathbb{C}$) given by the cocycle (a_{ij}). If $(f) \geqslant -D$ over U, then there exist the functions $f_i \in A(U_i \cap U_j)$ such that $f = f_i/\psi_j$ over $U_i \cap U_j$. Over $U_i \cap U_j \cap U$, we have therefore $f_i/\psi_i = f_j/\psi_j$, thus $f_i = a_{ij} f_j$. Therefore the family $\{f_i\}$ defines the holomorphic section f of the bundle E_D with required properties. Recall that the canonical bundle $K \to M$ is defined by the cocycle $k_{ij} := dz_j/dz_i$, where (U_i, z_i) is a map on M and dz_j/dz_i is the Jacobian of transformation of one map to another. Holomorphic section of the bundle K are simply holomorphic 1 forms on M and $\dim H^0(M, K) = p(M)$ is the genus of M. If (a_{ij}) is a cocycle of the bundle $E \to M$, then the adjoint bundle $E^* \to M$ has the cocycle $({}^t a_{ij}^{-1})$, thus the bundle $K \otimes E^*$ is defined by the cocycle $K_{ij} a_{ij}^{-1} : U_i \cap U_j \to \mathrm{Gl}(k, \mathbb{C})$. In particular, sections of the bundle $K \otimes E_D^*$, that is, the elements of the vector space $H^0(M, K \otimes E_D^*)$, are meromorphic forms Ω such that their divisors $(\omega) \geqslant D$. This shows an important relation between the first Chern number $c_1(E_D)$ and $\deg D$:

PROPOSITION. *Let $p(M)$ be genus of the surface M*

$$\deg D = \frac{1}{2\pi i} \int_M c_1(E_D) =: |c_1|(E_D).$$

Then the Riemann–Roch theorem can be formulated as follows

$$(R - R) \qquad \dim H^0(M, E_D) - \dim H^1(M, E_D) = |c_1|(E_D) + 1 - p(M).$$

Let K be a canonical bundle. Then

$$\dim H^1(M, E_D) = \dim H^0(M, K \times E_D^*),$$

that is, to the dimension of meromorphic 1 forms with multiplicity D; (R-R) takes the form

$$(R - R) \quad \dim H^0(M, E_D) - \dim H^0(M, K \times E_D^*) = |c_1|(E_D) + 1 - p(M).$$

We see therefore that in the case of Riemann surface, Hirzenbruch theorem is indeed equivalent to Riemann-Roch theorem! A natural question arises as to if there exists a relatively simple theorem which would cover both (R-R) and Gauss-Bonnet theorems. The answer is affirmative, such theorem holds (Riemann-Roch-Kotake.) Let M be a Riemann surface with genus $p(M)$, $K \to M$ canonical bundle over M, E holomorphic vector bundle over M with typical fiber $\mathbb{C}^k$. Denoting by $H^0(M,E)$ the space of holomorphic sections of the bundle $E \to M$ and by $E^* \to M$ the bundle adjoint to E, we have

$$(K) \quad \dim H^0(M,E) - \dim H^0(M, K \otimes E^*) = |c_1|(\det E) + k(1 - p(M)).$$

Taking in (K) $E = E_D$, we get (R-R) Let us recall that if we have a covering (U_i) of manifold M and when the vector bundle $E \to M$ is given by the cocycle $a_{ij} : U_i \cap U_j \to GL(k, \mathbb{C})$, then the line bundle $\det E \to M$ is given by the cocycle $\det a_{ij} : U_i \cap U_j \to \mathbb{C}$ and the hermitian structure of the bundle $E \to M$ is given by the family (A_j) of hermitian $k \times k$ matrices such that ${}^t g_{ik} A_j \overline{g_{ik}} = A_k$ on $U_i \cap U_j$. As a consequence, we have

$$\det A_j |\det g_{jk}|^2 = \det A_k \quad \text{on } U_i \cap U_j \ .$$

Thus the family $\det A = (\det A_j)$ defines hermitian structure on the bundle $\det E$. Since the functions $\det g_{jk}$ are holomorphic, $\{\partial\overline{\pi} \log(\det A_j)\} =: \partial\overline{\pi}\log(\det A_j)$ defines a closed form of the type $(1,1)$ which is nothing but the curvature from of the bundle $\det E$. We cannot, obviously, present here the proof of Kotake formula (K). Let us note only that it can be obtained with the help of heat transfer equations, which we discussed above (pages 58– 62.) One constructs two elliptic operators L, $\tilde{L}$ such that

$$(*) \qquad \dim H^0(M,E) - \dim H^0(M, K \otimes E^*)$$

$$= \dim E_0(L) - \dim E_0(\tilde{L}) = \dim(\ker L) - \dim(\ker \tilde{L})$$

and then on computes the difference of traces of corresponding Green functions of operators of heat transfer. When $dM = i/2h(z,\bar{z})dz \wedge d\bar{z}$ is the volume form of the surface M, then Kotake shows that the right hand side of $(*)$ has the form (Kotake formula)

$$(K) \qquad \frac{1}{2\pi i}\int_M d'd'' \log(\det A) + \frac{k}{4\pi i}\int_M d'd''\log h.$$

And we get the beautiful Kotake formula

$$(K') \qquad \dim H^0(M,E) - \dim H^0(M \otimes E^*)$$

$$\frac{1}{2\pi i}\int_M d'd''\log(\det A) + \frac{k}{4\pi i}\int_M d'd''\log h.$$

But

$$\frac{1}{2\pi i}\int_M d'd''\log h = \frac{1}{2\pi}\int_M \Omega = c_1(TM) = c_1(M);$$

where Ω is the curvature form of the bundle TM. Taking $E = 1 := M \times \mathbb{C}$ (trivial bundle), we obtain

$\dim H^0(M, K \otimes \mathbb{C}) = 1$ (because holomorphic functions on M are constant),

$$\dim H^0(M, K \otimes 1^*) = \dim H^0(M, K) = g(M.)$$

But $A = 1 \in \mathbb{C}$, and $\log\det A = 0$, and the left hand side of $(*)$ is equal

$$1 - p(M) = \frac{1}{4\pi}\int_M \Omega.$$

Finally

$$(G-B) \qquad 2(1 - p(M)) = \chi(M) = \frac{1}{2\pi}\int_M \Omega,$$

thus Gauss-Bonnet theorem. But we have

$$\frac{1}{2\pi i}\int_M d'd''\log(\det A) = |c_1|(\det E) = \int_M c_1(\det E),$$

and from $(*)$ and $(G-B)$, we finally obtain

$$(K-R-R) \quad \dim H^0(M, E) - \dim(M, K \otimes E^*) = |c_1|(\det E) + k(1 - p(M).)$$

In Part III we will present an ingenious proof of the (K-R-R) formula.

Now we can return to characteristic forms of real vector bundles.

5.2 Pontriagin forms. Pfaffian R^∇. Chern theorem once again

Characteristic forms of real vector bundles. Pontriagin forms. For complex bundles, invariant polynomials were understood as $GL(k, \mathbb{C})$ invariant polynomials . If E is a real bundle, then $E_{\mathbb{C}} := E \otimes \mathbb{C}$, that is, the complexification of E is already complex bundle. Introducing in fibers E_x the Euclidean metric $(\cdot|\cdot)_x$, we reduce the group $GL(k, \mathbb{R})$ down to the orthogonal group $O(k)$, and from now on we regard E as a Riemann bundle equipped with Riemann connection and local bases are orthonormal bases. Under these assumptions, the curvature matrix Ω is a skew-symmetric matrix of 2 forms (we made this assumption already while proving (C-G-B) theorem.) Since $E_{\mathbb{C}}$ is the complexisication of the real bundle E, then the natural isomorphism $\equiv E^*$ defined by Riemann metric, defines a $\mathbb{C}$-linear isomorphism of the bundles $E_{\mathbb{C}}$ and $E^*_{\mathbb{C}}$. Therefore (as a result of properties of Chern classes collected in the theorem) we have $c_j(E) = 0$ for odd j. Therefore locally

$$\det\left(1 + \frac{i}{2\pi}A\right) = \det\left(1 + \frac{i}{2\pi}{}^tA\right) = \det\left(1 - \frac{i}{2\pi}A\right)$$

when $A = -{}^tA$. Therefore, $c(A)$ is an even polynomial. In this way, we arrive at the definition of Pontriagin forms p, p_j:

$$p(E) = p(\Omega) := \det\left(1 + \frac{1}{2\pi}\Omega\right) = 1 + p_1(\Omega) + p_2(\Omega) + \ldots,$$

where $p_j(\Omega) = p_j(E)$ is a homogeneous function of order $2j$ in components of Ω. Thus for a real bundle $E \to M$, we have

$$p_j(E) = H^{4j}(M), \quad p_j(E) = (-1)^j c_j(E \otimes \mathbb{C}.)$$

This last equality is usually taken as the definition of Pontriagin form. Let $P(A)$ be an $O(k)$-invariant polynomial. Defining $P(\Omega)$ as above, for Riemann connection in the bundle E we obtain

PROPOSITION. *Any $O(k)$-invariant polynomial P such that $P(A) = P(gAg^{-1})$ for $g \in O(k)$; $A = -{}^tA$ is of the form $Q(p_1, \ldots, p_j)$ for some polynomial Q. Thus Pontriagin forms generate the set of characteristic forms $P(\Omega)$.*

As an example we present forms introduced by Hirzenbruch

$$L(x) = \prod_j \frac{x_j}{\tan hx_j} = 1 + \frac{1}{3}p_1 + \frac{1}{45}(7p_2 - p_1^2) + \ldots = \sum_j L_j(p_1, \ldots, p_j.)$$

Thus

$$L_0 = 1, \;\; L_1 = \frac{1}{3}p_1, \;\; L_2 = \frac{1}{45}(7p_2 - p_1^2), \;\; L_3 = \frac{1}{945}(62p_3 - 13p_1p_2 + 2p_1^3.)$$

$L(\cdot)$ is called the *Hirzennbruch L-polynomial*, $\hat{A}$, the (Hirzenbruch, or, in physical literature Dirac) genus is defined as follows

$$\hat{A}(x) = \prod_j \frac{x_j}{2\sin h(x_j/2)} := 1 + \frac{1}{24}p_1 + \frac{1}{5760}(7p_2 - p_1^2) + \ldots$$

These characteristic forms appear in the famous

5.3 Hirzenbruch signature theorem

The notion of signature of compact Riemann space of dimension $4l$ was earlier introduced by H. Weyl, even before the Hodge theory appeared. Using harmonic forms, we know that the bilinear form

$$B(\alpha, \beta) = \int_M \alpha \wedge \bar{\beta}$$

defines the quadratic form $\omega \to Q(\omega) := B(\omega, \omega)$. Signature of this form is called the *signature of manifold M* and is denoted by Sign M.

EXAMPLE. $M = M^4$. Then, similarly to (C-G-B), we have

$$\text{Sign } M = \frac{1}{12\pi^2}\int_M (\Omega_{12}^2 + \Omega_{13}^2 + \Omega_{14}^2 + \Omega_{23}^2 + \Omega_{24}^2 + \Omega_{34}^2.)$$

In general, we have beautiful

SIGNATURE THEOREM (Hirzenbruch) *If $M = M^{4k}$, then the signature*

$$\text{Sign } M = \int_M L_k(p_1(\Omega), \ldots, p_k(\Omega).)$$

This theorem, surprisingly, plays recently an important role in a quantum field theory, the theory of (super)strings.

It is not hard to construct the elliptic operator D, called also the Hirzenbruch-Hodge operator, such that ind D = Sign M. Thus both great theorems of Hirzenbruch, (R-R) and signature theorem, can be regarded as index theorems for some elliptic operators, similarly as Chern-Gauss-Bonnet theorem can be regarded as an index theorem for the operator $D = d + d^*$. Before we turn to the formulation of general index theorem of Atiyah and Singer, let us return once again to Chern theorem. There we had to do with Riemann manifold M of even dimension $2n$ and tangent bundle TM equipped with Riemann connection, in this case, Levi-Civita connection ∇. A central role was played by the form Pfaff(Ω) called the Euler form $e(M)$. It is easy to observe that Pfaff(A) is a $O(2n)$-invariant form for $-A = {}^tA = A^*$. Indeed, $\det A = (\text{Pfaff}A)^2$ and $\text{Pfaff}(gAg^*) = \text{Pfaff}(A)\det g = \text{Pfaff}(A)$. From general properties of invariant polynomials (Chern-Weyl theorem) we know that $d\text{Pfaff}(\Omega) = 0$ and thus Pfaff(Ω) represents the class $H^{2n}(M)$ of De Rham cohomology. The question arises what is a relation between nth Pontriagin form $p_n(\Omega)$ and the Euler class $e(M) = \text{Pfaff}(\Omega)$. The answer is given by

PROPOSITION. *Let $E = TM^{2n}$. Then*

$$e(M) := e(TM) = p_n(TM) = c_{2n}(TM \otimes \mathbb{C}.)$$

In particular for $n = 1$, thus M^2, we have $e(\Omega) = \frac{1}{4}(\Omega_{21} - \Omega_{12}) = \frac{1}{2\pi}K$.

In general

$$e(E) = \text{Pfaff}(\Omega^\nabla) = \text{Pfaff}(R^\nabla),$$

$$e(E) = p_n(\Omega.)$$

5.4 General index theorem (Atiyah-Singer)

In the case when M is a compact Riemann manifold of dimension m and when D is an elliptic operator $D : C^\infty(E) \to C^\infty(F)$, where $E, F \to M$ are vector bundles over M, Atiyah and Singer presented an ingenious construction of the vector bundle $\sigma(D) \to M$, called the symbol of the operator D such that we have the formula

$$(A - S) \qquad \text{index } D = \int_M (-1)^m \text{ch}(\sigma(D)) \text{td}(TM \otimes \mathbb{C}.)$$

We cannot present here the construction of the bundle $\sigma(D)$. There exist a number of proofs of Atiyah-Singer index theorem, and new proofs still appear. The most complete proof (except for original papers) is given in the monograph of the great expert in the problem Peter B. Gilkey *Invariance Theory. The Heat Equation and the Atiyah-Singer Index Theorem.* Recently the proofs appeared which made use of the methods of stochastic processes – this one actually could have expected because indeed equations of heat transfer are closely related with stochastic processes. Nowadays physicists present 'simple proofs' of index theorems. These are very stimulating for mathematicians: physical intuitions always fertilized mathematicians and *vice versa*! In this connection, the reader may find interesting the nice monograph by Michio Kaku, *Introduction to Superstrings*, Springer Verlag, 1988.

CHAPTER 6

Kähler Spaces. Bergman Metrics. Harish-Chandra-Cartan Theorem. Siegel Space (once again!)

On October 22, 1932, the twenty eight years old Privatdocent of Hamburg University, Erich Kähler, sent for publication a short (13 pages) paper entitled *Über eine bemerkenswerte Hermitische Metrik.* This modest paper, was to play an important role after the war, that was probably unexpected even for the author himself!

In order to better understand the notion of hermitian and Kähler metrics an complex manifold, let us recall the simple algebraic fact.

LEMMA. *let V be a complex vector space. Then the following sets are canonically isomorphic:*

(i) *hermitian forms h on V, that is, $h : V \times V \to \mathbb{C}$ such that*

$$h(ax_1 + bx_2, y) = ah(x_1, y) + bh(x_2, y),$$

$$h(x, y) = \overline{h(y, x)};$$

(ii) *quadratic forms g on V invariant with respect to multiplication by $i = \sqrt{-1}$*

$$q(ix, iy) = q(x, y);$$

(iii) *skew symmetric 2 forms ω on V invariant with respect to multiplication by i*

$$\omega(ix, iy) = \omega(x, y.)$$

The isomorphisms are given by

$$q = \operatorname{Re} h, \quad \omega = \operatorname{Im} h,$$

$$h(x,y) = q(x,y) - iq(ix,y),$$
$$h(x,y) = \omega(ix,y) + i\omega(x,y.)$$

Let us present now several equivalent definitions of the Kähler metric. If h is a hermitian metric on complex manifold M, that is, every fiber T_xM is equipped with hermitian (positive) scalar product $h_x = (\cdot|\cdot)_x$, where the function $x \to h_x$ is smooth, then writing $h = \operatorname{Re} h + i \operatorname{Im} h$ we see that if $g := \operatorname{Re} h$ is a Riemann metric on M and $ds^2 = (i/2)g_{i\bar{k}}dz_i d\bar{z}_k$, $g_{kj} = \overline{g_{i\bar{k}}}$, then $\omega : \operatorname{Im} h$ is a symplectic 2 form on M, $\omega = i/2(g_{jk}(dz_j \wedge d\bar{z}_k)$.

DEFINITION. A hermitian metric h on complex manifold M is *Kählerian* if the form $\omega := \operatorname{Im} h$ is closed, $d\omega = 0$.

From this definition we have immediate

COROLLARY. *On Riemann surface M^2 every hermitian metric is Kählerian because any 3 form, including $d\omega$ vanishes identically.*

EXAMPLE. On $\mathbb{C}^{n+1}$, the metric $h = dz_0 \otimes d\bar{z}_0 + \cdots + dz_n \otimes d\bar{z}_n =: (dz|dz)$ is Kählerian because $\operatorname{Im} \omega$ has constant coefficients.

EXAMPLE. The Fubini-Study metric on $P(\mathbb{C}^{n+1})$ given by the formula $h = \pi^* \tilde{h}$,

$$\tilde{h} := (z|z)^{-2}\{(z|z)(dz|dz) - (dz|z)(z|dz)\},$$

where $\pi : \mathbb{C}^{n+1} - \{0\} \to P(\mathbb{C}^{n+1})$ and $(\cdot|\cdot)$ are defined in Example above is Kählerian; this is, without doubts, the most important example.

PROOF. It is enough to show that $d\tilde{\omega} = 0$, where

$$\tilde{\omega} = \operatorname{Im} \tilde{h} = \frac{1}{2} i (z|z)^{-2}\{(z|z) dz \wedge d\bar{z} - [\bar{z}dz] \wedge [z d\bar{z}]\}.$$

To this end, we take $d^c := i(d'' - d')$ and check that

$$\tilde{\omega} = \frac{1}{4} dd^c \log(z|z) = \frac{1}{2} dd^c \log \|z\|.$$

□

This is, without doubts, the most important example of Kähler metric, as it follows from the following

THEOREM (CHOW). *Any complex submanifold of complex projective space $P(\mathbb{C}^{n+1})$ is algebraic (that is, is a set of zeros of a family of polynomials.)*

PROPOSITION. *Let X be a complex manifold equipped with Kähler metric h and let $j : M \hookrightarrow X$ be an imbedding (holomorphic.) Then $h_M : j^*h$ is a Kähler metric on M.*

PROOF. Let $\omega = \operatorname{Im} h$; if $j^*\omega = \omega_M$, then $\operatorname{Im} h_M = \omega_M$ and $d\omega_M = d(j^*\omega) = j^*d\omega = 0$. □

COROLLARY 1. *Any (complex) algebraic manifold possesses a natural Kähler metric induced by the Fubini-Study metric.*

COROLLARY 2. *Any complex submanifold $M \hookrightarrow \mathbb{C}^n$ (and thus any Stein manifold) is Kählerian.*

Already these examples and corollaries show the importance of Kähler spaces, that is, a manifold equipped with Kähler metric. The following theorem gives interesting characterizations of Kähler spaces.

THEOREM. *Let (M, h) be a complex hermitian manifold. Then the following conditions are equivalent.*

1. *h is Kählerian, that is $d(\operatorname{Im} h) = 0$;*
2. *in a neighborhood of any point $x \in M$ there exists a holomorphic map $(z_1, \ldots, z_n)$ such that the matrix $h_{jk} = h\left(\frac{\partial}{\partial z_j}, \frac{\partial}{\partial z_k}\right)$ differs from unit matrix only by small quantities of order two;*
3. *Levi-Civita parallel transport is a $\mathbb{C}$-linear mapping, that is,*

3′. *the operator of complex structure J has vanishing covariant derivative $\nabla J = 0$;*

4. *the metric h possesses Kähler potential p, that is,*

$$h_{j\bar{k}} = \frac{\partial p}{\partial z_j \partial \bar{z}_k}$$

thus h is plurisubharmonic on M;

5 *the only non-vanishing Christoffel symbol of the metric h is*

$$\Gamma^c_{ab} = h^{c\bar{c}} \frac{\partial h_{a\bar{c}}}{\partial z_b}.$$

COROLLARY 3. *Ricci tensor of Kähler metric* $R_{ab} = R^{\bar{c}}_{ab\bar{c}}$ *has the form*

$$(*) \qquad R_{a\bar{b}} = -\frac{\partial \log \det(h_{\mu\nu})}{\partial z_a \partial \bar{z}_{\bar{b}}},$$

and the curvature form $\Omega = R_{a\bar{b}} dz_a \wedge \bar{z}_{\bar{b}}$ *is closed.*

REMARKS. (a) Point 2. says that Kähler spaces regarded as Riemann spaces are particularly close to Euclidean spaces, in general Riemann spaces approximate Euclidean spaces only up to small quantities of order one.

(b) The existence of Kähler potential indicates a role of these spaces in complex analysis.

(c) We have already encountered many times expression (*) for curvature form in two dimensional case, that is, Riemann surface

$$R_{a\bar{b}} = -\frac{\partial(\log h)}{\partial z_a \partial \bar{z}_{\bar{b}}}$$

REMARK. On compact complex manifold we have Laplacians $\Delta_{p,q}$ acting on spaces $A^{(p,q)}(M)$ of forms of type (p,q). Let us denote by $b_{(p,q)}$ the dimension of the space of harmonic forms of type (p,q), that is, the forms $a_{IJ} dz^I \wedge d\bar{z}^J$, where $I = (p, |J|) = q$; in general $b_{(p,q)} \neq b_{(q,p)}$, but for Kähler spaces $b_{(p,q)} = b_{(q,p)}$!

Already in the above mentioned paper, Kähler observed close relation between his theory and the theory of relativity. As we know, an Einstein metric is a (pseudo)Riemannian metric g_{ij} such that

$$\mathrm{Ric} = \lambda g \quad \text{that is,} \quad R_{ij} = \lambda g_{ij}.$$

The hermitian metric h such that $\lambda g_{a\bar{b}} = R_{a\bar{b}}$ is called the *Kähler-Einstein* metric and plays an important role in the physical theories proposed recently: the theory of (super)strings. In connection with this, let me mention

6.1 Calabi hypothesis and Calabi-Yau spaces

Already in 1954, E. Calabi formulated an important hypothesis. Let M be a compact Kähler space with the form $\omega = i/2(g_{j\bar{k}} dz_j \wedge d\bar{z}_k)$ and let the form of the type $(1,1)$ $\sigma := \tilde{R}_{j\bar{k}} dz_j \wedge d\bar{z}_k$ be given, representing the first Chern class $[c_1(M)] \in H^2(M)$. Then on M there exists a Kähler metric $d\tilde{s}^2$ having $\tilde{R}_{j\bar{k}}$ as its Ricci tensor such that its fundamental form

$$\tilde{\omega} = i/2(\tilde{g}_{j\bar{k}} dz_j \wedge d\bar{z}_k)$$

is in the same cohomology class as ω.

In 1977, Shing-Tung Yau showed that this hypothesis is true. Moreover, he proved another claim by Calabi on existence of a unique Kähler–Einstein metric on Kähler manifold having a rich (ample) canonical line bundle $K = K(M)$. Both these claims reduces to solving the complex Monge-Ampére equation for real function φ:

$$(*) \qquad \det\left(g_{j\bar{k}} + \frac{\partial\varphi}{\partial z_j \partial\bar{z}_k} \right) = \det(g_{j\bar{k}}) e^{(c\varphi+F)}, \quad c = 0, 1;$$

where F is a given function such that $\int_M e^F = |M|$ for $c = 0$. Yau solved (even more general than) equation $(*)$, which is very highly non-linear, by introducing appropriate *a priori* estimates of a solution and its derivatives. As the celebrated L. Nirenberg said, in his *laudatio* in Warsaw in 1983 on the occasion of presenting Yau the Fields medal: 'The derivation of these estimates, though classical in spirit is a tour de force.' Yau theorems have a number of important consequences (also for physics), for example

(a) The only Kähler structure on complex projective space $P(\mathbb{C}^{n+1})$ is the classical one, that is the Fubini-Study metric. It is hard not to be satisfied with such a discovery: Fubini and Study found more than half a century ago the *only good* metric on the mother of all algebraic manifolds, the projective space.

(b) Solution of the *Severi conjecture* in affirmative: if a complex two dimensional surface (that is, M^4) is homotopic with $P(\mathbb{C}^3)$, then it is also biholomorphic equivalent to it.

(c) Yau (with Shoen) solved also in affirmative an outstanding and long lasting (since the formulation of the theory in 1917) problem in classical general relativity: the total energy of a finite distribution of masses and fields is positive.

I present here these impressive theorems also because they forcibly show that deep facts from global differential geometry can be sometimes obtained with the help of very complex investigations in the theory of partial differential equations, to cite Hermann Weyl, 'it is not sufficient to open eyes widely' (Weyl aimed this famous bitter remark at his contemporary philosophers.)

6.2 Bergman metrics on bounded domains

Bergman metrics are, as we know, yet another beautiful example of Kähler metrics, these are a natural generalization of Poincaré metric on unit disc in $\mathbb{C}$. Let O be a bounded domain in $\mathbb{C}^n$. As usual, let us denote by $L^2(O)$ the Hilbert space of functions square integrable with respect to the Lebesgue measure $d\mu$ on $\mathbb{C}^n = \mathbb{R}^{2n}$. Let $A^2(O)$ be a subset of $L^2(O)$ formed by functions holomorphic on O, that is $A^2(O) = A(O) \cap L^2(O)$. then, as we know (cf. *Analysis* II) for any compact subset $A \subset O$ there exists a constant $a(A) > 0$ such that $|f(z)| \leqslant a(A)\|f\|_{L^2(O)}$. It follows that on $A^2(O)$, convergence in the sense of Hilbert norm $\| \ \|$ results in uniform convergence, together with all derivatives, on any compact subset (with such a topology $A^2(O)$ is even a nuclear space with continuous imbedding $A^2(O) \hookrightarrow L^2(O)$. Thus $A^2(O) \hookrightarrow L^2(O) \hookrightarrow A(O)'$ is a Gelfand triple which has important consequences for spectral theory.) If $f_1, f_2, \ldots$ is an orthonormal basis in $A^2(O)$, then

$$(B) \qquad B_O(z,\bar{\zeta})B(z,\bar{\zeta}) := \sum_{n=0}^{\infty} f_n(z)\overline{f_n(z)}, \quad z, \bar{\zeta} \in O,$$

is a function on $O \times O$ which is holomorphic in z and antiholomorphic in the second variable, because the series (B) is uniformly convergent in any compact set $O \times O$.

DEFINITION. The function $B(\cdot,\cdot)$ is called the *Bergman kernel* and the hermitian metric

$$h_O(z,\bar{z}) = \sum_{j,k=1}^{n} \frac{\partial \log B(z,\bar{z})}{\partial z_j \partial \bar{z}_k} dz_j d\bar{z}_k$$

is called the *Bergman metric.*

This metric has several beautiful properties collected in

THEOREM. *Let O be a bounded domain in $\mathbb{C}^n$. Then*

(i) *h_O defines Kähler metric g on the manifold O: $g := \operatorname{Re} h$;*

(ii) *if $\varphi : O \to O_1$ is a holomorphism (that is holomorphic diffeomorphism) and if B and B_1 are Bergman kernels on O and O_1, then*

$$B(z,\bar{z}) = B_1(\varphi(z), \overline{\varphi(z)})|j(z,\varphi)|^2$$

where $j(z,\varphi)$ *is a Jacobian from* φ *to* z;

(iii) $f(z) = \int_O B(z,\zeta)\varphi(\zeta)d\mu(\zeta)$ *for any* $f \in A^2(O)$ *(reproducing property);*

(iv) B_O *does not depend on the choice of orthonormal basis* (f_j) *in* $A^2(O)$.

REMARKS. 1. Later Aronshajn observed that the inequality $|f(z)| \leqslant a(A)\|f\|$, from which (iii) and (iv) immediately follow, is fundamental for many functional spaces and on this basis he formulated his 'theory of reproductive kernels.'

2. The fact that O is bounded in $\mathbb{C}^n$ is an important assumption: we know that the only holomorphic functions on (whole) $\mathbb{C}^n$ are constants, thus for $O = \mathbb{C}^n$ Bergman kernels are meaningless.

3. Stefan Bergman created his theory in 1933, and thus not long after the paper of Kähler, in the famous paper *Über Kernfunktion eines Bereiches, ihr Verhalten am Rande*, J. Reine u. angew. Math. 169 (1933) 1–42 and ibid. 172 (1935), 89–128. Bergman was that time (similarly to Bochner) a professor at Berlin University.

6.3 Imbedding in projective spaces. Kodaira theorem

S. Kobayashi extended the notion of Bergman kernel to more general complex spaces in the following way:

Let M be a n-dimensional complex manifold and let H be a complex Hilbert space of holomorphic n forms f which are square integrable, to wit

$$\int_M i^{n^2} f \wedge \bar{f} < \infty, \quad (f|g) := \int_M i^{n^2} f \wedge \bar{g}.$$

It is said that H is *very ample* if the following two conditions are satisfied:

(a) for every $x \in M$ there exists $f \in H$ such that $f(x) \neq 0$ (thus elements of H split points x),

(b) if $(z_1, \ldots, z_n)$ is a map around x, then for any j there exists $h \in H$ such that $h^* dz_1 \wedge \cdots \wedge dz_n$ and $h(x) = 0$, $\frac{\partial h^*}{\partial z_j} \neq 0$. If now is a orthonormal basis in H, then

$$B_M \equiv B = b dz_1 \wedge \cdots \wedge dz_n \wedge d\bar{z}_1 \wedge \cdots \wedge d\bar{z}_n := \sum_{k=0}^{\infty} h_k \wedge \bar{h}_k$$

is a Bergman kernel in M.

The Bergman metric on M is again defined by

$$(*) \qquad ds^2 = g_{j\bar{k}} dz_j d\bar{z}_k, \quad \text{where } g_{j\bar{k}} := \frac{\partial \log B}{\partial z_j \partial \bar{z}_k}.$$

Such defined kernel B_M possesses all properties of classical Bergman function for bounded region: condition (a) guarantees that $B \neq 0$ everywhere, that is for $b > 0$, and condition (b) guarantees positive definiteness, thus ds^2 is hermitian metric. Kobayashi gives interesting geometric interpretation of conditions (a), (b).

Let for $x \in M$, $H(x)$ denotes $\{h \in H : h(x) = 0\}$. Condition (a) means that $H(x)$ is a hypersubspace in H, that is, it is a subspace of codimension one. As we know, a family of hypersubspaces forms a complex projective space isomorphic with the space of lines (that is, one dimensional subspaces) in H^*, H^* is a Hilbert space adjoint to H, thus $\{H(x), x \in M\}$ can be identified with the projective space $\mathbb{P}(H^*)$. We have therefore the mapping

$$M \ni x \xrightarrow{i} H(x) \in \mathbb{P}(H^*.)$$

Condition (b) says exactly that the mapping i is an imbedding (injection) $i : M \hookrightarrow \mathbb{P}(H^*)$ and the Bergman metric given by $(*)$ is nothing but the pull back i^* of the Fubini-Study metric on $\mathbb{P}(H^*)$.

This remark by Kobayashi shows the beauty and necessity of Bergman construction, it is not surprising that Bergman kernels play more and more important role in complex analysis, and recently in physics.

COROLLARY. *The space $A^2(O)$, with O being a bounded domain in $\mathbb{C}^n$ is very ample, indeed.*

Kodaira theorem on embedding in $\mathbb{P}(\mathbb{C}^{n+1})$. The notion of very ample space H introduced by Kobayashi is, without doubts modeled after the notion of *ample* holomorphic line bundle (HLB) $F \to M$ over complex compact manifold M which plays a fundamental role in famous proof of Kodaira theorem providing *inner* characterization of algebraic varieties. Let $s_0, \ldots, s_n$ be holomorphic sections of HWP $F \to M$ and let in every point $x \in M$, for at least one $j : s_j \neq 0 \in F$. If $\alpha : F_x \simeq \mathbb{C}$ is some isomorphism, then $\alpha s_0(x), \ldots, \alpha s_n(x) \neq 0 \in \mathbb{C}^{n+1}$. Since a is defined up to a (multiplicative) factor, we obtain in this way the mapping $M \ni x \to \psi(x) \in \mathbb{P}(\mathbb{C}^{n+1})$; ψ is thus a holomorphic mapping $\psi : M \to \mathbb{P}(\mathbb{C}^{n+1}$.

DEFINITION. The holomorphic line bundle $F \to M$ is *ample* if there exist a natural number k and holomorphic sections $s_0, \ldots, s_k$ of the bundle F^k (kth tensor power of the bundle F) such that $\psi \hookrightarrow \mathbb{P}(\mathbb{C}^{n+1}$ is a biholomorphic imbedding.

Therefore, if there exist an ample HLB over M, then M is an algebraic variety (Chow.) Kunihiko Kodaira presented in 1954 a simple criterion of ampleness of a bundle $F \to M$ and thus of the fact that M is algebraic.

THEOREM (KODAIRA). *($F \to M$ is ample) $\Longleftrightarrow$(the first Chern form $c_1(E)$ is cohomological with real 2 form $\varphi = ig_{j\bar{k}}dz_j \wedge d\bar{z}_k$ on M such that the hermitian form $g_{j\bar{k}}dz_jd\bar{z}_k$ is positive definite in any point.*

(Recall that if $[c_1(F)]$ has a integer valued representative, then the manifold M is called *Hodge manifold* and the integer valued metric $g_{j\bar{k}}$ is called *Hodge metric.* As an immediate corollary, we have the famous

KODAIRA THEOREM. *(The complex compact manifold M is algebraic) $\Longrightarrow$(M is Hodge.)*

Hodge spaces are very special compact Kähler spaces, because of the fundamental work of Kodaira (he was awarded for it (together with Serre) the Fields Medal in 1954, which was presented to him by Weyl), these spaces are called 'Kähler spaces of restricted type.'

Hans Grauert presented a completely different method (which does not make use from the theory of harmonic forms, that is, potential theory) of the proof of Kodaira theorem; this method works also for spaces which may have singularities, thus not being complex manifolds, and being known as the so called *complex spaces M*. After removal of singularities (which form an analytic subset $N \subset M$) $M - N$ is already holomorphic. We recommend the reader to read the wonderful review of Grauert presented at Stockholm Congress in 1962 *Die Bedeutung des Levischen Problems für die Analyse und Geometrie*, pp. 86–101.

6.4 Homogeneous complex spaces and bounded domains

We owe Felix Klein (1849–1925) the beginning of systematic investigations of homogeneous spaces (called often *Klein spaces*), that is the spaces M on which the group of automorphisms $G = \mathrm{Aut}(M)$ acts transitively: for every pair of points $x.x' \in M$, there exists an automorphism $g \in G$ such that $x' = gx$. In other words, the orbit Gx of any point $x \in M$ is the whole space M: $Gx = M$. Klein observed in his *Erlanger Program* of 1872 that investigation of homogeneous spaces is the most important task of geometry. There is no doubt that homogeneous spaces are particularly beautiful creations of mathematics, we have already mentioned some of them which are of particular importance, namely spaces of constant curvature, so called space forms, or symmetric spaces. If M is a complex manifold, then $\mathrm{Aut}(M)$ is the group $\mathrm{Hol}(M)$ of holomorphisms of the space M. If (M, g) is a Riemann space, then $\mathrm{Aut}(M, g) = \mathrm{Isom}(M, g)$ is a group of isometries of the metric space (M, g). From the property (ii) of Bergman kernels, we immediately have

PROPOSITION. *Let $M = O \subset \mathbb{C}^n$ be a bounded domain of general n-dimensional complex manifold having Bergman kernel (that is, for whose H is very ample.) Then Hol(M) $\subset$ Isom(M), if M is equipped with Bergman metric, any holomorphism is an isometry.*

Moreover, we have important

THEOREM A. *If the manifold M is homogeneous (in particular, if M is a homogeneous bounded domain in $\mathbb{C}^n$), then*

(1) *the space H is very ample, thus there exists a Bergman metric on M;*

(2) *the Ricci tensor $R_{j\bar{k}}$ of the metric $g_{j\bar{k}}dz_j d\bar{z}_k$ is non-degenerate, and moreover*

(3) *the Bergman metric is a Kähler-Einstein metric $R_{j\bar{k}} = ag_{j\bar{k}}$, $a =$ const.*

At this point we should present a number of deep and important facts concerning transformation groups, above all, the groups of isometries of Riemann spaces. The first is the Van der Waerden-van Dantzing theorem of 1924 and its generalization of 1939

MEYERS-STEENROD THEOREM. *Let (M, g) be a Riemann space. Then the connected group of isometries $G = Isom(M)$ is a Lie group (in the compact-open topology), and therefore is locally compact. For any $x \in M$, the group of isotropy (the stabilizer) $G_x = \{g \in G : gx = x\}$ is a compact subgroup of G. If the space M is compact, then G is compact as well.*

COROLLARY. *Let M be a complex compact manifold possessing Bergman kernel (for example, let H be very ample.) Then the group of holomorphisms $Hol(M)$ of the manifold M is a Lie group; the isotropy group of any point $x \in M$ is compact.*

PROOF. Since $\mathrm{Hol}(M)$ is a closed subgroup of $\mathrm{Isom}(M)$, we have the assertion. □

The very interesting case when point (3) of Theorem A holds is the following situation: on M a discrete group $\Gamma \subset \mathrm{Hol}(M)$ acts freely (that is, without fixed points.) In connection with this, we have

THEOREM B. *Let on complex manifold a discrete group $\Gamma \subset Hol(M)$ acts freely and such that the quotient space M/Γ is compact (e.g., when M is compact.) Then the Ricci tensor on M is non-degenerate.*

PROOF OF THEOREM A. Let $V = \mathrm{vol}(M)$ be a volume form on M, thus $V = v dz_1 \wedge \cdots \wedge dz_n \wedge d\bar{z}_1 \wedge \cdots \wedge d\bar{z}_n$ defined by Bergman metric. As we know

$$R_{j\bar{k}} = -\frac{\partial \log v}{\partial z_j \partial \bar{z}_k}.$$

If M is homogeneous, then the volume element is uniquely defined up to constant c, thus $v = cb$, where b is a Bergman density. We have therefore $R_{j\bar{k}} = -g_{j\bar{k}}$. □

PROOF OF THEOREM B. The forms B and V can be regarded as $2n$ forms on M/Γ because they are Γ-invariant. We have therefore the 2 forms

$$(*) \qquad \frac{1}{2\pi i} g_{j\bar{k}} dz_j \wedge d\bar{z}_k \quad \text{and} \quad -\frac{1}{2\pi i} R_{j\bar{k}} dz_j \wedge d\bar{z}_k$$

define the same class of de Rham cohomology belonging to $H^2(M/\Gamma)$, namely the first Chern class (form) $[c_1(M/\Gamma)]$. Taking the left expression in (*), we see that $c_1(M/\Gamma) \neq 0$. If however, $\det(R_{j\bar{k}}) = 0$, then from $c_1(M/\Gamma)^n = 0$, a contradiction. □

These considerations necessarily lead to the important class of spaces discovered, constructed, and thoroughly analyzed by E. Cartan, as we know these are

Symmetric spaces and their theory are a gigantic creation of one man, E. Cartan. A Riemann space (M, g) is locally symmetric if for any point $x \in M$ there exists a local reflection which is an isometry of some ball B_x. It turns out that this condition is equivalent to the fact that the curvature tensor is constant under parallel transport $\nabla R = 0$ – Cartan considered spaces with this condition first.

The space (M, g) is *(globally) symmetric* if for any point $x \in M$ there exists a global isometry $I_x \in \mathrm{Isom}(M)$. The important theorem holds

THEOREM. (1) *(M is locally symmetric)$\Longleftrightarrow$($\nabla R = 0$.)*

(2) *A simply connected complete locally symmetric space is symmetric.*

(3) *(M is symmetric)$\Longrightarrow$(M is homogeneous, and thus is of the form G/K, where $G =$ Isom(M) and K is a (compact) group of isotropy of some point $x \in M$.)*

REMARK. (ad (2).) The symmetry s_x with respect to x is constructed as follows: from point x one ejects a geodesics $t \to a(t, X)$ in direction X; then $s_x(m) := a(-t, X)$. Taking all $X \in T_xM$, we first obtain a local isometry. If (M, g) is complete, then, from Hopf-Rinow theorem it is geodesically complete, and extending arbitrarily these geodesics, we obtain the global isometry. Thus point (2) is proved.

As point (3) shows, there is very many symmetric spaces. Symmetric spaces are very closely related to semi-simple Lie groups. The semi-simple Lie groups were introduced and investigated in great details by great geometer Wilhelm Killing. For many years he was a professor and rector of Collegium Hosianum, the post-Jesuits college founded in Braniewo (German Braunsberg) by the Jesuit and later the cardinal Hosen. It is very interesting that in this catholic high school one of the greatest mathematicians of all times Weierstrass teached for many years. The pioneering research of Killing has been continued by E. Cartan and H. Weyl, who regarded his fundamental works on semi-simple groups (and algebras) of 1925 as his most important achievements.

As it is well known, the algebra LG of Lie group G is semi-simple if the Killing form $B(X,Y) := \mathrm{trace}(\mathrm{ad}\, X\ \mathrm{ad}\, Y)$ is non-degenerate (which is equivalent to the fact that LG is a direct sum of simple ideals.) If B is negative definite, then the group G is compact (Weyl.) It turns out that if a Lie group G is semi-simple and non-compact, then $\mathfrak{g} = LG$ can be decomposed into direct sum $\mathfrak{g} = \mathfrak{p} \oplus \mathfrak{h}$, where $B|\mathfrak{p}$ is positive definite and $B|\mathfrak{h}$ is negative definite. Then H is a compact subgroup generated by the subalgebra $\mathfrak{h}$ of the algebra $\mathfrak{g} = T_eG$ (unit element of G.) On the other hand, $B|\mathfrak{p}$ generates a Riemann metric on G/H and in this way G/H becomes a symmetric space. It can be shown that the curvature of the symmetric space G/H is given by the formula

$$K(X,Y) = -\|[X,Y]\|,$$

where $[X,Y]$ is the Lie bracket (of vector fields X, Y), and $\|\ \|$ is given by Killing form.

The reader can find the mastery exposition of the theory of symmetric spaces in the monograph[1] written in 1962 by, very young at the time, professor of MIT, Icelander Sigurdur Helgason. We also owe Helgason a beautiful theory of dualities of symmetric spaces, which is an extension and generalization of classical dualities on Lobachevski plane. This theory requires however a good knowledge of semi-simple Lie groups.

6.5 Symmetric spaces

$M = G/K$ where G is a non-compact semi-simple Lie group are called *symmetric spaces of non-compact type.* Let us return to complex manifolds with hermitian metric, for short, hermitian spaces. These spaces are called symmetric if any point $x \in M$ is a center of holomorphic isometry s_x, so that any point is an isolated point of $s_x \in \mathrm{Hol}(M)$, such that $s_x^2 = \mathrm{id}_x$. As we know already

FACT. Every symmetric hermitian manifold is Kählerian. As we know, every bounded symmetric domain $O \subset \mathbb{C}^n$ is a symmetric hermitian space of the non-compact type (if equipped with Bergman metric.) However, the surprising inverse theorem holds which was claimed by E. Cartan but fully proved only by Harish-Chandra (Cartan knew the proof only in particular

[1] *Differential Geometry and Symmetric Spaces*, Academic Press, 1962.

cases.)

HARISH-CHANDRA THEOREM ON BOUNDED DOMAINS. *Let M be a hermitian symmetric space of the non-compact type. Then M is (bi)holomorphically equivalent to a bounded domain O, that is, there exists a holomorphism $\varphi : M \to O$ of the space M and $O \subset \mathbb{C}^n$.*

PROOF of this magnificent theorem can be found in the cited Helgason's monograph (pp. 313–322.)

As we know, the symmetric bounded domains are the (unit) disc $\mathbb{D} \subset \mathbb{C}$; on D the group of holomorphisms acts transitively. The upper half-plane $\mathfrak{H}_1 = \{z \in \mathbb{C} : \operatorname{Im} z > 0\}$ is, seemingly, something completely different: it is a not bounded domain in $\mathbb{C}$, but we know that the Cayley transformation turns biholomorphically $\mathfrak{H}_1$ into $\mathbb{D}$.

Carl Ludwig Siegel, in connection with his investigations of 'symplectic geometry', gave a good n-dimensional analogue of the Lobachevski plane $\mathfrak{H}_1$, this is the famous and already known to us

Siegel half-space $\mathfrak{H}_n$; this is a complex manifold of dimension $\frac{1}{2}n(n+1)$ whose points are symmetric $n \times n$ matrices of the form $z = x + iy$, where x and y are real symmetric $n \times n$ matrices and y is positive definite $y > 0$.

Every symplectic matrix $g \in \mathrm{Sp}(2n, \mathbb{R})$ defines a holomorphism of the space $\mathfrak{H}_n$

$$T(g) : z \to (az + b)(cz + d)^{-1}.$$

The group $\mathrm{Sp}(2n, \mathbb{R})$ acts transitively on $\mathfrak{H}_n$ and its isotropy group of the point $p \in \mathfrak{H}_n$ is, as we know from general considerations, compact. Indeed, the isotropy group of the point $i1_n$ is nothing but the orthogonal group $\mathrm{SO}(2n)$. Thus

$$\mathfrak{H}_n \simeq \mathrm{Sp}(2n, \mathbb{R})/\mathrm{SO}(2n.)$$

It is not hard to show that $\mathfrak{H}_n$ is biholomorphic to the 'generalized Siegel disc' $\mathbb{D}^m$, $m = \frac{1}{2}n(n+1)$, where $\mathbb{D}^m = \{$ w-symmetric complex $n \times n$ matrix; $1_n - \bar{w}w > 0\}$, $\mathfrak{H}_n \ni z \to (1_n + iz)(1 - iz)^{-1} \in D^m$.

This mapping is a biholomorphism and it illustrates the general Harish-Chandra theorem.

As I have already mentioned, in 1926-27 Cartan succeeded in performing enormous work: he classified irreducible representations of symmetric spaces. Let us recall that symmetric space M can be (locally) represented in the following form $M = M_1 \times \ldots \times M_r$, where M_j, $j = 1, \ldots, r$ cannot be decomposed anymore, that is, they are irreducible. This classification reduces to the classification of some simple Lie algebras over $\mathbb{R}$. We have, among others, an important theorem saying that *simply connected symmetric hermitian space is of the form $M = M_0 \times M_- \times M_+$, where $M_0 = \mathbb{R}^k$, M_- is of compact, and M_+ of non-compact type, respectively.* In turn, such a space M can be represented as a product of irreducible hermitian symmetric spaces. Every irreducible hermitian symmetric spaces of non-compact type is of the form $M = G/K$, where G is a simple, connected, and non-compact Lie group and K is its maximal compact subgroup. The complete list of this spaces can be found in Helgason's book and also in the very good monograph of J. Wolf *Spaces of Constant Curvature.*

6.6 Spectral geometry

The Hodge theory shows that the topology (cohomology) of Riemann space is defined by harmonic forms $\mathcal{H}^k(M) \simeq H^k(M)$, $k = 0, \ldots, \dim M$. But harmonic k forms are zero eigen elements (vectors) of an appropriate Laplace-Weitzenböck-Hodge operator Δ_k on the space $\Lambda^k(M)$ of k forms.

The question, or the attempt to find some invariants of geometry of M, for example, curvatures, from the properties of spectrum $\mathrm{Spec}(\Delta_0)$ is aptly called spectral geometry. We see the first trial of this kind in Pythagorean school which, as gossip says, found a relation between the length of the string and the height of tone. Now we can see this from the spectrum of one-dimensional Laplacian $\Delta = -d^2/dx^2$ on the interval $M = [0, l]$, because if $-d^2u/dx^2 = \lambda u$, $u(0) = 0 = u(l)$, then $u_\lambda(x) = e^{i\lambda x}$, $\lambda = n/l\pi$, $n = 0, 1, \ldots$. Pythagorean influenced Plato in very substantial way. It is said that 'Pythagorean life' had as a goal to make the members of this brotherhood – order to hear the 'music', the 'harmony of spheres'. Maybe mathematics is just this 'harmony of spheres'? Pythagoreans teached three arts: arithmetic, geometry, and harmonics. And even now these three arts are main branches of mathematics: arithmetic (number theory), the mother of algebra, geometry, and harmonic analysis. The latter is, since Weyl, understood not only as the potential theory with the theory of harmonic forms, not only as the theory of Fourier transforms, but also as the general spectral theory and the theory of group representations. Perhaps this is the branch which puts together the whole mathematics: because our goal 'here' is to investigate and study representations – symbols of things? In this way spectral geometry is a descendant, the child of ancient Pythagorean dream: it joins together geometry with harmonics – the spectral theory of operators.

But the Laplace-Beltrami operator and the Weitzenböck operators Δ_p are the most natural operators on Riemann space!

Physics, statistical mechanics, theory of radiation are also predecessors of spectral geometry (Hurt.) Here the investigations by Weyl were also decisive: In 1910 the great Dutch physicist K. Lorentz posed it as a problem to show that the spectrum of black body radiation does not asymptotically depend on the shape of body M and only on its volume. Hilbert, who was present on the seminar, decided that this is enormously difficult problem. What a sensation it was, when not long after that the young student of Hilbert, H. Weyl, solved this problem in affirmative and, slightly later, the much most difficult one regarding vibrations of elastic body M^n.

Weyl formula (generalized). Let $0 \leqslant \lambda_1 \leqslant \lambda_2 \leqslant \ldots \to \infty$ be a sequence of eigenvalues of Δ_p, that is $\Delta_p \varphi_{\lambda_j} = \lambda_j \varphi_{\lambda_j}$, counted with multiplicities. Then

$$N_p(\Lambda) = \text{cardinal number of the set } \{l_k \leqslant \lambda\} \sim$$

$$\sim \binom{n}{p} \frac{|M^n|}{(4\pi)^{n/2}\Gamma(n/2+1)} \lambda^{n/2}.$$

Thus, the volume $|M^n|$ of the manifold M^n is defined in terms of the asymptotics of the spectrum $\mathrm{Spec}(\Delta, M)$ of the operator Δ on $M = M^n$. The Weyl formula is being proved in many ways and generalized for different elliptic operators. Perhaps the best insight into this kind of problems is provided by the theory of diffusion equation (heat transfer) and promoted by Marc Kac stochastic methods. The next major steps were the works of prematurely dead Indian mathematician Padoti and, already mentioned many times, Peter B. Gilkey. Let us present

PADOTI THEOREM (1970-71). 1. *Let two Riemann manifolds M, M_1 be isospectral, that is $\mathit{Spec}(\Delta, M) = \mathit{Spec}(\Delta, M_1)$. Then $|M| = |M_1|$ and* $\dim M = \dim M_1$.

2. *If, moreover $\mathit{Spec}(\Delta_p, M) = \mathit{Spec}(\Delta_p, M_1)$ for $p = 0, 1, 2$, then*

(a) *The scalar curvatures of M and M_1 are the same;*

(b) *If one of the manifolds is Einstein, then the second is as well;*

(c) *If one of the manifolds has constant sectional curvature, then the second has as well;*

(d) *There are some interesting equations for expressions built from components of curvature tensor R_{ijkl}.*

Gilkey obtained an analogous theorem independently. One introduces the notion of strong isospectrality of M and M_1 (the identity of spectra of wide class of 'natural' differential operators on M and M_1.) Then one obtains stronger consequences, for example, if M is locally symmetric ($\nabla R = 0$), then also M_1 is locally symmetric, or, if M is isometric to the standard sphere $S^n(r)$, then M_1 has this property as well.

For compact complex manifolds, we have

THEOREM (GILKEY). *If M, M_1 are complex manifolds and if $Spec(\Delta_{p,q}, M) = Spec(\Delta_{p,q}, M_1)$ for all p, q, then*

(a) *(M is Kähler)$\Longrightarrow$(M_1 is Kähler);*

(b) *($M = P(\mathbb{C}^n)$)$\Longrightarrow$($M = P(\mathbb{C}^n)$.)*

These beautiful theorems do not say however if the isospectrality implies isometricity of M and M_1. In 1964, using arithmetic of quadratic forms, J. Milnor constructed two tori of the same dimension (7) which are isospectral but not isometric. Later, with use rather elementary methods, an infinite number of isospectral pairs which are not isometric, were constructed.

P. Gilkey showed, by using the methods of heat transfer equations and the theory of invariants (e.g., Weyl theorem on tensor invariants with respect to orthogonal group) a number of theorems characterizing $\chi(M)$ or integrands in various index theorems.

Part II

General Structures of Mathematics

CHAPTER 1

Differentiable Structures. Tangent Spaces. Vector Fields

The notions in the title of this chapter are central for the whole of analysis, and, in particular, for what we are going to discuss below; here I restrict myself to presenting only fundamental definitions: those of a differential and analytical n-dimensional manifold $M = M^n$, vectors tangent to M, and tangent bundle TM. An analytical and connected manifold M^2 is called a *Riemann surface.* We have already discussed Riemann surfaces in part I, we will return to them later as well.

DEFINITION. A *differentiable (*analytic*) n-dimensional manifold* M^n is a pair $(M, \mathcal{A})$, where M is a Hausdorff space and $\mathcal{A} = \{(\mathcal{U}_j, h_j)\}_{j \in J}$ is a family $(\mathcal{U}_j, h_j)$ of maps (an atlas), that is, the following conditions are satisfied:

1. $(\mathcal{U}_j)_{j \in J}$ is an open covering of M, and $h_j : \mathcal{U}_j \to \mathcal{U}'_j$ are homomorphisms where $U'_j \subset \mathbb{R}^n$ are open neighborhoods (in the space $\mathbb{R}^n$).

2. $h_{ij} := h_i \circ h_j^{-1} : h_j(\mathcal{U}_i \cap \mathcal{U}_j) \to h_i(\mathcal{U}_i \cap \mathcal{U}_j)$ (h_{ij} are homomorphisms of domains in $\mathbb{R}^n$ and therefore we know what derivatives and differentiability means) are *differentiable* (analytic); the mappings h_{ij} are called *transition mappings.*

It follows from Kuratowski-Zorn lemma that there exists a *maximal atlas* $\mathcal{A}_m$ (it is not possible to add any new map to $\mathcal{A}_m$ without altering the property 2). This atlas is called the *differentiable (analytical) structure* on M. Since any differentiable (analytical) atlas on M can be extended to the maximal atlas $\mathcal{A}_m$ in essentially unique way, it is sufficient to give any atlas $\mathcal{A}$ on M in order to be able to talk about differentiable (analytical) structure.

REMARK 1. If only point 1. of the definition is satisfied, we talk about *topological manifold of dimension n*.

REMARK 2. In the definition, we could have replaced the space $\mathbb{R}^n$ with any Banach space E over $\mathbb{R}$; in that case, we would talk about *model manifold on (imitating) Banach space E*. Thus an n-dimensional manifold is a model manifold on $\mathbb{R}^n$.

REMARK 3. Using the notion of sheaf (which will be introduced later), one could present a more elegant definition of differentiable structure and, more generally, of a *space with a ring* (with an algebra). We will present this definition after introducing the notion of sheaf. This notion is very useful in many branches of analysis and algebraic geometry.

Historical remark. It seems that Bernhard Riemann was the first who carefully distinguished the topological, differentiable, and metric structures. He did that in his celebrated habilitation lecture *Über Hypothesen welche der Geometrie zugrunden legen* (10. 06. 1854). The notion of topological structure (Hausdorff 1914) we owe, in fact, Hermann Weyl. In his *Die Idee der Riemannischen Fläche* (1913) topology is defined by means of a basis of neighborhoods. The definition of differentiable manifold given above appeared for the first time in the fundamental monograph of Weyl devoted to the Einstein theory of gravity *Raum, Zeit, Materie* (1918).

It is well known that philosophers used to persistently claim – following Kant – that the 'real' space is the space $\mathbb{R}^3$ (or, more precisely, the Euclidean three dimensional metric space). I do not want to defend here this unacceptable statement (superstition?), but maybe there is a germ of truth in it? Indeed, we do not reject completely the Euclidean spaces $\mathbb{R}^n$, $n > 3$: general manifolds are *modeled* on $\mathbb{R}^n$! We operate much more firmly in some $\mathbb{R}^n$; maps and atlases help, make it possible even to define differentiation of functions and mappings: the 'creatures' living on M^n are locally defined by, easier for us to comprehend, creatures living in $\mathbb{R}^n$. Here we encounter some distant similarity with the process of abandoning the set of natural numbers $\mathbb{N}$ in construction of other worlds of numbers $\mathbb{R}$, $\mathbb{C}$, $\mathbb{H}$ and $\mathbb{R}^2$, $\mathbb{R}^3$, $\mathbb{R}^n$, ... We cannot reject our Euclidean tradition.

In Figure 1 we illustrate the notion of map and transition mapping.

Differential mappings $f : M \to N$ of smooth (i.e. differentiable) manifolds M and N. The maps reduce this problem to the notion of dif-

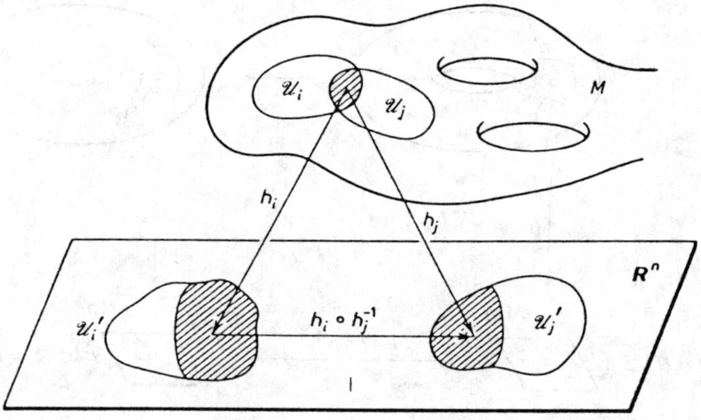

Fig. 1

ferentiability of mappings of regions in Euclidean (more generally, Banach) spaces as follows:

DEFINITION. Let M be modeled on $\mathbf{M}$ (for example $\mathbb{R}^m$), and N on $\mathbf{N}$ (for example $\mathbb{R}^n$). The mapping $f : M \to N$ is *differentiable at the point* $p \in M$ if for some maps $h : \mathcal{U} \to \mathcal{U}'$, $p \in \mathcal{U}$ and map $k : \mathcal{V} \to \mathcal{V}'$, $f(p) \in \mathcal{V}$ of the manifolds M and N, the superposition $k \circ f \circ h^{-1}$ is differentiable at the point $h(p) \in \mathcal{U}' \subset \mathbf{M}$. The mapping f is *differentiable* if it is differentiable at all points of $p \in M$.

Figure 2 illustrates this situation.

REMARK. The mapping $k \circ f \circ h^{-1}$ transforms the region $\mathcal{U}' \subset \mathbb{R}^m = \mathbf{M}$ into the region $\mathcal{V}'$ in the space $\mathbb{R}^n$ (in general, the Banach space $\mathbf{N}$), we know therefore what differential (tangent mapping), derivative, differentiability of this mapping mean. Of course, one can talk about differentiability of f in some subset $\mathcal{O} \subset M$.

A *germ at (or over)* $p \in M$ *of the mapping* $f : \mathcal{U} \to N$ defined in the neighborhood $\mathcal{U} \subset M$ of the point p, is a class f_p of mappings identical with f in some neighborhood $\mathcal{O}$ of p. More specifically, let f_1, f_2 be defined in some (in general, different) neighborhoods of $p \in M$,

$$f_1 : \mathcal{U}_1 \to N, \;\; f_2 : \mathcal{U}_2 \to N, \;\; p \in \mathcal{U}_1 \cap \mathcal{U}_2, \;\; f_1 = f_2 \text{ on } \mathcal{U}_1 \cap \mathcal{U}_2.$$

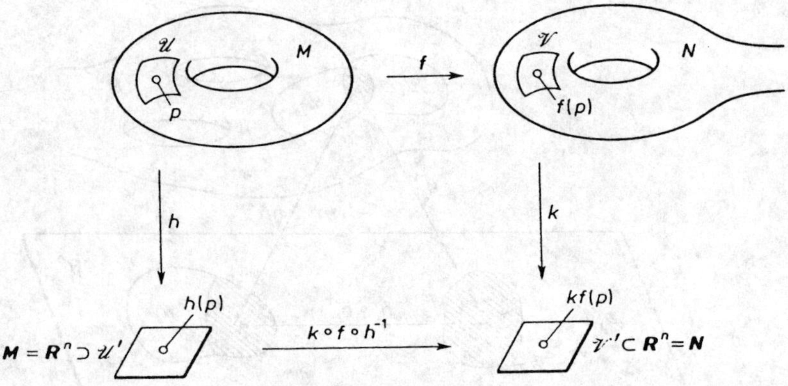

Fig. 2

Then we write $f_1 \overset{p}{\sim} f_2$. This is, obviously, an equivalence relation (because we can always take a smaller neighborhood of p). The class of this equivalence relation $\overset{p}{\sim}$ is denoted by f_p or $[\ ]_p$ and is called the *germ of mapping f at point p*, and f itself, a *representative of germ*. The notion of germ (of function, mapping, vector, tensor field, manifold, etc.) is very useful and was unknowingly used very early. The clear definition was probably first given by Hermann Weyl in *Die Idee der Riemannischen Fläche*.

Now we can talk about germs of differentiable mappings f at the point p.

Definition of tangent vector (first). Clearly, the set $\mathcal{E}_p(\mathcal{O}_p)$ of all germs of differentiable (holomorphic) functions, that is, the *stalk over p* is an $\mathbb{R}(\mathbb{C})$ ring and even an $\mathbb{R}(\mathbb{C})$ algebra if one defines operations on germs by operations on their representatives as follows

$$f_x \cdot g_x := [f \cdot g]_x, \quad f_x \pm g_x := [f \pm g]_x, \quad \lambda f_x := [\lambda f]_x.$$

DEFINITION (I). A *tangent vector at point p* of differentiable manifold $M = M^n$ is a linear mapping $X_p : \mathcal{E}_p \to \mathbb{R}$ satisfying Leibnitz rule

$$X_p(\varphi \cdot \psi) = X_p(\varphi)\psi + \varphi X_p(\psi), \quad \varphi, \psi \in \mathcal{E}.$$

The set T_pM of all tangent vectors at p is, in a natural way, a vector space over $\mathbb{R}$; we call this space the *tangent space to M at p*.

Since the germ f_p of the mapping $f : (M, p) \to (N, q)$ induces a homomorphism of $\mathbb{R}$-algebras

$$F^* : \mathcal{E}_p \to \mathcal{E}_q, \quad \varphi \to \varphi \circ f =: F^*\varphi, \quad q := f(p)$$

we have the *tangent mapping*

$$T_p f : T_p M \to T_q N, \quad X_p \to X \circ F^*.$$

In this way

$$T_p f(X)\varphi = X(\varphi \circ f).$$

The tangent mapping $T_p f$ is often called *differential* of f at p (and denoted by $d_p f$ or $(df)_p$) or *derivative* of f (and denoted $f'(p)$).

If with each point $x \in \mathcal{U} \subset M$ we associate tangent vector $X(x) \in T_x M$, we talk about the *vector field* X *over* $\mathcal{U}$. If $\mathcal{V} \subset \mathcal{U}$, then with every function $\varphi \in C^\infty(\mathcal{V})$ the vector field X associates the function $X(\varphi)$ on $\mathcal{V}$ by the rule

$$X(\varphi) : \mathcal{V} \to \mathbb{R}; \quad x \mapsto X(\varphi)(x) := X(x)\varphi_x.$$

If the function $X(\varphi)$ is differentiable, then the vector field is called *differentiable* itself.

If one denotes by TM the disjoint sum $\coprod_{x \in M} T_x M$ (called the *tangent bundle over* M) and by π the mapping $\pi : TM \to M$, given by the formula $\pi | T_x M = x$, then the vector field X over M is a mapping $X : M \to TM$, $X(x) \in T_x M$, that is,

$$\pi \circ X = \mathrm{id}_M.$$

In the language of vector bundles, this fact is expressed shortly as X *is a section of the tangent bundle* $TM \to M$. We will talk about vector bundles later. Then we will define topological and differentiable structures on TM and we will talk about continuous and differentiable sections. A smooth vector field over $\mathcal{U} \subset M$ is a differentiable section of the tangent bundle $\pi : TM \to M$ over the set $\mathcal{U}$. The set of smooth vector fields over M will be denoted by $\mathcal{X}(M)$ or $\mathcal{C}^\infty(TM)$ (more precisely, one should write $C^\infty(TM \xrightarrow{\pi} M)$, but our notation should not lead to misunderstandings). The reader will prove without difficulties the following

PROPOSITION 1. *The vector fields $\mathcal{X}(M)$ over M form an $\mathbb{R}$-vector space and even an $\mathcal{E}(M)$-algebra. We have*

1. $X(\varphi\psi) = X(\varphi) \cdot \psi + \varphi \cdot X(\psi)$ *(Leibnitz formula).*

2. $[X,Y](\varphi) := XY(\varphi) - YX(\varphi)$ *for* $\varphi \in \mathcal{E}(M)$ *and* $X, Y \in \mathcal{X}(M)$.
The commutator $[X,Y] := XY - YX$ *satisfies Jacobi identity:*
3. $[[X,Y],Z] + [[Z,X],Y] + [[Y,Z],X] = 0$.

Therefore, the vector space $\mathcal{X}(M)$ is equipped with bilinear mapping $[\cdot,\cdot] : \mathcal{X}(M) \times \mathcal{X}(M) \to \mathcal{X}(M)$ satisfying 3. and $[X,X] = 0$ and thus $\mathcal{X}(M)$ is a *Lie algebra.*

Below we will talk about Lie algebras very often. Here we would like only to mention the important notions (vector field, tangent bundle, germ of function, Lie algebra) in order to show the importance and necessity of the notion of tangent space. Since this is not a simple notion, we will present some other definitions of the tangent space T_xM.

As we know, if a vector space V is of finite dimension, then the dual space $V^* = L(V, \mathbb{R})$ is isomorphic with V. Since $\dim T_xM = \dim M$ (the dimension of the modelling space **M**), it is sufficient to define the *cotangent space* $(T_xM)^*$.

(Algebraic) definition of tangent space (second). Let $\mathfrak{m}_x$ be the space of germs vanishing at x: $\mathfrak{m}_x = \{f_x \in \mathcal{E}_x(M) : f(x) = 0\}$.

Let us recall the notions of ideal and maximal ideal.

DEFINITION. Let $\mathcal{R}$ be a commutative ring with unity (usually, we will denote operations in $\mathcal{R}$ by $+$ and $\cdot$). Then the subset $\mathfrak{u} \subset \mathcal{R}$ is an *ideal* in $\mathcal{R}$ if $\mathfrak{u}$ is a subgroup of the group $(\mathcal{R}, +)$ and $\mathfrak{u}\mathcal{R} \subset \mathfrak{u}$.

In any ring $\mathcal{R}$ there always exist two trivial ideals $\mathcal{R}$ and $\{0\}$. If $\mathfrak{u}$ is a subgroup in $\mathcal{R}$, then the quotient group $\mathcal{R}/\mathfrak{u} = \{r + \mathfrak{u} : r \in \mathcal{R}\}$ is the abelian group of layers $r + \mathfrak{u}$, but it is not, in general, a ring.

The reader will show the following simple theorem

THEOREM 2. *($\mathcal{R}/\mathfrak{u}$ is a ring)$\Longleftrightarrow$($\mathfrak{u}$ is an ideal).*

DEFINITION. An ideal $\mathfrak{p} \subset c\mathcal{R}$ is *prime* if

$$\forall_x \forall_y (xy \in \mathfrak{p}) \Longrightarrow x \in \mathfrak{p} \text{ or } y \in \mathfrak{p}.$$

An ideal $\mathfrak{m} \subset \mathcal{R}$ is *maximal* if it is not a subideal of any ideal different from $\mathcal{R}$.

The term prime ideal was introduced by Dedekind. Let $\mathbb{Z}$ denote, as usual, the ring of integers, then

$$(n\mathbb{Z} \textit{ is a prime ideal}) \Longleftrightarrow (n \textit{ is a prime number})$$

The question arises as to when a ring $\mathcal{R}$ is a field. The answer is

$$(*) \qquad (\mathcal{R} \textit{ is a field}) \Longleftrightarrow (\mathcal{R} \textit{ contains only trivial ideals}).$$

The following theorem characterizes maximal ideals

THEOREM 3. (a) *($\mathfrak{m}$ is a maximal ideal of the ring $\mathcal{R}$) $\Longleftrightarrow$($\mathcal{R}/\mathfrak{m}$ is a field)*

(b) *Let $\mathfrak{p}$ be an ideal in $\mathcal{R}$. Then*

$$(\mathfrak{p} \textit{ is a prime ideal}) \Longleftrightarrow (\mathcal{R}/\mathfrak{p} \textit{ does not have any divisors of zero}).$$

(c) *Every maximal ideal is prime.*

PROOF of (a) follows from (*), we leave (b) to the reader; (c) follows from (a) and (b).

PROPOSITION 4. (a) *$\mathfrak{m}_x$ is an ideal of the commutative ring $\mathcal{E}_x$, where*

$$f_x \cdot g_x := (fg)_x, \quad f, g \in \mathcal{E}.$$

(b) *$\mathfrak{m}_x$ is a maximal ideal of $\mathcal{E}_x$.*

PROOF. (a). is obvious because the sum of two functions vanishing at x vanishes at x, and the product of two functions, one of whose vanishes at x, vanishes at x.

(b). Let an ideal $\mathfrak{a} \subset \mathcal{E}_x$ contain the ideal $\mathfrak{m}_x$ and the function f such that $f(x) \neq 0$. Then $\mathfrak{a}$ contains (by definition of an ideal) all germs of the form $\mathfrak{m}_x + \mathbb{R}f = \mathcal{E}_x$. Indeed, let $g(x) = cf(x)$, $c \neq 0$ (we take $c = g(x)/f(x)$) and thus $g - cf \in \mathfrak{m}_x$, therefore $g \in \mathfrak{m}_x + \mathbb{R}f$. Thus $\mathfrak{a} = \mathcal{E}_x$ and the ideal $\mathfrak{m}_x$ is maximal. But every ideal contained in commutative ring $\mathcal{E}_x$ contains the whole of $\mathfrak{m}_x$. □

DEFINITION. A commutative ring $\mathcal{R}$ containing a single maximal ideal (different from 0 and $\mathcal{R}$) is called a *local ring*; usually the maximal ideal is denoted by $\mathfrak{m}(\mathcal{R})$ and the quotient field $\mathcal{R}/\mathfrak{m}$ is denoted by $\mathbf{k}(\mathcal{R})$.

Thus $\mathcal{E}_x$ is indeed a local ring (German *Stellenring*); this terminology resulted from the analysis of the ring $\mathcal{E}_x$, the stalk of germs over point x of manifold M. Similar situation arises in the case of the ring of germs of continuous functions $\mathcal{C}_x$, or holomorphic functions $\mathcal{O}_x$, where M is an analytic manifold.

DEFINITION (II). $(T_xM)^* := \mathfrak{m}_x/\mathfrak{m}_x^2$, where $\mathfrak{m}_x^2 := (\mathfrak{m}_x)^2$.

On the first sight, this definition does not have much to do with Definition (I), however the following lemmas clarify the situation.

A germ f_x is *stationary* if derivative (differential) of its representative f vanishes at x: $f'(x) = 0$ $((df)_x = 0)$.

LEMMA 5. *(f_x is stationary at x) $\Longleftrightarrow$($f - f(x) \in \mathfrak{m}_x^2$).*

PROOF. Since we have local situation, we can consider function f defined in a (convex) neighborhood $\mathcal{O} \subset \mathbb{R}^n$. From the mean value theorem we have

$$f(y) - f(0) = \int_0^1 \frac{\mathrm{d}}{\mathrm{d}t} f(ty)\, dt = \sum_{i=1}^n y_i g_i(y),$$

where $g_i(y) = \int_0^1 \partial_i f(ty)\, dt$. But $g_i(0) = 0 = \frac{\partial f}{\partial y_i(0)} = 0$ because f is stationary at zero. We have therefore

$$f = g_1 h_1 + \cdots + g_n h_n + f(x) \quad \text{where } f_i, g_i \in \mathcal{E}_x.$$

The reader will solve the following problems.

EXERCISES

1. $\mathfrak{m}_x^k$ is an ideal of germs $f \in \mathcal{E}_x$, whose derivatives vanish at x up to order $k-1$.

2. The Taylor series at zero defines homomorphisms of rings $\mathcal{E}_0(\mathbb{R}^n) \to \mathcal{R}[[x_1, \ldots, x_n]]$ (:= the ring of formal power series of n variables). The kernel of this homomorphism is $\mathfrak{m}_0^\infty := \bigcap_{k=1}^\infty \mathfrak{m}_0^k$.

DEFINITION $(II)'$. $\mathfrak{m}_x/\mathfrak{m}_x^2 =: (T_xM)^*$. The space dual to $\mathfrak{m}_x/\mathfrak{m}_x^2$ can be identified with linear mappings

$$(**) \qquad X : \mathcal{E}_x \to \mathbb{R} \text{ vanishing on } (\mathfrak{m}_x)^2\ .$$

vanishing on germs stationary at x by Lemma 5. We will denote this space by T_xM, as before.

LEMMA 6. *Every tangent vector X (that is, the vector satisfying $(**)$) has the property*

$$(D) \qquad X(\varphi\psi) = X(\varphi)\psi(x) + \varphi(x)X(\psi), \quad \text{for } \varphi, \psi \in \mathcal{E}.$$

PROOF. $f := (\varphi - \varphi(x))(\psi - \psi(x)) - \varphi(x)\psi(x) = \varphi\psi - \varphi(x)\psi - \varphi(\psi(x)$ is stationary at x. Thus $X(f) = 0$; this means that formula (D) holds. □

A linear functional (form) on commutative algebra $\mathcal{A}$ satisfying condition (D) is called *derivation* of algebra $\mathcal{A}$. Thus, the lemma above can be rephrased as

Every tangent vector X (in the sense of second definition) at point $x \in M$ is a derivation of the local algebra $\mathcal{E}_x$.

But the inverse theorem (to Lemma 6) holds as well, and thus we have

PROPOSITION 7. *Every derivation X of algebra $\mathcal{E}_x$ is a vector tangent to M at point x (in the sense of second definition), thus $X(f) = 0$ if f is stationary at x.*

Taking this proposition together with the previous lemma, we can formulate the following theorem.

THEOREM 8. *Both definitions of tangent space T_xM are equivalent.*

It remains only to present

PROOF OF PROPOSITION 7. Let us observe that $X(1) = X(1 \cdot 1) = X(1) \cdot 1 + 1 \cdot X(1) = 2X(1)$. Thus $X(c) = 0$ for any constant c. Let f be stationary at x. Then

$$f - f(x) = \sum_{i=1}^{n} g_i h_i, \quad \text{where } g_i, h_i \in \mathfrak{m}_x,$$

thus

$$X(f) = \sum_{i=1}^{n} (X(g_i)h_i(x) + g_i(x)X(h_i)) = 0.$$

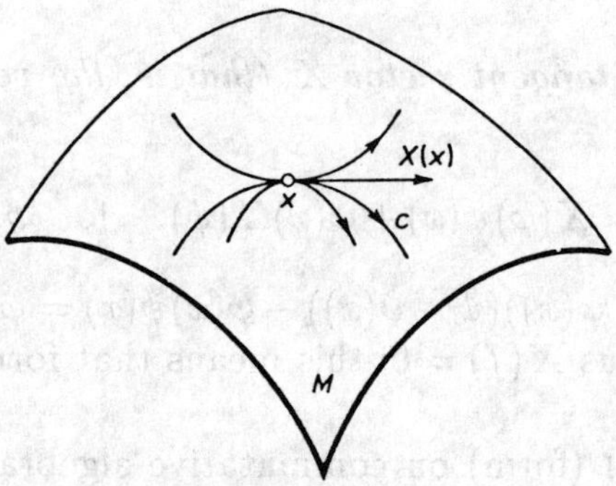

Fig. 3

□

Definition of tangent vector (third). Now we present 'geometric' definition (Fig. 3). Let $c : [0,1] \to M$ be a path on M starting from the point $x \in M$ (that is, it is a smooth mapping of some open neighborhood of the interval $[0,1]$ in $\mathbb{R}$ into M).

DEFINITION (III). A vector tangent to the curve c at the point $x = c(0)$ is an element $X = X_c \in T_xM$ defined by the formula

(1) $$X(f) := (f \circ c)'(0) = \frac{\mathrm{d}}{\mathrm{d}t} f \circ c(t)|_{t=0}, \quad f \in \mathcal{E}_x.$$

Clearly, formula (1) defines a tangent vector in the sense of Definition (I)

Two questions arise: 1. When two smooth paths c_1 and c_2 starting from the point x define the same tangent vector? In other words, when the velocities are equal? 2. Is every tangent vector X, that is, a derivations in $\mathcal{E}_x$, a vector tangent to some path? The answer to first question is immediate

(2) $$(f \circ c_1)'(0) = (f \circ c_2)'(0) \quad \text{for every } f \in \mathcal{E}_x.$$

We check that the identity above defines an equivalence relation in the set of paths starting at x – 'abstractly' speaking, a tangent vector $X \in T_xM$

is an equivalence class of paths with respect to the relation defined by (2). 'Geometrically' speaking, *two paths are equivalent* if they have the same velocity vector (that is, tangent vector) X at x.

The answer to 2. is given by the following

PROPOSITION 9. *Every vector $X \in T_xM$ is tangent to some curve on M.*

PROOF. Let $v \in \mathbb{R}^n$ (= $\mathbf{M}$, the model of manifold M). Let $\bar{c}(t) := tv$, $t \in [0,1]$. If $h : \mathcal{U} \to \mathcal{U}' \subset \mathbf{M}$ is a map in M (from an atlas defining differentiable structure of the manifold M) such that $h(x) = 0$, then $h^{-1} \circ \bar{c} =: c$ is a path on M starting from x and having the tangent vector X, whose representative is the triple (x, h, v). □

REMARK. At the end of the proof we made use of the fourth definition of tangent vector. We will use this one most often, for example below in the 'reconstruction of vector bundle from cocycle'. Let us shortly recall it.

Definition of tangent vector (fourth). Let $(\mathcal{U}_j, h_j)$, $j \in J$ be a differentiable atlas of manifold M modeled on a Banach space $\mathbf{M}$ (that is, $\mathbf{M} = \mathbb{R}^n$) and let

$$H_{ij} := h_i \circ h_j^{-1} := h_j(\mathcal{U}_i \cap \mathcal{U}_j) \to h_i(\mathcal{U}_i \cap \mathcal{U}_j).$$

Then, as we know, the derivatives of $a_{ij} := (h_i \circ h_j^{-1})'(x)$, $x \in \mathcal{U}_i \cap \mathcal{U}_j$ are smooth mappings (a_{ij} is an automorphism of M)

$$a_{ij} : h_j(\mathcal{U}_i \cap \mathcal{U}_j) \to \mathrm{GL}(\mathbf{M}) = \mathrm{Aut}(\mathbf{M}).$$

From the theorem on derivative of superposition, we immediately have

$$a_{ij} a_{ji} = a_{ii} = \operatorname*{id}_M,$$

$$(C) \qquad a_{ij} a_{jk} = a_{ik} \quad \text{over } \mathcal{U}_i \cap \mathcal{U}_j \cap \mathcal{U}_k \quad , i, j, k \in J.$$

We say that two triples are equivalent $(x, h_i, t) \sim (x, h_j, t')$, where $t, t' \in M$, if $t = a_{ij}(h_j(x))t'$, $x \in \mathcal{U}_i \cap \mathcal{U}_j$. It can be checked that $\sim$ is an equivalence relation by virtue of (C) which is called the *cocycle relation.*

DEFINITION (IV). The *tangent vector* X to M at point x is the equivalence class of triples (x, h_i, t) defined above. The triple (x, h_i, t) is called the *representative of the vector* X. The set of tangent vectors at $x \in M$ forms a vector space which is isomorphic (linearly) with the model space $\mathbf{M}$, is denoted by T_xM, and is called *tangent space to* M *at the point* $x \in M$.

EXERCISE. Show that definition (IV) is equivalent to the previous three definitions.

REMARK 1. This last definition is, from the operational point of view, the most useful one: the atlas $(\mathcal{U}_j, h_j)_{j\in J}$ is *explicitly* present here and it provides differential structure of manifold M. This definition *mutatis mutandis* is used for definition of fiber bundles and sections defined by specifying a cocycle of mappings corresponding to transitions from one map to another. The definition is classical: for $\mathbf{M} = \mathbb{R}^m$, a_{ij} is nothing but Jacobi matrix!

Definition (I) is, perhaps, the most modern: it seemingly does not make use of an atlas $\mathcal{A} = (\mathcal{U}_j, h_j)$, $j \in J$ but only of a presheaf $\mathcal{E} = \mathcal{E}(M)$ of smooth functions on M ($\mathcal{E}$ is called the *structural sheaf* of manifold M). But the atlas $\mathcal{A}$ is hidden in the definition of smooth functions $f : M \to \mathbb{R}$ ('there are no miracles in mathematics').

REMARK 2. In analysis and in algebraic topology one defines a structure (analytical, harmonic, algebraic) *in manifold* M by giving some structural sheaf, that is, some subsheaf of the sheaf of continuous functions on M as a family $\mathcal{S}_x$, $x \in M$ of some abelian groups.

In this way we arrive at the next important structure of modern mathematics, namely, the sheaf of germs and presheaf. These notions were developed by great mathematician Jean Leray when he was an internee in *Oflag.* Before the Second World War, Leray was a professor of mathematics in Nancy, he was a brilliant analyst and topologist. His pre-war works, written together with Jerzy Schauder, were highly acclaimed. In these works, they introduced methods of algebraic topology (theorem on degree of mapping, theorem on preservation of domain, due to Brouwer (1910–1911)) to analysis: the theory of differential and integral equations. Leray was captured as a reserve officer. During these hard times he began his important works that became a starting point of the theory of sheaves and cohomology (of manifold) valued in sheaves. He thought over the general theory of fiber spaces and spectral sequences. Also his great works generalizing the Riesz-Shauder theory are dated at this camp period. Some ideas of Leray have been later simplified and codified by Henri Cartan (son of Elie Cartan).

The notion of (pre)sheaf has its source in the theory of functions on

Riemann surfaces. A germ of holomorphic (or continuous, differentiable) function is an example of inductive limit. Therefore it is natural to present now the notion of inductive limit (and dual to it notion of projective limit) and the notion of (pre)sheaf. First, we present however the dual notion of projective limit of *projective system* of topological spaces which is slightly older and simpler (?). In topology this construction is encountered very often.

CHAPTER 2

Projective (Inverse) Limits of Topological Spaces

The notion of a system or projective family (often called earlier the inverse *spectrum*) is inseparable from the notion of Cartesian product of topological spaces and, at the same time, is its generalization. This notion appeared first in the work of Alexandrov of 1926 and was generalized by S. Lefschetz and codified in his monograph of 1952.

DEFINITION. *A projective family (inverse system)* of topological spaces is a family $\{X_i, P_{ij}, I\}$, where $(I, \prec)$ is a set ordered by relation $\prec$ (that is, for any pair of elements $i, j \in I$, there exists an element $k \in I$ such that $i \prec k$ and $j \prec k$); for any $i \in I$, X_i is a topological space such that

(1) $$P_{ij} : X_i \to X_j \quad \text{for } i \prec j \text{ is a continuous mapping}$$

(2) $$P_{ii} = \underset{X_i}{\mathrm{id}}, \quad i \in I,$$

$$P_{ik} = P_{ij} \cdot P_{jk} \quad \text{for any triple } i \prec j \prec k \ .$$

The *projective limit* $X_\infty = \lim\limits_{\leftarrow} P_{ij} X_j$ (denoted also $\lim\limits_{i \in I} \operatorname{proj} X_i$) is a subspace X_∞ of the product $\mathsf{X}_{i \in I} X_i$ formed by all (X_i), $i \in I$ (called *threads*) such that $P_{ij} x_j = x_i$, that is $P_{ij} \circ P_j = P_i$ for all $i \prec j$, where P_i is a restriction of projection

$$p_i : (x_j)_{j \in I} \to x_i \ \text{ to } \ X_\infty \qquad (\text{thus } P_i := p_i | X_\infty \to X_i).$$

EXAMPLE 1 (GENERAL). Let us assume that the set $(I, \prec)$ has the 'last' (greatest) element ω, that is, we have $i \prec \omega$ for all $i \in I$. The threads are now of the form

$$(x_i)_{i \in I} \quad \text{where } x_i = P_{i\omega}(x_\infty);$$

they are therefore described by their last element x_ω. Now $P_\omega : X_\infty \to X_\omega$ is a bijection, thus $X_\infty = \varprojlim X_i$ can be identified with X_ω.

Usually, we have to do with a situation described by

EXAMPLE 2 (RATHER GENERAL). Let X be a space (that is, vector space, abelian group, etc.) and let (X_i, τ_i) be a topological space with topology τ_i, $i \in I$. Let $f_i : X \to X_i$ be a mapping (linear, group homomorphism, etc.) Then on X there exists the poorest topology τ such that all mappings $f_i : (X, \tau) \to (X_i, \tau_i)$, $i \in I$ are (still) continuous (τ is an upper limit of the family of topologies $\{f_i^{-1}(\tau_i) : i \in I\}$. The topological space (X, τ) is called the *projective limit of the system* $\{(X_i, \tau_i), f_i, i \in I\}$.

EXAMPLE 3 (MOST ELEMENTARY). Let X_1 be a subspace of the topological space (X, τ) and let $p_1 : X_1 \to X$ be a canonical embedding. Then $\{(X_1, \tau|X_1), p_1\}$ is a (one element) projective family.

REMARK. It turns out that *any locally compact group contains an open subgroup being a projective limit of Lie groups.* This is the famous Yamabe theorem which became a decisive step in the proof of the fifth Hilbert problem: *Any locally compact group which is locally Euclidean (an thus is a topological manifold) is a Lie group.*

The following important theorem holds.

THEOREM 1. *1. The projective limit $X_\infty = \varprojlim P_{ij} X_j$ of Hausdorff spaces X_i is a closed subset of the product $\times_{i \in I} X_i$.*

2. If all spaces X_i are compact, then $\varprojlim P_{ij} X_j$ is a compact set.

PROOF. It is sufficient to prove 1., because 2. follows from 1. and Tichonov theorem (because a closed subset of compact space is compact).

Ad. 1. For any pair $i \prec j$

$$V_{ij} = \{x \in \times_{i \in I} X_i : P_{ij} \circ P_j(x) = P_i(x)\}$$

is a closed subspace of the Hausdorff space $\times_{i \in I} X_i$ because the mappings

$P_{ij} \circ P_j(x)$ and P_i are continuous. But the space $X = \bigcap_{i \prec j} V_{ij}$ is closed as an intersection of closed sets. □

REMARK. If the spaces X_i, $i \in I$ carry additional structure, e.g., they are (topological) groups, then one demands that the mapping P_{ij} be (continuous) homomorphisms of groups $X_i \to X_j$ for $i \prec j$. The structure of (topological) group is defined on $\underleftarrow{\lim} P_{ij} X_j$ in a natural way.

It is not hard to prove the following proposition characterizing projective limit:

PROPOSITION 2. *The projective limit* $\underleftarrow{\lim} P_{ij} X_j$ *of groups* X_i *is characterized, up to homomorphisms by the following property:*

For any group Y *and homomorphisms* $h_i : Y \to X_i$, $i \in I$ *such that* $P_{ij} \circ H_j = h_i$ *for all* $i \prec j$, *there exists a unique homomorphism* $h : H \to \underleftarrow{\lim} X_i$ *such that*

$$P_i \circ H = f_i \quad \text{for all } i \in I.$$

Chapter 3

Inductive Limits. Presheaves. Covering Defined by Presheaf

At this point we present shortly an important construction of topological space which is, in some sense, dual to projective limit (and is a generalization of disjoint union!)

Let again $(A, \prec)$ be an ordered set of a family of topological spaces E_α, $\alpha \in A$ and let continuous mappings be given

$$h_{\alpha\beta} : E_\alpha \to E_\beta \quad \text{for } \alpha \prec \beta;$$

moreover, let $h_{\alpha\alpha} = \mathrm{id}_{E_\alpha}$, $\alpha \in A$ and let the following conditions be satisfied

$$h_{\beta\gamma} \circ h_{\alpha\beta} = h_{\alpha\gamma} \quad \text{for } \alpha \prec \beta \prec \gamma.$$

Then $\{E_\alpha, h_{\alpha\beta}, A\}$ is called the *inductive family (simple family)*. (Observe that mappings $h_{\alpha\beta}$ are ordered in opposite direction than $P_{\alpha\beta}$ in the definition of projective family).

Now we form the *disjoint sum* $\tilde{E} := \coprod_{\alpha \in A} E_\alpha$ of spaces E_α, that is

$$\coprod_{\alpha \in A} E_\alpha = \bigcup_{\alpha \in A} (E_\alpha, \alpha), \quad \text{where } (E_\alpha, \alpha) := \{(x, \alpha) : x \in E_\alpha\},\ \alpha \in A.$$

Clearly, $x \to (x, \alpha)$ is a bijection of E_α on (E_α, α), moreover,

$$(E_\alpha, \alpha) \cap (E_\beta, \beta) = \emptyset \quad \text{for } \alpha \neq \beta.$$

When the sets E_α are pairwisely disjoint, then we can identify $\coprod_{\alpha \in A} E_\alpha$ with $\bigcup_{\alpha \in A} E_\alpha$ and this is the reason behind the name *disjoint union.* On $\coprod_{\alpha \in A} E_\alpha$ one defines now the (equivalence) relation $\sim$:

$$(x, \alpha) \sim (y, \beta),$$

if there exists such $\gamma \in A$, $\alpha, \beta \prec \gamma$ that

$$h_{\alpha\gamma}(x) = h_{\beta\gamma}(y).$$

It is immediate to check that $\sim$ is an equivalence relation.

The quotient space $\coprod_{\alpha \in A} E_\alpha / \sim =: E^\infty$ is called the *inductive limit* and is denoted by $\varinjlim h_{\beta\alpha} E_\alpha$ or $\lim_{\alpha \in A} \operatorname{ind} E_\alpha$.

We still must define topology on the set E^∞:

Let $h_\alpha : E_\alpha \to E^\infty$ be a canonical imbedding. Then

$$(*) \qquad h_\alpha \circ h_{\alpha\beta} = h_\alpha \quad \text{for } \alpha \prec \beta.$$

Since $E^\infty = \bigcup_{\alpha \in A} h_\alpha(E_\alpha)$, one can equip E^∞ in such the poorest (richest) topology that all mappings h_α, $\alpha \in A$ are (still) continuous. It is not known (in general) if E^∞ with such a topology is still a Hausdorff space, even if all E_α are Hausdorff spaces themselves.

The construction of inductive limit has arisen in connection with the notion of germ of mapping and is widely used in analysis, for example, in the Schwartz theory of distributions.

The reader will immediately show the following proposition.

PROPOSITION 1. *Let F be a topological space. Then $(f : \varinjlim h_{\alpha\beta} E_\alpha \to F$ is continuous$) \Longleftrightarrow (f \circ h_\alpha : E_\alpha \to F$ is continuous for all $\alpha \in A)$*

When E_α, $\alpha \in A$ are topological commutative groups (vector spaces), one demands the mappings $h_{\alpha\beta} : E_\alpha \to E_\beta$ be continuous homomorphisms. Then it is not hard to show that $\varinjlim h_{\alpha\beta} E_\alpha$ is an abelian group (a vector space).

The following important proposition holds.

PROPOSITION 2. *The inductive limit $\varinjlim (E_\alpha(h_{\alpha\beta}))$ of abelian groups E_α is characterized, up to an isomorphism, by the following property:*

For any abelian group H and arbitrary homomorphisms $f_\alpha : E_\alpha \to H$, $\alpha \in A$ such that $f_\beta h_{\alpha\beta} = h_\alpha$ (cf. $(*)$) there exists a unique homomorphism $f : \varinjlim E_\alpha \to H$ such that

$$f \circ h_\alpha = f_\alpha, \quad \alpha \in A.$$

EXAMPLE 1. Let A be a family of subsets of some (topological) space M ordered by relation $\alpha \prec \beta$ iff $a \subset \beta$. The mappings $h_{\alpha\beta} : \alpha \to \beta$ are inclusions. Then $\{\alpha, h_{\alpha\beta}\}$ is an inductive family.

Since all mappings $h_{\alpha\beta}$ are inclusions, $(x, \alpha) \sim (y, \beta)$ exactly when $x = y$. Denoting the inductive limit above by M^∞, we see that $h_\alpha : \alpha \to M^\infty$ is injective. If $x \in \alpha \cap \beta$, then $h_\alpha | \alpha \cap \beta = h_\beta | \alpha \cap \beta$. By

$$h(x) := h_\alpha(x) \quad \text{for } x \in \alpha,\ \alpha \in A$$

we denote the bijection

$$h : \bigcup_{\alpha \in A} \alpha \to M^\infty.$$

In this way we can identify $M^\infty := \lim\limits_{\overset{\alpha}{\to}} h_{\alpha\beta}\alpha$ with the set theoretical sum $\bigcup_{\alpha \in A} \alpha$.

EXAMPLE 2 (DUAL TO EXAMPLE 1). Let E be a space (for example, vector) and let (E_α, τ_α), $\alpha \in A$ be topological (vector) spaces and $f_\alpha : E_\alpha \to E$ be mappings (linear). Then, on the space E there exists such richest topology τ that the mappings $f_\alpha : (E_\alpha, \tau_\alpha) \to (E, \tau)$ are still continuous for any $\alpha \in A$. The space (E, τ) is called the *inductive limit* of the family $((E_\alpha, \tau_\alpha), \varphi_\alpha)$, $\alpha \in A$.

Germ of function. Stalk of sheaf as an example of inductive limit. The notion of inductive limit has grown, as we know, from the theory of analytic functions and Lie groups. In the theory of functions, we deal with functions defined in different regions of the complex plane $\mathbb{C}$ or Riemann surface X. The following definition of *germ of function at point a* (which has arisen most likely in connection with analytical continuation of germs of functions), is of paramount importance:

DEFINITION. Two functions f, g defined in some (usually different) neighborhoods $\mathcal{U}$, $\mathcal{V}$ of the point a belong to the same equivalence class (denoted by $\overset{a}{\sim} a$) if there exists a neighborhood $\mathcal{W} \ni a$ such that $\mathcal{W} \subset \mathcal{U} \cap \mathcal{V}$ and $f|\mathcal{W} = g|\mathcal{W}$. We denote this by $f \overset{a}{\sim} ag$, and this means that f and g are identical in some neighborhood of a. The (equivalence) class to which f belongs is called the *germ of function f at point a* (or *over a*) and denoted by f_a or $[f]_a$.

How this construction is related to the notion of inductive limit?

Let $(X, \mathcal{T})$ be a topological space with topology $\mathcal{T}$. The open sets belonging to $\mathcal{T}$ form a system ordered by inclusion $\mathcal{U} \prec \mathcal{V}$ if $\mathcal{U} \subset \mathcal{V}$. If with any $\mathcal{U} \in \mathcal{T}$ one associates a family of abelian groups $E_{\mathcal{U}}, \mathcal{U} \in \mathcal{T}$ and a family of homomorphisms $h^{\mathcal{U}}_{\mathcal{V}} : E_{\mathcal{U}} \to E_{\mathcal{V}}$ having the following properties

$$h^{\mathcal{U}}_{\mathcal{U}} = \underset{\mathcal{U}}{\mathrm{id}} \quad \text{for all } \mathcal{U} \in \mathcal{T} \ ,$$

$$h^{\mathcal{V}}_{\mathcal{W}} \circ h^{\mathcal{U}}_{\mathcal{V}} = h^{\mathcal{U}}_{\mathcal{V}} \quad \text{for } \mathcal{W} \subset \mathcal{V} \subset \mathcal{U},$$

then $(E_{\mathcal{U}}, h^{\mathcal{U}}_{\mathcal{V}}, \mathcal{T})$ is an inductive family. Often instead of $h^{\mathcal{U}}_{\mathcal{V}} f$, we write $f|\mathcal{V}$; the mappings $h^{\mathcal{U}}_{\mathcal{V}}$ are called *restrictions*, more precisely, *restriction homomorphisms.*

REMARK. In the general definition of inductive family we used the notation $h_{\mathcal{V}\mathcal{U}}$ instead of $h^{\mathcal{U}}_{\mathcal{V}}$.

DEFINITION. The above defined inductive family is called the *presheaf* of abelian groups over space X. If $a \in X$ and $\mathcal{U}$ runs through the family of neighborhoods of the point a, then

$$E_a := \underset{\mathcal{U} \ni a}{\underset{\rightarrow}{\lim}} E_{\mathcal{U}}.$$

is called the *stalk of the presheaf* $(E_{\mathcal{U}}, h^{\mathcal{U}}_{\mathcal{V}})$ at the point a. Thus $h_a : E_{\mathcal{U}} \to E_a$ associates with any element $f \in E_{\mathcal{U}}$ its equivalence class modulo $\overset{a}{\sim}$. Thus, $h_a(f) = f_a$, and therefore, the mapping h_a associates with an element f its germ f_a at a being an element of the stalk E_a at the point a.

DEFINITION. The *unbranched covering* $p : E \to X$ of the space X is the triple (E, p, X), where E, X are topological spaces and p is a local homomorphism.

We show that each presheaf defines some unbranched covering.

Covering space (cover) associated with a presheaf. Let X be a topological space and $\mathcal{F}$ a presheaf over X. Let $\mathcal{F}_x$, $x \in X$ be a family of all stalks called also the *fibers* defined by $\mathcal{F}$. Let us form the disjoint sum

$$|\mathcal{F}| := \coprod_{x \in X} \mathcal{F}_x \quad \text{of all fibers of the presheaf } \mathcal{F} \ .$$

Let $p : \mathcal{F} \to X$ be a mapping which associates with each germ $\varphi \in \mathcal{F}_x$ the point x, that is, p maps (the whole of) the fiber over x, $\mathcal{F}_x$, onto x. In other words, $p^{-1}(x) = \mathcal{F}_x$.

Now, let us define a basis of open sets (that is, a topology) on $|\mathcal{F}|$ as follows. Let $\mathcal{U} \subset X$ be an open set, $f \in \mathcal{F}(U)$ and let

$$[\mathcal{U}, f] := \{h_x(f) : x \in \mathcal{U}\} \subset |\mathcal{F}|.$$

The important theorem holds.

THEOREM 3. *The family of subsets* $[\mathcal{U}, f]$, *where* $\mathcal{U}$ *runs through open sets in* X, $f \in \mathcal{F}(\mathcal{U})$ *is a basis of some topology on* $|\mathcal{F}|$. *The mapping* $p : |\mathcal{F}| \to X$ *is locally topological, that is, the triple* $(|\mathcal{F}|, p, X)$ *is an unbranched covering of the space* X. *In this topology, each fiber* $\mathcal{F}_x$ *has the discrete topology (relative).*

We leave the proof to the reader (see also K. Maurin *Analysis part II*, Chapter XVI, 7).

Thus, with each presheaf $\{\mathcal{F}(\mathcal{U}), r^{\mathcal{U}}_{\mathcal{V}}\}$ over X we associate the unbranched covering $\{|\mathcal{F}|, p, X\}$ of the space X.

The question arises as to whether, also vice versa, with any unbranched covering $p : \mathcal{E} \to X$ one can naturally associate a presheaf?

The answer is affirmative and rather obvious, namely:

Let $(\mathcal{E}, p, X)$ be an unbranched covering, that is, the mapping $p : \mathcal{E} \to X$ is locally topological. Let $\mathcal{U}$, $\mathcal{V}$ be open subsets of X and $\mathcal{U} \subset \mathcal{U}$, $\mathcal{E}(\mathcal{U}) := p^{-1}(\mathcal{U}) \in \mathcal{E}$; let $f \in \mathcal{E}(\mathcal{U})$. We take $r^{\mathcal{U}}_{\mathcal{V}}|f := |f|\mathcal{V}$. Clearly $r^{\mathcal{U}}_{\mathcal{V}} : \mathcal{E}(\mathcal{U}) \to \mathcal{E}(\mathcal{V})$ satisfies the conditions

$$r^{\mathcal{U}}_{\mathcal{U}} = \mathrm{id}, \quad r^{\mathcal{V}}_{\mathcal{W}} r^{\mathcal{W}}_{\mathcal{U}} = r^{\mathcal{U}}_{\mathcal{V}} \quad \text{for } \mathcal{W} \subset \mathcal{V} \subset \mathcal{U},$$

and thus the inductive system $\{\mathcal{E}(\mathcal{U}), r^{\mathcal{U}}_{\mathcal{V}}\}$ forms a presheaf called the *canonical presheaf* $\Gamma(\mathcal{E})$ of the covering $p : \mathcal{E} \to X$. If we have two coverings $p_i : \mathcal{E}_i \to X$, $i = 1, 2$ of the space X, then the mapping $\varphi : \mathcal{E}_1 \to \mathcal{E}_2$ preserves fibers (is a mapping of coverings) if the diagram

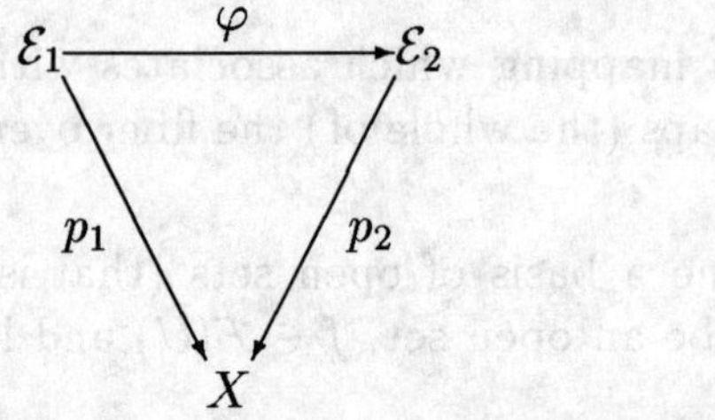

is commutative, that is $p_1 = p_2 \circ \varphi$. Indeed, then the fiber

$$\mathcal{E}_{1\,x} = p_1^{-1}(x) \xrightarrow{\varphi_x} p_2^{-1}(x) = \mathcal{E}_{2\,x}$$

and thus, $\varphi(\mathcal{E}_{1\,x}) \subset \mathcal{E}_{2\,x}$, $x \in X$. Since p_1, p_2 are locally topological, the mapping of coverings is locally topological as well and therefore is open (open sets are mapped into open sets). We showed therefore the following

COROLLARY 4. *Every mapping of coverings $\varphi : \mathcal{E}_1 \to \mathcal{E}_2$ with $\varphi \circ p_2 = p_1$ defines a mapping $\Gamma(\varphi)$ of presheaves $\Gamma(\mathcal{E}_1) \to \Gamma(\mathcal{E}_2)$, where $\Gamma(\varphi) = \{\varphi_{\mathcal{U}}, \mathcal{U} \in \mathcal{T}\}$ is a family of mappings*

$$\varphi_{\mathcal{U}} : \mathcal{E}_1(\mathcal{U}) \to \mathcal{E}_2(\mathcal{U}), \quad \textit{where } \ \mathcal{E}_i(\mathcal{U}) = p_i(\mathcal{U}),$$

$\mathcal{U}$ open in X, called the mapping of presheaves; $\{\mathcal{E}_1(\mathcal{U}), r_{1\mathcal{V}}^{\mathcal{U}}\} \to \{\mathcal{E}_2(\mathcal{U}), r_{2\mathcal{V}}^{\mathcal{U}}\}$.

Let us observe that the family $\{\varphi_{\mathcal{U}}\}$ is related to the family $\{r_{i\mathcal{V}}^{\mathcal{U}}\}$, $i = 1, 2$ as follows

$$\varphi_{\mathcal{U}} r_{1\mathcal{V}}^{\mathcal{U}} = r_{2\mathcal{V}}^{\mathcal{U}} \varphi_{\mathcal{U}}.$$

This leads to the following definition.

DEFINITION. Given the presheaves $\{\mathcal{E}_1(\mathcal{U}), r_{1\mathcal{V}}^{\mathcal{U}}\}$ and $\{\mathcal{E}_2(\mathcal{U}), r_{2\mathcal{V}}^{\mathcal{U}}\}$ over X, the family $\Phi = \{\Phi_{\mathcal{V}}^{\mathcal{U}}\}$ of mappings $\Phi_{\mathcal{U}} : \mathcal{E}_1(\mathcal{U}) \to \mathcal{E}_2(\mathcal{U})$ of abelian groups (vector spaces, rings, algebras, modules, etc) such that for all open sets $\mathcal{U}$, $\mathcal{V}$, $\mathcal{V} \subset \mathcal{U}$, $\Phi_{\mathcal{V}} r_{1\mathcal{V}}^{\mathcal{U}} = r_{2\mathcal{V}}^{\mathcal{U}}$ is called the *mapping of these presheaves* or the *presheaf mapping.*

LEMMA 5. *Every presheaf mapping $\Phi = \{\Phi_{\mathcal{U}}\}$ defines a mapping of coverings*

$$\check{\Gamma}(\Phi) : \check{\Gamma}(\mathcal{E}_1 \overset{p_1}{\to} X) \to \check{\Gamma}(\mathcal{E}_2 \overset{p_2}{\to} X)$$

(associated with this presheaf).

Obviously, the mapping $\{\Phi_\mathcal{U}\}$ is called *isomorphism, monomorphism, epimorphism* if for every $\mathcal{U}$, the mapping $\Phi_\mathcal{U}$ is a bijection, injection, or surjection. Similarly, if $\mathcal{E}_1 = \mathcal{E}_2$, $p_1 = p_2$, we can talk about automorphism. Analogously, we can talk about automorphisms of the covering $p : \mathcal{E} \to X$.

DEFINITION. The *automorphism of covering* $\varphi : \mathcal{E} \to \mathcal{E}$, $p = \varphi \circ p$ is a topological mapping of covering, it is customary to call it the *deck transformation.*

Authomorphisms of coverings (not necessarily unbranched) form a group which is denoted by $\mathrm{Aut}(\mathcal{E}/X)$ or, more precisely, $\mathrm{Aut}(\mathcal{E} \xrightarrow{p} X)$. We will deal with this object while discussing Riemann theory, where it plays a part analogous to the group of automorphisms $\mathrm{Aut}(\mathbf{L}/\mathbf{K})$ of extension $\mathbf{L} \supset \mathbf{K}$ of fields. Looking ahead a bit, we will present now an important notion.

DEFINITION. Let $\mathbf{L} \supset \mathbf{K}$ be an extension of fields (that is, $\mathbf{K}$ is a subfield of the field $\mathbf{L}$. Then an automorphism σ of the field $\mathbf{L}$ (that is, a mapping $\sigma : \mathbf{L} \to \mathbf{L}$ which is bijective and preserves all operations of $\mathbf{L}$) which leaves all elements of $\mathbf{K}$ invariant (that is, $\sigma|\mathbf{K} = \mathrm{id}_\mathbf{K}$) is called the *automorphism of extension* $\mathbf{K} \subset \mathbf{L}$. Clearly, such automorphisms form a group which is denoted by $\mathrm{Aut}(\mathbf{L}/\mathbf{K})$.

REMARK. Similarly one can talk about automorphisms of extension of rings $\mathcal{L} \supset \mathcal{K}$ and their group $\mathrm{Aut}(\mathcal{L} \supset \mathcal{K})$.

The reader has probably guessed already what was the relation between the group $\mathrm{Aut}(Y \xrightarrow{p} X)$ and the group $\mathrm{Aut}(L \supset p^*K)$ for rings of continuous functions $L := \mathcal{C}(Y)$ and $K := \mathcal{C}(X)$, where $p : Y \to X$ is a covering of topological spaces and p^* is a lift of the ring $\mathcal{K}$ to $\mathcal{L}$, namely

$$p^*(f) := f \circ p, \quad \text{thus } (p^*f)(y) := f(p(y)) \quad \text{for } f \in \mathcal{C}(X).$$

EXERCISES.

1. Show that for $\sigma \in \mathrm{Aut}(Y \xrightarrow{p} X)$ the association mapping $f \to \tilde{\sigma} f$ where $\tilde{\sigma} f := f \circ \sigma^{-1}$ is an automorphism of the ring $\mathcal{C}(Y)$.

2. Next, show that the mapping

$$\mathrm{Aut}(Y \xrightarrow{p} X) \ni \sigma \to \tilde{\sigma} \in \mathrm{Aut}(\mathcal{C}(Y))$$

is a homomorphism of groups.

3. Prove that the automorphism $\tilde{\sigma}$ preserves functions in the subring

$$p^*\mathcal{C}(X) = p^*\mathcal{K} \subset \mathcal{L} - \mathcal{C}(Y),$$

and that we have a homomorphism $\sigma \to \tilde{\sigma} = \mathcal{U}(\sigma)$ of groups

$$(*) \qquad \operatorname{Aut}(Y/X) \xrightarrow{U} \operatorname{Aut}(\mathcal{L} \supset \mathcal{K}).$$

In *Riemann theory of algebraic functions* we have a richer structure where we can talk not only about rings but also about function fields, and then (for compact X, Y), the mapping defined by formula $(*)$ is even an isomorphism of groups.

Sheaves (German *Garbe*, French *faisceau*). Let $\mathcal{S} = \{S(\mathcal{U}), r^{\mathcal{U}}_{\mathcal{V}}\}$ be a presheaf over X. As we know, by taking the inductive limit over each point $x \in X$, we can turn from the presheaf to the covering $p : |\mathcal{S}| \to X$, $|\mathcal{S}| = \coprod S_x$, $x \in X$. In turn, having arbitrary covering $\mathcal{S} \xrightarrow{p} X$, we had a natural mapping of the presheaf

$$\check{\Gamma} : \mathcal{S} \to \{S(\mathcal{U}), r^{\mathcal{U}}_{\mathcal{V}}\}.$$

One can form a superposition of these two mappings $\Phi : S \to \Gamma\check{\Gamma}(S)$, where $\Phi = \{\Phi(\mathcal{U})\}$ is a presheaf mapping. The question arises as to whether Φ is an isomorphism (of presheaves)? In other words:

1. When all $\Phi(\mathcal{U})$ are injective?

The answer is given by the following condition

I. Let $s, t \in S(\mathcal{U})$ be such that there exist an open covering $(\mathcal{U}_j)$, $j \in J$ of the set $\mathcal{U}$ such that if $r^{\mathcal{U}}_{\mathcal{U}_j} s = r^{\mathcal{U}}_{\mathcal{U}_j} t$ for all $j \in J$, then $s = t$.

2. When the injections $\Phi(\mathcal{U})$ are bijections?

The answer is given by the following condition

Let $\Phi(\mathcal{V}) : S(\mathcal{V}) \to \mathcal{S}(\mathcal{V})$ be injective for any open subset $\mathcal{V} \subset \mathcal{U}$;

$$(\Phi(\mathcal{U}) \text{ is a bijection}) \Longleftrightarrow (\text{condition II is satisfied}).$$

II. When $(\mathcal{U}_j)_{j \in J}$ is an open covering of $\mathcal{U}$ and $(s_j)_{j \in J}$, where $s_j \in S(\mathcal{U}_j)$ is a family for whose

$$r^{\mathcal{U}_j}_{\mathcal{U}_j \cap \mathcal{U}_k} s_j = r^{\mathcal{U}_k}_{\mathcal{U}_j \cap \mathcal{U}_k} s_k \quad \text{for all } j, k \in J \enspace ,$$

then there exists such $s \in S(\mathcal{U})$ that $r^{\mathcal{U}}_{\mathcal{U}_j} s = s_j$ for all $j \in J$. (The gluing condition: *from compatible pieces one can sew the whole*).

DEFINITION. The presheaf $\{S(\mathcal{U}), r^{\mathcal{V}}_{\mathcal{U}}\}$ satisfying conditions I and II is called the *sheaf over* X.

REMARK. For sheaves over X one can define naturally *algebraic operations* (depending on structure of fibers – stalks) and the notions of subsheaf and quotient sheaf.

COROLLARY 6. *The presheaf $\mathcal{E} = \mathcal{E}(X)$ of smooth functions on differentiable manifold X is a sheaf of local rings and even local algebras; each fiber $\mathcal{E}_x$ is an $\mathbb{R}$-algebra possessing a unique maximal ideal $\mathfrak{m}_x$ of functions vanishing at x.*

In the theory of analytic functions the important part is played – from the times of Weierstass and Riemann on – by the so-called *identity theorem* which can be phrased as follows.

THEOREM 7. *If we have two analytical functions f and g in the region $\mathcal{U} \subset \mathbb{C}$, that is if $f, g \in \mathcal{A}(\mathcal{U})$ which are identical in a neighborhood $\mathcal{U}(a)$ of a, $a \in \mathcal{U}$, then they are identical.*

PROOF is almost immediate. (cf. *Analysis II*, XVI, 7). Indeed, (*ex definitione*) f, g can be expanded in Taylor series convergent in some ball in a neighborhood of any point $z \in \mathcal{U}$. The formula for coefficients of the series has the form

$$k!a_k = \frac{\partial^{|k|}}{\partial z^k} f(z) = \frac{\partial^{|k|}}{\partial z^k} g(z) = k!b_k,$$

where $f(z) = \sum_k a_k z^k$, $g(z) = \sum_k b_k z^k$; thus $a_k = b_k$.

The identity theorem is a basis of a beautiful construction of analytical continuation of germ of holomorphic (analytical) function. The notion of presheaf systematizes this classical construction and highlights the important points. This leads immediately to the following definition:

DEFINITION. For a presheaf $\mathcal{F}$ over topological space X the *identity theorem* holds if the following condition is satisfied. If $\mathcal{U} \subset X$ is a *domain* (that is, an open and arcwise connected set) and if $f, g \in \mathcal{F}(\mathcal{U})$ are such that their germs over some $a \in \mathcal{U}$ are equal $[f]_a = [g]_a$, then $f = g$.

The following important theorem is immediate.

THEOREM 8. *Let X be a locally compact Hausdorff space, $\mathcal{F}$ a presheaf over X for whose the theorem on identities hold. Then the topological space $|\mathcal{F}|$ is a Hausdorff space.*

We recommend the reader to perform the simple proof as an exercise (cf. *Analysis II*, XVI, 7).

DEFINITION. Let X be a Riemann surface (more generally, an analytic manifold), $u : [0,1] \to X$ be a (continuous) curve in X with endpoints a and b: $u(0) = a$, $u(1) = b$. Then the germ $\psi \in \mathcal{A}_b$ is an *analytic continuation of the germ* $\varphi \in \mathcal{A}_a$ along the curve u if there exists a lift $\hat{u} : [0,1] \to |\mathcal{A}|$ of the curve u into a curve in the space of germs $|\mathcal{A}|$ (or, as some authors prefer, the covering space $|\mathcal{A}| \overset{p}{\to} X$) such that $\hat{u}(0) = \varphi$, $\hat{u}(1) = \psi$.

Roughly speaking, a curve in X with endpoints a, b can be lifted to curve $\hat{u} \subset |\mathcal{F}|$ with endpoints φ and ψ.

From the theorem on uniqueness of a lift (cf. *Analysis II*, XVI), it follows that *if there exists an analytic continuation along the curve u, then it is unique.*

In general, analytic continuations of a germ $\varphi \in \mathcal{A}$ along non homotopic curves with the same endpoints lead to *different* germs. The *Riemann–Weierstrass principle of analytic continuation* asserts that analytic continuations along homotopic curves lead to the same germ. This is the source of the magnificient theory due to Poincaré and was the starting point of a brilliant concept of Riemann – paraphrasing Hermann Weyl: the world of germs $|\mathcal{A}|$ even though infinite dimensional, has the stratification onto two dimensional surfaces, the connected components of germs $\varphi \in |\mathcal{A}|$. The Riemann surface $\mathcal{R}_\varphi$ of the germ $\varphi_0 \in \mathcal{A}_a$ is a connected component of this germ φ which, in the case of algebraic germs, should be completed by a *finite!* number of points. As it can be shown, the surprising and beautiful result follows.

THEOREM 9. *A Riemann surface $\mathcal{R}_\varphi$ of algebraic germ φ is compact and is an n-fold covering of the Riemann sphere $\bar{\mathbb{C}} = \mathbb{C} \cup \{\infty\}$, where n is a* *degree of an irreducible polynomial defining the germ φ.*

This, not too precise formulation is intended to encourage the reader to read the following chapter.

Above, we talked about rings and we said that $\mathcal{E}_x$ was a local algebra, that is a local ring in which multiplication by real numbers is defined. Since the notion of algebra is of fundamental importance in analysis and the whole of mathematics, in the following chapter we present the definition of algebra and many important examples of algebras: the famous algebras of quaternions of Hamilton, octonions of Cayley, algebras of Grassmann and Clifford. This algebras are irreplacible tools of modern physics. I will try to sweeten the dryness of information by historical remarks which show fascinating an usually painful process of birth of mathematical ideas.

CHAPTER 4

Algebras. Groups, Tensors, Clifford, Grassmann, and Lie Algebras. Theorems of Bott–Milnor, Wedderburn, and Hurwitz

In Frobenius theorem the important notion of algebra appeared. In this chapter present many important examples of algebras.

DEFINITION. A ring $\mathcal{A}$ which is at the same time a vector space over commutative field $\mathbf{K}$ and which satisfies the condition

$$(\alpha a)b = a(\alpha b) = \alpha(ab), \quad \text{for } \alpha \in \mathbf{K}, \ a, b \in \mathcal{A}$$

is called an *associative algebra* over $\mathbf{K}$ (in older terminology it was also called a *hypercomplex system.* If the associativity requirement is dropped, we have algebra an $\mathcal{A}$ over field $\mathbf{K}$. The *center* $Z(\mathcal{A})$ of the algebra $\mathcal{A}$ is $\{a \in \mathcal{A} : ax = xa \text{ for all } x \in \mathcal{A}\}$, that is, the set of commuting elements of $\mathcal{A}$.

As an example of associate algebras without divisors of zero over $\mathbb{R}$ one can give the fields $\mathbb{R}$, $\mathbb{C}$, and $\mathbb{H}$ and, according to Frobenius theorem, these are the only such finite dimensional algebras.

The theory of algebras is nowadays a large discipline with many important applications. It suffices to recall that *quantum mechanics* deals with operator algebras in Hilbert spaces. Of particular simplicity and beauty is the theory of *commutative Banach algebras* (with unity) called also *Gelfand theory*.

The most important finite dimensional algebras over **K** are *algebras of square matrices*, that is, the algebras of linear transformations of the space $\mathbf{K}^n$. Homomorphisms of algebras into such algebras are called (matrix) *representations* of these algebras. Let us present important examples of such situation.

EXAMPLES OF ALGEBRAS. A *group algebra* $G[\mathbf{K}]$ over (commutative) field **K** is a vector space over **K** with the basis $\{g \in G\}$, that is, it is a set of formal linear combinations

$$(1) \qquad a = \sum_{g \in G} \alpha_g g, \quad \text{where } \alpha_g \in \mathbf{K} \ ,$$

where only finite number of coefficients α_g differs from zero. The addition of elements $a \in G[\mathbf{K}]$ and the multiplication by $\alpha \in \mathbf{K}$ is defined componentwisely

$$\sum_g \alpha_g g + \sum_g \beta_g g = \sum_g (\alpha_g + \beta_g) g,$$

$$\lambda \sum_g \alpha_g g = \sum_g (\lambda \alpha_g) g.$$

The multiplication of elements, often denoted by $*$, is defined as follows

$$\left(\sum_g \alpha_g g\right) * \left(\sum_g \beta_g g\right) = \sum_g \gamma_g g, \quad \text{where } \gamma_g := \sum_s \alpha_s \beta_{s-g}.$$

The basis of algebra $G[\mathbf{K}]$ are sums with one term $1 \cdot g$ only.

REMARK. Group algebras were introduced (for finite groups) by Theodor Molien and Georg Frobenius and were used for investigations of representations of these groups.

For investigation of compact groups and (commutative) locally compact groups G, Hermann Weyl used systematically group algebra $G[\mathbf{K}]$. In that case, the summation over G is replaced by integration with respect to some invariant measure (integral) μ, the so-called *Hurwitz integral* (in the case of Lie group), and, more generally, *Haar integral* (in the case of arbitrary locally compact group). The formal linear combinations are replaced with continuous functions (more generally, locally μ-integrable ones) and it is customary to call the space $L^1(G,\mu)$ of these functions the *group algebra.* In the case of commutative group, the multiplication in $G[\mathbf{K}]$ is commutative and the algebra $L^1(G,\mu)$ is a commutative Banach algebra; we have

$$(f * h)(g) := \int_G f(s)h(s^{-1}g)\, d\mu(s).$$

The norm in $L^1(G,\mu)$ is, clearly, nothing but the integral

$$\int_G |f(g)|\, d\mu(g) =: |f|_{L^1(G,\mu)}.$$

The investigations of the theory of such algebras and their maximal ideals and its application to harmonic analysis on commutative locally compact groups established the fame of I. M. Gelfand and his school.

Tensor products of modules. Tensor algebra. The notion of tensor has been emerging rather slowly: as the etymology of the word indicates, this notion (and rather, the notion of tensor field) appeared first in physics, in the theory of elasticity: *tensio* in Latin means tension. The Hook law was formulated as follows: *ut tensio sic vis* (like tension, like force) – this was a linear relation between vector fields. Further examples from physics are tensor of inertia, tensors in Maxwell electrodynamics and so on. It seems that the algebraic, general notion of tensor came from H. Grassmann (*Lineare Ausdehnungslehre*) and was developed by Italian school of differential geometry. Codification of this notion appeared in Weyl's *Raum, Zeit, Materie* and it was popularized in his *Classical Groups.* The current abstract notion was in possession of mathematical community in thirties and was finally abstractly formulated by Claude Chevalley (1909 – 1985) (the co-founder of the Bourbaki brotherhood) in his work on Lie groups. Since we will often deal with tensor algebra and analysis in the later chapters, here we only shortly present definition of tensor product of modules.

DEFINITION. Let V_1 and V_2 be modules over a commutative and associative ring $\mathcal{R}$ with unity. A *tensor product of these $\mathcal{R}$-modules* is a $\mathcal{R}$-module $V_1 \otimes_{\mathcal{R}} V_2$ with the bilinear mapping

$$V_1 \times V_2 \to V_1 \otimes_{\mathcal{R}} V_2, \quad (v_1, v_2) \mapsto v_1 \otimes v_2 \in V_1 \otimes_{\mathcal{R}} V_2$$

possessing the universality property:

For any bilinear mapping

$$\beta : V_1 \times V_2 \to W,$$

where W is an arbitrary $\mathcal{R}$-module, there exists a unique linear mapping

$$b : V_1 \otimes_{\mathcal{R}} V_2 \to W$$

such that

$$\beta(v_1, v_2) = b(v_1 \otimes v_2), \quad v_1 \in V_1,\ v_2 \in V_2.$$

The question concerning existence and uniqueness immediately arises.

Existence (construction) of $V_1 \otimes_{\mathcal{R}} V_2$. The construction presented below is typically used for construction of new objects in algebra (we will encounter it also later, for example, in construction of Clifford and Grassmann algebras); it proceeds as follows.

Let F be a free $\mathcal{R}$-module generated by $V_1 \times V_2$ (thus, F is a set of formal linear combinations $\sum r_{ij}(v_i, v_j)$, $r_{ij} \in \mathcal{R}$). We consider a $\mathcal{R}$-submodule J formed by elements of the form

$$(v_1 + y, v_2) - (v_1, v_2) - (y, v_2),$$

$$(v_1, v_2 + z) - (v_1, v_2) - (v_1, z),$$

$$(J) \qquad (cv_1, v_2) - c(v_1, v_2),$$

$$(v_1, cv_2) - c(v_1, v_2),$$

where $v_1, y \in V_1$, $v_2, z \in V_2$, $c \in \mathcal{R}$.

Let

$$(T) \qquad V_1 \otimes_{\mathcal{R}} V_2 := F/J \quad \text{quotient module;}$$

we take

$$v_1 \otimes v_2 = (v_1, v_2) + J.$$

Thus the tensor product is $\mathcal{R}$-bilinear

$$(v_1 + y) \otimes v_2 = v_1 \otimes v_2 + y \otimes v_2,$$

$$v_1 \otimes (v_2 + z) = v_1 \otimes v_2 + v_1 \otimes z,$$

$$cv_1 \otimes v_2 = c(v_1 \otimes v_2),$$

$$v_1 \otimes cv_2 = c(v_1 \otimes v_2),$$

and we have the following isomorphism of $\mathcal{R}$-modules

$$\mathcal{R} \otimes_{\mathcal{R}} V \simeq V, \quad V_1 \otimes_{\mathcal{R}} V_2 \simeq V_2 \otimes_{\mathcal{R}} V_1,$$

$$(V_1 \otimes_{\mathcal{R}} V_2) \otimes_{\mathcal{R}} V_3 \simeq V_1 \otimes_{\mathcal{R}} (V_2 \otimes_{\mathcal{R}} V_3),$$

$$(\bigoplus_{i \in I} V_i) \otimes_{\mathcal{R}} W \simeq \bigoplus_{i \in I} (V_i \otimes_{\mathcal{R}} W)$$

for arbitrary $\mathcal{R}$-modules V, V_i, W.

If (x_i), $i \in I$, (y_j), $j \in J$ are bases in V_1, V_2, then $(x_i \otimes y_j)$, where $(i, j) \in I \times J$ is a basis of the module $V_1 \otimes_{\mathcal{R}} V_2$. In general, if V_1, V_2 are free $\mathcal{R}$-modules finitely generated (that is, finite dimensional vector spaces over field $\mathbb{R}$)

$$\dim(V_1 \otimes_{\mathcal{R}} V_2) = \dim V_1 \cdot \dim V_2.$$

The module $\bigotimes^p V := V \otimes_{\mathcal{R}} \ldots \otimes_{\mathcal{R}} V$ (p times) is called the *pth tensor power*. A *tensor algebra* $\mathcal{T}(V)$ of a $\mathcal{R}$-module V is the $\mathcal{R}$-module

$$\mathcal{T}(V) := \bigoplus_{p=0}^{\infty} \bigotimes^{p} V, \quad \text{where } V^0 := \mathcal{R},$$

and multiplication is defined to be the *tensor product.* Clearly, any $\mathcal{R}$-linear mapping of the $\mathcal{R}$-module V into associative $\mathcal{R}$ algebra $\mathcal{B}$ extends uniquely to the homomorphism of algebras $\mathcal{T}(V) \to \mathcal{B}$ mapping unity into unity. If V is a free module with a basis (v_i), $i \in I$, then $\mathcal{T}(V)$ is a free associative algebra having (v_i), $i \in I$ as its family of generators.

REMARK 1. Of course, the most important, and historically earliest, are tensor products of vector spaces (finite dimensional).

REMARK 2. In functional analysis (theory of functional spaces) we have to do with tensor products of infinite dimensional vector spaces, that is, $V_1 := \mathcal{C}(X)$, $V_2 := \mathcal{C}(Y)$, of continuous functions on compact spaces. In this case the vector space of continuous functions of two variables $\mathcal{C}(X \times Y)$ contains as a dense set (in appropriate topology)

$$V_1 \otimes_{\mathbb{R}} V_2 = \mathcal{C}(X) \otimes_{\mathbb{C}} \mathcal{C}(Y),$$

which is not complete as a linear space. The problem of introducing the appropriate topology on $V_1 \otimes_{\mathbb{R}} V_2$ arises then. It turns out that there does not exists any such natural topology – different kinds of topologies are needed. The systematic study of topology of tensor products of functional spaces (and, more generally, vector locally convex spaces) is due to very young Alexander Grothendieck. We will talk about these problems many times!

REMARK 3. As we know, algebraic operations on spaces of functions are defined by performing these operations on values of functions

$$(f_1 + f_2)(x) := f_1(x) + f_2(x).$$

If we have to do with functions valued in a vector space V, $f_i : X \to V$, $i = 1, 2$, then we can define the tensor product of such functions

$$(f_1 \otimes_{\mathcal{R}} f_2)(x) = f_1(x) \otimes_{\mathcal{R}} f_2(x).$$

More generally, this procedure is defined for the space of sections of vector bundles over manifold X, the so-called *vector and tensor fields.* From philosophical and historical point of view, it is very interesting to realize how tensors first appeared: these are the fields of physical objects, for example electromagnetic field, field of stresses of elastic continuum, spinorial fields in Dirac theory of electron and, more generally, in elementary particle physics, theory of gravity etc.

Clifford algebras. Let V be an n-dimensional vector space over field $\mathbf{K}$ and let Q be a quadratic form on V. With the form Q we associate the (symmetric) bilinear form

$$B(u, v) := Q(u + v) - Q(u) - Q(v).$$

REMARK. Quadratic forms can be defined also on modules, not only on vector spaces, which is important, for example, in number theory. We present the following definition

DEFINITION. A *quadratic form Q on $\mathcal{R}$-module X* is a function $Q : X \to \mathcal{R}$ such that for all $\alpha \in \mathcal{R}$

$$Q(\alpha x) = \alpha^2 Q(x), \quad \text{for } x \in X,$$

and the function $B(\cdot, \cdot)$ defined on $X \times X$ by the equation

$$B(x, y) := Q(u + v) - Q(u) - Q(v)$$

is bilinear over the ring $\mathcal{R}$.

EXAMPLE. If β is a bilinear form (infinitely symmetric), then, clearly, the function

$$Q(x) := \beta(x, x)$$

is a quadratic form with associated *symmetric* bilinear form

$$B(x, y) := \beta(x, y) + \beta(y, x).$$

The theory of quadratic forms is a gigantic branch of mathematics which develops fast since the times of Lagrange and Gauss till now (Milnor, Pfister, and others). We will mention impressive theorems of Minkowski, Hasse, C. L. Siegel later!

A *Clifford algebra* $\mathcal{C}(V)$ *over space* $V = (V, B)$ is an algebra $\mathcal{C}(V)$ and a $\mathbf{K}$-linear injection $\rho : V \to \mathcal{C}(V)$ such that for arbitrary $v \in V$

$$\rho(v)^2 = Q(v) \cdot 1.$$

Further, elements of $\rho(V)$ are multiplicative generators of the algebra $\mathcal{C}(V)$, that is, any element of $\mathcal{C}(V)$ is some (irreducible) linear combination of monomials of elements of $\mathcal{C}(V)$.

Of course, the problem of existence and uniqueness of the algebra $\mathcal{C}(V)$ arises. The affirmative answer is provided by the following theorem.

THEOREM 1. *The above problem of existence of the pair* $(\mathcal{C}, \rho)$ *has solution only up to an isomorphism, that is, if we have some pair* $(\mathcal{A}, \varphi)$, *where* $\mathcal{A}$ *is an algebra over* $\mathbf{K}$ *and* $\varphi : V \to \mathcal{A}$ *is an* $\mathbf{K}$*-homomorphism such that* $(\varphi(v))^2 = Q(v) \cdot 1$, *then there exists a unique homomorphism of algebras* $\psi : \mathcal{C} \to \mathcal{A}$ *such that* $\rho \circ \psi = \varphi$, *that is, the diagram*

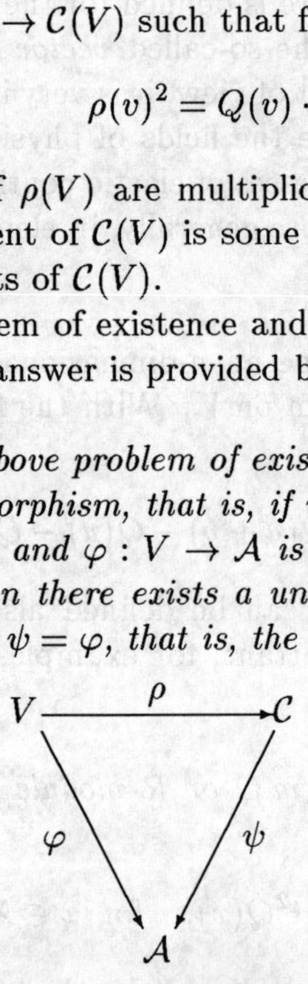

is commutative.

PROOF. The uniqueness follows from universality of the problem. The existence is proved as follows. Let us consider a tensor algebra over V

$$\mathcal{T}(V) := \bigoplus_{i=0}^{\infty} \mathcal{T}^i(V),$$

where $\mathcal{T}^0(V) := \mathbf{K}$ and $\mathcal{T}^i := V \otimes \cdots \otimes V$ (i times) for $i > 0$. Let $I(Q)$ be a two side ideal generated by elements of the form

$$t(v) = v \otimes v - Q(v) \cdot 1, \quad \text{where } v \in V$$

and 1 be the unity of the algebra $\mathcal{T}(V)$. Every element of $I(Q)$ can be represented in the form $\sum \lambda_i t(v) \mu_i$, where $\lambda_i, \mu_i \in \mathcal{T}(V)$, $v \in V$. Let $\mathcal{C}(V) := \mathcal{T}(V)/I(Q)$ and let $\rho : V \to \mathcal{C}(V)$ be a superposition of the monomorphism (embedding)

$$i : V \stackrel{\approx}{\rightarrow} \mathcal{T}^1(V) \subset \mathcal{T}(V)$$

and the projection

$$p : \mathcal{T}(V) \to \mathcal{C}(V) \ (:= \mathcal{T}(V)/I(Q)).$$

The pair $(\mathcal{C}(V), \rho)$ provides the solution of the problem, and therefore, the construction of Clifford algebra. Indeed, by virtue of universality of tensor algebras, the homomorphism $\varphi : V \to \mathcal{A}$ is such that the diagram

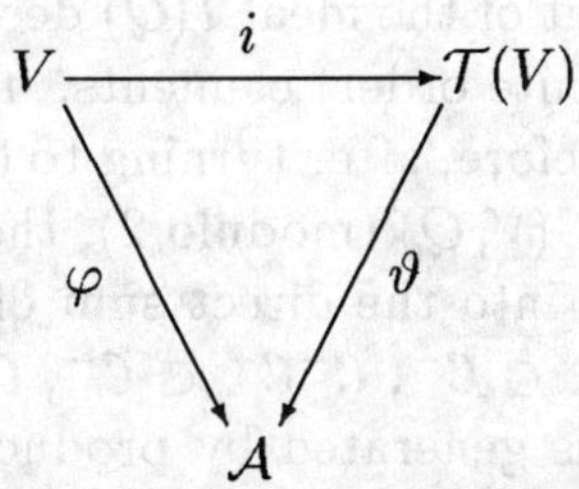

is commutative, with ϑ being a homomorphism of algebras.

Since $(\varphi(v))^2 = Q(v) \cdot 1$, the homomorphism ϑ vanishes on the ideal $I(Q)$ and therefore it defines the homomorphism we are looking for. □

COROLLARY 2. $\dim_{\mathbf{K}} \mathcal{C}(V) = 2^n$, *where* $n = \dim_{\mathbf{K}} V$.

PROOF. Taking $e_1, \ldots, e_n$ as a basis in the space V, we obtain the basis of $\mathcal{C}(V)$ as follows

$$e_0 := 1, \ \ e_1, \ldots, e_n, \ \ e_i e_j, \ i < j,$$

$$e_i e_j e_k, \; i < j < k,$$

$$\cdots\cdots\cdots$$

$$e_1 e_2 \cdots e_n, \quad \text{where } i, j, k \in \{0, \ldots, n\}.$$

REMARK 1. The Clifford algebra $\mathcal{C}(V)$ depends in an essential way on the quadratic form Q present in its definition, and therefore one often writes $\mathcal{C}(V, Q)$ or $\mathcal{C}(Q)$ instead of $\mathcal{C}(V)$. Usually, the mapping $\rho : V \to \mathcal{C}(V, Q)$ is an embedding under which one identifies elements of the space V with Clifford numbers, and therefore ρ is often omitted in the notation.

REMARK 2. If $e_1, \ldots, e_n$ is a basis of the space V, then in order to guarantee that the mapping $f : V \to \mathcal{A}$ into an algebra $\mathcal{A}$ defines a homomorphism of the algebra $\mathcal{C}(V, Q)$ into $\mathcal{A}$, it is necessary and sufficient that

$$f(e_i)f(e_j) + f(e_j)f(e_i) = 2Q(e_i, e_j) \cdot 1_{\mathcal{A}}.$$

Taking $f = \rho$, we see therefore that the products

$$e_{i_1} e_{i_2} \cdots e_{i_k}, \quad i_1 < i_2 \cdots < i_k$$

form a basis of the algebra $\mathcal{C}(V, Q)$, and indeed

$$\dim_{\mathbf{K}} \mathcal{C}(V, Q) = 2^{\dim_{\mathbf{K}} V}.$$

REMARK 3. Every element of the ideal $I(Q)$ decomposes into the sum of homogeneous (that is, of definite order) elements. The gradation of the tensor algebra $\mathcal{T}(V)$ defines therefore, after turning to the quotient $\mathcal{T}(V)/I(Q)$, the gradation of the algebra $\mathcal{C}(V, Q)$ (modulo 2), therefore the Clifford algebra $\mathcal{C} \equiv \mathcal{C}(V, Q)$ decomposes into the direct sum of two subspaces $\mathcal{C}^+$, $\mathcal{C}^-$, such that $\mathcal{C}^+\mathcal{C}^+ \subset \mathcal{C}^+$, $\mathcal{C}^+\mathcal{C}^- \subset \mathcal{C}^-$, $\mathcal{C}^-\mathcal{C}^+ \subset \mathcal{C}^-$, $\mathcal{C}^-\mathcal{C}^- \subset \mathcal{C}^+$. The vector space $\mathcal{C}^+$ (respectively, $\mathcal{C}^-$) is generated by products of even (respectively, odd) number of elements of the space V. The subspaces $\mathcal{C}^\pm$ play an essential role in the theory of *spinors*.

Algebra of quaternions as Clifford algebra. Let us consider the following important example.

EXAMPLE. Let $V = \mathbb{R}^2$, $Q(x_1, x_2) := -x_1^2 - x_2^2$. Then the Clifford algebra $\mathcal{C}(V, Q)$ has the basis $\{1, e_1, e_2, e_1e_2\}$ and is, therefore, $2^2 = 4$-dimensional and defined by relations

$$(*) \qquad e_1^2 = e_2^2 = -1, \quad e_1 e_2 = -e_2 e_1.$$

It is not hard to observe that the mapping $\mathcal{C}(V) \to \mathbb{H}$ given by

$$1 \to 1, \;\; e_1 \to i, \;\; e_2 \to k, \;\; e_1e_2 \to j$$

defines an isomorphism of $\mathcal{C}(V)$ into the algebra of quaternions $\mathbb{H}$.

Clifford algebras has become popular among physicists and mathematicians as a result of Dirac theory of (relativistic) electron. Let $V = M^{\mathbb{C}} := M + iM$ be a complexification of Minkowski space M ($\simeq \mathbb{R}^4$) with the metric $x_0^2 - (x_1^2 + x_2^2 + x_3^2)$ with respect to the orthonormal basis e_0, e_1, e_2, e_3, of the space M, being, at the same time a basis of $M^{\mathbb{C}}$.

PROPOSITION 3. *The Clifford algebra $\mathcal{C}(M^{\mathbb{C}})$ is isomorphic with the algebra of complex (4×4) matrices* $\mathrm{Mat}_4(\mathbb{C})$.

PROOF (exercise). Using the (spin) Pauli matrices

$$\sigma_\alpha, \quad \alpha = 0, 1, 2, 3$$

where

$$\sigma_0 = \begin{bmatrix} 1 & 0 \\ 0 & 1 \end{bmatrix}, \quad \sigma_1 = \begin{bmatrix} 0 & 1 \\ 1 & 0 \end{bmatrix}, \quad \sigma_2 = \begin{bmatrix} 0 & -i \\ i & 0 \end{bmatrix}, \quad \sigma_3 = \begin{bmatrix} 1 & 0 \\ 0 & -1 \end{bmatrix}$$

representing a basis of the field $\mathbb{H}$: $1 = \sigma_0$, $i = -i\sigma_1$, $k = -i\sigma_2$, $j = -i\sigma_3$, we construct the so-called *Dirac matrices*

$$\gamma_0 = \begin{bmatrix} \sigma_0 & -\sigma_0 \\ 0 & 1 \end{bmatrix}, \quad \gamma_k = \begin{bmatrix} 0 & \sigma_k \\ -\sigma_k & 0 \end{bmatrix}, \quad k = 1, 2, 3.$$

We check that the matrices γ_j satisfy the same relations in the algebra $\mathrm{Mat}_4(\mathbb{C})$ as the elements $\rho(e_j)$ do in the algebra $\mathcal{C}(M^{\mathbb{C}})$:

$$\gamma_0^2 = -\gamma_1^2 = \gamma_2^2 = \gamma_3^2 == -1, \quad \gamma_i\gamma_k + \gamma_k\gamma_i = 0, \;\; i \neq k.$$

Thus the $\mathbb{C}$-linear mapping $\sigma : M^{\mathbb{C}} \to \mathrm{Mat}_4(\mathbb{C})$ induces the homomorphism of algebras $\tau : \mathcal{C}(M^{\mathbb{C}} \to \mathrm{Mat}_4(\mathbb{C})$ such that $\tau\rho(e_j) = \gamma_j$.

We leave it to the reader as an exercise to check by direct computations that τ is surjective and since both algebra are sixteen dimensional, τ is an isomorphism. □

Historical remark. The discovery (construction) of algebras has its source in the quaternions theory of Hamilton. The dramatic history of discovery of quaternions is described in the beautiful book by van der Waerden

A History of Algebras (Springer, 1985), which main part consists of a wonderful essay on history of algebras. It was Hamilton who first defined complex numbers as pairs of real numbers. Understanding a role played by complex numbers in rotations of the plane $\mathbb{R}^2$ ($\simeq \mathbb{C}$), Hamilton searched for new numbers, which would play the same role in rotations of three-dimensional spaces. For triples (a, b, c) of real numbers he tried to define a multiplication which would preserve the norm. In analogy with complex numbers $a + ib$, Hamilton used to write his triplets as $a + ib + jc$. The failure in constructing an associative algebra with action in $\mathbb{R}^3$, of which Hamilton suffered for a long time, is explained (only) by Frobenius theorem. The sufferings of this famous October of 1843, are charmingly characterized in the letter which Hamilton wrote to his son Archibald not long before death in 1865: '...Every morning in the early of the above cited month, on my coming down to breakfast, your brother William Edward and yourself used to ask me: Well, Papa, can you multiply triples? Where to I was always to reply, with the sad shake of my head: No, I can only add and subtract them ...'. It was only on 16 October 1843 when quaternions were born: Hamilton writes further to his son (this matter was very scrupulately noted in his diary!), that when he was walking with his wife along the Royal Canal to the meeting of Academy – of whose he was a president – suddenly, the underground 'current of thoughts' surfaced and gave the answer to the question posed with such an effort for a long time: 'one should go to fourth dimension'. Hamilton writes: '...I pulled out on the spot a pocket-book, which still exists, and made an entry there and then. Nor could I resist the impulse – unphilosophical as it may have been – to cut with a knife on a stone of Brougham Bridge, as we passed it, the fundamental formula with the symbols i, j, k: $i^2 = j^2 = k^2 = ijk = -1$, which contains the solution of the Problem...'. The quaternions were born!

Pauli matrices in the theory of *spinning electron* have arisen from sufferings of Pauli trying to understand the anomalous Zeeman effect. An anecdote tells that a young men (Pauli) used to walk every day in the botanic garden in Copenhagen and one day a concerned lady asked him 'why are you walking so sad?'. He answered 'how one can be happy while investigating the anomalous Zeeman effect!' The works of Pauli contributed to discovery of Dirac of his famous equation of electron in 1928.

(Exterior) Grassmann algebra $\Lambda(V)$ over V. A *Grassmann algebra* $\Lambda(V)$ is a Clifford algebra $\mathcal{C}(V, Q)$, where Q is zero, that is, the identically vanishing quadratic form. Therefore, this algebra, usually denoted by $\Lambda(V)$

is generated by $e_1, e_2, \ldots, e_n$ satisfying the relations

$$e_i^2 = 0, \quad e_ie_j + e_je_i = 0 \quad \text{for } i \neq j\,.$$

Historical remark. When in January of 1853, Hamilton (1805– 1865) learned that Hermann Grassmann (1809–1877) had constructed his algebras in the monograph *Lineare Ausdehnungslehre*, he was scared that the unknown German from Stettin forestalled him in the discovery, which he considered to be the most important achievement of his life. Hamilton, as he himself confessed in a letter to his friend, devoured 100 pages of the work of Grassmann and realized, to his relief, that *Lineare Ausdehnungslehre* does not contain his quaternions. As these two last 'examples' show, the concepts of Grassmann and Hamilton are closely related to each other, they are elements of the corresponding Clifford algebras.

William Kongdom Clifford (1845–1879) was a deep researcher and a wide intellect: during his short life he enriched mathematics with many beautiful ideas: the famous *Clifford algebras*, which, due to the works of E. Cartan and H. Weyl, play a fundamental role in the theory of spinors and, in turn, in differential geometry and global analysis, and Lie group theory (spin structures, Dirac operators). Clifford was one of the first who understood and realized the importance of Grassmann's works. His work on Clifford numbers was published in 1878 under the symptomatic title *Applications of Grassmann's Extensive Algebra* (Am. J. Math. 1878, 1 (350–358)). The works of Clifford are very important for the theory of compact Riemann surfaces. Clifford was not only a popularizer of ideas of Riemann in England, the translator of collected works of Riemann into English, but also a brilliant continuator of his ideas; along with Riemann we can regard Clifford as a precursor of the idea of Einstein's theory of gravity that the (gravitational) mass influences curvature of the space-time. The interesting concept of his are also the so-called *Clifford parallels.* All works of Clifford are contained in one volume (similarly to works of Riemann).

Hermann Grassmann was a rare kind of a genius: not only he created the geometry of n-dimensional vector spaces, gave the most modern definition of the determinant, created exterior algebras, without whose the Cartan's theory of differential forms would have never been created, but also he obtained important results concerning Pfaff problem. This quite self-taught person from Stettin was one of the greatest European sanskritologists of his times:

in 1875 he wrote the first (German) dictionary to Rigveda and in the years 1877– 78 he performed the first translation of this holy book of Hinduism to German. The mathematical works of Grassmann had been appreciated much later than his works on the field of philology!

Algebra of octonions $\mathbb{O}$. Only several month (December 1843) after discovery of quaternions by Hamilton, John Graves found an eight dimensional algebra without divisors of zero, which, obviously, was not commutative, but was not associative, either. Its elements are called *octonions.* Graves (a friend of Hamilton's) published this result only in 1848; the same results has been independently obtained in 1845 by Arthur Cayley. We tell the unusual story of life of the latter in the first part of this book.

The discovery of octonions is again an example of independent realizations of ideas. The question arises as to whether there exist any other algebras without divisors of zero of dimensions different from 1, 2, 4, 8? The negative answer is provided by the famous theorem of 1958.

THEOREM 4 (BOTT–MILNOR). *The only finite dimensional algebras over $\mathbb{R}$ without divisors of zero are real and complex numbers, quaternions, and octonions:* $\mathbb{R}$, $\mathbb{C}$, $\mathbb{H}$, $\mathbb{O}$.

This theorem has been obtained by means of deep theorems of modern theory of vector bundles (theory of characteristic classes of Stiefel–Whitney). It turns out that the problem of existence of algebras without divisors of zero is closely related to the problem of *parallelization* of spheres S^k and projective spaces $P^n(\mathbb{R})$, that is, the problem of existence of smooth vector fields on these manifolds. According to the earlier theorem of Stiefel, it is known that if there exists a bilinear mapping $\mathbb{R}^n \times \mathbb{R}^n \to \mathbb{R}^n$ without divisors of zero, then the real projective space P^{n-1} is parallelizable. This theorem is much stronger than the one proved 60 years before by Hurwitz.

The great mathematician Adolf Hurwitz (1859–1919) showed in 1898 in the paper of major importance *Über die Composition der quadratischen Formen von beliebig vielen Variablen* (Nadt. Ges. der Wiss. Götingen, 309–316) that the identities $(a_1^2 + \cdots + a_n^2)(b_1^2 + \cdots + b_n^2) = c_1^2 + \cdots + c_n^2$ can hold only for $n = 1, 2, 4, 8$, and therefore the following theorem holds

THEOREM 5 (HURWITZ) *The only normed algebras over $\mathbb{R}$ are $\mathbb{R}$, $\mathbb{C}$, $\mathbb{H}$, and $\mathbb{O}$ for those dimensions of $\mathbb{R}^n$, for whose there exist algebras with division.*

The result of Hurwitz was obtained by topologists from Zürich, Stiefel and Eckmann in the course of investigations of some Lie algebras. As the prematurely deceased Italian mathematician Aldo Andreotti told me, from the work of Hurwitz one can deduce all fundamental linear equations of mathematical physics.

Historical remark. As we said already, the impulse for discovery of quaternions was to describe rotations of the space $\mathbb{R}^3$ (that is, elements of the orthogonal group $SO(3)$) in terms of a 'multiplication', similarly to the situation in $\mathbb{R}^2 \simeq \mathbb{C}$, where rotation can be obtained as the multiplication $z \to \omega z$ by $\omega \in \mathbb{C}$, $|\omega| = 1$. Indeed, after discovery of the algebra of quaternions $\mathbb{H}$, Hamilton – and independently Cayley – succeeded in describing any rotation in $\mathbb{R}^3$ in the form

$$u \to quq^{-1} \quad \text{where } u = x_1 j + x_2 k + x_3 j \in \mathbb{H},$$

$(x_1, x_2, x_3) \in \mathbb{R}^3$ and q is some quaternion, $q \in \mathbb{H}$. Thanks to Clifford numbers, that is, elements of the algebra $C(\mathbb{R}^n)$, this can be achieved for any n, that is, one can represent a rotation in $\mathbb{R}^n$ (and thus an element of the group $SO(\mathbb{R}^n) \equiv SO(n)$) in the form $u \to sus^{-1}$, where s is some invertible Clifford number. Several years after Clifford, in 1884, Rudolf Lipschitz (1832–1903) performed independently an analogous construction in the very important paper *Untersuchungen über die Summen von Quadraten* (Bonn 1884). The further investigations of these hypercomplex systems $\mathcal{A}$ we owe Theodor Molien (1861– 1941), who, in 1893, discovered two sided ideals in algebras $\mathcal{A}$.

THEOREM 6 (MOLIEN and slightly later (1894) CARTAN). *The simple, that is, not possessing any two sided ideals, algebras $\mathcal{A}$ are isomorphic with algebras of $n \times n$ real matrices if $\mathcal{A}$ has dimension n^2; $n \times n$ complex matrices if $\mathcal{A}$ has dimension $2n^2$; $n \times n$ quaternionic matrices if $\mathcal{A}$ has dimension $4n^2$.*

The general theory of algebras (over arbitrary fields) we owe Maclagan Wedderburn (1882–1943), who in 1908 generalized Molien– Cartan theorem

THEOREM 7 (MOLIEN–CARTAN). *A simple associative algebra with unity over arbitrary field $\mathbf{K}$ is isomorphic with an algebra with elements in the field (non commutative, in general) $\mathbf{D}$, such that $Z(\mathbf{D}) \supset \mathbf{K}$.*

This theorem, in turn, was extended in 1927 to some rings by Emil Artin (1898–1962), cf. van der Waerden, *Algebra*.

Another beautiful theorem concerning finite algebras was obtained by Wedderburn in 1905.

THEOREM 8 (WEDDERBURN). *Any finite ring with division is commutative (therefore, is a commutative field, that is, a Galois field).*

Lie algebras. A *Lie algebra* is characterized by the relation

$$ab + ba = 0 \quad \text{(skew symmetry)},$$

$$a \cdot bc + b \cdot ca + c \cdot ab = 0 \quad \text{(Jacobi identity)}.$$

Usually, the Lie product ab is denoted by the parenthesis (Lie bracket) $[a, b]$. Vector fields on manifolds are the most important examples of Lie algebras. The vector space T_eG of a Lie group G is, in a natural fashion, a Lie algebra – this is the reason for the name. We will keep returning to Lie algebras and groups many times. In this introductory chapter, we have just provided very brief information on algebras in order to draw reader attention on their importance in mathematics and physics.

If we look back at the part of history of mathematics related to the gigantic development of the theory of algebras, which started with a modest on the first sight achievement, the discovery, or construction, of the algebra of quaternions on 16 October 1843, perhaps, we find it more easy to agree with Hamilton, who regarded quaternions as his most beloved child?

Chapter 5

Fields and their Extensions

The general theory of extension of fields became possible only after creation of set theory by Dedekind, George Cantor (1845–1918), and Zermelo (1871–1953). It first appeared in the large, classical work of Ernst Steinitz (1871–1928) of 1910 *Algebraische Theorie der Körper*, Crelle Journal 137, 167–309. This was the time when Zemelo axiom of choice was still treated with suspicions. And it was the paper of Steinitz which became the most convincing argument in favour of this axiom or formulated later (equivalent) *Kuratowski–Zorn lemma.* In fact, Steinitz tries to excuse himself for using such a suspicious instrument.

The following important theorem holds.

THEOREM 1 (STEINITZ, 1910). 1. *Let* $\{\mathbf{K}_j\}$, $\psi \in J$ *be a (not empty) family of subfields of the field* $\mathbf{K}$. *Then* $\bigcap_{j\in J} \mathbf{K}_j$ *is also a subfield of* $\mathbf{K}$.

2. *For any field* $\mathbf{K}$ *there exist its smallest subfield, called the simple subfield of the field* $\mathbf{K}$ *(this subfield is the intersection of all subfields of* $\mathbf{K}$*).*

3. *If* B *is a subset of the field* $\mathbf{K}$, *then there exists the smallest subfield containing the set* B *(this is, obviously, the intersection of all subfields containing* B*). This subfield is called the subfield generated by* B.

4. *If* $\mathbf{L}$ *is an overfield of the field* $\mathbf{K}$: $\mathbf{L} \supset \mathbf{K}$ *and* $B \subset \mathbf{L}$, *then there exists the smallest field* $\mathbf{K}(B) \subset \mathbf{L}$ *generated by the set* $\mathbf{K} \cup B$. *In particular, if* $B = \{b_1, \ldots, b_n\}$, *then we write* $\mathbf{K}(b_1, \ldots, b_n)$ *instead of* $\mathbf{K}(B)$.

REMARK 1. In practise, one takes the set B to be disjoint from $\mathbf{K}$ and then we say that $\mathbf{K}(B)$ is an *extension* of $\mathbf{K}$ by the set B.

REMARK 2. $\mathbf{K}(b_1, \ldots, b_n)$ can be obtained by subsequent one-element extensions

$$\mathbf{K}(b_1, \ldots, b_n) = (\ldots((\mathbf{K}(b_1))(b_2))\ldots)(b_n).$$

REMARK 3. $\mathbf{K}(B)$ is formed by all rational relations of elements of the sets $\mathbf{K}$ and B: they are quotients of polynomials with coefficients in $\mathbf{K}$.

In what follows, we will only consider *commutative fields and rings (with a unit).*

Let $\mathbf{K}$ and $\mathbf{L}$ be fields; if $\mathbf{K} \subset \mathbf{L}$, we say that $\mathbf{L}$ is an *extension of the field* $\mathbf{K}$ and we often write $\mathbf{L}/\mathbf{K}$. In what follows, we will use both these notations. The dimension of the vector space $\mathbf{L}$ over $\mathbf{K}$, that is $\dim_{\mathbf{K}}(\mathbf{L}/\mathbf{K})$ will be denoted by $[\mathbf{L} : \mathbf{K}]$ and called the *degree of extension* $\mathbf{L}/\mathbf{K}$.

An element $a \in \mathbf{L}$ is *algebraic over* $\mathbf{K}$ if a is a root of some polynomial f with coefficients in $\mathbf{K}$, that is $f \in \mathbf{K}[x]$, $f(a) = 0$,

$$f(x) = \sum_{j=0}^{n} c_j x^j, \quad c_j \in \mathbf{K}.$$

If $c_n = 1$, then f is called a *normed* or *unitary* polynomial.

If $a \in \mathbf{L}$ is not an algebraic element over $\mathbf{K}$, then we say that it is transcendental over $\mathbf{K}$.

DEFINITION. The extension $\mathbf{L} \supset \mathbf{K}$ is *algebraic* if all elements of the space are algebraic over $\mathbf{K}$.

Below, the round brackets will always denote a *field* obtained by extension, the square brackets, for example $\mathcal{K}[x]$, denote a *ring* obtained by adding the variable x, that is, by forming *all polynomials* of the variable x with coefficients in the field $\mathbf{K}$.

We have the following simple but important theorem.

THEOREM 2. *If* $\mathbf{K}$ *is a subfield of the field* $\mathbf{L}$ *and* $a \in \mathbf{L}$, *then*

1. *(a is algebraic over $\mathbf{K}$)$\Longleftrightarrow$($[\mathbf{K}(a) : \mathbf{K}] < \infty$).*

2. *(a is algebraic over $\mathbf{K}$)$\Longleftrightarrow$(there exists an irreducible polynomial $f \in \mathcal{K}[x]$ such that $f(a) = 0$; the polynomial f is defined uniquely up to a constant factor).*

DEFINITION. The polynomial f from Theorem 2, 2. is called the *minimal polynomial* for a.

Characteristic of field K (char **K**). A field identical with its simple subfield is called a *simple field.* We have the following corollary.

COROLLARY 3. *Every simple field is isomorphic with*
(a) *the field* $\mathbb{Q}$ *of rational numbers or*
(b) *with a p-element field* $\mathbf{F}_p$, *where p is a prime number.*

PROOF. The smallest ring $\mathcal{R} \subset \mathbf{K}$ is isomorphic with the ring $\mathbb{Z}$ or $\mathbb{Z}/n\mathbb{Z}$ where n is a prime number, because (every) field **K** does not possess any divisors of zero. □

REMARK. Let us recall that *(the ring* $n\mathbb{Z}$ *is an ideal)*$\Longleftrightarrow$*(n is a prime number).*

DEFINITION. In the case a) of Corollary 3, char $\mathbf{K} = 0$; in the case b), char $\mathbf{K} = p$.

It can be expected that there exist fields of arbitrary prime characteristic. The reader will show without difficulties the following theorem.

THEOREM 4. 1. (char $\mathbf{K} = p$)$\Longleftrightarrow$(*p is the smallest natural number such that* $p \cdot 1 = 0$).
2.(char $\mathbf{K} = p \neq 0$)$\Longrightarrow$($(a+b)^p = a^p + b^p$; $(a-b)^p = a^p - b^p$; $(ab)^p = a^p b^p$ *for* $a, b \in \mathbf{K}$).

According to the following proposition, the most important for analysis are fields of characteristic 0.

PROPOSITION 5. *The fields* $\mathbb{Q}$, $\mathbb{R}$, $\mathbb{C}$, *and the fields* $\mathbb{Q}_p$ *(which is the complement of* $\mathbb{Q}$ *in the p-adic norm* $|\ |_p$*) all have characteristic 0.*

PROOF. $\mathbb{R}$, $\mathbb{C}$, and $\mathbb{Q}_p$ contain the simple field $\mathbb{Q}$. □

PROPOSITION 6. a) *Every finite field* **K** *is commutative and isomorphic to some field* $\mathbb{Z}/n\mathbb{Z}$.
b) *If* char $\mathbf{K} = p$ *and* **P** *is a simple field contained in* **K**, *then the field* **K** *contains* p^r *elements, where r is the dimension of the vector space* **K** *over* **P**, $r = \dim_{\mathbf{P}} \mathbf{K}$.

PROOF. There exists a bijection $f : \mathbf{K} \to \mathbf{P}^r$ given by the formula

$$(*) \qquad \mathbf{K} \ni x = \sum_{i=1}^{r} a_i k_i \to (a_1, \ldots, a_r) \in \mathbf{P}^r,$$

where $k_1, \ldots, k_r$ is a basis of the vector space $\mathbf{K}$ over $\mathbf{P}$. Denoting by $|B|$ the number of elements of the finite set B, from $(*)$ we have

$$|\mathbf{K}| = |\mathbf{P}^r| = p^r.$$

□

REMARK. For many years now it is customary to denote the extension $\mathbf{L} \supset \mathbf{K}$ and thus the overfield $\mathbf{L}$ of the field $\mathbf{K}$ by $\mathbf{L}/\mathbf{K}$. The reader will immediately see from the context if we have in mind an *extension* of the field $\mathbf{L}/\mathbf{K}$ or a quotient group (quotient space $\mathbf{L}/\mathbf{K}$).

EXAMPLE. $\mathbb{R}(\sqrt{-1}) = \mathbb{C}$; $[\mathbb{C} : \mathbb{R}] = 2$ and thus between $\mathbb{C}$ and $\mathbb{R}$ there is no interpolating field.

From now on we will only consider *commutative fields* (not stressing their commutativity). The reader will easily find out which facts hold also for noncommuatative fields.

The important relation between rational functions of one variable and transcendental elements is explained by the following theorem.

THEOREM 7. *If* $\mathbf{L} \supset \mathbf{K}$ *and* $a \in \mathbf{L}$ *is transcendental over (with respect to)* $\mathbf{K}$*, then:*

1. $[\mathbf{K}(a) : \mathbf{K}] = \infty$,
2. *the field* $\mathbf{K}(a)$ *is isomorphic with the field* $\mathbf{K}(x)$ *of rational functions of variable* x *with coefficients in the field* $\mathbf{K}$.

THEOREM 8. *Every finite extension is algebraic, that is,* ($[\mathbf{L} : \mathbf{K}] < \infty) \Longrightarrow (\mathbf{L} \supset \mathbf{K}$ *is algebraic over* $\mathbf{K}$).

PROOF. Let $[\mathbf{L} : \mathbf{K}] < \infty$, $a \in \mathbf{L}$. Then $\mathbf{K}(a) \subset \mathbf{L}$, and thus $[\mathbf{K}(a) : \mathbf{K}] \leqslant [\mathbf{L} : \mathbf{K}] < \infty$, and a is an algebraic element over $\mathbf{K}$. □

THEOREM 9. *Let* $\mathbf{L} \supset \mathbf{K}$. *Then the set* $\mathbf{Z} \subset \mathbf{L}$ *of all elements algebraic over* $\mathbf{K}$ *forms a field. This is the maximal algebraic extension of the field* $\mathbf{K}$

contained in **L**.

THEOREM 10. *(*$\mathbf{L} \supset \mathbf{K}$ *is an algebraic extension of the field* **K**, $\mathbf{M} \supset \mathbf{L}$ *is an algebraic extension of the field* **L***)*$\Longrightarrow$*(*$\mathbf{M} \supset \mathbf{K}$ *is an algebraic extension of the field* **K***).*

Algebraically closed field. An *algebraically closed field* **K** is such a field that for every polynomial $f \in \mathcal{K}[x]$ all its roots are contained in **K**.

EXAMPLE. *The field of complex numbers* $\mathbb{C}$ *is algebraically closed.* This is a formulation of fundamental theorem of algebra proved by Gauss (Gauss presented eight different proofs of this theorem, all of them are analytical, because, as a matter of fact, this is a theorem from analysis and topology, not from algebra)! Later we will present several (three, to be precise) proofs of this theorem. We have many equivalent conditions of algebraic closedness of a field.

THEOREM 11. *The following conditions are equivalent:*

1. *Every polynomial* $f \in \mathcal{K}[x]$ *such that* $\deg f > 0$ *has a root in* **K**.
2. **K** *is algebraically closed.*
3. *Every irreducible polynomial in* $\mathcal{K}[x]$ *is linear, that is,* $\deg f = 1$.
4. *(***L** *is an algebraic extension of* **K***)* $\Longrightarrow$*(*$\mathbf{L} = \mathbf{K}$*).*
5. *(*$[\mathbf{L} : \mathbf{K}] < \infty$*)*$\Longrightarrow$*(*$\mathbf{L} = \mathbf{K}$*).*

LEMMA 12. *(*$f \in \mathcal{K}[x]$, $\deg f \geqslant 0$*)*$\Longrightarrow$*(there exists* $\mathbf{L} \supset \mathbf{K}$ *such that* $f(\alpha) = 0$ *for some* $\alpha \in \mathbf{L}$*).*

By induction in degree of polynomial f, $\deg f$, we obtain the following corollary.

COROLLARY 13. *(*$f \in \mathbf{K}[x]$, $\deg f \geqslant 0$*)*$\Longrightarrow$*(there exists an extension* $\mathbf{L} \supset \mathbf{K}$ *such that* f *is a product of polynomials of degree* 1 *belonging to* $\mathcal{L}[x]$*).*

This leads to the natural definition.

DEFINITION. If $\mathbf{L} \supset \mathbf{K}$ and $f \in \mathbf{K}[x]$ is a product of linear polynomials belonging to $\mathbf{L}[x]$

$$f(x) = a_1(x - a_2) \cdots (x - a_n),$$

then the field $\mathbf{K}(a_1, \ldots, a_n) \subset \mathbf{L}$ is called the *field of decomposition* (German *Zerfallkörper*) of the polynomial f.

The question arises, given polynomial $f \in \mathbf{K}[x]$, how many fields of its decomposition exist? We can expect the answer in the form of the following theorem.

THEOREM 14. *Every two fields of decomposition $\mathbf{Z}_1$, $\mathbf{Z}_2$ of the polynomial $f \in \mathcal{K}[x]$ are $\mathbf{K}$-isomorphic, that is, there exists an isomorphism $\varphi : \mathbf{Z}_1 \to \mathbf{Z}_2$ such that $\varphi|_{\mathbf{K}} = \mathrm{id}_{\mathbf{K}}$ (that is, φ does not move the field $\mathbf{K}$).*

PROOF follows from the following simple observations

(A) Any isomorphism $\psi : \mathbf{K}_1 \to \mathbf{K}_2$ of fields can be extended to the isomorphism $\bar{\psi}$ of the ring of polynomials with coefficients in these fields $\bar{\psi} : \mathbf{K}_1[x] \to \mathbf{K}_2[x]$ by taking $\bar{\psi}(x) = x$. This condition defines $\bar{\psi}$ uniquely.

(B) If $f \in \mathbf{K}[x]$ is an irreducible polynomial and if in some extensions $\mathbf{L}_1 \supset \mathbf{K}$, $\mathbf{L}_2 \supset \mathbf{K}$, the polynomial has, respectively, the roots $a_i \in \mathbf{L}_i$, $i = 1, 2$, then there exists a $\mathbf{K}$- isomorphism $\varphi' : \mathbf{K}(a_1) \to \mathbf{K}(a_2)$ such that $\varphi'(a_1) = a_2$.

By induction in $\deg f$, we obtain the following important theorem.

THEOREM 15 (ON EXTENSION OF ISOMORPHISM). *If $\varphi : \mathbf{K}_1 \to \mathbf{K}_2$ is an isomorphism of fields, $\tilde{\varphi} : \mathbf{K}_1[x] \to \mathbf{K}_2[x]$ is its extension to an isomorphism of the ring of polynomials, and $\mathbf{Z}_i$ are fields of decomposition of the polynomials $f_i \in \mathbf{K}_i[x]$, $i = 1, 2$, then there exists an isomorphism $\psi : \mathbf{Z}_1 \to \mathbf{Z}_2$ which is an extension of φ.*

Theorem on identity (up to isomorphism) can be obtained from the theorem above by taking $\mathbf{K}_1 = \mathbf{K}_2 = \mathbf{K}$, $f_1 = f$, and $\varphi = \mathrm{id}_{\mathbf{K}}$.

COROLLARY 16. 1. *Any two algebraic closures of a field $\mathbf{K}$ are $\mathbf{K}$-isomorphic (we will denote them by* $\mathrm{alg}(\mathbf{K})$).

2. *Every $\mathbf{K}$-embedding $\varphi : \mathrm{alg}(\mathbf{K}) \to \mathrm{alg}(\mathbf{K})$ is an $\mathbf{K}$-automorphism.*

3. *($\mathbf{L} \supset \mathbf{K}$ is an algebraic extension)$\Longrightarrow$(every $\mathbf{K}$-embedding $\varphi : \mathbf{L} \to \mathrm{alg}(\mathbf{K})$ can be extended to a $\mathbf{K}$-automorphism $\psi : \mathrm{alg}(\mathbf{K}) \to \mathrm{alg}(\mathbf{K})$.*

THEOREM 17. *If $\mathbf{L} \supset \mathbf{K}$, where $\mathbf{L}$ is an algebraically closed field, then the following conditions are equivalent:*

1. *$\mathbf{L}$ is an algebraic extension of $\mathbf{K}$.*

2. *If an algebraically closed field* $\mathbf{M}$ *satisfies* $\mathbf{L} \supset \mathbf{M} \supset \mathbf{K}$, *then* $\mathbf{M} = \mathbf{L}$.

DEFINITION. The field $\mathbf{L}$ satisfying conditions 1. or 2. is thus the smallest algebraically closed field containing $\mathbf{K}$, therefore $\mathbf{L}$ is called the *algebraic closure of the field* $\mathbf{K}$ and denote it by $\bar{\mathbf{K}}$ or $\mathrm{alg}(\mathbf{K})$.

Separable algebraic extension. Perfect fields. Creator of the general theory of fields Ernst Steinitz (1871–1928) introduced an important notion (which, of course, was used earlier in particular cases) and called it, following the custom of these days, extension of 'first kind'; people used to talk about integral equations of first kind, abelian differentials of first kind, and so on. Only in 1930, Van der Waerden, in his *Modern Algebra* proposed the suggestive term *separable*. This term was immediately accepted.

DEFINITION. Let an element $a \in \mathbf{L}$ be an algebraic element over $\mathbf{K}$. If the minimal polynomial $f \in \mathbf{K}[x]$ does not possess multiple roots for a, then a is called the *separable* element over $\mathbf{K}$. The algebraic extension $\mathbf{L} \supset \mathbf{K}$ is *separable*, if all elements of $\mathbf{L}$ are separable over $\mathbf{K}$. Also, an irreducible polynomial, whose all zeros are separable is called a *separable* polynomial (its zeros are separated from each other).

DEFINITION. A field $\mathbf{K}$ is called *perfect* if any irreducible polynomial $f \in \mathbf{K}[x]$ is separable.

EXAMPLE 1. If characteristic of the field $\mathbf{K}$ equals 0 ($\mathrm{char}\, \mathbf{K} = 0$), then its every algebraic extension is separable and every irreducible polynomial $f \in \mathbf{K}[x]$ is separable.

EXAMPLE 2. Every field of characteristic 0 is perfect.

EXAMPLE 3. Every algebraic extension of a perfect field is separable.

EXAMPLE 4. Algebraically closed fields are perfect.

EXAMPLE 5. Finite fields, the so-called *Galois fields*, are perfect.

THEOREM 18 (ABEL THEOREM ON PRIMITIVE ELEMENT). *(*$\mathbf{L} \supset \mathbf{K}$ *is separable and finite)* $\Longrightarrow$ *(there exists* $\vartheta \in \mathbf{L}$ *such that* $\mathbf{K}(\vartheta) = \mathbf{L}$*);* ϑ *is called a primitive element.*

As it is easy to check, when $\mathbf{L} \supset \mathbf{K}$, $\mathbf{K}$- automorphisms form a group which is called the *group of automorphisms of extension* $\mathbf{L} \supset \mathbf{K}$ and is denoted by $\mathrm{Aut}(\mathbf{L}/\mathbf{K})$.

Evariste Galois was first to investigate the finite groups $\mathrm{Aut}(\mathbf{L}/\mathbf{K})$. This is the reason why sometimes $\mathrm{Aut}(\mathbf{L}/\mathbf{K})$ is denoted by $\mathrm{Gal}(\mathbf{L}/\mathbf{K})$; we reserve this notation to the case when $\mathbf{L}/\mathbf{K}$ is a Galois extension.

Theorem on primitive element was known to Niels Henrik Abel (1802–1829), but the explicit proof of this theorem was presented by Galois. This theorem simplifies investigations of finite extensions a lot; this is due to the simple form of the basis of $\mathbf{L}$ over $\mathbf{K}$

$$\sum_{k=1}^{[\mathbf{L}:\mathbf{K}]} a_k \vartheta^k,$$

by virtue of which one can easily consider structures and isomorphisms of these extensions; almost immediately we have the following useful theorem.

THEOREM 19. *Let the extension* $\mathbf{L} \supset \mathbf{K}$ *be finite and separable. Then*

$$|\,\mathrm{Aut}(\mathbf{L}/\mathbf{K})| = [\mathbf{L} : \mathbf{K}],$$

that is, the order of the group of automorphisms of the extension $\mathbf{L} \supset \mathbf{K}$ *is equal to the degree of this extension.*

Finally, we arrived at Galois extensions.

Galois extensions. An extension $\mathbf{L} \supset \mathbf{K}$ is called a *Galois extension* if $|\,\mathrm{Aut}(\mathbf{L}/\mathbf{K})| = [\mathbf{L} : \mathbf{K}]$; $\mathrm{Aut}(\mathbf{L}/\mathbf{K})$ is a *Galois group* of this extension and then we usually write $\mathrm{Gal}(\mathbf{L}/\mathbf{K})$ instead of $\mathrm{Aut}(\mathbf{L}/\mathbf{K})$.

In many branches of mathematics an important role is played by the notion of fixed point of a family G of mapping of space X, that is, such point $x_0 \in X$ that $g \cdot x_0 = x_0$ for all $g \in G$.

The set of fixed points of mapping G is often denoted by X^G. In particular, if $G \subset \mathrm{Aut}(\mathbf{L})$ and $\mathbf{L}$ is a field, then $\mathbf{L}^G$ is a field. Indeed, $0, 1 \in \mathbf{L}^G$; if $a, b \in \mathbf{L}^G$, then $a + b$, $-a$, ab, a^{-1} (if $a \neq 0$) also belong to $\mathbf{L}^G$.

Norms and traces. Let $G = \{g_1, \ldots, g_n\}$ be a finite group of automorphisms of the field $\mathbf{L}$. Let us define two mappings $\mathbf{L} \to \mathbf{L}^G$ as follows

(1) $\qquad N_G \equiv N : \mathbf{L} \to \mathbf{L}^G, \quad N(a) := g_1(a) \cdots g_n(a) \quad$ for $a \in \mathbf{L}$,

(2) $$\operatorname{tr}_G \equiv \operatorname{tr} : \mathbf{L} \to \mathbf{L}^G, \quad \operatorname{tr}(a) := g_1(a) + \cdots + g_n(a).$$

We check that these mappings indeed map into $\mathbf{L}^G$. For $g_j \in G$, we have

$$g_j(N(a)) = (g_j g_1)(a) \cdots (g_j g_n)(a) = N(a) \quad \text{because } \{g_j g_1, \ldots, g_j g_n\} = G,$$

$$g_j(\operatorname{tr}(a)) = (g_j g_1)(a) + \cdots + (g_j g_n)(a) = \operatorname{tr}(a).$$

DEFINITION. The mapping N_G is called *norm-like*, or shorter a *norm*, and the mapping tr_G is called *trace-like* or a *trace* for short.

From the definitions above, we have immediately

PROPOSITION 20. *$N_G(ab) = N_G(a)N_G(b)$, $\operatorname{tr}_G(\alpha a + \beta b) = \alpha \operatorname{tr}_G(a) + \beta \operatorname{tr}_G(b)$ for $a, b \in \mathbf{L}$, $\alpha, \beta \in \mathbf{L}^G$.*

REMARK. In the Galois theory, we have the following situation: $G = \operatorname{Aut}(\mathbf{L}/\mathbf{L}^G)$; denoting $\mathbf{K} = \mathbf{L}^G$, we see that G is a finite group of all $\mathbf{K}$-automorphisms of the field $\mathbf{L}$ and $|G| = [\mathbf{L} : \mathbf{K}]$. Then, one writes $\operatorname{tr}_{\mathbf{L}/\mathbf{K}}$ instead of tr_G and, similarly, $N_{\mathbf{L}/\mathbf{K}}$ instead of N_G; thus $\operatorname{tr}_{\mathbf{L}/\mathbf{K}} : \mathbf{L} \to \mathbf{K}$ $N_{\mathbf{L}/\mathbf{K}} : \mathbf{L} \to \mathbf{K}$, where the mappings are defined by (1) and (2).

Let us return to the general definition of trace and norm of an element $a \in \mathbf{L}$.

If we denote $a_j := g_j(a)$, we see that the trace $\operatorname{tr}(a)$ is the first symmetric fundamental polynomial; the norm $N(a)$ is the nth symmetric fundamental polynomial:

$$\operatorname{tr}(a) = s_1(a_1, \ldots, a_n) := a_1 + \cdots + a_n,$$

$$N(a) = s_n(a_1, \ldots, a_n) := a_1 \cdots a_n,$$

which polynomials are related to coefficients of the polynomial

$$\prod_{j=1}^{n}(z - a_j) = z^n + (-1)^1 s_1(a_1, \ldots, a_n) z^{n-1} + \cdots$$

$$\cdots + (-1)^n s_n(a_1, \ldots, a_n) = z^n + c_1 z^{n-1} + \cdots + c_n.$$

REMARK. A notion of the norm $N(a)$ of an element $a \in \mathbf{L} \supset \mathbf{K}$ will be useful in defining extensions of valuation of a norm on algebraic extensions of a complete field, the fundamental procedure in the theory of algebraic functions of one variable.

For some time now we have been moving in the circle of ideas of E. Galois, let us turn to sketching his theory now.

Normal extensions.

DEFINITION. An extension $\mathbf{L} \supset \mathbf{K}$ is called *normal* (sometimes, one shortly says that $\mathbf{L}$ is normal) if

1. is algebraic over $\mathbf{K}$;

2. every irreducible polynomial $W \in \mathbf{K}[t]$ which has a zero $a \in \mathbf{L}$ decomposes in $\mathcal{L}[t]$ into linear terms (that is, of order one).

The following useful criteria of normality hold. As we know, usually one extends fields by adding zeros of polynomials. And thus the following theorem holds.

THEOREM 21. *A field resulting from* $\mathbf{K}$ *by adding zeros of one, finitely many, or even infinitely many polynomials is a normal extension of the field* $\mathbf{K}$.

But the inverse theorem holds as well.

THEOREM 22. *Every normal extension* $\mathbf{L} \supset \mathbf{K}$ *is obtained by addition zeros of some set of polynomials; if* $\mathbf{L}$ *is a finite extension, it is sufficient to add zeros of (only) one polynomial.*

We leave the (simple) proof to the reader.

The term *normal* can be justified as follows: as we know, a subgroup $H \supset G$ is called *normal* or a *normal divisor* of G if for every $g \in G$

$$(*) \qquad gHg^{-1} \in H.$$

As we also know, the homogeneous space G/H is a group if and only if the group H is a normal divisor of the group G.

The following interesting theorem from Galois theory holds.

THEOREM 23. *(The intermediate field* $\mathbf{L} \supset \mathbf{Z} \supset \mathbf{K}$ *is a normal extension of* $\mathbf{K}$*)*$\Longleftrightarrow$*(the corresponding subgroup* H *is normal in* $\mathrm{Aut}(\mathbf{L}/\mathbf{K})$).

PROOF. $\Longrightarrow \mathbf{Z} \supset \mathbf{K}$ is normal, thus $g\mathbf{Z} = \mathbf{Z}$ for all $g \in \mathrm{Aut}(\mathbf{L}/\mathbf{K})$. From 4. of Theorem 11, we see that in this case $\mathrm{Aut}(\mathbf{L}/\mathbf{Z})$ is a normal divisor of the group $\mathrm{Aut}(\mathbf{L}/\mathbf{K})$.

If $H \subset \mathrm{Aut}(\mathbf{L}/\mathbf{K})$ is a normal divisor, then $\mathbf{L}^H = \mathbf{Z} \supset \mathbf{K}$ is a normal extension.

COROLLARY 24 $\mathrm{Gal}(\mathbf{L}/\mathbf{K})/H \simeq \mathrm{Gal}(\mathbf{Z}/\mathbf{K})$.

Now we present the general definition of Galois extensions which does not assumes finiteness of the extension; however, in the future we will consider finite extensions $\mathbf{L} \supset \mathbf{K}$ only.

DEFINITION. An extension $\mathbf{L} \supset \mathbf{K}$ is *Galois* if it is normal and separable.

Therefore, if $\mathbf{L} \supset \mathbf{K}$ is separable, then $\mathbf{L}$ is a Galois extension $\equiv \mathbf{L}$ is normal. Since the most important extensions are normal, some authors (that is, Van der Waerden) implicitly assume separability, and they identify Galois G-extensions with the normal ones.

THEOREM 25. *($\mathbf{L} \supset \mathbf{K}$ is Galois)$\Longleftrightarrow$(there exists a group $G \subset \mathrm{Aut}(\mathbf{L})$ such that $\mathbf{K} = \mathbf{L}^G$).*

COROLLARY 26. *If $\mathbf{K}$ is perfect (in particular, if $\mathrm{char}\mathbf{K} = 0$ or $\mathbf{K}$ is finite), then $\mathbf{L} \supset \mathbf{K}$ is Galois if and only if it is normal.*

In what follows, we will consider only *finite* Galois extensions and therefore the following theorem is very important.

THEOREM 27. *($\mathbf{L} \supset \mathbf{K}$ is finite and Galois)$\Longleftrightarrow$($\mathbf{L}$ is a field of decomposition of some polynomial $f \in \mathbf{K}[x]$ which has multiple roots).*

Therefore, we could regard the right hand side of Theorem 27 as a definition of (finite) Galois extensions.

We have yet another characteristic of finite Galois extensions which could be taken as an elegant definition; namely we have the following theorem.

THEOREM 28. *(A finite extension $\mathbf{L} \supset \mathbf{K}$ is Galois)*$\Longleftrightarrow$($|\mathrm{Aut}(\mathbf{L}/\mathbf{K})| = [\mathbf{L} : \mathbf{K}]$).

Thus, in what follows, while talking about Galois extensions $\mathbf{L} \supset \mathbf{K}$, we will keep in mind that the Galois group of this extension is finite and the number of its elements is equal to the degree of extension $\mathbf{L} \supset \mathbf{K}$.

After presenting these characterizations of Galois extensions and their Galois groups, we turn now to formulating the fundamental theorem of Galois theory.

Chapter 6

Galois Theory. Solvable Groups

If $\mathbf{L} \supset \mathbf{K}$ is an extension of $\mathbf{K}$, a decisive role is played by the group of automorphisms $\mathrm{Aut}(\mathbf{L}/\mathbf{K})$ of this extension, that is, a subgroup of automorphisms of the field $\mathbf{L}$ which preserve the field $\mathbf{K}$,

$$\mathrm{Aut}(\mathbf{L}/\mathbf{K}) = \{g \in \mathrm{Aut}(\mathbf{L}) : g|_{\mathbf{K}} = \underset{\mathbf{K}}{\mathrm{id}}\}.$$

DEFINITION. An extension $\mathbf{L} \supset \mathbf{K}$ is an *Galois extension* if it is

a) finite (that is, $[\mathbf{L} : \mathbf{K}] = n \leqslant \infty$),

b) $|\mathrm{Aut}(\mathbf{L}/\mathbf{K})| = [\mathbf{L} : \mathbf{K}]$ $(= \dim_{\mathbf{K}} \mathbf{L}/\mathbf{K})$.

Then, one often calls $\mathrm{Aut}(\mathbf{L}/\mathbf{K})$ a *Galois group* of the extension $\mathbf{L} \supset \mathbf{K}$ and denotes it by $\mathrm{Gal}(\mathbf{L}/\mathbf{K})$.

REMARK. If the group $\mathrm{Gal}(\mathbf{L}/\mathbf{K})$ is abelian (commutative), cyclic, solvable, then the extension $\mathbf{L} \supset \mathbf{K}$is called, respectively, *abelian (commutative), cyclic, solvable.*

THEOREM 1 (FUNDAMENTAL THEOREM OF GALOIS THEORY). *There exists a bijection $\mathcal{G}$ (that is, the one-to-one mapping) of the set $\mathcal{Z}$ of intermediate fields $\mathbf{L} \supset \mathbf{Z} \supset \mathbf{K}$ into the set $\mathcal{H}$ of subgroups $H \subset \mathrm{Aut}(\mathbf{L}/\mathbf{K})$ which preserves $\mathbf{Z}$, that is, $\mathbf{L}^H = \mathbf{Z}$. The mapping $\mathcal{G} : \mathcal{L} \to \mathcal{H}$ is defined by*

$$\mathcal{G}(\mathbf{Z}) := \mathrm{Aut}(\mathbf{L}/\mathbf{Z}). \tag{1}$$

The inverse mapping is the mapping $\mathcal{F} : \mathcal{H} \to \mathcal{Z}$ defined by $\mathcal{F}(H) := \mathbf{L}^H$, and thus, explicitly

$$\mathcal{F}(H) := \{g \in \mathrm{Aut}(\mathbf{L}) : G|_{\mathbf{Z}} = \underset{\mathbf{Z}}{\mathrm{id}}\} \equiv \mathbf{L}^H. \tag{2}$$

It is easy to check that $\mathcal{F}(H)$ *is a subgroup of* $\mathrm{Aut}(\mathbf{L}/\mathbf{K})$. *We have, clearly, the following statements:*

(3) $$(\mathbf{L} \supset \mathbf{Z}_1 \supset \mathbf{Z}_2 \supset \mathbf{K}) \iff (\mathrm{Aut}(\mathbf{L}/\mathbf{Z}_2) \subset \mathrm{Aut}(\mathbf{L}/\mathbf{Z}_1)),$$

(4) $$(g \in \mathrm{Aut}(\mathbf{L}/\mathbf{K})) \Longrightarrow \mathrm{Aut}(\mathbf{L}/g\mathbf{Z}) = g\,\mathrm{Aut}(\mathbf{L}/\mathbf{Z})g^{-1}.$$

PROOF (sketch). We will perform the proof in the case when $\mathbf{K}$ has characteristic zero; in this case the proof simplifies a bit. It is sufficient to show that for all intermediate fields $\mathbf{L} \supset \mathbf{Z} \supset \mathbf{K}$ and all subgroups $H \subset \mathrm{Aut}(\mathbf{L}/\mathbf{K})$, we have

(5) $$\mathbf{Z} = \mathbf{L}^{\mathrm{Aut}(\mathbf{L}/\mathbf{Z})} \quad \text{and} \quad H = \mathrm{Aut}(\mathbf{L}/\mathbf{L}^H).$$

These equalities will be showed if we prove that

(6) $$[\mathbf{L} : \mathbf{L}^H] = |H| \quad \text{for all subgroups } H \subset \mathrm{Aut}(\mathbf{L}/\mathbf{K})\,.$$

But, form the definition of Galois group, we have $\mathrm{Aut}(\mathbf{L}/\mathbf{L}^H) \supset H$, and thus

(7) $$[\mathbf{L} : \mathbf{L}^H] \geqslant |H|$$

This inequality follows from the following proposition.

PROPOSITION 2. *Let* $\mathbf{L} \supset \mathbf{K}$ *be a finite extension of the field* $\mathbf{K}$ *of characteristic* 0 *and let* $\mathbf{L} = \mathbf{K}(\vartheta)$. *Let* Ω *be a field of decomposition of the minimal polynomial for* ϑ *with respect to* $\mathbf{K}$. *If* $\mathbf{L} \subset \mathbf{M} \subset \mathbf{K}$ *and* φ *is some isomorphism* $\mathbf{M} \to \Omega$ *(of the field* $\mathbf{M}$ *into the field* Ω*), then there exists exactly* $[\mathbf{L} : \mathbf{M}]$ *extensions of the mapping* φ *to an isomorphism* $\mathbf{L} \to \Omega$.

PROOF OF PROPOSITION 2. Let $[\mathbf{L} : \mathbf{M}] = n$ and let $\vartheta_1, \ldots, \vartheta_n$ be zeros of a minimal polynomial for ϑ with respect to $\mathbf{M}$. The extensions $\varphi_j \supset \varphi$, $j = 1, \ldots, n$, we are looking for, are given by the formula

$$\varphi_j\left(\sum_{k=0}^{n-1} \alpha_k \vartheta^k\right) := \sum_{k=0}^{n-1} \varphi(\alpha_k)\vartheta^k \quad \text{for } , \alpha_0, \ldots, \alpha_{n-1} \in \mathbf{M}.$$

□

REMARK. The gist of the proof of proposition 2 is the fact that an irreducible polynomial of degree n in the field of decomposition has n different zeros, which for fields of characteristic $\neq 0$ is, in general, not true.

Now we can quickly complete the proof of Theorem 1 (fundamental theorem of Galois theory).

Let us consider the polynomial

$$f_h(x) = \prod_{h \in h} (x - h\vartheta),$$

where ϑ is a primitive element of the extension $\mathbf{L} \supset \mathbf{K}$; the polynomial $f_h(x)$ has coefficients belonging to the field $\mathbf{L}^h$ and the equality $\mathbf{L} = \mathbf{L}^H(\vartheta)$ holds. Thus $[\mathbf{L} : \mathbf{L}^H] \leqslant |H|$ holds, which together with inequality (7) proves the theorem. □

Historical remarks. Of course, Galois acted differently: at the beginning of XIX century the abstract notions of groups and fields did not exist. Recently we celebrated a centenary of the birth of the notion of extension of a field. The general notion of a group is due to von Dyck and H. Weber (1882). At those times mathematics was very concrete: people knew the group of permutation S_n (of n objects), groups of some concrete transformations, for example, the group of rotation of the plane; the extension of fields by addition of algebraic elements was known as well, but the general notions of fields and rings we owe Dedekind.

Galois investigated groups of polynomials and for these groups (polynomials) he developed his theory. The modern view on the Galois theory as a theory of (some) finite extensions of fields and automorphisms of these extensions we owe Dedekind again, who, as a young professor in Getingen and a friend of Riemann, presented this theory in this way in his lectures and then published it in his famous *XI Supplement* to lectures of Dirichlet on number theory which he edited. Further simplification and final codification of the Galois theory, we owe Artin, E. Noether, Van der Waerden, and, most of all, the friend and collaborator of Dedekind, Heinrich Weber (1842–1913). Galois investigated the problem of solvability of algebraic equations and also elliptic functions: he studied the papers of few years older N. H. Abel. The theorem on primitive elements was known already to Abel, and possibly even Lagrange, who certainly knew the important particular cases of the theorem, but we owe Galois the complete proof.

The way Galois looked at the problems of fields extensions and introduction of the Galois group of an extension was one of the greatest achievements of mathematical thought, so revolutionary that the contemporaries, in all other respects outstanding scientists like Cauchy and Poisson, were not able to comprehend these new and written in a telegraphic style notes of the brilliant youngster.

Galois group of an algebraic equation. The Galois group of an algebraic equation is the brilliant creation which is a direct predecessor of the group $\mathrm{Gal}(\mathbf{L}/\mathbf{K})$. We present here (following Helmut Koch) the definition which is very close to the original one.

Let $f \in \mathbf{K}[x]$ be a polynomial without multiple zeros in the field of decomposition $\mathbf{L}$ of the polynomial f (that is $\mathbf{L} \supset \mathbf{K}$ such that f over $\mathbf{L}$ decomposes into linear terms). Let $n = \deg f$ be the degree of f. Let now $y_1, \ldots, y_n$ be parameters and $\alpha_1, \ldots, \alpha_n$ zeros of f in $\mathbf{L}$. Next, let H be a ring of polynomials $h(y_1, \ldots, y_n) \in \mathbf{K}[y_1, \ldots, y_n]$ such that $h(\alpha_1, \ldots, \alpha_n) \in \mathbf{K}$; as we know from fundamental theorem on symmetric polynomials, H contains all symmetric polynomials.

DEFINITION (I). The group $G(f) = \{\pi \in S_n : h(\pi\alpha_1, \ldots, \pi\alpha_n) = h(\alpha_1, \ldots, \alpha_n)$ for all $h \in H\}$ is called the *Galois group of the polynomial* $f(x)$ or *of the equation* $f(x) = 0$.

In general, $G(f)$ is a proper (that is, $\neq S_n$) subgroup of the symmetric group S_n (of permutation of n elements). The following example shows however that it might happen that $G(f) = S_n$.

EXAMPLE. Let $f = x^n + a_1x^{n-1} + \cdots + a_n$, where $a_j := (-1)^j s_j$, s_j are elementary symmetric polynomials, that is

$$s_j = s_j(x_1, \ldots, x_n) := \sum_{i_1 < \cdots < i_k} x_{i_1} x_{i_2} \cdots x_{i_k}$$

of parameters $x_1, \ldots, x_n$ over some field $\mathbf{K}_0$, and the *ground* field is $\mathbf{K} = \mathbf{K}_0(s_1, \ldots, s_n)$. For the polynomial $h(y_1, \ldots, y_n) \in H$, the polynomial $h(x_1, \ldots, x_n) \in \mathbf{K}_0(x_1, \ldots, x_n)$ is symmetric, and thus $G(f) = S_n$.

The marvelous idea of Galois is that from the group $G(f)$ one can deduce many properties of zeros $\alpha_1, \ldots, \alpha_n$ of the polynomial f, for example, one can deduce the solution of the fundamental, at these times, problem concerning possibility of representing $\alpha_1, \ldots, \alpha_n$ in terms of radicals. The following *criterion of irreducibility* comes from Galois.

THEOREM 3 (GALOIS). *(The polynomial f is irreducible) $\Longleftrightarrow$ (the group $G(f)$ is transitive, that is, $G(f)$ acts transitively on the set $(\alpha_1, \ldots, \alpha_n)$ of* *zeros of the polynomial f).*

Proof. $\Longleftarrow$: Let $G(f)$ act transitively and let the polynomial f be an irreducible normed polynomial corresponding to α_1. Thus $f_1(\alpha_1) \in H$, and from definition of $G(f)$ we have, $f_1(\pi\alpha_1) = f_1(\alpha_1) = 0$ for all $\pi \in G(f)$. Thus $f = f_1$ (we write $\pi\alpha_1$ instead of $\alpha_{\pi 1}$).

$\Longrightarrow$(indirect). Let $G(f)$ not be a transitive group: let, for example, $\alpha_1, \ldots, \alpha_s$, $s < n$ be permuted by the group $G(f)$. Then fundamental symmetric polynomials of variables $\alpha_1, \ldots, \alpha_s$ are invariant with respect to all $\pi \in G(f)$, but the polynomial $(x - \alpha_1) \cdots (x - \alpha_s)$ has coefficients in $\mathbf{K}$ and therefore f is reducible, a contradiction. □

Let us recall the notion of transitivity of a group G of permutations of (arbitrary) set.

Definition. The group G of permutation of the set X acts *transitively* on X if for any two elements $x_1, x_2 \in X$ there exists a transformation $g \in G$ such that $g(x_1) = x_2$.

Remark. Above, we made use of the following theorem, coming from Galois, of course.

Theorem 4 (Galois). *Let for $g(y_1, \ldots, y_n) \in \mathbf{K}[y_1, \ldots, y_n]$.*

$$g(\pi\alpha_1, \ldots, \pi\alpha_n) = g(\alpha_1, \ldots, \alpha_n) \quad \text{for all } \pi \in G(f).$$

Then $g(\alpha_1, \ldots, \alpha_n) \in \mathbf{K}$.

We leave the proof to the reader as an exercise.

The notion of transitivity of groups of transformations was defined precisely by Camille Jordan in his monumental, 667 pages long work *Traité des substitutions et des équations algébraiques* of 1870; this is the first systematic book monograph on Galois theory. But the transitivity was, as have seen, known to Galois (and perhaps even Lagrange).

So what is the relation between the group $G(f)$ and the group $\mathrm{Aut}(\mathbf{L}/\mathbf{K})$? The expected answer is given by the following theorem.

Theorem 5. *Let $G = \mathrm{Aut}(\mathbf{K}(\alpha_1, \ldots, \alpha_n)/\mathbf{K})$. Then the mapping $\varphi : G \to G(f)$ (which with each element $g \in G$ associates the permutation of zeros $\alpha_1, \ldots, \alpha_n$ of the polynomial f caused by g) is an isomorphism!*

PROOF. Clearly, φ is an homomorphism $\varphi : G \to G(f)$. Let us now construct the inverse mapping φ_1 as follows. Let $\mathbf{K}(\vartheta) = \mathbf{K}(\alpha_1, \ldots, \alpha_n)$, let f_ϑ be a minimal polynomial for ϑ with respect to $\mathbf{K}$, and let $\psi \in \mathbf{K}[x_1, \ldots, x_n]$ be such polynomial that $\vartheta = \psi(\alpha_1, \ldots, \alpha_n)$ and let $\pi \in G(f)$. Then, from definition of $G(f)$, we have

$$f_\vartheta(\psi(\pi\alpha_1, \ldots, \pi\alpha_n)) = f_\vartheta(\psi(\alpha_1, \ldots, \alpha_n)) = f_\vartheta(\vartheta) = 0,$$

that is $\vartheta_\pi \psi(\pi\alpha_1, \ldots, \pi\alpha_n)$ is a zero of the polynomial f_ϑ.

The association $\vartheta \to \vartheta_\pi$ extends to the automorphism φ_1 of the field $\mathbf{K}(\vartheta)$ which does not move $\mathbf{K}$, thus $\varphi_1 \in G$, but $\varphi_1 \circ \varphi = \mathrm{id}_G$, $\varphi \circ \varphi = \mathrm{id}_{G(f)}$. □

We have shown the isomorphism $G(f) \simeq \mathrm{Aut}(\mathbf{K}(\alpha_1, \ldots, \alpha_n)/\mathbf{K})$.

Definition (I) of the group $G(f)$ of equation $f(x) = 0$, coming, in fact, from Galois, seems to be a bit heavy nowadays, but we have presented them for purpose in order to make the reader understand the way of thinking of the great creator of the theory. Today one uses, of course, the equivalent but simpler definition, as follows.

DEFINITION (II). A *Galois group of a polynomial* $f(x) \in \mathbf{K}[x]$ is the group $G(f) := \mathrm{Aut}(\mathbf{K}(\alpha_1, \ldots, \alpha_n)/\mathbf{K})$, where $\mathbf{K}(\alpha_1, \ldots, \alpha_n)$ is the field of decomposition of the polynomial f.

Clearly, every element $g \in G$ transform the set $\{\alpha_1, \ldots, \alpha_n\}$ onto itself, it defines a permutation $\pi_g \in S_n$ of roots of f. If such a permutation is known, then the automorphism $g_\pi \in G$ is defined; indeed, if $\pi\alpha_j = \alpha'_j$, $j = 1, 2, \ldots, s \leqslant n$, then every element of the field $\mathbf{K}(\alpha_1, \ldots, \alpha_n)$ regarded as a rational function $w(\alpha_1, \ldots, \alpha_n)$ turns into $w(\alpha'_1, \ldots, \alpha'_n)$. Thus, the group G_f of the polynomial f can be regarded as some group of permutations of its roots.

Now we introduce an important notion.

If the extensions $\mathbf{K}(\vartheta_1)$, $\mathbf{K}(\vartheta_2), \ldots \subset \mathbf{L}$ are such that $\mathbf{K}(\vartheta_1)$, $\mathbf{K}(\vartheta_2), \ldots$ are isomorphic: $\mathbf{K}(\vartheta_1) \simeq \mathbf{K}(\vartheta_1) \simeq \ldots$, then such *extensions* are called *associated (adjoint)* with respect to $\mathbf{K}$ and the quantities (elements) $\vartheta_1, \vartheta_2, \ldots$ which transform into each other under given isomorphisms are called *associated (adjoint)* quantities (elements). It is not hard to show the following theorem.

THEOREM 6. *Every two simple algebraic extensions* $\mathbf{K}(\alpha)$, $\mathbf{K}(\beta)$ *are* **K**-*equivalent, that is, there exists an isomorphism* $\varphi : \mathbf{K}(\alpha) \to \mathbf{K}(\beta)$ *which does not move* **K** *(that is,* $\varphi|\mathbf{K} = \mathrm{id}_{\mathbf{K}}$*) if* α *and* β *are zeros of the same in* $\mathbf{K}[x]$ *irreducible polynomial* f *(that is,* $f(\alpha) = f(\beta) = 0$*). Then there exists an isomorphism* $i : \mathbf{K}(\alpha) \to \mathbf{K}(\beta)$ *such that* $i(\alpha) = \beta$ *and* $i|\mathbf{K} = \mathrm{id}_{\mathbf{K}}$.

COROLLARY 7. *All zeros of irreducible polynomial* f *in* $\mathbf{K}[x]$ *belonging to the field* **L** *are adjoint to each other with respect to* **K**.

From these we have the following important theorem.

THEOREM 8. *(Two intermediate fields* $\mathbf{K} \subset \mathbf{Z}_1$, $\mathbf{Z}_2 \subset \mathbf{K}$ *are adjoint with respect to* **K***)*$\Longleftrightarrow$*(there exists* $g \in \mathrm{Aut}(\mathbf{L}/\mathbf{K})$ *such that* $g : \mathbf{Z}_1 \to \mathbf{Z}_2$*).*

Taking $\mathbf{Z}_j = \mathbf{K}(\alpha_j)$, we obtain the following useful corollary.

COROLLARY. *(*$\alpha_1, \alpha_2 \in \mathbf{L}$ *are adjoint with respect to* **K***)* $\Longleftrightarrow$*(there exists* $g \in \mathrm{Aut}(\mathbf{L}/\mathbf{K})$ *such that* $g\alpha_1 = \alpha_2$*).*

In this way, we once again obtain the Galois criterion of irreducibility of a polynomial f:

$$(\mathbf{K}[x] \ni f \text{ is irreducible}) \Longleftrightarrow (G_f \text{ is transitive}).$$

Since the number n of different adjoints of some element (quantity) $\alpha \in \mathbf{L}$ is equal to the degree of *irreducible* equation $f(x) = 0$ for α (that is $f(\alpha) = 0$), thus, if $n = 1$, α is a root of the linear equation $x - \alpha = 0$, therefore, $\alpha \in \mathbf{K}$. We showed therefore the following theorem (which we actually silently used in the first proof of transitivity of $\mathcal{G}(f)$).

THEOREM 9. $\alpha \in \mathbf{L}$ *is a fixed point of the Galois group* $\mathrm{Gal}\,(\mathbf{L}/\mathbf{K})$)$\Longrightarrow$($\alpha \in \mathbf{K}$).

One of the major successes of the Galois theory was the following beautiful theorem on solvability of algebraic equations (of arbitrary order) by radicals for $f(x) = 0$, where $f \in \mathbf{K}[x]$ and **K** has characteristic 0.

THEOREM 10 (GALOIS ON SOLVABILITY BY RADICALS). *An irreducible equation* $f(x) = 0$ *is solvable by roots (subsequent taking of the roots*))$\Longleftrightarrow$(*$G(f)$ is solvable*).

Solvable groups. The *commutator* of elements x and y of a group G is the expression $[x, y] : xyx^{-1}y^{-1}$. The *commutant* of a group G is

$$G' = G^{(1)} = [G, G] := \{[x, y] : x, y \in G\}.$$

It is easy to take commutant of the commutant G'; this will be denoted by $G^{(2)}$; $G^{(k+1)} := (G^{(k)})'$. It is not hard to prove the following lemma.

LEMMA 11. 1. $G \supset G^{(1)} \supset G^{(2)} \supset \cdots$;
2. *$G^{(k+1)}$ is a normal subgroup in $G^{(k)}$;*
3. *$G^{(k)}/G^{(k)}$ is abelian.*

It may happen that the descending sequence in 1. terminates at the group $\{e\} \supset G$ (consisting of the unit element only) for some m and then G is (*ex definitione*) *solvable*.

EXAMPLE 1. Abelian groups and, in particular, cyclic groups are solvable.

EXAMPLE 2. The groups S_n and A_n $(= S_n')$, the latter being the group of even permutations, called also the *alternating group*, are solvable for $n = 1, 2, 3, 4$.

Galois showed also the following famous theorem.

THEOREM 12 (GALOIS). (a) *If $H \subset G$ is a subgroup of G $H^{(k)} \subset G^{(k)}$ for $k = 1, 2, \ldots$, since the group A_5 is simple (it does not possess any nontrivial normal divisors), all the groups A_n for $n \geqslant 5$ are simple.*
(b) *Since $S_n' = A_n$, then S_n is not solvable for $n \geqslant 5$.*

In this way Galois proved the following famous theorem.

THEOREM 13 (ABEL–RUFFINI). *Equations of order five and higher are not (in general) solvable by radicals.*

Historical remark. The notion of solvable group (French *solvable*, German *ausflösbar*) was introduced by Galois. Ruffini presented a (not complete) proof of the theorem above in 1813 – but the proof of Ruffini has a flow in that it contains (not proved) hypothesis that for equation of order 5 all the roots are rational functions of roots themselves. Only the proof of Abel (1824) is complete. The proof of Galois of 1831 is very general and complete. The notes from 1829 were lost by Cauchy, and the ones from 1830 got lost by other academicians (Cauchy, Fourier). The famous work of 1831 was rejected by Poisson (also a member of the Academy) as being 'not understandable'!

CHAPTER 7

Ruler and Compass Constructions. Cyclotomic Fields. Kronecker–Weber Theorem

The ruler and compass constructions formed perhaps the most important momentum of the development of mathematics for hundreds and possibly even thousands of years. For Plato and his school only these constructions were perfect. The most famous are the following problems

1. The problem of trisection of the angle.
2. The problem of doubling of the cube.
3. The problem of quadrature of the circle.
4. The problem of rectification of the arc of circle.
5. The problem of construction of the regular n-gon.

These are the questions concerning the possibility of constructing the segments and points mentioned in points 1.–5. with the help of a ruler and compass alone.

The attempts to give an answer to these problem (similarly to the problem of solvability of an equation of order $n \geqslant 5$ by radicals) has led to creation of new branches of mathematics. Therefore we cannot shrug our shoulders or tap our forehead considering paranoid people who sometime ago tried to solve one of these famous five problem which are (in general) not solvable, as it had turned out later.

The Galois theory makes it possible to look at these problems anew and to find their solutions. This was a great success. Let us say few words about this new understanding and about the method itself. If we have given segments of the length 0,1, $a, b \in \mathbb{R}$, then, with the help of a ruler and compass, we can always construct the quantities $a+b$, $-a$, $a \cdot b$, $1/a$. Thus from 0 and 1 we can construct all numbers $x \in \mathbb{Q}$, and having numbers $a_0, \ldots, a_n$, we can construct every $y \in \mathbb{Q}(a_0, \ldots, a_n)$. In this natural way we are led to (finite) extensions of the field $\mathbb{Q}$. It is easy to observe that given $r > 0$, $r \in \mathbb{R}$, we can always construct $\sqrt{r}$. The subsequent addition of new elements (with the help of a ruler and compass) leads to the chain of extensions

$$(C) \qquad \mathbf{K}_0 := \mathbb{Q}(a_0, \ldots, a_n) \subset \mathbf{K}_1 \subset \cdots \subset \mathbf{K}_{j+1} \subset \cdots$$

where $\mathbf{K}_{j+1} := \mathbf{K}_j(\sqrt{r_j})$, $r_j \in \mathbf{K}_j$, $r_j > 0$, $j = 1.2. \ldots$

One can easily show the following theorem.

THEOREM 1. *($x \in \mathbb{R}$ can be constructed from $a_0, \ldots, a_n$ with the help of a ruler and compass)$\Longleftrightarrow$(x belongs to some $\mathbf{K}_j$ from the chain (C)).*

This theorem leads, with the help of Galois theory, to the following important theorem.

THEOREM 2. 1. *($x \in \mathbb{R}$ can be constructed from $a_0, \ldots, a_n \in \mathbb{R}$ with the help of a ruler and compass)$\Longleftrightarrow$($x \in \mathbf{L} \supset \mathbb{Q}(a_0, \ldots, a_n)$, where $\mathbf{L}$ is a Galois extension of order $2^m (= [\mathbf{L} : \mathbb{Q}(a_0, \ldots, a_n)])$ for some $m \in \mathbb{N}$).*

2. (x is a transcendental number)$\Longrightarrow$(x cannot be constructed with the help of a ruler and compass).

The proof of this theorem can be found in (almost) all textbooks on algebra and thus we will not present it here; we show however how one can apply this theorem to solve the five classical problems: what is important is the statement concerning the degree of extension 2^m!!!

Problem 2. Doubling of the cube. A cube of the edge = 1 is given, one needs to construct the cube of the edge = 2, so one needs to know how to construct $\sqrt[3]{2}$. But $\sqrt[3]{2}$ is an irreducible zero of the polynomial $x^3 - 2 \in \mathbb{Q}[x]$ over $\mathbb{Q}$, thus $\mathbb{Q}[\sqrt[3]{2}) : \mathbb{Q}] = 3$. Therefore, if $\sqrt[3]{2}$ would be possible to construct with the help of a ruler and compass, $\sqrt[3]{2}$ has to be an element of the field

$\mathbf{L}$ of order 2^m over $\mathbb{Q}$. But $\mathbb{Q} \subset \mathbb{Q}(\sqrt[3]{2}) \subset \mathbf{L}$ and

$$[\mathbf{L} : \mathbb{Q}] = [\mathbb{Q}(\sqrt[3]{2}) : \mathbb{Q}] \cdot [\mathbf{L} : \mathbb{Q}(\sqrt[3]{2})] = 3 \cdot p = 2^m$$

with p integer, thus 2^m must be divisible by 3, a contradiction. *The problem does not have any solution*!

Problem 1. Trisection of the angle leads to (zeros) of the polynomial

$$X \mapsto [X, *]$$

(also called the *adjoint mapping* of the algebra $\mathfrak{g}$.) The Killing form is an invariant. For $\phi \in \mathrm{Aut}(\mathfrak{g})$

$$\mathfrak{K}(\phi X, \phi Y) = \mathfrak{K}(X, Y)$$

holds.

THEOREM 3 (LINDEMANN THEOREM ON TRANSCENDENCE OF THE NUMBER π, 1882). *The number π is transcendental.*

Then the sensational result of Lindemann (1882) follows[1]

COROLLARY 4. *The squaring of the circle is impossible.*

The problem of construction of the regular n-gon has somewhat different character: we know that one can easily construct regular triangle, tetragon, hexagon, ... It was a great sensation when Gauss showed that one can construct a regular 17-gon.

The Galois theory provides the full answer to the question in the form of the following theorem.

THEOREM 5. *(It is possible to construct the regular n- gon)$\Longleftrightarrow$(n is a finite product of numbers from the set of powers of 2 and Fermat numbers).*

The known Fermat numbers (i.e. the primes $p_i = 2^{k_i} + 1$) are 3, 5, 17, 257, 65537, ... Therefore one can construct n-gons for

$$n = 3, 4, 5, 6, 8, 10, 12, 15, 16, 17, \text{(so-called, Gauss,)}\ 20 \ldots$$

[1]Carl Louis Ferdinand von Lindemann (1852–1939) was a teacher of Hurwitz and Hilbert in Köningsberg. His famous proof of transcendence of π was published in the short note *Über die Zahl π*, Math. Ann. **20** (1882), 213–225.

However, the n-gons which are missing in this sequence cannot be constructed, for example heptagon, 11-gon, 13-gon cannot be constructed with the help of a ruler and compass.

We will conclude this chapter with important examples, which by themselves are famous theorems.

EXAMPLES OF GALOIS EXTENSIONS

EXAMPLE 1. *Cyclotomic fields* played a role in the development of algebra and number theory which is hard to be overestimated. Let us consider $\mathbf{K} = \mathbb{Q}$, $\mathbf{L} = \mathbb{Q}(\zeta_m)$, where ζ_m is the root of order m of 1. Let us recall the definition and results. Let m be a natural number. Then ζ_m is called the *primitive mth root of unit* if ζ_m is such a complex number that $\zeta_m^m = 1$ (that is $\zeta_m = e^{2\pi i/m}$, this is the reason for the term *section of the circle – cyclotomy*) and $\zeta_m^j \neq 1$ for $1 \leqslant j \leqslant m$. Gauss considered the case when m is a prime number. The following important theorem holds.

THEOREM 6 (GAUSS). *For any natural number n which is prime with respect to m, the mapping $\zeta_m \to \zeta_m^n$ induces the automorphism $\mathbf{L} \overset{\text{on}}{\to} \mathbb{Q}$ which depends only on the classes of reminders n* mod m. *All these automorphisms form a group $G(m) \simeq (\mathbb{Z}/m\mathbb{Z})^\times$. The order of this group is finite and equal to $\varphi(m)$ (Euler function):*

$$\dim_{\mathbb{Q}} \mathbf{L} = [\mathbb{Q}(\zeta_m) : \mathbb{Q}] = \varphi(m) = |\operatorname{Aut}(\mathbb{Q}(\zeta_m)/\mathbb{Q}|,$$

and thus $\mathbb{Q}(\zeta_m) \supset \mathbb{Q}$ is a Galois extension with cyclic Galois group, and therefore abelian. The multiplicative group $(\mathbb{Z}/m\mathbb{Z})^\times$ is cyclic since it is a finite (sub)group of the multiplicative group of the field.

We 'showed' that the extension $\mathbb{Q}(\zeta_m) \supset \mathbb{Q}$ is cyclic (and therefore abelian). According to H. Hasse, in 1853 Leopold Kronecker already knew that every abelian extension $\mathbf{L} \supset \mathbb{Q}$ (that is, such that the group $\operatorname{Aut}(\mathbf{L}/\mathbb{Q})$ is abelian) is cyclotomic, in other words, he formulated an extraordinary claim that a cyclotomic extension containing the field $\mathbb{Q}$ contains all abelian extensions of $\mathbb{Q}$. This claim was, 33 years later, confirmed by Heinrich Weber, who proved the following famous theorem.

THEOREM 7 (KRONECKER–WEBER). *Every abelian extension $\mathbf{L} \supset \mathbb{Q}$ is contained in $\mathbb{Q}(\zeta)$, where ζ is some primitive root of* 1 *($\zeta = \sqrt[m]{1}$).*

This is one of the examples of brilliant mathematical intuition of great scientists.

EXAMPLE 2. Let again $\mathbf{L} = \mathbb{Q}(\zeta)$, where ζ is some primitive root of 1, and $a \in \mathbb{Q}(\zeta)$ be such that $\mathbf{M} := \mathbf{L}(\sqrt[m]{a})$ is of order m over $\mathbf{L}$, that is, the polynomial $x^m - a$ is irreducible over $\mathbf{L}$. Therefore, for any integer n, the mapping $\sqrt[m]{a} \to \zeta^n \sqrt[m]{a}$ induces an $\mathbf{L}$ automorphism of the field $\mathbf{M}$. The group $\mathrm{Aut}(\mathbf{M}/\mathbf{L} \simeq G(m)$ is cyclic and therefore, abelian.

EXAMPLE 3. Let p be a prime number, $\mathbf{F}_p := \mathbb{Z}/p\mathbb{Z}$, $q = p^m$, $m > 1$. Then $\mathbf{F}_q$ is a field with q elements, that is, a Galois field. The mapping $\Phi : \mathbf{F}_q \to \mathbf{F}_q$, where $\Phi(x) = x^p$, is an automorphism of the field $\mathbf{F}_q$; thus $\mathbf{F}_q$ coincides with the field of its units. All (integer) powers of Φ form a cyclic group of order m which is the Galois group $\mathrm{Gal}(\mathbf{F}_q/\mathbf{F}_p)$.

Since the notion of Galois extension is absolutely fundamental, we present here the third definition.

DEFINITION (III). The finite (algebraic) extension $\mathbf{L}/\mathbf{K}$ is a *Galois extension* if the following isomorphism of vector spaces (over $\mathbf{K}$) holds

$$\mathbf{L} \otimes_{\mathbf{K}} \mathbf{L} \simeq \mathbf{L} \oplus \ldots \oplus \mathbf{L} \quad \text{with } (\mathbf{L}/\mathbf{LK}) \text{ terms}) \ .$$

This definition is advocated by J. F. Pommaret in his interesting (and full of historical remarks) monograph *Lie Pseudogroups and Mechanics*, 1987. As it is well known, S. Lie dreamed of creating an analogue of Galois theory for differential equations.

EXERCISE. Prove equivalence of this definition with the preceding ones (the proof can be found on page 458 of the Pommaret book).

CHAPTER 8

Algebraic and Transcendental Elements

As we know, a complex number α is called an *algebraic number* if it is a zero of some polynomial

$$\alpha^m + a_1\alpha^{m-1} + \cdots + a_m = 0 \tag{1}$$

where the coefficients a_k are rational $a_k \in \mathbb{Q}$; α is an *integer algebraic number* if there exists a corresponding equation with *integer* coefficients $a_k \in \mathbb{Z}$.

If a complex number ϑ does not satisfy any equation of the form (1), then ϑ is a *transcendental number.* Thus ϑ is transcendental if it is not algebraic. The proof that some number is transcendental is usually a very difficult problem, and thus the proofs that e and π are transcendental, were considered to be a very important achievements.

THEOREM 1 (CHARLES HERMITE, 1873). *The number e is transcendental.*

THEOREM 1 (FERDINAND LINDEMANN 1882). *The number π is transcendental (cf. Theorem* 3).

Thus, *quadrature of the circle* is impossible by any geometric construction which makes only use of algebraic curves and surfaces! In particular, the quadrature of the circle with the help of a ruler and compass is impossible!

In this way, Liendemann (1852–1939) theorem concluded, in the negative, the problem of quadrature of the circle, the one of the oldest problems of mathematics. Nevertheless, there still exist maniacs who claim that they can, with the help of a ruler and compass, to construct the square of the area equal to the area of the disc $x^2 + y^2 \leqslant 1$. This a psychologically very interesting phenomenon: several years ago some, otherwise practically thinking businessman, kept approaching me claiming that he solved the problem of squaring of the circle. All arguments of the kind 'Why don't you read any textbook on algebra or the work of Liedemann (or the little book of C. L. Siegel *Transcendental Numbers* (1949)),' were being replied with 'But I just do know how to do this ...'.

The reader can find more information concerning transcendental numbers and the methods of proving transcendentality of some of numbers being values of particular functions satisfying differential equations in, that is, concerning the Gelfand–Schneider method, in Appendix to Chapter XVI of *Analysis*, part II.

CHAPTER 9

Weyl principle

In his charming book *Symmetry*, after presenting a basis of the Galois theory, Hermann Weyl writes 'There is a general principle, which is very fruitful in mathematics that *if you want to learn something about a mathematical object, investigate its automorphism group*'. I call this sentence the Weyl principle. Hermann Weyl used himself this principle in his work many times. Here, I will present some important examples.

Galois group. The Galois group of the extension $\mathbf{L} \supset \mathbf{K}$ of the field $\mathbf{K}$ made it possible to get deep insight into the structure of intermediate extensions of the field $\mathbf{K}$.

The group $\mathrm{Deck}(Y/X)$. The group $\mathrm{Deck}(Y/X)$ of covering mappings of the cover $Y \overset{p}{\to} X$ of the Riemann surface X by the surface Y is the group of automorphisms g of the manifold Y such that the fibers $p^{-1}(x)$ of the covering $p : Y \to X$ are preserved, that is, such g that $p \circ g = p$.

As we will see, the group $\mathrm{Deck}(Y/X)$ is isomorphic with the Galois group of th extension $\mathbf{L} \supset \mathbf{K}$, where $\mathbf{K} = \mathcal{M}(X)$ and, correspondingly, $\mathbf{L} = \mathcal{M}(Y)$ are the fields of meromorphic functions on X and Y, respectively. This group makes it possible to get deep insight into the structure of a Riemann surface.

A surface Y is a *Riemann surface* when it is a two dimensional manifold such that the transition mapping from one map to another are given by holomorphic functions, that is, convergent power series. The simplest Riemann surface (apart from the complex plane $\mathbb{C}$ with the atlas consisting of one map) is the so- called *Riemann sphere* $\bar{\mathbb{C}} := \mathbb{C} \cup \{\infty\}$ (which is homeomorphic with the two dimensional sphere $S^2 = \{x \in \mathbb{R}^3 : \|x\|^2 = 1\}$; one can easily find an atlas consisting of two maps. A holomorphic mapping

$f : Y \to \bar{\mathbb{C}}$ is (*ex defintione*) a meromorphic function. A pole of the function f is the point $y \in Y$ such that $f(y) = \infty$. Every meromorphic function is therefore a (holomorphic) covering of the Riemann sphere $f : Y \to \bar{\mathbb{C}}$. Riemann theory of algebraic functions is, as we will soon see, nothing but the theory of compact Riemann surfaces X.

Weyl group of compact Lie group G. A Lie group is a connected group being at the same time a differentiable manifold. Group actions are analytic. Such is, for example, the rotation group $\mathrm{SO}(\mathbb{R}^n)$, the Lorentz group, the group of Euclidean motions, the Poincaré group of special theory of relativity, and many others. (We will talk about Lie groups in the following chapters).

In any Lie group G, there exists a maximal (connected) abelian subgroup T, called a *maximal torus of the group* G. The group $T = T(G)$ is, in principle, unique because two maximal tori T_1 and T_2 are adjoint to each other, that is, there exists an inner automorphism $\sigma_g : G \to G$ (i.e., $\sigma_g x = gxg^{-1}$ for any $x \in G$) which transforms T_1 onto T_2. Moreover, *every point of a group belongs to some maximal torus* (this comprises the famous theorem of E. Cartan on *maximal tori*). And thus, the important structure for the group G is the pair $(G, T(G))$. Following the *Weyl principle* one should investigate automorphisms of the group G which do not change $T(G)$. Such automorphism group is called the *Weyl group* $W(G)$ *of the group* G and forms the basis of investigations of Lie groups. The group $W(G)$ is a finite group generated by some reflections. The discovery of the group $W(G)$ was a great achievement of Weyl which reminds the theory of platonic solids.

Let us consider the following example of Weyl group. Let $G = U(m)$ be the group of $m \times m$ unitary matrices, that is, the automorphisms of $\mathbb{C}^m$ preserving the natural scalar product in this space. The maximal torus $T = T(U(m))$ of the group $U(m)$ is the set of diagonal matrices

$$(*) \qquad \begin{bmatrix} e^{i\varphi_1} & & & & 0 \\ & \cdot & & & \\ & & \cdot & & \\ & & & \cdot & \\ 0 & & & & e^{i\varphi_m} \end{bmatrix}$$

and thus we have that $T = U(1) \times \cdots \times U(1)$, and indeed the torus T is a product of m circles, that is, the 'real' torus. The Weyl group of the group $\mathrm{U}(m)$ is therefore isomorphic with the group S_m of permutations of diagonal elements of the matrix (*).

The groups of Platonic solids. In the Euclidean space $\mathbb{R}^3$ there exists exactly five 'Platonic solids' that is, regular convex polyhedrons (inscribed) in the sphere S^3: *tetrahedron, hexahedron, octahedron, dodehedron, icosahedron* (the polyhedra with, respectively 4,6,8,12,and 20 faces). These polyhedra were known in antiquity, it is very interesting that the number of these regural figures is exactly five. The ancients, for example Platon in *Timaios* and Pythagoreans attached great importance to meditating these beautiful objects. There formed the basis of the philosophy of the nature of these days: they were associated with five elements: *fire, earth, air, cosmos (ether), water.* This doctrine (mediated by the great, wonderful *Comments to Euclid* by Proclos) played a fundamental role in Kepler's inspiration.

The subgroups of the group SO(3) of rotations of the Euclidean space preserving one of these five Platonic solids are called the *groups of motions of these solids.* These groups are finite. They are analogous to a Galois group of extension of fields or the group $\mathrm{Aut}(X)$ of automorphisms of a compact Riemann space X. The last sentence in the monumental work of Hermann Weyl *Die Idee der Riemannischen Fläsche* is the following: 'A closed Riemann surface of genus $p > 1$ admits only finite number of conformal transformations into itself', thus, the group $\mathrm{Aut}\, X$ is finite.

CHAPTER 10

Topology of Compact Lie Groups

H. Weyl started investigations of global properties of Lie groups in his classical work of 1925. These problems were noticed by E. Cartan, who proved the fundamental theorem on harmonicity of bi-invariant differential forms. This theorem in turn was an impulse of classical thesis of G. de Rham, which climaxed in the famous theorem on isomorphism of the cohomology groups $H^k(M,\mathbb{R})$ and de Rham groups $H^k_{dR}(M,\mathbb{R})$ for any smooth manifold M. The crowning achievement of this sequence is the famous Hodge – Kodaira theorem which says that on a compact manifold M in every class $[h] \in H^k_{dR}(M,\mathbb{R})$ there is exactly one harmonic form (i.e., such ω that $d\omega = d^*\omega = 0$. Therefore the de Rham group is isomorphic with the group of harmonic forms $H^k_{dR}(M,\mathbb{R}) \cong \mathcal{H}^k(M,\mathbb{R})$. Kodaira (and later de Rham and Bidal as well as A. Weil) showed also the famous theorem on harmonic decomposition:

$$L^{2,k}(M,\mathbb{R}) = \overline{d\Lambda^{k-1}(M,\mathbb{R})} \oplus \overline{d^*\Lambda^{k+1}(M,\mathbb{R})} \oplus \mathcal{H}^k(M,rz).$$

From Hodge – Kodaira theorem we have

COROLLARY 1 (CARTAN). *Let G be a compact Lie group with bi-invariant metric. Then a form ω is harmonic if and only if it is left- and right-invariant with respect to translations.*

Spheres and compact Lie groups. The simplest Lie group is S^1 (a circle). It can be seen that S^3 (three dimensional) can be equipped with structure of a compact Lie group[1]. What is the situation with even-dimensional spheres?

[1] It is sufficient to note that S^3 is homomorphic with the space of unit quaternions.

The answer is provided by

THEOREM 2 *The sphere* S^{2k}, $k = 1, 2, \ldots$ *is not a Lie group.*

This fact can be generalized and made more precise as follows

THEOREM 3 (HOPF) *Let G be a connected, compact Lie group. Then G has the same cohomologies as the Cartesian product of some odd-dimensional spheres whose dimension does not exceed the dimension of the manifold G:*

$$H^*(G, \mathbb{R}) \cong H^*\left(\bigtimes_j S^{2k_j - 1}, \mathbb{R}\right).$$

The particular case of classical groups was analysed in more details by R. Brauer. He found that

$$H^k(\mathrm{SU}(n), \mathbb{R}) = H^k(S^3, \mathbb{R}) \oplus H^k(S^5, \mathbb{R}) \oplus \ldots \oplus H^k(S^{2n-1}, \mathbb{R}),$$

$$H^k(\mathrm{Sp}(n), \mathbb{R}) = H^k(S^3, \mathbb{R}) \oplus H^k(S^7, \mathbb{R}) \oplus \ldots \oplus H^k(S^{4n-1}, \mathbb{R}).$$

These formulae were obtained by Brauer with the help of the theory of invariants. This is not surprising in view of the following theorem, which we already announced above.

THEOREM 4 (CARTAN–CHEVALLEY–EILENBURG). *For any connected, compact Lie group G*

$$H^k(G, \mathbb{R}) = \mathcal{H}^k(G) = \Lambda^k_{inv}(G), \quad k = 0, 1, 2, \ldots$$

Since left-invariant forms are defined by their values in unity of the group G, the vector spaces on the right hand side are in fact the spaces of G-invariant forms on the Lie algebra $\mathfrak{L}(G)$. Determination of these spaces is one of the fundamental problems of algebraic theory of invariants.

After this topological introduction, let us turn to our main topic.

Chapter 11

Representations of Compact Lie Groups

Compact Lie groups are 'perfect' entities in modern mathematics because:

(a) They are differentiable (and even analytical) manifolds of finite dimension and therefore can be analysed with the help of the most powerful tool of mathematics: ordinary and partial differential operators (equations);

(b) Denoting by $\mathfrak{L}(G)$ or $L(G) = T_e(G)$ the tangent space to G in unity e, we find that $L(G)$ can be identified with the set of left-invariant vector fields on G (or with the set of appropriate differential operators of order one). Thus $L(G)$ naturally becomes a Lie algebra: on $L(G)$ there exists a structure $[\cdot, \cdot]$ which satisfies the axioms of Lie algebra (see below);

(c) This rich algebraic structure, in turn, makes it possible to apply the powerful algebraic tools (theory of Lie algebras);

(d) An abstract (finite dimensional) Lie algebra $\mathfrak{g}$ defines (up to an isomorphism) a simply connected Lie group G such that $\mathfrak{L}(G) = \mathfrak{g}$. Therefore investigations of the Lie group G can be to large extent replaced by investigations of its Lie algebra $\mathfrak{g}$;

(e) Compactness of a Lie group G makes it possible to use the theory of representations of compact groups developed by Weyl and Peter. This theory is extraordinarily beautiful and simple: it reduces to large extend to the spectral theory of compact operators;

(f) And finally every complex, compact (connected) commutative Lie group is a torus. Moreover, as it was shown by E. Cartan, *every compact Lie group contains a maximal torus T (any other such a torus is conjugated $T_1 = gTg^{-1}$); and any point g_0 of the group G belongs to some maximal torus: $G = \bigcup_g gTg^{-1}$.* This fundamental Cartan theorem makes it possible to apply to T (and thus to the whole of G) the theory of representations of abelian groups: every representation of the torus T decomposes into the finite (and orthogonal) sum of unit representations (over $\mathbb{C}$), the so called *weights.* Thus the knowledge of these weights is the fundamental element of the theory of representations of compact Lie groups.

From these six elements H. Weyl was able to build a structure of extraordinary harmony and beauty, and he did not stop! He used the theory of semisimple Lie algebras created by W. Killing and perfected by E. Cartan, to crown his theory with the most beautiful chord: the theory of semisimple, compact Lie groups. Moreover using his brilliant invention, which he modestly called the 'unitary trick', H. Weyl obtained important theorems on representations of arbitrary (also non-compact) semisimple Lie groups.

In Weyl's considerations the fundamental role is played by the finite group $W(G) := N(T)/T$ (where $N(T) := \{g \in G; gTg^{-1} = T\}$ is a normalizator of T in G) nowadays called the *Weyl group.* This beautiful object made it possible for Weyl to geometrize a large part of the theory of representations of groups and the theory of semisimple Lie algebras. The Weyl group $W(G)$ acts on the Lie algebra $\mathfrak{L}(G)$ an on the Lie algebra $\mathfrak{L}(T)$ of the maximal torus T of the Lie group G in a natural way. The algebra $\mathfrak{L}(T)$ is called the *Cartan algebra* for the algebra $\mathfrak{L}(G)$. The Weyl group $W(G)$ is generated by the (finite) set of reflections with respect to hyperplanes contained in $\mathfrak{L}(G)$, s_j. Thanks to the fact that $\mathfrak{L}(G)$ is semisimple, one can define on it a natural scalar product $(\cdot,\cdot)$ called the Killing form, which gives the space $T_eG = \mathfrak{L}(G)$ the structure of Euclidean space. the Weyl group is therefore a group of symmetries of $\mathfrak{L}(G)$ generated by orthogonal reflections with respect to some hyperplanes related to weights. These hyperplanes decompose $\mathfrak{L}(G)$ into convex cones (with vertices at the point 0) called *Weyl chambers.* The Weyl group permutes Weyl chambers, that is it acts transitively and one-to-one on the set of Weyl chambers. Groups generated by reflections play a very important role in crystallography, and the theory of such groups is tied for ever with the name of H.M.S. Coxeter, who investigated them using very clear and suggestive method: Coxeter graphs and diagrams. Coxeter diagrams were slightly improved by Dynkin (see below), and they are called *Dynkin diagrams*, even though the credits should go to Coxeter!

And thus the theory of representations of compact Lie groups turned back to the starting point: the theory of symmetries , theory of crystals and Platonic solids (regular convex polyhedra). It is impossible not to feel a deep satisfaction realizing the royal way of great ideas from Plato to Weyl, the ideas that infiltrate the whole of mathematics and large areas of physics (in mathematics, for example, the theory of Riemann surfaces and abelian varieties; in physics – quantum theory.)

After this overview of Weyl's theory let us turn to more detailed discussion.

Lie algebra $\mathfrak{L}(G)$ of Lie group G. One-parameter subgroup of a Lie group G is a (differentiable) homomorphism $\xi : \mathbb{R} \to G$ of real axis (which is the simplest Lie group) into the group G. It can be easily checked that right shifts by the element $\xi(t)$, $t \in \mathbb{R}$ define left-invariant (that is, commuting with left translations) action of the group $\mathbb{R}$ on the manifold G. And vice versa, every left-invariant action $G \times \mathbb{R} \to G$ defines a one-parameter subgroup by the formula $\xi(t) = e \cdot t$. Moreover, to one-parameter subgroup $\xi : \mathbb{R} \to G$ corresponds the tangent vector $\xi_* \left(\frac{\mathrm{d}}{\mathrm{dt}}\right)_0 \in T_eG$. Vice versa, every tangent vector in T_eG defines a left-invariant vector field X on G. The integral curve γ_X of the field X going through the identity $e \in G$ is defined in a neighbourhood of the point $0 \in \mathbb{R}$ and it follows from uniqueness of the initial problem for ordinary differential equations that $\gamma_X(t+s) = \gamma_X(t)\gamma_X(s)$ for small s and t. It can be shown that γ_X uniquely extends to the one-parameter subgroup $\xi_X : \mathbb{R} \to G$.

We have therefore the following bijections of four objects:

$$\text{left-invariant action of } \mathbb{R} \text{ on } G$$
$$\updownarrow$$
$$\text{one-parameter subgroup in } G$$
$$\updownarrow$$
$$\{T_eG = \text{tangent vectors at } e \in G\}$$
$$\updownarrow$$
$$\{X_l(G) = \text{left-invariant vector fields on } G\}.$$

Each of these objects will be called *Lie algebra* of the group G. Lie algebra of the group G is denoted by $\mathfrak{L}(G)$. As a set of vector fields, Lie algebra

$\mathfrak{L}(G)$ is equipped with the Lie bracket $[X, Y] \in \mathfrak{L}(G)$ for every $X, Y \in= \mathfrak{l}(G)$ which satisfies the Jacobi identity

$$[[X,Y],Z]+[[Y,Z],X]+[[Z,X],Y]=0.$$

One defines the exponential mapping $\exp : \mathfrak{L}(G) \to G$ which maps every vector $X(e)$ to the point $\xi_X(1) \in G$ on the one-parameter subgroup ξ_X, that is, the point 'to which we arrive after unit of time.'

If $h : G_1 \to G_2$ is a homomorphism of Lie groups, then the mapping tangent to h at e (denoted by $T_e h$) maps the algebra $\mathfrak{L}(G_1)$ into $\mathfrak{L}(G_2)$; the diagram

$$\begin{array}{ccc} (T_eG_1) = \mathfrak{L}(G_1) & \xrightarrow{\;T_e h\;} & \mathfrak{L}(G_2) = (T_eG_2) \\ \Big\downarrow \exp & & \Big\downarrow \exp \\ G_1 & \xrightarrow{\;h\;} & G_2 \end{array}$$

is then commutative.

The most important group automorphisms are defined by elements $g \in G$ of the group itself; these are called inner automorphisms $a(g) : G \to G$, $a(g)x := gxg^{-1}$. Denoting the mapping tangent to $a(g)$ by $T_e a(g) : \mathfrak{L}(G) \to \mathfrak{L}(G)$, we have

DEFINITION 1. $\mathrm{Ad} : G \to \mathrm{Aut}(\mathfrak{L}(G))$ given by the formula $\mathrm{Ad}(g) = T_e a(g)$ is called the *adjoint representation* of the group G. A mapping tangent to Ad

$$\mathrm{Ad} : \mathfrak{L}(G) \to \mathfrak{L}\mathrm{Aut}(\mathfrak{L}(G)) = \mathrm{End}(\mathfrak{L}(G))$$

is a representation of the Lie algebra $\mathfrak{L}(G)$ of the group G in (additive) group of endomorphisms of this algebra (that is, in the general linear group associated with the vector space $\mathfrak{L}(G)$.)

Thus we avoided the unhandy notation

$$\mathrm{Ad} : \mathfrak{L}(G) \to \mathfrak{L}\mathrm{GL}(\mathfrak{L}(G)).$$

One can easily check the following

THEOREM 2.

(i) *A covariant tensor field* ψ *on* G *is bi-invariant if and only if*

$$\operatorname{Ad} g^* \psi_e = \psi_e$$

for all $g \in G$;

(ii) *A Lie group* G *has a bi-invariant volume form* vol *if*

$$|\det \operatorname{Ad}(g)| = 1$$

for all $g \in G$. *Such group is called em unimodular; a left-invariant Hurwitz integral (or measure) on such a group is right-invariant as well:*

$$\int_G f(g^{-1}x)d\mu(x) = \int_G f(xg)d\mu(x), \quad g \in G.$$

Let us introduce the following fundamental notion

DEFINITION 3. Let $\mathfrak{g}$ be the Lie algebra of the Lie group G. Then the bilinear form $\mathfrak{K}$ on $\mathfrak{g} \times \mathfrak{g}$

$$\mathfrak{K}(X,Y) = \operatorname{tr}(\operatorname{ad} X \operatorname{ad} Y)$$

is called the *Killing form* on $\mathfrak{g}$.

REMARK. The notion of Killing form can be introduced as it was done above for an abstract Lie algebra $(\mathfrak{g}, [\cdot,\cdot])$, defining the mapping $\operatorname{ad}_g : \mathfrak{g} \to \operatorname{End}(\mathfrak{g})$ by the formula

$$X \mapsto [X, *]$$

(also called the *adjoint mapping* of the algebra $\mathfrak{g}$.) The Killing form is an invariant. For $\phi \in \operatorname{Aut}(\mathfrak{g})$

$$\mathfrak{K}(\phi X, \phi Y) = \mathfrak{K}(X,Y)$$

holds.

DEFINITION 4. An (abstract) Lie algebra is *semisimple* if its Killing form is non-degenerate. A Lie group G is *semisimple* if its Lie algebra is semisimple.

It can be shown that for compact semisimple Lie group (over $\mathbb{C}$ or $\mathbb{R}$) the Killing form on $\mathfrak{L}(G)$ is (negative) definite, and thus one can define an inner product.

We have an important criterion for unimodularity.

THEOREM 5. *From each of the conditions above it follows that the Lie group G is unimodular*

(i) The set $\mathrm{Ad}(G)$ *is compact;*

(ii) The group G is compact;

(iii) The group G is abelian;

(iv) The group G is semisimple.

PROOF. (i) The group $\{|\det \mathrm{Ad}(g)|; g \in G\}$ is a compact subgroup of $\mathbb{R}_+$ ($\mathbb{R}_+$ is the multiplicative group of positive real numbers). Such subgroup consists of a single element 1, thus G is unimodular.

(ii) follows from (i).

(iii) is obvious since left invariance is in this case right invariance.

(iv) Every mapping Ad() preserves a non-degenerate Killing form $\mathfrak{H}$, and thus $(\det \mathrm{Ad}(g))^2 = 1$, $g \in G$.

Historical remarks. A brilliant and so simple idea of Hurwitz of introducing left (right) invariant volume forms by choosing an arbitrary n form ω_e at the point $e \in G$ and shifting it from the left with the help of $l_g x := gx$, $g \in G$ (and from the right with the help of $r_g x := xg$, $g \in G$) was of fundamental importance for investigations of Lie groups and theory of invariants. Hurwitz introduced his integral for the purpose of invariants theory in 1897!

If the group G is compact, then we normalize the bi-invariant Hurwitz–Haar measure dg such, as to satisfy

$$\int_G 1\, dg = 1.$$

Let us show a simple, but important application of the Hurwitz 'averaging method.'

THEOREM 6. *Let $U : G \to \mathrm{GL}(H)$ be a (continuous) representation of the compact group G in the Hilbert space $(H, (\cdot,\cdot))$. Then in H one can define a new inner product $(\cdot,\cdot)_1$ such that*

(i) *Operators* $U(g)$ *are unitary (that is, they preserve* $(\cdot,\cdot)_1$*);*

(ii) *The norms* $\|\cdot\|$ *and* $\|\cdot\|_1$ *are equivalent.*

In particular, every representation of a compact Lie group in Hilbert space is equivalent to unitary representation.

PROOF. The expression

$$(h|h')_1 := \int_G (U(g)h|U(g)h')\,dg$$

satisfies all the axioms of scalar product. For example, if $(h|h')_1 = 0$, then the non-negative continuous function $g \mapsto (U(g)h|U(g)h')$ must vanish identically. Taking $g = e$, we have $(h|h') = 0$ and thus $h = 0$.

Property (i) follows from invariance of dg. Indeed

$$(U(g_1)h|U(g_1)h')_1 = \int_G (U(g_1g)h|U(g_1g)h')\,dg =$$

$$= \int_G (U(g)h|U(g)h')\,dg = (h|h')_1.$$

In order to show (ii), let us first observe that for $c = \sup_{g\in G} \| U(g) \|$ the inequality $\| h \|_1^2 \leqslant c^2 \| h \|^2$ holds. Integrating over G the inequality $\| h \|^2 = \| U(g^{-1})U(g)h \|^2 \leqslant c^2 \| U(g)h \|^2$, we find $\| h \|^2 \leqslant c^2 \| h \|_1^2$.

In what follows we will assume that the representation (U, H) is unitary. Let us present a number of simple definitions and facts.

THEOREM 7 (WEYL, MASCHKE).

(i) *Every unitary representation of a compact Lie group* G *is semisimple that is, if* $H_1 \subset H$ *is a* U*-invariant space,* $U(g)H_1 \subset H_1$*, then the orthogonal completion* $H_2 = H_1^\perp$ *is also an invariant subspace, and thus the representation* (U, H) *decomposes into the simple (orthogonal) sum* $H_1 \oplus H_2$*;*

(ii) *Representation of a compact Lie group is semisimple.*

DEFINITION 8. Let V be a vector space and $S \subset \mathrm{Hom}(V, V)$. The set S is called

(i) *irreducible* if V does not possess a proper S-invariant subspace ;

(ii) *semisimple* if every S-invariant subspace possesses an S-invariant orthogonal complement;

(iii) *nilpotent* if $\dim V < \infty$ and V has a basis such that every element of S has the matrix (in this basis) of the form

$$\begin{bmatrix} 0 & & * \\ & \ddots & \\ 0 & & 0 \end{bmatrix}$$

(zeros on the diagonal and below, arbitrary elements over the diagonal);

(iv) *unipotent* if the set $\{s - I : s \in S\}$ is nilpotent

We say that the linear mapping is *semisimple, nilpotent, and unipotent* if the one-element set $S = \{s\}$ has one of these properties.

Similarly the representation (U, V) of the group G is called *irreducible, semisimple, nilpotent, and unipotent* if the set $\{U(g); g \in G\}$ possesses the respective property.

The following famous Weyl theorem is the climax of Weyl's theory of representations of Lie groups.

THEOREM 9 (WEYL THEOREM). *Every finite dimensional representation of semisimple Lie group is semisimple.*

PROOF. The proof of this deep theorem proceeds through the following steps.

(i) Every finite dimensional representation of a compact Lie algebra K is (as we know) semisimple;

(ii) The same holds for the algebra $\mathfrak{k} := \mathfrak{L}(K)$;

(iii) The same holds for the complexification $\mathfrak{k}_{\mathbb{C}} = \mathfrak{k} + i\mathfrak{k}$ of the algebra $\mathfrak{k}$;

(iv) This holds for every real Lie algebra $\mathfrak{g}$ such that $\mathfrak{g}_{\mathbb{C}} = \mathfrak{k}_{\mathbb{C}}$;

(v) 'Unitary trick': if $\mathfrak{g}$ is a real semisimple Lie algebra, then there exists a compact group K such that $\mathfrak{L}(K)_{\mathbb{C}} = \mathfrak{g}_{\mathbb{C}} = \mathfrak{k}_{\mathbb{C}}$.

Of course the point (v) is most difficult.

A *character* of (finite dimensional) representation U of the group G in the space H over field $\mathbb{K}$ is the function χ_U given by the formula

$$G \ni g \mapsto \chi_U(g) = \operatorname{tr} U(g) \in \mathbb{K},$$

where $\operatorname{tr} A$ denotes trace of the operator A. The following properties of characters follow immediately from the corresponding properties of a trace.

(i) Character is a constant function on classes of equivalent elements, that is

$$\chi_U(g_1 g g_1^{-1}) = \operatorname{tr}(U(g_1)U(g)U(g_1)^{-1}) = \operatorname{tr}(U(g)) = \chi_U(g);$$

(ii) $\chi_{U_1+U_2}(g) = \chi_{U_1}(g) + \chi_{U_2}(g)$;

(iii) $\chi_{U_1 \otimes U_2}(g) = \chi_{U_1}(g) \cdot \chi_{U_2}(g)$;

(iv) $\chi_{U^*}(g) = \chi_U(g^{-1})$.

Let us recall that if (e_j) is an orthonormal basis of the Hilbert space $(H, (\cdot|\cdot))$ and A is a linear and bounded operator such that the series

$$\operatorname{tr} A := \sum_j (Ae_j|e_j)$$

is absolutely convergent, then we say that A is of *trace class*. The number $\operatorname{tr} A$ is called the *trace of the operator*. In a finite dimensional space every operator is of trace class. The operator P_V of orthogonal projection on the subspace $V \subset H$ is of trace class if $\dim V < \infty$ and $\operatorname{tr} P_V = \dim V$.

If $A = A^*$ is a *Hilbert–Schmidt operator*, then the operator AA^* is of trace class because, by definition, $A = \sum_j \lambda_j P_j$, where $\sum_j |\lambda_j|^2 < \infty$, λ_j are eigenvalues of the operator A, and P_j are projectors on the eigensubspace corresponding to the eigenvalue λ_j:

$$Ah_j = \lambda_j h_j, \quad h_j \in P_j H.$$

The space of Hilbert–Schmidt operators possesses a structure of Hilbert space with the norm $\| A \|_{H-S}^2 := \sum_j |\lambda_j|^2$ and the inner product

$$(A, B)_{H-S} := \sum_j (Ae_j|Be_j).$$

On a compact Lie group G any integral operator with continuous kernel $a : G \times G \to \mathbb{R}$ is a Hilbert–Schmidt operator because the kernel is square integrable: $\int_{G\times G} |a(x,y)|^2 dxdy < \infty$, and this is a sufficient and necessary condition for an operator to be Hilbert–Schmidt.

It is easy to prove

PROPOSITION 10. *Let (U,H) be a unitary representation. Then $P_{H(G)} = \int_G (Ug)dg$ is a Hilbert–Schmidt operator and is an orthogonal projection on the subspace $H_0(G)$ consisting of vectors constant with respect to G:*

$$(h \in H_0(G)) \iff (U(g)h = h \text{ for every } g \in G).$$

From this and the commutation of the trace operation and integration we obtain

THEOREM 11 (SCHUR–WEYL). *If G is a compact group, then*

(i) every irreducible representation (U,H) of the group G is finite dimensional;

(ii) if (U_j, H_j), $j = 1,2$ are finite dimensional, unitary representations of the group G, then

$$\int_G \chi_{U_1} = dim_{\mathbb{C}} H_1(G),$$

where $H_1(G)$ is a subspace in H_1 consisting of all elements preserved by every mapping $U_1(G)$, $g \in G$;

(iii) $\int_G \overline{\chi_{U_1}} \chi_{U_2} = \dim \mathrm{Hom}_G(U_1, U_2)$ *of the space of* intertwining operators *, that is the operators $T : H_1 \to H_2$ such that*

$$TU_1(G) = U_2(G)T$$

for every $g \in G$;

(iv) if the representations U_1 and U_2 are irreducible, then

$$\int_G \overline{\chi_{U_1}} \chi_{U_2} = \begin{cases} 1, & \text{if } U_1 \cong U_2 \\ 0, & \text{if } U_1 \text{ is not equivalent to } U_2 \end{cases}$$

REMARK. Relation (iv) is called the *Schur–Weyl orthogonality relation.* Schur proved it for the group $O(n)$, and Weyl in the general case.

The climax of the theory of arbitrary compact groups (not necessarily Lie groups) is the famous

THEOREM 12 (PETER–WEYL). *Characters of irreducible representations form a complete orthonormal basis in the space $L^2(G)$.*

This can be phrased in the most elegant way as follows. Let $\hat{G}$ be a set of equivalence classes of irreducible, unitary representations of the group G; as we know $\hat{G}$ is parameterized by characters of irreducible representations $U_j : G \to \operatorname{Aut}(H_j)$. For every $j \in \hat{G}$ we have the alternative formulation

THEOREM 13 (PETER–WEYL). *The isomorphism*

$$L^2(G) \cong \bigoplus_{j\in\hat{G}} H_j \otimes H_j^* \cong \bigoplus_{j\in\hat{G}} \operatorname{Hom}(H_j, H_j)$$

holds.

The above isomorphism takes into account the action of the group G defined by translations: left action $l_g f(x) := f(g^{-1}x)$ corresponds to the action of U_j on H_j, and right action $r_g f(x) := f(xg)$ corresponds to the action of U_j on H_j^*.

This isomorphism can be also defined directly. For any $j \in \hat{G}$, let us consider the mapping $u_j : L^2(G) \to \operatorname{Hom}(H_j, H_j)$ given by the formula

$$U_j(f) = \int_G f(g)U_j(g)dg, \quad f \in L^2(G).$$

Then the mapping $f \mapsto (U_j(f))_{j\in\hat{G}}$ is a linear isomorphism, which becomes a natural isomorphism of Hilbert space if we renormalize the natural inner product on $H_j \otimes H_j^*$ by multiplying it by $\dim H_j$.

REMARK 1. It is interesting that in the Peter–Weyl theorem representatives of all equivalence classes of irreducible unitary representations appear, and thus all (irreducible) characters do.

REMARK 2. Theorem of Peter–Weyl opened a new chapter of mathematics – the abstract harmonic analysis (called also the general harmonic analysis.) In the case $G = T^1$ (and also if $G = T^1 \times T^1 \times \cdots \times T^1$) characters are exponential functions and the Peter–Weyl theorem is a fundamental theorem of classical theory of Fourier series.

Harmonic analysis on homogeneous spaces $X = G/K$ which started with works of E. Cartan and H. Weyl is related to the wonderful Peter–Weyl theorem in the following way.

Assume that the group G acts transitively on the space X. We can identify X with G/K, where $K = G_{x_0}$ is the isotropy group (stabilizer) of some point $x_0 \in X$, that is, $K = \{g \in G : g(x_0) = x_0\}$. Then we can identify $L^2(K)$ with the subspace in $L^2(G)$ consisting of functions which are preserved by (right) translations.

THEOREM 14 (WEYL).

(i) The isomorphism $L^2(X) \cong \bigoplus_{j\in\hat{G}} H_j \otimes (H_j^*)^K$ *holds. In this formula*

$$(H_j^*)^K = \{v^* \in H_j^* : v^* = U_j^*(k)v^*, \, k \in K\}$$

is the subspace of all K*-invariant vectors in* H_j^*.

(ii) G*-invariant subspaces can be characterized as the subspaces of the form*

$$\bigoplus_{j\in\hat{G}} H_j \otimes V_j,$$

where V_j *is, for every* j*, a subspace in* $(H_j^*)^K$.

If G is a Lie group, then $X = G/K$ is a smooth manifold and on X one can consider G-invariant linear differential operators D (for example, the Laplace operator on the sphere $S^2 = SO(3)/SO(2)$.) By taking the closure of such an operator we obtain a dense operator (in general not bounded), which we will denote by $D : L^2(X) \to L^2(X)$. The kernel of this operator $\ker D$ is an invariant space (in example above this is the space of *spherical harmonics.*) We find the isomorphism

$$\ker D = \bigoplus_{j\in\hat{G}} H_j \otimes V_j,$$

where $V_j = \ker D_j$ is a kernel of the mapping $D_j : (H_j^*)^K \to (H_j^*)^K$. this procedure extends to invariant systems of operators and differential equations.

As it is well known, the linear operators of classical analysis and physics are nothing but G-invariant operators for the corresponding groups and homogeneous spaces. The theory of such operators is, without doubts, the most important chapter of mathematical physics. Sometimes one must go beyond the theory of compact and abelian groups. This general circumstances are not totally under control yet, but in recent years a big progress has been done (see that is, the beautiful talk of W. Schmid on the Helsinki Congress of 1978.)

Let us turn to another results of the theory of representations of compact Lie groups. The major achievement of this theory after Weyl is, without doubts

The Borel–Weil–Bott–Konstant theorem. Let us start with some introductory remarks. It is not difficult to show that for a maximal torus T the space G/T possesses a (natural) structure of complex manifold on which G acts (through the left translations) as a group of holomorphic mappings. Moreover, any one dimensional representation $\sigma : T \to \mathbb{C}^*$ defines a homogeneous holomorphic line bundle $\mathcal{L}_\sigma \to G/T$. Since G acts on the bundle $\mathcal{L}_\sigma$, it acts on the cohomology groups $H^k(G/T, \mathcal{A}(\mathcal{L}_\sigma))$ of the sheaf of germs of holomorphic sections. According to the Hodge–Kodaira–Serre theory, any such group can be identified with the kernel of the G- invariant Laplace–Beltrami operator Δ on G/T which acts on the $(0, k)$ forms with coefficients in the bundle $\mathcal{L}_\sigma$. (Elements of this kernel are called harmonic $(0, k)$ forms with coefficients in the bundle $\mathcal{L}_\sigma$.) We encounter therefore the situation as in the case of G-invariant differential operators. The precise description of these cohomology groups is given by

THEOREM 15 (BOREL–WEIL–BOTT–KONSTANT).

(i) There exists an irreducible representation of the group G in the space $H^0(G/T, \mathcal{A}(\mathcal{L}_\sigma))$ (the space of holomorphic sections of the bundle $\mathcal{L}_\sigma$.) This is the representation U^σ induced by $\sigma \in \hat{T}$ corresponding to highest weight σ;[1]

(ii) For special σ all the spaces $H^k(G/T, \mathcal{A}(\mathcal{L}_\sigma))$ $k = 1, 2, ldots$ are zero. For the remaining σ there exist one and only one number $k = k(\sigma)$ such that $H^{k(\sigma)}(G/T, \mathcal{A}(\mathcal{L}_\sigma)) \neq \{0\}$. Moreover the representation of

[1]Definition of the representation of group G induced by representation σ of some subgroup $\Gamma \subset G$ will be given in the next chapter.

G in this space is irreducible.

(iii) All irreducible representations of the group G have the form as described in (ii).

REMARK. (a) Point (i) is usually called the Borel–Weil theorem. (b) The theorem above shows deep relation between complex analysis and irreducible representations of groups.

In order to explain in more details the relations between weights $\sigma \in \hat{T}$, we must turn back to the theory of Lie groups and Lie algebras. This is the subject of the next chapter.

CHAPTER 12

Nilpotent, Semimple, and Solvable Lie Algebras

Nilpotent and semisimple Lie algebras were introduced by W. Killing in 1888. Solvable (or integrable) algebras were introduced earlier by Lie in connection with his research on (algebraic) differential equations. Lie wanted to establish a theory for continuous transformation groups (of the space $\mathbb{R}^n$) which would be analogous to the Galois theory (and in the latter theory the decisive role is played by finite solvable groups.)

An endomorphism $N \in \mathrm{Hom}(V, V)$ of linear space V is nilpotent if some of its powers vanish $N^m = 0$; an algebra $\mathfrak{L}$ is *nilpotent* if $\mathfrak{L}^n = 0$ for some n ($\mathfrak{L}^0 = 1$, $\mathfrak{L}^1 = [\mathfrak{L}, \mathfrak{L}]$, ..., $\mathfrak{L}^j = [\mathfrak{L}, \mathfrak{L}^{j-1}]$.)

PROPOSITION 1 *Let $\mathfrak{L}$ be a Lie algebra. Then*

(i) ($\mathfrak{L}$ is nilpotent) $\Longrightarrow$(subalgebras of $\mathfrak{L}$ and homomorphic images of $\mathfrak{L}$ are nilpotent);

(ii) ($\mathfrak{L}/Z_{\mathfrak{L}}$ is nilpotent) $\Longrightarrow$($\mathfrak{L}$ is nilpotent), where $Z_{\mathfrak{L}}$ is the center of $\mathfrak{L}$, that is $\mathbb{Z}_{\mathfrak{L}} = \{z \in \mathfrak{L} : [x, z] = 0,\ x \in \mathfrak{L}\}$;

(iii) ($\mathfrak{L}$ is nilpotent) $\Longrightarrow$($Z_{\mathfrak{L}} \neq 0$);

(iv) ($\mathfrak{L}$ is nilpotent) $\Longrightarrow$($\mathrm{Ad}\, x$ is nilpotent for any $x \in \mathfrak{L}$).

Implication (iv) can be inverted.

THEOREM 2 (ENGEL). *If $\mathrm{Ad}\, x$ is nilpotent for any $x \in \mathfrak{L}$, then $\mathfrak{L}$ is nilpotent.*

There exists another, slightly different formulation of Engel theorem. Let V be a linear space of dimension n. The following sequence of subspaces

$$F = (V_{i_1} \subset V_{i_2} \subset \ldots \subset V_{i_s}), \quad \dim V_{i_j} = i_j$$

is called the *flag*.

EXAMPLE. Let F be a flag in V. The family of endomorphisms

$$\mathfrak{m}(F) = \{x \in \operatorname{End} V : x(V_{i_j} \subset V_{i_{j-1}}, \text{ for } j > 1\}$$

is an nilpotent subalgebra in the Lie algebra $\mathfrak{gl}(V) = \operatorname{Hom}(V, V)$. The matrices of endomorphisms from $\mathfrak{m}(F)$ are (in the appropriate basis) triangular matrices with zeros on the main diagonal; thus the algebra $\mathfrak{m}(F)$ is nilpotent.

THEOREM 3 (ENGEL). *If $\rho : \mathfrak{L} \to \operatorname{End} V$ is a representation of the algebra $\mathfrak{L}$ such that there exists a complete flag F in $V_i \subset V_j$ for every $x \in \mathfrak{L}$, then the algebra $\rho(\mathfrak{L})$ is nilpotent.*

We have the following generalization of Engel theorem.

THEOREM 4. *Let ρ be a finite dimensional representation of the algebra $\mathfrak{L}$ such that $\rho(x)$ is nilpotent for any $x \in \mathfrak{L}$. Then*

(i) $\rho(\mathfrak{L})$ is a nilpotent algebra;

(ii) $\mathfrak{L}/\ker\rho$ is a nilpotent algebra.

A class larger than nilpotent algebras is formed by

Solvable algebras. A sequence of derivations $\mathfrak{L}^{(k)}$, $k = 0, 1, 2, \ldots$ of the algebra $\mathfrak{L}$ is defined as follows

$$\mathfrak{L}^{(0)} := \mathfrak{L};\ \mathfrak{L}^{(1)} := [\mathfrak{L}^{(0)}, \mathfrak{L}^{(0)}];\ \ldots,\ \mathfrak{L}^{(k+1)} := [\mathfrak{L}^{(k)}, \mathfrak{L}^{(k)}].$$

The algebra $\mathfrak{L}$ is solvable if $\mathfrak{L}^{(k)} = 0$ for some k.

COROLLARY 5. *If $\mathfrak{L}$ is nilpotent, then $\mathfrak{L}$ solvable.*

PROPOSITION 6. *Let $\mathfrak{L}$ be an algebra. Then*

(i) ($\mathfrak{L}$ is solvable)$\Longrightarrow$(all the subalgebras and homomorphic images of the algebra $\mathfrak{L}$ are solvable);

(ii) (there exists a solvable ideal $\mathfrak{I} \subset \mathfrak{L}$ such that $\mathfrak{L}/\mathfrak{I}$ is solvable)$\Longrightarrow$($\mathfrak{L}$ is solvable);

(iii) (ideals $\mathfrak{I}_1$ and $\mathfrak{I}_2$ are solvable in $\mathfrak{L}$)$\Longrightarrow$(the ideal $\mathfrak{I}_1 + \mathfrak{I}_2$ is solvable).

COROLLARY 7. *In $\mathfrak{L}$ exists a maximal solvable ideal. It is called the radical of the algebra $\mathfrak{L}$ and is denoted by* Rad $\mathfrak{L}$.

DEFINITION 8. If $\mathfrak{L} \neq 0$ and Rad $\mathfrak{L} = 0$, we say that $\mathfrak{L}$ is semisimple (it does not contain solvable ideals).

COROLLARY 9. *If $\mathfrak{L}$ is not solvable, that is if $\mathfrak{L} \neq$ Rad $\mathfrak{L}$, then $\mathfrak{L}/$ Rad $\mathfrak{L}$ is semisimple.*

Now we present a number of theorems due to Lie which concern solvable algebras over $\mathbb{C}$. S. Lie called solvable algebras (and groups) 'integrable' as they play an important role in his theory of differential equations.

THEOREM 10 (LIE). *$\mathfrak{L}$ is solvable if and only if $[\mathfrak{L}, \mathfrak{L}]$ is nilpotent.*

THEOREM 11 (LIE). *Let $\mathfrak{L}$ be a solvable Lie algebra over $\mathbb{C}$ (or other algebraically closed field of characteristic 0.) Let $\rho : \mathfrak{L} \to \mathrm{Hom}(V, V)$ be a representation in finite dimensional space V over $\mathbb{C}$. Then all the matrices $\rho(x)$, $x \in \mathfrak{L}$ have a non-zero eigenvector v_1: $\rho(x)v_1 = \lambda(x)v_1$, $v_1 \neq 0$.*

COROLLARY 12 (LIE). *Assumptions as above. Then there exists a basis $v_1, v_2, \ldots, v_n$ of the space V such that all the endomorphisms $\rho(x)$, $x \in \mathfrak{L}$ are expressed as some upper triangular matrices, and thus $\mathfrak{L}$ stabilizes some flag in V.*

PROOF (based on Lie theorem). Let v_1 be as in Theorem 11. The space V_1 span by v_1 is ρ-invariant and ρ defines a representation ρ_1 of the algebra $\mathfrak{L}$ in the quotient space V/V_1. If $\dim V/V_1 \neq 0$ we apply the theorem to ρ_1 and obtain $v_2 \notin V_1$ such that $(v_2 + V_1 \in V/V_1$ is an eigenvector for all $\rho_1(x)$, $x \in \mathfrak{L}$. Following this route, we obtain the required basis $v_1, v_2, \ldots, v_n$.

This corollary makes it possible to construct ('general') solvable algebras. The following decomposition theorem holds.

THEOREM 13 (LEVI, 1905). *Every Lie algebra $\mathfrak{L}$ (over field of characteristic* 0*) can be decomposed into direct sum*

$$\mathfrak{L} = \operatorname{Rad}\mathfrak{L} + \mathfrak{s}$$

Let us observe that the algebra $\mathfrak{s}$ is isomorphic to the algebra $\mathfrak{L}/\operatorname{Rad}\mathfrak{L}$, and thus is semisimple. $\mathfrak{s}$ is called the *Levi algebra associated with* $\mathfrak{L}$..

REMARKS.

(a) The algebra $\mathfrak{s}$ in Levi decomposition is not uniquely defined; any two Lie algebras $\mathfrak{s}$ and $\mathfrak{s}'$ are conjugated in $\mathfrak{L}$ with respect to the adjoint group (that is $\mathfrak{s}' = \exp(\operatorname{ad} z)\mathfrak{s}$, where z runs through the maximal nilpotent ideal in $\mathfrak{L}$). Levi subalgebras are the maximal semisimple subalgebras of $\mathfrak{L}$.

(b) Levi decomposition is sometimes called Levi–Malcev or Levi–Whitehead decomposition.

(c) For groups we have the following analogue of Levi decomposition.

THEOREM 14. *If G is a connected and simply connected Lie group, then in G there exist* closed *simply connected Lie subgroups R and S such that R is the maximal solvable normal subgroup (normal divisor, invariant subgroup), and S is semisimple. Moreover $R \cap S = \{e\}$ and the mapping*

$$R \times S \ni (r, s) \mapsto rs \in G$$

is an analytic isomorphism. The decomposition $G = RS$ (or *$G = SR$) is also called the Levi decomposition.*

HISTORICAL REMARK. E.E. Levi (1883-1917) was a brilliant Italian mathematician (from Genoa), to whom we owe, apart from theorems on Lie groups and algebras, the construction of parametrix (that is the approximate fundamental solution) for elliptic equations, which was done independently of Hilbert (1907). The famous Levi problem (1911) was to find characterizations of domains of holomorphy in $\mathbb{C}^n$ in terms of the so called pseudo convectness (local property of the boundary.) This problem was solved by Bremermann, Norguet and Oka (1953/4) and, for arbitrary Stein manifolds, by Grauert (1959) (cf. Part Riemann and Complex Analysis.)

Reductive algebras. We say that the algebra $\mathfrak{L}$ is *reductive* if $\operatorname{Rad}\mathfrak{L} = Z_{\mathfrak{L}}$ (radical coincides with center of the algebra.)

PROPOSITION 15. *$\mathfrak{L}$ is reductive if and only if its adjoint representation $ad_{\mathfrak{L}}$ is reducible.*

If $\mathfrak{L}$ is reductive, then $\mathfrak{L}^{(1)} = [\mathfrak{L}, \mathfrak{L}]$ is semisimple and $\mathfrak{L} = Z_{\mathfrak{L}} + \mathfrak{L}^{(1)}$. Thus reductive algebras are direct sums of abelian and semisimple algebras. We say that the Lie group G is reductive if the algebra $\mathfrak{L}(G)$ is reductive. Compact Lie groups are reductive.

EXAMPLES 16.
1. Algebras $\mathfrak{gl}(n, \mathbb{R})$, $\mathfrak{gl}(n, \mathbb{C})$, $\mathfrak{u}(n)$ are reductive.
2. Algebras $\mathfrak{sl}(n, \mathbb{R})$, $\mathfrak{sl}(n, \mathbb{C})$, $\mathfrak{su}(n)$ are semisimple.

Reductive groups were introduced by E. Cartan in connection with his investigations of differential geometry. Let $\mathfrak{L}$ be a Lie algebra over the field $\mathbb{F}$. The Killing form $\mathcal{K} : \mathfrak{L} \times \mathfrak{L} \to \mathbb{F}$ is a bilinear mapping given by the formula $\mathcal{K}(x, y) := \operatorname{tr}(\operatorname{ad} x \cdot \operatorname{ad} y)$, where tr denotes the trace of a linear mapping. This form was introduced by Killing; its importance was appreciated in the famous Ph.D. thesis of Cartan. With the help of this form one could simply characterize important classes of algebras, and in the case when the algebra $\mathfrak{L}$ is semisimple, and can introduce on $\mathfrak{L}$ the Euclidean structure.

17 PROPERTIES OF KILLING FORM.
1. $\mathcal{K}(x, y) = \mathcal{K}(y, x)$ *(symmetry)*
2. $\mathcal{K}(\rho(x), \rho(y)) = \mathcal{K}(x, y)$ *for every automorphism ρ of the algebra $\mathfrak{L}$.*
3. *Operators* $\operatorname{ad} x$ *(for $x \in \mathfrak{L}$) are skew symmetric with respect to $\mathcal{K}$:*

$$\mathcal{K}([x, y], z) = cK(x, [y, z]), \quad -\mathcal{K}(\operatorname{ad} y \cdot x, z) = \mathcal{K}(x, \operatorname{ad} y \cdot z).$$

4. *Let* $\ker \mathcal{K} := \{x \in \mathfrak{L} : \mathcal{K}(x, y) = 0, \text{for all } y \in \mathfrak{L}\}$. *If* $\mathfrak{i}$ *is an abelian ideal, then* $\mathfrak{i} \subset \ker \mathcal{K}$.
5. $\ker \mathcal{K} = \{0\}$ *if and only if $\mathfrak{L}$ is semisimple.*
If the characteristic of $\mathbb{F}$ is infinite (that is, $\mathbb{F} = \mathbb{R}, \mathbb{C}$), then
6. $\operatorname{Rad}\mathfrak{L} = [\mathfrak{L}, \mathfrak{L}]^{\perp}$ *(orthogonal completion with respect to $\mathcal{K}$.)*
7. *$\mathfrak{L}$ is solvable if and only if $\mathcal{K}(\mathfrak{L}, [\mathfrak{L}, \mathfrak{L}]) = 0$.*

Points 5. and 7. are due to Cartan.

Corollary 18. *If $\mathfrak{L}$ is semisimple, then the Killing form defines the Euclidean structure on $\mathfrak{L}$. In that case we denote $\mathcal{K}(x, y)$ simply by (x, y). We will also identify H^* with H.*

Chapter 13

Reflections, Roots, and Weights. Coxeter and Weyl groups

Let $(V, (\cdot,\cdot))$ be an Euclidean space (over $\mathbb{R}$). We call *reflection* an element $s \in \mathrm{GL}(V)$ which leaves invariant all the points of some hypersurface S (that is, a subspace of codimension 1), and such that any vector v perpendicular to S (that is, $v \in S^{\perp}$) is moved by s into the opposite vector $s(v) = -v$. Reflection is an orthogonal mapping (it preserves the inner product.) Every non-vanishing vector $\alpha \in V$ defines the reflective plane L_α perpendicular to α ('the mirror') and the reflection s_α with respect to the plane L_α. We find that

$$s_\alpha(\beta) = \beta - \frac{2(\beta,\alpha)}{(\alpha,\alpha)}. \tag{1}$$

Let

$$<\beta;\alpha>:= \frac{2(\beta,\alpha)}{(\alpha,\alpha)}. \tag{2}$$

Therefore $s_\alpha(\beta) = \beta - <\beta;\alpha>$. In what follows we will assume that $\dim V < \infty$. In the theory of Lie algebras a very important role is played by finite systems of vectors (roots) which span V and satisfy the following conditions

(R1) The set $\Phi \subset V$ is finite and span V. Moreover $0 \notin \Phi$ (but in general Φ is not a basis in V.)

(R2) If $\alpha \in \Phi$, then the reflection s_α preserves Φ: $s_\alpha\Phi = \Phi$.

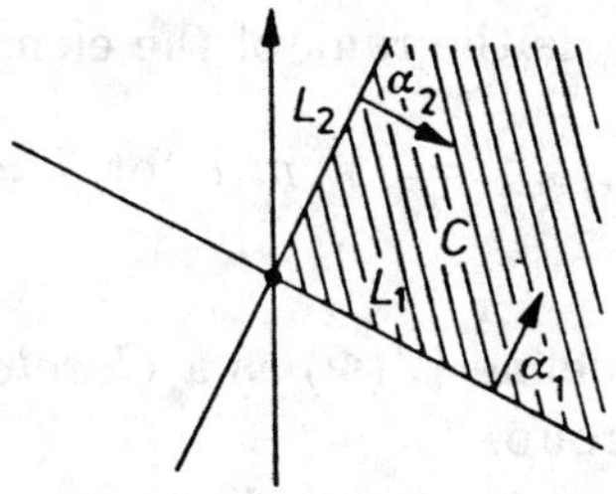

(R3) If $\alpha, \beta \in \Phi$, then $< \beta; \alpha > \in \mathbb{Z}$ (crystallographic condition.)

(R4) If $\alpha \in \Phi$ and $k\alpha \in \Phi$, then $k = \pm 1$.

DEFINITION 1. The set $\Phi \subset V$ satisfying R1–R3 is called the *root system* in V. We say that the root system Φ is *reduced* if it satisfies condition R4.

The systems of roots which will be considered in what follows are all reduced, and we will simply call them systems of roots instead of reduced systems of roots. The dimension dim V is called the *rank* of the root system.

Let Φ be a root system. Then the group A generated by all automorphisms of the space V preserving Φ is called the *group of automorphisms of the system* Φ and will be denoted by Aut(Φ). The subgroup $W(\Phi) \subset \text{Aut}(\Phi)$ generated by reflections s_α is called the *Weyl group* of the root system Φ. If $V = \bigoplus V_j$ (direct sum) and if Φ_j is a root system in V_j, $j = 1, 2, \ldots, l$, then $\Phi = \bigcup_{j=1}^{l} \Phi_j$ is a root system in V and the system Φ is called the *direct sum* of systems Φ_j.

We say that a system of roots is *irreducible* if Φ is not a direct sum of two non-empty root systems. The following fact holds: Every system of roots is a direct sum of some family of irreducible systems of roots. This decomposition is unique (modulo ordering of terms.) Connected components of the set $V - \bigcup_{\alpha \in \Phi} L_\alpha$ form convex cones, and are called *Weyl chambers*. The Weyl group permutes the chambers (that is, on the set of chambers, the Weyl group acts transitively and without fixed points.) The closure $\bar{C}$ of any chamber C is a fundamental region of the group $W(\Phi)$. Let $L-1, L_2, \ldots, L_r$ be walls of the chamber C (see the Figure.) Then $\alpha_1, \alpha_2, \ldots, \alpha_r$ is a basis of the space V and is called the *basis of the system of roots defined by the Weyl chamber* C.

The Weyl group $W(\Phi)$ is generated by reflections s_{α_j}, $j = 1, 2, \ldots, r$. Moreover

(*) $\left(s_{\alpha_j} s_{\alpha_k}\right)^{m_{jk}} = I$, m_{jk} is the rank of the element $s_{\alpha_j} s_{\alpha_k}$;

(**) $m_{jj} = 1$ (because $s_{\alpha_j} s_{\alpha_j} = I$), m_{jk} for $j \neq k$ may take the values $2, 3, \ldots$.

Relations (*) and (**) define $W(\Phi)$ as a *Coxeter group* (ex definitione!). Thus $W(\Phi)$ is a Coxeter group.

The chamber C is a convex cone, and thus it defines partial ordering in V such that a vector is non-negative if it is a linear combination with non-negative coefficients of the base vectors $\alpha_1, \alpha_2, \ldots, \alpha_r$. In this way every root is either negative, or positive: $\Phi = \Phi^+ \cup \Phi^-$ and its coordinates in the basis $\alpha_1, \alpha_2, \ldots, \alpha_r$ are integers. The subgroup $\Gamma(\Phi)$ in the (abelian) group V is a lattice, that is,

$$\Gamma(\Phi) = Z\alpha_1 \oplus Z\alpha_2 \oplus \ldots \oplus Z\alpha_r.$$

This is a discrete subgroup, invariant with respect to $W(\Phi)$. Elements of this subgroup are called *root weights* of the system of roots Φ. Observe that $\Gamma(\Phi)$ is a subgroup in the group of parallel translations of the space V. The semidirect product $W_a(\Phi) := W(\Phi) \times_s \Gamma(\Phi)$ is called the *affine Weyl group* of the system Φ. The group $W_a(\Phi)$ is a discrete group of transformations of the space V generated by reflections with respect to hyperplanes $L_{\alpha,k} := \{v \in V :< \alpha, v >= k\}$, where $\alpha \in \Phi$, $k \in \mathbb{Z}$. The quotient space $V/W_a(\Phi)$ is compact, and is a simplex if the system of roots is irreducible. In the latter case $W_a(\Phi)$ is called the *crystallographic group.*

Crystallographic groups are discrete subgroups Γ in the group $R(n)$ of all moves in the Euclidean space E^n, such that the space $R(n)/\Gamma$ is compact. Investigation of these groups is related to (flat) ornaments in $n = 2$, and to crystals in $n = 3$. At the end of nineteenth century, Fedorov (1889–1890) and, independently, A.M. Schoenfliess (1891) classified spacial crystallographic groups. (We say that two crystallographic groups are equivalent if they are adjoined in the group of affine transformation of the space.)

THEOREM 2 (FEDOROV–SCHOENFLIESS). *If we consider affine transformations preserving orientation, in E^3 there exists 230 (equivalence classes of) crystallographic groups. If, however, we allow for arbitrary affine transformations, we have in E^3 exactly 219 (equivalence classes of) crystallographic groups.*

In 1900 D. Hilbert posed the following problem (the so called 18 Hilbert problem): Prove that for any n there exists only a finite number of (equivalence classes of) crystallographic groups. This problem was solved in the positive by L. Bieberbach in 1910. The original proof of Bieberbach was soon simplified by Frobenius (1911).

THEOREM 3 (BIEBERBACH).

(i) *Every crystallographic group in E^n contains n linearly independent translations; the group of linear parts (see remark below) is finite.*

(ii) *Two crystallographic groups in E^n are equivalent if they are equivalent as abstract groups.*

(iii) *For any n there exists a finite number of (equivalence classes of) crystallographic groups.*

REMARK. As we know, motions in the space E^n are transformations of the form $g(x) = Ax + b$, where $A \in SO(n)$, $b \in \mathbb{R}^n$; A is called the *linear part*, and b the *translational part* of the mapping g.

Zassenhaus showed in 1948 that every "abstract space group" (see below) is isomorphic with some crystallographic group.

DEFINITION 4. The group R is called an *abstract space group* of rank n if there exists an exact sequence of groups $0 \to T \to R \to P \to 1$, where T is a free, abelian group of rank n, and P is a finite group acting faithfully on T.

REMARK. The problems related to crystallographic groups had great influence on the development of mathematics, not only of the group theory (Coxeter, Weyl groups.) One can mention such branches as number theory, theory of abelian and automorphic functions, differential geometry. There is a relation between crystallographic groups and so called Euclidean space forms, that is, flat compact Riemann spaces. Every crystallographic group Γ is a fundamental group $\pi_1(M)$ of a connected, flat, compact Riemann manifold M ($M = \mathbb{R}^n/\Gamma$)[1].

The notion of the system of roots can be generalized to the case when a linear space V does not possess any (natural) Euclidean structure.

GENERALIZATION 5. The system $\Phi \subset V$ is called a root system if

[1] Cf. J.A. Wolf, *Spaces of Constant Curvature*, McGraw–Hill, New York, 1967.

(i) Φ generates V and $0 \notin \Phi$,

(ii) For every $\alpha \in \Phi$ there exists $\alpha^* \in V^*$ such that $\alpha^*(\alpha) = 2$ and $s_\alpha \in \operatorname{End}(V)$ given by the formula $s_\alpha(v) = v - \alpha^*(v)\alpha$,

(iii) $n(\alpha, \beta) := \beta^*(\alpha) \in \mathbb{Z}$ for every $\alpha, \beta \in \Phi$.

The endomorphism s_α of the space V is called the *reflection* with respect to the hyperplane $\ker \alpha^*$. (Obviously $s_\alpha(\alpha) = -\alpha$.) The set $\Phi^* = \{\alpha^* : \alpha \in \Phi\}$ is a root system in the space V^* with the property that $\alpha^{**} = \alpha$ for every $\alpha \in \Phi$. We say that Φ^* is a root system in V^* dual to Φ.

In the case of a semisimple Lie algebra (that is, for $\mathfrak{L}(G)$, G semisimple and compact), V is taken to be $V = \mathfrak{L}(T)^*$, where T is a maximal torus in G, that is, $\mathfrak{L}(T)$ is a Cartan algebra in $\mathfrak{L}(G)$, and the roots are linear functions on Cartan algebra.

13.1 Weights of representations of Lie algebra

Let $\mathfrak{g}$ be a Lie algebra, and V a vector space over $\mathbb{C}$ (or other algebraically closed field of characteristic 0) and let $\rho : \mathfrak{g} \to \operatorname{Hom}(V, V)$ be a representation of $\mathfrak{g}$ in V. The element $\gamma \in \mathfrak{g}^*$ for which there exists the vector $v \in V$, $v \neq 0$ and the number $n \in \mathbb{N}$ such that for every $h \in \mathfrak{g}$

$$(*) \qquad (\rho(h) - \gamma(h) I_v)^n v = 0$$

is called the *weight of representation* ρ. The vectors v which satisfy $(*)$ form the *weight space* of V associated with the weight γ. If $\mathfrak{g}$ is a Cartan algebra of Lie algebra $\mathfrak{L}$, and $\rho = \operatorname{ad}_{\mathfrak{g}}$ is an adjoint representation of $\mathfrak{g}$ in $\mathfrak{L}$, the weight is called a *root of* $\mathfrak{L}$ *with respect to* $\mathfrak{g}$. The associated weight spaces are called *root spaces*.

13.2 Classification of root systems. Coxeter diagrams

Let $\alpha_1, \alpha_2, \ldots, \alpha_r$ be a basis of a given root system Φ. The matrix n_{ij}, $i, j = 1, 2, \ldots, r$, where

$$n_{ij} = < \alpha_i, \alpha_j > = 2\frac{(\alpha_i, \alpha_j)}{(\alpha_j, \alpha_j)}$$

is called the *Cartan–Killing matrix* of the root system Φ. On the diagonal $n_{jj} = 2$, and n_{ij}, $i \neq j$ may take the values $0, -1, -2, -3$. The Cartan–Killing matrix does not depend on the choice of a basis in Φ (modulo permutation of indices.) Two root systems with identical Cartan–Killing matrices are isomorphic.

DEFINITION 6. *Coxeter graph* of the root system Φ is a graph, whose vertices are elements $\alpha_1, \alpha_2, \ldots, \alpha_r$ forming a basis of the root system Φ, and vertices α_i and α_j are connected by $n_{ij} \cdot n_{ji}$ edges (that is, by 1, 2, 3 or 0 lines.)

THEOREM 7 (COXETER). *The Coxeter graph for Φ is connected if and only if the root system Φ is irreducible.*

Dynkin enriched the information provided by Coxeter graph: if simple roots α_i, α_j are not perpendicular and if they have different length, then on two or three edges connecting the vertex i with the vertex j one place the symbol ¿ (arrow head) directed from the longer to the shorter root (vertex). Such modified graph is often called the *Dynkin diagram.* In some cases over each vertex one place a number proportional to the square of the length of the given root.

THEOREM 8 (COXETER). *The full catalogue of non-isomorphic, irreducible root systems given by bases of root systems can be represented (by Coxeter–Dynkin diagrams) as follows (r is a number of roots in the basis):*

Cartan–Killing matrices

$$A: \begin{bmatrix} 2 & -1 & 0 & & & & & & 0 \\ -1 & 2 & -1 & 0 & & & & \cdot & 0 \\ 0 & -1 & 2 & -1 & 0 & \cdot & \cdot & \cdot & 0 \\ \cdot & \cdot & \cdot & \cdot & \cdot & \cdot & \cdot & \cdot & \cdot \\ 0 & 0 & 0 & 0 & \cdot & \cdot & \cdot & -1 & 2 \end{bmatrix}$$

$$B: \begin{bmatrix} 2 & -1 & 0 & & & & & & 0 \\ -1 & 2 & -1 & 0 & \cdot & \cdot & & & 0 \\ \cdot & \cdot & \cdot & \cdot & \cdot & \cdot & \cdot & \cdot & \cdot \\ 0 & 0 & 0 & \cdot & \cdot & \cdot & -1 & 2 & -2 \\ 0 & 0 & 0 & \cdot & \cdot & \cdot & 0 & -1 & 2 \end{bmatrix}$$

$$C: \begin{bmatrix} 2 & -1 & 0 & & & & & & 0 \\ -1 & 2 & -1 & 0 & \cdot & \cdot & \cdot & \cdot & 0 \\ \cdot & \cdot & \cdot & \cdot & \cdot & \cdot & \cdot & \cdot & \cdot \\ 0 & 0 & 0 & \cdot & \cdot & \cdot & -1 & 2 & -2 \\ 0 & 0 & 0 & \cdot & \cdot & \cdot & 0 & -2 & 2 \end{bmatrix}$$

$$D: \begin{bmatrix} 2 & -1 & 0 & & & & & & 0 \\ -1 & 2 & -1 & 0 & \cdot & \cdot & \cdot & & 0 \\ \cdot & \cdot & \cdot & \cdot & \cdot & \cdot & \cdot & \cdot & \cdot \\ 0 & 0 & & \cdot & \cdot & -1 & 2 & -1 & 0 & 0 \\ 0 & 0 & & \cdot & \cdot & \cdot & -1 & 2 & -1 & -1 \\ 0 & 0 & & \cdot & \cdot & \cdot & 0 & -1 & 2 & 0 \\ 0 & 0 & & \cdot & \cdot & \cdot & 0 & -1 & 0 & 2 \end{bmatrix}$$

$$E_6: \begin{bmatrix} 2 & 0 & -1 & 0 & 0 & 0 \\ 0 & 2 & 0 & -1 & 0 & 0 \\ -1 & 0 & 2 & -1 & 0 & 0 \\ 0 & -1 & -1 & 2 & -1 & 0 \\ 0 & 0 & 0 & -1 & 2 & -1 \\ 0 & 0 & 0 & 0 & -1 & 2 \end{bmatrix}$$

$$E_7: \begin{bmatrix} 2 & 0 & -1 & 0 & 0 & 0 & 0 \\ 0 & 2 & 0 & -1 & 0 & 0 & 0 \\ -1 & 0 & 2 & -1 & 0 & 0 & 0 \\ 0 & -1 & -1 & 2 & -1 & 0 & 0 \\ 0 & 0 & 0 & -1 & 2 & -1 & 0 \\ 0 & 0 & 0 & 0 & -1 & 2 & -1 \\ 0 & 0 & 0 & 0 & 0 & -1 & 2 \end{bmatrix}$$

$$E_8: \begin{bmatrix} 2 & 0 & -1 & 0 & 0 & 0 & 0 & 0 \\ 0 & 2 & 0 & -1 & 0 & 0 & 0 & 0 \\ -1 & 0 & 2 & -1 & 0 & 0 & 0 & 0 \\ 0 & -1 & -1 & 2 & -1 & 0 & 0 & 0 \\ 0 & 0 & 0 & -1 & 2 & -1 & 0 & 0 \\ 0 & 0 & 0 & 0 & -1 & 2 & -1 & 0 \\ 0 & 0 & 0 & 0 & 0 & -1 & 2 & -1 \\ 0 & 0 & 0 & 0 & 0 & 0 & -1 & 2 \end{bmatrix}$$

$$F_4: \begin{bmatrix} 2 & -1 & 0 & 0 \\ -1 & 2 & -2 & 0 \\ 0 & -1 & 2 & -1 \\ 0 & 0 & -1 & 2 \end{bmatrix}$$

$$G_2: \begin{bmatrix} 2 & -1 \\ -3 & 2 \end{bmatrix}$$

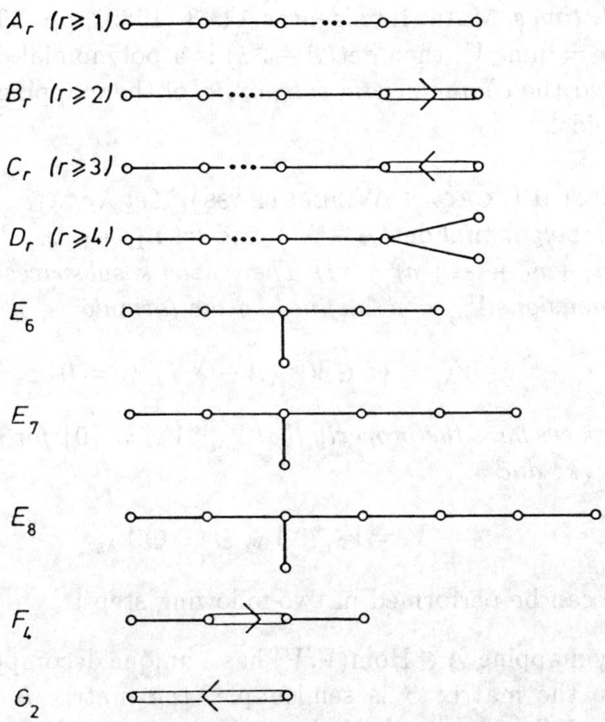

Coxeter–Dynkin diagrams

13.3 Relation with semisimple complex Lie algebras

As we have mentioned already many times, the classical works of H. Weyl (1925-1926) and investigations of regular polyhedra which started with Pythagoreans and Plato were the momentum for Coxeter's research. Semisimple algebras (over $\mathbb{C}$) were classified first by W. Killing (1888/1889), a pupil of Weierstrass. Generalizing the fundamental notions of linear algebra of an eigenvector and eigenspace of linear mapping $A : V \to V$, he introduced the notions of root and root space. Let us recall these former definitions, assuming that V is a vector space over the field $\mathbb{C}$. A vector $v \in V$, for which there exists $\lambda \in \mathbb{C}$ and $n \in \mathbb{N}$ such that $(A - \lambda I)^n v = 0$ is called the *root vector of* A belonging to λ. The set of root vectors belonging to λ forms a root subspace V_λ (cf. A.J. Coleman *The greatest mathematical paper of all times*, Math. Inteligencer **11**, 3 (1989), pp. 29-38.)

When $r = \dim_{\mathbb{C}} V$, then $\det(A - \lambda I)$ is a polynomial of degree r in variable λ, called the *characteristic polynomial* of the mapping A. The following theorem holds.

THEOREM 9 (JORDAN–WEIERSTRASS). *Let $\lambda_1, \lambda_2, \ldots, \lambda_k$ be different roots of the polynomial* $\det(A - \lambda I)$ *and let $n_1, n_2, \ldots, n_k$ be their multiplicities ($n_1 + n_2 + \cdots + n_k = r$.) There exist k subspaces $V_{\lambda_1}, V_{\lambda_2}, \ldots, V_{\lambda_k}$ $\subset V$ of dimensions $V_{\lambda_j} = n_j$ defined by the formula*

$$V_{\lambda_j} = \{v \in V : (A - \lambda_j I)^{n_j} v = 0\}.$$

These subspaces have the property that $V_{\lambda_i} \cap V_{\lambda_j} = \{0\}$ for $i \neq j$, $AV_{\lambda_j} \subset V_{\lambda_j}$, $j = 1, 2, \ldots, k$, and

$$(*) \qquad V = V_{\lambda_1} \oplus V_{\lambda_2} \oplus \cdots \oplus V_{\lambda_k}.$$

PROOF can be performed in two following steps:

(i) Every mapping $A \in \mathrm{Hom}(V, V)$ has a unique decomposition $A = S + N$, where the matrix S is semisimple, the matrix N is nilpotent, and $SN = NS$ (the so-called Jordan decomposition).

(ii) We take the decomposition of V onto eigenspaces of the operator S – this is exactly equation $(*)$. From properties of Jordan decomposition it follows that $AV_{\lambda_j} \subset V_{\lambda_j}$ and $\dim V_{\lambda_j} = n_j$.

Jordan decomposition is also of fundamental importance for classification of Lie algebras: it shows the necessity of decomposition of algebras into semisimple and nilpotent ones, which are, in some sense, complementary.

GENERAL EXAMPLES 10.

1. The matrix S is semisimple, if (after appropriate change of the basis) it can be diagonalized:

$$S = \text{Diag}\{\lambda_1, \lambda_2, \ldots, \lambda_k\}.$$

2 The matrix N is nilpotent if it can be expressed in the form of an upper triangular matrix:

$$N = \begin{bmatrix} 0 & & * \\ & \ddots & \\ 0 & & 0 \end{bmatrix}.$$

In his Ph.D. thesis (1894), E. Cartan brought out a class of subalgebras in a semisimple Lie algebra.

DEFINITION 11. Let $\mathfrak{l}$ be a semisimple Lie algebra. We say that $\mathfrak{h} \subset \mathfrak{l}$ is a *Cartan subalgebra* if it is the maximal abelian subalgebra and all its elements are semisimple (and thus can be simultaneously diagonalized.)

Let now $\mathfrak{g} = \mathfrak{L}(G)$ be Lie algebra of the semisimple group G. The Killing form $\mathcal{K}$ on $\mathfrak{g}$ is non-degenerate, and since G is compact, $\mathcal{K}$ is negative-determinate.

Let $\mathfrak{h}$ be the maximal (abelian) subalgebra in $\mathfrak{g}$, such that all the operators $\text{ad}_{\mathfrak{g}}\, x$, $x \in \mathfrak{h}$ are semisimple. (They are semisimple, and thus can be simultaneously diagonalized.) In other words, $\mathfrak{h}$ is a Cartan subalgebra for $\mathfrak{g}$. For linear form (functional) $\alpha^* \in \mathfrak{h}^*$ we consider the space

$$L_\alpha = \{X \in \mathfrak{g} : [h, X] = \alpha^*(h)X,\ h \in \mathfrak{h}\}.$$

If $L_\alpha \neq \{0\}$, then α is called the *root of the algebra* $\mathfrak{g}$ with respect to Cartan subalgebra $\mathfrak{h}$, and L_α is called the *root space.*

THEOREM 12 (KILLING, CARTAN). *The set of roots Φ of a semisimple algebra (with respect to any Cartan subalgebra) is a reduced root system.*

We have therefore in our disposal the developed theory of Coxeter group, Weyl groups, Cartan matrix, Coxeter graphs, etc. It turns out that to connected Coxeter graphs (with crystallographic condition) there correspond, in the case of semisimple Lie algebras, simple complex Lie algebras[2], and the

[2]That is, semisimple Lie algebras having exactly two ideals.

correspondence is one-to-one. Therefore there exist four infinite sequences of simple Lie algebras, and moreover five "exceptional algebras" E_6, E_7, E_8, F_4, G_2. The four infinite sequences correspond to algebras of the classical Lie groups. Thus we have:

- A_n — $\mathfrak{sl}(n+1, \mathbb{C})$ — the algebra of the group $\mathrm{SL}(n+1, \mathbb{C})$ (special linear) formed by the $(n+1) \times (n+1)$ matrices of determinant 1. This is a set of endomorphisms of the space $\mathbb{C}^{n+1}$ with vanishing trace.
- B_n — $\mathfrak{so}(2n+1, \mathbb{C})$ — the algebra of the group $\mathrm{SO}(2n+1, \mathbb{C})$ (special orthogonal). This is an algebra of $(2n+1) \times (2n+1)$ skew symmetric matrices.
- C_n — $\mathfrak{sp}(2n, \mathbb{C})$ — the algebra of the group $\mathrm{SP}(2n, \mathbb{C})$ (symplectic). This is a set of $(2n) \times (2n)$ matrices preserving the symplectic form.
- D_n — $\mathfrak{so}(2n, \mathbb{C})$.

As we know, to a Lie algebra there corresponds a unique simply connected Lie group. In our table only the groups $\mathrm{SO}(2n+1, \mathbb{C})$ are not simply connected. The universal covering of the group $\mathrm{SO}(2n+1, \mathbb{C})$ is the group $\mathrm{Spin}(2n+1, \mathbb{C})$ (the multiplicity of the covering is equal 2.) This latter group was discovered by E. Cartan in 1909, many years before the discovery of Dirac of his famous equation of 'spinning electron.'

The idea of the proof of Cartan theorem – as it is always in the case of great ideas – is beautiful and simple. Let G be a compact Lie group. As we know already, in G there exists a maximal torus of dimension r. All maximal tori are adjoined to each other, and a maximal torus passes through every point of the group. Therefore

$$G = \bigcup_{g \in G} gTg^{-1}.$$

The torus T (being an abelian algebra) has its own Lie algebra $\mathfrak{L}(T)$, which is r-dimensional and commutative. Thus for any compact Lie group there exists an associated Cartan algebra $\mathfrak{L}(T)$; the Lie algebra of the maximal torus $T \subset G$. The dimension of $\mathfrak{L}(T)$ is called the *rank of the group* G. The exponential mapping $\exp : \mathfrak{L}(T) \to T$ is a homomorphism and covering. After identifying $\mathfrak{L}(T)$ with $\mathbb{R}^r$, the kernel $\ker(\exp) \subset \mathbb{R}^r$ is a discrete subgroup, and therefore a lattice $\Gamma \subset \mathbb{R}^r$. The (adjoint) representation $\mathrm{Ad} : T \to \mathfrak{L}(T)$

decomposes (in view of Schur lemma) into irreducible unitary representations $U_j : T \to S^1$, $U_j(t) = e^{2\pi i \vartheta_j(t)}$: the image of $t \in T$ under U_j is a rotation of the plane $\mathbb{C}$ by the angle $2\pi\vartheta_j(t)$. The numbers $\vartheta_j(t)$ are defined modulo $\mathbb{Z}$. Therefore the functions $\vartheta_j : \mathfrak{L}(T) \to \mathbb{R}$ can be extended onto the covering $\mathfrak{L}(T)$ of the torus T. Let us denote these functions $\alpha_j : \mathfrak{L}(T) \to \mathbb{R}$. These are linear functions, $\alpha_j \in \mathfrak{L}(T)^*$, and on the lattice $\Gamma := \exp^{-1}(1)$ they take integer values; thus α_j are roots. The family $\{\pm\alpha_j\}$ is the root system of the algebra $\mathfrak{L}(G)$, also called the root system of the group G, and denoted $\Phi(G)$.

EXAMPLE 13 For $G = U(n)$, the maximal torus is the following set of diagonal matrices:

$$T = \text{Diag}\left(e^{2\pi i x_1}, \ldots, e^{2\pi i x_1}\right), \quad x_j \in \mathbb{R}.$$

It can be shown that the roots are differences $x_i - x_j$, $i \neq j$.

As we know, a root system Φ is (modulo isomorphism) independent of the choice of Cartan subalgebra $\mathfrak{h} \subset \mathfrak{g}$. More precisely, this is a subject of Witt theorem:

THEOREM 14 (WITT). *Let $(\mathfrak{g}, \mathfrak{h}, \Phi)$, $(\mathfrak{g}', \mathfrak{h}', \Phi')$ be complex semisimple Lie algebras, where $\mathfrak{h}$, $\mathfrak{h}'$ are Cartan subalgebras, and Φ, Φ' the corresponding root systems. Let $A : \mathfrak{h} \to \mathfrak{h}'$ be an isomorphism which induces bijection Φ on Φ'. Then A extends to the isomorphism $\tilde{A} : \mathfrak{g} \to \mathfrak{g}'$ of Lie algebras.*

Thus Φ defines the algebra $\mathfrak{g}$, is its 'skeleton'.

THEOREM 15 (VAN DER WAERDEN). *With any root system Φ there is associated at most one Lie algebra.*

THEOREM 16 (WITT, 1941). *With any root system Φ there is associated one and only one Lie algebra.*

EXAMPLE 17. Let $\mathfrak{g} = \mathfrak{sl}(n, \mathbb{C})$ and let $\mathfrak{h}$ be a Cartan algebra consisting of diagonal matrices with trace 0. Let E_{ij} be the matrix having one at the place (i, j), and all other entries equal zero. Let e_j be the linear functional mapping the diagonal matrix into jth entry on the diagonal, to wit

$$\text{Diag}(\lambda_1, \ldots, \lambda_p, \ldots, \lambda_n) \stackrel{e_j}{\to} \lambda^p, \quad J = 1, 2, \ldots, n.$$

Then for every $h \in \mathfrak{h}$ we have

$$[h, E_{ij}] = ((e_i - e_j)h) \cdot E_{ij},$$

and thus $e_i - e_j$ is a root of $\mathfrak{g}$ with respect to $\mathfrak{h}$, as long as $i \neq j$. Since

$$\mathfrak{g} = \mathfrak{h} + \sum_{i \neq j} \mathbb{C}E_{ij},$$

$e_i - e_j$, $i \neq j$ are all roots. A basis of the root system is, for example, $\alpha_j := e_j - e_{j+1}$, $1 \leqslant j \leqslant n-1$. Thus the space V has dimension $n-1$ (over $\mathbb{R}$.) It can be easily checked that the reflection s_{α_j} in $\alpha_j^{\perp}$ exchanges α_j and α_{j+1} leaving all other base vectors invariant. Thus the Weyl group is isomorphic with the symmetric group Σ_n. It can be similarly shown that the Weyl group for $\mathfrak{su}(n)$ is also isomorphic with Σ_n; *ditto* for $\mathfrak{u}(n)$.

Historical remark. Van der Waerden proved his uniqueness theorem already in 1933. Witt theorem is by far more complicated. Witt proved this theorem in 1941 in his classical paper which we have already cited. In this paper simple proofs of famous Cartan theorems are also presented:

(a) on the fact that irreducibility of a root system Φ is equivalent with simplicity of the algebra $\mathfrak{L}$;

(b) on unique decomposition of a semisimple algebra into simple algebras corresponding to decomposition of Φ into irreducible (simple) root systems.

The work of Witt completed geometrization of the classification of (semi) simple Lie algebras. Many years later (1966), Serre gave the a priori proof of Witt theorem, not citing (and therefore not knowing) the Witt's work.

To end of this chapter we present the famous

Weyl integral formula. Let G be a compact connected Lie group, T its maximal torus, $\chi_1, \ldots, \chi_n$ characters of T. Then for any continuous, invariant function on G (that is $f(gxg^{-1}) = f(x)$, $g \in G$),

$$\int_G f(g) d\mu(g) = \frac{1}{|W|} \int_T f(t) \prod_{j=1}^{n} (\chi_j(t) - 1)\, d\nu(t),$$

where μ, ν are normalized Haar measures on G and T respectively, and W is the Weyl group.

EXAMPLE. $G =$SU(2), T is a set of matrices of the form

$$t(\varphi) = \begin{pmatrix} e^{i\varphi} & 0 \\ 0 & e^{-i\varphi} \end{pmatrix}, \quad |W| = 2, \ \ d\nu = d\varphi/2\pi,$$

$$\chi_1(t(\varphi)) = e^{2i\varphi} \quad \chi_2,(t(\varphi)) = e^{-2i\varphi},$$

$$\prod_{j=1}^{2} \chi_j(t(\varphi) - 1) = 4\sin^2\varphi.$$

Chapter 14

Covariant Differentiation. Parallel Transport. Connections

Parallel transport along curves (connection in fiber bundle). Since M is a connected differentiable manifold, it is arcwise connected as well, and we can define a product of paths as follows (Fig. 1).

Let

$$\gamma_1 : [0,1] \to M, \quad \gamma_1(0) = x, \ \gamma_1(1) = y,$$
$$\gamma_2 : [0,1] \to M, \quad \gamma_2(0) = y, \ \gamma_2(1) = z.$$

Then we define the curve $\gamma_1\gamma_2 : [0,1] \to M$ as follows

$$\gamma_1\gamma_2(t) := \begin{cases} \gamma_1(2t), & 0 \leqslant t \leqslant \frac{1}{2}, \\ \gamma_2(2t-1), & \frac{1}{2} \leqslant t \leqslant 1, \end{cases}$$

$$\gamma_1^{-1}(t) := \gamma_1(1-t).$$

Definition. Let $E \to M$ be a bundle with typical fiber $\mathbf{E}$. Then, with any path $\gamma : [a,b] \to M$ in the base M one can associate the isomorphism P_γ of fibers $E_{\gamma(a)} \to E_{\gamma(b)}$ such that

1. P_γ continuously depend on the path γ (for compact topology),
2. $P_{\gamma_1\gamma_2} = P_{\gamma_1} \circ P_{\gamma_2}$, $P_{\gamma^{-1}} = P_\gamma^{-1}$,
3. P_γ is independent of the parametrization of the path.

We say that in the bundle $E \to M$ the *parallel transport* along paths in M is given, or, in other words, that we have *connection* in the bundle.

The most important situation arises if we consider a vector bundle (for example, the tangent bundle $TM \to M$); then we talk about parallel transport of vectors along curves. In what follows, we will assume that the bundle

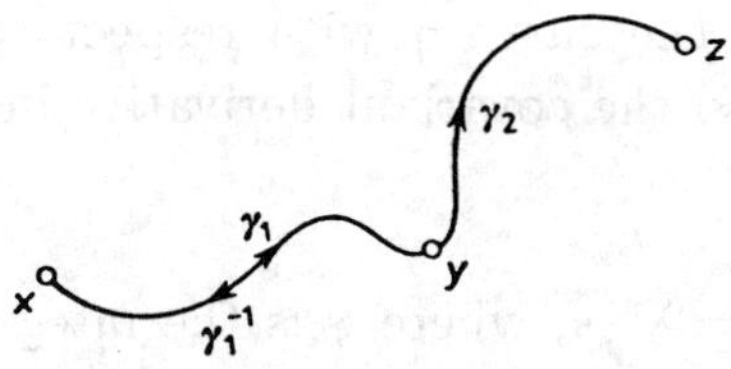

Fig. 1

$\pi : E \to M$ is a vector bundle. We will show that, by integrating linear differential equations, connection will lead us to so-called *covariant derivative.*

As we know from the theory of linear differential equation, the equation

$$(*) \qquad \frac{\mathrm{d}e}{\mathrm{d}t} = A(t)e, \quad \text{where } t \mapsto A(t) \in L(\mathbf{E}) = L(\mathbf{E};\mathbf{E})$$

which satisfies Lipschitz condition, defines 'evolution', that is, the one- parameter family of (linear) isomorphisms U_t of the Banach space $\mathbf{E}$ such that $U_{t_1+t_2} = U_{t_1} \cdot U_{t_2}$, $U_{-t} = U_t^{-1}$, $U_0 = \mathrm{id}_{\mathbf{E}}$.

Since the fibers E_x are isomorphic with the typical fiber, the space $\mathbf{E}$, and since the bundle $\mathbf{E}|\mathcal{O}$ is locally isomorphic to the product $\mathcal{O} \times \mathbf{E}$, where $\mathcal{O}$ is some subset of the model $\mathbf{M}$ of the manifold M, local trivializations make it possible to reduce the problem to integration of the ordinary differential equations $(*)$.

From connection (parallel transport) to covariant derivative. Let $s \in C^\infty(E)$ be a smooth section of the vector bundle $E \to M$ and let $\gamma \to \|_\gamma$ be a connection in E (that is, a parallel transport along smooth curves $\gamma : [0,1] \to M$). The curve $\gamma : [0,1] \to M$ defines the one-parameter family $s_t := s(\gamma(t)) \in E_{\gamma(t)}$, $t \in [0,1]$. Let us parallelly transport the vectors s_t along γ to the point $x = \gamma(0)$; it this way we obtain the one-parameter family $\tilde{s}_t \in E_x \simeq \mathbf{E}$, $0 \leqslant t \leqslant 1$ of vectors belonging now to the single fiber E_x over x, which carries the structure of a Banach space ($\simeq \mathbf{E}$). We therefore know what the derivative of such family

$$\lim_{t\to 0} \frac{\tilde{s}(t) - s(\gamma(0))}{t} =: \nabla_{\dot{\gamma}} s(x)$$

exactly means.

This expression is called the *covariant derivative* of the section s (of the bundle $E \to M$) along the curve γ with respect to the given connection $\|_\gamma \equiv P_\gamma$. This defines also the covariant derivative in direction of the vector field $X \in \mathcal{X}(M)$:

DEFINITION. $\nabla_X s := \nabla_{\dot\gamma} s$, where γ is the integral curve of the field X starting from the point $x = \gamma(0)$. If $\nabla_{\dot\gamma} s = 0$, then we say that the section s is *parallel along* γ or that it is *horizontal along* γ.

This last term will be explained later. We immediately check that such defined operation ∇_X has the properties collected in the following proposition

PROPOSITION 1. *Let* $\|$ *be a linear connection in the vector bundle* $E \to M$. *Then* ∇_X *defines the* $\mathbb{R}$*-bilinear mapping*

$$\nabla : \mathcal{X}(M) \times C^\infty(E) \to C\infty(E), \quad \nabla(X, s) := \nabla_X(s)$$

with the properties

$$\nabla_X(fs) = X(f)s + f\nabla_X s, \quad f \in C^\infty(M), \;\; s \in C^\infty(E), \tag{1}$$

$$\nabla_{fX} s = f\nabla_X s. \tag{2}$$

In the proof of formulae (1) and (2) one uses only the definition and, in fact, one repeats the elementary proof of Leibnitz formula for derivative of a product of functions.

REMARK 1. Formula (1) is often written in the form

$$\nabla_X(fs) = X(f) \otimes s + f\nabla_X s, \tag{1*}$$

Thus every linear connection in the bundle $E \to M$ defines the mapping ∇ which is mentioned in the proposition. This leads us to the following definition.

DEFINITION Let $E \to M$ be a vector bundle. The bilinear mapping $\nabla : \mathcal{X}(M) \times C^\infty(E) \to C^\infty(E)$ denoted by $\nabla_X s := \nabla(X, s)$ with the properties (1), (2) is called the *covariant derivative (or differential)* in the bundle

$E \to M$.

Proposition 1 can be phrased as follows:

Every parallel transport along curves (that is, a linear connection in the bundle E) defines covariant derivative in the bundle $E \to M$.

REMARK. In the *finite* dimensional situation, that is, when both the dimensions of the manifold M and the typical fiber $\mathbf{E}$ are finite, it makes sense to regard ∇ as the mapping

$$\nabla : C^\infty(E) \to C^\infty(T^*M \otimes E) \simeq C^\infty(\mathrm{Hom}(TM, E)) \tag{3}$$

satisfying the Leibnitz rule

$$\nabla(fs) = df \otimes s + f\nabla_X s. \tag{3*}$$

However, in the infinite dimensional situation, tensor products of Banach, and even Hilbert spaces do not carry Banach or Hilbert structure: the product $H_1 \otimes H_2$ of two infinite dimensional Hilbert spaces is *never* complete in any pre-Hilbert structure. One must therefore complete the algebraic product $H_1 \otimes H_2$ in different (not equivalent!) norms which leads to different topologically complete tensor products. We will talk about these products later – here we just wanted to stress the essential difference between finite and infinite dimensional cases.

As we have shown, linear connection defines covariant derivative. The claim arises that also vice versa, covariant derivative defines parallel transport along curves, so we turn to the next subject

From covariant derivative to parallel transport.

THEOREM 2. *Let* $\dim M$, $\dim E < \infty$. *Then the covariant derivative defines the parallel transport along curves, that is, the linear connection in the vector bundle* $E \to M$.

More precisely, for any smooth curve $\gamma : [0,1] \to M$ *there exists* $\|_\gamma : E_{\gamma(0)} \to E_{\gamma(1)}$ *being a (linear) isomorphism of fibers. In particular, from each point* $e_0 \in E_{\gamma(0)}$ *there starts the smooth curve* $\tilde\gamma : [0,1] \to E$ *such that* $\pi\tilde\gamma = \gamma$ *and* $\tilde\gamma(0) = e_0$. *The curve* $\tilde\gamma$ *is a horizontal section of the bundle* $E \to M$ *along the curve* γ.

PROOF. A local trivialization of the bundle E is the isomorphism

$$E|\mathcal{O} = \pi^{-1}(\mathcal{O}) \simeq \mathcal{O} \times \mathbf{E}.$$

Therefore, the definition of covariant derivative in this trivialization leads to the linear differential equation of the form

$$\nabla_X s(x) = Ds_x \cdot X_x + \Gamma_\varphi(\varphi(p)) \cdot (X_x, s_x)$$

where

$$\Gamma_\varphi(\cdot) : \varphi(\mathcal{O}) \to L(M, \mathbf{E}; \mathbf{E}).$$

The problem of existence of linear connection reduces therefore to the problem from the theory of linear differential equations

$$\frac{\mathrm{d}y}{\mathrm{d}t} = -\Gamma(\varphi(\gamma(t), (\varphi \circ \gamma)'(t), y(\varphi(t)), \quad y(0) = y, \tag{4}$$

and thus to the linear equation of the form

$$\frac{\mathrm{d}e}{\mathrm{d}t} = A(t)e, \quad \text{where } t \to A(t) \in L(\mathbf{E}). \tag{5}$$

As we know from the theory of differential equations, (5) defines a (local) one-parameter family of linear isomorphisms $U_t : \mathbf{E} \to \mathbf{E}$:

$$U_{t_1+t_2} = U_{t_1} \circ U_{t_2}, \quad U_{t^{-1}} = U_t^{-1}, \quad U_0 = \underset{\mathbf{E}}{\mathrm{id}}.$$

Returning to the bundle $E|\mathcal{O} \to \mathcal{O} \subset M$, we have the correspondence of $\gamma : [0,1] \to M$ with the isomorphisms

$$\|_\gamma : E_{\gamma(0)} \to E_{\gamma(1)}$$

possessing the required properties. □

Holonomy (monodromy) group. Let us return to the general notion of connection in the fiber bundle $\pi : E \to M$ with typical fiber $\mathbf{E}$. Let us consider closed curves, that is, loops with the beginning and the end at the point $x \in M$. The parallel transport along such loop γ defines an isomorphism of the fiber E_x over x. We have therefore the mapping P of the set of loops

$$\Omega(M, x) \ni \gamma \to P_\gamma \in \mathrm{Aut}(E_x) \simeq \mathrm{Aut}(\mathbf{E})$$

into some subgroup Φ_x of the group of automorphisms of the fiber E_x. This group is called the *holonomy group* (of connection P) at the point x. Recently, one often uses the older term (introduced for the case of finite groups by C. Jordan in 1870), the *monodromy group* at x.

REMARK 1. In the case when $E = \tilde{M} \to M$ is an unbranched covering of the manifold M, the connection P associates with every class of homotopical loops $[\gamma] \in \pi_1(M; x)$ the mapping $\sigma([\gamma]) = P_\gamma$ because the homotopic curves $\gamma_1 \sim \gamma_2$ provide the same mapping $P_{\gamma_1} = P_{\gamma_2}$. Thus σ is a homomorphism of the fundamental group $\pi_1(M; x)$ in the group of automorphisms of the fiber $E_x = \tilde{M}_x$; its image $\sigma(\pi_1(M; x)) = \Phi_x$ is called the *monodromy of the fiber* E_x or the *discrete group of holonomy* because every fiber of the covering $\tilde{M} \to M$ is discrete. We have here a very close relation with the group $\mathrm{Deck}(\tilde{M} \to M)$!

REMARK 2. The mentioned representation $\sigma : \pi_1(M; x) \to \Phi_x$, often called the *monodromy representation* appeared in Riemann's lectures on (linear ordinary) differential equations held in winter semester 1856/57, and thus a year before the publication of his fundamental work on hypergeometric equations *Beiträge zur Theorie der durch Gausssche Reihe* $F(\alpha, \beta, \gamma, x)$ *dastellbaren Funktionen*, 1857, Werke, 63–87. It is hard to overestimate the richness and depth of ideas of this paper and also its influence on the subsequent development of mathematics. The hypergeometric equation (of Gauss – it was known already to Euler!) is the simplest linear ordinary differential equation with three singular points of its coefficients – the simplest in the sense that, in general, such equations may have *more* than three singular points – and, if the singular points exist at all, there must be at least three of them. The works and lectures of Riemann (which contained much more than the subsequent publications: the remaining notes were left forgotten for several dozen years in the archive of the Göttingen University) used to open new branches of mathematics. There one encounters the inverse problem – partially solved by Riemann himself – *For a given group of linear transformations contained in* $\mathrm{GL}(n, \mathbb{C})$ *find the differential equation with regular singularities, whose monodromy group is the given group.* This problem posed by Riemann in 1857 and becoming famous as the 21 Hilbert Problem was in full solved by Helmut Röhrl 100 years later, in 1957. We will say more about a history of the solving this Riemann–Hilbert problem later.

All this is very nice – the reader may say – but do connections exist? We must therefore show the existence of covariant derivative and parallel transport.

Existence of covariant derivative and parallel transport. We will show this fact in three steps.

Step I. When the bundle E is trivial, that is, when $E = \mathcal{U} \times \mathbf{E}$, where $\mathcal{U}$ is an open subset of some Banach space M, then the derivative ∇ is the ordinary derivative because the section $s \in C^\infty(\mathcal{U} \times \mathbf{E})$ can be identified with the mapping $s : \mathcal{U} \to \mathbf{E}$.

Step II. When $(\Phi, \varphi, \mathcal{U})$ is a trivialization of the restriction $E|\mathcal{U}$, then this trivialization reduces the problem to step I: in $E|\mathcal{U}$ we have therefore the connection $\nabla_{\mathcal{U}}$.

Step III. General case. Let $(\mathcal{U})_i$, $i \in I$ be a covering of M making the trivialization of every restriction $E|\mathcal{U}_i$, $i \in I$ possible. Let (α_i), $i \in I$ be a decomposition of unity related to the covering $(\mathcal{U}_i)$, $i \in I$ and let $\nabla_i = \nabla_{\mathcal{U}_i}$ be the derivative on $E|\mathcal{U}_i$.

We define derivative on $E \to M$ by the formula

$$\nabla s := \sum_i \alpha_i \nabla_i s_i, \quad \text{where } s_i := s|\mathcal{U}_i.$$

□

We proved therefore the following theorem.

THEOREM 3. *On every vector bundle $E \to M$ there does exist a covariant derivative.*

It is even more simple to show the existence of Riemann metric on any bundle $E \to M$ with typical fiber $\mathbf{E}$ being an Euclidean space (or, more generally, Hilbert space) $(\mathbf{E}, (\cdot|\cdot)$. We leave this as an exercise to the reader.

The existence of parallel transport along curves follows immediately from the theorem above and the discussion at the beginning of this chapter.

We have therefore the following corollary.

COROLLARY 4. *On every vector bundle $E \to M$ there does exist a parallel transport along curves (that is, a linear connection).*

Chapter 15

Remarks on Rich Mathematical Structures of Simple Notions of Physics Based on Example of Analytical Mechanics

Let us pause for a moment to take a bird-eye view at our laborious but necessary climbing. This will make us realize one more time the seemingly paradoxical but almost necessary way of development of mathematical ideas (or, perhaps, all ideas?).

Many fundamental notions of mathematics were born in relation with physics: *mathematics is the language (Logos) of physics.* The interesting phenomenon of this chapter of development of ideas, which mathematics is itself, is that involved notions requiring a number of mathematical structures used to arise *earlier* than simple notions, which required poorer mathematical structures. And so, the notions of velocity, acceleration, force, stress – and thus some tensor fields have arisen before the notions of vectors and tensors. It was for mathematical correctness, to make it possible to define these elementary notions generally and precisely, to make sense of the statement that *acceleration is derivative of velocity*, that one must define covariant derivative, and ought to have in the disposal the notion of *connection in vector bundle.* In order to be able to talk about smoothness of a vector field on a (differential) manifold M, one must introduce a differentiable structure in the family of tangent space $TM = \coprod_{x\in M} T_xM$, so that at the end (or rather at the beginning) one can say what a tangent vector at the point $x \in M$ really is. One also needs the notion of differentiable manifold, and thus we have at least eight levels high architecture:

1. Differentiable manifold M.
2. Tangent space T_xM at (each) point of M.
3. The space of tangent bundle $TM = \coprod_{x\in M} T_xM$.
4. Differential structure on TM.
5. Smooth surjection (projection) $\pi : TM \to M$.
6. Section s of the bundle $TM \to M$, that is the mapping $s : M \to TM$ such that $s(x) \in T_xM$.
7. Connection ∇ in the tangent bundle $TM \to M$.
8. Acceleration (along curve) $c :]0,1[\to M$ as $\nabla\dot{c}$ is therefore only the eighth level of this construction.

Therefore, the notion so fundamental and elementary for physics as acceleration and the statement as fundamental as the *Second Newton Law* providing relation between force and acceleration requires all these eight levels. And what about the theory of elasticity and the theory of electromagnetism of Maxwell, Hertz, and Minkowski or the Einstein's theory of gravity!

One can easily understand the sorrow, bitterness, and even rage of older generation of physicists that 'youngsters make simple things complicated', these people were not taught the notions and structures of modern mathematical analysis and it was impossible for them to catch up with the new development later. It was already Poincaré who convincingly showed (almost a hundred years ago) that physics requires analysis on differentiable manifolds, that the phase space of a system of material points is nothing but the cotangent bundle $T^*M \to M$; both he and E. Cartan (and also Lie) realized that Hamiltonian mechanics (*implicitly*) makes use of the symplectic structure of cotangent bundle $T^*M \to M$: if $(x,p) = (x^1,\ldots,x^n,p_1,\ldots,p_n)$ are local coordinates in the space T^*M, where n is the dimension of the manifold M, then the (canonical) symplectic form on T^*M is the differential 2 form

$$\omega := \sum_{i=1}^{n} dx^i \wedge dp_i \tag{1}$$

The manifold T^*M is orientable and its volume form

$$\omega^n = \omega \wedge \omega \wedge \ldots \wedge \omega \tag{2}$$

is invariant with respect to the vector field X_H on T^*M, called the *Hamiltonian vector field* whose action in local coordinates has the from

$$\frac{\mathrm{d}x^i}{\mathrm{d}t} = \frac{\partial H}{\partial p_i}, \quad \frac{\mathrm{d}p_i}{\mathrm{d}t} = -\frac{\partial H}{\partial x^i}, \quad i = 1, \ldots, n, \tag{3}$$

where the function $H = H(x,p)$ has usually the form

$$H(x,p) = \frac{1}{2}\sum_{i,j=1}^{n} g^{ij}(x)p_i p_j + U(x). \tag{4}$$

In turn, $(g^{ij}) = (g_{ij})^{-1}$ is a Riemann metric on T^*M defining the canonical isomorphism of spaces

$$T_xM \leftrightarrow T_x^*M = (T_xM)^*.$$

Thus $H \in C^\infty(T^*M)$, $U \in C^\infty(M)$.

The *kinetic energy* of this system of n degrees of freedom is the positive definite quadratic form of velocities

$$T(x,\dot{x}) = \frac{1}{2}\sum_{i,j=1}^{n} g_{ij}(x)\dot{x}^i\dot{x}^j, \quad \text{where } \dot{x}^i := \frac{\mathrm{d}x^i}{\mathrm{d}t} \tag{5}$$

which defines the Riemann metric (g_{ij}) on TM (or, as one use to say traditionally, on M).

The *Lagrange function (Lagrangian)* is the function

$$L := T - U, \tag{6}$$

where $U \in C^\infty(M)$ is the potential energy of the system. Therefore, a mechanical system (in the Lagrange approach) is the triple (M, g, U) and the principle of least action says that the motion of the system proceeds along the trajectory $c : [t_1, t_2] \to M$ being the extremal of the action (action functional)

$$\Phi(c) := \int_{t_1}^{t_2} L(\dot{c})dt, \quad \text{where } \dot{c}(t) \in T_{c(t)}M, \quad t \in [t_1, t_2]. \tag{7}$$

The local coordinates x^i of the trajectory of the system (M, g, U) satisfy therefore the Euler–Lagrange equations

$$\frac{\mathrm{d}}{\mathrm{d}t}\frac{\partial L}{\partial \dot{x}^i} = \frac{\partial L}{\partial x^i}, \quad i = 1, \ldots, n. \tag{8}$$

Physicists use to call $\frac{\partial L}{\partial x^i}$ the *generalized force* (which is just some one form); $\frac{\partial L}{\partial \dot{x}^i} := p_i$ are called the *coordinates of generalized momentum.* Thus we achieved the 'mechanical' interpretation of local coordinates (x^i, p_i) of the cotangent bundle T^*M.

The transition form Lagrangian to Hamiltonian approach is made with the help of the famous Legendre transformation. It is not hard to show that the Hamilton function (*Hamiltonian*) can be interpreted as the total energy of the system $E := T + U$ (one must perform Legendre transformation). One shows almost immediately that $X_H(H) = 0$, thus H is a first integral of system of Hamilton equations (3), or, in other words, a first integral of the vector field X_H. The equation $X_H(H) = 0$ expresses therefore the law of *conservation of mechanical energy.* Every integral curve of the field X_H lies on the level surface $H(x, p) = h$; the constant h is the total energy of the trajectory because $H(x, p) = E(x, \dot{x})$.

The local one-parameter group defined by X_H preserves the symplectic form ω on T^*M, and it is therefore a (local) one-parameter group of symplectic transformations of the symplectic manifold (T^*M, ω). Earlier, these symplectic transformations were also-called canonical transformations. One should recall at this point that the general theory of symplectic manifolds (V, ω), that is, $2n$-dimensional differentiable manifolds equipped with a differentiable two form ω of maximal rank (ω is non-degenerate, its matrix is invertible at any point), has naturally grown from the Hamiltonian optics and mechanics. The importance of these notions and theories related to them is enormous in modern mathematics: one should recall such fast developing and important branches of mathematics as the theory of dynamical systems, ergodic theory, and the theory of hyperbolic equations (generalizations of wave equation) which comprise independent fields of research and whose investigators use to forgot quite often about sources of these theories.

For ages, motions of cellestial bodies fascinated people (cf. the title of the Copernicus work). Every respectable mathematician starting from the times of Newton and Kepler till Poincaré was engaged in research in celestial mechanics. The problem of the 'system of the world', that is, the stability of the Solar System caused anxiety for great minds of the times of racionalisnm. The famous works of Lagrange, Poisson, and most of all Dirichlet on the theory of *three body problem* (Sun – Earth – Moon) worked in terms of the power series. Under the influence of Weierstrass and Mittag-Leffler, the king of Sweden, Oscar II funded a price for investigation of convergence of power series in t (time) for the problem of N bodies for all values of the parameter $t \in \mathbb{R}$. In 1889 this price was awarded to Poincaré who had showed that *no quantitative method was capable of solving this problem.* This was the final blow at quantitative methods and the end of some era, and, at the same time, the beginning of qualitative methods (among others, topological), so much more powerful. Poincaré started a new era of differential geometry, analysis, and topology. In Part I, we have already mentioned some wonderful fruits of these investigations (theorems of Poincaré, Lepschetz, Hopf, Chern). These remarks are intended to encourage the reader: mathematics is great, beautiful, and very much alive!

Chapter 16

Tangent Bundle TM. Vector, Fiber, Tensor and Tensor Densities, and Associate Bundles

Let us return, encouraged by what was said in Chapter 15, to the theory of vector bundles! Since the notion of tangent bundle TM of differentiable manifold is perhaps the most fundamental notion of modern analysis (which is today to large extent the analysis on manifolds) and the since tangent bundle is the 'mother of vector bundles' and, more generally, fiber bundles, we will present once again the construction of the bundle $TM \to M$ without defining tangent spaces T_xM. This construction can be extended – virtually without changes – to the case of general fiber bundles.

Thus, let $\varphi_j : \mathcal{O}_j \to \mathcal{O}'_j$, $j \in J$ be an atlas of the smooth manifold M modelled on the (Banach) space $\mathbf{F}$. Instead of the group of linear automorphisms of $\mathbf{F}$, $\mathrm{GL}(\mathbf{F})$ we will simply write $\mathrm{Aut}(\mathbf{F})$. From the definition of atlas, we know that

$$\varphi_i \circ \varphi_j^{-1} : \varphi_j(\mathcal{O}_i \cap \mathcal{O}_j) \to \varphi_i(\mathcal{O}_i \cap \mathcal{O}_j), \quad i, j \in J$$

are diffeomorphisms, and thus the derivative $(\varphi_i \circ \varphi_j^{-1})'(x) \equiv d_x(\varphi_i \circ \varphi_j^{-1})$ at the point $x' = \varphi_j(x)$ is an element of the group $\mathrm{Aut}(\mathbf{F})$; we will denote this element by $g_{ij}(x)$. We immediately check the equality

$$g_{ij}(x) \cdot g_{jk}(x) = g_{ik}(x), \quad x \in \mathcal{O}_i \cap \mathcal{O}_j \cap \mathcal{O}_k.$$

Thus we have the differentiable mappings

$$g_{ij} : \mathcal{O}_i \cap \mathcal{O}_j \to \mathrm{Aut}(\mathbf{F}), \tag{1}$$

such that

$$(2) \qquad g_{ij} \cdot g_{jk} = g_{ik} \quad \text{over } \mathcal{O}_i \cap \mathcal{O}_j \cap \mathcal{O}_k \quad \text{(cocycle relation).}$$

The family of mappings (g_{ij}), $i, j \in J$ satisfying relation (2) is called the *cocycle of tangent bundle* of the manifold M. This term is justified by the following theorem.

PROPOSITION 1 (ON RECONSTRUCTION). *Let $(\mathcal{O}_j)$, $j \in J$ be an open covering of the (smooth) manifold M modelled over $\mathbf{F}$; let $g_{ij} : \mathcal{O}_j \cap \mathcal{O}_i \to$ Aut$(\mathbf{F})$ satisfy (2). Then there exists a smooth vector bundle $\pi : E \to M$ with the differentiable atlas*

$$(3) \qquad \{h_j : E_{\mathcal{O}_j} :\equiv \pi^{-1}(\mathcal{O}_j) \overset{\simeq}{\to} \mathcal{O}_j \times \mathbf{F},\ j \in J\}$$

such that g_{ij} are transition mappings of this atlas.

This bundle is called the *tangent bundle* of the manifold M and is denoted by TM, or, more precisely, $(TM, \pi, \mathbf{F})$, or, even more precisely, $(TM, \pi, (g_{ij}), \operatorname{Aut}\mathbf{F})$. The space T_xM is isomorphic with $\mathbf{F}$ and is called the *tangent space* at point x or the *fiber* over $x \in M$.

PROOF. Let us take $E' = \bigcup_j \mathcal{O}_j \times \mathbf{F} \times \{j\} \subset M \times \mathbf{F} \times J$, with E' equipped with the induced topology of the space $M \times \mathbf{F} \times J$; J has discrete topology. We introduce on E' the equivalence relation as follows

$$(x, t, i) \sim (x', t', j) \Leftrightarrow x = x', \quad t = g_{ij}(x)t'.$$

The equality $g_{ij} \cdot g_{jk} = g_{ik}$ shows that the relation $\sim$ is an equivalence relation. Now, we equip $E = E'/\sim$ with the quotient topology, and we take $k : E' \to E$ to be a canonical mapping. Since the relation $\sim$ is consistent with the projection $E' \to M$, $(x, t, i) \to x$, we obtain the continuous mapping $\pi : E \to M$. Of course, $\pi^{-1}(x) \simeq \mathbf{F}$ and we have

$$(4) \qquad E_{\mathcal{O}_j} := \pi^{-1}(\mathcal{O}_j) = k(\mathcal{O}_j \times \mathbf{F} \times \{j\}),$$

and moreover, the restriction $k|\mathcal{O}_j \times \mathbf{F} \times \{j\} \to E_{\mathcal{O}_j}$ is an homeomorphism. The linear maps

$$(5) \qquad h_j : E_{\mathcal{O}_j} \to \mathcal{O}_j \times \mathbf{F}$$

are defined as inverses of the superposition of the restriction $k|\mathcal{O}_j \times \mathbf{F} \times \{j\}$ with the identification

$$\mathcal{O}_j \times \mathbf{F} \times \{j\} \simeq \mathcal{O}_j \times \mathbf{F}.$$

It follows from the construction above that for

$$(6) \qquad \varphi_{ij} = h_i \circ h_j^{-1} : \mathcal{O}_i \cap \mathcal{O}_j \times \mathbf{F} \to \mathcal{O}_i \cap \mathcal{O}_j \times \mathbf{F}$$

we have

$$\varphi_{ij}(x,t) = (x, g_{ij}(x)t) \quad \text{for } (x,t) \in \mathcal{O}_i \cap \mathcal{O}_j \times \mathbf{F},$$

and thus g_{ij} are the *transition mappings* related to the atlas (h_j), $j \in J$ of the bundle E. □

The bundle $E \to M$ is differentiable (smooth) if $g_{ij}, i, j \in J$ are differentiable (smooth), in that case, we say also that the atlas $A = (h_j)$, $j \in J$ is differentiable (smooth). Two atlases A and A' of the bundle $E \to M$ are *consistent* if $A \cap A'$ is a differentiable atlas. This relation is an equivalence relation. The equivalence class of differentiable atlases is called the *differentiable structure* (*linear* if $\mathbf{F}$ is a Banach space).

REMARK 1. A tangent vector v at the point $x \in M$ is an equivalence class $[(x,t,j)]$, where

$$(8) \qquad (x,t,i) \sim (x,t',j) \quad \text{if } t = g_{ij}(x)t'.$$

REMARK 2. A vector field (*tangent*) on an open set $\mathcal{O} \subset M$ is a smooth mapping $s_{\mathcal{O}} : \mathcal{O} \to E$ such that $\pi \circ s_{\mathcal{O}} = \mathrm{id}_{\mathcal{O}}$; it is represented therefore by the family of mappings

$$(9) \qquad s_i : \mathcal{O}_i \cap \mathcal{O} \to \mathbf{F}, \quad i \in J,$$

such that $s_i(x) = g_{ij}(x) f_j(x)$ for all $x \in \mathcal{O}_i \cap \mathcal{O}_j \cap \mathcal{O}$.

REMARK 3. Among others, the advantage of this formulation of vector fields on manifold – due, in principle, to Norman Steenrod – over the classical one is that, the vectors tangent to M were organized into the differentiable manifold $TM := E$ and therefore the methods of topology and analysis can be readily applied.

REMARK 4. On the set of all smooth (or continuous) vector fields $C^\infty(\mathcal{O})$ over $\mathcal{O} \subset M$ one can introduce, in a natural way, a restriction mapping, and in this way one obtains the sheaf C_E^∞ of smooth (or continuous C_E) sections of the bundle E.

REMARK 5. If $\mathbf{F} = \mathbb{R}^n$, then $C_E^\infty(\mathcal{O})$ is isomorphic with the space of all families (f_j), $j \in J$, $f_j \in C^\infty(\mathcal{O}_j \cap \mathcal{O})^n$ such that (9) holds.

We said already that the tangent bundle $TM \to M$ is the 'mother of all fiber bundles', among others, the tensor bundles over M. Before presenting their construction, let us define in general terms a fiber bundle over M with typical fiber $\mathbf{F}$, with $\mathbf{F}$ being a topological space and then present the construction of such bundles from a given cocycle of transition mappings.

DEFINITION. The *bundle of vector spaces*, or, in short, the *vector bundle* with typical fiber $\mathbf{F}$ over the topological space (differentiable manifold) M is the quadruple $E, \pi, M, \mathbf{F})$, where E is a topological space and $\pi : E \to M$ is a smooth surjection. Every fiber $E_x := \pi^{-1}(x)$ is a vector space isomorphic (linearly!) with the given vector Banach space $\mathbf{F}$ called the *typical fiber*. The axiom of *local triviality* holds: For every point $x \in M$ there exists an open neighbourhood $\mathcal{O} \ni x$ together with the homomorphism

$$h : E_{\mathcal{O}} := \pi^{-1}(\mathcal{O}) \to \mathcal{O} \times \mathbf{F}$$

with the following properties

1. h preserves fibers, that is the following diagram is commutative

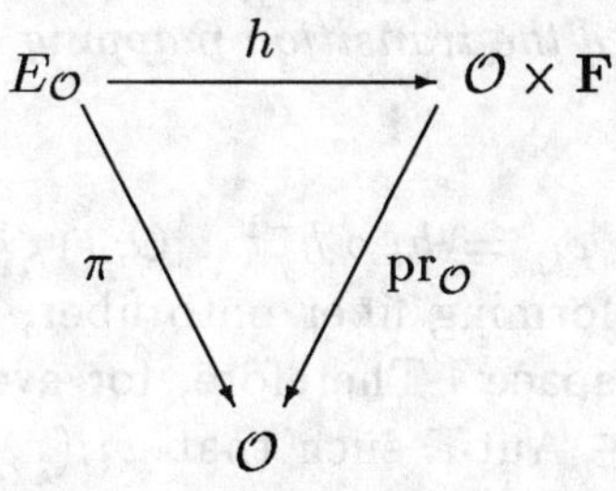

2. For every $x \in \mathcal{O}$, the mapping $h|E_x : E_x \to \xi \times \mathbf{F} \simeq \mathbf{F}$ is an isomorphism of Banach spaces.

The mapping $h : E_{\mathcal{O}} \to \mathcal{O} \times \mathbf{F}$ is called the *linear map* of E over $\mathcal{O}$. If $(\mathcal{O}_j)$, $j \in J$ is an open covering of M and $h_j : E_{\mathcal{O}_j} \to \mathcal{O}_j \times \mathbf{F}$ are corresponding linear maps, then the family (h_j), $j \in J$ is called the *atlas (of the bundle)* E. If $\dim \mathbf{F} = 1$ we talk about *line bundle*. The number $\dim \mathbf{F}$ is often called the *rank* of E.

A bundle $E \to M$ is *trivial*, if there exists a map $h : E \to M \times \mathbf{F}$.

The following simple theorem holds.

THEOREM 2 (ON EXISTENCE OF COCYCLE OF TRANSITION MAPPINGS). *Let* $(E, \pi, M, \mathbf{F})$ *be a vector bundle and* (h_j) *an atlas on* E

$$h_j : E_{\mathcal{O}_j} \to \mathcal{O}_j \times \mathbf{F}, \quad j \in J.$$

Then there exists a unique continuous mapping

$$g_{ij} : \mathcal{O}_i \cap \mathcal{O}_j \to \operatorname{Aut} \mathbf{F}$$

such that for the mappings

$$\varphi_{ij} := h_i \circ h_j^{-1} : \mathcal{O}_i \cap \mathcal{O}_j \times \mathbf{F} \to \mathcal{O}_i \cap \mathcal{O}_j \times \mathbf{F}$$

we have

$$\varphi_{ij}(x, f) = (x, g_{ij}(x) f) \quad \textit{for all } (x, f) \in \mathcal{O}_i \cap \mathcal{O}_j \times \mathbf{F}.$$

Over the point $x \in \mathcal{O}_i \cap \mathcal{O}_j \cap \mathcal{O}_k$*, the cocycle relation*

$$g_{ij} g_{jk} = g_{ik}, \quad i, j, k \in J$$

holds (from this relation, $g_{ii} = \mathrm{id}$, $g_{ij} = g_{ji}^{-1}$*).*

The mapping g_{ij} *is called the transition mapping (from the map* h_i *to the map* h_j*).*

PROOF. The mapping $\varphi_{ij} = h_i \circ h_j^{-1} : \mathcal{O}_i \cap \mathcal{O}_j \times \mathbf{F} \to \mathcal{O}_i \cap \mathcal{O}_j \times \mathbf{F}$ is a homeomorphism transforming fiber onto fiber, and on each fiber it is an isomorphism of vector space. Therefore, for every $x \in \mathcal{O}_i \cap \mathcal{O}_j$ there exists the mapping $g_{ij}(x) \in \operatorname{Aut} \mathbf{F}$ such that $\varphi_{ij}(x, f) = (x, g_{ij}(x) f)$. The continuity of the mapping

$$\mathcal{O}_i \cap \mathcal{O}_j \ni x \to g_{ij} \in \operatorname{Aut} \mathbf{F}$$

follows from the fact that φ_{ij} is a homeomorphism. The relation $g_{ij} g_{jk} = g_{ik}$ follows from the corresponding relation for φ_{ij}. □

REMARK 1. Taking in the definitions above any manifold or even topological space $\mathbf{F}$ in place of the Banach space $\mathbf{F}$, we obtain a fiber bundle with typical fiber $\mathbf{F}$. Of course, in this case one does not assume linearity of the isomorphisms $h|E_x$.

REMARK 2. Particularly important are bundles with typical fiber $\mathbf{F}$ being a topological group G. These are the so-called *principal bundles.* Then, the group G acts from the right on the bundle space E

$$E \times G \ni (z, g) \to z \cdot g \in E$$

and fibers E_x can be regarded as orbits

$$z \cdot G = \{p \in E : p = z \cdot g,\ g \in G\}.$$

We will devote two chapters to investigation of fiber bundles. Here we only note that the so-called *homogeneous spaces* $G \to G/H$ is in a natural way a principal bundle with typical fiber H, where H is a subgroup of the group G.

Without any changes in the proof, the theorem on reconstruction of fiber bundle from given cocycle holds. Since this theorem is basic for construction of many bundles we will present (one more time) its formulation and then the construction of tensor bundles.

THEOREM 3 (ON RECONSTRUCTION OF FIBER BUNDLE FROM TRANSITION COCYCLE). *Let G be a subgroup of the group* $\mathrm{Aut}(\mathbf{F})$ *of continuous (smooth ...) automorphisms of a topological space (differentiable manifold* $\mathbf{F}$*) understood as a group acting on* $\mathbf{F}$ *from the left. Let* $(\mathcal{O}_j)$, $j \in J$ *be a covering of the manifold* M. *With every set* $\mathcal{O}_i \cap \mathcal{O}_j$ *we associate a continuous (smooth ...) mappings*

$$g_{ij} : \mathcal{O}_i \cap \mathcal{O}_j \to G$$

and let

$$g_{ij} g_{jk} = g_{ik}, \quad \text{over } \mathcal{O}_i \cap \mathcal{O}_j \cap \mathcal{O}_k \quad i, j, k \in J$$

hold.

Then there exists a (unique up to isomorphism) bundle (E, π, M, G) *and a continuous (smooth ...) atlas*

$$\{h_j : E_{\mathcal{O}_j} \to \mathcal{O}_j \times \mathbf{F},\ \ j \in J\}$$

of the bundle E whose transition mappings are given by g_{ij}. Every mapping $s_j : \mathcal{O}_j \cap \mathcal{O} \to \mathbf{F}$ which represents sections $s \in C_E^\infty(x)$ satisfies

$$s_i(x) = g_{ij}(x)s_j(x) \quad \textit{for } x \in \mathcal{O}_i \cap \mathcal{O}_j \cap \mathcal{O}.$$

PROOF. Word for word as in the reconstruction for the tangent bundle TM. □

This theorem makes it possible to construct important vector bundles, for example tensor bundles on manifold M from the tangent bundle TM and cotangent bundle T^*M. The prescription is as follows. Let $\mathbf{F}_\alpha$ be a finite dimensional vector space of dimension r_α; then one can construct the direct sum $\mathbf{F}_1 \oplus \mathbf{F}_2$, the adjoint space $\mathbf{F}_{\alpha_2} := \mathbf{F}^*_{\alpha_1}$, the tensor product $\mathbf{F}_1 \otimes \mathbf{F}_2$, the exterior product $\mathbf{F}_1 \wedge \mathbf{F}_2$, the space of tensors

$$\mathbf{F}^r_s := \underbrace{\mathbf{F} \otimes \cdots \otimes \mathbf{F}}_{r} \otimes \underbrace{\mathbf{F}^* \otimes \cdots \otimes \mathbf{F}^*}_{s}$$

the Clifford algebra Cliff($\mathbf{F}$) over $\mathbf{F}$, the space of (anti)symmetric tensors, an so on. Each of these operations can be applied to the tangent bundle TM, cotangent bundle T^*M, and so on and one obtains the corresponding vector bundles, for example, the tensor product of the bundles $(E_1, \pi_1, M, \mathbf{F}_1) \otimes (E_2, \pi_2, M, \mathbf{F}_2)$ which is denoted for short as $E_1 \otimes E_2$, or $E_1 \wedge E_2$. The question what is the form of the cocycles of transition mappings is answered in the following table.

Vector bundle	Cocycle	Rank of bundle
A	$g_{ik} = \begin{bmatrix} a_{11}(x) & \dots & a_{1n}(x) \\ \dots & \dots & \dots \\ a_{m1}(x) & \dots & a_{mn}(x) \end{bmatrix}$	n
B	$h_{ik} = \begin{bmatrix} b_{11}(x) & \dots & b_{1n}(x) \\ \dots & \dots & \dots \\ b_{m1}(x) & \dots & b_{mn}(x) \end{bmatrix}$	m
$A \oplus B$	$g_{ik} \oplus h_{ik}(x) = \begin{bmatrix} g_{ik}(x) & 0 \\ 0 & h_{ik}(x) \end{bmatrix}$	$n + m$

Vector bundle	Cocycle	Rank of bundle
A^*	$g^*_{ik}(x) = {}^t g_{ki}(x) = {}^t g_{ik}(x)^{-1} = \check{g}_{ik}(x)$	n
$A \otimes B$	$g_{ik} \otimes h_{ik}(x) = (a_{\mu\nu}(x) \cdot b_{\kappa\lambda}(x)),$ where (ν, κ) is the index of row (μ, λ) is the index of column $(\nu, \mu) = 1, \ldots, n;$ $(\kappa, \lambda) = 1, \ldots, m$	$n \cdot m$
$\mathrm{Hom}(A, B)$	$h_{ik} \otimes \check{g}_{ik}(x)$	$n \cdot m$
$\Lambda^p A$	$\Lambda^p g_{ij}(x) = (c_{i_1 \ldots i_p, j_1 \ldots j_p}(x))$ where $i_1 \ldots i_p$ is the index of row $j_1 \ldots j_p$ is an index of column $1 \leqslant i_1 < \cdots < i_p \leqslant n$ $1 \leqslant i_1 < \cdots < i_p \leqslant n$ $c_{i_1 \ldots i_p, j_1 \ldots j_p} := \det(a_{i_\nu j_\kappa})$ $\nu, \kappa = 1, \ldots, p$	$\begin{pmatrix} n \\ p \end{pmatrix}$
TM $\dim M = n$	$g_{ik}(x) = (a^\mu_\nu)^n,$ where $a^\mu_\nu = \frac{\partial(y_1, \ldots, y_n)}{\partial(x_1, \ldots, x_n)}$ $g_{ik}(x) = (\varphi_i \circ \varphi_k)'(x),$ $\varphi_i \circ \varphi_k^{-1}(x) = y_\nu(x_1, \ldots, x_n),$ $\nu = 1, \ldots, n$	n
$T^*M := (TM)^*$		n
$\varkappa(M) = \Lambda^n T^* M$ $= \det(T^* M)$ $= (\det(TM))^*$	If $x^1_{\mathcal{O}_i}, \ldots, x^n_{\mathcal{O}_i}$ are coordinates in $\mathcal{O}_i$, then $g_{ij} = \frac{\partial(x^1_{\mathcal{O}_i}, \ldots, x^n_{\mathcal{O}_i})}{\partial(x^1_{\mathcal{O}_j}, \ldots, x^n_{\mathcal{O}_j})}$	1
$\det TM$	$h_{ij} = \frac{\partial(x^1_{\mathcal{O}_j}, \ldots, x^n_{\mathcal{O}_j})}{\partial(x^1_{\mathcal{O}_i}, \ldots, x^n_{\mathcal{O}_i})}$	1

REMARK. In the table, $\det A := \Lambda^r A$, $r = \mathrm{rank} A$ and is called the *deter-*

minant of bundle A or *determinant bundle.* The bundle $\varkappa(M)$ is called the *canonical bundle of manifold.* Both these bundles are line bundles.

Tensors (more precisely, sections of vector bundles) are the most important objects of classical physics and geometry. From the point of view of applications, tensor bundles are the most important bundles, indeed. These are obtained from the 'mother of all bundles' the tangent bundle (TM, π, M) as follows. Let the differentiable manifold M be n dimensional, that is, modelled on the space $\mathbb{R}^n$. The tangent bundle $TM \to M$ is a vector bundle with typical fiber $\mathbb{R}^n$ and cocycle a_{ij} being the Jacobi matrix (derivative) of the mapping $\varphi_i \circ \varphi_j^{-1}$, where $\varphi_i : \mathcal{O}_i \to \mathbb{R}^n$, $i \in J$ is an atlas of the manifold M. Obviously we have the cocycle relation

$$(*) \qquad a_{kj} \cdot a_{ji} = a_{ki} \quad \text{for } x \in \mathcal{O}_i \cap \mathcal{O}_j \cap \mathcal{O}_k.$$

Let G be a topological (Lie) group of transformations of the space $\mathbf{F}$ and let $\rho : \mathrm{GL}(\mathbb{R}^n) \to G$ be a continuous (differentiable) homomorphism. Let us take

$$(**) \qquad g_{ij}(x) := \rho a_{ij}(x).$$

It follows from $(*)$ that g_{ij}, $i, j \in J$ is a cocycle, and therefore from reconstruction theorem it follows that there exists a fiber bundle $E \to M$ over M with the typical fiber $\mathbf{F}$, group G, and cocycle (g_{ij}) denoted $(E, \pi, M, \mathbf{F}, (g_{ij}))$. This bundle is called the *tensor bundle of type ρ over manifold M* and its sections $s : M \to E$ are called the *tensor fields of type ρ over manifold M.* We owe this general approach to tensor fields Charles Ehresmann (1940's) and Norman Steenrod who in 1951 wrote the first monograph on the theory of fiber bundles, the famous *The Topology of Fiber Bundles.* It is this book, where the unified theory of topological fiber bundles and connections in them was introduced for the first time. This book played an important role in the development of modern topology and differential geometry. It is written very clearly and understandably!

We owe Hermann Weyl the notion of tensor densities, which are the correct mathematical objects to describe physical quantities like field strengths. Following Steenrod, we will show how these quantities can be described in the language of tensors of type ρ.

DEFINITION. Let $a \in \mathrm{GL}(n, \mathbb{R}) := \mathrm{Aut}(\mathbb{R}^n)$; $\breve{a}$ is an *inverse matrix* of the transposed matrix ${}^t a$ (earlier $\breve{a}$ was often called *contragradient matrix*). Let us denote by $|a| = \det(a)$ the determinant of the matrix a. The fiber $\mathbf{F}$ has the coordinates ($\dim \mathbf{F} = s + t$)

$$f^{\alpha_1 \dots \alpha_s}_{\beta_1 \dots \beta_t}, \quad \alpha_i, \beta_j = 1, \dots, n,$$

and let $w \in \mathbb{R}$. The linear representation $\rho : \mathrm{GL}(n, \mathbb{R}) \to G$ is defined by the formula

$$(\rho(a) \cdot f)^{\alpha_1 \dots \alpha_s}_{\beta_1 \dots \beta_t} = |a|^w a^{\alpha_1}_{\gamma_1} \cdots a^{\alpha_s}_{\gamma_s} f^{\gamma_1 \dots \gamma_s}_{\delta_1 \dots \delta_t} \check{a}^{\delta_1}_{\beta_1} \cdots \check{a}^{\delta_t}_{\beta_t}$$

where, on the right hand side we employed the Einstein's summation convention, that is, all repeating upper and lower indices are summed from 1 to n. Tensor fields of type ρ are called *s-contravariant and t-covariant densities of weight w*. If $w = 1$, then a tensor density t is called *Weyl density*. If $w = 0$ and $s = 0$, we have fields of t-linear forms or t-covariant tensors. The most important in this class are the totally antisymmetric (or totally symmetric) fields. Differential t forms are totally antisymmetric t-covariant tensor fields.

The field of symmetric bi-covariant tensors which are positive definite is the *Riemann metric tensor*. We encountered such tensors already many times. Another tensors of great importance for mathematics and physics are *curvature tensors* very closely related with curvature forms in vector bundles, in particular in the tangent bundle TM; in the latter case one shortly talks about curvature tensor of manifold M.

As Theorem 3 on reconstruction clearly shows, the *most important* ingredient of a vector bundle with typical fiber $\mathbf{F}$ and cocycle (g_{ij}) is, in fact, the G-cocycle (g_{ij}) whose values belong to the group $\mathrm{Aut}\,\mathbf{F}$, the group of automorphisms of the typical fiber, or, more generally, to some subgroup $G \subset \mathrm{Aut}(\mathbf{F})$. In the Steenrod construction a representation ρ of the group $\mathrm{Aut}(\mathbb{R}^n)$

$$\rho : \mathrm{Aut}(\mathbb{R}^n) \to G$$

appeared.

It is clear that we have here two bundles:

I. A bundle whose typical fiber is a topological group H and which is defined by the cocycle (h_{ij}) with values in H; this is the so-called *H-principal bundle $P \to M$*.

II. Any representation, that is, the homomorphism $\rho : H \to G$ leads to a new bundle $P(\rho)$, whose cocycle is (ρh_{ij}). Thus, an H- principal bundle gives rise, generates with the help of the representation ρ, the whole family of bundles $P(\rho)$ called the bundles *associated with the bundle P*.

Ehresmann codified the procedure of turning from generic principal bundle to the bundles associated with it by the representation ρ. The germ of the idea of principal bundle was the great idea of Cartan of *bundles of frames* which he called *'méthode de repére mobile'* – the method of moving frame *(repére)*. We will return to the notions of principal bundles, bundles of frames, and associate bundles in the following chapters. Here I would like only to stress the courage and imagination of the creators of these constructions during the time when these ideas evolved and were taking the final shape:

1. Detachment of the tangent space T_xM from the ground M (tangent vectors are not some arrows touching the manifold M at the point $x \in M$).

2. Organization of the family $(T_xM)_{x\in M}$ into the tangent bundle $TM \to M$.

3. Detachment of the G-principal bundle P from the bundle TM (more generally, vector bundle).

4. Formation of the family $P(\rho)$ of associated bundles, $\rho \in \{$ representation of the group G in some representation space $\}$.

CHAPTER 17

G-spaces. Group Representations

Let G be a group and P a topological space (differentiable manifold). We say that G *acts on* P *from the right* if there exists a continuous (differentiable) mapping

$$\varphi : P \times G \to P$$

such that for all $a \in G$ the mapping

$$R_a := \varphi(\cdot, a) : P \to P$$

is a homomorphism (diffeomorphism) of P and

$$\varphi(\varphi(p, a), b) = \varphi(p, ab), \quad \text{thus } R_{ab} = R_a \circ R_b.$$

If e is a unit of G, then $R_e = \mathrm{id}_P$. For simplicity, we will use the simplified notation

$$R_a(p) = p \cdot a \quad \text{and thus } p \cdot (ab) = (p \cdot a) \cdot b.$$

Correspondingly, one defines the *left action* of G on P

$$\psi : G \times P \to P, \quad \psi(ab, p) = \psi(a, \psi(b, p)).$$

We write ψ in place of $\psi(a, p) \equiv L_a p \equiv a \cdot p$. Of course, $R_a L_b = L_b R_a$ (the left and right actions commute). Every right action defines the left action in a natural way, to wit

$$g \cdot x := x \cdot g^{-1}.$$

The *orbit*, more precisely, the *G-orbit* of the point p is the set $p \cdot G = \{p \cdot a : a \in G\}$ in the case of right action; for the left action, the definition is analogous. In what follows, we will only talk about right actions.

If the orbit of some point is the whole space, that is, if $p_0 \cdot G = P$, then, of course, this property holds for every $p \in P$ and we say that the group *acts transitively.* In other words

$$\forall_{p_1, p_2 \in P} \exists_{g \in G} p_2 = p_1 \cdot g.$$

Clearly, G acts transitively on each orbit.

The group G acts *freely* if

$$(p \cdot a = p \quad \text{for some} \quad p \in P) \Longrightarrow (a = e)$$

thus, no transformation R_a possesses fixed points for $a \neq e$. The group G *acts effectively* if

$$(R_a = \underset{P}{\mathrm{id}}) \Longrightarrow (a = e),$$

that is

$$(p \cdot a = p \quad \text{for all} \quad p \in P) \Longrightarrow (a = e)$$

The effectiveness of action of the group G can be therefore formulated as follows: observing that the mapping

$$G \ni g \stackrel{\rho}{\to} R_g \in \mathrm{Aut}(P)$$

is a homomorphism of the group G in the group of automorphisms (that is, permutations) of the set P, we say that this homomorphism $g \to R_g$ is a representation of the group G in the (representation) space P. Of course, the kernel $\ker \rho$ of this homomorphism is a normal subgroup of G. We have therefore the following definition.

DEFINITION. The action $\rho : G \to \mathrm{Aut}(P)$ is *effective* if $\ker \rho = \{1\}$, that is, if the kernel of the homomorphism ρ is trivial.

The group G acts *freely* on P if for every $x \in P$ the stabilizer $G_{(x)} = \{1\}$; thus no automorphism R_g, $g \neq 1$ possesses fixed points.

We have, of course, (free) $\Longrightarrow$(effective).

EXERCISE. Show that $\ker \rho = \bigcap_{x \in P} G_{(x)}$.

Usually, the set P is a topological space and even a differentiable manifold, also the group G is topological (for example, discrete) or even a Lie group, which means that G is a differentiable manifold itself (of finite dimension) and the group action is differentiable.

DEFINITION. If we have left (right) action of the group G on the space P, then the pair (P, G) is called the *left (right) G space.* If the action of G is transitive, then the pair (P, G) is called the *homogeneous space.*

Earlier, G spaces were called Klein spaces (Felix Klein, 25.04.1849–22.06.1925, was the first to define and investigate these spaces (cf. his famous *Erlangen Program* of 1872).) The *Erlangen Program* was an inauguration lecture of very young – 23 years old – professor and it revolutionized the contemporary geometry, unifying it to some extent. Vulgarizing and simplifying things somehow, one could say that the *Erlangen Program* posed it as the main goal of geometry to investigate G spaces, invariants of G transformations, and transformation groups themselves. In spite of the fact that the most important spaces, that is, Riemann spaces, are not, in general, G spaces, the theory of fiber bundles and, most of all, the theory of principal fiber bundles could be regarded as a continuation of the Klein's Program. Perhaps the most important class of G spaces is introduced by gigantic theory of group representations.

DEFINITION. Let V be a topological space (one most often considers a vector space V.) The homomorphism

$$\rho : G \to \mathrm{Aut}(V)$$

(that is $\rho(ab) = \rho(a)\rho(b)$) is called the *representation of the group* G in the space V; V is called the *representation space* of the group G. Therefore, the representation is the triple (ρ, G, V). For applications, the most important are *linear representations*: V is some vector space. Usually, such representations are called *group representations.* Of course, the representation (ρ, G, V) defines (or even is itself) a left action of the group G

$$G \times V \to V$$

$$g \cdot v := \rho(g)v.$$

If $G = \mathbb{R}$, then $\rho(t)$, $t \in \mathbb{R}$ is called the one-parameter group of transformations, in other words, $(\rho(t), t \in \mathbb{R})$ is a representation of 'time', that is, the group $\mathbb{R}$ in the space V.

The theory of group representations is not a single tool in investigations of groups (following the ancient rule 'if you want to to understand the essence of some notion, investigate as it acts in different situations', that is, 'investigate its representations'), but became an irreplacible and more and more often used technique of modern physics and chemistry, and one of the most powerful unifying principles saving the modern mathematics from breaking into disconnected parts. Soon after quantum mechanics was created (Heisenberg, 1925, Schrödinger, 1926, Dirac, 1926–27), Hermann Weyl, and, independently, Eugen Wigner in 1927 introduced the theory of group representations into quantum theory. (Weyl formulated the Heisenberg commutational relations in terms of group representations, avoiding in this way difficult problems related to domains of unbounded operators in Hilbert spaces. Wigner started investigations of representations of Lorentz and Poincaré groups.) The fundamental work of Weyl of 1927 *Gruppentheorie und Quantummechanik* was extended in 1928 to the wonderful monograph (of the same title) with unbelievable richness of ideas.

One can look at the evolution of a (quantum) physical system as at a one-parameter group of unitary transformations $\rho(t)$, $t \in \mathbb{R}$ of a Hilbert space $V = (V, (\cdot|\cdot))$ of possible states of the system. In 1931, Marshall Stone showed that every such group is generated (uniquely defined) by a self-adjoint operator (usually not bounded!) in the space V, more precisely,

$$\rho(t) = \exp(itH), \quad t \in \mathbb{R}.$$

This famous Stone theorem was already suggested by Weyl in his *Gruppentheorie und Quantummechanik*!

After this digression showing not only a role of transformation groups and also unity of mathematics and physics, let us turn to further examples, each of whose is a starting point of a large theory.

THEOREM 1. *Let G be a Lie group, $H \subset G$ its closed subgroup; H acts on G from the right: $G \times H \to G$, $g \to g \cdot h$. The (homogeneous) space G/H is a differentiable manifold. The fibration $\pi : G \to G/H$, where π is a projection $g \to g \cdot H := \{gh : h \in H\}$ possesses as fibers the H-orbits: $\pi^{-1}(g) = g \cdot H$ (the typical fiber is H, of course).*

The proof can be found in any textbook on the Lie group theory or differential geometry.

REMARK. The space of orbits G/H is often called *homogeneous space.* This terminology is justified by the following proposition.

PROPOSITION 2. *Let M be a homogeneous space, that is, on the manifold a Lie group G acts transitively (from the left). Let $m \in M$ and let $G_{(m)} := \{g \in G : g \cdot m = m\}$ be an isotropy group of the point m. Then the mapping*

$$f_m : G/G_{(m)} \to M \quad \textit{where } gG_{(m)} \to g \cdot m$$

is a diffeomorphism.

PROOF. The mapping f_m is well defined, and is an injection; the assumption of transitivity of the action of G guarantees that this mapping is a surjection, too. It remains to show that f_m is an *immersion*, that is, that its differential Tf_m is injective at every point. But f_m is *G-equivariant*, that is $f_m(g \cdot m) = g \cdot f_m(m)$ and thus has constant rank. Since f_m is an injection, it is an immersion, by virtue of the implicit function theorem. □

DEFINITION I. Let G be a Lie group acting from the right on the manifold P, $P \times G \to P$. The quadruple (P, π, M, G) is a *smooth fiber bundle* over M with typical fiber G.

More precisely:

1. $M \simeq P/G$ is the space of G-orbits. Thus every fiber $\pi^{-1}(m)$ is a G-orbit.

2. P is locally trivial (with typical fiber $\simeq G$); every point of the base M possesses a neighbourhood $\mathcal{O}$ such that there exists the equivariant diffeomorphism φ called the *map of the bundle.*

3. $\varphi : \pi^{-1}(\mathcal{O}) \to \mathcal{O} \times G$, $\varphi(p \cdot b) = \varphi(p) \cdot b$, where $\mathcal{O} \times G$ is a right G space with the group G acting on the second term $(m, a) \cdot b := (m, ab)$.

Chapter 18

Principal and Associated Bundles

Particularly important, and in some sense conceptually natural, are the bundles having as their typical fiber the group G itself, that is the bundles (P, π, M, G), where the group G acts on itself by right multiplication

$$G \ni g \to ga \in G, \quad a \in G.$$

Thus, if we have a G-fiber bundle $(E, \pi, M, \mathbf{F}) \simeq E_g$ defined by the cocycle $g = (g_{ij})$ of transition mappings, in a unique way we can associate with this bundle the principal bundle (P, π, M, G) having the same cocycle $g = (g_{ij})$, $i, j \in J$. This bundle is called the *principal bundle associated with bundle* E_g. We will see in a moment that also vice versa, having a principal bundle (P, π, M, G) and representation $\rho : G \to \mathbf{F}$ of the group G in the space $\mathbf{F}$, one can construct in a natural way the bundle $P \times_\rho \mathbf{F}$ over M, having the typical fibre $\mathbf{F}$. The construction goes as follows:

In the Cartesian product $P \times \mathbf{F}$ we define an equivalence relation: we denote $\rho(g)f$ by $g \cdot f$ and we have

$$(z \cdot g, f) \sim (z, g \cdot f), \quad z \in P,\ f \in \mathbf{F} \tag{1}$$

or, if you like, $(z, f) \cdot g := (z \cdot g, g^{-1} \cdot f)$,

$$P \times_\rho \mathbf{F} := (P \times \mathbf{F})/ \sim = (P \times \mathbf{F}/G. \tag{2}$$

Let $q : P \times \mathbf{F} \to (P \times \mathbf{F})/ \sim$ denote the canonical projection, that is,

$$q(z, f) := (z, f)/G.$$

We have the commutative diagram

$$\begin{array}{ccc} P \times \mathbf{F} & \xrightarrow{\quad q \quad} & P \times_\rho \mathbf{F} \\ \Big\downarrow{\scriptstyle \pi_P} & & \Big\downarrow{\scriptstyle \pi} \\ P & \xrightarrow{\quad p \quad} & M \end{array}$$

It follows from this construction that the bundle $(P \times_\rho \mathbf{F}, \pi, M)$ has the typical fibre $\mathbf{F}$, that is $\pi^{-1}(m) = \mathbf{F}$. Such bundles is called the *bundles associated with the principal bundle* (P, p, M, G) defined by the representation $(\rho, \mathbf{F})$ of the group G in $\mathbf{F}$.

The notion of principal bundle is very important; for this reason we present two more definitions of this notion.

Two maps $\varphi_j : \pi^{-1}(\mathcal{O}_j) \to \mathcal{O}_j \times G$, $j = 1, 2$ give the transition map over $\mathcal{O}_1 \cap \mathcal{O}_2 \equiv \mathcal{O}_{12}$ of the following form

$$\text{(3)} \quad \mathcal{O}_{12} \times G \xrightarrow{\varphi_1^{-1}} \varphi^{-1}(\mathcal{O}_{12}) \xrightarrow{\varphi_2} \mathcal{O}_{12} \times G, \quad (m, a) = (m, g_{12}(m) \cdot a),$$

where $g_{12} : \mathcal{O}_{12} \to G$ is defined by

$$\text{(3)} \qquad \varphi_2 \varphi_1^{-1}(m, a) = (m, g_{12}(m) \cdot a), \quad m \in \mathcal{O}_{12}\ a \in G.$$

The quadruple (P, π, M, G) is called the *G–principal bundle* over M .

REMARK 1. In definition above it should be noticed that G acts on P *from the right*, and the transition mappings $g_{12}(m)$ act on G *from the left.*

REMARK 2. Condition (3) implies that $\pi(p \cdot a) = \pi(p)$, $p \in P$, $a \in G$, and indeed the G-orbit of the point p is the fiber $\pi^{-1}(\pi(p))$ containing the point p.

REMARK 3. It follows from condition (3) that the action of G is free: $G_{(p)} = \{e\}$ for any $p \in P$. In particular, the orbits are submanifolds contained in P.

REMARK 4. Let us recall that a smooth fiber bundle over M is the quadruple $(P, \pi, M, \mathbf{F})$, where P, M, $\mathbf{F}$ are manifolds called, respectively, *bundle space*, *base manifold*, and *typical fiber*; $\pi : P \to M$ is a smooth projection on M. There exists an open covering $(\mathcal{O}_j)$ of the base manifold M and the family of commutative diagrams

$$\begin{array}{ccc} \pi^{-1}(\mathcal{O}_j) & \xrightarrow[\simeq]{\varphi_j} & \mathcal{O}_j \times \mathbf{F} \\ & {\scriptstyle\pi}\searrow \quad \swarrow & \\ & \mathcal{O}_j & \end{array}$$

(φ_j is a diffeomorphism). The family φ_j is called the *atlas of the bundle* $(P, \pi, M, \mathbf{F})$.

REMARK 5. We have $\varphi_{ij} := \varphi_i \varphi_j^{-1} : \mathcal{O}_i \cap \mathcal{O}_j \times \mathbf{F} \to \mathcal{O}_i \cap \mathcal{O}_j \times \mathbf{F}$, $\varphi_{ij}(m, f) = (m, g_{ij} f)$ for all $(m, f) \in \mathcal{O}_i \cap \mathcal{O}_j \times \mathbf{F}$.

The mappings g_{ij} are called *transition mappings* associated with the atlas (φ_j). We check that

$$(*) \qquad g_{ij} g_{jk} = g_{ik} \quad \text{(the so called \textit{cocycle relation})},$$

$g_{ij}(m) \in \mathrm{Aut}(\mathbf{F})$ holds.

Now we show the following important theorem.

THEOREM 1 (ON RECONSTRUCTION)

1. *Let M be a manifold, $(\mathcal{O}_j)_{j \in J}$ an open covering of M, and let*
$$g_{ij} : \mathcal{O}_i \cap \mathcal{O}_j \to \mathrm{Aut}(\mathbf{F})$$
satisfies the cocycle relation $(*)$. *Then there exists the smooth fiber bundle $(P, \pi, M, \mathbf{F}$ and the smooth atlas $\{\varphi_j = \pi^{-1}(\mathcal{O}_j) \to \mathcal{O}_j \times \mathbf{F}, j \in J\}$ whose transition mappings are g_{ij}.*

2. *If the smooth mappings $s_i : \mathcal{O}_i \to \mathbf{F}$ such that $s_i(x) = g_{ij}(x) s_j(x)$ for $x \in \mathcal{O}_i \cap \mathcal{O}_j$ are given, then there exists a (smooth) section $s : M \to P$ such that $\varphi_i(s(x)) = (x, s_i(x))$ for $x \in \mathcal{O}_i$.*

PROOF. *Ad 1.* Let

$$P' = \bigcup_{j \in J} \mathcal{O}_j \times \mathbf{F} \times \{j\} \subset M \times \mathbf{F} \times J$$

be equipped with the topology induced by the topology of $M \times \mathbf{F} \times J$ (J has a discrete topology.) On P' we introduce the equivalence relation

$$(x, f, i) \sim (x', f', j) \Leftrightarrow x = x',\ f = g_{ij} f'.$$

The cocycle relation shows that $\sim$ is indeed an equivalence. We take $P = P'/\sim$, equipping P with quotient topology. Let $k : P \to P'$ be the canonical projection. Since the relation $\sim$ is consistent with the projection $P' \to M$, the continuous mapping $\pi : P \to M$ is therefore induced. The fiber is obviously $\pi^{-1}(x) = \mathbf{F}$ (if $\mathbf{F}$ is a vector space, group, then the fibers $\pi^{-1}(x)$ are also vector spaces, groups etc.) We have

$$P_{\mathcal{O}_i} := \pi^{-1}(\mathcal{O}_i) = k(\mathcal{O}_i \times \mathbf{F} \times \{i\}),$$

and

$$(**) \qquad k|\mathcal{O}_i \times \mathbf{F} \times \{i\} \to P_{\mathcal{O}_i}$$

is a homeomorphism. The maps $\varphi_j : P_{\mathcal{O}_i} \to \mathcal{O}_i \times \mathbf{F}$ are defined as inverses of the homeomorphisms $(**)$ composed with the identification

$$\mathcal{O}_i \times \mathbf{F} \times \{i\} \simeq \mathcal{O}_i \times \mathbf{F}.$$

It follows from the construction above that the transition mappings defined by the atlas (φ_j), $j \in J$ are g_{ij}.

Ad 2. Obvious. □

DEFINITION (I). When $\mathbf{F}$ is a one dimensional vector space, then $(P, \pi, M, \mathbf{F})$ is called the *line bundle.*

DEFINITION (II). Let G be a Lie group. The *smooth principal bundle* with the structural group G is the pair $(\mathcal{P}, T)$, where:

1. $\mathcal{P} = (P, \pi, M, \mathbf{F})$ is a smooth fiber bundle,
2. $T : P \times G \to P$ is an action of G on P from the right,
3. $\mathcal{P}$ possesses the atlas $\psi_j := \varphi_j^{-1} : \mathcal{O}_j \times \mathbf{F} \to \mathcal{O}_j \times \mathbf{F}$, $j \in J$ such that

$$\psi_j(x, ab) = \psi_j(x, a) \cdot b, \quad x \in \mathcal{O}_j,\ a, b \in G.$$

(Above we denoted $T(z, a)$ by $z \cdot a$.)

Let us make some important remarks.

REMARK 1. (3)$\Longrightarrow p(z \cdot g) = p(z)$, $z \in P$, $g \in G$; moreover the action of T is *free* ($G_z = \{e\}$.) In particular, the G-orbits are submanifolds of the manifold P. Sometimes we will denote these orbits by $G_x := p^{-1}(x)$, $x \in M$.

REMARK 2. Thus $z \cdot G$ is a fiber containing z:

$$p^{-1}(p(z)) = z \cdot G.$$

We have an important characterizations of a principal bundle which we present in the form of the following theorem.

THEOREM 2. *Let $\mathcal{P} = (P, \pi, M, \mathbf{F})$ be a smooth fiber bundle. Next, let $T : P\times \to P$ be a smooth, free, right action whose orbits are identical with fibers of the bundle. Then $(\mathcal{P}, T)$ is a principal bundle.*

PROOF. Let $(\mathcal{O}_j)$, $j \in J$ be an open covering of M such that there exist sections $s_j : \mathcal{O}_j \to P$ $(p \circ s_j = \mathrm{id}_{\mathcal{O}_j})$. Define

$$\psi_j : \mathcal{O}_j \times G \overset{\simeq}{\longrightarrow} p^{-1}(\mathcal{O}_j)$$

by taking

$$\psi_j(x, g) := s_j(x) \cdot g, \quad g \in G.$$

Then $(\psi_j, \mathcal{O}_j)$, $j \in J$ is an atlas of the bundle $\mathcal{P}$ satisfying 3, because

$$\psi_j(x, ab) = s_j(x) \cdot ab = (s_j(x) \cdot a) \cdot b = \psi_j(x, a) \cdot b.$$

□

DEFINITION (III). The *principal bundle* is a fiber bundle (P, p, M, G) satisfying assumptions of the above theorem.

The existence of global sections is a very restrictive property of the bundle. The following theorem holds.

THEOREM 3. *The principal bundle $\mathcal{P} = (P, p, M, G)$ is trivial) $\Longleftrightarrow$($\mathcal{P}$ possesses a smooth global section).*

PROOF. $\Longleftarrow$ Let $\mathcal{P}$ possesses a section s over the open set $\mathcal{O} \subset M$. Then s defines the homomorphism $\varphi : \mathcal{O} \times G \to P$ of principal bundles defined by the formula

$$\varphi(x,g) = s(x) \cdot g \quad x \in \mathcal{O},\ g \in G,$$

which can be regarded as an isomorphism of a trivial bundle with the restriction $P_{\mathcal{O}} := p^{-1}(\mathcal{O})$ of the bundle P to $\mathcal{O}$. In particular, if $\mathcal{O} = M$, we get the thesis.

$\Longrightarrow$ Obvious. □

In the theory of Riemann surfaces (that is, connected, one dimensional, holomorphic complex manifolds) an important role is played by line bundles $E \to M$ over M, the so called holomorphic bundles, whose fibers are one dimensional: $\dim_{\mathbb{C}} E_x = 1$.

The following, quite simple, theorem holds.

THEOREM 4. *Let M be a Riemann surface, $p : E \to M$ a holomorphic vector bundle over M of rank n (the typical fiber $\mathbf{F}$ has dimension $n = \dim_{\mathbb{C}} \mathbf{F}$.) Then the following conditions are equivalent:*

1. *The bundle E is holomorphically trivial.*
2. *There exist n globally holomorphic sections $f_1, f_2, \ldots, f_n$ of the bundle E such that over any $x \in M$ the vectors $f_1(x), f_2(x), \ldots, f_n(x) \in E$ are linearly independent.*
3. *The cocycle (g_{ij}) of the transition mappings decomposes, that is, there exist holomorphic $(n \times n)$-matrices*

$$g_i \in \mathrm{GL}(n, \mathcal{A}(\mathcal{O}_i))$$

(that is, their elements are holomorphic functions on $\mathcal{O}_i \subset M$, where $(\mathcal{O}_i)$, $i \in I$ is a covering of the Riemann surface M), such that

$$g_{ij} = g_i g_j^{-1}, \quad \text{over } \mathcal{O}_i \cap \mathcal{O}_j,\ i, j \in I.$$

The proof of this theorem is not difficult and can be found in many textbooks. As an example, we present the proof of

(1) $\Longrightarrow$ (2). We have therefore the map $h : E \to M \times \mathbb{C}^n$. Let $e_1, \ldots, e_n$ be a canonical basis in $\mathbb{C}^n$, and $f_1, \ldots, f_n$ sections of E such that

$$h(f_k(x)) = (x, e_k) \quad \text{for all } x \in M,\ k = 1, \ldots, n.$$

Then all f_k are holomorphic and linearly independent at any $x \in M$. □

Historical remark. From the point of view of the history of ideas, it is interesting to observe that sections of vector bundles appeared earlier than vectors: first people dealt with vector fields, for example, velocity, which is a vector field, as force is a covector field – and thus some section of the cotangent bundle $T^*M \to M$. Similarly, tensor fields appeared earlier than tensors: the so called *stress tensor* or *deformation tensor* are tensor fields (that is, a section of some tensor bundle!) over M. From this it could be seen how long and tough was going from Galileo, Kepler, and Newton, from the early days of dynamics, hydrodynamics, and the theory of elasticity to the modern mathematical structures: differential manifold, vector bundle (and general) fiber bundle of Lie group, transformation group, principal bundle, connections in vector and principal bundles, parallel transport .... The milestones on this road were built by great mathematicians and physicists: Riemann, F. Klein, Poincar'e, Grassmann, Lie, Weyl, H. Hopf, E. Cartan, Steenrod, Chern, just to mention only the greatest ones.

Vector fields on a manifold M, that is, sections of the bundle TM, are of great importance not only for physics, but also make it possible to get insight into global structure of the manifold M, its topology. One of the most beautiful and most important phenomenon in this area (which we have already discussed in Part I of this book) is the famous relation between indices of the vector field X on compact manifold M and its Euler characteristic.

THEOREM 5 (POINCAR'E). *Let M be a compact surface (oriented) of genus p_M, and let V be a vector field on M. With any critical point of the field $V \not\equiv 0$ (that is, the point $x_0 \in M$ such that $V(x_0) = 0$) one can associate the integer* index $V(x_0)$ *called the index of the field V in x_0 (with non-critical points we associate* index $V(x) = 0$.*) Then the equality*

$$\sum_{x \in M} \text{index } V(x) = 2(1 - p_M)$$

holds.

REMARK. Since the field has only a finite number of critical points, the left hand side is a finite sum.

Chapter 19

Induced Representations and Associated Bundles

The most important source of group representations are the so-called *induced representations*, the notion which was introduced explicitely for *finite* groups at the end of previous century by Frobenius. The case of *compact* groups was treated systematically in the classical monograph of André Weil (1938-39), and the general theory for the case of arbitrary locally compact groups was developed by George Mackey in 1950s. It is interesting that the main impulse for the Mackey's work was the famous work of (the physicists!) E. Wigner from 1939 on representations of the Poincaré group. Since I devoted a separate monograph to the theory of group representations and there is a number of excellent textbooks devoted to this subject, in this chapter I will present only two definitions of induced representation, the second of whose will be based on the concept of associated bundle.

DEFINITION (I). Let K be a subgroup of the group G and let (ρ, V) be a representation of the group K in the (vector) space V. Let $\mathcal{H}^\rho$ be a space of mappings $f : G \to V$ such that

$$f(g_1 h) = \rho(h) f(g_1), \quad h \in K,\ g_1 \in G.$$

The mappings $U^\rho(g) : \mathcal{H} \to \mathcal{H}$ defined by the formula

$$U^\rho(g) f(x) := f(g^{-1}x), \quad f \in \mathcal{H}^\rho,\ x \in G$$

define the representation $(U^\rho, \mathcal{H}^r)$ of the group G in the space $\mathcal{H}^\rho$, called the *representation induced by representation* (ρ, V) of the subgroup $K \subset G$.

We check that U^ρ is a representation of the group G. Indeed

$$U^\rho(gg')f(x) = f((gg')^{-1}x) = f(g'^{-1}g^{-1}x) = (U^\rho(g')f)(g^{-1}x) =$$

$$= U^\rho(g)U^\rho(g')f(x);$$

thus $U^\rho(gg') = U^\rho(g)U^\rho(g')$, and therefore U^ρ is a homomorphism of the group G in the group of transformations of the space $\mathcal{H}^\rho$.

DEFINITION (II). Let us consider the associated bundle $P \times_\rho V$ and K-principal bundle $(P, p, G/K)$. Let $M = G/K$. Let

$$H^\rho = \{\text{sections } f \text{ of the bundle } P \times_\rho V \to M\}.$$

The formula

$$(U^\rho f)(x) = f(g^{-1}x), \quad \text{where } x \in G/K$$

defines the *induced representation of the group* K in the space V.

EXERCISE. Compare these two definitions of induced representations.

EXAMPLE 1. The most important example is the so called *(left) regular representation.* The inducing representation is, in this case, the identity

$$\rho(k) = \mathrm{id}_V, \quad k \in K,$$

$$U^\rho(g)f(x) = f^{-1}(x), \quad x \in G$$

EXAMPLE 2. A bit more general: let X be a G-homogeneous space and

$$L_g x = g \cdot x, \quad U(g)f(x) := f(g, x)$$

is also often called the *regular representation.* Usually (if G is not finite), the space $\mathcal{H}^\rho$ is not a space (manifold) of finite dimension. But then one makes additional restrictive assumptions concerning the group G and its subgroup K:

1. The group G is locally compact;

2. The homogeneous space G/K is locally compact, the space V is either finite dimensional or a Hilbert space.

Then on the space G/K one constructs a left invariant measure (integral) μ and considers the space of sections of the associated bundle which are square integrable. This space is, in a natural way, a Hilbert space

$$\mathcal{H}^\rho = L^2(P(\rho) \to G/K; \mu).$$

Then the induced representation $(U^\rho, \mathcal{H}^\rho)$ is unitary.

Representations of this type are natural tools of complex analysis, and also of quantum mechanics.

EXAMPLE 3. When G is a compact Lie group, the homogeneous space G/K often carries the structure of complex manifold. Then it is useful to consider holomorphic sections of the holomorphic line bundle $L \to G/K$. The beautiful theorems of A. Borel, R. Bott, A. Weil, and B. Konstant give criteria for irreducibility of induced representations. Of course, these theorems require good knowledge of the theory of compact Lie groups, in particular, the theory of semisimple Lie groups and algebras.

CHAPTER 20

Vector Bundles and Locally Free Sheaves

Let $p : E \to M$ be a vector bundle of rank $r = \dim \mathbf{F} < \infty$. The smooth sections of the bundle E over $\mathcal{U} \subset M$ form the vector space $C^\infty(\mathcal{U}, E)$; let us recall that the section s over $\mathcal{U}$ is the mapping $s : \mathcal{U} \to M$ such that $p \circ s = \mathrm{id}_{\mathcal{U}}$. Obviously, the family $\mathcal{U} \to C^\infty(\mathcal{U}, E)$, where $\mathcal{U}$ runs through open sets in the manifold M possesses the structure of a sheaf. If $\mathcal{V} \subset \mathcal{U}$, then there exists a mapping $\rho^{\mathcal{V}}_{\mathcal{U}} : (\mathcal{U}, E) \to (\mathcal{V}, E)$ of restriction of sections over $\mathcal{U}$ to sections over $\mathcal{V}$. This sheaf is denoted by $\mathcal{C}^\infty_E$ and $(\mathcal{U}, E)$ by $\mathcal{C}^\infty_E(\mathcal{U})$ in order to be in agreement with notations used in the definition of sheaves.

Let us observe that the sheaf $\mathcal{C}^\infty_E$ is a sheaf of modules over the ring $C^\infty(M)$ of smooth functions. It is useful, in this context, to consider the sheaf $\mathcal{C}^\infty(M)$ which we denoted by $\mathcal{E}$ or $\mathcal{E}(M)$ with canonic restrictions. Every stalk $\mathcal{C}^\infty_{E,x}$ over $x \in M$ is a module over the local ring $\mathcal{C}^\infty_x(M) = \mathcal{E}_x$.

Since, for trivializations $E|\mathcal{U} \xrightarrow{\sim} \mathcal{U} \times \mathbb{R}^r$, $\mathcal{C}^\infty_E(\mathcal{U}) \simeq \mathcal{E}(\mathcal{U})^r$ (direct sum of r copies of $\mathcal{E}(\mathcal{U})$), such sheaves are called *locally free sheaves of rank r*. And thus we have the following theorem.

THEOREM 1. *The association $E \to \mathcal{C}^\infty_E$ is a one- to-one relation between vector bundles (of finite rank) and locally free sheaves of finite rank (both categories of objects are considered up to an isomorphism!)*

PROOF. It remains to associate with any locally free sheaf $\mathcal{F}$ of rank r over connected manifold M a vector bundle $E \to M$ of rank r. Let $(\mathcal{U}_i)$ be a covering of M such that $\mathcal{F}|\mathcal{U}_i$ is free and

$$h_i : \mathcal{F}|\mathcal{U}_i \to \mathcal{E}(\mathcal{U}_i)^{r_i}$$

is the corresponding isomorphism. Then

$$H_j \circ h_i^{-1} : \mathcal{E}(\mathcal{U}_i \cap \mathcal{U}_j)^{r_i} \to \mathcal{E}(\mathcal{U}_i \cap \mathcal{U}_j)^{r_j} \tag{1}$$

is an isomorphism of sheaves of modules. Since the space M is connected, all the numbers r_i are the same. We take $r_i = r$. But every endomorphism of a sheaf of modules $\mathcal{E}(\mathcal{U})^r$ is given by the $r \times r$ matrix $g(\mathcal{U})$, and thus the isomorphism (1) defines the $r \times r$ matrix g_{ij} whose elements are smooth functions ($\mathbb{R}$ valued) over $\mathcal{U}_i \cap \mathcal{U}_j$. We immediately check that the matrices g_{ij} satisfy the cocycle relation and thus they can be regarded as transition mappings of some smooth vector bundle $E \to M$ of rank r. The reader will check that $\mathcal{C}_E^\infty \simeq \mathcal{F}$. □

In this way the association $E \leftrightarrow \mathcal{C}_E^\infty$ makes it possible to relate with every homomorphism of vector bundles a homomorphism of locally free sheaves over $\mathcal{E}$. Using the terminology of the category theory, we have to do with equivalence of two categories of objects: finite rank vector bundles and locally free sheaves of finite rank. In this way, one can construct, for example, with the help of algebraic operations direct sum, tensor product, exterior product, homomorphisms, etc. on vector bundles, the corresponding objects of sheaf theory. However, the associated structures are completely different, as showed in the following warning.

WARNING! The fiber E_x over x of the bundle $E \to M$ is a vector space $\mathbf{F}$ and thus an object completely different from the stalk $\mathcal{C}_{E,x}^\infty$ of a sheaf over $x \in M$! This is clearly shown in the following example of a trivial bundle

$$E = M \to \mathbb{R}, \quad \text{thus } E_x = \mathbb{R},$$

$$\mathcal{C}_E^\infty = \mathcal{C}^\infty(M) = \mathcal{E}(M)$$

thus the stalk $\mathcal{E}_x$ is a local ring of germs of function at x (having discrete topology!)

So what is the relation between the fiber E_x of the bundle $E \to M$ and the stalk of the sheaf $\mathcal{C}_E^\infty$? The answer is

$$E_x \simeq (\mathcal{C}_E^\infty)_x / \mathfrak{m}_x (\mathcal{C}_E^\infty)_x. \tag{2}$$

As we know, the fiber $(\mathcal{C}_E^\infty)_x$ is a local ring, and thus it possesses a unique maximal ideal $\mathfrak{m}_x(\mathcal{C}_E^\infty)_x$.

In the case of tangent bundle $E = TM$, relation (2) takes the form known to us from the (algebraic) definition of tangent vectors and covectors, to wit

$$TM_x \simeq (\mathfrak{m}_x/\mathfrak{m}_x^2)^*$$

or, for cotangent bundle, $(T^*M)_x \simeq \mathfrak{m}_x/\mathfrak{m}_x^2$.

REMARK 1. The sets $\mathfrak{m}_x/\mathfrak{m}_x^2$ and $(\mathfrak{m}_x/\mathfrak{m}_x^2)^*$ are called, respectively, Zariski cotangent and Zariski tangent spaces at x.

REMARK 2. The sheaf theory has its sources in complex analysis and is now indispensable in the theory of complex spaces. Complex analysis works with locally free sheaves rather than with vector bundles.

CHAPTER 21

Axiom of Covering Homotopy

If $p : X \to Y$ is a locally topological covering, then p satisfies the so called

AXIOM (OF COVERING HOMOTOPY). *For arbitrary polyhedron W (that is, finite cell space) and for arbitrary homotopy*

$$f : W \times [0,1] \to Y \tag{1}$$

(f is, obviously, continuous) and for an arbitrary continuous mapping $g_0 : W \times \{0\} \to X$ such that

$$pg_0(x,0) = f(x,0) \tag{2}$$

there exists a continuous mapping $g : W \times [0,1] \to X$ satisfying

$$pg = f, \quad \text{and } g(x,0) = g_0. \tag{3}$$

REMARK 1. The spaces X, Y must not be Hausdorff spaces.

REMARK 2. If f, g satisfy (3), then we say that homotopy g covers homotopy f (for the covering p).

Therefore, the axiom of covering homotopy states that if the mapping g_0 covers $f(\cdot, 0)$, then the *full* homotopy f is being covered by some homotopy g.

This property is also phrased slightly different by saying that the *axiom of lifting homotopy to polyhedra* holds.

For our purposes this second formulation is perhaps more useful because connection in fiber bundle is just lifting of curves. We will show that fiber bundles possess the property of lifting homotopy to polyhedra: indeed, the following theorem holds.

THEOREM 1. *Let $p : E \to M$ be a fiber bundle. Then p satisfies the axiom of covering homotopy.*

PROOF. STEP I. Let us assume that $E = M \times \mathbf{F}$ is a trivial bundle. Then the covering mapping g_0 can be written in the form

$$g_0(x, 0) = (f(x, 0), h(x, 0)).$$

Taking

$$g_0(x, t) = (f(x, t), h(x, t)), \quad t \in [0, 1],$$

we obtain covering homotopy.

STEP II. Let now $W_0 \subset W$ be a cell space, and let on this space the homotopy

$$g_0 : W_0 \times [0, 1] \to E = M \times \mathbf{F}$$

be defined such that it covers homotopy f. In other words we have a given covering mapping on the subspace

$$(W_0 \times [0, 1]) \cup (W \times \{0\}) \subset W \times [0, 1].$$

We must extend the mapping g_0 to the covering mapping

$$g : W \times [0, 1] \to E.$$

Let us observe that it is sufficient to construct the extension of the mapping g_0 on a single cell $\mathbf{D}_i^k \times [0, 1] \subset W \times [0, 1]$.

The restricted mapping g_0 defines a covering mapping on the subset $S := (\mathbf{D}_i^k \times \{0\}) \cup (S_i^{k-1} \times [0, 1])$ having the form of a full 'cup' with the bottom $\mathbf{D}_i^k \times \{0\}$ and the wall $S_i^{k-1} \times [0, 1]$ (cf. Fig. 1)

We transform the tube $\mathbf{D}_i^k \times [0, 1]$ topologically into itself such that the 'cup' S transforms into the 'bottom' $\mathbf{D}_i^k \times \{0\}$ (cf. Fig. 1). In this way the problem of extension of the covering mapping g_0 from the set S onto the whole of $\mathbf{D}_i^k \times [0, 1]$ is equivalent with the construction of extension of covering mapping with the set $\mathbf{D}_i^k \times \{0\}$ onto the whole of $\mathbf{D}_i^k \times [0, 1]$. But this step was already performed.

LAST STEP. Now we make a sufficiently fine decomposition of the polyhedron W and the interval $[0, 1] = I$ into finite number of intervals I_l, such

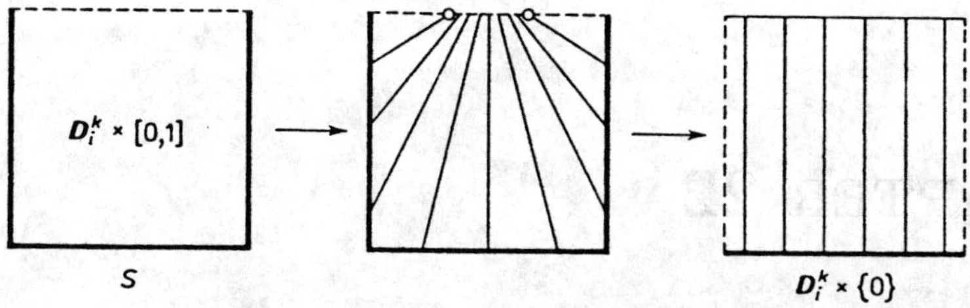

Fig. 1

that under mapping f, the product $\mathbf{D}^n \times I_l$ of the cell $\mathbf{D}^n \subset W$ and I_l maps into one trivialization of the covering $p : E \to M$. Then we have

$$f|\mathbf{D}^n \times I_l : \mathbf{D}^n \times I_l \to \mathcal{U} \subset M$$

and the mapping g maps $\mathbf{D}^n \times I_l$ into

$$p^{-1}(\mathcal{U}) = E|\mathcal{U} \overset{\sim}{\to} \mathcal{U} \times \mathbf{F}$$

which is a trivial bundle, for whose the construction of covering homotopy was made already. □

And thus, the objects of our utmost interest satisfy axiom of covering homotopy, that is the axiom of lifting homotopy to polyhedra.

CHAPTER 22

Serre Fibering. General Theory of Connection. Corollaries

In his (classical now) Ph.D. thesis *Homologie singuliére des espaces fibrés. Applications*, Ann. of Math. 54 (1951), 425–501, which, at that time, electrified the mathematical world, very young Jean-Pierre Serre (born in 1926) turn the things upside down: He takes the axiom of covering homotopy as the definition of a fiber space. In section 2 of chapter II of this wonderful work, he writes as follows: 'In order to find properties of the spectral sequence of homology of spaces ... we make use only of the *theorem on lifting homotopy to polyhedra*. And we take it for the definition ...' and the definition follows which today is regarded as a fundamental one in the theory of fiber spaces.

DEFINITION. The triple (E, p, M), where E, M are topological spaces and $p : E \to M$ is a continuous surjection is called the *Serre fibration* if the axiom of covering homotopy (that is the axiom of lifting homotopy to polyhedra) holds.

We see that examples of Serre fibrations are

1. Fiber bundles (this is what Theorem 1 of previous chapter says);
2. Fiber spaces of Hurewicz–Steenrod (the proof in the Steenrod's book);
3. Principal fiber bundles;
4. The spaces of paths $\tilde{\Omega}(X, x_0) \to X$ (as we will see in a moment, these spaces are not covered by the classification 1–3).

As these large classes of fiber spaces show, the class of Serre fiberings is very wide, and, what is more important, natural – as the beautiful theorems of Serre (from the work cited above) clearly show. One obtains new theorems in classical calculus of variations: *existence of infinitely many geodesics connecting two points of Riemann manifold M and existence of infinitely many geodesics in Riemann manifold M which transversally cross two given submanifolds contained in M.*

Our goal is much more modest: we will show that

a) Serre fibering is characterized by the possibility of lifting curves from the base M to the fiber space $E \to M$.

b) If that base M is locally contractible, then the Serre fibration $p : E \to M$ is a fiber bundle.

c) And most of all, the Serre approach makes it possible to look from a different perspective at the theory of connection in fiber bundles and at analytical extension of a germ of meromorphic function along a path.

As we saw, a linear connection, that is a parallel transport along curves $c : [0,1] \to M$ is nothing but a lift of the curve c to the curve $\tilde{c} : [0,1] \to E$ (in the bundle space) with given starting point $\tilde{c}(0) \in p^{-1}(x_0)$, $p\tilde{c} = c$, $c(0) = x_0$. We take this geometric property of parallel transport as the definition of connection in Serre fibering.

DEFINITION. Let $p : X \to Y$ be a continuous surjection, with Y, X being arbitrary topological spaces. Let us denote by $\Omega(Y,p,X)$ the set of pairs (c,y) with $c \in X^I$, $y \in Y$ such that c is a curve in X starting from the point $p(y)$, that is

$$\Omega(Y,p,X) = \{(c,y) \in X^I \times Y : c(0) = p(y)\},$$

where X^I is equipped with the topology of compact convergence and $\Omega(Y,p,X)$ with the relative topology in the product $X^I \times Y$. Similarly, Y^I is equipped with the compact topology of continuous mappings $I \to Y$. The *connection in the fibering* $p : Y \to X$ is the mapping

$$\Gamma : \Omega(Y,p,X) \to Y^I$$

with the following properties

C.1. Γ is continuous;

C.2. for an arbitrary pair $c,y) \in \Omega(Y,p,X)$, $\tilde{c} := \Gamma(c,y)$ is a curve in Y starting from the point y: $\tilde{c}(0) = y$;

C.3. $p\tilde{c} = c$, that is, $\tilde{c}$ is a lift of c.

Not in every Serre fibration a connection exists, however, we will show that the existence of a connection implies the property of lifting homotopy:

THEOREM 1. *If in the triple (Y, p, X) there does exist a connection Γ, then $p : Y \to X$ is a Serre fibering.*

PROOF. Actually, we will show even little more. Let W be a topological space (not necessary cellular) and let $F : W \to Y$ be continuous and H be a homotopy of the mapping $p \circ F : W \to X$. With the help of connection Γ we lift the homotopy H (into Y). For $w \in W$ we define $H_w \in H^I$ (which is a path in X) by the equation

$$H_w(t) := H(w, t).$$

Let $K : W \times I \to Y$ be given by

$$K(w, t) := \Gamma(H_w, F(w))(t).$$

From the property C.2 we have that $K(w, 0) = F(w)$; from the property C.3 it follows that $p(K(w, t)) = H_w(t) = H(w, t)$. It is sufficient to show that the mapping $K : W \times I \to Y$ is continuous. But from Bourbaki theorem (see below) we know that it is sufficient to show continuity of the mapping $G : W \to Y^I$ given by the formula

$$G(w) := \Gamma(H_w, F(w)).$$

From Bourbaki theorem and continuity of H it follows that $w \to H_w$ is continuous, but $w \to F(w)$ is continuous by assumption and Γ is a continuous mapping in both its arguments (property C.1). Therefore, G is continuous, and thus continuity of K is also proved.

THEOREM 2 (BOURBAKI). *Let X be a topological space and $\tilde{\Omega}(X, x_0)$ a space of paths $c : [0, 1] \to X$ starting at x_0 equipped with the compact topology. Let W be a topological space (not necessarily Hausdorff). For $F : W \to \tilde{\Omega}(X, x_0)$ to be continuous, it is sufficient and necessary that the mapping $G : W \times I \to X$ given by the formula*

$$G(w, t) := F(w)(t), \quad w \in W,\ t \in I = [0, 1]$$

be continuous.

PROOF (simple) can be found in Bourbaki or in the monograph by C. Teleman *Elemente de topologie si Varietati differentiabile*, Bucarest, 1961.

Using Bourbaki theorem, we show that the space $\tilde{\Omega}(X, x_0)$ is contractible.

THEOREM 3 (SERRE). *The space $\tilde{\Omega}(X, x_0)$ is contractible.*

PROOF. We define the mapping $h : \tilde{\Omega}(X, x_0) \times I \to \tilde{\Omega}(X, x_0)$ which with every path $c \in \tilde{\Omega}(X, x_0)$ and with every $t \in I$ associates the path $c_t := c(t, s)$, $s \in I$. Therefore, $h(c, t) := c_t$. Using Bourbaki theorem and taking $W = \tilde{\Omega}(X, x_0) \times I$, $F = h$, we will show that h is continuous. Thus, the mapping $G : W \times I \to X$ has the form

$$G((c, t), s) = G(w, s) = F(w)(s) = h(c, t)(s) = c_t(s) = c(ts).$$

Taking $t' = u(t, s) = ts$, we see that the mapping $(t, s) \to c(t')$ is continuous being a superposition of continuous mappings

$$(t, s) \to u(t, s), \quad t' \to c(t').$$

Thus the mapping $G : (c, (t, s)) \to c(ts)$ is continuous as well, because we know from Bourbaki theorem that the mapping $(c, t') \to c(t')$ is continuous, and G is a superposition of continuous mappings

$$(c, (t, s) \to (c, u(t, s)), \quad (c, t') \to c(t').$$

Therefore, a continuous mapping h is a continuous deformation of a (constant) h_0 which with very path c associates a constant path $h_0(c) : I \to \{x_0\}$. Thus h 'contracts $\tilde{\Omega}(X, x_0)$ to the point $\{x_0\}$.' □

DEFINITION. The *standard Serre fibering* is the fibering $p : \tilde{\Omega}(X, x_0) \to X$, where $p(c) := c(1)$, that is, with every path $c : [0, 1] \to X$ with the staring point x_0 one associates its endpoint. The fiber $p^{-1}(x)$ over x is the space $\Omega(X, x_0, x)$ of path with starting point x_0 and endpoint x; the fiber $p^{-1}(x)$ *is not finite dimensional.* Let us sketch the following interesting proposition.

PROPOSITION 4 (SERRE). *In a standard Serre fibering, connection does exist, it is therefore a Serre fibering.*

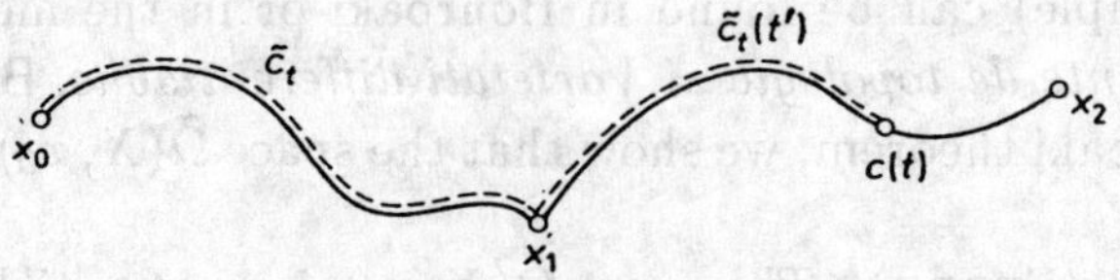

Fig. 1

PROOF (main line of reasoning). Let c be a path in X leading from x_1 to x_2: $c(1) = x_1$, $c(2) = x_2$. What the sentence 'x_1 is covered by y' really means? It means exactly that there exists a path c_1 starting at x_0 and ending at x_1, that is such that $c_1(0) = x_0$, $c_1(1) = x_1$. The lift $\tilde{c}$ (that is, covering) of the path $c \in X^{[0,1]}$ is a path in $\tilde{\Omega}(X, x_0)$, that is, a one-parameter family of paths $\tilde{c}_t$ starting at x_0 of the following form

$$\tilde{c}_t \in \tilde{\Omega}(X, x_0)$$

$$\tilde{c}_t(t') = \begin{cases} c_1(t') & \text{for} \quad 0 \leqslant t' \leqslant 1 \\ c(t') & \text{for} \quad 1 \leqslant t' \leqslant t \end{cases}$$

One can introduce the parameter $t' := 1/t + t$, and then t' runs through the interval $[0, 1]$). In Fig. 1, the dashed line denotes the lifted curve $\tilde{c}_t$

$$\Gamma(c, c_1)(t) := \tilde{c}_t = c_1 \circ c|_{[1,t]},$$

where

$$c_1 \circ c(s) = \begin{cases} c_1(2s) & \text{for} \quad 0 \leqslant s \leqslant \frac{1}{2} \\ c(2s-1) & \text{for} \quad \frac{1}{2} \leqslant s \leqslant 1. \end{cases}$$

Clearly, $p\Gamma(c, c_1) = c$, $\tilde{c}_0 = \Gamma(c, c_1)(0) = c_1$. It remains to show continuity of the mapping Γ; in order to do that one must make use of Bourbaki theorem.□

As a further consequence of the existence of connection in the triple (Y, p, X), we show local triviality of Serre fiberings.

Let us recall that the triple (Y, p, X) is a fiber bundle if it is locally trivial, that is, if every point x of the base X possesses a neighbourhood $\mathcal{O}$

and if there exists the topological space $\mathbf{E} = \mathbf{E}_0$ called the typical fiber (for $\mathcal{O}$) such that

$$p^{-1}(\mathcal{O}) =: Y|\mathcal{O} \simeq \mathcal{O} \times \mathbf{E}$$

The important theorem holds.

THEOREM 5. *If the fibering (Y, p, X) possesses a connection Γ and if the base X is locally contractible – for example, if X is a manifold – then (Y, p, X) is locally trivial, that is, it is a fiber bundle.*

PROOF. Let x be a point in X and let $\mathcal{O}$ be a contractible (to x) neighbourhood of x. For the assumption, it does exist

$$h : \mathcal{O} \times [0,1] \to \mathcal{O}$$

such that

$$h(x', 0) = x', \quad x' \in \mathcal{O},$$

$$h(x', 1) = x, \quad h(x, t) = x \quad \text{for } t \in [0,1].$$

With every $x' \in \mathcal{O}$ we associate a path $c_{x'}$ starting at x and ending at x': $c_{x'}(t) := h(x', 1-t)$, $t \in [0,1]$. Let us consider the mapping

$$\varphi : \mathcal{O} \times E_x \to p^{-1}(\mathcal{O}),$$

where $E_x := p^{-1}(x)$, given by

$$\varphi(x', e) := \Gamma(c_{x'}, e)(1).$$

We will show that φ is topologically 1. a bijection, 2. continuous. Indeed, φ is a superposition of two continuous mappings

$$(x', e) \to (x', e, 1) \quad \text{and} \quad (x', e, t) \to \Gamma(c_{x'}, e)(t).$$

The continuity of the second mapping is being proved as follows. From the Bourbaki theorem, the continuity of the mapping $x' \to c_{x'}$ follows. From the property C.3, the mapping $(x', e) \to \Gamma(c_{x'}, e)$ is continuous as a superposition of two continuous mappings.

It remains to show that φ^{-1} is continuous as well. It associates with every point $y' \in p^{-1}(\mathcal{O})$ the pair (x', y'), where $x' = p(y')$ and $y = \Gamma(\tilde{c}_x, y')(1)$. The mapping $y' \to p(y') = x'$ is continuous, similarly, is $y' \to y$ as a superposition of the continuous mappings $y' \to x' \to c_{x'} \to \tilde{c}_{x'}$,

$$(y', c) \to (y', c, 1), \quad (y', c, t) \to \Gamma(c, y')(t).$$

□

Corollary 6. *If the base X is contractible and connection does exist in (Y, p, X), then the bundle (Y, p, X) is trivial.*

Proof. In the theorem one takes $\mathcal{O} = X$. □

CHAPTER 23

Homology. Cohomology. de Rham Cohomology

In mathematics one often encounters the following situation. Let a sequence of abelian groups (modules) ... be given

(1*) $$\{C^n, n \in \mathbb{Z}\}$$

together with homomorphisms $d_n : C^n \to C^{n+1}$ (called *differentials* or *coboundary homomorphisms*) for whose

(2*) $$d_{n+1}d_n = 0 \quad \text{(zero group) for all } n \ .$$

Then $K = \{C^n, d_n, n \in \mathbb{Z}\}$ is called the *(co)chain complex.*

Since $\operatorname{im} d_{n-1} \subset \ker d_n$ we can form the quotient group

(3*) $$H^n(K) := \ker d_n / \operatorname{im} d_{n-1}$$

called the *nth cohomology group of the complex K*.

The dual notion is the complex of chains of abelian groups C_n

(1) $$K_0 = \{C_n, \partial_n, n \in \mathbb{Z}\};$$

we assume that $C_n = 0$ for $n < 0$.

Let us consider the homomorphisms $\partial_n : C_n \to C_{n-1}$ satisfying the relation

(2) $$\partial_n \partial_{n+1} = 0 \quad \text{for all } n \ .$$

Since $\operatorname{im} \partial_{n+1} := \partial_{n+1}(C_n) \subset \ker d_n$ we can form the quotient group

$$H_n(K_0) := \ker \partial_n / \operatorname{im} \partial_{n+1} \tag{3}$$

called the *nth homology group of the complex* K_0.

Now we will present an example which happen to be fundamental for the whole of mathematics.

de Rham complex. Let X be a differential manifold, $\mathcal{E}^{(r)} = \mathcal{E}^{(r)}(X)$ be a $\mathbb{R}$ vector space of differential forms of order r, $d_r : \mathcal{E}^{(r)} \to \mathcal{E}^{(r+1)}$ be the standard exterior derivative. As we know $d_{r+1} d_r = 0$ and thus $\{\mathcal{E}^{(r)}, d_r\} = K$ is a cochain complex; cohomology groups of this complex are called *de Rham cohomologies*:

$$H^r_{dR}(X) := \ker d_r / \operatorname{im} d_{r-1}. \tag{4}$$

Of course, $H^r_{dR}(X) = 0$ if $r > \dim X$. de Rham groups are, clearly, vector spaces over $\mathbb{R}$ and they may be of infinite dimension, however, the following important theorem holds.

THEOREM 1 (DE RHAM). *If X is a compact manifold, then* $\dim H^r_{dR} < \infty$.

The proof is not simple, it requires some acquittance with the theory of elliptic operators.

Along with exterior differentiation, the most important operation which can be performed on differential forms is integration. E. Cartan was led to the operation d by classical theorems of integral calculus (multiple integrals); Gauss, Green, Ostrogradski, and, most of all, Stokes theorems.

With any r dimensional simplex σ_r in oriented manifold X, that is, with any (smooth) mapping $\sigma_r : I^r \to X$, where $r = \dim X$ and with any form $\overset{r}{\varphi} \in \mathcal{E}^r$ with support in the image of σ_r we associate the real number

$$\int_{\sigma_r} \overset{r}{\varphi}$$

called the *integral of the form* $\overset{r}{\varphi}$. The r *chain* c_r is a formal combination of r simplexes

$$c_r = \sum_i \sigma^i_r, \quad n_i \in \mathbb{Z}, \quad \text{thus } c_r = \bigoplus \mathbb{Z}\sigma_r;$$

$$\partial_r \sigma_j := \sum_{k=0}^{r} (-1)^k \epsilon_k \sigma_j^k,$$

where σ_j^k denotes some $(r-1)$ simplex ('kth face' of σ_j) and $\epsilon_k = \pm 1$. It can be checked that $\partial_r \partial_{r+1} = 0$.

An integral of $\overset{r}{\varphi}$ over chain c_r is, of course, defined by linearity

$$< c_r, \overset{r}{\varphi} >= \sum_i \int_{\sigma_i} \overset{r}{\varphi}.$$

The surprising fact holds, whose proof is given in *Analysis* part II, and which, even for the such great mathematician as René Thom is – as he himself confesses – still a mystery. This is the following theorem.

THEOREM 2 (POINCARÉ–STOKES). *The formula*

$$\int_{\partial c_r} \overset{r-1}{\varphi} =< \partial c_r, \overset{r-1}{\varphi} >=< c_r, d\overset{r-1}{\varphi} > \quad (= \int_{c_r} d\overset{r-1}{\varphi}) \tag{5}$$

holds.

Formula (5) is a basis of mathematical analysis and classical field theory in physics; its beauty is unusual: it shows that the operation d of exterior derivative is dual to the operation ∂ of formation of boundary. As we will see later, for arbitrary abelian group A, and even the sheaf of abelian groups over manifold X, one can introduce a cohomology group (Cech) $H^r(X, A)$. The famous de Rham theorem states the there is an *isomorphism of vector spaces*

$$H^k(X, \mathbb{R}) \simeq H^k_{dR}(X, \mathbb{R}) \quad \text{for } k = 0, 1, \ldots, \dim X. \tag{6}$$

Let us now present the fundamental theorem

Theorem on exact sequence of cohomologies. The sequence of homomorphisms of groups

$$\cdots \to G_k \overset{\varphi_1}{\to} G_{k-1} \overset{\varphi_{k-1}}{\to} G_{k-2} \to \cdots$$

is exact in the term G_{k-1} if $\operatorname{im} \varphi_k = \ker \varphi_{k-1}$; if the sequence is exact in every term, it is called exact.

DEFINITION. The complex $K = \{C_1^n, d_n\}$ is a *subcomplex* of the complex $K = \{C^n, d_n\}$ if the groups C_1^n are subgroups of the groups C^n and $d_n(C_1^n) \subset C_1^{n+1}$. In such situation the differentials can be turned into the group $C_2^n := C^n/C_1^n$ to obtain the so-called *quotient complex* $\{C_2^n, d_n\} =: K_2$ denoted by K/K_1. We obtain the exact sequence of complexes

$$0 \to K_1 \to K \to K_2 \, (= K/K_1) \to 0 \tag{7}$$

Let us now form an exact sequence of cohomology groups of (above) complexes as follows

$$H^n(K) := \ker d_n / d_n(C^{n-1}), \quad H^n(K_1) := (\ker d_n \cap C_1^n)/d_n(C_1^{n-1}). \tag{8}$$

Since $C_1^n \subset C^n$, $d_n(C_1^{n-1}) \subset d_n(C^{n-1})$, by associating with the group element $\ker d_n \cap C_1^{n-1}/d_n(C_1^{n-1})$ its leaf with respect to the larger subgroup $d_n(C^{n-1})$, we obtain the homomorphism

$$i_n : H^n(K_1) \to H^n(K). \tag{9}$$

Analogously, by making use of the homomorphism $C^n \to C_2^n = C^n/C_1^n$, in obvious way we obtain the homomorphism

$$j_n : H^n(K_1) \to H^n(K_2). \tag{10}$$

But there exist also less obvious homomorphism

$$\delta_n : H^n(K_2) \to H^{n+1}(K_1)$$

(reminding the homomorphism ∂ in the sequence of homology group) which we construct as follows.

Let $u \in H^n(K_2)$; to it corresponds an element $y \in \ker d_n$ in the group C^n/C_1^n. Let us take its preimage $\bar{y}$ in the group C^n. Since $dy = 0$ in C^{n+1}/C_1^{n+1}, $d\bar{y} \in C_1^{n+1}$, and thus $d\bar{y} \in \ker d_{n+1}$. We check that the leaf $d\bar{y} + d_n C_1^n$ defines an element of the group $H^{n+1}(K_1)$ depending only on the initial element u (and not on auxiliary elements y and $\bar{y}$). In this way we get the homomorphism

$$\delta_n : H^n(K_2) \to H^{n+1}(K_1). \tag{11}$$

Tracing back the construction of the homomorphisms i_n, j_n, and δ_n, we obtain the following theorem.

THEOREM 3. (ON EXACT SEQUENCE OF COHOMOLOGIES). *Let*

$$0 \to K_1 \to K \to K_2 (= K/K_1) \to 0$$

be an exact sequence of chain complexes. Then the sequence

(12)

$$\cdots \overset{j_{n-1}}{\to} H^{n-1}(K_2) \overset{\delta_{n-1}}{\to} H^n(K_1) \overset{i_n}{\to} H^n(K) \overset{j_n}{\to} H^n(K_2) \overset{\delta_n}{\to} H^{n+1}(K_1) \overset{i_{n+1}}{\to} \cdots$$

is exact.

Chapter 24

Cohomology of Sheaves. Abstract de Rham Theorem

Cohomology groups of the space X with values in the sheaf $\mathcal{F}$ (of abelian groups) over X were defined by Leray following the earlier definition of Čech. Later this cohomology was codified and perfected by H. Cartan and J.-P. Serre. Here we will briefly sketch the leading ideas; for details we send the reader to many excellent textbooks and monographs.

Let X be a paracompact space, that is a manifold having countable basis of neighborhoods, or, more generally, being a countable sum of compact sets. As we know, in that case, for any covering $\mathfrak{U} = (\mathcal{U})_j$, $j \in J$ there exists a corresponding partition of unity, and in the case of smooth manifold X, an (arbitrary) differentiable partition of unity.

Let $\mathcal{F} = \{\mathcal{F}(\mathcal{U})\}$ be a sheaf of abelian groups. Let us recall: $\mathcal{U}$ runs through open subsets of the space X, $\mathcal{F}(\mathcal{U})$ is an abelian group (for example, vector space) and we have restriction homomorphisms $r^{\mathcal{V}}_{\mathcal{U}} : \mathcal{F}(\mathcal{U}) \to \mathcal{F}(\mathcal{V})$ for $\mathcal{V} \subset \mathcal{U}$ satisfying the standard relations. The system $\{\mathcal{F}(\mathcal{U}), r^{\mathcal{V}}_{\mathcal{U}}\}$ is called the *presheaf* and if the two further axioms (of sewing and uniqueness) are satisfied, it is called the *sheaf of abelian groups.*

For any multiindex $i = (i_0, i_p) \in J^{p+1}$, we assume

$$\mathcal{U}(i) = \mathcal{U}(i_0, \ldots, i_p) = \mathcal{U}_{i_0} \cap \mathcal{U}_{i_1} \cap \cdots \cap \mathcal{U}_{i_p}.$$

In order to define cohomology, we must characterize the groups $C^p(\mathfrak{U}, \mathcal{F})$.

Definition. The *p cochain* of the covering $\mathfrak{U}$ valued in the sheaf $\mathcal{F}$ is the mapping which associates with every multiindex $i \in J^{p+1}$ a section $c_i \in \Gamma(\mathcal{U}(i), \mathcal{F})$ such that c_i is alternating, that is, it is a skew symmetric function of i (it changes sign when two indices are altered); $\Gamma(\emptyset, \mathcal{F}) := 0$ – a one element group.

(1) $$C^p(\mathfrak{U}, \mathcal{F}) = \{\text{set of all } p \text{ chains } \mathfrak{U} \text{ valued in sheaf } \mathcal{F}\}$$

is, clearly, and abelian group. Now we must define the mapping

(2) $$\delta^p : C^p(\mathfrak{U}, \mathcal{F}) \to C^{p+1}(\mathfrak{U}, \mathcal{F}), \quad \delta^p(c)_i = \sum_{j=0}^{p+1} (-1)^j c_{i_0 \ldots \hat{i}_j \ldots i_{p+1}},$$

where $\hat{i}_j$ means that the index i_j was omitted. We will often write δ instead of δ^p, if it does not lead to any misunderstandings. We have

(3) $$\delta^{p+1}\delta^p = 0.$$

Now, following the general recipe, we form the abelian group

(4) $$H^p(\mathfrak{U}, \mathcal{F}) := \ker \delta^p / \operatorname{im} \delta^{p-1},$$

called the *pth cohomology group of covering* $\mathfrak{U}$ *valued in sheaf* $\mathcal{F}$.

$Z^p(\mathfrak{U}, \mathcal{F}) := \ker \delta^p$ is called the *group of p cocycles.* $B^p(\mathfrak{U}, \mathcal{F}) := \operatorname{im} \delta^{p-1}$ is called the *group of p boundaries.*

Since $c_{i_0} - c_{i_1} = 0$ over $\mathcal{U}_{i_0} \cap \mathcal{U}_{i_1}$ for all i_0, i_1, then (by definition of sheaf!) it follows the existence of $f \in \Gamma(X, \mathcal{F})$ whose restriction to $\mathcal{U}(i)$ is c_i; f is a global section of $\mathcal{F}$. Thus

(5) $$H^0(\mathfrak{U}, \mathcal{F}) \simeq \Gamma(X, \mathcal{F}) \quad \text{group of global sections of } \mathcal{F}.$$

In order to remove dependence of cohomology groups on the covering $\mathfrak{U}$, one checks that at if $\mathfrak{B} > \mathfrak{U}$ is a refinement of $\mathfrak{U}$, then we have a (unique) homomorphism $\sigma(\mathfrak{B}, \mathfrak{U}) : H^p(\mathfrak{U}, \mathcal{F}) \to H^p(\mathfrak{B}, \mathcal{F})$ which makes it possible to turn to the inductive limit

(6) $$H^p(X, \mathcal{F}) := \lim_{\underset{\mathfrak{U}}{\to}} H^p(\mathfrak{U}, \mathcal{F}),$$

called the *pth cohomology group of the space* X valued in sheaf $\mathcal{F}$. This makes it possible to obtain the following theorem which is analogous to the theorem on exact sequence of homotopies (see below).

THEOREM 1 *If X is paracompact and*

$$0 \to \mathcal{F} \to \mathcal{G} \to \mathcal{H} \to 0$$

is an exact sequence of of sheaves of abelian groups over X, then

$$0 \to H^0(X,\mathcal{F}) \to H^0(X,\mathcal{G}) \to H^0(X,\mathcal{H}) \to H^1(X,\mathcal{F}) \to$$
$$\to H^0(X,\mathcal{G}) \to H^0(X,\mathcal{H}) \to \ldots$$

is exact.

REMARK. Of course, taking inductive limit is an inconvenient operation, and therefore the *acyclic coverings* are important.

DEFINITION. An open covering $\mathfrak{U} = (\mathcal{U}_j)$, $j \in J$ is *acyclic* with respect to the sheaf $\mathcal{F}$ if $H^q(\mathcal{U}(i_0,\ldots,i_p),\mathcal{F}) = 0$ for all $p \geqslant 0$, $q \geqslant 1$.

For such coverings the following useful theorem holds.

THEOREM 2 (LERAY LEMMA). *If $\mathfrak{U}$ is a (locally finite) covering of X which is acyclic with respect to $\mathcal{F}$, then there exist the isomorphisms*

$$H^q(\mathfrak{U},\mathcal{F}) \simeq H^q(X,\mathcal{F}), \quad q = 0,1,2,\ldots$$

It is important to know for which sheaves of abelian groups $\mathcal{W}$ on paracompact space X, the cohomology groups $H^q(X,\mathcal{W})$, $q \geqslant 1$ vanish. Such are, for example, *fine sheaves.*

The definition of such sheaves is modeled on partition of unity.

DEFINITION. Let $\mathcal{F}$ be a sheaf over a paracompact X. Then the sheaf $\mathcal{F}$ is *fine* if for any locally finite open covering $\mathfrak{U} = (\mathcal{U}_j)$, $j \in J$ of the space X there exists a family $(h_i)_{i\in I}$ of homomorphisms $h_i : \mathcal{F} \to \mathcal{F}$ such that

1. For every $i \in I$ there exists an open set $A_i \subset X$ such that $A_i \subset \mathcal{U}_i$ and $h_i(\mathcal{F}_x) = 0$ for $x \notin A_i$.
2. $\sum_{i\in I} h_i =$ identity.

THEOREM 3. *Let $\mathcal{F}$ be a fine sheaf over paracompact space X. Then $H^q(X,\mathcal{F}) = 0$ for $q \geqslant 1$.*

PROOF. For $q \geqslant 1$, we define the homomorphism

$$k^q : C^q(\mathfrak{U}, \mathcal{F}) \to C^{q-1}(\mathfrak{U}, \mathcal{F})$$

such that for the co-differentials $\delta^q : C^q(\mathfrak{U}, \mathcal{F}) \to C^{q+1}(\mathfrak{U}, \mathcal{F})$

$$k^{q+1}\delta^q + \delta^{q-1}k^q = \text{identity}. \tag{7}$$

Then, of course, $H^q(\mathfrak{U}, \mathcal{F}) = 0$ for $q \geqslant 1$. Let us construct such k^q. Let $f \in C^q(\mathfrak{U}, \mathcal{F})$, then the cochain $k^q f$ associates with any multiindex $(i_0, \ldots, i_{q-1})$ the section $k^q f(i_0, \ldots, i_{q-1})$ of the sheaf $\mathcal{F}$ over $\mathcal{U}(i_0, \ldots, i_{q-1})$. For any $i \in I$, let $t(i, i_0, \ldots, i_{q-1})$ be a section of $\mathcal{F}$ over $\mathcal{U}(i_0, \ldots, i_{q-1})$ which is equal to $h_i(f(i, i_0, \ldots, i_{q-1}))$ over the smaller set $\mathcal{U}(i, i_0, \ldots, i_{q-1})$ and vanishes outside this set. Let

$$(k^q f)(i_0, \ldots, i_{q-1}) := \sum_{i \in I} t(i, i_0, \ldots, i_{q-1}).$$

We check that $k^{q+1}\delta^q f + \delta^{q-1}k^q f = f$ which completes the proof. □

Now we present several examples of fine sheaves.

Examples of fine sheaves. Let $\mathfrak{U} = (\mathcal{U}_i)_{i \in I}$ be a locally finite covering of a paracompact manifold X and let $(\varphi_i)_{i \in I}$ be a smooth partition of unity corresponding to the covering $\mathfrak{U}$.

EXAMPLE 1. Let $\mathcal{E}$ be a sheaf of germs of smooth functions on X. We define

$$h_i : \mathcal{E} \to \mathcal{E}, \quad \text{where } h_i(f) := \varphi_i f, \quad \text{for } f \in \mathcal{E},\ i \in I.$$

We have therefore the epimorphism of presheaf $(\mathcal{F}(\mathcal{U}))$ and thus also the epimorphism of sheaf $\mathcal{F}$. Clearly, these homomorphisms satisfy points 1. and 2. of Definition. Therefore, $\mathcal{E}$ is a fine sheaf. The proof in the case of the following Examples 2. and 3. is similar.

EXAMPLE 2. $\mathcal{E}^{(q)}$ being a sheaf of germs of smooth differential q forms over X is fine.

EXAMPLE 3. More generally, the sheaf of smooth germs of smooth vector bundle $E \to X$ of finite rank is fine.

For sheaves $\mathcal{E}^{(q)}$ of Example 2 the following important theorem holds.

THEOREM 4 (POINCARÉ LEMMA). *The sequence*

$$0 \to \mathbb{R} \xrightarrow{d} \mathcal{E}\,(=\mathcal{E}^{(0)}) \xrightarrow{d} \mathcal{E}^{(1)} \xrightarrow{d} \cdots \xrightarrow{d} \mathcal{E}^{(q)} \xrightarrow{d} \mathcal{E}^{(q+1)} \to \cdots$$

is exact (d is the exterior derivative and $\mathbb{R}$ *denotes the constant sheaf).*

PROOF. Since $dd = 0$, it is sufficient to show that (for a sufficiently small neighbourhood X) for q form ω on $\mathcal{U}$ such that $d\omega = 0$, there exists the $(q-1)$ form ρ on some neighbourhood $\mathcal{V} \subset \mathcal{U}$ such that $d\rho = \omega$. But this is just the classical Poincaré lemma (proved in *Analysis* part II). □

REMARK 1. This theorem was known already to Voltera.

REMARK 2. An analogous theorem holds for the sequence of complex forms on complex manifold X and is in that case called the *Grothendieck–Dolbeaut lemma.*

Now we can show abstract de Rham theorem.

THEOREM 5 (ABSTRACT DE RHAM THEOREM; H. CARTAN). *Let*

$$(R) \qquad 0 \to \mathcal{F} \xrightarrow{h} \mathcal{F}_0 \xrightarrow{h^0} \mathcal{F}_1 \xrightarrow{h^1} \mathcal{F}_2 \xrightarrow{h^2} \cdots \xrightarrow{h^{p-1}} \mathcal{F}_p \xrightarrow{h^p} \cdots$$

be an exact sequence of sheaves over paracompact space X such that

$$(7) \qquad H^q(X, \mathcal{F}_p) = 0 \quad \text{for } g \geqslant 1$$

(then (R) is called a resolution of the sheaf $\mathcal{F}$*), for example, let* $\mathcal{F}_p$ *be fine sheaves. The sequence (R) defines the sequence*

$$(8) \qquad 0 \to \Gamma(X, \mathcal{F}) \xrightarrow{h_*} \Gamma(X, \mathcal{F}_0) \xrightarrow{h_*^0} \cdots \xrightarrow{h_0^{p-1}} \Gamma(X, \mathcal{F}_p) \xrightarrow{h_*^p} \cdots$$

Since $h_*^{p+1} h_*^p = 0$, $\{\Gamma(X, \mathcal{F}_p), h_*^p\}$ *is a chain complex.*

Then

$$(9) \qquad \begin{array}{ll} H^q(X, \mathcal{F}) \simeq \ker h_*^q / \operatorname{im} h_*^{q-1} & q \geqslant 1, \\ H^0(X, \mathcal{F}) \simeq \ker h_*^0 \end{array}.$$

PROOF. Since $\ker h_*^0 \simeq \Gamma(X, \mathcal{F}) \simeq H^0(X, \mathcal{F})$, the case $q = 0$ is proved. Let $\mathcal{K}_p := \ker h^p$. Therefore, the sequence (R) defines the exact sequence of sheaves

$$(10) \qquad 0 \to \mathcal{K}_p \to \mathcal{F}_p \xrightarrow{h^p} \mathcal{K}_{p+1} \to 0 \quad \text{for } p \neq 0.$$

Since $H^q(X, \mathcal{F}_p) = 0$ for $q \geqslant 0$, the exact sequence (R) gives the isomorphism

$$H^{q-1}(X, \mathcal{K}_{p+1}) \simeq H^q(X, \mathcal{K}_p), \quad p \geqslant 2. \tag{11}$$

But $\mathcal{K}_0 = \mathcal{F}$, and thus, using (11) several times, we get

$$H^1(X, \mathcal{K}_{q-1}) \simeq H^q(X, \mathcal{F}), \quad q \geqslant 1. \tag{12}$$

For $p = q - 1$, the exact sequence (10) contains the exact sequence

$$H^0(X, \mathcal{F}_{q-1}) \overset{h_*^{q-1}}{\to} H^0(X, \mathcal{K}_q) \to H^1(X, \mathcal{K}_{q-1}) \to 0. \tag{13}$$

Since $H^0(X, \mathcal{K}_q) = \ker h_*^q$, $H^0(X, \mathcal{K}_{q-1}) = \Gamma(X, \mathcal{F}_{q-1})$, the isomorphism (9) follows from (12) and (13). □

As an immediate consequence of the above theorem and Poncaré lemma, we get the following theorem.

THEOREM 6 (DE RHAM).

$$\begin{aligned} H^q(X, \mathbb{R}) &\simeq H^p_{dR}(X, \mathbb{R}) \quad q \geqslant 1, \\ H^0(X, \mathcal{F}) &\simeq \ker(\mathcal{E}^{(0)} \overset{d}{\to} \mathcal{E}^{(1)}) \end{aligned} \tag{9}$$

On the left hand side we have pth cohomology group with values in $\mathbb{R}$ (constant sheaf, and on the right hand side the pth de Rham cohomology group. □

REMARK. Analogous theorem holds in the complex case.

CHAPTER 25

Homotopy Group $\pi_k(X, x_0)$. Hopf Fibering. Serre Theorem on Exact Sequence of Homotopy Groups of a Fibering

Surely, not in all Serre fiberings $p : X \to Y$ a connection exists, even though we have lifting of curves from the base Y to curves in the bundle space of given starting point x_0: in order to do this, it suffices to take as the polynomial W the one point set $\{x_0\}$. But let us stress *the connection in the bundle* $p : X \to Y$ *is something more that the possibility of lifting (arbitrary) curves; the operation (of lifting)* Γ *must be continuous as well!*

As we mentioned already, Serre defined the class of his fibering for purpose of the homotopy theory (his Ph.D. thesis opened a new chapter of this theory!) Even though we cannot prove the fundamental theorems of Serre, I could not resist formulating one of them. In order to do that, we must first briefly define a notion of the homotopy group and Hopf fibering.

Let X be a topological space with distinguished point x_0. By the *k-dimensional* (or *kth) group* $\pi_k(X, x_0)$ *of the space* (X, x_0) we understand the set of homotopically equivalent mappings of the k-dimensional sphere S^k into X with the distinguished point $s_0 \in S^k$ which is mapped into x_0. The group $\pi_1(X, x_0)$ is the fundamental group (of Poincaré) and, in general, is not commutative.

Obviously, the mapping $p : E \to M$ induces the homomorphism

$$p_* : \pi_k(E, y_0) \to \pi_k(M, x_0), \quad k = 1, 2, \ldots$$

The following theorem holds.

THEOREM 1 (ON EXACT SEQUENCE OF FIBERINGS, SERRE).
Let $p : E \to M$ be a Serre fibering (that is, a fiber bundle), $x_0 \in M$, $y_0 \in p^{-1}(x_0) =: F$, and let $j : F \to E$ be a canonical embedding.
Then there exists a homomorphism of homotopy groups

$$\partial : \pi_k(M, x_0) \to \pi_{k-1}(F, y_0),$$

such that the sequence of homomorphisms

$$\ldots \to \pi_k(F, y_0) \xrightarrow{j_*} \pi_k(E, y_0) \xrightarrow{p_*} \pi_k(M, x_0) \xrightarrow{\partial}$$

$$\xrightarrow{\partial} \pi_{k-1}(F, y_0) \xrightarrow{j_*} \pi_{k-1}(E, y_0) \xrightarrow{p_*} \pi_{k-1}(M, x_0) \to \ldots \to \pi_1(E, y_0) \xrightarrow{p_*} \pi_1(M, x_0)$$

is exact.

REMARK. Let us recall that a sequence of homomorphisms of groups

$$\ldots \to G_k \xrightarrow{\varphi_k} G_{k-1} \xrightarrow{\varphi_{k-1}} G_{k-2} \to \ldots$$

is *exact in the term* G_{k-1} if $\operatorname{im} \varphi_k = \ker \varphi_{k-1}$; if the sequence is exact in all terms, then it is *exact.*

PROOF of existence of the homomorphism ∂. Let $\varphi : I^k \to M$ be a representative of the element $[\varphi]$ of the group $\pi_k(M, x_0)$. Instead of the sphere S^k we can consider the multicube I^k whose boundary is mapped into the point x_0.

Since $I^k = I^{k-1} \times I$ and since the boundary $\partial I^k \xrightarrow{\varphi} x_0$, $\varphi(I^{k-1} \times \{1\}) = x_0$.

We will regard φ as a homotopy of the mapping of the multicube I^{k-1}. Let $\psi(u, 0) \equiv y_0$, $u \in I^{k-1}$. Therefore the mapping $\psi(\cdot, 0)$ covers the mapping $\psi(\cdot, 0)$. As a consequence of the axiom on covering homotopies, we extend the mapping ψ into $I^{k-1} \times I$ and ψ covers φ. In particular, $\psi(u, 1)$ covers $\varphi(u, 1) \equiv 1$ and thus the mapping $\varphi(\cdot, 1) I^{k-1} \to E$ and $\psi(\cdot, 1)$ maps E into the fiber $F(= p^{-1}(x_0))$. Since $\varphi(\partial I^{k-1} \times I) = x_0$, we can choose the covering ψ such that $\psi(\partial I^{k-1} \times I) = y_0$. Therefore $\psi(\cdot, 1)$ maps the multicube I^{k-1} into the fiber F and it boundary ∂I^{k-1} in the distinguished point y_0, and thus $\psi(\cdot, 1)$ represents an element of the group $\pi_{k-1}(F, y_0)$. One must still check if, when φ and φ' are homotopic, then the corresponding ψ and ψ' (which we have constructed above) are homotopic as well.

We cover the homotopy Φ connecting φ with φ' with the homotopy Ψ connecting ψ with ψ'. Thus the correspondence $\varphi \to \psi$ correctly defines the mapping $\partial : \pi_k(M, x_0) \to \pi_{k-1}(F, y_0)$. We leave it to the reader to check the additivity of ∂.

The proof of exactness in the term $\pi_k(F, y_0)$. Let $\varphi : I^{k+1} = I^k \times I \to M$ represent an element of the group $\pi_{k+1}(M, x_0)$ and let $\psi : I^k \times I \to E$ cover φ. Let the restriction $\psi(\cdot, 1)$ represent the element $\partial([\varphi]) \in \pi_k(F, y_0)$. The element $j_*\partial([\varphi])$ is, therefore, represented by the mapping $\psi(u, 0)$, that is, $j_*\partial([\varphi]) = 0$. And vice versa, If $\varphi : I^k \times I \to E$ is a homotopy connecting $\varphi(u, 1) \subset F$ and the constant mapping $\varphi(u, 0)$, then, from the definition of ∂, the mapping $\varphi(\cdot, 1)$ represents the element $\partial([p\varphi])$, thus $\ker j_* \subset Im\partial$.

We leave the proof of exactness in the terms $\pi_k(E, y_0)$ and $\pi_k(M, x_0)$ to the reader. □

EXAMPLE 1. Let us consider the simplest situation, namely the covering $\mathbb{R} \to S^1$ with the fiber $\mathbb{Z}$.

Since $\pi_k(\mathbb{R}) = 0$, $k \geqslant 0$, $\pi_k(\mathbb{Z}) = 0$, $k \geqslant 1$, we have

$$\to \pi_k(\mathbb{Z}) \to \pi_k(\mathbb{R}) \to \pi_k(S^1) \to \ldots \to \pi_1(S^1) \to \pi_0(\mathbb{Z}) \to \pi_0(\mathbb{R}) \to \pi_0(S^1),$$

thus

$$\ldots \to 0 \to 0 \to \pi_k(S^1) \to \ldots \to 0 \to 0 \to \pi_1(S^1) \to \mathbb{Z} \to 0 \to 0.$$

Therefore $\pi_1(S^1) = \mathbb{Z}$, $\pi_k(S^1) = 0$ for $k > 1$.

This is indeed a very simple example of application of the powerful theory.

But let us consider the famous example of H. Hopf, who first showed the nontrivial groups of homotopy of S^2, $\pi_2(S^2) = \mathbb{Z}$ and $\pi_3(S^2) = \mathbb{Z}$.

EXAMPLE 2. *Hopf fibering* $p : S^3 \to S^2$ *with the fiber* S^1. The exact homotopy sequence has the form

$$\ldots \to \pi_3(S^1) \to \pi_3(S^3) \to \pi_3(S^2) \to$$
$$\to \pi_2(S^1) \to \pi_2(S^3) \to \pi_2(S^2) \to$$
$$\to \pi_1(S^1) \to \pi_1(S^3) \to \pi_1(S^2).$$

Therefore, using the previous example $\pi_3(S^1) = \pi_2(S^1) = \pi_2(S^3) = \pi_1(S^3) = \pi_1(S^2)$, and for the sequence above we have

$$\to 0 \to \mathbb{Z} \to \pi_3(S^2) \to 0 \to 0 \to \pi_2(S^2) \to \mathbb{Z} \to 0 \to 0.$$

From exactness of this sequence, we have

$$\pi_2(S^2) = \pi_1(S^1) = \mathbb{Z},$$

$$\pi_3(S^2) = \mathbb{Z}.$$

Let us observe that the generator of the group $\pi_3(S^2)$ is represented by the Hopf fibering $p : S^3 \to S^2$ itself.

Indeed, it follows from exactness of the sequence that $\pi_3(S^3) \xrightarrow{p_*} \pi_3(S^2)$. Since generators of the group $\pi_3(S^3)$ are represented by the identity mapping $\varphi : S^3 \to S^3$, $P_*[\varphi] = [p\varphi] = [p]$.

Let us apply the theorem on exact homotopy sequence to the standard Serre fibering $p : \tilde{\Omega}(X, x_0) \to X$, taking in this theorem $E = \tilde{\Omega}(X, x_0)$, $X = M$, and the projection p associates with the curve $c : [0, 1] \to X$ with the starting point $x_0 = c(0)$ its end point $x = c(1)$. The fiber is therefore the space $\Omega(X, x_0)$ of loops c with beginning and end at the point x_0: $c(0) = c(1) = x_0$. Since the space $\tilde{\Omega}(X, x_0)$ is contractible, all its homotopy groups vanish $\pi_n(\tilde{\Omega}(X, x_0)) = \pi_n(E) = 0$, $n = 1, 2, \ldots$. Therefore we arrive at the exact sequences

$$0 \xrightarrow{p_*} \pi_n(X, x_0) \xrightarrow{\partial} \pi_{n-1}(\Omega(X, x_0)) \xrightarrow{j_*} 0, \quad n \geqslant 0.$$

But the exactness of this sequence ($\operatorname{im} p_* = \ker \partial$, $\operatorname{im} \partial = \ker j_*$) means that ∂ is a bijective homomorphism (injective surjection) and thus ∂ is a group isomorphism. Therefore we proved the following theorem.

THEOREM 2 (SERRE). *If X is an arcwisely connected space, then*

$$\pi_n(X, x_0) \simeq \pi_{n-1}(\Omega(X, y_0)), \quad n \geqslant 1$$

where y_0 is, for example, the constant path $I \to x_0$.

REMARK. The theorem above may be used to define inductively the nth homotopy group $\pi_n(X, x_0)$ of the space X as the $(n-1)$st homotopy group of the space of loops $\Omega(X, y_0)$

$$\pi_n(X, x_0) =: \pi_{n-1}(\Omega(X, y_0), y_0).$$

This is what some authors indeed do: this formulation is somehow advantageous, for example it is easier to prove commutativity of the group $\pi_n(X, x_0)$, $n > 1$. However, the standard definition of the group $\pi_n(X, x_0)$ as a set of classes of homotopically equivalent mappings of spheres $(S^n, s_0) \to (X, x_0)$ employed here, was historically the first.

CHAPTER 26

Various Benefits of Characteristic Classes (Orientability, Spin Structures). Clifford Groups, Spin Group

In Part I of this book we introduced various characteristic classes. Some of their applications concern existence of particular structures on a differential manifold. These are, in general, difficult problems. As we know, the tangent bundle $TX \to X$ ($\dim X = n$) is (*ex definitione*) a $\mathrm{GL}(n, \mathbb{R})$-bundle; let us recall that if the group $\mathrm{GL}(n, \mathbb{R})$ of the bundle TX can be reduced to some Lie subgroup $G \subset \mathrm{GL}(n, \mathbb{R})$, that is, if the transition matrices $g_{ij}(x) \in G$, $x \in \mathcal{U}_i \cap \mathcal{U}_j$, then we say that this reduction is a G-structure.

EXAMPLE 1. An $O(n)$-structure is nothing but the Riemann structure.

EXAMPLE 2. For $n = 2m$ let us consider the group $\mathrm{GL}(m, \mathbb{C})$ as a subgroup of the group $\mathrm{GL}(n, \mathbb{R})$. Then the $\mathrm{GL}(m, \mathbb{C})$-structure is called the *almost complex structure.*

Obviously, a complex manifold possesses a almost complex structure. The inverse theorem *does not hold.*

THEOREM 1 (FRÖLICHER–NIJENHUIS). *An integrable almost complex structure (that is, such that their torsion tensor vanishes) is complex.*

This theorem is very deep. A somehow simplified proof was given by

Hörmander in his monograph *Introduction to Complex Analysis in Several Variables.*

EXAMPLE 3. The sphere $S^2 \simeq \mathbb{P}^1(\mathbb{C})$ possesses a complex structure, but the six dimensional sphere S^6 possesses (only?) an almost complex structure and it was not known for a long time if this structure is integrable – this was considered to be a scandal (by mathematicians!)

The almost complex structure on S^6 is introduced with the help of octonions on S^7 (cf. Steenrood.)

The following theorem is highly nontrivial.

THEOREM 2 (A. BOREL–J.-P. SERRE). *Among the even dimensional spheres (only these have a chance to possess an almost complex structure) only S^2 and S^6 allow for the existence of complex structures.*

As we know, along with Riemann surfaces the most important compact complex spaces are projective spaces $\mathbb{P}^m(\mathbb{C})$.

THEOREM 3 (CHOW). *Compact complex submanifolds in $\mathbb{P}^m(\mathbb{C})$ are algebraic, that is, they are zero loci of a (finite) family of polynomials.*

Orientability of X and the first Whitney class $w_1(X)$. The Stiefel–Whitney classes for $O(n)$ bundles $E \to X$ are the cohomology classes

I_W $w_i(E) = w_i \in H^i(X, \mathbb{Z}_2)$, $0 \leqslant i \leqslant n$.

For these classes Hirzenbruch gave axioms which are similar to the axioms of Chern classes.

II_W *(naturalness)* $w(f^*E) = f^*w(E)$, where $w := w_0(E) + \cdots + w_n(E)$ is the *total Whitney class*;

III_W $w(E \oplus E') = w(E) \cdot w(E')$;

IV_W *normalization condition.*

When $E = TX$, then the class $w_i(TX)$ is denoted for short $w_i(X)$ and is called the *Whitney class of the manifold X.*

We have the exact sequence of groups

$$1 \to SO(n) \to O(n) \overset{\det}{\to} \mathbb{Z}_2 \to 1$$

and the exact sequence of cohomologies, the interval of whose

$$w_1 = (\det)_* : H^1(X, O(n)) \to H^1(X, \mathbb{Z}_2)$$

defines the first Stiefel–Whitney class $w_1(E)$ of the vector bundle $E \to X$.

From the exact sequence of cohomologies we obtain the following proposition

PROPOSITION 4. *The group $O(n)$ of the bundle $TX \to X$ (more generally, the vector bundle $E \to X$) reduces to $SO(n)$; therefore the manifold X (the bundle $E \to X$) is orientable if and only if $w_1(X) = 0$ (more generally, $w_1(E) = 0$).*

It is much more difficult to prove the theorem of Borel–Hirzenbruch which provides the sufficient and necessary condition for existence of spinor structure.

Existence of spinor structure on the manifold X ($\equiv w_1(X) = 0$). In this problem one has to lift the structure group $SO(n)$ of the Riemann manifold X (that is, the bundle $TX \to X$) to the spin group $Spin(n)$ which is a double covering of $SO(n)$. The possibility of performing such a lift is called the *existence of spinor structure*, or, for short, the *spin structure.*

THEOREM 5 (BOREL–HIRZENBRUCH). *On a differentiable and orientable manifold X (that is, $w_1(X) = 0$) the spinor structure exists if and only if the second Whitney class vanishes; $w_2(X) = 0$.*

The (very brief) sketch of the proof.

The starting point is now the exact sequence of groups

$$1 \to \mathbb{Z}_2 \to Spin(n) \to SO(n) \to 1$$

Let the tangent bundle τ be an element of the set of cohomologies $H^1(X, SO(n))$. It can be shown that there exists an exact sequence of cohomology sets with distinguished elements and with the coboundary mapping

$$\delta : H^1(X, SO(n)) \to H^2(X, \mathbb{Z}_2),$$

such that $\delta(\tau) = w_2(X)$. Therefore, the tangent bundle $\tau : TX \to X$ is associated with a $Spin(n)$ principal bundle (and thus X possesses a spinor structure) if and only if $w_2(X) = 0$. In order to define the group $Spin(n)$, we must turn back to Clifford algebras.

The group $Spin(n)$. Let Q be a non-degenerate quadratic form on the vector space V over the field $\mathbf{k}$ (char $\mathbf{k} \neq 2$) of dimension $n \geqslant 3$. The special orthogonal group of the space (V, Q) will be denoted by $SO(n)$ or, more precisely, $SO(Q)$. Let $\mathcal{C} = \mathcal{C}(Q)$ be the Clifford algebra of the pair (V, Q); let $\mathcal{C}^+$ (resp. $\mathcal{C}^-$) be the subspace of $\mathcal{C}$ generated by products of even (resp. odd) number of elements of the space V. Let β be an antiautomorphism of the algebra $\mathcal{C}$ defined by the formula $\beta(v_1 v_2 \ldots v_j) = v_j \ldots v_2 v_1$. The injection $V \hookrightarrow \mathcal{C}$ makes it possible to define the Clifford group.

The Clifford group is $G = \{s \in \mathcal{C} : s \text{ is invertible in } \mathcal{C} \text{ and } sVs^{-1} = V\}$; $G^+ := G \cap \mathcal{C}^+$ is called the *special Clifford group.*

DEFINITION. $Spin(n) = Spin(Q) = \{s \in G^+ : s\beta(s) = 1\}$. A linear representation σ of the group $Spin(n)$ in the space V is

$$\sigma(s) \cdot v := svs^{-1}, \quad s \in Spin(n), v \in V,$$

and is called its *vector representation.* It turns out that

$$\sigma(Spin(Q)) = SO(Q), \quad \text{and} \quad \ker \sigma = \{\pm 1\}.$$

If $\mathbf{k} = \mathbb{R}$ and the quadratic form Q is positive or negative definite, then the group $Spin(n)^{\mathbb{R}}$ of real points of the algebraic group $Spin(n)$ is called the *spin (spinor) group.* This group is a connected and simply connected compact Lie which is a double covering of the group $SO(n, \mathbb{R}) = SO(n)$. The following isomorphisms hold

$$Spin(3)^{\mathbb{R}} \simeq SU(2), \quad Spin(4)^{\mathbb{R}} \simeq SU(2) \times SU(2),$$

$$Spin(5)^{\mathbb{R}} \simeq Sp(4), \quad Spin(6)^{\mathbb{R}} \simeq SU(4).$$

Let $\mathbf{C}$ be a complexification of the Clifford algebra $(\mathbb{R}^n, Q)$ with $Q = -(x_1^2 + \cdots + x_n^2)$. Then $\mathbf{C}$ possesses an irreducible representation in a vector space S of dimension $2^{[n/2]}$ over $\mathbf{C}$ which defines a representation of the group $Spin(n) \subset \mathbf{C}$ in the space S. Every spinor structure on the manifold X defines the vector bundle $\pi_S : S(X) \to X$ with typical fiber S called the *spinor bundle.* The Levi–Civita connection on X defines connection on $S(X)$ in a canonical way. On the space $C^\infty(S(X))$ of smooth sections of the bundle $S(X) \to X$ called the *spinor fields* acts the linear differential operator of order one D, called the *Dirac operator* and given by the formula

$$Du = \sum_{j=1}^{n} s_j \cdot \nabla_{s_j} u \quad \text{(where $\cdot$ denotes multiplication in algebra } \mathbf{C}) \ .$$

If $Du = 0$, then u is called the *harmonic spinor (harmonic spinor field).*

After this digression let us return to spinor structures. Let $P \to X$ be an $SO(n)$ principal bundle associated with the tangent bundle $TX \to X$. Let $\tilde{P}$ be a $Spin(n)$ principal bundle over X (with the fiber $Spin(n)$ being a double covering of the fiber $SO(n)$). We have therefore the diagram

$$\begin{array}{ccc} \tilde{P} & \xrightarrow{\ \pi\ } & P \\ & \searrow \qquad \swarrow & \\ & X & \end{array}$$

Thus π is a double covering. Since connected coverings of a manifold Y are in one-to-one correspondence with classes of subgroups of the fundamental group $\pi_1(Y)$, it is not hard to see that double coverings of the bundle P correspond to elements of the group $\mathrm{Hom}(H_1(P), \mathbb{Z}_2) = H^1(P, \mathbb{Z}_2)$ which shrink to the nontrivial element of the group $H^1(SO(n), \mathbb{Z}_2) = \mathbb{Z}_2$. Indeed we have the exact (this is not obvious) sequence of groups

$$(*) \quad \ldots \to H^1(X, \mathbb{Z}_2) \xrightarrow{\pi^*} H^1(P, \mathbb{Z}_2) \xrightarrow{i^*} \underbrace{H^1(SO(n), \mathbb{Z}_2)}_{\mathbb{Z}_2} \xrightarrow{\delta} H^2(X, \mathbb{Z}_2) \to \ldots$$

In other words, every spinor structure on X corresponds to the unique element $a \in H^1(P, \mathbb{Z}_2)$ such that $i^*a = 1$. As a result of exactness of the sequence $(*)$, such element exists only if and only if $w_2(X) = 0$.

Moreover, if $i^*a = 1 = i^*b$, then $b - a \in \ker i^* = \mathrm{im}\, \pi^*$, and vice versa. Since for dimension 0 the mapping i^* is surjection and π^* is bijection for dimension 1, if $w_2(X) = 0$, then *the number of different spin structures on X is equal to the number of elements of the group $H^1(X, \mathbb{Z}_2)$.*

EXAMPLE. Let X be a Riemann surface of genus p. As we know $w_2(X)$ is represented by Euler form which, in turn, is the product $\chi(X) \cdot \mathrm{vol}\, X$ of the Euler number and the volume form. But as we know $\chi(X) = 2(1 - p)$

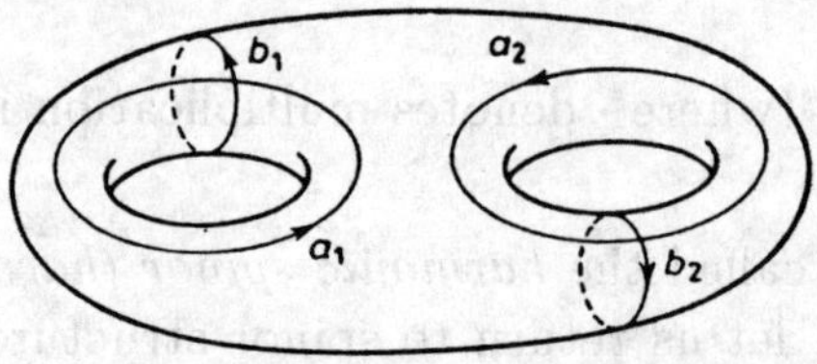

and thus $w_2(X) = 0$ (because it is a class with values in $\mathbb{Z}_2$) and X possesses spinor structures. But in the theory of Riemann surfaces one shows that the group $H_1(X, \mathbb{Z}_2)$ has $2p$ generators (cycles $a_1, \ldots, a_p, b_1, \ldots, b_p$, cf. the figure below.) Thus the rank of the group $H_1(X, \mathbb{Z}_2)$ is $2p$ and it possesses $2^{2p} = 4^p$ elements. We showed therefore the following proposition

PROPOSITION 6. *A Riemann surface X of genus p has 4^p different spinor structures.*

We must already finish our tale about characteristic classes of vector bundles and various benefits resulting from using them. The reader certainly became convinced that these classes not only form an important part of modern differential geometry and analysis, but are also necessary for investigations of existence and for findings some basic structures of manifolds.

CHAPTER 27

Divisors and Line Bundles. Algebraic and Abelian Varieties

As in the theory of Riemann surfaces, divisors play a decisive role in global analysis of complex manifolds of higher dimensions. Here the situation is completely different: There are many divisors and meromorphic functions on a Riemann surface, however, there may be no global meromorphic functions and divisors on an arbitrary complex manifold X of complex dimension greater than 1. The exceptional roles are played in this context by projective algebraic varieties X in the complex projective space $i : X \to \mathbb{P}^n$.

Let us recall here the famous

THEOREM 1 (CHOW). *Every compact complex submanifolds in $\mathbb{P}^n(\mathbb{C})$ is a set of zeros of a family of polynomials (and thus is a projective algebraic variety) in the classical sense.*[1]

We will illustrate the statements above on the example of the theory of complex tori. There exist tori (of dimension greater than 1, of course) on which there do not exist any meromorphic functions different from the constant ones, and which, therefore, cannot be imbedded into the projective space $\mathbb{P}^n(\mathbb{C})$.

DEFINITION 2. Let U_i be an open covering of the complex manifold X.

[1]The classical definition is the following. $X = \{z \in \mathbb{P}^n(\mathbb{C}) : Q_i(z_0, \dots, z_n) = 0, i = 1, 2, \dots, r\}$, where $Q = (Q_1, \dots, Q_r)$ is a homogeneous system of polynomials of $n + 1$ variables and $z_0, \dots, z_n$ are homogeneous coordinates $z \in \mathbb{P}^n(\mathbb{C})$, is called the *projective algebraic manifold.*

We say that the family $(U_i, f_i)_{i\in I}$, where f_i is meromorphic over U_i, defines the *divisor* D if $\frac{f_i}{f_k} \in A^*(U_i \cap U_k)$ for every i, k such that $U_i \cap U_k \neq \emptyset$. Another family (U_i', f_i') defines the same divisor D if $\frac{f_i'}{f_k'} \in A^*(U_i' \cap U_k')$.[2] With the meromorphic function $g \in M(X)$ we associate the divisor (g) represented by (U_i, g_i). Divisors of (globally defined) meromorphic functions are called *principal.* Divisors form, in a natural way, the abelian group denoted by $\mathrm{Div}(X)$. The following simple relations hold.

$$(f \cdot g) = (f) + (g), \quad (\frac{f}{g}) = (f) - (g),$$

$$((f) \geqslant 0) \Longleftrightarrow (f \in A^*(X)).$$

Divisors D_1 and D_2 are (linearly) equivalent (we write $D_1 \sim D_2$) if $D_1 - D_2$ is a principal divisor.

Similarly to the one dimensional case, divisor can be represented as a formal (locally finite) linear combination of analytical hypersurfaces $D = \sum_k p_k \cdot V_k$, $\dim_{\mathbb{C}} V_k = n - 1$, $p_k \in \mathbb{Z}$ (and this was, of course, the initial definition!) If $f \in M(X)$, then its divisor $(f) = (f)_\infty - (f)_0$, where $(f)_\infty$ is the *polar divisor* and $(f)_0$ the *zero divisor*, and they say what is the multiplicity of poles (zeros) of the function f. In the case of Riemann surfaces $\dim_{\mathbb{C}} V_k = 0$ and divisors are combinations of integer-valued points, and thus there is many of them. In 1949, A. Weil observed that the language of line bundles simplifies the theory of divisors very much.

PROPOSITION 3 (A. WEIL). *With every divisor $D \in Div(X)$ one can associate the line bundle $L_D \to X$. If D is given by the family $U_i, f_i)$, then $g_{ik} := f_i/f_k \in A^*(U_i \cap U_k)$ is a cocycle of the bundle L_D. Moreover*

(i) $(D \sim D') \Longleftrightarrow (L_D \simeq L_{D'})$

(ii) *Every holomorphic section s of the bundle L_D can be identified with a meromorphic function on X, and thus with an element of the space $L(D)$.*

DEFINITION 4. The *Chern class* $c(D)$ of the divisor D is the Chern class of the bundle associated to this divisor $c(D) := c(L_D)$.

Projective embeddings. Kodaira theorem. For more then a hundred years (since the times of Riemann and Weierstrass) the greatest mathematicians investigated problems related to the possibility of embedding of

[2]Let us recall that $A^*(U)$ is the set of functions holomorphic on U and not taking the value 0 on this set.

holomorphic compact manifolds in $\mathbb{P}^n(\mathbb{C})$. The final solution was found by Japanese mathematician K. Kodaira, who worked at Princeton at the time.

THEOREM 5 (KODAIRA, 1954). *X is a projective manifold if and only if there exists a positive line bundle $L \to X$. In such a case there exist k_0 such that for all $k \geqslant k_0$ the holomorphic imbedding $i_{L^k} : X \to \mathbb{P}^n$ can be defined as follows: there exist holomorphic sections of the line bundle $L^k \to X$, $s_0, \ldots, s_n$ and the holomorphic injection $i_{L^k}(x) := [s_0^V, \ldots, s_n^V] \in \mathbf{P}^n$, $x \in V$.*[3]

(The function s^V corresponds to the section S over the trivializing neighbourhood V.)

There is yet another equivalent intristic condition (suggested by A. Weil) (as we know, only Kähler manifolds can be embedded in $\mathbb{P}^n$.)

THEOREM 6 (KODAIRA 1954). *The compact Kähler manifold X is a projective submanifold if it possesses a Hodge metric, that is, such Kähler metric H that $\operatorname{Im} H$ is an integer, $\operatorname{Im} H \in H^2_{dR}(H, \mathbb{Z})$ (in other words, the periods of H are integers.)*

REMARK. The main step in the proof of Theorem 6 was performed by Kodaira and generalized by Nakano.

THEOREM 7 ON VANISHING (KODAIRA–NAKANO). *Let X be a complex manifold of dimension n, $L \to X$ be a positive line bundle. Then*

$$H^q(X, A^p(L)) = 0 \quad \text{for } p + q > n.$$

By dualization we obtain

COROLLARY. *If $L \to X$ is a negative line bundle. Then*

$$H^q(X, A^p(L)) = 0 \quad \text{for } p + q < n.$$

Theorems on vanishing of cohomology groups are of utmost importance in analysis (Cousin problems), algebraic and differential geometry, and, recently, also in physics.

---

[3]If the bundle L is defined by the cocycle g_{ij}, then the bundle L^k is defined by the cocycle $h_{ij} = (g_{ij})^k$.

The surprising for the first sight Kodaira condition (positivity of $L \to X$) is rather natural; one is led to this condition by the theta function theory which was initiated by Riemann and continued by such giants as Weierstrass, Klein, Poincaré, Lefschetz, Kähler, Siegel, Andreotii, and others.

One dimensional tori $\mathbb{C}/\Gamma$ (that is connected Riemann surfaces of genus $p = 1$) and meromorphic functions on them comprise the theory of elliptic functions, the theory to which many volumes long monographs were devoted at the turn of XIX and XX centuries. This theory is becoming popular again nowadays. It was already Jacobi (one of the founders of the theory of elliptic functions), who observed that every elliptic function $f_\Gamma \in M(\mathbb{C})$ can be represented as a quotient ϑ_1/ϑ_2 of two holomorphic functions $\vartheta_i \in A(\mathbb{C})$ which can be represented by rapidly convergent power series, the so called ϑ-series. The inverse problem (see below) led Riemann and Weierstrass to investigations of functions of p complex variables, where p is the genus of the Riemann surface S. It was the brilliant idea of Riemann to introduce the functions ϑ of p variables, called by him 'ϑ-Reihen'; these are the power series of the form[4]

$$\vartheta_\tau(z) = \sum_{m \in \mathbb{Z}^g} \exp\left(m^T \tau m + 2m^T z\right),$$

where τ is a complex $(g \times g)$ matrix satisfying the relations $\tau^T = \tau$, $\operatorname{Im} \tau > 0$ (τ has positive definite imaginary part.) The series (ϑ) is uniformly convergent on every compact subset of the space $\mathbb{C}^p$ and thus it represents an entire function.

Riemann discovered that for any Riemann surface of genus p there exists the associated (complex) p-torus $\operatorname{Jac}(S) = \mathbb{C}^p/\Gamma$, where Γ is a discrete subgroup in $\mathbb{C}^p$. This torus is called the *Jacobian of the surface* S. Riemann had an intuition that investigations of the Jacobian make it possible to get a deep insight into the very structure of the surface S: its properties are reflected by the structure of $\operatorname{Jac}(S)$. Since $\mathbb{C}^p/\Gamma$ is an abelian group, it possesses a number of properties which the initial surface S does not have at

[4]For $p = 1$ this formula gives

$$Q_3(z|\tau) = \sum_{m=-\infty}^{\infty} e^{\pi i(m^2\tau + 2mz)}.$$

all. The group Γ is called the *group of periods* of S and is defined as follows.

The cohomology group $H_1(S, \mathbb{Z})$ has a basis represented by $2g$ one dimensional cycles $\alpha_1, \ldots, \alpha_{2p}$. Taking as a basis in the space $h^1_{dR}(S, \mathbb{R})$ Abel differentials $\omega^1, \ldots, \omega^g$, we obtained the so called *period matrix* whose columns correspond to the periods on the cycle I_k

$$\Omega = (\pi^j_k) = \left(\int_{\alpha_k} \omega^j \right), \quad j = 1, \ldots, g, \; k = 1, \ldots, 2g.$$

Let us recall that $\omega^1, \ldots, \omega^g$ form a basis of the space $A^{0,1}(S)$ of holomorphic one forms. It can be shown that the following theorem holds

THEOREM 8 (RIEMANN) *The vectors* $\gamma_k = \pi^1_k, \ldots, \pi^p_k)$, $k = 1, \ldots, 2p$ *(columns of the matrix* (π^j_k)*) are R-linearly independent and span the discrete group* $\Gamma := \mathbb{Z}\gamma_1 + \cdots + \mathbb{Z}\gamma_{2p}$. *This group is called the group of periods of the surface S and is denoted by* $Per(\omega^1, \ldots, \omega^g)$*:*

$$\mathrm{Per}(\omega^1, \ldots, \omega^p) \left\{ \left(\int_C \alpha_1, \ldots, \int_C \alpha_g \right) : C \in H_1(S, \mathbb{Z}) \right\}.$$

In what follows we will denote by π the canonical projection on the Jacobian $\mathbb{C}^p \xrightarrow{\pi} \mathbb{C}^p/\Gamma$.

In order to be able to formulate a number of fundamental theorems in a simpler way, let us introduce some mappings and notations. We show how some basic theorems on analysis on Riemann manifolds can be phrased in the language of Jacobi tori.

The canonical mapping $\kappa : S \to Jac(S)$ is defined by $S \ni x \to \left(\int_{x_0}^x \alpha^1, \ldots, \int_{x_0}^x \alpha^g \right)$ modulo the (period) group Γ, where the reference point x_0 is fixed. We immediately extend the mapping κ to the divisor group $Div(S)$ by taking $\kappa(D) = \sum r_j \kappa(x_j)$ for $D = \sum r_j x_j$. We use the notation $\kappa^r : S^r \to Jac(S)$, $0 < r \leqslant g$, where $\kappa^r(x_1, \ldots, x_r) := \sum_{j=1}^r \kappa(x_j)$, $W_r := \kappa^r(S^r)$,

$Div_0(S)$ is the group of divisors of degree 0 (deg $D := \sum r_j = 0$),
$Div_g(S)$ is the group of principal divisors ($Div_g(S) = \{(f) : f \in M(S)\}$).

THEOREM 9 (ABEL). *Let* $D, D_1 \geqslant 0$, deg $D =$ deg D_1. *Then*

$$(D_1 \sim D \;\text{ thus }\; D_1 = D + (f)) \Longleftrightarrow (\kappa(D) = \kappa(D_1) \in Jac(S)).$$

DEFINITION 10. The *Picard group* of surface S is the quotient

$$Pic(S) := Div_0(S)/Div_g(S).$$

Since the Abel theorem is one of the most important theorems of global complex analysis, let us phrase it in slightly different form.

THEOREM 11 (ABEL). *Let $D \in Div_0(S)$ and let c be such 0-chain that its image $\partial c = D$. Then the mapping*

$$(*) \qquad \varphi : Div_0(S) \to Jac(S)$$

where $\varphi = \kappa | Div_0(S)$ has the kernel $\ker \varphi = Div_g(S)$.

The *inverse Jacobi problem* is to check if the injection $j : Div_0(S)/Div_g(S) \to Jac(S)$ (induced by φ) on the quotient $Pic(S)$ is a surjection.

This problem was posed by Jacobi (in different form, of course – the notion of Riemann surface did not exist at the time) and after important preliminary investigations by Göppel and Rosenheim, was fully solved by Riemann (1857) and Weierstass in his lectures '*Über die Theorie von Abelschen Transzendenten*' (1896) with the help of the ϑ function.

THEOREM 12 (RIEMANN–WEIERSTRASS). *The inverse Jacobi problem has a solution. More precisely, the mapping $j : Pic(S) \to Jac(S)$ is an isomorphism; moreover $W_p = Jac(S)$.*

The modern proof (without ϑ function) is presented, for example, in *Analysis, Part II.* The crowning achievement of all the wonderful ideas, born as a result of problems with solving the inverse problem, is the famous Torelli–Andreotti–Weil theorem which was formulated in 1914 by Torelli and precisely proven by A. Andreotti (1952, 1958) and A. Weil (1957). The simplest (but still quite complex) proof is due to H.H. Martens. Since this theorem is so important and so beautiful, we will present to different formulations of it.

DEFINITION 13. The *polarized abelian variety* is the pair $(X, [\omega])$, where $X = \mathbb{C}^p/\Gamma$ is a torus which can be holomorphically imbedded into some projective space $\mathbb{P}^n$ and ω is the class defined by Hodge form. Two polarized abelian varieties $(X, [\omega])$ and $(X', [\omega'])$ are *isomorphic* if there exist a biholomorphic mapping $f : X \to X'$ such that $[f^*\omega'] = [\omega]$.

Now we can present the first version.

THEOREM 14 (TORELLI). *Let S, S' be compact Riemann surfaces of genus $p \geqslant 1$. If their polarized Jacobians are isomorphic, then S and S' are isomorphic (that is, biholomorphically equivalent):*

$$((Jac(S), [\omega_S]) \simeq (Jac(S'), [\omega_{S'}])) \Longleftrightarrow (S \simeq S').$$

REMARK. In the case $p = 1$, this statement is the classical fact; indeed we have

PROPOSITION 15. *Let S be of genus 1 (a torus). Then $S \simeq Jac(S)$. More precisely, $S \ni x \to \kappa(x) := \int_0^x \omega \pmod{\Gamma} \in Jac(S)$ is an isomorphism.*

REMARK. The mapping κ is given by the elliptic integral of first kind

$$\kappa(x) = \int_0^x \frac{dz}{\sqrt{W(z)}} \pmod{\Gamma} \in \mathbb{C}/\Gamma = Jac(S),$$

where $z : S \to \mathbb{P}^1$ is a meromorphic function (coordinate) on S and W is a polynomial of order 3 or 4. Then $f : z \circ \kappa^{-1} \circ p_1$ (where $p_1 : \mathbb{C} \to \mathbb{C}$) is a biperiodic meromorphic function mapping $\mathbb{C}$ into $\mathbb{R}$, that is, the so called elliptic function. It was a brilliant discovery of Abel (1821) and Jacobi (1822) that by turning to inverse function, the studies of elliptic integrals are reduced to elliptic functions. Similarly, the function inverse to the integral

$$\arcsin x = \int_0^x \frac{dz}{\sqrt{1 - z^2}} \pmod{2k\pi}$$

is 1-periodic and satisfies the beautiful addition theorem. The important step towards discovery of elliptic functions was investigation of elliptic integrals in complex domains: only then the harmony of elliptic and abelian functions, for example, the addition theorem $j([D_1] + [D_2]) = j([D_1]) + j([D_2])$ which contains addition of Abel integrals ($p = 1$), could have been revealed.

CHAPTER 28

General Abelian Varieties and Theta Function

$Jac(S)$ is a very important but quite special example, which, at the same time, is an algebraic variety. From Kodaira embedding theorem, one easily infers the following fact known already to Frobenius.

THEOREM 1. *The necessary and sufficient condition for the torus* $X = \mathbb{C}/\Gamma$ *to be an algebraic variety is that there exists a hermitian form on* $\mathbb{C}^p$, $H : \mathbb{C}^p \times \mathbb{C}^p \to \mathbb{C}$ *having the following properties*

$$(F-R)_I \qquad \begin{array}{l} \text{(i) } H \text{ is positively definite } (\operatorname{Re} H > 0) \\ \text{(ii) } A = \operatorname{Im} H : \Gamma \times \Gamma \to \mathbb{Z}. \end{array}$$

Conditions above are called the *Frobenius–Riemann conditions*, and the form $A = \operatorname{Im} H$ is called the *Riemann form*. The pair $(\mathbb{C}/\Gamma, H)$, where H is a Riemann form, is often called the *polarized abelian variety.* Because of the importance of conditions $(F-R)_I$, we will present these conditions once again in the matrix form, which can be obtained once the appropriate bases are taken.

$(F-R)_{II}$. Let Ω be a $(p \times 2p)$ matrix (of periods) of the lattice Γ; their columns $\gamma_1, \ldots, \gamma_{2p}$ form a $\mathbb{Z}$-basis of the lattice Γ. The torus $\mathbb{C}^p/\Gamma$ is an abelian variety if and only if there exists a integer valued, skew symmetric, non degenerate $(2p \times 2p)$ matrix D (the principal matrix corresponding to Ω) such that the following condition hold

$$(F-R)_{II} \qquad \Omega D \Omega^T = 0, \quad \sqrt{-1}\Omega D \bar{\Omega}^T > 0.$$

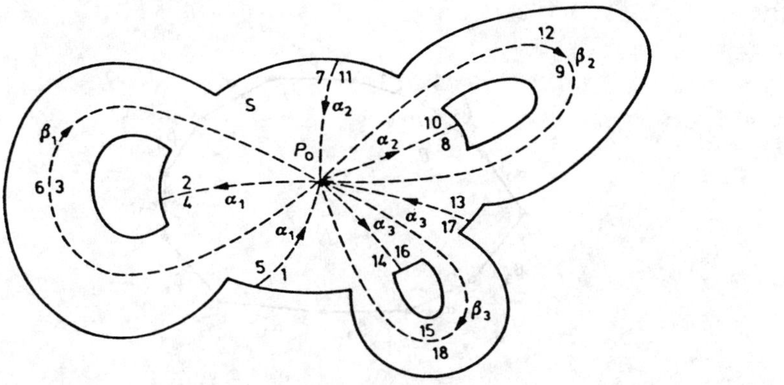

Fig. 1

Frobenius investigated the forms of the matrix A in details. The fundamental theorem of this theory is

THEOREM 2 (FROBENIUS). *Let A be a skew symmetric, non degenerate, bilinear form on a free $\mathbb{Z}$-module Γ and let $A : \Gamma \times \Gamma \to \mathbb{Z}$. Then Γ has an A-orthogonal $\mathbb{Z}$-basis $v_1, \ldots, v_p, v_1', \ldots, v_p'$ such that $A(v_j, v_j') = \epsilon_j$, $j = 1, \ldots, p$, where $\epsilon_1, \ldots, \epsilon_p \in \mathbb{N}$ and $\epsilon_j | \epsilon_{j+1}$ (ϵ_j divides ϵ_{j+1}.) Thus, in the basis above A has the form*

$$M_A = \begin{bmatrix} 0 & \Delta_\epsilon \\ -\Delta_\epsilon & 0 \end{bmatrix}, \quad \text{where } \Delta_\epsilon = \mathrm{diag}[\epsilon_1, \ldots, \epsilon_p].$$

Thus $\det A = [\epsilon_1 \cdots \epsilon_p]^2$, $\Gamma = \Gamma_1 \oplus \Gamma_2$, where Γ_1 is span by $v_1, \ldots, v_p$, and Γ_2 by $v_1', \ldots, v_p'$; Γ_1 and Γ_2 are A-isotropic.

REMARK. The basis above is called the *Frobenius basis* and the numbers $\epsilon_1, \ldots, \epsilon_p$ are called the *elementary divisors* of the form (matrix) A. The submodule Γ_k is A-isotropic if $A(v, w) = 0$ for $v, w \in \Gamma_k$. The number $\mathrm{Pf}(A) := (\det A)^{1/2} = \epsilon_1 \cdots \epsilon_p$ is called the *Pfaffian* of the form A.

EXAMPLE 3. The torus $Jac(S)$ has a canonical polarization. The surface S is topologically equivalent to the sphere with p handles. In Fig. 1 we presented three handles (in clockwise order). The surface S can be cut along $2g$ closed curves $\alpha_1, \beta_1, \ldots, \alpha_p, \beta_p$ all starting from the point p_0; in this way we obtain the polygon π called the *normal form* of the surface S (the surface S itself can be obtained by identification of the corresponding

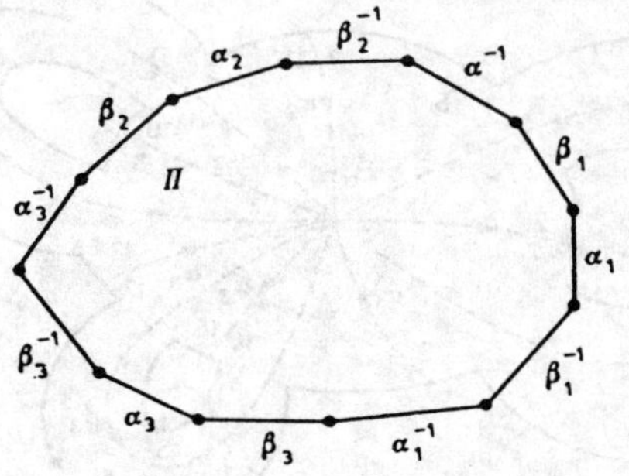

Fig. 2

faces of this polygon). Walking along the boundary of the normal form, as the successive numbers in Fig. 1 indicate, and having the interior of the the polygon on the left hand side, we successively encounter the following curves

$$\alpha_1^{-1}, \beta_1^{-1}, \alpha_1, \beta_1, \alpha_2^{-1}, \beta_2^{-1}, \alpha_2, \beta_2, \alpha_3^{-1}, \beta_3^{-1}, \alpha_3, \beta_3, \ldots$$

Thus the normal form is as presented in Fig. 2. With no loss of generality, one can assume that the period matrix has the form $\Omega = (I_p, \tau)$. The $(F - R)_I I$ conditions take the form $\tau^T = \tau$, $\operatorname{Im} \tau > 0$ (cf. Springer, *Introduction to Riemann Surfaces*, Addison Wesley, Reading, 1957.)

The example below shows that for $p \geqslant 2$ the tori are not, in general, abelian varieties (they cannot be polarized), and thus there do not exist any non constant meromorphic functions on them.

EXAMPLE 4. $p = 2$. Let the torus has the following period matrix

$$\Omega = \begin{bmatrix} 1 & 0 & \sqrt{-2} & \sqrt{-5} \\ 0 & 1 & \sqrt{-3} & \sqrt{-7} \end{bmatrix}.$$

Let A be a (4×4) non degenerate antisymmetric matrix with rational entries such that $\Omega A \Omega^T = 0$. Then the element in the first row and the second column of this matrix is given by the formula

$$a_{12} + a_{13}\sqrt{-3} + a_{14}\sqrt{-7} - a_{23}\sqrt{-2} - a_{24}\sqrt{-5} + a_{34}\sqrt{15} - a_{24}\sqrt{15} = 0.$$

It follows that

$$a_{12} = a_{13} = a_{14} = a_{23} = a_{24} = 0 \quad (\text{because } a_{ij} \in \mathbb{Q}).$$

But $A^T = -A$ and therefore $a_{21} = a_{31} = a_{41} = a_{32} = a_{43} = 0$; thus $\det A = 0$ which contradicts the assumption that $\det A \neq 0$.

DEFINITION 5. The set of complex $g \times g$ symmetric matrices τ such that $\operatorname{Im} \tau > 0$ (positive definiteness) forms an open convex set called the *Siegel half-space.* We will denote this set by $\mathfrak{H}_g$.

REMARK. $\mathfrak{H}_1$ is just the upper half-plane in $\mathbb{C}$ which we have already encountered in non Euclidean geometry and which forms the covering manifold of hyperbolic Riemann surface (that is, such Riemann surfaces that their genus $p > 1$.) C.L. Siegel (1896–1980), one of the greatest modern mathematicians, is the author of the theory of automorphic functions and modular functions of dimension $g > 1$. He contributed a lot to the theory of quadratic forms, analytical number theory, celestial mechanics, symplectic geometry. We will present some of the fundamental results of Siegel in a moment.

28.1 Theta functions

Let $(\mathbb{C}^p/\Gamma, H)$ be a polarized abelian variety. The *theta function of type* (H, ρ) is a holomorphic function ϑ on $\mathbb{C}^p$ which for every pair $z \in \mathbb{C}^p$, $\gamma \in \Gamma$ satisfies

$$\vartheta(z+\gamma) = \rho(\gamma) \exp[\pi H(z,\gamma) + \frac{1}{2}\pi H(\gamma,\gamma)]\vartheta(z) = u_\gamma(z) \cdot \vartheta(z)$$

where the function $\rho : \Gamma \to S^1$ ($\{z \in \mathbb{C} : |z| = 1\}$) is called the *semicharacter* and satisfies

$$(sc) \qquad \rho(\gamma_1, \gamma_2) = \rho(\gamma_1)\rho(\gamma_2) \exp(i\pi A(\gamma_1, \gamma_2)),$$

with $A = \operatorname{Im} H$. (From the definition of polarization $A(\gamma_1, \gamma_2) \in \mathbb{Z}$.)

REMARK. The function $u_\gamma(\cdot)$ is often called the *automorphy factor* of the function ϑ. We have

$$(CA) \qquad u_{\gamma_1\gamma_2}(z) = u_{\gamma_1}(\gamma_2(z))u_{\gamma_2}(z) \quad \text{for } \gamma_1, \gamma_2 \in \Gamma,\ z \in \mathbb{C}^p,$$

where $\gamma(z) := z + \gamma$. Condition (sc) is necessary for (CA). In the theory of functions, relation of the form $f(\gamma(z)) = u_\gamma(z) f(z)$ defines the so called *automorphic forms.* A quotient of two such forms f_1, f_2 with the same automorphy factor satisfies $(f_1/f_2)(\gamma(z)) = (f_1/f_2)(z)$ $\gamma \in \Gamma$ and is therefore an automorphic function (with respect to Γ.) In the case of the ϑ functions of the type (H, ρ) their quotient is a meromorphic function such that $g(z+\gamma) = g(z)$, $\gamma \in \Gamma$ and thus an abelian function. Abelian functions are functions on the torus $\mathbb{C}^p/\Gamma$. Frobenius computed the dimensions of the spaces of ϑ functions. We have

THEOREM 6 (FROBENIUS). *Let $(\mathbb{C}^p/\Gamma, H)$ be a polarized abelian variety. Then the vector space of ϑ function of the type (H, ρ) has the dimension*

$$(\det \operatorname{Im} H)^{1/2} = \epsilon_1 \cdots \epsilon_g,$$

where ϵ_k are divisors of the form $\operatorname{Im} H$.

This theorem can be turned into the more geometrical form.

THEOREM 7 (WEIL). *Let us construct the line bundle $L(H, \rho) \to \mathbb{C}^p/\Gamma$ as follows: in the set $\mathbb{C}^p \times \mathbb{C}$ we introduce the the equivalence relation*

$$(*) \qquad (z+\gamma, t) \sim (z, u_\gamma, t).$$

Then $(\mathbb{C}^p \times \mathbb{C})/\sim$ is a holomorphic line bundle over $\mathbb{C}^p/\Gamma$ denoted by $L(H, \rho)$ and with the ϑ functions of type (H, ρ) we associate holomorphic sections $\tilde{\vartheta} \in A(L(H, \rho))$ of the bundle $L(H, \rho)$.

The reader sees it that relation $(*)$ is exactly the functional equation for ϑ function

$$\vartheta(z+\gamma) = u_\gamma(z) \cdot \vartheta(z).$$

Frobenius theorem now takes the form

$$\dim_{\mathbb{C}} A(L(H, \rho)) = \sqrt{\det \operatorname{Im} H} = \operatorname{Pfaff}(\operatorname{Im} H).$$

Bundles over the torus $\mathbb{C}^p/\Gamma$ are precisely described by

THEOREM 8 (APPELL–HUMBERT–WEIL).

(i) *Every bundle on the torus $\mathbb{C}^p/\Gamma$ is of the form $L(H, \rho)$, where H is a hermitian form on $\mathbb{C}^p$ such that $A = \operatorname{Im} H$ is integer valued on $\Gamma \times \Gamma$.*

(ii) $(L(H_1, \rho_1) \simeq L(H_2, \rho_2)) \Longleftrightarrow (H_1 = H_2, \rho_1 = \rho_2)$.

(iii) *The Chern class of the bundle* $L = L(H, \rho)$ *is represented by* $\operatorname{Im} H$.

(iv) *(the bundle* $L(H, \rho) \to \mathbb{C}^p/\Gamma$ *is positive)*$\Longleftrightarrow$*(H is a Riemann form.)*

The precursor of the Kodaira theorem on projective embedding is the famous theorem of 1921.

THEOREM 9 (LEFSCHETZ). *Let* $X = \mathbb{C}^p/\Gamma$ *have a polarization* H *and let* $L(H, \rho) \to X$ *be a positive line bundle over* X*; let us denote this bundle by* L. *Then* $L^3 = L(3H, \rho^3)$ *defines the holomorphic embedding* $i_L : X \to \mathbb{P}^{N-1}$, *where* $N = 3^p \cdot \epsilon_1 \cdots \epsilon_p$.

This embedding is given by fundamental sections of the space $A(L^3)$, that is, ϑ functions of the type $(3H, \rho^3)$.

REMARK. As we know the projective coordinates on the space $\mathbb{P}^{N-1}$ are quotients $x_1 : x_2 : \ldots : x_N$ and thus the embedding is given by abelian functions. It can be seen from the proof of Lefschetz theorem that the line bundle $L(H_1, \rho_1)$ induces an embedding in $\mathbb{P}^{N-1}$ if the principal divisors of the Riemann form H are not less than 3. The simplest example is $p = 1$.

Embedding of elliptic curve $p = 1$. Let the torus $\mathbb{C}/\Gamma$ have the basis of periods $(1, \tau)$, that is $\Gamma = \{n + m\tau; n, m \in \mathbb{Z}\}$. We consider the form $H(z, s) = \dfrac{1}{\operatorname{Im} \tau} z \cdot \bar{s}$ and we check that H satisfies the Frobenius–Riemann conditions

$$H(1, 1) = \operatorname{Im}(\tau, \tau) = 0, \quad \operatorname{Im} H(\tau, 1) = -\operatorname{Im} H(1, \tau) = 1.$$

Biperiodic meromorphic functions are called *elliptic functions*. The function with fundamental periods $(1, \tau)$ is called the *Weierstrass functions* $\wp$; it has double poles in the lattice points $n + m\tau$ and is given by the formula

$$\wp(z) = \frac{1}{z^2} + \sum_{(0,0) \neq (n,m) \in \mathbb{Z}^2} \left[\frac{1}{(z - n - m\tau)^2} - \frac{1}{(n + m\tau)^2} \right].$$

The Weierstrass functions satisfies the following *Weierstrass differential equation* (the coefficients of this equation are defined on page 323)

$$(W) \qquad (\wp')^2 = 4\wp^3 - 60 G_4 \wp - 140 G_6.$$

The mapping $\mathbb{C} \to \mathbb{P}^2$ given by $z \to (1, \wp(z), \wp'(z))$ (in projective coordinates) induces an isomorphism of the torus $X = \mathbb{C}/\Gamma$ with the flat curve given by third order equation of the form $z_0 z_1^2 = 4z_1^3 + az_0^2 z_1 + bz_0^3$ which can be obtained from (W).

The case $g = 2$. The next simplest case is the embedding in $\mathbb{P}^{N-1}$ where $N = 3 \cdot 3$; this is the embedding $i_{L^3} : X \to \mathbb{P}^8$, where the manifold X of dimension 2 has the principal polarization $\epsilon_1 = \epsilon_2 = 1$.

Theta functions and divisors If $\pi : \mathbb{C}^p \to \mathbb{C}^p/\Gamma$ is the canonical projection, then every positive divisor D on $\mathbb{C}^p/\Gamma$ defines the positive divisor $\pi^*(D)$ on $\mathbb{C}^p$: if (V_i, f_i) is a representative of the divisor D, then $(\pi^{-1}(V_i), f_i \circ \pi)$ is, *ex definitione* a representative of the divisor $\pi^*(D)$. The following important theorem holds.

THEOREM 10 (RIEMANN). *Let θ be a ϑ function on $\mathbb{C}^p$ and let $0 \geqslant D \in Div(\mathbb{C}^p/\Gamma)$ satisfy $\pi^*(D) = (\theta)$. Then the mapping $\theta \mapsto D$ is a surjective homomorphism of the (multiplicative) subgroup of ϑ functions of the space $\mathbb{C}^p$ onto the (additive) subgroup of positive divisors on $\mathbb{C}^p/\Gamma$. The kernel of this homomorphism is the group of trivial ϑ functions, that is the functions of the form $z \to \exp(W(z))$, where W is a polynomial of order 2 on $\mathbb{C}^p$.*

Now we can look at Jacobi tori from the more general viewpoint. $Jac(S)$ has a natural principal polarization defined by condition $(\epsilon_1 = \epsilon_2 = \ldots = \epsilon_p = 1)$. Thus we have *single* ϑ function, this is the *Riemann ϑ function* θ_τ. The corresponding divisor $\Theta \in \mathrm{Div}(\mathbb{C}^p/\Gamma)$ which provides the canonical polarization $\{\Theta\}$ of the torus $\mathbb{C}^p/\Gamma$.

In order to be able to phrase theorems on functional fields, let us collect some simple facts from the theory of fields.

28.2 Strictly transcendental extensions. Transcendental degree

Let K be a subfield of the field L; the extension $L \supset K$ of the field K is called *strictly transcendental* if there exists a set $B \subset L$ which is algebraically independent with respect to the field K, such that $L = K(B)$. Thus every strictly transcendental extension of the field K is K isomorphic with the field of rational functions $K(B)$, where B is some set of variables. Every maximal subset of the field L which is algebraically independent with respect to the field K is called the *transcendental basis of L with respect to K*. It turns out that all such transcendental bases have the same cardinality; it is called *transcendental dimension (or degree)* of the field L with respect to the field K and is denoted by $\dim \operatorname{tr}_K L$ (or $\operatorname{tr} \deg_K L$. An arbitrary extension $\tilde{K}$ of the field K can be performed in two steps: first one obtains the strictly transcendental extension $K(B)$, and then one takes algebraic extension of the field $K(B)$. If $B = \{b_1, \ldots, b_g\}$, then $\tilde{K}$ is called the *algebraic functional field of g transcendental variables* $(b_1, \ldots, b_g)$. The fields of functions of one variable are particularly well known. As we will see in a moment, the theory of abelian functions leads to the fields of functions of g variables. In fact, the field of elliptic functions, or, more generally, the fields of automorphic functions were the first examples of functional fields. Investigations of these fields was an important momentum in development of the theory of fields.

Chapter 29

Theorems on Algebraic Dependence

Let Γ be a lattice in $\mathbb{C}$, then the elliptic Γ periodic functions form a field E_Γ; indeed, meromorphic functions (on $\mathbb{C}/\Gamma$) form a field. Riemann and Weierstrass knew it already that every two functions $f_0, f_1 \in E_\Gamma$ are algebraically dependent, moreover, Weierstrass showed that $E_\Gamma = \mathbb{C}(\wp, \wp')$, that is, every elliptic Γ periodic function is a rational function of the Weierstrass function $\wp$ and its derivative. Similar theorem was later proved for automorphic function of one variable. As we know the problem of uniformization of compact Riemann surface S led F. Klein and H. Poincaré to creation of the theory of automorphic functions.

Let $\pi : \tilde{S} \to S$ be a universal covering for S ($\tilde{S}$ is a simply connected Riemann surface.) It turns out that $\tilde{S}$ is biholomorphically equivalent to the Riemann sphere $\hat{\mathbb{C}} := P(\mathbb{C})$, or the plane $\mathbb{C}$, or the half plane $\mathfrak{H}$ (for $g > 1$, $\tilde{S} \sim \mathfrak{H}$ always), moreover $S = \mathfrak{H}/\Gamma$, where for Γ one can take the group of transformations of the covering. Thus functions from $\mathcal{M}(S)$ (meromorphic on S) can be identified with functions from $\mathcal{M}_\Gamma(\tilde{S})$ which are meromorphic on $\tilde{S}$ and Γ invariant, that is $f(\gamma z) = f(z)$, $\gamma \in \Gamma$, and therefore with Γ automorphic functions on $\tilde{S}$. On a compact Riemann surface we have again

THEOREM 1 *Every two functions in $\mathcal{M}(S)$ are algebraically dependent. There exist two functions $f_1, f_2 \in \mathcal{M}(S)$ such that every function $f \in \mathcal{M}(S)$ is a rational function of f_0 and f_1, that is, $\mathcal{M}(S) = \mathbb{C}(f_0, f_1)$.*

This fact can be expressed shortly as follows: The set of Γ automorphic functions is a field of algebraic functions of $p = 1$ variable.

Now we present a number of impressive theorems of C.L. Siegel who developed a general procedure of proving theorems on algebraic dependence of automorphic functions.

THEOREM 2 (SIEGEL–THIMM (1939)). *Let X be a compact complex manifold of dimension p. Then every p meromorphic functions $f_1, \dots, f_p \in \mathcal{M}(X)$ are algebraically dependent. Moreover, tr $\deg_{\mathbb{C}} \mathcal{M}(X) \leqslant p = \dim X$.*

By merging this theorem with the fact that on a abelian variety $\mathbb{C}^p/\Gamma$ we have p meromorphic functions which are algebraically (and even analytically) independent, we obtain

COROLLARY 3 (POINCARÉ). *The function field K_Γ of Γ periodic abelian functions on $\mathbb{C}^p$, where the torus $\mathbb{C}^p/\Gamma$ is an algebraic variety (the field $\mathcal{M}(\mathbb{C}^p/\Gamma)$) is a finite extension of the field $\mathbb{C}(f_1, \dots, f_p)$, where $f_1, \dots, f_p$ is an arbitrary set of algebraically independent abelian functions.*

Algebraic dependence of modular functions of p variables. As we saw, for the case of algebraic torus $\mathbb{C}^p/\Gamma$ the theorem hold which described the situation precisely. The general theory of automorphic functions of p variables was created by C.L. Siegel. As the theory of the ϑ functions of p variables has shown it already, this theory is by far more complex and difficult than the case of one complex variable: we saw that already the very existence of automorphic functions is not always guaranteed (Riemann–Frobenius relation in the case of abelian functions!) Similarly to the role played in the theory of one variable by $\mathfrak{H}_1$, the fundamental role in the theory of p complex variables is played by the *Siegel half space* $\mathfrak{H}_p = \{Z\colon (p \times p)$ symmetric matrices with $\operatorname{Im} Z > 0\}$. It can be shown that $\mathfrak{H}_p$ is a contractible region in $\mathbb{C}^n$ $n = p(p+1)/2$. Similarly to the case $p = 1$, automorphisms of the space $\mathfrak{H}_p$ form a group which is isomorphic with the group $Sp(2n, \mathbb{R})$, that is the group of real $(2n \times 2n)$ matrices T such that $T^T JT = J$, where $J = \begin{bmatrix} 0 & 1_n \\ -1_n & 0 \end{bmatrix}$. The following important theorem holds.

THEOREM 4 (SIEGEL). *Let $\mathfrak{H}_p$ be the Siegel upper half plane. Then $Sp(2p, \mathbb{R})$ acts transitively in biholomorphically on $\mathfrak{H}_p$ as follows $Z \to TZ :=$*

$\frac{AZ+B}{CZ+D}$, where $T = \begin{bmatrix} A & B \\ C & D \end{bmatrix}$. The group $K = Sp(2p, \mathbb{R}) \cap O(2p)$ is the group preserving the point $\sqrt{-1}1_p$. K is a maximal compact subgroup in $Sp(2p, \mathbb{R})$; all maximal compact subgroups in $Sp(2p, \mathbb{R})$ are conjugate with K (by inner automorphisms). The correspondence $K \ni T \to C\sqrt{-1} + D \in U(g)$ defines a topological isomorphism.

In the theory of elliptic functions ($p = 1$) a very important role was played by the modular group Γ_1, the group of integer valued (2×2) matrices with determinant 1: two one dimensional tori with periods $(1, \tau)$ and $(1, \tau')$, where $\tau, \tau' \in \mathfrak{H}_1$ are biholomorphically equivalent if and only if $\tau' = g\tau$, with g being an element of the modular group. The *modular functions* are the meromorphic functions on $\mathfrak{H}_1$ which are Γ_1 invariant, and thus they are automorphic functions. They form therefore an algebraic functional field as well.

We owe Siegel the very interesting and deep theory of modular functions of p variables.

DEFINITION 5. The *modular Siegel group* $\Gamma_p := Sp(2p, \mathbb{Z})$ is the group of integer valued symplectic matrices.

The following theorem of utmost importance holds.

THEOREM 6 (SIEGEL–BAILY). *The field of Γ_p invariant, meromorphic functions on $\mathfrak{H}_g$ (the so called modular Siegel functions) is an abelian functional field of $p(p+1)/2$ variables, thus for these functions theorem on algebraic dependence holds.*

REMARK 1. In the initial version Siegel used the additional assumption that these functions are to be a product of modular forms (like abelian functions,which are products of ϑ functions.) The deep theory of compactification of the space $\mathfrak{H}_p/\Gamma_p$ let Baily and Satake to prove that this assumption is not necessary. Later, in the joint beautiful paper, Andreotti and Grauert, by introducing the notions of pseudoconcaveness for the region X and the discrete group Γ of automorphisms of X proved the flowing important

THEOREM ON FINITENESS (ANDREOTTI–GRAUERT) 7. *The Γ automor-*

phic functions form the space of finite dimension.

The pseudoconcaveness property of the group Γ makes it possible not to assume the compactness of the the orbits X/Γ, which property Siegel must have usually assumed. It turns out that the modular Siegel group Γ_p is pseudoconcave which gives the theorem on algebraic dependence of modular functions.

REMARK 2. The group G of transformations of the space X acts *transitively* if for every two points $x_1, x_2 \in X$, there exists $g \in G$ such that $g \cdot x_1 = x_2$ (in other words for every two points there exists an orbit of G passing through them.)

To complete this point we will present an important notion of the automorphic form which we *implicite* encountered already several times. In the theory of elliptic functions the so called modular forms appear; these are, for example, the Eisenstein series G_4 and G_6 which appear in the equation of the Weierstrass function $\wp$.

DEFINITION 8. Let U be a complex manifold and Γ a subgroup of automorphisms of U, let G be a group acting on the (vector) space V. The *automorphy factor* σ of the group Γ is the mapping $\sigma : U \times \Gamma \to \mathbb{C}$ such that $\sigma(\cdot, \gamma) \in A(U)^*$:

$$(*) \qquad \sigma(x, \gamma_1 \cdot \gamma_2) = \sigma(\gamma_2 x, \gamma_1)\sigma(x, \gamma_2), \quad \text{for } x \in U,\ \gamma_1, \gamma_2 \in \Gamma$$

(one often writes $\sigma_\gamma(x) \equiv \sigma(x, \gamma)$).

The *automorphic form* (of type σ) is the function $f : X \to V$ (taking values in the vector space V) such that

$$f(\gamma \cdot x) = \sigma_\gamma(x) \cdot f(x).$$

Observe that for any integer k, σ^k is also an automorphy factor. The space of automorphic forms $f \in A(U)$ such that $f(\gamma \cdot x) = \sigma_\gamma^k(x) \cdot f(x)$ will be denoted by $A(\Gamma)_k$.

Now we present some examples.

EXAMPLE 1. $U = \mathbb{C}^p$, Γ is a lattice in $\mathbb{C}^p$, $\gamma \cdot z = z + \gamma$, and $\sigma_\gamma(z)$ is the automorphism factor of the ϑ functions theory. The automorphic forms

$A(\Gamma)_k$ (with values in $\mathbb{C}$) are ϑ functions of the type (H, ρ)

$$\sigma_\gamma(z) := \rho(\gamma)\exp(\pi H(z,\gamma) + \frac{1}{2}\pi H(\gamma,\gamma)).$$

EXAMPLE 2. U is a compact domain in $\mathbb{C}^n$ with Γ being a discrete subgroup of the group of biholomorphic automorphisms such that U/Γ is a compact variety. Let $\sigma(x,\gamma) = j(x,\gamma)$ be the Jacobian of the transformation $\gamma : U \to U$ at the point x. Then for every bounded and holomorphic function f on U, the *Poincaré series* of natural weight $k \geqslant 2$

$$\mathcal{P}_{f,k} = \sum_{\gamma\in\Gamma} f(\gamma(x))j(x,\gamma)^k, \quad k \text{ integer and} \geqslant 2$$

is compctly convergent and represents an automorphic form of weight k (with values in $\mathbb{C}$)

$$\mathcal{P}_{f,k}(\gamma(z)) = j(z,\gamma)^{-k}\mathcal{P}_{f,k}(z).$$

The form $\mathcal{P}_{f,k}$ is called the *Poincaré series of weight* k.

It is easy to observe that given above-mentioned assumptions, U is a region of holomorphy: the Bergmann function tends to infinity when z tends to the boundary of U. (The variety $X = U/\Gamma$ satisfies assumptions of Kodaira theorem; it is therefore an algebraic manifold if Γ acts without fixed points.)

EXAMPLE 3. $U = \mathfrak{H}_g$ is the Siegel upper half plane, $\Gamma = \Gamma_g$ is the Siegel modular group. For $Z \in \mathfrak{H}_g$ we have

$$\gamma \cdot Z = \begin{bmatrix} A & B \\ C & D \end{bmatrix} Z = \frac{AZ+B}{CZ+D}, \quad \gamma = \begin{bmatrix} A & B \\ C & D \end{bmatrix} \in \Gamma.$$

We take $\sigma_\gamma(Z) = J_\gamma(Z)$. Then $A(\Gamma_g)$ are Siegel *modular forms.*

EXAMPLE 4. Taking in Example 3 $U = \mathfrak{H}_1$, $\Gamma = SL_2(\mathbb{Z})$, $\sigma_\gamma(z) = J_\gamma(z)$, $g = 1$, we obtain the classical model of modular forms (of one complex variable) of weight k

$$f(\gamma \cdot z) = (cz+d)^{-2k} f(z), \quad \gamma = \begin{bmatrix} a & b \\ c & d \end{bmatrix} \in SL_2(\mathbb{Z}).$$

If $\Gamma = \Gamma(\omega_1, \omega_2)$, $\tau := \omega_1/\omega_2 \in \mathfrak{H}_1$, then

$$G_k(\omega_1,\omega_2) = \sum_{0\neq m,n\in\mathbb{Z}} (m\omega_1 + n\omega_2)^{-k}$$

is called the *Eisenstein series.* We take $g_2 = 60G_4$, $g_3 = 140G_6$, $\Delta = g_2^3 - 27g_3^2$, $j = 12^3 g_2^3/\Delta$.

THEOREM 9
(i) $G_k(\tau) := G_k(1, \tau)$ *is a modular form of weight* k.
(ii) $j(\tau) := j(1, \tau)$ *is a modular function holomorphic on* $\mathcal{H}_1$ *and having a simple pole at* ∞.
(iii) *The modular functions on* $\mathfrak{H}_1$ *form the field* $\mathfrak{M}$ *generated by* j; $\mathfrak{M} = \mathbb{C}(j)$.

The function j called the *elliptic modular function* was discovered by Dedekind in 1877 and by Klein a year later. With the help of this function Picard proved his famous

THEOREM 10. *Every holomorphic function on the complex plane* $\mathbb{C}$ *which is not constant, takes every complex value with the possible exception of at most one value.*

The importance of the function j in the theory of elliptic functions lies, among others, in the following fact. Let S be a Riemann surface isomorphic with the torus $\mathbb{C}/L$ and denote by j_A the number $j(\tau)$ where $(1, \tau)$ generates L. Let S' be another such surface. Then

THEOREM 11. *(the surfaces* S *and* S' *are isomorphic)*$\Longleftrightarrow$*(*$j_S = j_{S'}$*)*$\Longleftrightarrow$*(*$L \simeq L'$*, where* L, L' *are lattices defining* S *and* S'*, respectively)*$\Longleftrightarrow$*(if* $(1, \tau)$, $(1, \tau')$ *generate* L *and* L'*, then* $\tau' = \gamma\tau$*, where* $\gamma \in Sp(2, \mathbb{Z})$*)*

We can present yet another version of this theorem.

THEOREM 12 (TORELLI). *Let* S *and* S' *be Riemann surfaces of genus* $p \geqslant 1$ *and let* $(\pi i 1_p, A)$ *and* $(\pi i 1_p, A')$ *be period matrices with respect to canonical bases. Then* $(S \simeq S')$$\Longleftrightarrow$*(there exists* $\gamma \in Sp(2p, \mathbb{Z})$ *such that* $(\pi i 1_p, A)\gamma = (\pi i 1_p, A')$*, where* $A' = \pi i B^{-1} C$*).*

Roughly speaking, period matrices are modularly equivalent. We can once again dwell upon the problem of

Projective embedding with the help of automorphic forms.

Let U be a complex manifold and $A(\Gamma)_k$ a finite dimensional (vector) space of automorphic forms. Let $f_0, f_1, \dots, f_N$ be a basis in $A(\Gamma)_k$. We define the holomorphic mapping $U \ni z \to (f_0(z), f_1(z), \dots, f_N(z)) \in \mathbb{C}^{N+1}$. If $f_0, f_1, \dots, f_N$ do not vanish simultaneously at any point $z \in V$, we obtain (by turning to projective coordinates) holomorphic mappings $U \to \mathbb{P}^N(\mathbb{C})$. Since for every $\gamma \in \Gamma$, the points z, γz have the same image in $\mathbb{P}^N$, by denoting by U/Γ the space of Γ orbits in U, we obtain the continuous mapping $U/\Gamma \to \mathbb{P}^N(\mathbb{C})$. Let us present the condition guaranteeing that U/Γ is again a complex manifold.

DEFINITION 13. The group Γ acts on U *properly discontinuously* if for every compact $C \subset U$ the set $\{\gamma \in \Gamma : \gamma \cdot \mathbb{C} \cap C \neq \emptyset\}$ is finite.

It is easy to see that in that case the space U/Γ is a locally compact space. If, moreover, Γ acts without fixed points, U/Γ is a complex manifold and $U \to U/\Gamma$ is its covering.

Now, if the mapping $f : U/\Gamma \to \mathbb{P}^N(\mathbb{C})$ and its derivative are injective, then f is called the *projective embedding* by automorphic functions. Observe that the change of basis $f_0, f_1, \dots, f_N$ of the space $A(\Gamma)_k$ is equivalent to some projective mapping in $\mathbb{P}^N(\mathbb{C})$. Employing this general procedure in Example 1 above, we obtain Lefschetz theorem on projective embedding. We see that it is important to have a criterion of properly discontinuous action of the group Γ. This criterion is the subject of the following theorem.

THEOREM 14 (SIEGEL). *Let G be a locally compact group, K its compact subgroup, Γ a discrete subgroup of G. Then Γ acts on the quotient space $U := K\backslash G$ in the properly discontinuous way.*

The crown example is due to Siegel again.

EXAMPLE 15 $G = Sp(2p, \mathbb{R})$, $K = Sp(2p, \mathbb{R}) \cap O(2p)$, Γ is some discrete subgroup of $Sp(2p, \mathbb{R})$, for example, the modular group. As we know $\mathfrak{H}_p = G/K$ and thus the modular group Γ acts freely on $\mathfrak{H}_p$; then $\mathfrak{H}_p$ is a universal covering of the quotient space $\mathfrak{H}_p/\Gamma$ (since, as we know, $\mathfrak{H}_p$ is connected!)

EXAMPLE 16 $G = \mathbb{C}^p$, $K = O$, Γ is a lattice in $\mathbb{C}^p$. $O\backslash G/\Gamma = \mathbb{C}^p/\Gamma$ is a complex manifold, the torus and $\mathbb{C}^p$ is its universal covering.

Part III

The Idea of the Riemann Surface

To Hermann Weyl, the great mathematician,
physicist, and philosopher

1 Introduction

The present part is central for the whole of this book and its title is borrowed from the celebrated monograph of the young Hermann Weyl *Idee der Riemannschen Fäsche.* (Of course, our presentation does not have much in common with the Weyl's classic, however in this way I want to express admiration and gratitude to the greatest, most influential master of mathematics of our century.)

Among a number of wonderful ideas we owe Riemann, the idea of Riemann surface is, without doubts, the most beautiful, everlasting, intensively developing, unifying and fertilizing a number of other ideas, penetrating the whole body of mathematics, and, in turn, many branches of physics. It is a language, and thus an organ of cognition and creation, a co-originator of *kosmos noethos*, the world of ideas. This idea originated from the germ of algebraic functions, or elliptic integrals (differentials) and their local inverses: elliptic functions. Riemann, as any great creator, was very concrete: his surfaces were connected with their mother, the Riemann sphere $S^2 = \mathbb{P}^1$ with an umbilical cord - a meromorphic function (here, algebraic) - which were created from its germ by the process of analytical continuation. Remember, at that time the notion of topological space was not known. Therefore Riemann cuts his surfaces M with $2p$ cuts and in this way he can spread it over the plane $\mathbb{C}$

One of our goals is to present the proof of Riemann-Roch theorem. Riemann was a great analyst: he regarded the theory of functions on his surfaces X as a theory of the operator

$$d'' : C^\infty(A^{0,0}(X)) \to C^\infty(A^{0,1}(X)),$$

or, using the cohomology language, of the mappings

$$d'' : H^0(X) \to H^1(X).$$

For that reason he investigated the equation $d''u = \omega$ for given $(0,1)$ form ω and $d''u = 0$. The Laplace operator on Riemann surface has the form

$$d'd''f = \Delta f.$$

The theory of harmonic functions (and forms) is often called the 'potential theory". It can be said, therefore, that Riemann investigated the potential theory on his surfaces. This is the reason why he was interested in Dirichlet principle which characterizes harmonic entities as critical points of the corresponding Dirichlet energy integral (cf. Part 'Riemann and Calculus of Variations'.).

A Riemann surface is nothing but holomorphic atlas $(\mathcal{U}_j, z_j)$, $j \in J$, where $\mathcal{U}_j$ are domains in $\mathbb{C}$ related by holomorphic transition functions. Thus complex analysis on Riemann surfaces (and, more general, on complex manifolds) X works with the atlas $\mathcal{U}_i$ and the corresponding 1 and 2 cycles.

Our first goal will be to formulate a simple corollary of Cauchy formula.

THEOREM. *Let $\mathcal{U}$ be a domain in $\mathbb{C}$. An integral operator P with the Cauchy kernel $(\zeta - z)^{-1}$ is a* fundamental solution *of the operator $d'' : C^\infty(A^{0,0}(\mathcal{U})) \to C^\infty(A^{0,1}(\mathcal{U}))$, $Pd'' : C^\infty(A^{0,0}(\mathcal{U})) \to C^\infty(A^{0,0}(\mathcal{U}))$,*

$$P\omega = \frac{1}{2\pi i}\int_{\mathcal{U}} \frac{d\zeta}{\zeta - z} \wedge \omega(\zeta) = \frac{1}{2\pi i}\int_{\mathcal{U}} d' \log(\zeta - z) \wedge \omega.$$

This means that the equations

$$d''P\omega = \omega \quad \text{and} \quad Pd''f = f$$

are satisfied for compactly supported sections in $\mathcal{U}$.

REMARK. On a differentiable manifold M the spaces of (p, q) forms are denoted by $A^{p,q}(M)$. Therefore for operators d'', P we can write

$$A^{0,0}(Y) \underset{P}{\overset{d''}{\rightleftarrows}} A^{0,1}(Y)$$

Therefore P is (in some sense) left and right inverse of d''.

PROOF follows from Cauchy formula. If $u \in C^1(\bar{\mathcal{U}})$, we have

$$u(z) = \frac{1}{2\pi i}\int_{\partial\mathcal{U}} \frac{u(\zeta)}{\zeta - z} d\zeta + \int_{\mathcal{U}} \frac{d''u}{\zeta - z} d\zeta \wedge d\bar{\zeta}, \quad z \in \mathcal{U}.$$

Conversely, if $\omega = \varphi d\bar{z}$, where $\varphi \in C_0^k(\mathcal{U})$, $k \geqslant 1$, then $d''P\omega = \omega$, that is, $\frac{\partial u}{\partial \bar{z}} = \varphi$, where $u(z) := \frac{1}{2\pi i}\int_{\mathcal{U}} \frac{\varphi d\bar{\zeta}}{\zeta - z} d\zeta$.

$(\zeta - z)^{-1}$ is called the Cauchy kernel.

2 Fredholm–Noether operators. Parametrices

In the global situation, the atlas $\{(\mathcal{U}_j, z_j)\}$ on X, we have for every coordinate map $(\mathcal{U}_j, z_j)$ the corresponding Cauchy kernel $(\zeta_j - z_j)^{-1}$ and the local operator P_j. We cannot construct a global fundamental solution for the operator d'' on a compact Riemann surface for a holomorphic vector bundle $E \to M$, acting on (p, q) forms with values in E: $d'' : A^{0,1}(X) \to A^{0,1}(X)$, but using adequate partitions of unity we can construct a *parametrix*, that is, an approximate fundamental solution of d'' – more precisely

DEFINITION. Let $A : C^\infty(F_1) \to C^\infty(F_2)$ be a (linear) elliptic operator, where F_1, F_2 are differentiable vector bundles on compact manifold X. The *parametrix* P is a (pseudodifferential) operator $P : C^\infty(F_2) \to C^\infty(F_1)$ such that

$$PA = 1 - S_1, \quad AP = 1 - S_2,$$

where 1 is the identity, and S_j are integral operators with smooth kernels (therefore trace class)

$$S_j = \int_X s_j(x, x)dx < \infty, \quad j = 1, 2.$$

REMARK. Clearly S_j are compact operators. If we consider Banach spaces H_1, H_2 obtained by completion of $C^\infty(F_j)$, $j = 1, 2$, then

DEFINITION. Let $A \in L(H_1, H_2)$, and H_1, H_2 be Banach spaces. Then the linear continuous operators $Q_j \in L(H_2, H_1)$ such that $Q_1A - 1$ and $AQ_2 - 1$ are compact are called parametrices (left and right) of A. The operator $A \in L(H_1, H_2)$ with two sided parametrices is called Fredholm-Noether operator.

THEOREM (FRITZ NOETHER). *(A is a Fredholm-Noether operator)* $\Longleftrightarrow$ *(*dim ker L *and* dim coker L *are finite dimensional).*

Therefore the index of A, index$A$ = dim ker A – dim coker A *is well defined.*

Before we present (a quite simple) proof, let us make some historical remarks and recall important points of the F. Riesz theory.

HISTORICAL REMARKS. Fritz Noether (born in 1884), the younger brother of Emmy Noether was the son of Max Noether, one of the founders of modern algebraic geometry. We owe Fritz Noether the notion of index of an operator. It is (historically) *unjustified* to call Noether operators the Fredholm operators. The Russian school is - in this respect - correct: they call Fredholm-Noether operators *Noether operators.* But, of course, the name of Ivar Fredholm is not completely out of place: We owe Fredholm the theory of linear integral operators (with kernels with weak singularities.) Such operators are compact; this was observed by Hilbert, the founder of spectral theory of compact operators. Hilbert used himself the language of continuous bilinear forms which is equivalent to the language of continuous operators; compact forms (operators) were called by Hilbert completely continuous.

The notion and the name 'parametrix' is due to Hilbert (1904-5). Almost at the same time this notion was introduced and applied by brilliant Italian mathematician E.E. Levi. We owe E.E. Levi fundamental investigations on the field of the theory of several complex variables (cf. Part 'Riemann and Complex geometry'.)

REMARK. Usually the Fredholm–Noether operator $A \in L(H_1, H_2)$ for Banach spaces H_1, H_2 is *defined* by the following properties:

1. $\ker A$ is finite dimensional;
2. $\operatorname{coker} A$ is finite dimensional (that is, $\dim H_2/AH_1 < \infty$);
3. AH_1 is closed in H_2.

But 3. follows from 1. and 2., therefore we can dispense of 3, and the theorem of Noether gives the equivalence of both definitions.

EXERCISE. Prove that (1. and 2.) $\Longrightarrow$3.

We will denote by $F(H_1, H_2)$ the set of Fredholm-Noether maps. Before we turn to the proof of Noether theorem, we will present the famous theorem of F. Riesz

THEOREM (F. RIESZ). *Let H be a Banach space, K a compact operator on H. Then the operator $A := 1 - K$ is a Fredholm-Noether operator.*

PROOF. $\ker A$ is finite dimensional: Since $1|_{\ker A} = K|_{\ker A}$, it follows that $1|_{\ker A}$ is compact $\Longrightarrow \dim\ker A < \infty$. Since the Banach space X is finite dimension, it follows that the identity 1_X is compact.

Let us denote by H' the dual Banach space, that is, $H' = L(H, \mathbb{C})$ and by T' the dual (adjoint) operator. It is obvious that (K is compact) $\Longrightarrow$(K' is compact).

2. We prove that $\operatorname{coker} A$ is of finite dimension. $(H/(A(H)))' \simeq \ker A' = \ker(1 - K')$. Therefore, by 1., $H/A(H)'$ is of finite dimension $\Longrightarrow H/A(H)$ is of finite dimension.

DEFINITION. If $A \in F(H_1, H_2)$, then $\dim\ker A - \dim\operatorname{coker} A$ is called the index of A .

COROLLARY. *($K : H \to H$ is compact) $\Longrightarrow$($\operatorname{ind}(1 - K) = 0$).*

We have the following useful theorem

THEOREM (F. RIESZ). *If $A \in F(H_1, H_2)$ is Fredholm-Noether, then*
(a) $A' \in F(H_2', H_1')$;
(b) $\ker A' = (\operatorname{coker} A)'$;
(c) $\operatorname{coker} A' = (\ker A)'$;
(d) $\operatorname{ind} A' = -\operatorname{ind} A$.

PROOF. From the exact sequence

$$0 \to \ker A \to H_1 \xrightarrow{A} H_2 \to \operatorname{coker} A \to 0$$

it follows that the sequence

$$0 \leftarrow (\ker A)' \leftarrow H_1' \xleftarrow{A} H_2' \leftarrow (\operatorname{coker} A)' \leftarrow 0$$

is exact; therefore (a)-(d) follow. □

PROOF OF NOETHER THEOREM. $\Longrightarrow$Since $Q_1 A - 1$ is compact, $\ker Q_1 A$ is of finite dimension; hence $\dim\ker A < \infty$ because $\ker A \subset \ker Q_1 A$. If $AQ_2 - 1$ is compact, then $AQ_2(H_2)$ has a finite codimension; therefore $A(H_1)$ is of finite codimension, because $A(H_1) \supset AQ_2(H_2)$.

$\Longleftarrow$ Let $\dim\ker A, \dim\operatorname{coker} A < \infty$. Denote by V and W subspaces such that

$$H_1 \simeq \ker A \times V, \quad H_2 \simeq A(H_1) \times W.$$

Then $A|_V : V \to A(H_1)$ is bijective. Denote by $Q : H_2 \to H_1$, defined by $Q|_{A(H_1)} := (A|_V)^{-1}$ which is continuous by the Banach theorem on inverse operator and $Q|_W = 0$. Therefore the map $1-QA$ is a projection onto ker A, whence compact; and $1-AQ$, being a projection on W is compact as well. □

Thus we have the set $F(H_1, H_2)$ of Fredholm-Noether operators which have index: these are such mappings that there exists a parametrix $P \in L(H_1, H_2)$, where

$$(*) \qquad PA = 1 - S_1, \quad AP = 1 - S_2,$$

S_j are trace class (there exist the function $\operatorname{tr} : S_j \to \mathbb{Z}, j = 1, 2$). We have a quite elementary but very useful

THEOREM (TRACE FORMULA FOR THE INDEX). *Let A satisfy* $(*)$. *Then* $\operatorname{ind} A = \operatorname{tr} S_1 - \operatorname{tr} S_2$.

PROOF can be found in the treatise of Hörmander (vol. 3, 19.1.4).

Now we are prepared to present the proof of Riemann-Roch theorem (R-R theorem) which we understand as an index theorem for the operator

$$d'' : H^0(E \otimes A^{0,0}) \to H^0(E \otimes A^{0,1})$$

acting between the spaces of sections of bundles of forms valued in the holomorphic vector bundle E of rank r, $E \to M$, over compact Riemann surface of genus p_X:

$$\operatorname{ind} d'' = \dim H^0(X, \mathcal{O}(X)) - \dim H^1(X, \mathcal{O}(X)) =$$

$$(R-R) \qquad = c_1(\det E) - \frac{n}{2} c_1(K),$$

where $K = K_X$ is the canonical bundle of X, and $c_1(E)$ denotes the first Chern class of a line bundle $E \to X$. We know that $c_1(K_X) = 2(p_X - 1)$, and that, by Kodaira-Serre duality,

$$\dim H^1(X, \mathcal{O}(X)) = \dim H^0(X, \mathcal{O}(K \otimes E^*)).$$

In order to be able to apply the abstract theorems of index theory, the Cauchy-Riemann operator $d'' = A$ should be a *continuous* mapping of Hilbert spaces $d'' : H_1 \to H_2$. Since d'' is a linear operator of order 1, for H_2 we take, of course, the Sobolev space $W_1^2(A^{0,1} \otimes E)$ of first order derivatives of sections of the bundle $A^{0,1} \otimes E$; $H_1 = W_0^2(A^{0,0} \otimes E)$.

We know that the problem resides in the construction of a good parametrix P of the operator d'' and in computation of the traces $\operatorname{tr} S_1$, $\operatorname{tr} S_2$. Now we present a beautiful proof of Riemann–Roch theorem due to Siebners.

3 Proof of Riemann–Roch theorem

STEP I: CONSTRUCTION OF THE PARAMETRIX P OF THE OPERATOR d''. The Cauchy formula gives the local fundamental expansion of P_j in the map $\mathcal{U}_j$ of the atlas $(\mathcal{U}_j)$ on the surface X. On the diagonal $\zeta_j = z_j$ the Cauchy kernel $(\zeta_j - z_j)^{-1}$ has a singularity, and the parametrix P is to provide integral operators S_1, S_2 with continuous kernel. The construction we are going to present has a general character: for a differential operator of arbitrary order on manifolds one constructs a fundamental solution of the 'frozen' operator having constant coefficients. Then one takes appropriate partitions of unity; more precisely:

Let (φ_j) be a smooth partition of unity subordinate to the covering $(\mathcal{U}_j)$: $spt\, \varphi_j \subset \mathcal{U}_j$ $0 \geqslant \varphi_j \geqslant 1$, $\sum \varphi_j = 1$. Let now $\alpha_j \in C^\infty(X)$ be such that $spt\, \alpha_j \subset \mathcal{U}_j$ and $\alpha_j(x) = 1$ for $x \in spt\, \varphi_j$.

The defining cocycle l_{ij} of the holomorphic vector bundle $E \to M$ is given by $l_{ij} \in \mathrm{GL}(n, \mathcal{O}(\mathcal{U}_i \cap \mathcal{U}_j))$ and is an $n \times n$ invertible matrix of holomorphic functions on $\mathcal{U}_i \cap \mathcal{U}_j$. The section $s \in H^0(E \otimes A^{0,p}, \mathcal{U}_j)$ is represented by an n-tuple of elements of $H^0(A^{0,p}, \mathcal{U}_j)$. The local operators P_j are acting component-wisely on n-tuples of $(0,1)$ forms. Now we can define the parametrix P for the operator d'':

$$P := \sum \varphi_j P_j : C^\infty(E \otimes A^{0,1}) \to C^\infty(E \otimes A^{0,0}).$$

We have to compute $Pd''f$ and $d''P\omega$. Since $\alpha_j d'' f = d''(\alpha_j f) - d''\alpha_j \cdot f$,

$$Pd''f = \sum \varphi_j P_j(d''(\alpha_j f) - d''\alpha_j \cdot f) = \sum \varphi_j \alpha_j f - \sum \varphi_j P_j f d''\alpha_j =$$

$$f - \sum \varphi_j P_j f d''\alpha_j = f - S_1 f.$$

But $d''\alpha_j = 0$ in a neighborhood of the diagonal (x,x), therefore S_1 not only has a smooth kernel, but also $\operatorname{tr} S_1 = 0$.

Similarly,

$$d''P\omega = d''\sum \varphi_j P_j \alpha_j \omega = \sum d''\varphi_j (P_j \alpha_j \omega) + \sum \varphi_j d''(P_j \alpha_j \omega) =$$

$$= \sum \varphi_j \alpha_j \omega + \sum P_j (a_j \omega) d''\varphi_j = \omega - S_2 \omega,$$

where

$$S_2\omega := -\sum d''\varphi_j P_j(\alpha_j\omega) = \sum d''\varphi_j (P_j - P_i)\alpha_i\omega - \sum d''\varphi_i (P_j \alpha_i \omega).$$

But $\alpha_i = 1$ on *spt* $d''\varphi_i$, hence, in a neighborhood of the diagonal,

$$\sum d''\varphi_j (P_j \alpha_i \omega) = P_j \sum (d''\varphi_i)\omega = P_j \omega d'' \sum \varphi_j = 0;$$

therefore

$$\mathrm{tr}(\sum d''\varphi_i P_j \alpha_i \omega) = 0.$$

By the trace formula for the index, we have proved the following interesting

PROPOSITION (L.M. SIEBNER–R.J. SIEBNER). $P := \sum \varphi_j P_j \alpha_j$.

1. $Pd''f = f - S_1 f$ and $d''P\omega = \omega - S_2\omega$, where S_1 and S_2 are integral operators with smooth kernels.

2. Moreover, $\mathrm{tr}\, S_1 = 0$ and $\mathrm{tr}\, S_2 = \mathrm{tr} \sum_i d''\varphi_i (P_i - P_j)\alpha_i$.

COROLLARY I. *(a) d'' is Fredholm–Noether;*

(b) $\mathrm{ind}(d'') = \mathrm{tr} \sum_i d''\varphi_i (P_i - P_j)\alpha_i$, where P_j is the fundamental solution on $\mathcal{U}_j$ given by the Cauchy kernel.

COROLLARY II (FINITENESS THEOREM). *If $E \to X$ is a holomorphic vector bundle on a compact Riemann surface X, then the cohomology spaces $H^0(X, E)$ and $H^1(X, E)$ are finite dimensional $\mathbb{C}$ vector spaces.*

We will see that the Riemann–Roch theorem follows from the formula (*b*) above. But before we turn to the proof of this theorem, let us recall the definition of line bundles and their Chern functions c_1.

The *canonical bundle* $K = K_X$ is the most important line bundle on Riemann surface X: it is T^*X, the holomorphic cotangent bundle of X. Its holomorphic sections $\omega \in H^0(K)$ are holomorphic $(1,0)$ forms called Abelian differentials of first kind. The sheaf $\mathcal{O}(X)$ of holomorphic sections of K is denoted by Ω, $\Omega(X)$, or Ω_X. The line bundle is defined by the transition functions $k_{ij} \in \mathcal{O}(\mathcal{U}_i \cap \mathcal{U}_j)$, holomorphic and nowhere zero on $\mathcal{U}_i \cap \mathcal{U}_j$ such that

$$dz_j = k_{ij} dz_i, \quad \text{that is }, k_{ij} = \frac{\mathrm{d}z_j}{\mathrm{d}z_i} \quad \text{on } \mathcal{U}_i \cap \mathcal{U}_j.$$

Locally, $\omega \in \Omega(\mathcal{U})$ is given by $\omega_i = f_i dz_i$, where f_i are holomorphic functions on $\mathcal{U}_i \cap \mathcal{U}$, $f_i \in \mathcal{O}(\mathcal{U}_i \cap \mathcal{U})$, $f_i = f_j \frac{\mathrm{d}z_j}{\mathrm{d}z_i} = k_{ij} f_j$. If $(\{\mathcal{U}_i, z_i)\}$ is a holomorphic atlas of X with transition functions $z_i = z_{ij}(z_j)$, then the Jacobian $J(z_{ij}) = \left|\frac{\mathrm{d}z_j}{\mathrm{d}z_i}\right|^2 = |k_{ij}|^2$.

If M is a complex manifold of complex dimension m, then the canonical bundle K_M is defined as $\Lambda^m T^*(M)$, the m-th exterior power of the $\mathbb{C}$ cotangent bundle T^*M of M.

The determinant bundle $\det E$ of a vector bundle $E \to X$ given by the cocycle (l_{ij}) is defined by the cocycle $\det l_{ij}$. It is a line bundle.

On a Riemann surface X, for any $r \in \mathbb{Z}$ we have the r-th tensor product $K^{\otimes r}$, where $K^{-1} = K^*$. Elements of $\mathcal{O}(K^{\otimes r})$ are called *holomorphic differentials* of order r; locally they are given by transition functions $\left(\frac{\mathrm{d}z_j}{\mathrm{d}z_i}\right)^r$. In Teichmüller theory of utmost importance are *holomorphic quadratic differentials*; these are elements of $\mathcal{O}(K^{\otimes 2})$. It follows from the Riemann–Roch theorem that $\dim_{\mathbb{C}} H^0(\mathcal{O}(K_X^{\otimes 2})) = (3p-3)$, $p > 1$, where p is the genus of X[1]. (This fact is of paramount importance in Teichmüller theory!)

Chern classes $c(K_X)$ and $c(\det E)$. For the canonical bundle K_X given by the cocycle (k_{ij}) there are positive functions $k_j \in C^\infty(\mathcal{U}_i)$ such that $d' \log k_{ij} = d' \log k_j - d' \log k_i = d' \log \frac{k_j}{k_i}$. Similarly, for the bundle $\det E$, where E is given by (l_{ij}), there exist positive functions $l_i \in C^\infty(\mathcal{U}_i)$ such that $d' \log l_{ij} = d' \log l_j - d' \log l_i = d' \log \frac{l_j}{l_i}$. The corresponding Chern forms and Chern numbers are given by

$$c(K_X) = \int_X d''d' \log k_j \quad \text{and} \quad c(\det E) = \int_X d''d' \log l_j.$$

More generally, for any line bundle $E \to X$ given by the cocycle (f_{ij}) and for a family (r_j) of non-vanishing C^∞ functions $r_i : \mathcal{U}_i \to \mathbb{C}$ such that for all $z \in \mathcal{U}_i \cap \mathcal{U}_j$, $r_i(z) = r_j(z)|f_{ji}(z)|^2$, $\varphi := \frac{1}{2\pi\sqrt{-1}} d'd'' \log r_i \in A^{1,1}(\mathcal{U}_i)$, $i \in I$ is a well defined global $(1,1)$ form on X and we have

[1]This follows immediately from $R-R$, because $c(K) = 2*(p-1)$, and $c(K^{\otimes r}) = rc(K)$: $\dim_{\mathbb{C}} H^0(X, \mathcal{O}(c(K^{\otimes r})) = c(K^{\otimes r}) - (p-1) = (2r-1)(p-1)$

DEFINITION.

$$(Ch) \qquad c(F) := \int_X \varphi = \frac{1}{2\pi\sqrt{-1}} \int_X d'd'' \log r_i$$

is the Chern number of the line bundle $F \to X$. The function $F \to c(F) \in \mathbb{Z}$ defined on C^∞ line bundles (LB) on X is called the *Chern function* . It has the following nice properties:

THEOREM. *Let $c : LB \to \mathbb{Z}$ be the Chern function; then*
1. $c(F_1 \otimes F_2) = c(F_1) + c(F_2)$;
2. $c(F^*) = -c(F)$;
3. $c(K_X) = 2(p_X - 1)$, *where p_X is the genus of X,* $p_X := \dim H^0(X, \mathcal{O}(K))$;
4. If F_D is a line bundle defined by the divisor D, then $c(F_D) = \deg D$;
5. $c(K^r) = rc(K)$;
6. $\dim H^0(X, \mathcal{O}(K^r)) = (2r-1)(p_X - 1)$, *where* $K^r \equiv K^{\otimes r}$.

REMARK. For a vector bundle E of rank $n \geqslant 1$, $c(E) := c(\det E)$.

PROOF. 1. and 2. follow immediately from (Ch). 3.–5. follow from the Riemann-Roch theorem: put $E = K$, then $\dim_{\mathbb{C}} H^0(X, \mathcal{O}(K)) - \dim_{\mathbb{C}} H^0(X, K \otimes K^*) = g_X - 1 = c(K) - \frac{1}{2}c(K) = \frac{1}{2}c(K)$. Therefore $c(K) = 2(p_X - 1)$. 6. For $r > 1$, $H^1(X, \mathcal{O}(K^r)) = 0$, thus $\dim H^0(X, \mathcal{O}(K^r)) = c(\det K^r) - \frac{1}{2}c(K) = rc(K) - \frac{1}{2}c(K) = \frac{2r-1}{2}c(K) = (2r-1)(p_X - 1)$. □

Similar formulas hold for compact Kähler manifold M. Let us recall some important notions. The hermitian structure induces a Riemann metric, which defines the (torsion free) Levi-Civita connection in TM. There exists a natural isomorphism between the *real* tangent bundle TM and the holomorphic tangent bundle $T'M$ ($TM^{\mathbb{C}} = T'M \oplus T''M$).

The complex structure defines connection compatible with the hermitian metric and with the holomorphic structure on $T'M$. There are several equivalent definitions of Kähler manifold (we use here the summation convention):

DEFINITION. 1. M is Kähler if these two connections agree.

If $g_{a\bar{\beta}} dz^\alpha \otimes d\bar{z}^\beta$ is a hermitian metric on M, then $\omega := \frac{\sqrt{-1}}{2} g_{a\bar{\beta}} dz^\alpha \wedge d\bar{z}^\beta$ is called the *Kähler form*;

2. M is Kähler if $d\omega = 0$;

3. There is a function F such that $g_{\alpha\bar{\beta}} = \frac{\partial^2 F}{\partial z^\alpha \partial z^{\bar{\beta}}}$, that is

$$\omega = \frac{\sqrt{-1}}{2} d' d'' F.$$

F is called the *Kähler potential*;

4. At each point $z_0 \in M$ one can introduce holomorphic normal coordinates, that is

$$g_{\alpha\bar{\beta}}(z_0) = \delta_{\alpha\bar{\beta}}, \quad \frac{\partial g_{\alpha\bar{\beta}}}{\partial z^k}(z_0) = 0 = \frac{\partial g_{\alpha\bar{\beta}}}{\partial z^{\bar{l}}}(z_0)$$

for all α, β, k, l. The Ricci form (tensor) is obtained as

$$R_{\alpha\bar{\beta}} dz^\alpha \wedge dz^{\bar{\beta}} = -d' d'' \log(\det g_{\alpha\bar{\beta}}),$$

and the Chern form (class) is proportional to the Ricci form

$$c(M) = c_1(M) = -\frac{\sqrt{-1}}{2\pi} d' d'' \log(\det g_{\alpha\bar{\beta}}) = \frac{\sqrt{-1}}{2\pi} R_{\alpha\bar{\beta}} dz^\alpha \wedge dz^{\bar{\beta}} =$$
$$= -\frac{\sqrt{-1}}{2\pi} d' d'' \log\left(\det \frac{\partial^2 F}{\partial z^\alpha \partial z^{\bar{\beta}}}\right).$$

Since every Riemann surface is Kählerian, we obtain the previous relations.

After these remarks let us return to the beautiful proof of Riemann-Roch theorem due to Siebners.

PROOF OF RIEMANN-ROCH THEOREM (SIEBNER- SIEBNER). As we know we have the local fundamental solution

$$P_j \omega := \frac{1}{2\pi\sqrt{-1}} \int_{\mathcal{U}_j} d' \log(\zeta_j - z_j) \wedge \omega(\zeta_j).$$

We have already seen the importance of the operator $(P_i - P_j) =: \Theta_{ij}$; the following lemma describes this operator.

LEMMA. *Let ρ be a global $(0,1)$ form with coefficients in $E \to M$ having support in $\mathcal{U}_i \cap \mathcal{U}_j$. Then near the diagonal*

$$Q_{ij}\rho = (P_i - P_j)\rho =$$

$$\int_{\mathcal{U}_i \cap \mathcal{U}_j} \left\{ l_{ij}^{-1}(z_j) l'_{ij}(z_j) - \frac{1}{2} I \left(\frac{\partial}{\partial \zeta_j} \right) \log k_{ij}(z_j) + O(\zeta_j - z_j) \right\} d\zeta_j \wedge \rho_j(\zeta_j).$$

Here I denotes the identity matrix and l'_{ij} is the derivative.

PROOF. In terms of the j basis of sections of E the representation of $P_i\rho$ near the diagonal is

$$(P_i\rho)_j = l_{ij} \int_{\mathcal{U}_i \cap \mathcal{U}_j} d' \log(\zeta_i - z_i) \wedge \rho_i(\zeta_i) =$$

$$= \int_{\mathcal{U}_i \cap \mathcal{U}_j} l_{ij}^{-1}(z_j) l_{ij}(\zeta_j) \left(d' \log(\zeta_i - z_i) - \frac{1}{2} k_{ij}(z_j) + O(\zeta_j - z_j) \right) \wedge \rho_j(\zeta_j).$$

Here we use the expansion

$$\zeta_i - z_i = (k_{ji}^{1/2}(z)(\zeta_j - z_j) + O(|\zeta_j - z_j|^2),$$

whence,

$$d'(\zeta_j - z_j) = \frac{1}{2} \log k_{ji}(z_j) + d' \log(\zeta_j - z_j) + O(\zeta_j - z_j), \quad k_{ji} = k_{ij}^{-1}.$$

Moreover, near the diagonal

$$l_{ij}(\zeta_j) = l_{ij}(z_j) + l_{ij}(z_j) + O(|\zeta_j - z_j|^2),$$

therefore

$$l_{ij}^{-1} l_{ij}(\zeta_j) = I + l_{ij}^{-1}(z_j) l'_{ij}(z_j) + O(|\zeta_j - z_j|^2).$$

Therefore we have

$$(P_i\rho)_j = \int_{\mathcal{U}_i \cap \mathcal{U}_j} \left\{ d' \log(\zeta_i - z_i) I + \frac{1}{2} d' \log k_{ij}(z_j) I + \right.$$

$$\left. + \, l_{ij}^{-1}(z_j) l'_{ij}(\zeta_j) d\zeta_j + O(\zeta_j - z_j) \right\} \wedge \rho_j(\zeta_j).$$

Subtracting $(P_j\rho)_j = \int_{\mathcal{U}_i \cap \mathcal{U}_j} d' \log(\zeta_i - z_i) \wedge \rho_j(\zeta_j)$ we obtain the formula for $Q_{ij}\rho$. □

Now the proof of the Riemann–Roch theorem follows quickly. For any global section ω of $E \otimes A^{0,1}$ we have $\omega = \sum \varphi_j \omega$ (φ_j is a partition of unity.) We have to evaluate the trace $\operatorname{tr} Q_{ij}$ of the operator Q given by

$Q\omega := \sum_{ij} d''\varphi_j Q_{ij}\alpha_j\varphi_i\omega$. Applying the Lemma to $\rho = \alpha_j\varphi_j\omega$ and using the elementary formula

$$\operatorname{tr} A^{-1}\left(\frac{\mathrm{d}}{\mathrm{d}z}\right) A = \frac{\mathrm{d}}{\mathrm{d}z}\log\det A,$$

we obtain the desired Riemann-Roch formula for $\operatorname{ind} d'' = \operatorname{tr} Q$ as follows

$$\operatorname{tr} Q = \sum_{ij}\int_{\mathcal{U}_i\cap\mathcal{U}_j} \alpha_j\varphi_i d''\varphi_j d'\log\det l_{ij} - \frac{n}{2}\sum_{ij}\int_{\mathcal{U}_i\cap\mathcal{U}_j} \varphi_i d''\varphi_j d'\log k_{ij} =$$

(since $d'\log k_{ij} = d'\log k_j - d'\log k_i$, $d'\log l_{ij} = d'\log l_j - d'\log l_i$, $\sum d''\varphi_j = d''\sum\varphi_j = 0$, and since $\alpha_j = 1$ on the support of $d''\varphi_j$)

$$\operatorname{tr} Q = \sum_{ij}\int_{\mathcal{U}_i\cap\mathcal{U}_j} \varphi_i d''\varphi_j d'\log\det l_j - \frac{n}{2}\sum_{ij}\int_{\mathcal{U}_i\cap\mathcal{U}_j} \varphi_i d''\varphi_j d'\log k_j.$$

But

$$\varphi_i d''\varphi_j d'\log k_j = d''(\varphi_i\varphi_j d'\log k_j) - \varphi_j d''\varphi_i d'\log k_j - \varphi_i\varphi_j d''d'\log k_j,$$

and the corresponding formula holds for l_j; thus, by virtue of the Stokes formula we have

$$\operatorname{tr} Q = \sum_{ij}\int_{\mathcal{U}_i\cap\mathcal{U}_j} \varphi_i\varphi_j d''d'\log\det l_j - \frac{n}{2}\sum_{ij}\int_{\mathcal{U}_i\cap\mathcal{U}_j} \varphi_i\varphi_j d''d'\log k_j +$$

$$+\sum_{ij}\int_{\partial(\mathcal{U}_i\cap\mathcal{U}_j)} \varphi_i\varphi_j\left(\frac{n}{2}d'\log k_j - d'\log l_j\right).$$

But on $\partial(\mathcal{U}_i\cap\mathcal{U}_j)$ either φ_i or φ_j vanishes, therefore

$$\operatorname{ind} d'' = \int_X d''d'\log l_j - \frac{n}{2}\int_X d''d'\log k_j = c(\det E) - \frac{n}{2}c(K).$$

□

Let us present a number of important corollaries of the Riemann-Roch theorem.

The Riemann theory would not be interesting, had it not for existence of nontrivial meromorphic functions on compact Riemann surfaces. Thus the first term $\dim H^0(X, \mathcal{O}_D)$ in the Riemann-Roch formula

$$(R-R) \qquad \dim H^0(X,\mathcal{O}_D) - \dim H^1(X,\mathcal{O}_D) = \deg D - (p-1)$$

plays the decisive role. Riemann himself was only able to prove the inequality $\dim H^0(X,\mathcal{O}_D) \geqslant \deg D - (g-1)$. The equality $(R-R)$ is due to the pupil of Riemann, Gustav Roch (1839-1866). Let us recall that $\mathcal{O}_D(X) = \{f \in \mathcal{M}(X) : \mathrm{ord}_x(f) \geqslant -D(X)$ for every $x \in X\}$. Since divisors with arbitrary $\deg D$ exist, the spaces $H^0(X,\mathcal{O}_D)$ may have arbitrary (but always finite) dimension.

From $(R-R)$ it follows

PROPOSITION. *For every holomorphic line bundle, the number*

$$\chi(E) = \dim H^0(X,\mathcal{O}_D) - \dim H^1(X,\mathcal{O}_D) - c(E) \quad (= c(K_X))$$

is independent of E, that is, it is the same for any line bundle.

But also *vice versa*: $(R-R)$ follows from the Proposition. One often proves $(R-R)$ by proving the Proposition first.

In complex analysis a very important role is played by *vanishing theorems*, also called Theorems B because the famous Cartan-Serre theorem characterizing Stein manifolds was called Théorem B (cf. part V.)

Before formulating these theorems let us introduce the frequently used notation for dimensions of the spaces $H^q(X,\mathcal{O})$: $i(D) = \dim H^1(X,\mathcal{O}_D) = \dim H^0(X,K_D) = l(K-D)$, $l(D) := \dim H^0(X,\mathcal{O}_D)$, even though, from the pedagogical point of view, one should reduce to the minimum the amount of new notations (it is easy to forget them.) The number $i(D)$ is the *index of specialty* of divisor D. The elementary fact holds.

THEOREM. *For compact X and $D \in \mathrm{Div}(X))$ with $\deg D < 0$ $H^0(X,\mathcal{O}_D) = 0$, hence $l(D) = 0$.*

PROOF (BY CONTRADICTION). $\emptyset \neq H^0(X,\mathcal{O}_D) \ni f \neq 0$. Then $(f) \geqslant -D$, and therefore $\deg(f) \geqslant -\deg D > 0$; this contradicts the fact that $\deg(f) = 0$.

COROLLARY. *$i(D) = p - 1 - \deg D$ if $\deg D < 0$.*

THEOREM B (VANISHING THEOREM). *If $\deg D > 2p-1$ $(= \deg K)$, then*

(a) $H^1(X,\mathcal{O}_D) = 0$;

(b) $l(D) = \deg D + 1 - p$.

PROOF. $(\deg K = 2p-2) \Longrightarrow (\deg(K-D) < 0) \Longrightarrow (i(D) = l(K-D) = 0)$ by the preceding theorem.

REMARK. For divisors with $\deg D = 2p-2$, Theorem B does not hold: $H^1(X, \mathcal{O}_K) \simeq \mathbb{C}$ and $\deg K = 2p-2$.

COROLLARY. *For compact* X, $H^1(X, \mathcal{M}) = 0$, *where* $\mathcal{M}$ *is a sheaf of meromorphic functions.*

PROOF. Let $\xi \in H^1(X, \mathcal{M})$ be a cohomology class which is represented by the cocycle $(f_{ij}) \in Z^1((\mathcal{U}_i), \mathcal{M})$. Passing to the refinement of the covering $(\mathcal{U}_i)$ of X, we can assume that the total number of poles of all f_{ij} is finite. Then there exists a divisor D with $\deg D > 2p-2$ such that $(f_{ij}) \in Z((\mathcal{U}_i), \mathcal{O}_D)$. By Theorem B the cocycle (f_{ij}) is cohomologous to zero relative to the sheaf $\mathcal{O}_D$, and thus relative to the sheaf $\mathcal{M}$. □

4 The fundamental theorem for compact surfaces

The question of great importance is if any holomorphic line bundle has meromorphic sections? The answer to the positive is provided by another vanishing theorem which so important that the great master of complex analysis Robert C. Gunning does not hesitate to call it *fundamental.*

THEOREM (FUNDAMENTAL THEOREM FOR COMPACT SURFACES). *For a compact surface* X *the following statements hold.*

(a) $H^1(X, \mathcal{M}^*) = 0$;

(b) Every holomorphic line bundle $E \to M$ *has a nontrivial meromorphic section;*

(c) Every line bundle $E \to M$ *is equivalent to a line bundle* L_D *of divisors.*

REMARK. (a), (b), and (c) are equivalent assertions.

Let us recall that on an *arbitrary* complex manifold X the sheaf of germs of divisors $\mathcal{D} := \mathcal{M}^*/\mathcal{O}^*$ and $\mathrm{Div}(X)) := H^0(X, \mathcal{D})$ (global sections of $\mathcal{D}$) are divisors. The abelian group of divisors $\mathrm{Div}(X)$ is additive. Sections of $\mathcal{O}^*$ are nowhere vanishing holomorphic functions $f \in \mathcal{O}(X)$; $\mathcal{O}^*$ is a subsheaf of $\mathcal{M}^*$. From the exact sequence

$$0 \to \mathcal{O}^* \to \mathcal{M}^* \to \mathcal{D} \to 0$$

one obtains the exact cohomology sequence (we write $\mathcal{O}^*(X) = H^0(X, \mathcal{O}^*)$, and so on)

$$(*)\ \ 0 \to \mathcal{O}^*(X) \to \mathcal{M}^*(X) \xrightarrow{\psi} \mathrm{Div}(X)) \xrightarrow{\eta} H^1(X, \mathcal{O}^*) \to H^1(X, \mathcal{M}^*) \to 0.$$

PROOF OF THE THEOREM. From the exactness of the sequence it follows the equivalence of (a), (b), and (c). If $H^1(X, \mathcal{M}^*) = 0$, then η is surjective., thus every line bundle $L \in H^1(X, \mathcal{O}^*)$ is of the form $\eta(D) = L_D$.

(A line bundle is $\eta(D)$ for some D) $\Longleftrightarrow (\eta(D) = L_D$ has a nontrivial meromorphic section).

We are done if we prove that for every line bundle E there exists L_D such that $H^0(X, \mathcal{M}(E \otimes L_D)) \neq 0$. We will show even more, namely that for any line bundle there exists L_D such that $H^0(X, \mathcal{O}(E \otimes L_D)) \neq 0$. But the latter can be easily achieved by taking D such that $\deg(E \otimes L_D) = c(E) + c(L_D) > c(K) = 2(p-1)$.

FURTHER REMARKS. It turns out that also on *non compact* Riemann surface X, $H^1(X, \mathcal{M}^*) = 0$ (thus $H^1(X, \mathcal{M}^*) = 0$ for arbitrary Riemann surface.) But this fact is difficult to show; it amounts of the famous Behnke-Stein theorem that on a non compact Riemann surface X every divisor $D \in \mathrm{Div}(X)$ is a divisor of some meromorphic function $f \in \mathcal{M}^*(X)$.

Thus on compact surface X instead of investigating divisors one can investigate line bundles; these notions are equivalent. The group $\psi\mathcal{M}^* \subset \mathrm{Div}(X)$ is the group of *principal divisors* (divisors of meromorphic functions.) The quotient group $\mathrm{Div}(X)/\psi\mathcal{M}^*(X) = \mathrm{Div}(X)/\operatorname{im}\psi$ is a group of divisor classes $\simeq H^1(X, \mathcal{O}^*)$ (the group of (isomorphy) classes of holomorphic line bundles.)

$$(D_1 \simeq D_2) \Longleftrightarrow (L_{D_1} \simeq L_{D_2}).$$

The reader wonders perhaps that if we are talking about Theorem B there must be some Theorem A. Indeed, *Theorems A* concern sheaves $\mathcal{F}$ and vector bundles $L \to X$ and assert that there are *many* global sections, or that holomorphic global sections generate over arbitrary $x \in X$ a fiber L_x (stalk $\mathcal{F}_x$.) Theorems A and B were proved in the 1950s (for X – Stein and $\mathcal{F}$ – analytic coherent sheaf) by Cartan and Serre (cf. Part 'Riemann and Complex Geometry'.) Let us now present

THEOREM A (FOR COMPACT RIEMANN SURFACES). *If D is a divisor on X with $\deg D > 2p - 1$, then for any $x \in X$ there exists a holomorphic section $s \in H^0(X, L_D)$ of the line bundle $L_D \to X$ such that $s(x) \neq 0$. (Equivalently, there exists a divisor D' linearly equivalent to D ($D' \sim D$) not containing x in its support).*

COROLLARY A. *Let $L \to X$ be a holomorphic line bundle on X with $\deg L > 2p$. Then*

(a) If $x, y \in X$, $x \neq y$, then there exists $s \in H^0(X, L)$ such that $s(x) = 0$, $s(y) \neq 0$ (that is, there are many holomorphic sections of X);

(b) If $x \in X$, there exists $s \in H^0(X, L)$ such that $\operatorname{ord}_x(s) = 1$.

PROOF (AFTER R. NARASIMHAN). Let s_x be the standard section of $L(\{x\})$ with $\operatorname{Div}(s_x) = x$. The mapping $A : H^0(X, L(D - x)) \to H^0(X, L(D))$, $f \to f \otimes s_x$ is *not* surjective since (by $(R - R)$)

$$\dim H^0(X, L(D - x)) = \deg(D - x) + 1 - p < \deg D + (1 - p) = \\ = \dim H^0(X, L(D)),$$

because $\deg(D - x) > 2p - 2$. The image of A, $\operatorname{Im} A$ consists of sections of $L(D)$ vanishing at x. Consider next the line bundle $L \otimes L(-x)$. It follows from theorem A that there exists $s' \in H^0(X, L \otimes L(-x))$ with $s'(y) \neq 0$. Take $s = s' \otimes s_x$, where s_x is the standard (holomorphic) section of $L(x)$. If $x \neq y$, $s(y) \neq 0$, $s(x) = 0$; if $x = y$, then $\operatorname{ord}_x(\operatorname{Div} s_x) = 1$. □

5 Embedding of Riemann surfaces

Since the times of Euler and Gauss differential geometry have dealt with surfaces 'lying' in $\mathbb{R}^3$. Algebraic geometry deals with algebraic varieties, 'loci of a family of polynomials'. It was early realized that one should consider *points at infinity*, in this way the notion of the projective space $\mathbb{P}$ (first the projective plane) was established. As we know the theory of compact Riemann surfaces appeared from the theory of algebraic functions and the problems of integration of algebraic differentials (elliptic and hyperelliptic integrals: Galois, Gauss, Abel, Jacobi, Riemann, Weierstrass, and many others.) Only the development of analysis and topology made it possible to spell precisely naive intuitions concerning embeddings.

Theorem A makes it possible to prove important *imbedding theorem* which asserts that every compact Riemann surface can be embedded in some

$\mathbb{P}^N$. The strategy is as follows. Let $p = p_X$ be the genus of X; take any holomorphic line bundle $L(D) = L \to X$ with $\deg L = \deg D > 2p$ and let $N = \dim H^0(X, L) - 1 = \deg D - 1$. Let $s_0, s_1, \ldots, s_N$ be a basis of $H^0(X, L)$. If $y \in X$, take a neighborhood $\mathcal{U}$ of y and the section $\sigma \in H^0(\mathcal{U}, L)$ such that $\sigma(x) \neq 0$ for $x \in \mathcal{U}$. By Theorem A the sections $s_0, s_1, \ldots, s_N$ cannot have common zeros. Therefore $i_L(x) := \left(\frac{s_0(x)}{\sigma(x)} : \cdots : \frac{s_N(x)}{\sigma(x)}\right) \in \mathbb{P}^N$ is a well defined map $i_L : X \to \mathbb{P}^N$. We have the following

THEOREM (EMBEDDING THEOREM). *If $\deg L < 2p$, then $i_L : X \to \mathbb{P}^N$ is a holomorphic embedding of X into $\mathbb{P}^N$.*

PROOF. (a) i_L is injective. Let $x \neq y$ and as in Corollary A choose $s \in H^0(X, L)$ with $s(x) = 0$ and $s(y) \neq 0$. If $s = a_0 s_0 + \cdots + a_N s_N$, then $i_L(x)$ is in the hypersurface $a_0 z_0 + \cdots + a_N z_N = 0$ and $i_L(y)$ is outside this hypersurface; therefore i_L is injective.

(b) The differential (tangent map) of i_L is injective. For given $x \in X$, as in Corollary A choose a section $s \in H^0(X, L)$ such that $\text{ord}_x(s) = 1$, and let k be such that $s_k(x) \neq 0$ and $s = a_0 s_0 + \cdots + a_N s_N$. Thus

$$\frac{s}{s_k} = a_k + \sum_{j \neq k} \frac{s_j}{s_k}.$$

The functions $f_j = \frac{s_j}{s_k}$, $j \neq k$ form inhomogeneous coordinates for $i_L(y)$ for y in a neighborhood of x: $i_L(y) = (f_0(y), \ldots, 1, f_{k+1}(y), \ldots, f_N(y))$. Since $\text{ord}_x\left(\frac{s_j}{s_k}\right) = 1$, we have $df_j(x) \neq 0$ for at least one $j \neq k$. The proof is complete. □

We recall that the line bundle $L \to X$ is *ample* if for some positive integer $m > 0$ the map i_{L^m} (where $L^m := L^{\otimes m}$) is an embedding of X into some projective space $\mathbb{P}^N$. Thus we have

COROLLARY. *A holomorphic line bundle L on compact X is ample if* $\deg L > 0$.

Clearly the number N is not the smallest possible one. It is much harder to show that $N = 3$ suffices:

THEOREM. *Every compact surface X can be holomorphically imbedded in $\mathbb{P}^3$.*

Chow theorem says that every compact complex manifold which can be imbedded holomorphically in $\mathbb{P}^N$ is algebraic. Thus compact Riemann surfaces are algebraic, or as algebraic geometers use to say *X is a smooth algebraic curve in $\mathbb{P}^N$*. Open, that is, non compact Riemann surfaces cannot be imbedded in $\mathbb{P}^N$ but they are Stein manifolds and can be regarded as closed analytic submanifolds of some $\mathbb{C}^M$. There is a famous

CONJECTURE. *Any open Riemann surface can be properly embedded in* $\mathbb{C}^2$.

Since the canonical bundle $K_X = K$ is an extremely important line bundle, the mapping $i_K : X \to \mathbb{P}^{p-1}$, where $p = p_X$ is the genus of X, must be particularly interesting. $i_K(x) := (\omega_1 : \cdots : \omega_p) \in \mathbb{P}^{p-1}$, where the one forms $\omega_1, \ldots, \omega_p$ form a basis of $H^0(X, K)$. The map i_K is called the *canonical map* of X and $i_K(X) \subset \mathbb{P}^{p-1}$ is the *canonical curve* of X. We know from Corollary A that i_K is injective if for any $x \neq y$ there exists a holomorphic form ω such that $\omega(x) = 0$ and $\omega(y) \neq 0$; i_k is an immersion if for any $x \in X$ there exists ω with $\mathrm{ord}_x(\omega) = 1$. Thus

$(i_K$ is an embedding$) \Longleftrightarrow($for any x, y, $\dim H^0(X, K - x - y) < \dim H^0(X, K - x) = p - 1)$

But by $(R-R)$, $\dim H^0(X, K - x - y) = p - 3 + \dim H^0(X, x + y)$, therefore

$(\dim H^0(X, K - x - y) < \dim H^0(X, K - x)) \Longleftrightarrow (\dim H^0(X, x + y) = 1)$

We have proved

THEOREM. *The canonical map i_K fails to be an embedding if and only if there exists a meromorphic function $f : X \to \mathbb{P}^1$ having only two poles, that is if X is a 2-sheeted branched covering of the Riemann sphere $\mathbb{P}^1$.*

6 Hyperelliptic surfaces. Hyperelliptic involutions

Such surfaces, called *hyperelliptic* are distinguished among compact Riemann surfaces. They can be - as we will see in a moment - of arbitrary genus. The theorems above can be phrased as follows

(The canonical map of X fails to be an embedding) $\Longleftrightarrow$ (X is hyperelliptic).

We can expect that hyperelliptic surfaces have properties different from curves that are not hyperelliptic: indeed in the case of Torelli theorem or in the Teichmüller theory the case of hyperelliptic surfaces must be analyzed separately. Immediate questions arise:

1. Why these surfaces are called hyper*elliptic*. In other words, what is the relation with ellipticity?

2. Are there compact surfaces which are not hyperelliptic?

Ad. 1. As we know everything started with elliptic integrals, that is with integrals of the form $\int \frac{dz}{\sqrt{P(z)}}$, where $P(z)$ is a polynomial of third or fourth degree. As we know from the Riemann-Hurwitz formula, if the covering $f : X \to \mathbb{P}^1$ with total branching order b is n-sheeted, then

$$p_X = \frac{1}{2}b - n + 1.$$

Let X be hyperelliptic, thus $n = 2$ and $p = b/2 - 1$. For example, let $f : X \to \mathbb{P}^1$ be the Riemann surface of $\sqrt{P(z)}$, where $P(z) = (z - a_1) \cdots (z - a_k)$ with all roots a_k distinct. The genus p of X is $p = [(k-1)/2]$, where $[c]$ denotes the largest integer smaller than c. We have the basis of holomorphic forms on X:

$$\omega_j = \frac{z^{j-1}dz}{\sqrt{P(z)}}, \quad 1 \leqslant j \leqslant [(k-1)/2],$$

where z denotes the meromorphic function $f : X \to \mathbb{P}^1$.

Exercise. Prove that ω_j are holomorphic. Take local coordinates at critical points $a_1, \ldots, a_k$.

In this way we answered the first question.

Ad. 2. This problem troubled already Riemann! In the course of answering this question Riemann has encountered the famous moduli problem: he counted the number of 'parameters' (moduli) describing a hyperelliptic surface of genus p: locally one needs $2p - 1$ of them; for an arbitrary surface of genus $p > 1$, one needs $3p - 3$ parameters, and thus for $p \geqslant 3$, $(3p - 3) - (2p - 1) = p - 2 \geqslant 1$. Thus not every surface of genus $p \geqslant 3$ is hyperelliptic. This is the famous Riemann count which has led, much later, Teichmüller to his famous theory.

PROPOSITION. *Every surface of genus $p = 2$ is hyperelliptic.*

Thus one must look for non-hyperelliptic objects among surfaces of genus $p \geqslant 3$.

The proof of the proposition follows from the fact that if $p = 2$ and $x \in X$ is a Weierstrass point (see below), then there exists a meromorphic function which is holomorphic on $X - \{x\}$ having at x a pole of order 2. Or even simpler form Theorem A: Take $D := 2x$, then $\deg(K + 2x) \geqslant 4 = 2p$. The existence follows from Theorem A.

REMARK. Such meromorphic functions are called abelian differentials of the second kind. Similarly one proves the existence of a meromorphic form having poles of order 1 *only* at two points $x_1 \neq x_2$. The current terminology is as follows: meromorphic differentials are called *abelian*; abelian holomorphic differentials are of the *first kind*; while abelian differentials with zero residue are of the *second kind*; general differentials which may have residues ($\neq 0$) are of the *third kind.* Thus it follows from Theorem A

PROPOSITION. *On every compact Riemann surface there exist abelian differentials of second and third kind.*

7 Weierstrass points. Wronskian

The following question arises in relation with the considerations above. As we know when X is compact then the field of meromorphic functions $\mathcal{M}(X)$ defines the surface completely. If X is of genus p, then for every $x \in X$ there exists $f \in \mathcal{M}(X)$ which has a pole only at x and this pole is of order $\geqslant p+1$. Thus particularly interesting are points x for whose there exists a function $f \in \mathcal{M}(X)$ the only pole of order $\geqslant p$. The problem posed by Weierstrass is if such points (called the *Weierstrass points*) exist and how they can be characterized, is, perhaps, most important in the context of hyperelliptic X. If X is of genus $p > 1$, how many there are Weierstrass points? These questions arose in the course of construction of the impressive theory of hyperelliptic curves (and integrals.)

DEFINITION. Given $x \in X$, $p = p_X$, let $(\mathcal{U}, z)$ be a coordinate at x with $z(x) = 0$. x is a Weierstrass point if there exists $f \in \mathcal{M}(X)$ and constants $a_0, \ldots, a_{p-1}$ not all equal zero such that

1. f is holomorphic on $X - \{x\}$;
2. $f - \sum_{j=0}^{p-1} \frac{a_j}{z^{j+1}}$ is holomorphic at x.

For existence of such f it is sufficient (and necessary) that $\operatorname{res}_x\left(\sum_{j=0}^{p-1} \frac{a_j}{z^{j+1}}\omega\right) = 0$ for all holomorphic forms ω on X. Let $\omega_1, \ldots, \omega_p$ be a basis of $H^0(X, \Omega)$ and let $\omega_k = f_k dz$ on $\mathcal{U}$. If $f_k = \sum_{j=0}^{\infty} f_{k,j} z^j$, then $f_{k,j} = \frac{1}{j!} f_k^{(j)}(0)$, therefore the residuum $\operatorname{res}_x\left(\sum_{j=0}^{p-1} \frac{a_j}{z^{j+1}}\omega_k\right) = \sum_{j=0}^{p-1} \frac{a_j}{j!} f_k^{(j)}(0)$, $k = 1, \ldots, p$. We have thus another characterization of Weierstrass points.

THEOREM. *($x \in X$ is a Weierstrass point) $\Longleftrightarrow$(equations $\sum_{j=0}^{p-1} a_j f_k^{(j)}(0) = 0$, $k = 1, \ldots, p$ have a nontrivial solution $(c_1, \ldots, c_{p-1})$) $\Longleftrightarrow$($\det(f_k^{(j)}(0)) = 0$), $1 \leqslant k \leqslant p$, $0 \leqslant j \leqslant p-1$.*

The last condition makes use of the notion of Wronskian

$$W(f_1, \ldots, f_g)(z) := \det(f_k^{(j)}(z)), \;\; 1 \leqslant k \leqslant p, \;\; 0 \leqslant j \leqslant p-1.$$

COROLLARY. *x is a Weierstrass point if and only if $W(f_1, \ldots, f_g)(x) = 0$.*

Now we can give a global definition of Wronskian on X as a section of the line bundle $K^N \to X$, where $N = \frac{1}{2}p(p+1)$ and $K = K_X$ is the canonical bundle of X. Indeed, let $(\mathcal{U}_i, z_i)$ be an open covering of X, then the transition maps $\kappa = \frac{\mathrm{d}z_j}{\mathrm{d}z_i}$ define K and the assignment $W_i := W(f_{1,i}, \ldots, f_{g,i})$ $(\omega_k = f_{k,i} dz_i)$ satisfies $W_i = \kappa_{ij} W_j$ on $\mathcal{U}_i \cap \mathcal{U}_j$.

Therefore W_i define a holomorphic section $W \in H^0(X, K^N)$ of the line bundle K^N. N is the number of equations which are obtained for $\omega_k = f_k dz = \tilde{f}_k d\tilde{z}$ on $\mathcal{U} \cap \tilde{\mathcal{U}}$: we have $f_k = \psi \tilde{f}_k$, where $\psi := \frac{d\tilde{z}}{dz}$. By induction

$$\frac{d^m f_k}{dz^m} = \psi^{m+1} \frac{d^m \tilde{f}_k}{d\tilde{z}^m} + \sum_{\mu=0}^{m-1} \varphi_{m\mu} \frac{d^\mu \tilde{f}_k}{d\tilde{z}^\mu},$$

where $\varphi_{m\mu} \in \mathcal{O}(\mathcal{U} \cap \tilde{\mathcal{U}})$ and are independent of k.

If we take $W_z(\omega_1, \ldots, \omega_p) := W(f_1, \ldots, f_p)$, we find

$$W_z(\omega_1, \ldots, \omega_p) = \left(\frac{\mathrm{d}\tilde{z}}{\mathrm{d}z}\right)^N W_{\tilde{z}}(\omega_1, \ldots, \omega_p).$$

Thus we have a global version of the famous

THEOREM (WEIERSTRASS). *There is a non zero section W of the line bundle K_X^N, where $N = \frac{1}{2}p(p+1)$ such that the zeros of W are exactly Weierstrass points of X.*

Since for $p > 1$, $\deg(\operatorname{Div} W) = \deg K^N = N \deg K = (p-1)p(p+1)$,

COROLLARY. *1. There are at most $(p-1)p(p+1)$ Weierstrass points; 2. If $p_X > 1$ there exist Weierstrass points.*

A positive integer w is a *gap* for $x \in X$ when there do not exist any meromorphic function f such that $f \in \mathcal{O}(X - \{x\})$ with $\operatorname{ord}_x f = w$. The reader may be willing to prove the following

WEIERSTRASS GAP THEOREM. *Let X has genus $p > 0$ and let $x \in X$. Then there exists exactly p gaps for x: $1 = w_1 < w_2 < \cdots < w_p \leqslant 2p - 1$.*

PROOF. Put $l_n := l(nx)$, then by Theorem B, $l_n = n + 1 - p$ for all $n > 2p - 1$. Observe that $l_n \leqslant l_{n+1}$ and that w is a gap for x if $l_w = l_{w-1}$. When one passes from l_n to l_{n+1}, the increase of at most 1 is possible; thus such increase takes place exactly $p-1$ times. Since $l_n = l_{n-1} + 1$ for $n > 2p$, there are no gaps $\geqslant 2p$. Since there are $2p - 1$ steps on the way from 1 to g, the case $l_n = l_{n-1}$ takes place exactly p. □

8 Hyperelliptic involution

Let X be hyperelliptic and $f : X \to \mathbb{P}^1$ be a function of degree 2 ($\deg(\operatorname{Div} f) = 2$.) We can assume that $f^{-1}(\infty)$ consists of two distinct points (if it does not, we compose f with an automorphism of $\mathbb{P}^1$). Denote by $C = C(f)$ the set of critical points of f and let $B := f(C)$; thus we have $B \subset \mathbb{P}^1 - \{\infty\} = \mathbb{C}$. If $c \in C$, then $\operatorname{ord}_c(f) = 2$, hence $f^{-1}(b)$ consists of one point if $b \in B$. B is called the *branch locus* of f. Now define the map $\tau_1 : X - C \to X - C$ such that $\tau_1(x) \neq x$ with $f(x) = f(\tau_1(x))$. Extend τ_1 to the holomorphic map $\tau : X \to X$ by setting $\tau(c) = c$ for $c \in C$. Clearly $\tau^2 = \mathrm{id}_X$. τ is called the *hyperelliptic involution* or *sheet interchange*. We recall the Riemann–Hurwitz formula for $f : X \to Y$: $\chi(X) + |B(f)| = \deg(f)\chi(Y)$, where $\chi(M) = 2 - 2p_M$. Since $\chi(\mathbb{P}^1) = 2$, we have $B(f) = 2 \cdot 2 + 2(p-2) = 2p + 2$.

We have proved the $\Longrightarrow$part of

PROPOSITION (HURWITZ). *Let X have genus p. Then*

(X is hyperelliptic) $\Longleftrightarrow$(There exists a holomorphic involution $\tau \in$ Aut(X), $\tau^2 = 1$ that fixes $2p+2$ points).

PROOF. $\Longleftarrow$ let τ be a holomorphic involution with $2p+2$ fixed points. Consider the subgroup $\Gamma \subset \mathrm{Aut}(X)$ of order 2 generated by τ and the 2-sheeted covering $f : X \to X/\Gamma$ which is branched at $2p-2$ fixed points of τ. The Riemann–Hurwitz formula implies that $2p+2 = 2p-1+\deg(f)\eta(X/\Gamma)$, $4 = 2\eta(X/\Gamma)$, therefore $\eta(X/\Gamma) = 2$ and $p_{X/\Gamma} = 0$ (that is, $X/\Gamma = \mathbb{P}^1$). Thus on X there lives a meromorphic functions of degree 2. □

COROLLARY. *On hyperelliptic X of genus p there exists exactly $2p+2$ fixed points of the holomorphic involution $\tau \in$ Aut(X) and each of these is a Weierstrass point of X.*

EXERCISE 1. Let $f : X \to \mathbb{P}^1$ be of degree 2, then f is isomorphic to the canonical map $i_{K_X} : X \to i_{K_X}(X) \subset \mathbb{P}^1$. Prove that f is unique up to an automorphism of $\mathbb{P}^1$, that is *two functions of degree 2 differ only by Möbius transformation*

$$f \to \frac{af+b}{cf+d}, \quad a, b, c, d \in \mathbb{C},\ ad - bc \neq 0.$$

EXERCISE 2. Show that conversely:

Any Weierstrass point on hyperelliptic X is a branch point of, essentially unique, map $X \to \mathbb{P}^1$ of degree 2 – the canonical map.

Hint: Identify X with the surface $w^2 - (z - z_1)\cdots(z - z_{2p+1}) = 0$ with distinct z_i.

We are sorry to have to leave the domain of hyperelliptic surfaces: this was and still is a wide test field of mathematics. Such surfaces are very rich (they can be of arbitrarily high genus) and have interesting properties. Using them as examples we can better understand such fundamental notions (which, in fact, were born here) as 'branch point', 'genus', 'Weierstrass point', 'holomorphic involution'. On the other hand, they are even more interesting, because they are not (completely) typical: they posses conformal involution (because they are a 2 folded covering of the Riemann sphere, which is quite exceptional.) Next the canonical bundle is not very ample. For them a beautiful Clifford theorem holds.

9 Clifford theorem

Clifford theorem is quite deep and we cannot present here its proof; we will content ourselves with its formulation. Let us recall that divisor $D = \sum D(x)$ is *effective* if all $D(x) \geqslant 0$, $X \in X$; we write $D \geqslant D'$ if $D - D'$ is effective. It is customary to write

$$\mathcal{L}(D) := H^0(X, \mathcal{O}(X)) = H^0(X, L_D).$$

Two divisors are *linearly equivalent* if $L_D \simeq L_{D'}$, and one writes in this case $D \sim D'$. D is principal if $D = \mathrm{Div}(f) \equiv (f)$ for some $f \in \mathcal{M}(X)$. The *complete linear system* of D (or L_D), denoted by $|D|$ is the set of effective divisors linearly equivalent to D. More generally, any linear subspace of $|D|$ is called the linear system of D (or L_D). We write $\dim |D| = h^0(D)$. Given two meromorphic functions f, g,

$$((f) = (g)) \Longleftrightarrow (\text{there exists a nonzero constant } \lambda \text{ such that } f = \lambda g.)$$

We have thus a $1-1$ correspondence between D and $\mathbb{P}\mathcal{L}(D) = \mathbb{P}(H^0(X, \Lambda_D))$. Therefore $|D|$ is a projective space.

THEOREM (CLIFFORD). *Let $D \geqslant 0$ and $i(D) \equiv H^1(X, L_D) > 0$. Then*
1. $\dim |D| \leqslant \frac{1}{2} \deg D$;
2. If $\dim |D| = \frac{1}{2} \deg D$, *then either* $D = 0$, *or* $D \sim K_X$, *or* X *is hyperelliptic.*

10 Riemann bilinear relations. Abel–Jacobi map

We came now to the one of the most magnificent idea of Riemann: his wonderful 'bilinear relations' that lead, on the one hand, to the complex p dimensional torus $\mathbb{C}^p/\Lambda$ called the Jacobian of compact Riemann surface X and denoted $\mathrm{Jac}\,(X)$, and, on the other, to the marvellous holomorphic function ϑ, the famous Riemann theta function and to the hypersurface $\Theta \subset \mathrm{Jac}\,(X)$ being a divisor of zeros $\{z : \vartheta(z) = 0\}$.

These discoveries not only made it possible for Riemann to see the problems of his predecessors Abel and Jacobi (Abel theorem and Jacobi inversion problem) from completely new perspective, but also gave their surprising general solutions. This approach of Riemann is so fascinating that it overshadowed the achievements of his predecessors and these great mathematicians are often forgotten in lectures and monographs.

Let us recall the construction of Jacobian Jac(X) of a Riemann surface of genus p. Let $a_1, \ldots, a_p, b_1, \ldots, b_p$ be a *symplectic* basis of the space $H_1(X, \mathbb{Z})$ of cocycles, that is $[a_j, a_k] = 0 = [b_j, b_k]$, $[a_j, b_k] = -[b_k, a_j] = d_{jk}$, where $[\ ,\ ]$ is the intersection number of the loops representing the classes a_j, b_k, $j, k = 1, \ldots, p$. This means that the intersection matrix is of the form

$$\begin{pmatrix} 0 & 1_p \\ -1_p & 0 \end{pmatrix}.$$

If $\omega_1, \ldots, \omega_p$ form a basis of the space Ω^1 of holomorphic differential forms, then the $p \times 2p$ matrix

$$\Omega := \left(\int_{a_i} \omega_j, \int_{b_i} \omega_j \right)$$

is called the *period matrix* of X. Already Riemann observed that one can choose the basis $\omega_1, \ldots, \omega_p$ (canonical basis) such that Ω has the form $\Omega = (1_p, Z)$, where $Z \in \mathfrak{H}_p$, that is Z is a symmetric $p \times p$ matrix with positive definite imaginary part: $Z = {}^tZ$, $\operatorname{Im} Z > 0$; $\mathfrak{H}_p$ is the Siegel upper half plane.

Given a point $\tau \in \mathfrak{H}_p$, consider the lattice $\Lambda = \Lambda_\tau \subset \mathbb{C}^p$ generated by the columns of the $p \times 2p$ matrix $(1_p, \tau)$ and form the complex torus $\mathrm{Jac}_\tau(X) := \mathbb{C}^p/\Lambda_\tau$. In what follows we will omit the index $\tau = Z$ and we will write $\mathrm{Jac}\,(X) = \mathbb{C}^p/\Lambda$. $\mathrm{Jac}\,(X)$ can be intristically defined as follows. Let V be the dual of $H^0(X, \Omega^1)$, $V = \Omega^1(X)^*$; Riemann defines now the linear form on V by

$$H^0(X, \Omega^1) \ni \omega \to \int_\gamma \omega \in \mathbb{C}, \quad \text{if } \gamma \in H_1(X, \mathbb{Z}).$$

Taking a_j, b_k, and ω_j as above, it can be easily checked that in this way we obtained an injection p of $H_1(X, \mathbb{Z})$ into a lattice in $H^0(X, \Omega^1)^*$. We have $\mathrm{Jac}\,(X) = H^0(X, \Omega^1)^*/p(H_1(X, \mathbb{Z}))$. We define the *Abel–Jacobi map* $A : X \to \mathrm{Jac}\,(X)$ as follows

$$A(x) := \left(\int_{x_0}^{x} \omega_1, \ldots \int_{x_0}^{x} \omega_p \right), \quad \text{mod } \Lambda\ ,$$

where all integrals are taken along the same curve l connecting the base point x_0 with x. If l_1 is another curve connecting x_0 with x, then $l_1 - l = \gamma$ with $\gamma \in H_1(X, \mathbb{Z})$, therefore $\int_{l_1} \omega_k = \int_l \omega_k + \int_\gamma \omega_k$ for all k.

We know how important is existence of an ample line bundle $L \to M$ on a compact manifold M: it makes a projective imbedding of $M \to \mathbb{P}^N$ in some projective space $\mathbb{P}^N$ possible.

Polarization of a complex torus M is a datum of an ample line bundle $L \to M$. Such a pair (M, L) is a polarized abelian variety. The polarization is *principal* if $\dim H^0(M, L) = 1$, that is if L possesses (up to a constant) one holomorphic section. In such case we say that (M, L) is a principally polarized abelian variety. As we will see $\mathrm{Jac}\,(X)$ is a principally polarized abelian variety.

But not every principally polarized abelian variety is a Jacobian of a compact Riemann surface X. The famous Schottky problem (partially solved by Schottky) is to give a criterion which would determine if a given principally polarized abelian variety is a Jacobian.

I will tell later on a solution, or solutions of this difficult problem; let me stress here a problem which is philosophically interesting: it is not always clear (even in mathematics!) if a problem has been solved, in other words, what one should expect of a solution. For example, the famous work of the Japanese mathematician T. Shiota *Characterization of Jacobian varieties in terms of soliton equations*, Invent. Math. **83** (1986), 333–382 which proves the claim of Novikov that Jacobians can be characterized by the Kadomcev–Pietashvili equation (of the theory of plasma!) is considered to be a solution of the Schottky problem. But some distinguished mathematicians are not fully satisfied; and thus H.M. Farkas writes: 'While the Schottky problem was solved (by Shiota, KM), it, in my opinion, has not been solved in the spirit of Schottky. Since I personally believe that a solution in Schottky spirit will yet be found ...'

But let us return to Riemann and Jacobians. In what follows I will sometimes denote the period matrix $Z \in \mathfrak{H}_p$ by τ.

The *Riemann theta function*

$$\vartheta(z, \tau) := \sum_{m \in \mathbb{Z}^p} \exp 2\pi i (\frac{1}{2} {}^t m \tau m + {}^t m z)$$

is defined for all $z \in \mathbb{C}^p$ and $\tau \in \mathfrak{H}_p$. In the coordinate form

$$\vartheta(z_1, \ldots, z_p, \tau) = \sum_{m_1, \ldots, m_p \in Z} \exp \pi i \left(\sum_{j,k=1}^{p} \tau_{jk} m_j m_k + 2 \sum_{j=1}^{p} m_j z_j \right),$$

or

$$\vartheta(\zeta, Z) = \sum_{m \in Z^p} \exp \pi i \,(< m, Zm > +2 < m, \zeta >).$$

It is easy to see that ϑ is a holomorphic function on $\mathbb{C}^p \times \mathfrak{H}_p$. The series above is normally convergent on $\mathbb{C}^p$ thanks to $\operatorname{Im} Z > 0$.

This is one of the most important and most wonderful functions in mathematics it is almost omnipresent: as Pierre Cartier observed recently it plays a quite important role in quantum mechanics.

As we will see in a moment, the Riemann theta function and its natural generalizations can be regarded as sections of some line bundles $L(H,\rho)$ on complex tori. This is a consequence of the following quasi periodicity of ϑ.

$$\vartheta(\zeta + n + {}^t m\tau, \tau) = \exp 2\pi i \left(-\frac{1}{2}{}^t m\tau m - {}^t m\zeta \right) \vartheta(\zeta, \tau) \tag{1}$$

for $n, m \in \mathbb{Z}^p$; therefore $n + {}^t m\tau \in \Lambda_\tau$. For fixed $\tau \in \mathfrak{H}_g$ the quasi periodicity (1) shows that zeros of ϑ are well defined modulo Λ_τ. In this way we have a hypersurface defined on $\operatorname{Jac}(X)$, the famous *theta divisor*

$$\Theta = \Theta_\tau = \{z \in \operatorname{Jac}(X) : \vartheta(z,\tau) = 0\}. \tag{2}$$

We know about close relation between divisors and line bundles. And indeed every principal polarization of M can be obtained by a Θ divisor.

THEOREM. *Any principally polarized abelian variety (M, L) is of the form $\mathbb{C}^p/\Lambda_\tau, \Theta_\tau)$ for some $\tau \in \mathfrak{H}_p$.*

Two principally polarized abelian varieties (M, L) and (M', L') are *isomorphic* if there exists a biholomorphic map $f : M \to M'$ such that $f^* : L' \to L$. The *Siegel modular group* plays a decisive role in the theory of principally polarized abelian varieties. We have

THEOREM (SIEGEL). *If $Z, Z' \in \mathfrak{H}_p$, then principally polarized abelian varieties corresponding to Z and Z' are isomorphic if and only if there exists*

$$\begin{pmatrix} A & B \\ C & D \end{pmatrix} \in \operatorname{Sp}(2p, \mathbb{Z})$$

such that

$$Z' = \frac{AZ + B}{CZ + D}.$$

11 Linear bundles on complex tori: Appel–Humbert theorem

As we know there is a one-to-one correspondence between hermitian forms H on a complex vector space V and real alternating forms E on V satisfying the identity $E(ix, iy) = E(x, y)$, $x, y \in V$. This correspondence is given by the formulas

$$E(x, y) = \operatorname{Im} H(x, y),$$

$$H(x, y) = E(ix, iy) + iE(x, y).$$

If we have a complex torus $\mathbb{C}^p/\Lambda$, then the form E (or H) on $\mathbb{C}^p$ is the *Riemann form* with respect to Λ if

1. E is alternating;
2. E takes integral values on $\Lambda \times \Lambda$ ($E(\Lambda, \Lambda) \subset \mathbb{Z}$);
3. $(x, y) \to E(ix, y)$ is symmetric and positive.

Analogously, we can call a hermitian form on $\mathbb{C}^p$ a Riemann form with respect to Λ if its imaginary part $E := \operatorname{Im} H$ is a Riemann form with respect to Λ. $\sqrt{E}$ is called the Pfaffian of E. We have the famous

THEOREM (FROBENIUS). *Let E be an alternating, non degenerate bilinear (i.e., symplectic) form on a free $\mathbb{Z}$ module Λ. Then there exists a symplectic basis $e_1, v_1, \ldots, e_p, v_p$ for Λ with respect to E, and Λ is an E-orthogonal direct sum $\Lambda = [e_1, v_1] \oplus \cdots \oplus [e_p, v_p]$ of 2 dimensional modules $[e_j, v_j]$ such that $E(e_j, v_j) = d_j$ is an integer > 0 and $d_1|d_2, \ldots, d_{p-1}|d_p$ (d_j divides d_{j+1} for $1 \leqslant j < g$.) Therefore the matrix E with respect to this Frobenius basis has the form*

$$(E(e_i, v_j)) = \left(\begin{array}{ccccccc} & & & \vdots & & & d_1 \\ & 0 & & \vdots & & & \vdots \\ & & & \vdots & d_p & & \\ \cdots & \cdots & \cdots & \cdots & \cdots & \cdots & \cdots \\ & & -d_p & \vdots & & & \\ & \vdots & & \vdots & & 0 & \\ -d_p & & & \vdots & & & \end{array}\right)$$

and $\mathit{Pfaff} E = d_1 \cdots d_p$.

We know the importance of line bundles on manifolds; thus we are happy to have a theorem which we owe Appel and Humbert which we present in the form given by A. Weil

THEOREM (APPEL–HUMBERT). *The isomorphism classes of line bundles on* $M = \mathbb{C}^p/\Lambda$ *are in one-to-one correspondence with the pairs* (H, ρ), *where* H *is hermitian on* $\mathbb{C}^p$ *and integral on* $\Lambda \times \Lambda$ *and* $\rho : \Lambda \to S^1$ *is a quasi character of* Λ, *that is,* $|\rho(\lambda)| = 1$ *and* $\rho(\lambda_1)\rho(\lambda_2) = \rho(\lambda_1 + \lambda_2)e^{\pi i E(\lambda_1,\lambda_2)}$.

The line bundle corresponding to the pair (H, ρ) is denoted by $L(H, \rho)$ and is obtained as $(\mathbb{C}^p \times \mathbb{C})/\sim$, where

$$(z, \xi) \sim \left(z + \lambda, \rho(\lambda)e^{\pi H(z,\lambda)+\frac{\pi}{2}H(\lambda,\lambda)}\xi\right), \quad z \in \mathbb{C}^p,\ \xi \in \mathbb{C},\ \lambda \in \Lambda.$$

The function $\lambda \to e_\lambda := \rho(\lambda)e^{\pi H(z,\lambda)+\frac{\pi}{2}H(\lambda,\lambda)}$ is a cocycle on the group Λ with values in $H^0(\mathbb{C}^p, \mathcal{O}^*(\mathbb{C}^p))$. The Chern class $c(L(H, \rho))$ of this line bundle is equal to $\operatorname{Im} H = E \in H^2(M, \mathbb{Z})$. If the cocycles $\{e_\lambda^{(j)}\}$ correspond to the pairs (H_j, ρ_j), then $\{e_\lambda^{(1)}, e_\lambda^{(2)}\}$ correspond to the pair $(H_1 + H_2, \rho_1\rho_2)$; therefore we have

COROLLARY 1. $L(H_1, \rho_1) \otimes L(H_2, \rho_2) \simeq L(H_1 + H_2, \rho_1\rho_2)$.

COROLLARY. *Holomorphic sections of* $L(H, \rho)$ *correspond to holomorphic functions* ϑ *on* $\mathbb{C}^p$ *such that*

$$\vartheta(z + \lambda) = e_\lambda(z)\vartheta(z) = \rho(\lambda)e^{\pi H(z,\lambda)+\frac{\pi}{2}H(\lambda,\lambda)}\vartheta(z), \quad \lambda \in \Lambda.$$

Such functions are called the ϑ-functions of the type (H, ρ).

Frobenius computed dimension of the space of such functions.

THEOREM. *If* H *is positive definite, then*

$$\dim H^0(M, L(H, \rho)) = \mathit{Pfaff}\, E,$$

where $E = \operatorname{Im} H$, $M = \mathbb{C}^p/\Lambda$.

Therefore, if $\operatorname{Pfaff} E = 1$, then $\mathbb{C}^p/\Lambda$ is a principally polarized abelian variety because the matrix E has all elementary divisors $d_1 = \cdots = d_p = 1$.

The crowning of efforts of Riemann, Frobenius, Poincaré is the famous theorem which we owe one of the founders of modern topology, S. Lefschetz

THEOREM (LEFSCHETZ). *Let $M = \mathbb{C}^p/\Lambda$, H be a hermitian forma on $\mathbb{C}^p$ such that* $\operatorname{Im} H$ *is integral on $\Lambda \times \Lambda$; let ρ be a quasi character $\rho : \Lambda \to S^1$ satisfying assumptions of the preceding theorem. Then the following three conditions are equivalent.*

1. *H is positive definite (and thus is a Riemann form);*
2. *For an integer $k \geqslant 3$ the space of holomorphic sections of the bundle $L^{\otimes k}(H,\rho)$ defines an imbedding of M into some $\mathbb{P}^N$;*
3. *For every subtorus $N \subset M$ there exists an integer p and a section s of the bundle $L^{\otimes p}(H,\rho)$ together with two points $x_1, x_2 \in M$, $x_1 - x_2 \in N$ such that $s(x_1) = 0$, $s(x_2) \neq 0$.*

PROOF. 2. $\Longrightarrow$3. is obvious. The most important implication is 1. $\Longrightarrow$2. which is usually called Lefschetz theorem. We present the proof for $k = 3$ (the case $k > 3$ can be proved analogously.) Since we identify sections of the bundle $L(H,\rho)$ with the ϑ functions of the type (H,ρ) and $L^{\otimes k}(H,\rho) \simeq L(kH,\rho^3)$, if ϑ is a section of the bundle $L(H,\rho)$, then for any pair of points $a, b \in \mathbb{C}^p$ the function $\vartheta(z-a)\vartheta(z-b)\vartheta(z+a+b)$ is a section of the bundle $L^3(H,\rho)$. Observe that for arbitrary z_0 this section σ does not vanish at z_0: we take the points $a, b \in \mathbb{C}^p$ such that $\vartheta(z_0 - a) \neq 0$, $\vartheta(z_0 - b) \neq 0$, $\vartheta(z_0 + a + b) \neq 0$. Then the proof proceeds as in the case of Corollary of Theorem A: Let $\vartheta_1, \ldots, \vartheta_d$ be a basis in the space of holomorphic sections of the bundle $L^{\otimes 3}$; we define the mappings $i_{L^3} : M \to \mathbb{P}^d$ which in homogeneous coordinates have the form $i_{L^3}(\pi(z)) := (\vartheta_1(z), \ldots, \vartheta_d(z)) \in \mathbb{P}^d$, where $z \in \mathbb{C}^p$ and $\pi : \mathbb{C}^p \to \mathbb{P}^p/\Lambda$ is the canonical projection. Using the above observation concerning sections σ, we prove that i_{L^3} is injective. □

12 ϑ-functions. The great Riemann theorems: 'Abel theorem', 'Jacobi inversion', and 'ϑ divisor theorem'

But let us return to the treatise of Riemann of 1857 *Theorie der Abelschen Functionen.* In the first part of this work, and later in his lectures, Riemann formulates and proves the theorem which he calls Abel theorem. As we know, for existence of a meromorphic function f with given zeros and poles $x_1, \ldots, x_k$, $y_1, \ldots, y_k$ on a compact Riemann surface X it is necessary that the sum of residua of the function f^* be equal zero, in other words $\deg f = 0$. But since X may be not simply connected this condition is not be sufficient. Riemann gives the solution to this problem as follows.

THEOREM (ABEL'S THEOREM OF RIEMANN). *Let D be a divisor on X of degree 0. Then*
$(D \sim 0$ (i.e., $D = (f)$ for some f)) $\Longleftrightarrow A(D) = 0$ in Jac (X)).

Thus the assertion of the theorem is the following. Let $x_1, \ldots, x_k$, $y_1, \ldots, y_k$ be points of X with $x_i \neq y_j$ for all i, j. The necessary and sufficient condition for existence of a meromorphic function f with $\sum x_i$ as its divisor of zeros and $\sum y_j$ as its divisor of poles is that

$$\sum_{j=1}^{k} \int_{x_0}^{x_j} \vec{\omega} \equiv \sum_{j=1}^{k} \int_{x_0}^{x_j} \vec{\omega} \mod \Lambda, \quad \vec{\omega} = (\omega_1, \ldots, \omega_p).$$

We see how much more compact is the formulation in the language of Jac (X) and Abel–Jacobi mapping A.

Abel himself did not formulate the theorem in this form. He was mainly interested in *addition theorems* for integrals of algebraic functions (like addition theorems for trigonometric functions.) As R. Narasimhan writes: 'It seems to have been Riemann who first recognized the relevance of Abel's work to the problem of constructing functions with given zeros and poles. This relationship was pursued further by A. Clebsch, who made several beautiful geometric applications of Abel's theorem.'

We can extend, in a natural way, the mapping A into the kth Cartesian product $X^d = X \times \cdots \times X$, $d > 1$. Let us denote by $W^k \subset$ Jac (X) the set of points in Jac (X) of the form $\sum_{j=1}^{k} A(x_j)$, $x_1, \ldots, x_k \in X$. Similarly, denoting by $X_k = Sym_k X$ the k-fold symmetric product of X we can define, the map $A_d = A : X_d \to$ Jac (X), $x_1 + \cdots + x_d \to A(x_1 + \cdots + x_d) := A(x_1) + \cdots + A(x_d)$. Denote $W_d := A_d(X_d)$. The famous Θ theorem of Riemann asserts

THEOREM (RIEMANN Θ THEOREM). *There exists a point $\kappa \in$ Jac (X) such that $W_{p-1} = \Theta - \kappa$, moreover $2\kappa = A(K_X)$, where K_X is the canonical divisor in X; κ is called the Riemann vector.*

In other words, the theta divisor Θ consists, up to translation, of points in Jac(X) of the form $A(x_1) + \cdots + A(x_{p-1})$.

Another great achievement of Riemann was the complete solution of the famous

Jacobi inversion problem. The works of Abel, in particular his Abel theorem suggested to Jacobi the following inverse problem: Let $\omega_1, \ldots, \omega_p$ be a basis of $\Omega^1(X)$ – the space of holomorphic 1 forms. Given any p complex numbers $z_1, \ldots, z_p$ find $x_1, \ldots, x_p \in X$ such that $\sum_{j=1}^{p} \int_{x_0}^{x_j} \omega_{jk} = z_k$, $k = 1, \ldots, p$.

Jacobi solved this problem only in the simplest case $p = 2$. This problem was considered to be the major question of complex analysis in 30s through 50s of XIX century. Young Weierstrass decided to devote his live to finding its solution. And the solution was found by Riemann in his fundamental papers and lectures on ϑ function. Weierstrass solved the inverse problem a bit later by other means.

Let us denote by $\mathrm{Div}_0(X) \subset \mathrm{Div}(X)$ the group of divisors of degree 0 and by $\mathrm{Div}_P(X) \subset \mathrm{Div}(X)$ the group of principal divisors. The quotient group

$$\mathrm{Pic}(X) := \mathrm{Div}(X)/\mathrm{Div}_P(X)$$

is called the *Piccard group.* The restricted Piccard group is

$$\underset{0}{\mathrm{Pic}}(X) := \underset{0}{\mathrm{Div}}(X)/\underset{P}{\mathrm{Div}}(X).$$

Since $\mathrm{Div}(X)/\mathrm{Div}_0(X) = \mathbb{Z}$, we have the exact sequence

$$0 \to \underset{0}{\mathrm{Pic}}(X) \to \mathrm{Pic}(X) \to \mathbb{Z} \to 0.$$

the Riemann Abel theorem asserts that the kernel of the Abel–Jacobi map $A : X \to \mathrm{Jac}\,(X)$ is equal to $\mathrm{Div}_P(X)$. Hence, by passing to the quotient we obtain the injective mapping $j : \mathrm{Pic}_0(X) \to \mathrm{Jac}\,(X)$. The inverse problem is if this map is injective. The answer is yes. This is the famous Riemann theorem:

Riemann solution of the inverse problem of Jacobi:
For any compact Riemann surface X the map

$$j : \underset{0}{\mathrm{Pic}}(X) \to \mathit{Jac}\,(X)$$

is a surjection and even an isomorphism.

An elegant proof is given in the Forsters book; it does not make use of Riemann theta function and its divisor Θ, but makes use of the Riemann–Roch theorem. The Riemann solution of the problem is given by a series of his theorems.

THEOREM III. *Let $e \in Jac(X)$ and assume that $A(X) = W^1$ is* not *entirely contained in $\Theta + e$. Then $W_1 \cap (\Theta + e)$ consists of exactly p points (counted with multiplicities.) In other words, if the section $x \to \vartheta(A(x) - e) \not\equiv 0$, then it has exactly p zeros.*

THEOREM IV. *If $e \in Jac(X)$ and $W_1 \not\subset \Theta + e$ and $x_1, \ldots, x_p$ are the points at which $\vartheta(A(x) - e)$ vanishes, then*

$$\sum_{j=1}^{p} A(x_j) = e - \kappa,$$

where κ is a constant independent of e. Moreover, there are no other p-tuple of points $y_j \in X$ for whose

$$\sum_{j=1}^{p} A(y_j) = e - \kappa.$$

But in his magnificent work Riemann does not stop here. He investigates the geometry of the divisor Θ an, in particular, its singular points. This is the subject of the following famous

RIEMANN SINGULARITY THEOREM. *Singular points of Θ are those points $\sum_{j=1}^{p} A(x_j) - \kappa$ for whose the divisor $D = \sum_{j=1}^{p} x_j$ is special (that is, $h^0(D) \geqslant 1$).*

This theorem was extended in 1973 by G. Kempf the case of W_k, $2 \leqslant k \leqslant p-2$ and made even more precise: Kempf considers the tangent cone on $T_a(\mathrm{Jac}\,(X))$, where a is a singular point of Θ.

The great and mighty idea of Riemann surfavce will be developped in all following Parts: On Teichmüller approach to Riemann Moduli Problem, on complex geometry, and last but not least, on number theory.

Part IV

Riemann and Calculus of Variations

CHAPTER 1

Introduction

Some geometrical (circle, sphere, straight line, ...) and physical (soap bubbles) entities are distinguished from a family of similar objects by the fact that they minimalise or maximalise some quantity (length, area, energy) and they have fascinated people (not only mathematicians) since antiquity. The philosophical and theological background for existence of such 'perfect' entities was given by Leibnitz in his *Theodicea* and his reach correspondence. And thus it is perhaps not strange at all that it was Leibnitz who, in the letter to König, many years before Maupertuis, presented first the principle of least action.

The critics (e.g., Voltaire) of the Leibnitz's *Theodicea* did not understand, or did not want to understand the depth of the Leibnitz concepts, mistaking them with dull optimism, which does not see monstrosity of the empirical reality around us. But Leibnitz was a child of a terrible war which has been devastating Central Europe for 30 years (he was born in Leipzig in 1646; he lost his father at the age of 5.) This, perhaps the most original and most universsal mind of the epoch was a realistically minded politician and a man able to solve important technical problems: he saved, for example, a silver mine in Harz Mountains from flood; he was the organizer of many Academies and the spiritual father of the Petersburg Academy (Peter the Great sought contact with Leibnitz, and took from him some advises.) Leibnitz was was one of the greatest philosophers of all times: he had strength to built such a splendid pieces of work as *Monadologie* and *Theodicea*.

And thus Leibnitz is not only the founder (alongside with Newton) of calculus who introduced the notation of differential df and integral $\int f$, but also of the calculus of variations, that is, the 'differential calculus' of functionals, the functions

$$e : \mathcal{F} \to \mathbb{R}$$
$$f \to \int e(f)$$

defined on the functional spaces $\mathcal{F}$ ($e(f)$ is the energy density.) The bravery of Leibnitz concepts takes the breath away: it must be remembered that, in this times, the notion of a function and of topology of function spaces was not known – people knew several 'functions' defined by power series, and did not analyze their convergence; this was a very modest beginning indeed, but it was heavy loaded with the energy waiting for being formed to the final shape by great mathematicians, physicists, and astronomers.

In the case of the Leibnitz works we can see how, behind a great scientific theory, stands the great philosophy, to say it better: some great philosophically – theological concept incarnates itself in (great) scientific theories, an later, perhaps, even in technology. At this point I cannot resist the remark that Leibnitz was a grand, grand father of computing machines.

The attempts – usually awkward and not convincing – to reconcile the evil of the world with God's goodness and almightness are, in my view, fated to fail: some wise men said 'who preaches almightness of God, makes ground for atheism.' How Leibnitz solves this dilemma? His answer is that God had chosen from the innumerable *possible* worlds the best possible, but that the *perfect world is not possible.* (Perhaps this impossibility was due to the (hidden) existence of *autonomous* evil – not explicitly expressed by Leibnitz?) This infinite multitude (of possible worlds) can only be conceived by an infinite understanding, the monad of monads. This was, for Leibnitz, the proof of God's reality, His existence. The best possible world is distinguished by a *pre-established harmony* between the kingdom of nature on the one hand, and the heavenly kingdom of grace and freedom on the other. (This is a deep interpretation of the old hermetic principle!) Through this harmony the effective canons unite with the purposive canons. Thus bodies move due to their own internal laws in accordance with the thoughts and ideas, desires of (their) souls. (This is the fundamental conception of a monad: a body, a stone is a monad, is a totality of its history and relations of the whole reality, the 'illusion' of primitive causality arises when one picks one or few elements of this totality of aspects and features of monads. In this way, the contradiction between the predetermination of the empirical world following strict laws (of nature) and constantly experienced spontaneity and freedom of the individual is removed. (Without firm structures and laws freedom is impossible – one has to discern between arbitrariness and freedom!)

The best possible world must 'obey' specific laws since an ordered, structured world is better than the chaotic one. This proves – for Leibnitz – the necessity of natural world. Or, put it in modern terms: one cannot separate 'world' from natural laws, an electron from Dirac equation. The contents of the natural laws is only determined in a moral sense: they must satisfy the criteria of beauty and simplicity in this best of all possible worlds (Platonic identity of truth and beauty!)

This leads Leibnitz to variational principles: if a physical process did not yield an extremal value for a particular action, or energy, integral, the world could be improved, and therefore, would not be the best possible one. Consequently, Leibnitz uses beauty and simplicity of variational principles as the 'proof' of his pre-established harmony.

The most famous principle is the *Dirichlet principle* which Riemann brilliantly used in his theory of Riemann surfaces: in the proof of Riemann–Roch theorem and in Riemann mapping theorem. In fact, the Riemannian Dirichlet principle was accepted by Riemann without proof and precise formulation, which was noted by Weierstrass. Of course, Riemann realized this gaps, but he did not doubt the truth of his theorems based on the Dirichlet principle. Here we encounter an unusual phenomenon – characteristic for mathematics, but contradicting popular ideas concerning the very idea of mathematical research. The loopholes in the Riemann's proofs, together with deep convictions that his thesis are true resulted in creation of a number of extremely fruitful theories leading to Riemann thesis in another way. Let us recall the most famous ones: the 'alternating method' of Schwarz and Neumann and the related 'methode de baloyage' (sweeping out) method of Poincaré about which I will say more later; the method of integral equations of Fredholm; the method of parametrix of Hilbert and E.E. Levi. And, at the end, the 'rescuing' the Dirichlet principle by Hilbert (1900) which was the beginning of the so-called direct methods of calculus of variations and one of the main sources of the modern functional analysis. We will devote a lot of attention to the latter in context of the fundamental problem of the theory of minimal surface — the 'Plateau problem.'

As we know, Riemann saw his theory as a potential theory (theory of functions and harmonic forms) on his surfaces. The theory of minimal surfaces on which Riemann worked simultaneously with Weierstrass led to very important maps, harmonic maps, $u : M \to N$ of Riemann manifolds defined by extremum of an action integral. Presently the theory of harmonic maps is a gigantic discipline, having a great number of applications not only in mathematics.

I will gladly discuss some basic theorems of this theory also because they turned out to be the most natural (at least in the Riemann spirit) approach not only to the Plateau problem, but also to the, posed by Riemann, extremely important 'moduli problem,' the problem of parametrization of equivalent classes of compact Riemann surfaces of genus $p > 1$. This problem led O. Teichmüller to formulation of an extremely reach theory called nowadays the Teichmüller theory. Teichmüller was able to present a construction of $\dim_{\mathbb{R}} 6(p-1)$-dimensional manifold, called by A. Weil the Teichmüller space $\mathcal{T}_p$, and being a (universal) covering of the Riemann space of moduli $\mathcal{M}_p$: $\mathcal{M}_p = \mathcal{T}_p/\Gamma_p$. The Teichmüller $\mathcal{T}_p$ has become the object of intensive investigations. It turns out that it possesses extremely interesting geometry and complex structure: it is a Stein manifold, $\dim_{\mathbb{C}} \mathcal{T}_p = 3(p-1)$, and it has a Kähler metric. The 'Riemannian' approach to this problems was advertised by Antony Tromba and started with research of Eells and others. This approach makes use of some analogues of the energy integral and the corresponding harmonic maps.

As we know, already in 1918, Hermann Weyl, trying for the first time to unify Einstein theory of gravity and Maxwell's electrodynamics, formulated gauge theories, being the theory of linear connections on differetiable manifolds. Not knowing the theory of connections in principal bundles, Yang and Mills formulated the theory which, as it turned out, plays a basic role in the theory of elementary particles. The Yang–Mills theory in which the critical points of the Yang–Mills functional are connections turned out not only very interesting by its own, but it has become an important tool of a number of mathematical theories (e.g., differential topology.) An important, independently developing, beautiful mathematical theory has been created yet again. In this theory the moduli problem appears as well.

1.1 General criteria for existence of minimizers of functionals

The general fact guaranteeing existence of a minimum of a continuous function is provided by the famous

WEIERSTRASS THEOREM. *On a compact space W each continuous function $E : W \to \mathbb{R}$ attains its infimum and supremum.*

For infimum it suffices the lower semicontinuity of E.

Let us recall the notion of semicontinuity:

$E : X \to \mathbb{R}$ is lower semicontinuous (l.s.c.) if for every $a \in \mathbb{R}$ $X_a = \{x \in X : E(x) \leqslant a\}$ is closed.

An equivalent definition is that E is l.s.c. at x_0 if

$$\lim_{x \to x_0} \inf E(x) = E(x_0).$$

E is l.s.c. on X if it is l.s.c. at every $x_0 \in X$.

Briefly, a continuous function is both l.s.c. and upper s.c. on X.

For the direct methods of calculus of variations we have the generalized

WEIERSTRASS THEOREM. *Let X be compact and $E : X \to \mathbb{R}$ be l.s.c. Then*

1. *E attains its minimum (there exists $x_0 \in X$ such that $f(x_0) \leqslant f(x)$) for all $x \in X$.*
2. *E is bounded from below.*

There is another, useful version of this theorem.

WEIERSTRASS THEOREM′. *Let X be a Hausdorff space and $E : X \to \mathbb{R} \cup \{\infty\}$. Suppose that for any $a \in \mathbb{R}$*

$$X_a = \{x \in X : E(x) \leqslant a\}$$

is compact. Then

1. *E is bounded from below.*
2. *There exists $x_0 \in X$ such that $E(x_0) = \inf E(X)$.*

We know that a ball in a Hilbert (or general Banach) space H is compact if and only if $\dim H < \infty$.

In order to develop direct methods of calculus of variations, for example, construction of a minimal sequence $x_n \to x_0$, $n \to \infty$, where $E(x_0) = \inf E(H)$ one has to use *weakly* convergent sequences:

DEFINITION. A sequence (x_n) in a (separable) Hilbert space H is weakly convergent to $x_0 \in H$ if for each $l \in H$, $(x_n|l) \to (x_0|l)$, $n \to \infty$.

Here are

Important examples of weakly l.s.c. (w.l.s.c.) functions.

1. Let $(H, (\cdot|\cdot)$ be a Hilbert space; then the function $E(x) :=\| x \|^2= (x, x)$ is w.l.s.c., and more generally

2. Let $b : H \times H \to \mathbb{R}$ be a continuous bilinear symmetric form on H such that $b(x, x) \geqslant 0$. Then
$$E(x) = b(x, x)$$
is w.l.s.c. on H.

   Proof. Let $x_n \to x$ weakly.
$$0 \leqslant b(x_n - x, x_n - x) = b(x_n, x_n) - b(x, x) - 2b(x, x_n - x).$$
   By F. Riesz representation there exists such $l \in H$ that
$$b(x, x_n - x) = (l|x_n - x) \to 0 \quad n \to \infty.$$
□

As a corollary we obtain important

THEOREM. *The Dirichlet integral* $E(f) = \int |\nabla f|^2$ *is w.l.s.c. on the Sobolev space* $W_1^2(\mathbb{D}, \mathbb{R})$. □

Therefore the Dirichlet integral is bounded from below and attains its lower bound on a ball in $W_1^2(\mathbb{D}, \mathbb{R})$.

REMARK. Similar fact holds for a generalized energy integral for maps $u : M \to N$, where M, N are Riemann manifolds.

There is an important relation between

1.2 Convexity and weak lower semi continuity

PROPOSITION. *Let* $E : (H, (\cdot, \cdot) \to \mathbb{R} \cup \{\infty\}$ *be continuous and* convex, *that is, for every* $x, y \in H$ *and* $0 \leqslant t \leqslant 1$
$$E(tx + (1 - t)y) \leqslant tE(x) + (1 - t)E(y).$$

Then E is w.l.s.c.

PROOF follows from Banach–Saks theorem: Let $x_n \to x$ *weakly*, then by Banach–Saks $x^N := N^{-1}\sum_{n=1}^N x_n \to x$ *strongly* in H for $N \to \infty$. Hence, by continuity and convexity of E, we have $E(x) = \lim_{N\to\infty} E(x^N) \leqslant \lim_{N\to\infty} N^{-1}\sum_{n=1}^N x_n \leqslant \liminf_{n\to\infty} E(x_n)$. □

This Proposition encompasses both Examples above.

CHAPTER 2

The Plateau Problem

2.1 Coercity of Dirichlet integral

We saw that the Dirichlet integral $\mathcal{D}$ is w.l.s.c., but it has another important property called *coercivity* which plays an important role in solution of the Plateau problem.

DEFINITION. Let M be a subset of a separable Hilbert space H. Then $E : M \to \mathbb{R}$ is *coercive* if for any sequence (x_n) in M, $E(x_n) \to \infty$ if $\| x_n \| \to \infty$.

An important criterion for existence of a minimum is

THEOREM. *Let M be a subset of a separable Hilbert space H which is weakly closed. Let $E : M \to \mathbb{R}$ be 1. sequentially w.l.s.c., and 2. coercive.*

Then there exists a minimizer $x_0 \in M$ of E, that is,

$$E(x_0) = \inf E(M).$$

PROOF. Let $a := \inf E(M) \geqslant -\infty$ and let $(x_n) \subset M$ be a sequence such that $E(x_n) \to a$ as $n \to \infty$ By coerciveness of E the set (x_n) is bounded, and therefore weakly relatively compact. Let $x_n \to x_0$ be a weakly convergent subsequence. Since M is weakly closed $x_0 \in M$. By 1.,

$$a \leqslant E(x_0) \leqslant \operatorname{liminf}_{n\to\infty} E(x_n) = a.$$

hence $E(x_0) = \inf E(M)$. □

We see how fascinatingly simple are proofs of these general theorems guaranteeing existence of the minimizer. The proofs for the case of concrete functionals of calculus of variations present different level of difficulty.

Before we turn to the formulation of the Plateau problem, let us introduce an important class of surfaces spanning a Jordan curve γ in $\mathbb{R}^d$ $S(\gamma) = \{f \in W_1^2(\mathbb{D}, \mathbb{R}^d) : f|_{\partial\mathbb{D}} \in C^0(\partial\mathbb{D}, \mathbb{R}^d)$ is a weakly monotone parametrization of $\gamma\}$.

PROPOSITION. *The Dirichlet functional*

$$\mathcal{D}(f) = \int_{\mathbb{D}} |\nabla f|^2$$

is coercive in $S(\gamma)$.

PROOF. For $f \in W_1^2(\mathbb{D}, \mathbb{R}^d)$ with $f|_{\partial\mathbb{D}} \in L^\infty(\partial\mathbb{D}, \mathbb{R}^d)$ denote by $\| \ \|_0^2$ the $L^2 = W_0^2$ norm. Then by the Sobolev inequality $\| f \|_0^2 \leqslant c \| \nabla f \|_0^2 + \| f \|_{0,\partial\mathbb{D}} \leqslant c\mathcal{D}(f) + \| f \|_{L^\infty(\partial\mathbb{D})}$. Therefore for each $f \in C(\gamma)$, $\| f \|_1^2 \leqslant c\mathcal{D}(f) + c_0(\gamma)$, where $c_0(\gamma)$ is independent of f; hence for $\| f_n \|_1 \to \infty$, $\mathcal{D}(f_n) \to \infty$. □

Much more difficult is to prove the weak closedness. $S(\gamma)$ is *not* weakly closed in $W_1^2(\mathbb{D}, \mathbb{R})$ because of

LEMMA. *The weak closure of each orbit* $\mathrm{Aut}(\mathbb{D}) \circ f = \{f \circ g : g \in \mathrm{Aut}(\mathbb{D})\}$, $f \in W_1^2(\mathbb{D}, \mathbb{R})$ *contains the constant map.*

2.2 The Rado–Douglas solution of Plateau problem

Let γ be a C^1 curve in $\mathbb{R}^n$ and $\bar{\mathbb{D}} \subset \mathbb{R}^n$ be a closed unit disc. The classical (naive) Plateau problem is to minimize the area integral

$$\mathcal{A} = \int_{x^2+y^2<1} \left(\det \nabla u \cdot {}^t\nabla u\right)^{1/2} dx dy =$$

$$= \int_{x^2+y^2<1} \left(|u_x|^2 |u_y|^2 - (u_x u_y)^2\right)^{1/2} dx dy$$

in the class of all differentiable mappings $u : \bar{\mathbb{D}} \to \mathbb{R}^n$ such that

$$u|_{\partial\mathbb{D}} : \partial\mathbb{D} \to \gamma \tag{1}$$

is a (weakly) monotone, orientation preserving parametrization.

However $\mathcal{A}$ is *invariant under arbitrary changes of parameters.* It was observed by Lagrange that if a smooth surface is locally area-minimizing for a fixed boundary γ, then the mean curvature of that surface vanishes.

Following Riemann, Weierstrass, Schwarz, and Darboux, one introduces isothermal coordinates (conformality!):

$$|u_x|^2 - |u_y|^2 = 0 = u_x u_y. \tag{2}$$

This linearizes the Euler–Lagrange equations for least area: namely they reduce to Laplace equation

$$\Delta u = 0. \tag{3}$$

Now we can formulate

Problem of Plateau. Find solution u of equations (2) and (3) with the Plateau boundary condition (1).

For *unknotted* curves Garnier (1926) was able to prove the existence of solutions of (1) – (3) with the help of the methods of complex analysis (Riemann theory of hypergeometric equation was an important tool!) But general solution was found by direct methods of calculus of variations by Tibor Rado (in 1930) and later by Jesse Douglas (1931). In their procedure the following reductions-observations are crucial.

1. One replaces the complicated area functional $\mathcal{A}$ with simpler Dirichlet energy functional (this was Riemann's idea!)

$$\mathcal{E}(u) := \frac{1}{2}\int_{x^2+y^2<1} |\nabla u|^2 dx dy =$$

$$= \frac{1}{2}\int_{x^2+y^2<1} \left(|u_x|^2 + |u_y|^2\right) dx dy. \tag{4}$$

By definition, we have

$$\mathcal{A}(u) \leqslant \int_{x^2+y^2<1} \left(|u_x|^2 |u_y|^2\right)^{1/2} dx dy \leqslant$$

$$\leqslant \frac{1}{2}\int_{x^2+y^2<1} \left(|u_x|^2 + |u_y|^2\right) dx dy = \mathcal{E}(u),$$

with equality holding if and only if u is conformal: (2) is satisfied then. In this way we achieved the first drastic reduction of the symmetries of the problem: replace $\mathcal{A}$ with $\mathcal{E}$; then all differentiable parametrizations are reduced to *conformal reparametrizations.* This is because the energy (Dirichlet) functional $\mathcal{E}$ is (only) conformally invariant $\mathcal{E}(u) = \mathcal{E}(u \circ g)$, for all $g \in \mathcal{G}$, where $\mathcal{G}$ is the conformal group of all Möbius transformations of the disc $\mathbb{D} = \{z : |z| < 1\}$:

$$\mathcal{G} = \{g : z \to g(z) = e^{i\varphi}\frac{a+z}{1-\bar{a}z} \quad \text{with } a \in \mathbb{C}, |a| < 1, 0 < \varphi < 2\pi\}.$$

But the Möbius group $\mathcal{G}$ acts non compactly: for any $u \in C(\gamma)$, the orbit $\{u \circ g : g \in \mathcal{G}\}$ accumulates weakly also at constant functions (see below.)

2. We can get rid of the conformal invariance of $\mathbb{D}$ by making use of the famous Hilbert *three points condition* (1900). For any oriented triple $\exp(i\varphi_1)$, $\exp(i\varphi_2)$, $\exp(i\varphi_3)$, $0 \leqslant \varphi_1 < \varphi_2 < \varphi_3 < 2\pi$, there exists a unique $g \in \mathcal{G}$ such that

$$g(\exp(2\pi ik/3)) = \exp(i\varphi_k), \quad k = 1, 2, 3.$$

Now we can turn to the final reduction, obtaining a function space admissible for the energy functional $\mathcal{E}$.

3. Fix the parametrization ψ of γ (we assume that ψ is a diffeomorphism $\psi : \partial\mathbb{D} \to \gamma$), and let

$$C^\times(\gamma) = \{u \in C(\gamma) : u(\exp(2\pi ik/3)) = \psi(\exp(2\pi ik/3)), k = 1, 2, 3\}$$

equipped with W_1^2 topology.

We see that for every $u \in C(\gamma)$ there exists $g \in \mathcal{G}$ with $u \circ g \in C^\times(\gamma)$. Now we have to prove the following fundamental

THEOREM (DOUGLAS, RADO). *For any C^1 embedded curve γ there exists a minimizer u of the energy integral in $C(\gamma)$.*

The main steps of proof are the following. First we have the crucial proposition

GENERALIZED POINCARÉ INEQUALITY (POINCARÉ, FRIEDRICHS). *For* $u \in C^{\times}(\gamma)$,

$$\int_{\mathbb{D}} |u|^2 \leqslant c \int_{\mathbb{D}} |\nabla u|^2 + \int_{\partial\mathbb{D}} |u|^2 \leqslant c\mathcal{E}(u) + c(\gamma),$$

which is a consequence of

LEBESGUE–COURANT LEMMA (LEBESGUE 1907, COURANT 1936). *The set* $C^{\times}(\gamma)$ *is weakly closed in* W_1^2.

PROOF OF THE THEOREM. $\mathcal{E}$ is coercive on $X = C^{\times}(\gamma)$ in $W_1^2(\mathbb{D}, \mathbb{R}^n)$. Moreover, $\mathcal{E}$ is weakly lower semicontinuous on $W_1^2(\mathbb{D}, \mathbb{R}^n)$. By general functional analysis criteria, $\mathcal{E}$ attains its minimum in $C^{\times}(\gamma)$, which, by conformal invariance equals to the minimum on $C(\gamma)$. □

Thus we have 'only' to prove Proposition.

PROOF OF PROPOSITION (after Struwe). We use the fixed parametrization g which associates with every $u \in C^{\times}(\gamma)$ a continuous map $\xi : \mathbb{R} \to \mathbb{R}$ such that

$$g(\exp(i\xi(\varphi))) = u(\exp(i\varphi)), \quad \xi(0) = 0.$$

By (1) such functions are continuous, monotone, and $\xi - \mathrm{id}$ is 2π periodic. $\xi(2\pi k/3) = 2\pi k/3$ for all $k \in \mathbb{Z}$ by Hilbert three point condition. Let us denote by V the set of such functions ξ, that is $V = \{\xi : \mathbb{R} \to \mathbb{R}$ with ξ monotone, $\xi(\varphi + 2\pi) = \xi(\varphi) + 2\pi$, and $\xi(2\pi k/3) = 2\pi k/3$ for all $\varphi \in \mathbb{R}, k \in \mathbb{Z}\}$. Clearly V is convex. Let now u_n be a sequence in $C^{\times}(\gamma)$ with associated $\xi \in V$ such that $u_n \to u$ *weakly* in $W_1^2(\mathbb{D})$. But every ξ_n is monotone and $0 \leqslant \xi(\varphi) \leqslant 2\pi$, for $\varphi \in [0, 2\pi]$. Therefore, some sequence $\xi_n \to \xi$ almost everywhere on $[0, 2\pi]$, thus almost everywhere on $\mathbb{R}$ (be periodicity.) Now we have to consider two cases:

1. ξ is continuous;

2. ξ is discontinuous at some φ_0.

Case 1. x is continuous; therefore it follows from monotonicity that $\xi_m \to \xi$ uniformly on $\partial\mathbb{D}$. Therefore $u|_{\partial\mathbb{D}}$ is continuous and satisfies (1), that is $u \in C^{\times}(\gamma)$ and Proposition is proved.

Case 2. Is much more difficult. Now the Lebesgue–Courant lemma enters the stage. We will show that this case is impossible.

Assumption 2. leads to the following situation: There exists $\epsilon > 0$ *independent* of φ_1, φ_2 such that

$$u_n(\exp(i\varphi)) - u_n(\exp(i\psi)) \geqslant \epsilon > 0, \tag{5}$$

for all $\varphi \in I_1$, $\psi \in I_2$ if $n \geqslant n_0(\varphi_1, \varphi_2)$ is large enough; $I_1 = \{\varphi \in I_0 : \varphi \leqslant \varphi_1\}$, $I_2 = \{\varphi \in I_0 : \varphi \geqslant \varphi_2\}$, $I_0 := (2\pi(k-1)/3, 2\pi(k+1)/3)$, with $k \in \mathbb{Z}$ such that $|\varphi_0 - -2\pi k/3| < \frac{\pi}{2}$.

Equality (5) contradicts the Lebesgue–Courant lemma which asserts that elements $u_n \in W_1^2$ have, uniformly in $n \in \mathbb{N}$ arbitrary small oscillations on balls $B_{\rho n}$. □

This completes the proof of Rado–Douglas solution of Plateau problem. For completeness we present

Proof of Lebesgue–Courant lemma. Let $f \in W_1^2(\bar{\mathbb{D}}, \mathbb{R}^d)$ with $\mathcal{E}(f) = K$. Let $0 < \delta < 1$, $x_0 \in \bar{\mathbb{D}}$. Then there exists r with $\delta < r < \sqrt{\delta}$ for whose $f|_{\partial B(x_0,r)\cap\bar{\mathbb{D}}}$ is absolutely continuous and

$$|f(x_1) - f(x_2)| \leqslant (8\pi K)^{1/2} \log\left(\frac{1}{\delta}\right)^{-1/2} \tag{6}$$

for all $x_1, x_2 \in \partial B(x_0, r) \cap \bar{\mathbb{D}}$.

Proof. Since f is an element of the Sobolev space W_1^2, it follows that for almost all r, $f|_{\partial B(x_0,r)\cap\bar{\mathbb{D}}}$ is absolutely continuous. Taking polar coordinates (ρ, φ) centered at x_0, we have (be Schwarz inequality)

$$|f(x_1) - f(x_2)| \leqslant \int_0^{2\pi} |\partial_\varphi f(x)| d\varphi \leqslant (2\pi)^{1/2} \left(\int_0^{2\pi} |\partial_\varphi x|^2 d\varphi\right)^{1/2}. \tag{7}$$

But the energy $\mathcal{E}(f; \partial B(x_0, r) \cap \bar{\mathbb{D}})$ of f on $\partial B(x_0, r) \cap \bar{\mathbb{D}}$ equals

$$\frac{1}{2}\int_{\mathbb{D}} \left(|\partial_\rho f|^2 + \frac{1}{\rho^2}|\partial_\varphi f|^2\right) \rho d\rho d\varphi.$$

Therefore there exists $\delta < r < \sqrt{\delta}$ with

$$\int |_{\partial B(x_0,r)\cap\bar{\mathbb{D}}} \frac{1}{\rho}|\partial_\varphi f|^2 d\rho d\varphi \leqslant \frac{2K}{\int_\delta^{\sqrt{\delta}} \frac{1}{\rho} d\rho} = \frac{2K}{\log \delta^{-1}}. \tag{8}$$

Then (5) follows. □

EXERCISE 1. Prove the Poincaré inequality: For a bounded domain $\mathcal{U}$ in $\mathbb{R}^n$ of diameter d and $v \in W^2_{0,1}(\mathcal{U}) := W^2_1(\mathcal{U})-$ closure of $C^\infty_0(\mathcal{U})$,

$$\int_{\mathcal{U}} |v|^2 dx \leqslant d^2 \int_{\mathcal{U}} |\nabla v|^2 dx.$$

Hint: take $\mathcal{U} \subset [0,d] \times \mathbb{R}^{n-1} =: P$, $v \in C^\infty_0(\mathcal{U}) \subset C^\infty_0(P)$. Apply the mean value theorem and Schwarz inequality.

EXERCISE 2. Prove the generalized Poincaré inequality. An elegant proof can be found in Struwe, *Variational Methods*, Springer 1996, pp. 240-241.

Let us now present an interesting variational characteristics of conformal mappings: conformal mapping is an extremal deformation. Indeed:

PROPOSITION. *Let $\mathcal{U}$ be a domain in $\mathbb{R}^2$ and let $f \in W^2_1(\mathcal{U}, \mathbb{R}^d)$. If for any differentiable family of diffeomorphisms*

$$h_t : \bar{\mathcal{U}} \to \bar{\mathcal{U}}_t, \quad h_0 = \mathrm{id}$$

the 'variation of independent variables' vanishes, that is, if

$$\frac{\mathrm{d}}{\mathrm{d}t}\mathcal{D}(f \circ h_t^{-1}; \mathcal{U}_t)|_{t=0} = 0,$$

then f is conformal.

PROOF. Let $\varphi = (\varphi^1, \varphi^2) \in C^1(\mathcal{U}, \mathbb{R}^2)$ be such that $|t| \parallel \nabla\varphi \parallel_{L^\infty} <$ 1 and $h_t := \mathrm{id} + t\varphi : \mathcal{U} \to \mathcal{U}_t := h_t(\mathcal{U})$. Since $\nabla h_t = \mathrm{id} + t\nabla\varphi$, h_t are diffeomorphisms $\bar{\mathcal{U}} \to \bar{\mathcal{U}}_t$. But

$$(h_t^{-1} \circ h_t = \mathrm{id}) \Longrightarrow (\nabla h_t^{-1} \circ h_t = (\nabla h_t)^{-1} = \mathrm{id} - t\nabla\varphi + o(t^2).$$

Thus (u, v are coordinates in $\mathbb{R}$)

$$\det(\nabla h_t) = 1 + t(\varphi^1_u + \varphi^2_v) + o(t^2).$$

Therefore

$$2\mathcal{D}(f \circ h_t^{-1}; \mathcal{U}_t) = \int_{\mathcal{U}_t} |\nabla(f \circ h_t^{-1})|^2 = \int_{\mathcal{U}_t} |\nabla f \cdot (\nabla h_t^{-1}) \circ h_t|^2 \det(\nabla h_t) =$$

$$= \int_{\mathcal{U}} \{|\nabla f|^2 - 2t(|f_u|^2\varphi_u^1 + |f_v|^2\varphi_u^2 + f_u \cdot f_v(\varphi_u^2 + \varphi_v^1) +$$

$$t(|\nabla f|^2 + \varphi_u^1 + \varphi_v^2) + o(t^2)\}.$$

Thus $t \mapsto \mathcal{D}(f \circ h_t^{-1}; \mathcal{U}_t)$ is differentiable at $t = 0$ and

$$0 = \frac{\mathrm{d}}{\mathrm{d}t}\mathcal{D}(f \circ h_t^{-1}; \mathcal{U}_t)|_{t=0} = -\frac{1}{2}\int_{\mathcal{U}}[(|f_u|^2 - |f_v|^2)(\varphi_u^1 - \varphi_v^2) + 2f_u \cdot f_v].$$

If we now introduce complex coordinates in $\mathbb{C} \simeq \mathbb{R}^2$, $w := u + iv$, $\varphi := \varphi^1 + i\varphi^2$, and define

$$\Phi := (\varphi_u - if_v)^2 = |f_u|^2 - |f_v|^2 - 2if_u \cdot f_v = (\partial_w f)^2,$$

where $\partial_w := \frac{\partial}{\partial u} - \frac{\partial}{\partial v}$, $\partial_{\bar{w}} := \frac{\partial}{\partial u} + \frac{\partial}{\partial v}$, then the integrand in square parenthesis above equals $\mathrm{Re}(\Phi) \cdot \partial_{\bar{w}}\varphi$, and we have finally

$$\frac{\mathrm{d}}{\mathrm{d}t}\mathcal{D}(f \circ h_t^{-1}; \mathcal{U}_t)|_{t=0} = -\frac{1}{2}\int_u \mathrm{Re}(\Phi \cdot \partial_{\bar{w}}\varphi).$$

This last integral can vanish for all $\varphi \in C^1(\bar{\mathcal{U}}, \mathbb{R}^2)$ if and only if $\Phi(w) \equiv 0$ in $\mathcal{U}$, that is, if f is conformal. □

Here is an important property of the function $\Phi = (\partial_w f)^2$.

PROPOSITION. *Let $\mathcal{U}$ be domain in $\mathbb{C}$ and assume that $f \in W_1^2(\mathcal{U}, \mathbb{R}^d)$ is harmonic, then $\Phi := (\partial f)^2 \equiv (f_u - if_v)^2$ is a holomorphic function of $w = u + iv$.*

PROOF is immediate since $\Delta = \bar{\partial}\partial = -\partial\bar{\partial}$. Therefore

$$\partial\Phi = 2\bar{\partial}\partial f \cdot \partial f = 2\Delta f \cdot \partial f = 0.$$

□

REMARK. We see the fundamental role played by the function $\Phi = (\partial f)^2$. If Φ is holomorphic (that is, if $\bar{\partial}\Phi = 0$), then $\Phi(dw)^2$ is called the holomorphic quadratic differential. If X is a Riemann surface, holomorphic quadratic differentials are sections of $K_X^{\otimes 2}$ and from Riemann–Roch theorem it follows directly

THEOREM. *For Riemann surface of genus $p > 1$*

$$\dim_{\mathbb{C}} H^0(X, K_X^{\otimes 2}) = 3(p-1),$$

that is, the real dimension of holomorphic quadratic differentials is equal $6(p-1)$ or $H^0(X, K_X^{\otimes 2}) \simeq \mathbb{R}^{6(p-1)}$.

This observation is fundamental for the Teichmüller approach to the Problem of Moduli: the famous Teichmüller theorem asserts that $\mathcal{T}_p \simeq \mathbb{R}^{6(p-1)}$ since the cotangent space at each point of $\mathcal{T}_p$ is isomorphic to $H^0(X, K_X^{\otimes 2})$, the space of quadratic holomorphic differentials on X.

Riemann saw clearly a close relation between

2.3 Riemann mapping theorem and Plateau problem

If we take $d = 2$, from Douglas–Rado theorem we obtain the Riemann mapping theorem. Let $\mathcal{U}$ be a domain in $\mathbb{C}$ bounded by a Jordan curve γ of class C^1: $\partial\mathcal{U} = \gamma$. Then there exists a map $f : \bar{\mathbb{D}} \to \bar{\mathcal{U}}$ which is conformal in $\mathcal{U}$ and $f|_{\partial\mathbb{D}} : \partial\mathbb{D} \to \gamma$ is homeomorphic.

PROOF. Since the solution $f \in W_1^2(\mathbb{D}, \mathbb{R})$ of the Plateau problem ($f(\mathbb{D}) = \mathcal{U}$) is conformal and since f maps $\partial\mathbb{D}$ monotonously onto γ, f_z, as it follows from Cauchy integral formula, does not vanish in $\mathbb{D}$.

REMARK. The classical, strong form of Riemann, where $\mathcal{U}$ is *any* simply connected domain such that $\mathcal{U} \neq \mathbb{C}$ is now being proved by Koebe method (simplified by Carathéodory) which is variational as well. This method makes use of the following property of conformal maps (Schwarz–Pick):

Denote by $d = d_{\mathbb{D}}$ the Poincaré (hyperbolic) metric on $\mathbb{D}$ given by $ds^2 = \frac{4|dz|^2}{(1-|z|^2)^2}$. Then for every pair $x_1, x_2 \in \mathcal{U}$ and for a conformal map $\mathcal{U} \overset{onto}{\to} \mathbb{D}$, $d(f(x_1), f(x_2)) \geqslant d(u(x_1), u(x_2))$, where u is any biholomorphism of $\mathcal{U}$ onto $u(\mathcal{U}) \subset \mathbb{D}$, $u(\mathcal{U}) \neq \mathbb{D}$.

In this way one constructs a sequence $f_n : \mathcal{U} \to \mathcal{U}_n := f_n(\mathcal{U})$ of the conformal maps exhausting $\mathbb{D}$, to wit

$$f_k(\mathcal{U}) \subsetneq f_{k+1}(\mathcal{U}) \subsetneq \ldots \subsetneq \mathbb{D},$$

and $f := \lim f_n$ gives the biholomorphism $f : \mathcal{U} \to \mathbb{D}$.

EXERCISE. 1. Prove this construction (which is given in any textbook on function theory.)

2. Prove the SCHWARZ–PICK LEMMA

For any biholomorphic map $h : \mathbb{D} \to \mathbb{D}$

$$\frac{|h'(z)|}{1 - |h(z)|^2} \leqslant \frac{1}{1 - |z|^2}, \quad z \in \mathbb{D}.$$

Moreover, if the equality holds at one point, then h is a biholomorphic automorphism of $\mathbb{D}$, that is, $h \in \mathrm{Aut}(\mathbb{D})$ and the equality holds for all points of $\mathbb{D}$.

3. Geometric form of 2. Denote by

$$d(z_1, z_2) := \inf_c \int_c \frac{2|dz|}{1 - |z|^2},$$

(c being rectifiable curves in $\mathbb{D}$ connecting z_1 and z_2.) Then the Schwarz–Pick lemma has the beautiful form

For every holomorphic map $h : \mathbb{D} \to \mathbb{D}$

$$d(h(z_1), h(z_2)) \leqslant d(z_1, z_2), \quad z_1, z_2 \in \mathbb{D}.$$

If the equality holds for some pair z_1, z_2, then $h \in \mathrm{Aut}(\mathbb{D})$.

4. Prove Riemann mapping theorem in the following steps:

(i) There exists a biholomorphic map of $\mathcal{U}$ onto *part* of $\mathbb{D}$.

(ii) Let $\mathcal{F}$ be a family of holomorphic injections $u : \mathcal{U} \to \mathbb{D}$ such that $u(0) = 0$ and $u'(0) > 0$. Then there exists $f \in \mathcal{F}$ with greatest $f'(0)$ (by (i) we can assume that $\mathcal{U} \subset \mathbb{D}$.) (Hint: Use the Montel–Vitali theorem on compactness of the bounded sets in Fréchet metric.)

(iii) The map f of (ii) is a surjection of $\mathcal{U}$ onto $\mathbb{D}$.

2.4 Representation formulas for minimal surfaces. Enneper–Weierstrass theorem. Scherk surface

Probably the first who found such (local) representation in the case $d = 3$ was the great French geometer, the cofounder of Ecolle Politechnique, Gaspard Monge (1746–1818). Another representations we owe Weierstrass and Riemann (lectures of 1861/62), H.A. Schwarz, Darboux (1842–1917) in his celebrated four-volume monograph *Lecons sur la théorie générale des surfaces et sur les applications géometriques du calcul infinitesimal* (1887–1896), and many others. These formulas made it possible to construct a number of minimal surfaces and to gain a lot of experience, but the solution of the Plateau problem was beyond the reach on nineteenth century mathematics.

As we know, in the theory of surfaces in $\mathbb{R}^3$ the decisive role was played by the Gauss mapping $G : M \to S^2$. It can be expected that for minimal surfaces this mapping will possess especially beautiful properties. And indeed, when M is a Riemann surface immersed in $\mathbb{R}^3$ and if we regard the Gauss mapping as $G : M \to \mathbb{P}^1(\mathbb{C})$, then the following theorem holds.

THEOREM. *(A Riemann surface immersed in $\mathbb{R}^3$ is minimal)$\Longleftrightarrow$(Gauss map $G : M \to \mathbb{P}^1(\mathbb{C})$ is holomorphic).*

We will show that this theorem holds also for Riemann surfaces immersed in $\mathbb{R}^d$, $d \geqslant 3$. But first we must define the (generalized)

Gauss map of a surface immersed in $\mathbb{R}^d$

Let $\mathrm{Gr}_0(2, \mathbb{R}^d)$ be the set of oriented 2-planes in $\mathbb{R}^d$ which contain the origin 0; we can regard $\mathrm{Gr}_0(2, \mathbb{R}^d)$ as a subset of the projective space $\mathbb{P}^{d-1}(\mathbb{C})$:

For each $P \in \mathrm{Gr}_0(2, \mathbb{R}^d)$ take a positively oriented basis (X, Y) such that

$$(*) \qquad |X| = |Y|, \quad X \cdot Y = 0,$$

and assign the point

$$\Phi(P) := \pi(X - iY)$$

where $\pi : \mathbb{C}^d - \{0\} \to \mathbb{P}^{d-1}(\mathbb{C})$ is the canonical projection

$$(w_1, \ldots, w_d) \to (w_1 : w_2 : \cdots : w_d) := \{(cw_1, \ldots, cw_d) : c \in \mathbb{C} - \{0\}\}.$$

Clearly, the point $\Phi(P)$ is contained in the quadric

$$\mathcal{Q}_{d-2}(\mathbb{C}) := \left\{(w_1, \ldots, w_d) : w_1^2 + \cdots + w_d^2 = 0\right\} \subset \mathbb{P}^{d-1}(\mathbb{C}).$$

This follows from the fact that $(X-iY)\cdot(X-iY) = X\cdot X - 2i(X\cdot Y) - Y\cdot Y = 0$. The map Φ: $\mathrm{Gr}_O(2,\mathbb{R}^d) \to \mathcal{Q}_{d-2}(\mathbb{C})$ is bijective, and we can identify both manifolds.

Now consider a surface $x = (x_1,\ldots,x_d) : M \to \mathbb{R}^d$ immersed in $\mathbb{R}^d$. For each $p \in M$ the oriented tangent plane T_pM is canonically identified with an element of $\mathrm{Gr}_O(2,\mathbb{R}^d)$ after the parallel translation $p \to 0$.

DEFINITION. The *Gauss map* G of a surface M immersed in $\mathbb{R}^d$ is defined as

$$G : M \to \mathcal{Q}_{d-2}(\mathbb{C}), \quad p \to \Phi(T_pM).$$

For positively oriented isothermal coordinates (u,v) the vectors $X = \frac{\partial x}{\partial u}$, $Y = \frac{\partial x}{\partial v}$ provide a positive basis satisfying $(*)$. Therefore, the Gauss map is locally given by

$$G = \pi(X - iY) = \left(\frac{\partial x_1}{\partial z},\ldots,\frac{\partial x_d}{\partial z}\right), \qquad z = u + iv.$$

We can write $G = (\omega_1 : \cdots : \omega_d)$ with globally defined holomorphic 1-forms $\omega_j := \partial x_j \equiv \frac{\partial x_j}{\partial z}\,dz$.

Now we have

THEOREM. *(A Riemann surface $x : M \to \mathbb{R}^d$ is minimal)$\Longleftrightarrow$(The Gauss map $G : M \to \mathbb{P}^{d-1}(\mathbb{C})$ is holomorphic).*

PROOF. $\Longrightarrow$Assume that M is minimal. Then

$$\frac{\partial}{\partial \bar z}\left(\frac{\partial x}{\partial z}\right) = \frac{1}{4}\Delta x = 0,$$

and therefore $\frac{\partial x}{\partial z}$ is holomorphic, and hence G is holomorphic.

$\Longleftarrow$Let G be holomorphic. The problem is local. For a holomorphic coordinate z, take $f_j := \frac{\partial x_j}{\partial z}$, $j = 1,\ldots,d$. We can assume that f_d has no zeros. Since f_j/f_d is holomorphic, we have

$$(**) \qquad \frac{1}{4}\Delta x_j = \frac{\partial x_j}{\partial z \partial \bar z} = \frac{\partial}{\partial \bar z}\left(\frac{f_j}{f_m} f_m\right) = f_j \frac{1}{f_m}\frac{\partial f_m}{\partial \bar z}.$$

Put $\frac{1}{f_d}\frac{\partial f_m}{\partial \bar z} := h_1 + ih_2$ with real valued functions h_1 and h_2; taking the real part of both sides of $(**)$ we see that

$$\Delta x = 2\left(\frac{\partial x}{\partial u} h_1 + \frac{\partial x}{\partial v}\eta_2\right) \in T_pM.$$

Since for each $X \in T_pM$, $\Delta x \cdot X = 0$, we have $\Delta x \cdot \Delta x = 0$. Therefore $\Delta x = 0$, and hence M is minimal. □

By definition, a surface (with Riemann metric) is *flat* if and only if its Gaussian curvature vanishes identically, thus, if and only if the Gauss map G is constant.

The reader will prove the following

PROPOSITION. *(A minimal surface immersed in $\mathbb{R}^d$ is flat (that is, its Gauss map is constant))$\Longleftrightarrow$(M lies in a plane).*

For $d = 3$, as a corollary to the Theorem, we have the classical

PROPOSITION. *(A surface M immersed in $\mathbb{R}^3$ is minimal)$\Longleftrightarrow$(The classical Gauss map $G : M \to \mathbb{P}^1(\mathbb{C})$ is holomorphic, i.e., G is meromorphic).*

In his lectures Riemann proved a theorem equivalent to the celebrated *Enneper–Weierstrass representation of a minimal surface.*

THEOREM (ENNEPER–WEIERSTRASS). *For holomorphic coordinate $z = x + iv$, G is represented as $G = (\omega_1 : \omega_2 : \omega_3) = (f_1 : f_2 : f_3)$, where $\omega_j = f_j dz = \partial x_j$, $j = 1, 2, 3$. Set $\omega = \omega_1 - i\omega_2 \not\equiv 0$, $g := \frac{f_3}{f_1 - if_2}$.*
Then we have

(1) $\omega_1 = \frac{1}{2}(1 - g^2)\omega$, $\omega_2 = \frac{i}{2}(1 + g^2)\omega$, $\omega_3 = g\omega$;

(2) $x_1 = \operatorname{Re} \int_{z_0}^{z} (1 - g^2)h + const$
$x_2 = \operatorname{Re} i \int_{z_0}^{z} (1 + g^2)h + const$
$x_3 = \operatorname{Re} \int_{z_0}^{z} gh + const.$

(3) *(the holomorphic form ω has zero of order $2k$) $\Longleftrightarrow$(g has a pole of order k).*

PROOF. Consider $\omega = hdz$. Clearly $gh = f_3$. Since $f_1^2 + f_2^2 + f_3^2 = 0$, we have

$$\frac{1}{2}(1 - g^2)h = f_1, \quad \frac{i}{2}(1 + g^2)h = f_2.$$

On the other hand

$$2 \operatorname{Re} \int_{z_0}^{z} f_j(\zeta)d\zeta = \int_{z_0}^{z} \frac{\partial x_j}{\partial u} du + \frac{\partial x_j}{\partial v} = x_j(z) - x_j(z_0), \quad j = 1, 2, 3.$$

Scherk surface

Therefore (2) is proved; (1) and (3) are obvious. □

The beautiful example of this representation is provided by the famous Scherk surface given by

$$x_1 = \mathrm{Re} \int_0^z (1-z^2)\varphi(z)$$

$$x_2 = \mathrm{Re}\, i \int_0^z (1+z^2)\varphi(z)$$

$$x_3 = \mathrm{Re} \int_0^z 2z\varphi(z)$$

with

$$\varphi(z) = \frac{2}{1-z^4} = \frac{2}{(1+z)(1-z)(z+i)(z-i)}$$

on the domain $\mathbb{C} - \{\pm 1, \pm i\}$. The spherical Gauss image of the Sherk surface omits exactly four points of $S^2 = \mathbb{P}^1$: the points ± 1 and $\pm i$.

2.5 Minimal surfaces and value distribution theory

As we saw, due to the intermediation of the Gauss map, the theory of minimal surfaces is extremely closely related to the theory of meromorphic functions and in turn with the large and constantly developing 'value distribution theory' which could be found in the classical theorems of Casorati–Weierstrass and Picard. This theory will be forever connected with the name of great Finnish mathematician Rolf Nevanlina (1895–1980), the founder of famous Finnish school of function theory, and his brilliant pupil Lars V. Ahlfors (1907–1994). This relations are described in the nice monograph of Japanese researcher Hirotaka Fujimoto *Value Distribution Theory of the Gauss Map of Minimal Surfaces in* $\mathbb{R}^n$. I made use of this book while writing the previous sections. In 1988 Fujimoto obtained the following important

THEOREM (FUJIMOTO). *The number of exceptional values of the Gauss map of a non-flat, complete, minimal surface immersed in* $\mathbb{R}^3$ *is at most 4 (four).*

□ The number four is the best possible: we saw that the Gauss map of the Scherk surface omits precisely four points of $\mathbb{P}^1$.

We leave now minimal surfaces, and return to the Plateau–Douglas problem in connection with Teichmüller theory.

Harmonic maps are solutions of natural variational problems of differential geometry (and physics): these are critical or stationary points of the energy (physicists use the term 'action') $E(u) = E(u, g, \gamma)$ of the maps $u : (N, \gamma) \to (M, g)$ of Riemann manifolds with metric tensors $\gamma_{\alpha\beta}$ and g_{ij}. The energy $E(u)$ of u is given by

$$E(u) = \frac{1}{2}\int_M |du|^2 dM$$

where $dM = *1_M$ is the volume element on M, and the norm $|\cdot|$ in the integrand is the tensor product norm on $T^*N \otimes u^{-1}TM$.

If $(x^1, \ldots, x^n)$ and $(u^1, \ldots, u^m)$ are, respectively, local coordinates on N and M, then $\frac{1}{2}|du|^2 \equiv e(u)$ is called sometimes the energy density (in the physical literature the models defined below are called sigma-models) of u

and is given by

$$e(u) = \frac{1}{2}\gamma^{\alpha\beta}(x) g_{ij}(u) \frac{\partial u^i}{\partial x^\alpha}\frac{\partial u^j}{\partial x^\beta}.$$

Let $\gamma = \det(\gamma_{\alpha\beta})$, and Γ^i_{jk} be Christoffel symbols on M. If u is of class C^2, $E(u) < \infty$, and u is a critical point of $E(\cdot)$ (with respect to variations that vanish on ∂N if $\partial N \neq \emptyset$), then u is called *harmonic* and satisfies the corresponding Euler– Lagrange equations:

$$(E-L) \qquad \frac{1}{\sqrt{\gamma}}\frac{\partial}{\partial x^\alpha}\left(\sqrt{\gamma}\gamma^{\alpha\beta}\frac{\partial}{\partial x^\beta}u^i\right) + \gamma^{\alpha\beta}\Gamma^i_{jk}\frac{\partial}{\partial x^\alpha}u^j\frac{\partial}{\partial x^\beta}u^k = 0$$

This is a nonlinear elliptic system of m partial differential equations, where the principal part (of second order) is the Laplace–Beltrami operator on N and is, therefore, of the form of a divergence, while the nonlinearity is quadratic in gradients of the solution.

The differential du can be regarded as a section of the bundle $T^*N \otimes u^{-1}TM$:

$$e(u) = \frac{1}{2}(du|du)_{T^*N\otimes u^{-1}TM}.$$

Therefore $e(u)$ is the trace of the pullback, via u, of the metric tensor g_{ij} of M. Therefore $e(u)$, and also $E(u)$ are independent of the choice of coordinates, and thus intristically defined.

LEMMA. *u is harmonic if*

$$(1) \qquad \operatorname{tr}(\nabla du) =: \tau(u) = 0,$$

where ∇ is the covariant derivative in the bundle $T^*N \otimes u^{-1}TM$.

PROOF follows from the important formula

$$\tau^k(u) = (\operatorname{trace}(\nabla du))^k =$$

$$(2) \qquad \gamma^{\alpha\beta}\left(\frac{\partial u^k}{\partial x^\alpha \partial x^\beta} - {}^N\Gamma^\gamma_{\alpha\beta}\frac{\partial u^k}{\partial x^\gamma} + {}^M\Gamma^k_{ij}\frac{\partial u^i}{\partial x^\alpha}\frac{\partial u^j}{\partial x^\beta}\right).$$

COROLLARY. *(E-L) and (1) are equivalent.*

We have an important observation for harmonic maps of Riemann surfaces ($\dim_{\mathbb{R}} N = \dim_{\mathbb{R}} M = 2$):

Let N and M be (Riemann) surfaces, then we have conformal metrics

$$\sigma^2 dz d\bar{z} = \sigma^2(dx^2 + dy^2), \quad z = x + iy;$$

$$\rho^2 du d\bar{u} = \sigma^2(du_1^2 + du_2^2), \quad u = u_1 + iu_2.$$

Then the Laplace–Beltrami operator on N is given by

$$\Delta = \frac{1}{\sigma^2}\frac{\partial}{\partial z}\frac{\partial}{\partial \bar{z}},$$

and in these coordinates (E–L) takes the form

$$\Delta u = \frac{1}{\sigma^2} u_{z\bar{z}} + \frac{1}{\sigma^2}\frac{2\rho_u}{\rho} u_z u_{\bar{z}} = 0.$$

We have

PROPOSITION. *If $u : N \to M$ is a harmonic map, then*

(*i*) $$\varphi := (|u_x|^2 - |u_y|^2 - 2iu_x \cdot u_y)dz = \\ = 4\rho^2 u_z u_{\bar{z}} dz^2$$

is a harmonic quadratic differential.

(ii) ($\varphi = 0$)$\Longleftrightarrow$(u is conformal).

COROLLARY. *Since $\varphi = 0$ is (by Liouville theorem) the only holomorphic quadratic differential on S^2, any harmonic map $u : S^2 \to M$ is either conformal or anticonformal.*

REMARK. We see importance of holomorphic quadratic differentials in the theory of minimal surfaces. It follows from Riemann–Roch theorem that for Riemann surface X of genus $p \geqslant 2$

$$\dim_{\mathbb{C}} H^0(X, K_X^{\otimes 2}) = 3(p-1).$$

We know that on a Riemann surface X holomorphic functions (and forms) are harmonic; this was a basis of the Riemann theory of complex analysis: for him this theory was the potential theory, that is, the theory of harmonic functions on X. We know that Riemann surfaces are particular examples of Kähler manifolds. The following theorem shows once more that Kähler theory is a natural extension of Riemann complex analysis.

THEOREM (LICHNEROWICZ). *For Kähler manifolds N and M every holomorphic map (in particular, holomorphic form) $u : N \to M$ is harmonic.*

PROOF. The (E–L) equation for the case of Kähler manifolds takes the form

$$\tau^i(u) = \gamma^{\alpha\bar{\beta}} \left(\frac{\partial u^i}{\partial z^\alpha \partial \bar{z}^{\bar{\beta}}} + \Gamma^i_{jk} \frac{\partial u^j}{\partial z^\alpha} \frac{\partial u^k}{\partial z^{\bar{\beta}}} \right).$$

For holomorphic normal coordinates on N at z_0, the Christoffel symbols vanish at z_0. Therefore for holomorphic u, that is, such that $u^i_{,z\bar{\beta}} = 0$, $i = 1, \ldots, m$, $\tau^i(u) = 0$, whence u is harmonic. □

REMARK. For Kähler manifold the complex and Riemann structures are comparable.

REMARK. Holomorphic maps between arbitrary (*non* – Kählerian) manifolds are *not* harmonic in general.

Harmonic maps occur in many different situations in geometry, analysis, and physics:

(a) If $M = \mathbb{R}$, τ is the Laplace–Beltrami operator on (N, γ), and harmonic maps are harmonic functions;

(b) If $\dim N = 1$ (for example if $N = S^1$), harmonic maps are geodesics parametrized by arc length;

(c) For $\dim N = 2$, harmonic maps are minimal immersions; in particular if $M = \mathbb{R}^d$, we have minimal surfaces in a parametric form;

(d) In general, if $u : N \to M$ is an isometric immersion, $u^*\gamma = g$, then *(u is harmonic)$\Longleftrightarrow$(u is a minimal immersion)*;

(e) If (N, γ), (M, g) are Kähler and $u : N \to M$ is holomorphic, then u is harmonic;

(f) If $\dim N = 2$ and M is a symmetric space, then harmonic maps are known to physicists as *non-linear sigma models.* Of particular interest is the case $N = S^2$, $M = \mathbb{P}^n(\mathbb{C})$, and sigma models are prototypes of non- abelian field theories: they have similar properties to – far more difficult – four dimensional Yang–Mills fields (conformal invariance, existence of instantons and topological charges.) For $M = \mathbb{P}^n(\mathbb{C})$, such sigma models occur in the studies of ferromagnetics;

(g) $\dim N = 2 = \dim M$, these are conformal maps of two dimensional domains;

(h) Perhaps the simplest approach to Teichmüller theory and Riemann moduli problem is through harmonic maps between Riemann surfaces of some genus $p \geqslant 2$. This beautiful theory was developed by Eells and Earle, Tromba and Fisher, M. Wolf, Jost, and others.

A fundamental problem in the theory of harmonic maps is to answer the question:

UNder which conditions a given map can be deformed to a harmonic map? or, in other words *Which homotopy classes contain harmonic maps? Are such harmonic maps unique?*

The answer is provided by

THEOREM (EELLS–SAMPSON). *If the tangent space of M has non-positive sectional curvature, the answer is affirmative.*

This answer is provided by the heat flow method for harmonic maps (Eells–Sampson, Hartman, Jost.)

Before turning to heat transfer equation, we present

2.6 Some properties of harmonic maps. Theorems of Eells–Sampson, Hartman, and corollaries

Let us recall some elementary notions and formulas. Let $h : M \to \mathbb{R}$ be a function (of class C^2) on Riemann manifold M, and denote by ∇_Y the covariant (Levi–Civita) derivative in direction of Y. Then the Hessian , or the second fundamental form of h at point $x \in M$ is defined by

$$\nabla^2 h(x)(Y, Z) := (\nabla_Y \operatorname{grad} h(x), Z), \tag{1}$$

where Y, Z are vectors tangent at x.

If $u \in C^2(N, M)$, $v \in C^2(M, V)$ are mappings between Riemann manifolds, $h \in C^2(M, \mathbb{R})$, then the following,

Riemann chain rules hold: Denote by $\Delta = \Delta_N$ the Laplace–Beltrami operator on N, and let (e^α) be an orthonormal frame on N. Then

$$\Delta(h \circ u) = \nabla^2 h(u_{e^\alpha}, u_{e^\alpha}) + ((\operatorname{grad} h) \circ u | \tau(u))_M. \tag{2}$$

More generally,

$$\Delta(v \circ u) = \nabla dv(u_{e^\alpha}, u_{e^\alpha}) + (dv) \circ u \circ \tau(u). \tag{2'}$$

In particular, if u is harmonic, that is, if $\tau(u) = 0$,

$$\Delta(h \circ u) = \nabla^2 h(u_{e^\alpha}, u_{e^\alpha}). \tag{3}$$

In local coordinates

$$\Delta(h \circ u) = \gamma^{\alpha\beta} \nabla^2 h(u_{x^\alpha}, u_{x^\beta}). \tag{4}$$

PROPOSITION. *(i) If $h \in C^2(M, \mathbb{R})$ is strictly convex on M and $u : N \to M$ is harmonic, then $h \circ u$ is a* subharmonic *functions on N.*

(ii) Suppose that N is compact (possibly with boundary $\partial N \neq \emptyset$) and $u : N \to M$ is harmonic. If there exists a strictly convex function on $u(N)$, then u is constant.

(iii) If $v : M \to V$ is totally geodesic (that is, if $\nabla dv = 0$) and u is harmonic (then $v \circ u$ is harmonic), then u is a constant map.

PROOF. (i) From (3) it follows that $\Delta(h \circ u) > 0$, and this is definition of subharmonicity.

(ii) From the maximum principle for subharmonic functions it follows that $h \circ u$ is constant; since the Hessian of h is definite, it follows from (3) that u =const.

(iii) Follows immediately from $2'$. □

The archetype of Eells–Sampson–Hartman theorem is the famous theorem of Hilbert. This is the first theorem on global differential geometry of manifolds of arbitrary dimension.

THEOREM (HILBERT). *(i) Let (M, g) be a compact Riemann manifold. Then every free homotopy class of closed loops in M contains closed geodesic.*

(ii) Any two (not necessarily distinct) points of M can be connected by a geodesic belonging to given homotopy class.

These geodesics are obtained (as harmonic maps $u : S^1 \to M$) by minimizing the energy $E(c)$ of all maps $c : S^1 \to M$ in given homotopy class.

Hilbert theorem opened a new chapter of global differential geometry and was a starting point of direct methods of calculus of variations. Hilbert used the Dirichlet integral $E(c) = \int_0^1 |dc|^2$. Eells and Sampson made use of the 'method of heat equation': the energy $E(u_t)$ of family of mappings $U(\cdot, t) : N \to M$ converges to harmonic map u_∞, $u_t \to u_\infty$, $t \to +\infty$. This follows from the important theorem due to Eells and Sampson (improved by Hartman):

THEOREM (EELLS–SAMPSON). *Consider the 'heat equation', that is, the parabolic system*

$$(P) \qquad \frac{\partial u(x,t)}{\partial t} = \tau(u(x,t)), \quad \textit{for } x \in N, t \geqslant 0,$$

with the initial condition

$$u(x,0) = v(x), \quad x \in N,$$

where $v : N \to M$ is a given *mapping. Then,*

(a) The energy $t \to E(u(\cdot, t))$ is a decreasing function of t.

(b) If M and N are compact and M has nonpositive sectional curvature, then $\frac{\mathrm{d}^2}{\mathrm{d}t^2} E(u(\cdot, t)) \geqslant 0$ (t-convexity of E.)

Moreover,

(c) A solution for (P) exists for all $t \geqslant 0$ and as $t \to \infty$ it converges uniformly to the harmonic map u_∞. In particular, any map $v \in C^3(N, M)$ is homotopic to harmonic map.

We have immediately

COROLLARY. *Any continuous map $v : N \to M$ is homotopic to harmonic map if M has nonpositive curvature.*

PROOF . (a)

$$\frac{\mathrm{d}}{\mathrm{d}t}E(u(\cdot,t)) = \frac{\mathrm{d}}{\mathrm{d}t}\frac{1}{2}\int |du|^2 = \int \left(\frac{\partial}{\partial t}du|du\right) =$$

$$= \int d\left(\frac{\partial}{\partial t}u|du\right) = -\int \left(\frac{\partial}{\partial t}u|\tau(u)\right) = -\int \left|\frac{\partial}{\partial t}u\right|^2 \leqslant 0.$$

(b)

$$\frac{\mathrm{d}^2}{\mathrm{d}t^2}E(u(\cdot,t)) = \int_N \left|\nabla\frac{\partial u}{\partial t}\right|^2 - \int_N \left(R^M\left(du\cdot e_\alpha, \frac{\partial u}{\partial t}\right) du\cdot e_\alpha | \frac{\mathrm{d}u}{\mathrm{d}t}\right) \geqslant 0$$

since M has nonpositive sectional curvature (and thus the second integral is negative.)

(c) Of course, this point is difficult, one makes use of subtle inequalities in order to be able to apply the theory of Julius Schauder.

The necessary estimates are based on the formula

$$(*) \qquad \Delta_N e(u(x,t)) - \frac{\partial e(u,t)}{\partial t} =$$

$$= |\nabla du(x,t)| + (du\cdot Ric^N(e^\alpha)|du\cdot e^\alpha) - \left(R^M(du\cdot e^\alpha, du\cdot e^\beta)du\cdot e^\alpha|du\cdot e^\beta\right),$$

where Ric^N is the Ricci tensor on N, R^M is the curvature tensor on M and e^α is the orthonormal frame on N.

If $R^M \leqslant 0$, from (*) we obtain

$$(**) \qquad \Delta_N e(u(x,t)) - \frac{\partial e(u,t)}{\partial t} \geqslant -ce(u).$$

The (strong) maximum principle gives the pointwise bound for $e(u(x,t))$ in terms of $E(u(\cdot,t))$. Since $t \to E(u(\cdot,t))$ is decreasing, one obtains necessary estimates:

LEMMA 1. *If $u(x,t)$ solves the parabolic problem (P) for $t \in [0,T)$, and M has nonpositive sectional curvature, then for every $\alpha \in (0,1)$*

$$|u(\cdot,t)|_{C^{2+\alpha}(N,M)} + \left|\frac{\partial u}{\partial t}(\cdot,t)\right|_{C^\alpha(N,M)} \leqslant c_2.$$

The constant c_2 depends on α, T, the initial value $v = u(\cdot, 0)$, and the geometry of N and M.

LEMMA 2. *If M has nonpositive sectional curvature, then solution for (P) exists for all $t \geqslant 0$.*

PROOF. One easily proves that for all $t \in [0, T)$ and $x \in N$

$$\left|\frac{\partial u(x,t)}{\partial t}\right| \leqslant \sup_{x \in N}\left|\frac{\partial u(x,0)}{\partial t}\right|,$$

which shows that the set of $T \in [0, \infty)$ with the property that the solution exists for all $t \in [0, T)$ is open and nonempty. Lemma 1 implies the this set is closed as well, and thus it is the whole of $[0, \infty)$. □

The last step is provided by

LEMMA 3.*If $u(x,t)$ remains in a bounded subset of M, then there exists a sequence $t_n \to \infty$, for which $\frac{\partial u(x,t_n)}{\partial t}$ converges uniformly in x as $t_n \to \infty$.*

PROOF. By virtue of the $C^{2+\alpha}$ bounds for $u(\cdot, t)$ of Lemma 1, we can assume that $u(\cdot, t_n)$ converges, as $t_n \to \infty$, uniformly to the harmonic map u_∞. One proves that $u(\cdot, t)$ converges uniformly to u_∞ as $t \to \infty$. □

REMARK. This simplified and improved version the Eells–Sampson proof is due to Philip Hartman, and it appeared *Canadian J. Math.* **19** (1967), 673–687.

We owe Hartman the beautiful

UNIQUENESS THEOREM FOR HARMONIC MAPS (HARTMAN). *Let $u_1, u_2 : N \to M$ be two homotopic harmonic maps from N to nonpositively curved M. For fixed $x \in N$, let $s \to f(x, s)$ be the unique geodesic from $u_1(x)$ to $u_2(x)$ in the homotopy class determined by the homotopy between u_1 and u_2, and let $s \in [0, 1]$ be the parameter proportional to arc length. Then*

1. For each $s \in [0, 1]$ is a harmonic map with $E(u(\cdot, s)) = E(u_1) = E(u_2)$.

2. The length of the geodesic $f(x, \cdot)$ is independent of x. Thus any two harmonic maps can be joined by a parallel family of harmonic maps of equal energy.

3. If, moreover, M has negative *sectional curvature, then the harmonic map $u : N \to M$ is unique in its homotopy class, unless it is constant or it maps N into a closed geodesic.*

In the last case the non-uniqueness can be only due to rotations of the geodesic.

In Teichmüller theory one is interested in compact Riemann surfaces X of negative curvature, equipped with different metrics. Therefore the following corollary is of paramount importance.

COROLLARY. *Let $N = M = X$ be compact Riemann manifold, and let two metrics g and g_0 on X have negative (sectional) curvature.*

Then there exists a unique *smooth harmonic map $u(g) : (X, g) \to (X, g_0)$ which is homotopic to identity. Moreover u is a diffeomorphism, and $g \to u(g)$ is C^∞-smooth.*

REMARK. There are other methods of proving existence of harmonic maps: by Hildebrandt–Kaul–Widman (1977), Karin Uhlenbeck (1970), and by Shoen–Uhlenbeck (1982).

The reader will try to show existence of solution of the Dirichlet boundary problem for equation $\frac{\partial u}{\partial t} = \tau(u)$:

THEOREM (HAMILTON 1975). *Assumption for M as above; N is compact with non empty boundary $\partial N \neq \emptyset$; $v : N \to M$ is a continuous map, $u(x, 0) = v(x)$, $x \in N$, $u(y, t) = v(y)$, $y \in \partial N$.*

Then (P) has a smooth solution $u(x, t)$ for $t \geqslant 0$. As $t \to \infty$, $u(\cdot, t) \to u_\infty(\cdot)$, where u_∞ is harmonic, unique, homotopic to v, with the same boundary values on ∂N.

We now give important corollaries to the Eells–Sampson–Hartman (E–S–H) theorem.

COROLLARY 1. (Hilbert theorem on geodesics above.)

COROLLARY 2 (PREISSMANN THEOREM 1943). *If M is compact of negative sectional curvature, then every abelian subgroup of $\pi_1(M)$ is cyclic.*

PROOF. Let a and b be commuting elements of $\pi_1(M)$. Then the homotopy between ab and ba makes it possible to construct a map $v : T^2 \to M$, where T^2 is the two dimensional torus. By (E–S–H) theorem, v is homotopic to the harmonic map $u : T^2 \to M$, and $u(T^2)$ is contained in a closed geodesic. Therefore a and b are homotopic to the multiple of this closed geodesic. □

COROLLARY 3. *((M, g) is of negative curvature)$\Longrightarrow$(Each isometry of M is homotopic to the identity, and the isometry group of M is discrete).*

PROOF follows from Hartman theorem since isometries are harmonic.

Now we prove important properties of harmonic maps that follows from the formula $(*)$ for $\Delta e(u)$.

THEOREM. *Let N be compact with $Ric^N \geqslant 0$, and suppose that the sectional curvature of M is nonpositive. Let $u : N \to M$ be harmonic. Then*

1. u is totally geodesic (that is, $\nabla du \equiv 0$) and $e(u) = \mathrm{const}$.

2. If Ric^N is positive at one point $x_0 \in N$, then $u = \mathrm{const}$.

3. If the sectional curvature of M is negative, then u is either constant or it maps N onto closed geodesic.

PROOF. 1. Since N is compact (and $\partial N = \emptyset$), $\int_N \Delta e(u) dN = 0$, therefore, the integral over the right hand side of $(*)$ vanishes. Since, by assumption, the integral is pointwisely nonnegative, it has to vanish identically. Therefore $\nabla du \equiv 0 \Longrightarrow u$ is totally geodesic. But, as we saw, $\Delta e(u) = 0$ $\Longrightarrow e(u)$ is harmonic on M, and since N is compact, $e(u) =$const.

2. ($Ric^N(x_0)$ is positive definite)$\Longrightarrow$($R^N_{\alpha\beta}(x_0) u^i_{x^\alpha} u^i_{x^\beta} = 0$)$\Longrightarrow$($e(u)(x_0) = 0$)$\Longrightarrow$($e(u) \equiv 0$ (by 1))$\Longrightarrow u =$const).

3. Since M has a *negative* sectional curvature, by similar reasoning we see that $\dim(du(T_x N)) \leqslant 1$ for each $x \in N$. If $\dim(du(T_x N)) = 0$, then $u =$ const, otherwise u is totally geodesic, and it maps N onto closed geodesics.

COROLLARY 4. *(M is compact and of nonpositive curvature)$\Longrightarrow$ ($\pi_m(M)= 0$ for $m > 1$).*

PROOF. Let $v : S^m \to M$; then by (E–S–H) v is homotopic to harmonic map $u : S^m \to M$. Since $Ric^{S^m} > 0$ for $m > 1$, it follows from 1. of the theorem that u =const. □

COROLLARY 5. *Let M be a Riemann surface of genus $p > 1$. Then every free homotopy class of closed loops on M contains a unique closed (simple) geodesic.*

PROOF. Since the genus of M is larger than 1, we can equip M with a metric of negative curvature (even constant curvature equal -1) by uniformization theorem. Then the assertion follows from Hartman theorem. □

CHAPTER 3

Teichmüller Theory. Riemann Moduli Problem

> **Motto.** Riemann's classical problem of moduli is not a problem with a single aim, but rather a *program* to obtain maximum information about a whole complex of questions which can be viewed from several different angles.
>
> Lars V. Ahlfors

This motto is the first sentence of the beautiful opening address at the International Congress of Mathematicians held in Stockholm in August, 1962 *Teichmüller Spaces* given by the great expert on the Riemann moduli problem. Lars V. Ahlfors was the first winner of the Fields medal (in 1936, together with J. Douglas).

The above words of the great scientist characterize perfectly gigantic efforts and tremendous results of the greatest mathematicians which originated from the short remark of Riemann in his classical treatise on algebraic functions (that is, compact Riemann surfaces.) Riemann suggests that the set (space) of equivalence classes $\mathcal{M}_p$ of surfaces of genus $p > 1$ could be parametrized by $3(p-1)$ complex parameters (or $6(g-1)$ real ones) which he called *moduli.*

During the next 150 years this short remark by Riemann has been transformed into the great, rich, and beautiful mathematical theory (which is far from being completed.) The first who became fascinated by this remark was Felix Klein, the tireless propagator of Riemannian ideas (and the cofounder, along with Poincaré, of the theory of automorphic functions) and his pupil, collaborator, and son-in-law, Robert Fricke.

As a result of the race against Poincaré , Klein (according to his own words) has lost his creative strength. His work was continued by R. Fricke, the men of incredible diligence and erudition: he was the leading expert in the theory of modular and automorphic functions of his times; the author of wonderful textbooks, monographs, and review articles. It is Robert Fricke whom we owe the first solution of the Riemann moduli problem, hidden in the second volume of the monograph *Vorlesungen über die Theorie der automorphen Functionen.*

In this work Fricke introduces the notion of marked Riemann surface, and he constructs the space $\mathcal{F}_p$ (of equivalence classes) of marked Riemann surfaces of genus $p \geqslant 2$, now called the *Fricke space*, and he showed that $\mathcal{F}_p \simeq \mathbb{R}^{6(p-1)}$.

Oswald Teichmüller undoubtedly did not know Fricke's work and he introduces a different space, $\mathcal{T}_p$, called today the *Teichmüller space.* Teichmüller showed that this space has a structure of differentiable manifold (over $\mathbb{C}$) of dimension $3(p-1)$ (in 1939, the precise definition of complex manifold did not exist yet.)

Thus we have the famous

TEICHMÜLLER THEOREM. *$\mathcal{T}_p$ is a cell, that is, it is homeomorphic (even diffeomorphic) to $\mathbb{R}^{6(p-1)}$.*

Much later the bijection between the Fricke space $\mathcal{F}_p$ and the Teichmüller space $\mathcal{T}_p$ was proved. (Cf. the excellent textbook of Inayoshi and Taniguchi.)

The first edition of Fricke's monograph appeared before the Uniformization Theorem was proved (Koebe, Poincaré, 1907.) But the work of Fricke played a very important role in the subsequent development of the theory due to Jacob Nielsen and Werner Fenchel (cf. J. Nielsen *Collected Mathematical Papers* Vols. 1 and 2, Birkhäuser, 1986.) We will talk about important Fenchel–Nielsen coordinates on $\mathcal{T}_p$ later; they play important role in, considered by physicists, the theory of (super)strings.

A new chapter in the moduli theory was opened by the work of brilliant, tragic Osfald Teichmüller (work of 1939 and posthumous of 1944), who looked at the problem in a new way: with the aid of quasiconformal mappings $f: M \to M'$, that is, the mappings satisfying Beltrami equation:

$$\frac{\partial f}{\partial \bar{z}} = \mu \frac{\partial f}{\partial z}.$$

Quasiconformal mappings were introduced by Camillo, Herbert Grötzsch already in 1928 in a geometrical way as follows: for any differentiable map $f: Q \to Q'$, $w = f(z)$, its *dilatation at point* $z \in Q$ is defined by

$$K_f(z) := \frac{|f_z(z)| + |f_{\bar{z}}(z)|}{|f_z(z)| - |f_{\bar{z}}(z)|},$$

and its *dilatation* is $K[f] := \sup_{z \in Q} K_f(z)$.

Differentiable, orientation preserving homeomorphisms $w = f(z)$ between Riemann surfaces with bounded dilatation are called *quasiconformal*.

3.1 Teichmüller metric

Teichmüller introduces a *marking* (in the sense of Fricke) in the following way: Let M_0 be a fixed Riemann surface. Variable surfaces M_i have to be equipped with an orientation preserving homeomorphism f_i from M_0 to M_i. Two such marked surfaces (M_1, f_1) and (M_2, f_2) are identified if a biholomorphic map from M_1 to M_2 takes f_1 to a homeomorphism homotopic to f_2.

DEFINITION. The Teichmüller distance $d(P, Q)$ of two marked Riemann surface is defined in terms of dilatation

$$d_T(P, Q) := 2^{-1} \log \inf_f \{K[f] : \ f : P \to Q \text{ is quasiconformal}\}.$$

The function d_T is indeed a distance:
1. It is symmetric since $K[f] = K[f^{-1}]$;
2. The triangle inequality follows from the fact that

$$K[f_1 \circ f_2] \leqslant K[f_1] \cdot K[f_2];$$

3. Finally, $(d_T(P, Q) = 0) \Longrightarrow (P = Q)$.

Teichmüller defines topology on $\mathcal{T}$ by his distance and he proves that

THEOREM. *$(\mathcal{T}, d_T)$ is a complete metric space.*

3.2 The analytic structure of the Teichmüller space $\mathcal{T}_p$

Teichmüller introduces the notion of analytic family of compact Riemann surfaces, he was a predecessor of the Kodaira–Spencer theory of deformations of complex structures. Then the cotangent spaces $T_P^*\mathcal{T}_p$ are isomorphic with the space A_2 of *holomorphic quadratic differentials:*: $A_2(g) = H^0(M, K_M^{\otimes 2})$, where $K_M = \Omega_M$ is a canonical bundle of the surface M. Since, by Riemann–Roch theorem, $\dim_{\mathbb{C}} H^0(M, K_M^{\otimes 2}) = 3(p-1)$, Teichmüller obtains his famous

TEICHMÜLLER THEOREM. $\dim_{\mathbb{C}} \mathcal{T}_p = 3(p-1)$, *and equipped with* $d_T(\cdot,\cdot)$ *this space is a complete Finsler space.*

3.3 The moduli space

Now the moduli space, $\mathcal{M}_p$, is easy to identify. Let P and Q be points of $\mathcal{T}_p$ representing marked Riemann surfaces M_j equipped with homeomorphisms $f_j : M_j \to M_0$, $j = 1, 2$. Any isomorphism of M_1 and M_2, in the abstract sense, defines a homotopy class of homeomorphisms of M_0. The group Γ_p of such homotopy classes is called the *Teichmüller modular group.* It acts on $\mathcal{T}_p$ in such a way that its orbits consist exactly of equivalence classes of marked Riemann surfaces (since it operates on Teichmüller markings.) The quotient $\mathcal{T}_p/\Gamma_p$ is the Riemann moduli space $\mathcal{M}_p$.

One proves that the action of Γ_p is holomorphic. Since Γ_p has fixed points, the moduli space $\mathcal{T}_p/\Gamma_p$ has singularities: $\mathcal{M}_p$ is a *complex space* (Grauert.)

This is a very impressive construction. Of course there were some loopholes and Teichmüller himself realized it very well. Rigorous proofs were given much later by Ahlfors (1960-61) and Bers (1960). The subsequent developments was strongly influenced by remarks and conjectures of André Weil (1958). He introduced another metric on $\mathcal{T}_p$ following works of H. Petersson who presented his famous *scalar product* on automorphic forms in 1949.

Since the cotangent space $T_Q^*\mathcal{T}_p$ is a space of holomorphic quadratic differentials, it is equipped with the Petersson scalar product. Therefore it is natural to introduce this product on the tangent bundle $T\mathcal{T}$, called now the Petersson–Weil product $(\cdot|\cdot)_{PW}$ (cf. Part 'Riemann and Number Theory.) Weil conjectured that the PW-metric is hermitian and Kählerian and he hoped that it is complete.

Ahlfors proved that $(\cdot|\cdot)_{PW}$ is indeed Kählerian and even has nonpositive Ricci and (holomorphic) sectional curvatures. More precisely, the bound on curvature was later obtained independently by Wolpert and Tromba. It reads

$$Ric^{\mathcal{T}_p} \leqslant \frac{-1}{2\pi(p-1)}.$$

However, the PW metric is *not* complete!

All these beautiful results were obtained by hard analysis of elliptic equations.

Now, in the spirit of this book, we turn to another approach to Teichmüller theory.

Chapter 4

Riemannian Approach to Teichmüller Theory. Harmonic Maps and Teichmüller Space

Before we turn to working out our program, let us recall some classical facts from the theory of compact *two dimensional* manifolds, to stress the existence of an abyss between C^∞ structures and the complex ones. In the two dimensional case this abyss is much deeper than between topological and differentiable structures.

The classical statement says that the genus of a surface characterizes its topology (that is, two surfaces M and N are homeomorphic if and only if they have the same genus p.) Moreover

THEOREM. *(M, N are compact oriented C^∞ manifolds of the same genus)$\Longrightarrow$(M and N are diffeomorphic).*

This shows that the classification of compact, connected 2-manifolds up to homeomorphism is the same as the classification up to diffeomorphism, and thus the C^∞ does not provide us with any finer classification. Riemann knew that the situation changes dramatically once complex structure is introduced; this was the source of the problem of moduli.

Let us recall the notion of complex structure on M. It is such an atlas $c := \{\mathcal{U}_j, \varphi_j\}$ that $\varphi_i \circ \varphi_j^{-1}$ are holomorphic maps. If (M, c) is a Riemann surface and $f : M \to M$ is a *diffeomorphism*, the pullback f^*c of c by f is the *new* complex structure (atlas) $f^*c := \{(f^{-1}(\mathcal{U}_j), \varphi_j \circ f)\}$. Clearly, $(\varphi_j \circ f) \circ (\varphi_j \circ f)^{-1} = \varphi_i \circ f \circ f^{-1} \circ \varphi_j^{-1} = \varphi_j \circ \varphi_j^{-1}$. (M, c) and (M, f^*c)

are holomorphically equivalent, whereas (in general) $f : (M, c) \to (M, c)$ is *not* a holomorphic equivalence. Hence the classification of Riemann surfaces should be such that (M, c) and (M, f^*c) are in the same equivalence class.

Denote by $\mathcal{D} = Diff^+(M)$ the group of orientation preserving diffeomorphisms of M, and by $\mathcal{C}$ the set of all complex structures on M. Then we have a natural action of $\mathcal{D}$ on $\mathcal{C}$ (from the right) by pullback:

$$\mathcal{C}(M) \times Diff^+(M) \to \mathcal{C}(M), \quad (c, f) \to f^*c.$$

DEFINITION. The *Riemann moduli space* of M denoted by $\mathcal{M}$ or $\mathcal{R}(M)$ is the space of $\mathcal{D}$ orbits $\mathcal{M} := \mathcal{C}(M)/\mathcal{D}(M)$.

Thus $\mathcal{M}_p$ is the set of Riemann surfaces of genus p which are *not* holomorphically equivalent.

The following examples were known to Riemann.

EXAMPLE 1. $\mathcal{M}_0 = \{\text{point}\}$, since any Riemann surface homeomorphic to sphere is isomorphic to the Riemann sphere $\mathbb{P}^1(\mathbb{C})$.

The second example is much more interesting (elliptic case) and requires some knowledge of elliptic functions.

EXAMPLE 2. $\mathcal{M}_1 = \mathbb{C}$. Any Riemann surface of genus 1 is homeomorphic to torus $T^1 = \mathbb{C}/\Lambda$. Two such tori E_Λ and $E_{\Lambda'}$ are biholomorphic if and only if there exists an automorphism $z \to az + b$ of $\mathbb{C}$ carrying the lattice Λ to Λ'. If the lattice Λ is generated by $\lambda_1, \lambda_2 \in \mathbb{C}$ and Λ' by λ'_1 and λ'_2, this will be the case if and only if $\tau = \lambda_1/\lambda_2$ and $\tau' = \lambda'_1/\lambda'_2$ are related by

$$\tau' = \frac{\alpha\tau + \beta}{\gamma\tau + \delta} \quad \begin{pmatrix} \alpha & \beta \\ \gamma & \delta \end{pmatrix} \in SL(2, \mathbb{Z}).$$

Since we can take τ lying in the upper half plane $\mathfrak{H}_1 = \{z \in \mathbb{C} : \operatorname{Im} z > 0\}$, $\mathcal{M}_1$ is the set of isomorphism classes of elliptic curves, and can be identified with $\mathfrak{H}_1/\mathrm{SL}(2, \mathbb{Z})$. The set of moduli is the fundamental domain F of the group $\mathrm{SL}(2, \mathbb{Z})$, called the modular group (see Figure)

$$F = \left\{ z \in \mathfrak{H}_1 : -\frac{1}{2} \leqslant \operatorname{Re} z < \frac{1}{2}, |z| \geqslant 1 \text{ if } \operatorname{Re} z \leqslant 0, |z| > 1 \text{ if } \operatorname{Re} z > 0 \right\}.$$

Thus $\mathcal{M}_1 \simeq F \simeq \mathbb{C}$.

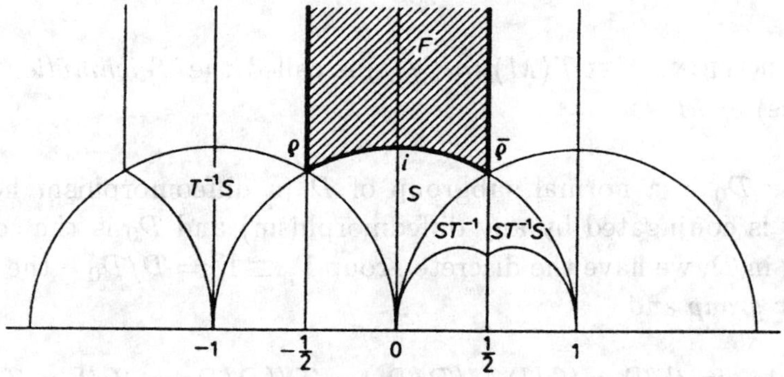

The modular group

We have therefore the Riemann observation (and conjecture):

$$\dim_{\mathbb{C}} \mathcal{M}_p = \begin{cases} 0, & g = 0 \\ 1, & g = 1 \\ 3(p-1), & g > 1 \end{cases}$$

This last example is of great interest as it is a sort of pilot one: we encountered the basic features of the moduli problems (this holds as well for the case of objects different than Riemann surfaces, for example instantons – the self dual connections.)

What we have here is *infinite* dimensional space $\mathcal{C}$ of geometrical structures (in our case, conformal structures) on M (torus in our case) and the infinite dimensional group $\mathcal{D}$ acting on the space $\mathcal{C}$, but the space of orbits $\mathcal{C}/\mathcal{D}$ is *finite* dimensional and its points (orbits) are called moduli.

The structure of the moduli space $\mathcal{M}_p$ for $p > 1$ is mysterious, it is still not completely understood.

The breakthrough in understanding of $\mathcal{C}/\mathcal{D}$ is due to Osfald Teichmüller (1939, 1943): he introduced the subgroup $\mathcal{D}_0$ of $\mathcal{D}$ of such diffeomorphisms in $Diff^+(M)$ which are *homotopic* to identity. There is a fact due to Reinhold Baer (and Jacob Nielsen):

PROPOSITION (BAER, NIELSEN,1928) *1. Suppose that $h_0, h_1 \in \mathcal{D}$ are homotopic. Then they are also* isotopic, *that is, there exists a homotopy h_t, $t \in [0,1]$ which is a diffeomorphism for each t.*

2. $\pi_0(\mathcal{D}) = \mathcal{D}/\mathcal{D}_0$. ($\pi_0(X)$ is, by definition, the number of connected components of X.)

DEFINITION. $\mathcal{T} \equiv \mathcal{T}(M) := \mathcal{C}/\mathcal{D}_0$ is called the *Teichmüller space* (moduli space) of M.

Since $\mathcal{D}_0$ is a normal subgroup of $\mathcal{D}$ (a diffeomorphism homotopic to identity is conjugated by any diffeomorphism) and $\mathcal{D}_0$ is the component of identity in $\mathcal{D}$, we have the discrete group $\Gamma_p \equiv \Gamma := \mathcal{D}/\mathcal{D}_0$ – the *Teichmüller modular group* and

$$\mathcal{M}_p := \mathcal{C}/\mathcal{D} = (\mathcal{C}/\mathcal{D}_0)/(\mathcal{D}/\mathcal{D}_0) = \mathcal{T}/(\mathcal{D}/\mathcal{D}_0) = \mathcal{T}/\Gamma = \mathcal{T}_p/\Gamma_p.$$

Thus the Teichmüller space $\mathcal{T}$ covers the Riemann moduli space $\mathcal{M}$.

In this chapter we will investigate geometry of the Teichmüller space with the aid of methods of the theory of harmonic maps $u : (M, \gamma) \to (M, g)$ between the Teichmüller surfaces (M, γ) and (M, g). But first let us present

PROOF OF PROPOSITION. 1. We put hyperbolic metric g on M. We can minimize the energy in the isotopy class of h_i, $i = 0, 1$, obtaining the harmonic diffeomorphism u_i isotopic to h_i. Since h_0 and h_1 are homotopic, by uniqueness of harmonic map $u_0 = u_1$.

2. It follows from 1. that $\mathcal{D}_0$ is the connected component of identity in $\mathcal{D}$. □

Before we turn to the heart of the 'Riemannian harmonic road' to Teichmüller theory, let us make some

Observations and identifications. We will use the expressions complex structure, conformal structure, holomorphic structure exchangeable, because, from the uniformization theorem, we have important

OBSERVATION. For each conformal structure on M there exists a unique hyperbolic metric g (obtained from the Poincaré metric on $\mathfrak{H}_1 \simeq \mathbb{D}$) with constant curvature $K(M) = -1$.

In the following we will frequently identify conformal (complex) structure with the corresponding hyperbolic metric.

As we saw for every diffeomorphism $v : M \to M$ and a conformal metric g on M, $v : (M, v^*g) \to (M, g)$ is conformal, and it is legitimate to identify the structure v^*g with g.

The starting point are two following theorems.

THEOREM A (SAMPSON 1978). *1. Given any conformal structure (M, g), for any other hyperbolic structure (M, γ) there exists a* unique *harmonic map*

$$u(g, \gamma) : (M, g) \to (M, \gamma)$$

which is homotopic to identity.

2. $u(g, \gamma)$ is a diffeomorphism, and $\varphi(u) := \rho^2 u_z \bar{u} dz^2$ is a holomorphic quadratic differential, where z is a conformal coordinate on (M, g), and the hyperbolic metric γ is locally represented as $\rho^2 du d\bar{u}$.

3. $(\varphi(u) \equiv 0) \Longleftrightarrow (u$ is conformal$) \Longleftrightarrow (g = \gamma)$. Therefore we have a natural injective map

$$\Phi(g) : \mathcal{T}_p \to \mathcal{Q}(g) := H^0(M, K_M^{\otimes 2}),$$

where $\mathcal{Q}(g)$ is the space of holomorphic quadratic differentials on (M, g).

REMARK 1. We already encountered the mapping $\Phi = \Phi(g)$ in the proof of the solution of Plateau Problem, where the equivalence 3. was crucial.

REMARK 2. The next theorem was proved by Michael Wolf, a student of Wolpert in 1987, under assumption that dim $\mathcal{T}_p$ is known *a priori.* This is an awkward and strong assumption and it was dropped in 1991 by Jost in his beautiful monograph.

THEOREM B (WOLF, 1987, JOST, 1991). *For any g the map $\Phi(g)$: $\mathcal{T}_p \to \mathcal{Q}(g)$ is surjective, and hence bijective.*

As a consequence of Wolf–Jost and Riemann–Roch theorems we obtain the Teichmüller asserting that $\mathcal{T}_p$ is diffeomorphic to $\mathbb{R}^{6(p-1)}$.

The map $\Phi(g)$ makes it possible to introduce a natural differentiable structure on the Teichmüller space $\mathcal{T}_p$; in order to make this structure canonical, that is, independent of g, one has to show that the transition mappings $\Phi(g_2) \circ \Phi(g_1)^{-1}$ are differentiable for any two hyperbolic g_1, g_2

Let us now present the idea of proofs of these marvelous theorems.

Sampson theorem follows from theorems of Eells–Sampson and Shoen–Yau.

For a fixed source metric g on M one can choose a unique metric $\tilde{\gamma}$ on any $\mathcal{D}_0$ orbit such that $(M, g) \to (M, \tilde{\gamma})$ is harmonic. The *energy density*

$e(v)$ of the mapping $z \to v(z)$ is

$$e(v) = \mathcal{H} + \mathcal{L}$$

where

$$\mathcal{H} = \gamma(v(z))|v_z|^2/g(z) \qquad \text{and } \mathcal{L} = \gamma(v(z))|v_{\bar{z}}|^2/g(z).$$

The energy $E(v)$ on the space of such mappings equals

$$E(v) = \int_M e(z)g(z)dzd\bar{z},$$

and the Euler–Lagrange equation for harmonic maps u is

$$\tau(u) = u_{z\bar{z}} + \left(\frac{\gamma_u}{\gamma} \circ u\right) \cdot u_z u_{\bar{z}} = 0.$$

We have therefore the following characteristic of the decomposition of the pullback of the metric γ under u:

$$u(\gamma)dud\bar{u} = \varphi dz^2 + e(u)gdzd\bar{z} + \bar{\varphi}(d\bar{z})^2,$$

where the quadratic differential

$$\varphi dz^2 := gu_z u_{\bar{z}} dz^2$$

is holomorphic.

By theorem B any differential $\varphi \in \mathcal{Q}(g)$ represents a hyperbolic metric $m(\varphi)$. M. Wolf introduced the energy

$$E(\varphi) := E(u(g, m(\varphi))$$

which depends on the *target* metric $\gamma = m(\varphi)$ on (M, γ).

THEOREM C (WOLF). *The (Wolf) energy $E(\varphi)$ is a* proper *exhausting function on $\mathcal{T}_p$, that is,*

$$E_c := \{\varphi \in \mathcal{Q}(g) : E(\varphi) \leqslant c\}$$

is compact for any $c \in \mathbb{R}$ ($E(\varphi) \geqslant 0$.)

REMARK. We know that the energy of a map $u : (M, g) \to (M, \gamma)$ depends on both metrics g and γ, $E(v) = E(v, g, \gamma)$. Tromba considered previously the energy $E(v, g)$ with fixed *target* metric γ, and he obtained a theorem similar to Theorem C. Then Wolf considered the energy $E(\varphi) = E(v, \gamma)$ with fixed *source* metric which has better properties. Jost used Wolf's energy and simplified the proof.

Both energies possess similar wonderful properties: as we will see $E(\varphi)$ is *plurisubharmonic* (after introducing a complex structure on $\mathcal{T}_p$) and is a $d'd''$ potential for the Kählerian Petersson–Weil metric.

Theorem A is proved as follows: since $\log \mathcal{H}$ satisfies an elliptic equation, the injectivity of $\Phi(g)$ in the Sampson theorem reduces to maximum principle which holds for solutions of elliptic equations.

In theorem B, the surjectivity of $\Phi(g)$ follows from the properties of Φ which are (as noted by Georg Schumacher (1990)) ultimately the properties of the energy $E(\varphi)$.

THEOREM D (WOLF). *The energy $E(\varphi)$ is an exhaustion function on Teichmüller space $\mathcal{T}_p$, that is,*

$$E(\varphi) : \mathcal{T} \to \mathbb{R}^+$$

is a proper map.

THEOREM E (WOLF, TROMBA). *1. The energy E is a strongly plurisubharmonic function.*

2. E has a unique minimum, and this is its only critical point.

We recall now the very important notion of

Plurisubharmonic function (p.s.h.) Let G be a domain in $\mathbb{C}^n$; then the C^2 function $p : G \to \mathbb{R}$ is (strictly) plurisubharmonic if for every $z \in G$ the Levi form

$$L(p) := \sum_i \frac{\partial^2 p}{\partial z_i \partial \bar{z}_i} dz_i d\bar{z}_i \quad (\equiv d'd''p)$$

has the signature $L(p)(z) \geqslant 0$ (resp. > 0.)

Since on a complex manifold transition maps are holomorphic, the signature of $d'd''p$ is independent of the coordinate system and thus the notion of plurisubharmonic function is well defined on every complex manifold X. This definition can be weakened; one can talk about continuous (and even upper semicontinuous) functions. The following observation is obvious

LEMMA. *If X is a compact complex manifold, then every plurisubharmonic function on X is constant.*

PROPOSITION *($p : G \to \mathbb{C}$ is plurisubharmonic)$\Longleftrightarrow$(the function $\mathbb{C} \ni t \to p(z + tw)$ is subharmonic for arbitrary $z, w \in \mathbb{C}^n$ in the part of $\mathbb{C}^n$ where it is defined).*

This is the reason for the term *pluri*subharmonic.

Since we would like to investigate another properties of the space $\mathcal{T}$, let us pause for a moment to make some

Observations on important spaces in mathematics and physics.

The history of mathematics (and physics as well) shows the most important role in analysis and geometry is played by

Three types of spaces

1. The Euclidean spaces $E^n = (\mathbb{R}^n, (\cdot, \cdot))$;
2. $\mathbb{C}^n$;
3. Complex projective spaces $\mathbb{P}^n(\mathbb{C}) = \mathbb{P}(\mathbb{C}^{n+1})$.

It is clear that (closed) submanifolds of these spaces and the structures on them (metric, differential, complex) should also play an important role.

Thus we obtain three natural types of manifolds:

1'. Riemannian manifolds;

2'. Stein manifolds (Remmert, Bishop);

3'. Algebraic (Chow's theorem.)

One would like to characterize this fundamental objects in an intristic way (and *not* as submanifolds.) Therefore one obtains the following manifolds:

1". Riemannian manifolds (X, g);

2". Stein manifold X as a complex manifold equipped with a strictly plurisubharmonic exhausting function $p : X \to \mathbb{R}$;

3". Hodge manifold = Kählerian manifold with rational first Chern class (Kodaira theorem.)

How should we understand the appearance of the Kähler and Kodaira manifolds? A hermitian Bergman metric on a X is a pullback of the Fubini–Study metric on $\mathbb{P}(\mathbb{C}^n)$. But the Fubini–Study is Hodge, therefore A. Weil conjectured that every compact Hodge manifold is algebraic. This conjecture was proven in the famous Kodaira theorem. Kählerian manifolds with negative (holomorphic) sectional curvature (and similarly for negative Ricci curvature) are geodesically convex: any two points can be connected by a single geodesics.

It turns out that the Teichmüller space $\mathcal{T}_p$ possesses all the above properties: it is a complex manifold of dimension $3(p-1)$, it is a Stein manifold (because E is strictly plurisubharmonic and exhausting), it has a Kähler metric: for any $g \in \mathcal{T}(M)$ there exists a single critical point of the energy E, and the Hessian $\nabla^2 E(g)$ is given by

$$\nabla^2 E(g) = (\cdot|\cdot)_{PW}.$$

The energy E on $\mathcal{T}_p$ is the $d'd''$ potential for the Petersson–Weyl metric at the reference point. Therefore $(\cdot|\cdot)_{PW}$ is Kählerian.

Of course, the original definition of Stein manifold was completely different. It developed in connection with investigations of *domain of holomorphy* in the theory of functions of several complex variables: the domain of holomorphy $X \subset \mathbb{C}^n$ is a domain (open, connected set) for which there exists at least one holomorphic function $f : X \to \mathbb{C}$ which cannot be extended as a holomorphic function through any boundary point of X. There is the following, classic remark of Riemann.

THEOREM (RIEMANN). *Every domain $X \subset \mathbb{C}^1$ is a domain of holomorphy.*

The reader will prove this theorem as an exercise (cf. Part 'Riemann and Complex Geometry'.)

For several complex variables the situation is drastically different. It was a great 'sensation' when Fritz Hartogs constructed a domain $H \subset \mathbb{C}^2$ for which *every* holomorphic function can be extended across ∂H . It was a fundamental and difficult theorem to find characteristics (possibly geometrical) of the domain of holomorphy.

THEOREM (HARTOG'S CONTINUITY THEOREM, 1906). *Let $n \geqslant 2$, G be a domain in $\mathbb{C}^{n-1}$, U and open non empty subset of G; $r_1, r_2 \in \mathbb{R}$, such that $0 < r_1 < r_2$. Let*

$$S := \{z \in \mathbb{C} : |z| < r_2\}, \quad R := \{z \in \mathbb{C} : r_1 < |z| < r_2\}.$$

then every *holomorphic function in $X' := (U \times S) \cup (G \times R)$ can be holomorphically extended onto $X = G \times S$.*

PROOF.

$$\tilde{f}(z) := \frac{1}{2\pi i}\int_{|w|=\rho}\frac{f(z_1,\ldots,z_{n-1},w}{w-z_n}dw,$$

where $r_1 < \rho < r_2$, is a holomorphic extension of f onto $G \times \{z_n \in \mathbb{C} : |z_n| < \rho\}$, and $\tilde{f} = f$ on $U \times \{z_n \in \mathbb{C} : |z_n| < \rho\}$ (Cauchy theorem.) The identity theorem gives holomorphic extension of f onto X. □

A complex manifold X is *holomorphically convex* (h.c.) if for any discrete sequence $\{x_n\}_1^\infty$ there exists a holomorphic function $f : X \to \mathbb{C}$ such that $|f(x_n)| \to \infty$ as $n \to \infty$.

EXERCISE 1. Prove that (h.c.) is equivalent to the more 'geometric' condition: for each compact $K \subset X$ the set (holo-convex hull of K) $\hat{K} := \bigcap_{f\in\mathcal{O}(X)}\{x \in X : |f(x)| \leqslant \sup|f(K)|\}$ is compact.

EXERCISE 2. If we consider linear functions only, we obtain conv$K$, the convex hull of K.

EXERCISE 3. Each open subset $X \subset \mathbb{C}$ is holo-convex.

REMARK. Exercise 3 suggests (together with Riemann theorem) that (h.c.) is important for characterization of domain of holomorphy. Indeed we have the following important

THEOREM. *(A domain $X \subset \mathbb{C}$ is a domain of holomorphy)$\Longleftrightarrow$(X is holo-convex).*

In 1951 Karl Stein introduced the following important class of complex manifolds.

DEFINITION. A complex manifold X (with countable topology) is a Stein manifold if

1. it has 'many' globally holomorphic functions: for every two distinct points $a, b \in X$ there exists a holomorphic function $f : X \to \mathbb{C}$ such that $f(a) \neq f(b)$ (one says that holomorphic functions on X repair points);
2. X is holomorphically convex.

REMARK. K. Stein called his manifolds holomorphically complete. It is a deep theorem of Grauert–Narashimhan showing that both definition (holomorphically complete vs. holomorphically convex) are equivalent. No wonder this theorem is difficult to prove: what the existence of a single p.s.h. exhaustive function has to do with holomorphic completeness?

Perhaps even more astonishing is the following

THEOREM (BISHOP, REMMERT, NARASHIMHAN). *(X is Stein)*$\Longleftrightarrow$*(X is a closed complex submanifold of some* $\mathbb{C}^N$ *).*

REMARK 1. $N = 2\dim_{\mathbb{C}} X + 1$ suffices.

REMARK 2. There are corresponding theorems even for complex spaces.

We have seen that $\mathcal{T}_p$ is a Stein manifold. This important theorem was proved first (in a quite different way) by Bers and Ehrenpreis (Bull. AMS, **70** (1964), 761–764.)

The role of plurisubharmonic functions in characterization of domains of holomorphy was clearly recognized by the great Japanese Kioshi Oka who proved that $X \subset \mathbb{C}^2$ is a domain of holomorphy if and only if X is Stein, that is, if on X lives a p.s.h. function.

The famous Levi problem was solved in full by Hans Grauert in 1958 (Ann. Math. **68** (1958), 460–472.)

We talk about the P–W scalar product for such a long time that it is time to give its definition. In 1949 H. Petersson presented his scalar product which, in the case of holomorphic quadratic differentials $\varphi, \psi \in \mathcal{Q}(g)$ has the form

$$(\varphi|\psi)_P := \int_M \frac{\varphi\bar{\psi}}{g^2} g dz d\bar{z}.$$

Since $T_g^*\mathcal{T}(M) = \Phi(g)$ this defines a harmonic form in each tangent space of $T_g^*\mathcal{T}$. This hermitian metric on $\mathcal{T}(M)$ was introduced by André Weil in 1958, therefore it is called the Petersson– Weil metric.

We see that 'Riemannian approach to Teichmüller theory' is crystal clear and natural: harmonic map $u : (M, g) \to (M, \gamma)$ gives the fundamental bijection $\Phi(g) : \mathcal{T}(M) \to \mathcal{Q}(g)$, from this $\mathcal{T}_p \simeq \mathbb{R}^{6(p-1)}$ which makes it possible to introduce a canonical *differentiable* structure on $\mathcal{T}(M)$ with $\Phi(g_1) \circ \varphi(G_2)^{-1}$ being transition maps of the atlas. Everything else follows from the properties of the Dirichlet integral. But let us do not deceive ourselves, even though the problem is conceptually clear, the necessary proofs are far from being trivial. Already the introduction of a complex structure on $\mathcal{T}(M)$ is not easy. This was done by Teichmüller in his posthumous paper.

The *almost* complex structure on $\mathcal{T}$ is given immediately in the canonical way: we identify the cotangent space of $\mathcal{T}_p$ at (M,γ) with $\mathcal{Q}(\gamma)$, and define $\mathcal{J} : \mathcal{Q}(\gamma) \to \mathcal{Q}(\gamma)$, $\mathcal{J}^2 = -\text{id}$. From the fundamental paper of Newlander–Nirenberg *Complex analysis coordinates in almost complex manifolds*, Ann. Math. **65** (1957),391–404, we know the criterion for integrability of $\mathcal{J}$:

THEOREM (NEWLANDER, NIRENBERG). *Vanishing of the Nijenhuis tensor*

$$N(\mathcal{J})(X,Y) := [\mathcal{J}X, \mathcal{J}Y] - \mathcal{J}[\mathcal{J}X, Y] - \mathcal{J}[X, \mathcal{J}Y] - [X,Y]$$

for any smooth vector fields X, Y is sufficient for integrability of $\mathcal{J}$.

THEOREM. *For the canonical almost complex structure $\mathcal{J}$ on $\mathcal{T}_p$ the Nijenhuis tensor $N(\mathcal{J}) = 0$. Therefore $\mathcal{T}_p$ possesses the complex structure of a complex manifold.*

REMARK. Of course, Teichmüller obtained this theorem by quite different means, his holomorphic family of Riemann surfaces. His proof was not complete and was completed by in a very ingenious way by Lipman Bers in 1960.

REMARK. On Riemann surface each almost complex structure is integrable and therefore complex. Since the problem is local, consider $M = \mathbb{R}^2$ as a manifold with ordinary Cartesian coordinates (x,y). For every $q \in M$, the endomorphism of T_qM given by

$$\mathcal{J}(q) : a\left(\frac{\partial}{\partial x}\right)_q + b\left(\frac{\partial}{\partial y}\right)_q \to -b\left(\frac{\partial}{\partial x}\right)_q + a\left(\frac{\partial}{\partial y}\right)_q, \quad a, b \in \mathbb{R}$$

has the square $\mathcal{J}^2 = -\text{id}$. The tensor field $q \to \mathcal{J}(q)$ is an almost complex structure on M whose Nijenhuis tensor vanishes $N(\mathcal{J}) \equiv 0$. Once more a deep abyss between $\dim_{\mathbb{C}} = 1$ and $\dim_{\mathbb{C}} > 1$ is clearly seen.

REMARK. A proof of Newlander–Nirenberg theorem (by Kohn method) is given in the Hörmander's monograph *An Introduction to Complex Analysis in Several Variables*, 1966. There are other proofs of the Newlander–Nirenberg theorem. None of them is simple, this theorems expresses a very deep fact from the theory of elliptic operators.

Let us return to Teichmüller theory. As noted by Wolpert, the space $\mathcal{T}(M)$ equipped with the Petersson–Weil metric is not complete. But we have

THEOREM. *The energy is P-W-convex (that is, any two points of $\mathcal{T}(M)$ can be joined by a geodesics.) Thus*

(E is P-W-convex) $\Longrightarrow$(E is strictly plurisubharmonic and exhausting) $\Longrightarrow$($\mathcal{T}(M)$ is a Stein manifold).

As we know from Hopf–Rinow theorem if a Riemann manifold is complete it is also *geodesically* complete: every geodesic can be infinitely continued. True, $\mathcal{T}(M)$ is not P-W-complete, but we have a sort of remedy in form of the beautiful Wolpert theorem for which the Riemannian method gives simpler proof.

THEOREM (WOLPERT). *Any two points in Teichmüller space $\mathcal{T}_p$ can be joined by a unique W-P-geodesics.*

PROOF (TROMBA). Since E is geodesically convex and proper, any two points in compact balls $E^{-1}[0,r]$, $r > 0$ can be joined by a unique geodesic. Since $\mathcal{T}_p = \bigcup_{r>0} E^{-1}[0,r]$, the result follows.

PROPOSITION. *E has a unique minimum and this is the only critical point.*

PROOF. That E has a minimum follows from its properness.

(Strict P-W-convexity)$\Longrightarrow$(all critical points are non-degenerate minima).

Should E have more than one critical point, then, by Morse theory, there has to be another critical point which is not a minimum; contradiction. □

REMARK. This argument is called the *mountain point principle* and it follows from Morse inequalities.

The non-compactness of the P-W-metric has an interesting consequence, the so called

Thurston earthquakes: if a geodesics in $\mathcal{T}_p(M)$ is defined for only finite interval of time it must crash into a surface of convergence. This phenomenon can be described in the following way.

THEOREM. *Let the class* $[g] \in \mathcal{T}(M)$ *and let* $c(t,[g]), t^-[g] < t < t^+[g]$ *be a P-W geodesic in* $\mathcal{T}(M)$ *with initial point* $[g] = c(0,[g])$, *where* $(t^-[g], t^+[g])$ *is the maximal domain of definition. If* $t^+[g] < \infty$, *then for any sequence* $t_n \to t^+[g]$, *a non-trivial geodesics on the surface* M, $c(t_n[g])$ *is shrinking in length to zero. In such a case one says that* $(M, c(t_n,[g]))$ *is developing a node.*

In order to be able to formulate the beautiful theorems of Wolpert and Tromba on upper bounds of the P-W curvatures of $\mathcal{T}_p$, we have to tell about

4.1 Hermitian hyperbolic geometry of Kobayashi

To start with, we have to introduce some fundamental notions in hermitian geometry.

As we know the most important simply connected Riemann surface is the disc $\mathbb{D} = \mathbb{D}_1 \simeq \mathfrak{H}_1$ with the Poincaré metric

$$ds^2_{\mathbb{D}} = \frac{4|dz|^2}{(1-|z|^2)^2} = \left(\frac{2|dz|}{(1-|z|^2)}\right)^2.$$

The Gaussian curvature $K^{\mathbb{D}} = -1$. For the disc $\mathbb{D}_a := \{z \in \mathbb{C} : |z| < a\}$, we have

$$ds^2_{\mathbb{D}_a} = \frac{4a^2|dz|^2}{A(a^2-|z|^2)^2}, \quad A > 0,$$

and $K^{\mathbb{D}_a} = -A$. Therefore we have distance in $\mathbb{D}$ given by

$$\rho(z_1, z_2) := \inf_c \int_c \frac{2|dz|}{1-|z|^2}.$$

We have the following fundamental corollary of Schwarz lemma (in Pick formulation.)

Every holomorphic map $f : \mathbb{D} \to \mathbb{D}$ *satisfies*

1. $\rho(f(z_1), f(z_2)) \leqslant \rho(z_1, z_2)$, $z_1, z_2 \in \mathcal{D}$. *Hence*
2. *every biholomorphism (automorphism) of* $\mathbb{D}$ *is a* ρ*-isometry.*

In general, the Gaussian curvature $K(h)$ of the Riemannian metric $h(z)^2|dz|^2$, $h(z) > 0$ is given by

$$K(h) = -\frac{4}{h^2} d'd'' \log h.$$

The foundation and the source of hyperbolic geometry is the Ahlfor's lemma which, on the first sight, seems to be only a mild modification if the Schwarz lemma.

AHLFORS LEMMA. *If M is a one dimensional Kähler manifold with the metric ds_M^2 whose Gaussian curvature K^M is bounded by the negative constant $-B$,*

$$K^M \leqslant -B < 0,$$

then every holomorphic $f : \mathbb{D}_a \to M$ satisfies

$$f^* ds_M^2 \leqslant \frac{A}{B} ds_{\mathbb{D}_a}^2.$$

COROLLARY (SCHWARZ–PICK LEMMA). If $f : \mathbb{D} \to \mathbb{D}$ is holomorphic, then

$$f^* ds_{\mathbb{D}}^2 \leqslant ds_{\mathbb{D}}^2.$$

PROOF. In Ahlfors lemma put $M = \mathbb{D}$, $a = 1$, $A = B = 1$. □

We have another formulation of Ahlfors lemma.

AHLFORS LEMMA. *Given a pseudometric $2h(z)dzd\bar{z}$ on $\mathbb{D}$ whose Gaussian curvature is $k(h) \leqslant -1$,*

$$h(z) \leqslant 4(1 - |z|^2)^{-2}.$$

There is a natural generalization of the Ahlfors lemma to arbitrary dimension due to Kobayashi.

THEOREM (KOBAYASHI). *Let M be m dimensional hermitian manifold whose holomorphic sectional curvature $K_s^M \leqslant -B < 0$ (B is constant.) Then for every holomorphic $f : \mathbb{D}_a \to M$ we have*

$$f^* ds_M^2 \leqslant \frac{A}{B} ds_{\mathbb{D}_a}^2.$$

We prove only Ahlfors lemma since the proof of Kobayashi theorem is the same if one makes use of the inequality for sectional curvatures of a submanifold $M' \subset M$

$$K_s^{M'} \leqslant K_s^M$$

which will be proved below.

PROOF OF THE AHLFORS LEMMA. Let u be a non negative function on $\mathbb{D}_a$ defined by $f^*(ds_M^2) := u ds^2_{\mathbb{D}_a}$. We are done if we prove that $u \leqslant \frac{A}{B}$.

Let us consider two situations:

1. u may attain its maximum on $\partial\mathbb{D}_a$, but we can reduce this case to the case when u attains its maximum in $\mathbb{D}_a$ as follows. Let $0 < t < a$ and denote by ds_t^2 the metric on $\mathbb{D}_t$. Let $z_0 \in \mathbb{D}_a$; we can choose such t that $z_0 \in \mathbb{D}_t$. Define $u_t \geqslant 0$ by $f^*(ds_M^2) := u_t ds_t^2$. Then $u_t(z_0) \to u(z_0)$ as $t \to a$. It is therefore sufficient to prove that $u_t \leqslant A/B$ on $\mathbb{D}_t$. If $hdzd\bar{z} = f^*(ds_M^2)$, then h is bounded on the compact set $\tilde{\mathbb{D}}$, the closure of $\mathbb{D}_t$. But

$$\frac{4t^2}{A(t^2 - |z|^2)^2} \to \infty$$

on $\mathbb{D}_t$, hence u_t approaches 0 on $\partial\mathbb{D}_t$ and attains its maximum in $\mathbb{D}$.

2. Let u attain its maximum at $z_0 \in \mathbb{D}_a$. If $u(z_0) = 0$, then $u \equiv 0$ and we are done. Assume that $u(z_0) > 0$; hence f is biholomorphic in an open neighborhood $\mathcal{O}$ of z_0. We can use the coordinate map z of $\mathbb{D}_a$ as a local coordinate in $f(\mathcal{O})$. If we write $ds_M^2 = 2hdzd\bar{z}$ on $f(\mathcal{O})$, then $f^*(ds_M^2) = 2hdzd\bar{z}$ on $\mathcal{O}$. If we write $ds^2_{\mathbb{D}_a} = 2gdzd\bar{z}$, then $u = h/g$, and the Gaussian curvatures are

$$K^M = -\frac{1}{h} d'd'' \log h, \quad K^{\mathbb{D}_a} = -A \equiv \frac{1}{g} d'd'' \log g.$$

But $K^M \leqslant -B$ (by assumption) and

$$d'd'' \log u = d'd'' \log h - d'd'' \log g = -K^M h - Ag \geqslant Bh - Ag.$$

Since u attains its maximum at z_0, $d'd'' \log u \leqslant 0$. Therefore

$$(0 \geqslant Bh - Ag) \Longrightarrow A/B \geqslant h/g = u(z_0) \geqslant u(z).$$

□

PROOF OF THE KOBAYASHI THEOREM is the same as of the Ahlfors lemma: we can assume that u defined by $u ds^2_{\mathbb{D}_a} := f^*(ds_m^2)$ attains its maximum at $z_0 \in \mathbb{D}_a$ and that $u(z_0) > 0$. Then f is non degenerate in some $\mathcal{O} \in z_0$. Thus $f(\mathcal{O})$ is a one dimensional *submanifold* of M; its sectional curvature, which is a Gaussian curvature, $\leqslant -B < 0$ (by the theorem on sectional holomorphic curvature of a submanifold.) The rest of the proof is the same as in the case of Ahlfors lemma. □

Now we have to define

Ricci curvature and holomorphic sectional curvatures. If $\pi : E \to M$ is a holomorphic vector bundle of rank r over complex manifold M of dimension m, and g is a hermitian inner product on E, then for a basis of local holomorphic sections of E, $e_1, \ldots, e_r$, $g_{\alpha\bar{\beta}} := g(e_\alpha, \bar{e}_\beta)$. Let ∇ be a hermitian connection on $E \to M$, and $R = R^\nabla$, the curvature of ∇. The components $K^\alpha_{\beta i\bar{j}}$ of the curvature R in the local coordinate system $z^1, \ldots, z^m$ of M and the corresponding vector fields $Z_i := \frac{\partial}{\partial z^i}$, $\bar{Z}_j := \frac{\partial}{\partial \bar{z}^i}$ are given by

$$K^\alpha_{\beta ij} e_\alpha = R(Z_i, Z_j) e_\beta.$$

The Ricci curvature *Ric* is given by

$$K_{i\bar{j}} = K^\alpha_{\alpha i\bar{j}} = -\frac{\partial^2 \log G}{\partial z^i \partial \bar{z}^j}, \quad G := \det g.$$

Given an element s^α_x of $E_x = \pi^{-1}(x)$, $x \in M$, we consider the hermitian form at $x \in M$

$$K_s := -K_{\alpha\bar{\beta}i\bar{j}} s^\alpha \bar{s}^\beta dz^i d\bar{z}^j, \quad \text{where } K_{\alpha\bar{\beta}i\bar{j}} := \frac{\partial^2 g_{\alpha\bar{\beta}}}{\partial z^i \partial \bar{z}^j}.$$

Now we are ready to present an important

DEFINITION. 1. If K_s is negative (positive) definite for every non zero s, then the hermitian bundle $\pi : E \to M$ is said to have *negative* (*positive*) curvature.

2. If the form $K_{i\bar{j}} dz^i d\bar{z}^j$ is negative (positive), then the hermitian bundle $\pi : E \to M$ is said to have *negative* (*positive*) Ricci curvature.

We have very useful

THEOREM (KOBAYASHI–GOLDBERG). *If $E' \to M$ is a holomorphic subbundle of $E \to M$, that is, if $E'_x \subset E_x$ is a subspace for every x, then*

$K'_s \leqslant K_s$, that is, $K_s - K'_s$ is a positive semidefinite form.

PROOF. We can take orthonormal local coordinates $e_1, \ldots, e_r$ in $E \to M$ at x (thus $g(e_\alpha, \bar{e}_\beta) = \delta_{\alpha\beta}$) such that $e_1, \ldots, e_{r'}$ are sections of E' at x. Hence

$$-(K'_{\alpha\bar{\beta}i\bar{j}})_x = -(K_{\alpha\bar{\beta}i\bar{j}})_x - \sum_{\rho=r'+1}^{r} \left(\frac{\partial g_{\alpha\rho}}{\partial z^i} \frac{\partial g_{\rho\bar{\beta}}}{\partial \bar{z}^j} \right)_x$$

□

COROLLARY. *If $M' \subset M$ is a holomorphic submanifold and M has* negative *holomorphic sectional curvature, then M' also has negative sectional curvature.*

4.2 Hyperbolic complex analysis

This term means the study of holomorphic mappings into *negatively* curved hermitian manifolds M. For the tangent bundle $E := TM \to M$ one introduces *holomorphic bisectional curvature* as follows: take any two unit tangent vectors $s, t \in TM$, then

$$K_s(t, \bar{t}) := -K_{i\bar{j}k\bar{l}} s^i \bar{s}^j t^i \bar{t}^j$$

is called holomorphic bisectional curvature determined by s and t. In particular $K_s(s, \bar{s})$ is called *holomorphic sectional curvature* of M *determined by* s and is denoted by K_s^M or $K^M(s)$.

DEFINITION. A hermitian manifold is *negatively curved* if $K^M(s) \leqslant -A < 0$ for some positive constant A.

EXAMPLE. Of course the disc $\mathbb{D}$ is negatively curved with $A = 1$. Every Riemann surface with $p > 1$ is negatively curved as well.

Sometimes negatively curved complex manifolds are called *hyperbolic.* But now, after fundamental studies of Kobayashi and his school hyperbolicity is an attribute of a wider class of hermitian manifolds, namely such that the Kobayashi pseudodistance is a distance. What we are interested in is

4.3 Hyperbolicity of the Teichmüller space

In 1960 Lars Ahlfors proved that Petersson–Weil metric is Kählerian. Shortly afterwards, in 1961, he proved

THEOREM (AHLFORS). *1. The Ricci curvature of P–W metric is negative.*

2. The holomorphic bisectional curvature and the scalar curvature of $\mathcal{T}_0$ are both negative.

Much sharper results were obtained (by other methods) independently by Wolpert and Tromba in 1986.

THEOREM (TROMBA, WOLPERT). *1. The holomorphic (bi)sectional curvature of Teichmüller space is bounded from above by*

$$K_s^{\mathcal{T}_p} \leqslant -\frac{1}{2\pi(p-1)} < 0 \quad \textit{since } p > 1\ .$$

2. The same bound holds for the Ricci curvature $Ric^{\mathcal{T}_p}$.

COROLLARY. *The Teichmüller space $\mathcal{T}_p$ is hyperbolic.*

4.4 Kobayashi pseudodistance. Kobayashi hyperbolic spaces

A generalization of Poincaré metric to higher dimensional manifolds was introduced by Constantin Carathéodory already in 1932. But more elastic pseudodistance was introduced by Soshichi Kobayashi in 1967 in the following way:

Let M be a complex manifold. Given two points $x, y \in M$, we set $d_M^1(x, y) := \inf \rho(a, b)$, where the infimum is taken over all points $a, b \in \mathbb{D}$ such that there exists a holomorphic $h : \mathbb{D} \to M$ with $h(a) = x$, $h(b) = y$; ρ denotes the Poincaré distance in $\mathbb{D}$. For any positive integer n put

$$d_M^n(x, y) := \inf \sum_{i=1}^{n} d_M^1(x_{i-1}, x_i),$$

where the infimum is taken over all points $x_0, \ldots, x_n \in M$ with $x_0 = x$, $x_n = y$. Clearly

$$d_M^{n+1}(x, y) \leqslant d_M^n(x, y), \quad x, y \in M.$$

DEFINITION. The Kobayashi pseudodistance d_K^M (or simply d_K if we consider only one M) is defined by

$$d_K^M(x, y) := \lim_{n\to\infty} d_M^n(x, y).$$

It is obvious that $d_K^M : M \times M \to \mathbb{R}$ is continuous and has all properties of pseudodistance:

$d_K^M(x, y) \geqslant 0$

$d_K^M(x, y) = d_K^M(y, x)$ (symmetry)

$d_K^M(x, z) \leqslant d_K^M(x, y) + d_K^M(y, z)$ (triangle inequality.)

If $d_K^M(x, y)$ is a distance, that is, if $(d_K^M(x, y) = 0) \Longleftrightarrow (x = y)$, then d_K^M is called *Kobayashi distance* and (M, d_K^M) is called *hyperbolic manifold.* We have obvious facts.

EXAMPLE 1. $d_K^{\mathbb{D}} = \rho$ – Poincaré distance.

EXAMPLE 2. $d_K^{\mathbb{C}} \equiv 0$ since bounded holomorphic functions are constant (Liouville theorem.)

EXAMPLE 3. $(M_1, M_2$ are hyperbolic$) \Longrightarrow (M_1 \times M_2$ is hyperbolic with product metric.)

Kobayashi distance is invariant with respect to $\mathrm{Aut}(M)$ and has the *distance decreasing property*:

PROPOSITION. *If M and N are complex manifolds and $f : M \to N$ is holomorphic, then*

$$d_K^N(f(x_1), f(x_2)) \leqslant d_K^M(x_1, x_2), \quad x_1, x_2 \in M.$$

In particular, if M is hyperbolic (and thus d_K^M is a distance) then any biholomorphic $f \in \mathrm{Aut}(M)$ is an isometry.

The reader may be willing to prove some of the following properties of the Kobayashi distance.

THEOREM (KOBAYASHI). *1. Negatively curved hermitian manifold is hyperbolic (in the sense of Kobayashi.)*

2. If $\tilde{M} \to M$ is an (unbranched) covering of a complex manifold M, then

(M is hyperbolic)$\Longleftrightarrow$($\tilde{M}$ is hyperbolic).

Therefore all Riemann surfaces M_p of genus $p >$ are hyperbolic since the universal covering $\tilde{M}_p \simeq \mathcal{D}$.

3. The following generalization of Picard theorem holds

Let M be a hyperbolic manifold; then each holomorphic map $f : \mathbb{C}^n \to M$ is constant.

PROOF OF 3. Since $d_K^{\mathbb{C}^n} \equiv 0$ we have $d_K^M(f(x), f(y)) = d_K^{\mathbb{C}^n}(x, y) = 0$ and thus $f(x) = f(y)$.

As we know Klein and Poincaré proved that the automorphism group $\mathrm{Aut}(M)$ of a compact Riemann surface of genus $p > 0$ is finite. We have a corresponding fact for hyperbolic manifolds.

THEOREM (KOBAYASHI). *For compact hyperbolic M the group* $\mathrm{Aut}(M)$ *is finite.*

Submanifolds. Let X be a complex submanifold of Y or let $f : X \to Y$ be holomorphic and injective. Then
(X is hyperbolic)$\Longrightarrow$(Y is hyperbolic).
Therefore, since the discs $\mathbb{D}_r$ are hyperbolic,
1. Polidiscs are hyperbolic;
2. Bounded domains in $\mathbb{C}^n$ as subsets of products of $\mathbb{D}^K$ are hyperbolic.
Recently hyperbolic manifolds play an important role in number theory.

Let us return to Teichmüller theory. We have encountered two interesting metrics on Teichmüller space: the original metric of Teichmüller d_T which is Finslerian and complete, but not hermitian, and the P–W metric which is Kählerian, but not complete. Now we have constructed the third invariant metric on $\mathcal{T}_p$, the Kobayashi metric which, we denoted d_K. What are relations between these metrics? The answer is given by Royden (1971–74), and now we will talk about these deep results.

4.5 Invariant metrics of Teichmüller space

Perhaps the most surprising discovery of Royden (which would certainly please Teichmüller) is the following.

THEOREM (ROYDEN). *The Teichmüller metric d_T coincides with the Kobayashi metric d_K, $d_T = d_K$. Hence the Teichmüller metric is characterized by*

$$d_T(x, y) = \inf \rho(x, y),$$

where the infimum is taken over all holomorphic maps $h : \mathbb{D} \to \mathcal{T}_p$ with $h(a) = x$, $h(b) = y$, $a, b \in \mathbb{D}$, $x, y \in \mathcal{T}_p$

We saw that the Teichmüller modular group Γ_p acts on $\mathcal{T}_p$ as a group of d_T isometries. But conversely

THEOREM (ROYDEN, 1971). *1. Any biholomorphic map between domains in $\mathcal{T}_p$ which is a d_T isometry is introduced by an element of Γ_p.*

2. Any biholomorphic map $f : \mathcal{T}_p \to \mathcal{T}_p$ is induced by an element of the modular group Γ_p. For $p > 2$, $\mathrm{Aut}(\mathcal{T}_p) = \Gamma_p$. But for $p = 2$ we have

3. $\mathrm{Aut}(\mathcal{T}_p) = \Gamma_p/\mathbb{Z}_2$.

The following observation is due to Georg Schumacher.

PROPOSITION (SCHUMACHER 1990). *If $p > 2$ the automorphism group of the modular space $\mathcal{M}_p$ consists only of the identity* $\mathrm{Aut}(\mathcal{M}_p) = \mathrm{id}$.

REMARK. For $p = 2$ the statement does not hold since $\mathcal{M}_2 = \mathbb{C}^3/\mathbb{Z}_5$ has many automorphisms.

In the Bers approach to Teichmüller theory, the following objects are of great importance.

4.6 Harmonic Beltrami differentials on (M, g)

Let the metric g be represented $\lambda^2 dz d\bar{z}$ and $\psi \in \mathcal{Q}(g)$ be a holomorphic quadratic differential. Then, since $\bar{\psi}_z = 0$

$$\mu := \frac{\bar{\psi}}{\lambda^2}$$

satisfies Beltrami equation

$$(B) \qquad \mu_z + \frac{2\lambda_z}{\lambda}\mu = 0.$$

Solutions of (B) are called *harmonic Beltrami differentials* and the space of these differentials is denoted by $H(g)$. Bers proved that the tangent space at $[g]$

$$T_{[g]}\mathcal{T} = H(g).$$

In Riemannian approach to Teichmüller theory this fact is almost obvious since the space $H(g)$ and $\mathcal{Q}(g)$ are mutually dual and the dual pairing is given by

$$\mathcal{Q}(g) \times H(g) \ni (\varphi, \mu) \to$$

$$(D) \qquad \to <\varphi, \nu> = \int_M \varphi\psi = \int \varphi \frac{\bar{\psi}}{\lambda^2}\lambda^2 dz \otimes d\bar{z}.$$

This pairing corresponds to the following Serre pairing:

If Θ_M is the sheaf of germs of holomorphic vector fields on M and $H^1(M, \Theta_M)$ is the first Dolbeault space of cohomology classes of d'' closed $(0,1)$ forms with values in the holomorphic tangent bundle of M. Then every class in $H^1(M, \Theta_M)$ is, by virtue of Hodge theorem, represented by a single harmonic representant $H^1(M, \Theta_M) = H(g)$: a harmonic Beltrami differential and the pairing (D) is the Serre duality

$$(S) \qquad H^0(M, K_M^{\otimes 2}) \times H^1(M, \Theta_M) \to \mathbb{C}.$$

Since $\mathcal{Q}(g)$ is the *cotangent* space of $\mathcal{T}(M)$ at $[g]$, $H(g)$ is the tangent space $T_{[g]}\mathcal{T}(M)$ and Bers theorem is proved.

The first cohomology group $H^1(M, \Theta_M)$ is called the *space of infinitesimal deformations (of complex structures)* of M. Let (M, c) be a complex structure c of a Riemann surface M given by the atlas $c : \{\mathcal{U}_j, z_j\}$, where $D_j = z_j(\mathcal{U}_j)$ are domains in $\mathbb{C}$. The identification of D_j and D_k is given by the biholomorphic map from the open set $D_{kj} = z_k(\mathcal{U}_j \cap \mathcal{U}_k) \subset D_k$ onto $D_{jk} = z_j(\mathcal{U}_j \cap \mathcal{U}_k) \subset D_j$.

A *deformation* (M_1, c_1) of (M, c) is regarded as a gluing of some domains D_j by different identification $f_{jk}(\cdot, s)$ being a biholomorphic map from D_{kj} onto D_{jk} with the parameter $s = (s_1, \ldots, s_n)$ such that $f_{jk}(z_k, 0) = z_{jk}(z_k)$. If all f_{jk} are holomorphic (or C^∞ functions), we have holomorphic (resp. differentiable) family $S \ni s \to (M_s, c_s)$ of Riemann surfaces. This notion was introduced by Teichmüller in his posthumous paper of 1944.

We consider now *infinitesimal* deformation M_s, and for simplicity (of writing) we consider only one parameter deformations: $n = 1$. Take the derivative of $f_{jk}(z_k, t)$ with respect to s at $s = 0$. We can regard it as a holomorphic vector field on $\mathcal{U}_j \cap \mathcal{U}_k$ which we denote by

$$\vartheta_{jk} = \frac{\partial f_{jk}(z_k, 0)}{\partial s}\frac{\partial}{\partial z_j}, \quad z_k = z_{kj}(z_j).$$

Since on $\mathcal{U}_j \cap \mathcal{U}_k$ we have $f_{jk}(f_{kj}(z_l, s), s)) = z_j$,

$$\vartheta_{jk} + \vartheta_{kj} = 0 \quad \text{on } \mathcal{U}_j \cap \mathcal{U}_k.$$

Similarly, the relation $f_{jl}(z_l, s) = f_{jk}(f_{kl}(z_l, s), s))$ on $\mathcal{U}_j \cap \mathcal{U}_k \cap \mathcal{U}_l$ gives

$$\vartheta_{jk} + \vartheta_{kl} + \vartheta_{lj} = 0 \quad \text{on } \mathcal{U}_j \cap \mathcal{U}_k \cap \mathcal{U}_l.$$

Therefore $\vartheta = \{\vartheta_{jk}\}$ is a cocycle and defines an element $[\vartheta] \in H^1(M, \Theta_M)$, the first cohomology group with coefficients in Θ_M, the sheaf of germs of holomorphic vector fields on the Riemann surface M.

We see that $[\vartheta]$ represents the derivative of the complex structure c_s of the (M_s, c_s) with respect to s at $s = 0$.

Of course, the notion of complex holomorphic family $\mathcal{X} \to X_s$ of complex manifolds of arbitrary dimension is introduced in the same way.

The theory of deformations of complex structures is a huge and difficult domain of complex analysis which was developed by Kodaira, Spencer, Kuranishi, Grauert, Nirenberg and many others. In this theory an important role is played by the Kodaira–Spencer map $\rho_{\mathcal{H}} \equiv \rho : T_0(S) \to H^1(M, \Theta_X)$ which we have constructed for the case of compact Riemann surfaces.

The existence of deformations of complex structure for compact complex manifold $X = X_0$ is difficult problem. We will present a solution due to Kodaira, Nirenberg, and Spencer (1958) based on the theory of elliptic differential equations. This result was independently obtained by Grauert who used entirely different approach.

Let (S, s_0) be a complex manifold of dimension m (with the base point s_0 which, for simplicity, will be denoted by 0 below, and X a complex manifold of dimension n. Let $s = (s_1, \ldots, s_m)$ be complex coordinates around 0. Denote by $\mathcal{X} = \{X_s : s \in S\}$ a smooth family of compact complex manifolds X_s which are diffeomorphic to $X \simeq X_0$. We sawthat $\mathcal{X}$ can be defined (or understand) as a family of complex structures c_s defined on the same differentiable manifold X.

For any tangent vector v of S at 0 $(= s_0)$, $v = \sum_{\alpha=1}^{m} v^\alpha \frac{\partial}{\partial s_\alpha} \in T_0S$ we constructed

$$v\varphi(0) = \sum_{\alpha=1}^{m} v^\alpha \frac{\partial \varphi(s)}{\partial s_\alpha}$$

where $\varphi(s)$ is a family of $(0,1)$ forms

$$\varphi_j^\beta(z, s) = \sum_{\nu=1}^{n} \varphi_{j\bar{\nu}}^\beta(z, s) d\bar{z}^\nu, \quad \text{on } \mathcal{U}_j \subset X,$$

where $\{\mathcal{U}_j\}$ is a (locally finite) covering of X. Denote by $\rho_{\mathcal{X}}(v) \in H_{d''}^{0,1}$ the d'' cohomology class of $v\varphi(0)$. In view of the Dolbeaut isomorphism $H_{d''}^{0,1}$ can be canonically identified with $H^1(X, \Theta)$, where $\Theta = \Theta_X$ is the sheaf over X of germs of holomorphic vector fields.

Thus we have the linear map

$$\rho_{\mathcal{X}} : T_0(S) \to H^1(X, \Theta).$$

called the Kodaira–Spencer map.

The following important theorem gives sufficient condition for existence of a family of deformations $c_s = (X_s, c_s)$, $s \in S$.

THEOREM (KODAIRA–NIRENBERG–SPENCER, GRAUERT 1957). *Let X be a complex manifold and $\Theta = \Theta_X$ the sheaf of (germs of) holomorphic vector fields on X. If $H^2(X, \Theta) = 0$, then there exists a smooth family $\mathcal{X} = \{(X_s, c_s) : s \in S\}$ of deformations X_s of X such that the Kodaira–Spencer map $\rho_{\mathcal{X}} : T_0(S) \to H^1(X, \Theta)$ is an isomorphism.*

In order to be able to formulate theorems of the theory of deformations of complex structures, we must introduce a number of new notions. We can look at the holomorphic (smooth) family $\mathcal{X} = \{X_s, s \in S\}$ of compact manifolds, called the deformation of X, considering, for example, the triple (X, π, S), where $\pi : X \to S$ is a holomorphic, surjective map whose fibers $X_s := \pi^{-1}(s)$ are our manifolds X_s. We say that the family is over S or is parametrized by S.

If $f : S' \to S$ is a holomorphic map and $\mathcal{X} = (X, \pi, S)$ is any family, then the fibered product $X \times_S S' =: X'$ defined by

$$X \times_S S' = \{(x, s') \in X \times S' : \pi(x) = f(s')\}$$

is, in a natural way, a family $\mathcal{X}' = (X', \pi', S')$ over S':

$$\pi' : X' \to S', \quad (x, s') \to s',$$

and is called the *pullback of $\mathcal{X}$ by f*.

If $\mathcal{X}_i = (X_i, \pi_i, S_i)$, $i = 1, 2$ is a pair of families, then a (homo)morphism from $\mathcal{X}_1$ to $\mathcal{X}_2$ consists of a pair of holomorphic maps $G : X_1 \to X_2$ and $f : S_1 \to S_2$ such that $\pi_2 g = f \pi_1$. There is always a homomorphism from the pullback to the original family. There is a theorem due to Grauert and Fischer on local triviality of families which tells that

THEOREM (GRAUERT–FISCHER). *(A smooth family of compact complex manifolds is locally trivial)$\Longleftrightarrow$(all fibers are locally isomorphic).*

One considers germs of deformations identifying two deformations over (S, s_0) if they coincide in a neighborhood of s_0.

DEFINITION. A deformation $\mathcal{X} = (X, \pi, S)$ of $V \equiv X_{s_0}$ with base point is *(locally) complete* if (locally) every deformation $\mathcal{X}' = (X', \pi', S')$ of V with base point s'_0 is obtained as a pullback of $\mathcal{X}$ by a suitable holomorphic map $f : S' \to S$ with $f(s'_0) = s_0$. If f is always *uniquely* determined by $\mathcal{X}'$, the deformation is *(locally) universal.* A deformation $\mathcal{X} = (X, \pi, S)$ of V with base point s_0 is called *versal* if for any $\mathcal{X}'$ as above there is an f whose derivative $df(s'_0)$ is uniquely determined.

REMARK. Universality of a deformation is too strong a property: there are many manifolds V which have *no* universal deformation. Therefore the weaker property, versality, was introduced. But one has to allow for bases S with singularities: S should be only a *complex space.* The main theorem of deformation theory due to Kuranishi asserts the existence of versal deformations for compact complex spaces. We formulate this theorem for manifolds.

THEOREM (KURANISHI). *Every compact complex manifold has a versal deformation. Such family is called* Kuranishi family.

REMARK. Generalization of this theorem to complex spaces is due to Grauert; cf. Part 'Riemann and Complex Geometry'.

Kuranishi proved his magnificent theorem first in 1962, and then a simpler and improved version in 1964. That last proof uses the Kodaira–Nirenberg–Spencer result and strong elliptic systems of differential equations.

The Kodaira–Nirenberg–Spencer theorem can now be formulated in the following way.

If $H^2(V, \Theta_V) = 0$, then V admits smooth versal deformations.

There is a characterization of smooth families.

THEOREM (KODAIRA–SPENCER, 1958). *(A smooth family is complete)* $\Longleftrightarrow$ *(The Kodaira–Spencer map is surjective).*

Let us return to Teichmüller theory.

DEFINITION. The *universal Teichmüller curve* is the fiber bundle π :

$\tilde{\mathcal{T}}_p \to \mathcal{T}_p$ over $\mathcal{T}_p$, where the fiber over $g \in \mathcal{T}_p$ is conformally equivalent to (M, g).

THEOREM. *The universal Teichmüller curve* $\tilde{\mathcal{T}}$ *is a complex manifold of dimension* $\dim_{\mathbb{C}} \tilde{\mathcal{T}} = 3p - 2$. *In the chart* $\mathcal{Q}(g)$ *the fiber over* γ *can be identified with the fiber over* g *via the harmonic map*

$$u(g, \gamma) : (M, g) \to (M, \gamma).$$

Now we take $\mathcal{X} = \tilde{\mathcal{T}}$ and $S = \mathcal{T} = \mathcal{T}(M)$. The basis of the smooth family (X, π, S) of deformations of the Riemann surface M is, of course, the Teichmüller space $(\mathcal{T}(M), 0)$ itself with the base point $s_0 = 0$. As we know, the space $H^1(M, \Theta_M)$ of infinitesimal deformations of (M, g) can be identified with the space $H(g)$ of harmonic Beltrami differentials, which in turn can be identified with the tangent space $T_0(\mathcal{T}(M))$ of the Teichmüller space $\mathcal{T}$ at the base point $0 = [g]$. With these identifications, the Kodaira–Spencer map ρ is an isomorphism and we have

THEOREM. *The Kodaira–Spencer map* $\rho_{\tilde{\mathcal{T}}} = \rho : T_0(\mathcal{T}(M)) \to H^1(M, \Theta_M)$ *is bijective.*

Recall that on $\mathcal{T}(M)$ we introduced the complex structure as multiplication by $\sqrt{-1}$ in each $\mathcal{Q}(g)$, the cotangent space at g. This complex structure on $\mathcal{T}_p$ induced a complex structure on $\mathcal{T}_p \times \mathfrak{H}_1$ (where $\mathfrak{H}_1$ is the upper half plane) with the following interesting property. By uniformization theorem, if $g \in \mathcal{T}_p$ and if (M, g) is regarded as the quotient space $\mathfrak{H}_1/\Gamma(g)$, where $\Gamma(g)$ is a discrete subgroup of $\mathrm{Aut}(\mathfrak{H}_1)$, then multiplication by $\sqrt{-1}$ on $\mathcal{Q}(g) \times \mathfrak{H}_1$ is $\Gamma(g)$ invariant.

4.7 Wolpert formulas for Petersson–Weil form

To be able to present the beautiful Wolpert formula for Kählerian P–W form of the P–W metric and corresponding formulas for smooth families of compact Kähler, we have to introduce an operation which is of great importance in de Rham cohomology.

Fiber integral. Let $f : E \to B$ be a smooth fiber bundle with compact typical fiber of dimension n. There is a canonical homomorphism denoted by $\int_{E/B}$ or $f_* : A^l_{cv}(E) \to A^{l-n}_{cv}(B)$ such that for all l-forms α with compact support in vertical direction, that is, such that the restriction to each fiber $f^{-1}b$ has compact support,

$$\int_E \alpha \wedge f^*\beta = \int_B \left(\int_{E/B} \alpha \right) \wedge \beta$$

or, in other notation,

$$\int_E \alpha \wedge f^*\beta = \int_B f_*\alpha \wedge \beta$$

for all differential forms on B.

The map $\int_{E/B}$ is called the fiber integral (integration along the fiber) commutes with exterior differentiation

$$d_B \int_{E/B} \alpha = (-1)^n \int_{E/B} d_E \alpha.$$

This important formula shows that the fiber integration descends to the cohomology $H^\bullet_{cv}(E)$ and we can use the same notation

$$F_* \equiv \int_{E/B} : H^\bullet_{cv}(E) \to H^{\bullet - n}_{cv}(B).$$

We have an analog of the Poincaré lemma for cv-supports.

PROPOSITION (POINCARÉ LEMMA FOR VERTICAL SUPPORTS). *Integration along the fiber defines an isomorphism*

$$f_* H^*_{cv}(B \times \mathbb{R}^n) \tilde{\to} H^{*-n}(B).$$

This is a special case of

THEOREM (THOM ISOMORPHISM). *If the vector bundle $f : E \to B$ over manifold B of finite type is orientable, then*

$$H^*_{cv}(E) \simeq H^{*-n},$$

where n is the rank of E. This isomorphism is given by $\int_{E/B}$.

REMARK. The manifold M of dimension n has a *good cover* if all finite intersections $\mathcal{U}_{\alpha_0} \cap \ldots \cap \mathcal{U}_{\alpha_k}$ are diffeomorphic to $\mathbb{R}^n$. A manifold which has a *finite* good cover is said to be of *finite type*. Clearly every manifold has a good cover. Every compact manifold is of finite type.

Let us recall for a Kähler manifold (X, g) the Kähler form ω_X in local holomorphic coordinates $(z^1, \ldots, z^n)$ can be written as

$$\omega_X = \omega = \sqrt{-1} g_{\alpha\bar{\beta}} dz^\alpha \wedge d\bar{z}^{\bar{\beta}}.$$

If $R_{\alpha\bar{\beta}}$ is the Ricci tensor, then

$$Ric(\omega_X) = -\sqrt{-1} d'd'' \log \det(g_{\alpha\bar{\beta}}) = \sqrt{-1} R_{\alpha\bar{\beta}} dz^\alpha \wedge d\bar{z}^{\beta}$$

and the first Chern form

$$c_1(X) = \frac{\sqrt{-1}}{2\pi} R_{\alpha\bar{\beta}} dz^\alpha \wedge d\bar{z}^{\bar{\beta}} = -\frac{\sqrt{-1}}{2\pi} d'd'' \log \det(g_{\alpha\bar{\beta}}).$$

In 1986–87 Scott Wolpert achieved a magnificent unification of algebraic geometry and analytic approach to Teichmüller theory. In his work the Petersson–Weil form ω_{PW} on $\mathcal{T}_p(M)$ plays a central role. Its class in $H^2(\mathcal{M}_p, \mathbb{R})$ generates this space, and $\frac{1}{2\pi^2}[\omega_{PW}]$ is rational and extends to an element of $H^2(\bar{\mathcal{M}}_p, \mathcal{Q})$, where $\bar{\mathcal{M}}_p$ is a compactification of the moduli space $\mathcal{M}_p$.

The P–W form extends as a $(1,1)$ current (of a corresponding line bundle) which is the Chern form of *continuous* hermitian metric on this bundle. The smoothing of the metric shows the positivity of this bundle on $\bar{\mathcal{M}}_p$. In this way Wolpert proved that $\bar{\mathcal{M}}_p$ is embedded in the complex projective space $\mathbb{P}^N(\mathbb{C})$. This was first proved by Knudsen and Mumford in 1976, of course, with the help of quite different methods (algebraic geometry.) Thus we have

THEOREM (KNUDSEN, MUMFORD, WOLPERT). *The compactification $\bar{\mathcal{M}}_p$ of the moduli space $\mathcal{M}_p$, $p > 1$ is a projective variety.*

Let $\pi : \tilde{\mathcal{T}}_p \to \mathcal{T}_p$ be a universal Teichmüller curve which in a natural way defines a hermitian line bundle. Denote by $c_1(\tilde{\mathcal{T}}_p/\mathcal{T}_p, g)$ its Chern form. Its square $c_1^2(\tilde{\mathcal{T}}_p/\mathcal{T}_p, g)$ is a $(2,2)$-form whose fiber integral

$$\int_{\tilde{\mathcal{T}}_p/\mathcal{T}_p} c_1^2(\tilde{\mathcal{T}}_p/\mathcal{T}_p, g) \in A^{1,1}(\mathcal{T}_p).$$

In 1986 Wolpert proved that this is the desired Petersson–Weil form on Teichmüller space.

THEOREM (WOLPERT).

$$(W1) \qquad \int_{\tilde{\mathcal{T}}_p/\mathcal{T}_p} c_1^2(\tilde{\mathcal{T}}_p/\mathcal{T}_p, g) = \omega_{PW}(\mathcal{T}_p).$$

REMARK. This construction of ω_{PW} descends to $\mathcal{M}_p = \mathcal{T}_p/\Gamma_p$.

The compactifying divisor $\mathcal{D} := \bar{\mathcal{M}}_p - \mathcal{M}_p$ is a union $\mathcal{D} = \mathcal{D}_0 \cup \cdots \cup \mathcal{D}_{[p/2]}$ where the generic points of $\mathcal{D}_0$ correspond to Riemann surfaces of genus $p-1$ and two punctures, and for $j > 0$, $\mathcal{D}_j$ are surfaces consisting of two components of genus j and $p-j$, respectively, with one puncture on each of them. This compactification is due to Mumford (1977) and Knudsen (1983).

The symplectic structure of $\mathcal{T}_p$ was expressed by Wolpert in a surprisingly simple way in terms of Frenckel–Nielsen coordinates.

THEOREM (WOLPERT). *Let $0 < l_j < \infty$, $-\infty < \tau_j < \infty$ for $j = 1, \ldots, 3(p-1)$ be the Frenchel–Nielsen coordinates on $\mathcal{T}_p(M)$. Then the Petersson–Weil form is*

$$(W2) \qquad \omega_{PW} = -\sum_{j=1}^{3(p-1)} d\tau_j \wedge dl_j.$$

Proof of this formula cannot be simple, of course. If $\{c_j\}$, $j = 1, \ldots, 3(p-1)$ is a system of decomposing curves on the Riemann surface M (they are simple closed geodesics) and $t \to l_j(t)$ is a geodesic length function on the geodesic $c_j : l_j = l_{c_j}$, we obtain

COROLLARY. *For every simple closed geodesic c on M*

$$\omega_{PW}\left(\frac{\partial}{\partial l_c}, \cdot\right) = d\tau_c.$$

It follows directly from $(W2)$ that π_{PW} possesses a differentiable extension to $\bar{\mathcal{M}}_p$. On the other hand, the Petersson–Weil forms singular near the compactifying divisor $\bar{\mathcal{M}}_p - \mathcal{M}_p$ with respect to the complex structure of the moduli space and the extension of ω_{PW} onto $\bar{\mathcal{M}}_p$ is of class $L^2(\bar{\mathcal{M}}_p)$.

4.8 Generalization to higher dimensions

Is it possible to obtain a generalization of Teichmüller theory to compact Kähler manifolds of arbitrary dimension? There are principal differences between $\dim_{\mathbb{C}} = 1$ and $\dim_{\mathbb{C}} > 1$. An important theory was created — and is still in progress — by Georg Schumacher and his Japanese collaborators. This theory is deeply connected with Quillen theory of determinant bundles. Schumacher and Fujiki in their huge work obtained a fiber integral formula of Wolpert type for Kähler–Einstein manifolds.

Let X be a Kähler manifold with Kähler form ω_X and Ricci flat form $Ric(\omega_X)$. The Kähler form is called *Kähler–Einstein* if

$$Ric(\omega_X) = k\omega_X,$$

where k is a real constant which can be normalized to $-1, 0, 1$. In the sequel we will assume that $k = -1$.

Generalized Petersson–Weil metric can be defined as follows. Let $(X, g_{\alpha\bar{\beta}})$ be a compact Kähler–Einstein manifold of constant non positive curvature. Let $f : \mathcal{X} \to (S, s_0)$ be universal deformation of X. By means of the Kodaira–Spencer map $\rho = \rho_{\mathcal{X}}$ we can identify the tangent space $T_{s_0}(S)$ of S at s_0 with the space $H^1(X, \Theta_X)$ of infinitesimal deformations; Θ_X is the sheaf of germs of vector fields.

According to the Calabi–Yau theorem, on every fiber $\mathcal{X}_1 = f^{-1}(1), 1 \in S$ of our family there exists a *unique* Kähler form $\omega_{\mathcal{X}_1}$ with $k = -1$. The relative volume forms $g(s)$ constitute a hermitian metric on the anti canonical bundle $K^{-1}_{\mathcal{X}/S}$. The $(2\pi/k)$-fold of the Chern form is (now) a real locally $d'd''$ exact $(1,1)$ form $\omega_{\mathcal{X}}$ on $\mathcal{X}$ of class C^∞. The Kähler– Einstein condition imply that all restrictions

$$\omega_{\mathcal{X}} | \mathcal{X}_1 = \omega_{\mathcal{X}_s}, \quad s \in S.$$

If n is the (complex) dimension of the family $f : \mathcal{X} \to S$, the $(n+1)$-fold power $c_1^{n+1}(\mathcal{X}/S, g)$ of the first Chern form c_1 of $K^{-1}_{\mathcal{X}/S}$ equipped with the metric $g(\cdot)$ is a real $(1,1)$-form (precisely as in the classical case $n = 1$.) We have the following magnificent generalization of the Wolpert formula (W).

THEOREM (FUJIKI–SCHUMACHER, 1988). *The generalized Petersson–Weil form on the base of the universal family $\mathcal{X} \to S$ of Kähler–Einstein manifold of Ricci curvature $k = -1$ can be represented as a fiber integral*

$$\omega_{PW} = -\frac{2\pi^{n+1}}{k^n(n+1)!}\int_{\mathcal{X}/S} c_1^{n+1}(\mathcal{X}/S, g).$$

In particular ω_{PW} possesses locally (also around singular poles) a $d'd''$ potential of class c^∞.

REMARK. For $n = 1$, X a Riemann surface of genus $p > 1$, the family $\tilde{\mathcal{T}}_p \to \mathcal{T}$, we have exactly the first Wolpert formula $(W1)$.

Let us return to Teichmüller theory.

4.9 Metrics on Teichmüller space (general remarks)

Above we concentrated on the Petersson–Weil which turned out to be Kählerian, but, in fact, Teichmüller himself introduced another metric which is not Riemannian but only a Finsler metric. As was shown by Wolpert, the Petersson–Weil metric is not complete which was initially considered a sad fact: the completeness of a space is very much desired (this is why the non complete space of rational numbers $\mathbb{Q}$ is being completed to the space of real numbers.) It was considered a very important achievement of analysis when it was shown that the Lebesgue functional space $L^p(X, \mu)$ are complete and one can present many more examples of this kind.

It was an important event, with far reaching consequences, when, at the turn of centuries Hermann Minkowski introduced in a finite dimensional vector space $\mathbf{F}$ a non-Euclidean norm $|\ |$. Minkowski constructed his norms in a geometrical way: Every (bounded) symmetric and convex set $W \subset \mathbf{F}$ defines a seminorm $|f|_W := \sup\{t \geqslant 0 : tf \in W\}$. The Finsler space is a differential manifold X whose tangent bundle is equipped with Minkowski metric: every tangent space T_xX possesses a norm $|\ |_x$, and $x \to |\ |_x$ is smooth. Riemann was already talking about such metrics in his habilitation lecture *Über Hypothesen welche der Geometrie zugrunde liegen.* Did this lecture inspired Minkowski?

In his thesis of 1914 written under supervision of Carathéodory, Paul Finsler developed this weighty sentence of Riemann into a beautiful theory, known today as Finsler geometry. Stefan Banach generalized the concept of Minkowski norm into infinite dimensional vector spaces. Hilbert spaces are related to Banach spaces in the same way as Euclidean spaces are related to Minkowski ones.

But let us return to Osfald Teichmüller. His work is a treasury and a source of many fruitful concepts; let us only note that Teichmüller observed that the cotangent space $T^*_{[g]}\mathcal{T}(M_0)$ is the space $\mathcal{Q}(g)$ of holomorphic quadratic differentials on fixed M_0. Every holomorphic (= conformal) structure on M_0 is given by the Beltrami differential μ on M_0. We denoted by $H(g)$ the space of harmonic differentials on (M, g), and we identified $H(g)$

with the tangent space $T_g\mathcal{T}_p$. The Riemann surface given by $= \mu$ is denoted by M_μ. Two structures μ and ν are said to be equivalent if there exists a conformal mapping from M_μ onto M_ν which is homotopic to identity. We know that $\mathcal{T}_p(M)$ is, by definition, $\mathcal{C}(M)/Diff_0(M)$. The precise definition of the space of Beltrami differentials will be given later, here it suffices to know that this space is dual to $\mathcal{Q}(g)$, where the dual pairing $\mathcal{Q}(g) \times H(g)$ or $\mathcal{Q}(g) \times H(g) \to \mathbb{C}$ is given by

$$< q|\mu >:= \int_M q\mu, \quad q \in \mathcal{Q}(M),\ \mu \in H(g).$$

Hence the norm on the tangent space to $\mathcal{T}_p$ is given by duality

$$(*) \qquad \| \mu \| = \sup\{< q, \mu >: \| q \| \leqslant 1\}.$$

The most natural norm on $\mathcal{Q}(M) \equiv \mathcal{Q}(g)$ is the L^1 norm

$$(**) \qquad \| q \|_1 := \int_m |q|.$$

DEFINITION. The infinitesimal metric which the norms $(*)$ or $(**)$ induce on $\mathcal{T}_p$ is the Teichmüller metric denoted by ds_T, and the corresponding distance is denoted by d_T.

Following the beautiful article by H.L. Royden *Invariant metrics on Teichmüller space* (1974), we can introduce Petersson–Weil metric as follows.

A conformal metric $\lambda|dz|$ on M gives rise to the hermitian product on $\mathcal{Q}(M)$ by setting

$$(H) \qquad (q_1|q_2)_\lambda := \int_M q_1\bar{q}_2\lambda^{-2}.$$

Let $\| q \|_\lambda$ be the corresponding norm. From (H), $(**)$, and Schwarz inequality we obtain

$$\| q \|_1 \leqslant \| q \|_\lambda \ Area_\lambda(M),$$

where $Area_\lambda(M)$ is the area of M in the λ metric. For the Poincaré metric on M we have $Area_\lambda(M) = 4\pi(p-1)$. The hermitian metric on $\mathcal{T}_p$ dual to the norm $\| \ \|_\lambda$ is clearly the Petersson–Weil metric denoted by ds_{PW}. Thus we obtained an interesting relation between both metrics.

PROPOSITION (ROYDEN, 1974).

$$ds^2_{PW} \leqslant 4\pi(p-1)ds^2_T, \quad d_{PW} \leqslant \sqrt{4\pi(p-1)}d_T.$$

4.10 The period map. Royden theorems

As we know the Riemann period matrix $Z = (Z_{ij})$ is of paramount importance in the theory of ϑ-functions, Jacobians, etc. Let us recall some definitions. Let M be a Riemann space of genus $p \geqslant 2$; choose the *canonical* basis of the homology $H_1(M, \mathbb{Z})$ $a_1, \ldots, a_p, b_1, \ldots, b_p$ and holomorphic differentials $\omega_1, \ldots, \omega_{2p}$ normalized by

$$\int_{a_j} \omega_k = \delta_{jk}.$$

The matrix $Z = (Z_{ij})$ defined by

$$Z_{kj} := \int_{b_j} \omega_k$$

is called the *period matrix* of M. This is a symmetric $p \times p$ matrix with *positive definite imaginary part* Y_{jk} given by

$$Y_{jk} = \frac{1}{2\sqrt{-1}} \int_M \omega_k \bar{\omega}_j,$$

where here and in the sequel we will write $\omega_k \bar{\omega}_j$ for $\omega_k \wedge \bar{\omega}_j$. The Siegel upper half space $\mathfrak{H}_p$ consists of all symmetric $p \times p$ matrices with positive imaginary part.

DEFINITION. The mapping $\varphi : \mathcal{T}_p \to \mathfrak{H}_p$ given by $\varphi(M) := Z$ is called the *period map*; this map is holomorphic.

On the Siegel space $\mathfrak{H}_p$ there exists the only invariant hermitian metric constructed by Siegel (and independently by Bergman) which is a natural generalization of the Poincaré metric on the half plane $\mathfrak{H}_1$:

$$ds^2_{SB} := \frac{1}{4}\operatorname{trace}(Y^{-1} dZ Y^{-1} \overline{dZ}), \quad \text{where } Y := \operatorname{Im} Z.$$

If we make an infinitesimal variation of the conformal structure on M by means of a Beltrami differential μ, the variation of the period matrix Z is given by the formula

$$dZ_{ij} = \int_M \omega_i \omega_j \mu,$$

where ω_i, ω_j are normalized Abelian differentials. For $Y_{jk} = \operatorname{Im} Z = \frac{1}{2\sqrt{-1}} \int_M \omega_k \bar{\omega}_j$, and $C = (C_{jk}) := Y^{-1/2}$ put $u_j := 2^{-1/2} C_{jk} \omega_k$; then u_j are Abelian differentials, and we have

$$\frac{1}{\sqrt{-1}} \int_M u_i \bar{u}_j = \frac{1}{2\sqrt{-1}} C_{ik} \int_M \omega_k \bar{\omega}_l C_{jl} = (Y^{-1/2} Y Y^{-1/2})_{ij} = \delta_{ij}.$$

Therefore (u_j) is an orthonormal basis with respect to the scalar product

$$(u_i | u_j) := \frac{1}{\sqrt{-1}} \int_M u_i \bar{u}_j.$$

We have further

$$(Y^{-1/2} dZ Y^{-1/2})_{ij} = C_{ik} dZ_{kl} C_{lj} = C_{ik} \int_M \omega_k \omega_l \mu C_{ij} = 2 \int_M u_i u_j \mu,$$

whence $\frac{1}{2} dZ_{ij} = \int_M u_i u_j \mu$. If we define the Bergman kernel form for the space of Abelian differentials

$$k(x,y) := \sum u_j(x) \overline{u_j(y)},$$

we obtain, using the fact that $ds^2_{SB} = \frac{1}{4} \operatorname{trace}(dZ' \overline{dZ'})$, for the Siegel–Bergman metric the interesting formula

$$ds^2_{SB} = \int_M \int_M \mu(x) k(x,y) \overline{\mu(y)}.$$

We have the following

Theorem (Royden). *Let $\varphi : \mathcal{T}_p \to \mathfrak{H}_p$ be the period map, and let ds_{SB} be the Siegel–Bergman metric on $\mathfrak{H}_p$. Then the pullback to $\mathcal{T}_p$ of this metric is*

$$(\varphi^* ds_{SB})^2 = \int_M \int_M \mu(x) k(x,y) \overline{\mu(y)},$$

where $k(x,y)$ is the Bergman kernel form for the space of Abelian differentials with the scalar product

$$(u,v) = \frac{1}{\sqrt{-1}} \int_M u \bar{v}.$$

We have yet another metric on the Teichmüller space $\mathcal{T}_p$, the pullback $\varphi^* ds_K$ of the Kobayashi metric ds_K on the Siegel space $\mathfrak{H}_p$:

$$\varphi^* ds_K = \max_{\|u\|=1} \left| \int_M u^2 \mu \right|.$$

But $\| u \| = 1 \Longrightarrow \int |u|^2 \leqslant 1$, and we have

$$\varphi^* ds_K = \max_{\|u\|=1} \left| \int_M u^2 \mu \right| \leqslant \max_{\|u\|=1} \left| \int_M \mathcal{Q} \mu \right| = ds_T$$

where ds_T is the Teichmüller metric on $\mathcal{T}_p$. But Royden proved that for $\mathcal{T}_p$ the Teichmüller metric coincides with the Kobayashi metric. Thus we obtain the interesting

PROPOSITION (ROYDEN, EFGRAFOV–POSTNIKOV). *Any holomorphic map of* $(\mathcal{T}_p, dz_T)$ *into* $(\mathfrak{H}_p, ds_k)$ *is distance decreasing.*

4.11 The period map and Torelli theorems

One of the formulations of the famous Torelli theory asserts the following. We know that the period matrix Z depends on the basis: if we take another basis, then the period matrix Z transforms by Siegel symplectic modular group $\Gamma_p = \mathrm{Sp}(2p, \mathbb{Z}) : Z' = (AZ + B)(CZ + D)^{-1}$, where A, B, C, D are $p \times p$ matrices with integer entries. Thus to any compact Riemann surface (therefore to any algebraic function) there corresponds one point of the moduli space $\mathcal{A}_p := \mathcal{H}_p / \Gamma_p$. Similarly the Riemann moduli space $\mathcal{M}_p$ is obtained from Teichmüller space $\mathcal{T}_p$ dividing it by Teichmüller modular group Γ_p, and the period map descends to the moduli space $\tilde{\varphi} : \mathcal{M}_p \to \mathcal{A}_p$. Therefore the following properties of the period map $\tilde{\varphi}$ are of paramount importance; this is the famous theorem conjectured by Torelli.

THEOREM (TORELLI). *The period map* $\tilde{\varphi} : \mathcal{M}_p \to \mathcal{A}_p$ *is injective.*
In classical wording
If the period matrices of two Riemann surfaces are on the same $\mathrm{Sp}(2p, \mathbb{Z})$ *orbit in* $\mathfrak{H}_p$, $p > 1$, *then these two surfaces are biholomorphically (conformally) equivalent.*

REMARK. The original proof of Torelli (1913) contained so many loopholes, that it cannot be regarded as a real proof. However nobody doubted that Torelli theorem is correct. The original formulation (clearly equivalent to the one above) is as follows

THEOREM (TORELLI). *Let* M, N *be Riemann surfaces of genus* $p > 1$ *and let* $\Theta(M)$ *and* $Q(N)$ *be theta divisors of the corresponding Jacobians* $\mathrm{Jac}(M)$, $\mathrm{Jac}(N)$, *respectively.*

If there exists a biholomorphism $\psi : \mathrm{Jac}(M) \to \mathrm{Jac}(N)$ *such that* $\psi^*(\Theta(N)) = \Theta(M)$, *then* M *and* N *are biholomorphic.*

REMARK. The pair $(\mathrm{Jac}(M), \Theta(M))$ is called the principally polarized Jacobian. All proofs of Torelli theorem construct a Riemann surface from the principally polarized Jacobian and show that this is possible in an essentially single way.

For many years the Torelli theorem was as a sleeping beauty from the tale of Grimm brothers. The prince who woke her up was great Italian mathematician Aldo Andreotti in the series of papers starting in 1952. André Weil became interested in the Torelli theorem thanks to his contacts with Andreotti. Weil regarded Andreotti's proofs as not sufficiently systematic and, in his work of 1957, he presented the first fully correct proof, working in the case of arbitrary field characteristic. A number of other proofs followed. The proof which is considered today to be the most simple and elementary is due to Henrik Martens (1963), and it appeared in his PhD thesis written under supervision of L. Bers at Courant Institute *Torelli's Theorem and a Generalization for Hyperbolic Surfaces* Com. Pure App. Math. **16** (1963), 97–110. The Martens proof became a classic and is being reproduced in reviews and monographs, for example in R. Narashimhan's *Compact Riemann Surfaces*, Birkhäuser, 1992, of which I made use while writing the present exposition. (Cf. the lovely book of E. Reyssat *Quelques Aspects des Surfaces de Riemann*, Birkhäusen, 1992.)

At this point I must end the tale on Teichmüller theory, but I cannot resist to mention relation between Teichmüller theory and Plateau–Douglas problem. This will be the subject of the next chapter.

CHAPTER 5

Teichmüller theory and Plateau–Douglas problem

Already in early nineteen-thirties Douglas posed the problem of minimal surfaces of higher genus $p > 1$. The problem of existence in Euclidean space of minimal surfaces of higher genus spanning given contour has been considered by Douglas, Courant, and Shifman. Douglas papers were severely criticized by A. Tromba who in several papers written in collaboration with T. Tomi gave a solution via Teichmüller theory.

Another formulation and solution of the Plateau–Douglas problem in Euclidean space and even on Riemann manifold is given in the monograph of Jürgen Jost *Two–Dimensional Geometric Variational Problem*, J. Wiley, 1991. In this section we will try to formulate Tromba's approach.

Formulation of the problem. Given a smooth compact surface S with k boundary curves $c_1, \ldots, c_k$, let $\gamma = (\gamma_1, \ldots, \gamma_k)$ be a system of k disjoint, oriented Jordan curves in $N = \mathbb{R}^d$.

If $u : S \to N$ is a map of class C^1 the area is

$$A(u, S) := \int_S |\nabla u| ds,$$

and the energy, or Dirichlet, integral $E(u, g)$, where g is a Riemann metric on S is given by

$$E(u, g) := \frac{1}{2} \int_S g^{\alpha\beta} < u_{x^\alpha} | u^{x^\beta} > dv_g,$$

where (x^1, x^2) are local coordinates on S. We know that the energy $E(u, g)$

depends only on the conformal class of g, that is,

$$E(u, g) = E(u, \lambda g)$$

for any positive function λ on S; E is only invariant under diffeomorphism f of S:

$$E(u \circ f, f^* g) = E(u, g).$$

We know that $u : S \to \mathbb{R}^d$ is harmonic if and only if for any system of conformal coordinates on S defined by $\varphi : S \to \mathbb{R}^2$, the pullback $\mathcal{U} := u \circ \varphi^{-1}$ is harmonic in the classical sense, that is, $\Delta U = 0$. We recall that the map $u : S \to N = \mathbb{R}^d$ is conformal if and only if for any system $\varphi : S \to \mathbb{R}^2$ of conformal coordinates $w := x^1 + ix^2$ on (S, g), the pullback $U = u \circ \varphi^{-1}$ satisfies

$$(U_w | U_w) = 0$$

that is, if $U(w) = (U^1(w), \ldots, U^d(w))$, then

$$(U^1_w)^2 + \cdots + (U^d_w)^2 = 0.$$

To treat Plateau–Douglas problem with the help of variational methods, it is an obvious idea to treat as natural domains of our mappings not closed Riemann surfaces, but oriented Riemann surfaces with boundaries. Therefore, we have to construct the *Teichmüller space* for surfaces with k boundary components, each diffeomorphic to the (oriented) unit circle.

Schottky double. It was an ingenious idea of Friedrich Schottky to handle manifolds with boundaries by associating with the manifold M an oriented, closed (that is, without boundary) manifold denoted by $2M$ in the following way. Let $\partial M = c_1 \cup \cdots \cup c_k$ (disjoint sum.) Let (M', c') be such that there is an isometry $i : (M, c) \to (M', c')$, and (M', c') has opposite orientation to (M, c). Identifying each $m \in \partial M$ with $i(m)$, we obtain a closed surface denoted by $2M := M \cup M'$ (disjoint sum), and on $2M$ we introduce a complex coordinate system $2c$. $(2M, 2c)$ is called the *Schottky double* of (M, c). It follows from construction that $2M$ is equipped with the orientation preserving involution i ($i^2 = \mathrm{id}$) with fixed points $c_1 \cup \cdots \cup c_k$.

REMARK. The similar construction, also due to Schottky makes it possible to construct for every compact *non orientable* surface Σ without boundary an oriented *compact* surface Σ' of constant curvature.

In what follows we will consider only oriented compact surfaces $2M$ without boundary with the C^∞ involution i, $i^2 = \mathrm{id}$ such that the set of fixed points of i consist of k disjoint curves.

We have therefore the following

DEFINITION. The Teichmüller space $\mathcal{T}(2M)$ is defined to be the quotient space $\mathcal{C}^s/\mathcal{D}_0$, where $\mathcal{D}_0$ is now the space of those C^∞ diffeomorphisms of $2M$ which are homotopic to identity, i symmetric, and map each half of $2M$ into itself. We will *not* consider all metrics on $2M$, but only those for whose the involution i is an *isometry.* Such surfaces has been called by Tromba *symmetric Riemann surfaces.*

The theory of Teichmüller spaces $\mathcal{T}(2M)$ for symmetric Riemann surfaces is constructed as in the classical case, that is, for the case of Riemann surfaces without boundaries. Again a fundamental role is played by the space $\mathcal{Q}(g)^s$ of holomorphic quadratic differentials, which are now assumed to be *real* on ∂M. Thus, from thc Riemann–Roch theorem we obtain

PROPOSITION. $\dim_{\mathbb{R}} \mathcal{Q}(g)^s = -3\chi(M) = 6p - 3 + 3k.$

We have

MODIFIED TEICHMÜLLER THEOREM FOR SYMMETRIC SURFACES. *Let $2M$ be a symmetric Riemann surface with symmetry i as above. The Teichmüller space $\mathcal{T}(2M)$, where M has genus p is a C^∞ manifold of dimension $-3\chi(M) = 6p - 6 + 3k$, where k is the number of boundary curves $c_1, \ldots, c_k$ of M. The cotangent space to $\mathcal{T}(2M)$ at $[g]$ can be identified with the real part of holomorphic quadratic differentials $\mathcal{Q}(g)^s$ which are real on ∂M.*

Unoriented minimal surfaces. For the proof of unoriented minimal surfaces of higher genus — a vast, interesting, and important family of minimal surfaces — one needs Teichmüller theory for unoriented surfaces S. By Schottky construction, one passes to a $\mathbb{Z}_2$ cover S' of S. Therefore one has to work with metrics g on S which satisfy an additional $\mathbb{Z}_2$ symmetry; such metrics are denoted by $Met'_{-1}(S)$. We obtain the Teichmüller space $\mathcal{T}(S') = Met'_{-1}(S)/\mathcal{D}_0(S)$ and the corresponding generalization of Teichmüller theorem.

After this remark let us return to oriented minimal surfaces S of genus $p > 1$.

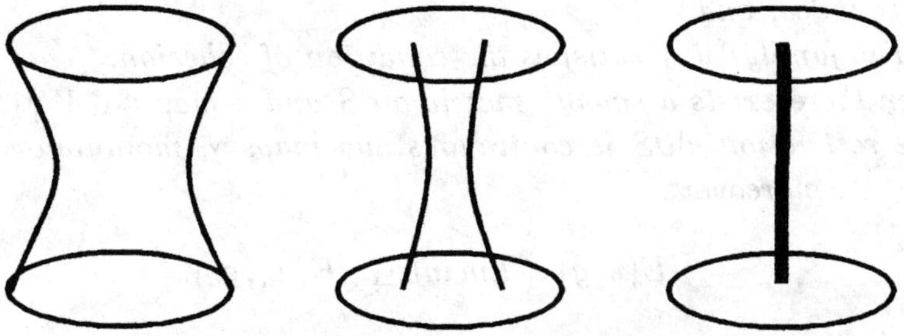

Condition of cohesion

In order to exclude degeneration of minimal surfaces $u_n \to u_0$, $n \to \infty$ which can take place in the case $\partial S \neq \emptyset$ (cf. Figure), Richard Courant has introduced the natural *condition of cohesion*, called by Schoen and Yau *incompressibility condition*:

A family (for example, a sequence) of smooth maps $u : S \to N \equiv \mathbb{R}^d$ satisfies the *condition of cohesion* (*incompressibility condition*) if there is a lower bound of the length of images under any $u \in \mathcal{F}$ of all homotopically nontrivial closed loops in S.

REMARK. In general it is difficult to check the condition of cohesion, and therefore several authors replaced it by other condition (cf. Jost's solution of the Plateau–Douglas problem.)

Now we ready to formulate the fundamental result of the theory of Tromba and his collaborators giving a solution of Plateau–Douglas problem.

THEOREM (TOMI–TROMBA, 1988). *Let S be a compact smooth surface which is not simply connected and has $k \geqslant 1$ boundary components $\gamma_1, \ldots, \gamma_k$, with the genus of Schottky double $p > 1$. Let $c_1, \ldots, c_k$ be pairwise disjoined, rectifiable Jordan curves in $\mathbb{R}^d$ $(= N.)$ Further, let (u_n, g_n), $n = 1, 2, \ldots$ be a sequence of maps $u_n : S \to \mathbb{R}^d$ of class $C^0 \cap W_1^2(S, \mathbb{R}^d)$ which maps γ_i monotonously onto c_i, $i = 1, 2, \ldots$, and g_n are smooth metrics on S. Suppose that*

1. *The energy (Dirichlet) integrals $E(u_n, g_n)$ and* $\sup(u_n(S)$ *are uni-*

formly bounded; and

2. The family $\{u_n\}$ *satisfies the condition of cohesion.*

Then there exists a smooth metric on S *and a map* $u \in W_1^2(S, \mathbb{R}^d)$ *such that the restriction* $u|\partial S$ *is continuous and maps* γ_i *monotonously onto* c_i, $i = 1, 2, \ldots$; *moreover*

$$E(u, g) \leqslant \liminf_{n\to\infty} E(u_n, g_n).$$

In the case when S *is oriented and all* u_n *map* γ_i *onto* c_i, $i = 1, 2, \ldots$ *with a prescribed fixed orientation,* u *can be chosen to map* γ_i *onto* c_i *in the orientation preserving way.*

Of course, we cannot even present a draft of the proof of this magnificent achievement. Let us only stress the fundamental role which is played in the problem of moduli, that is, in the theory of space $\mathcal{M}_p = Met_{-1}/Diff^+$, by the famous

MUMFORD COMPACTNESS THEOREM (1971).

First version. *Let* M *be a closed, connected, smooth surface, and let* $\{g_n\}$ *be a sequence of smooth metrics of curvature* -1, *that is,* $g_n \in Met_{-1}(M)$ *such that all their closed geodesics are bounded in length by* $l_0 > 0$, *independent of* n. *Then there exist smooth diffeomorphisms* f_n *of* M ($f_n \in Diff^+(M)$) *such that there exists a subsequence* $\{g_m\}$ *of* $\{g_n\}$, $f_m \in Diff^+(M)$, *and* $g \in Met_{-1}(M)$ *satisfying* $f_m^* g_m \xrightarrow{C^\infty} g$.

Second version. *Let* $\{\Gamma_n\}$ *be a sequence of discrete subgroups of* $\mathrm{Aut}(\mathfrak{H}_1)$ *with compact quotients* $\mathfrak{H}_1/\Gamma_n := (M, g_n)$. *Suppose that all nontrivial closed geodesics are bounded in length by* $l_0 > 0$. *Then there exists a subsequence* $\{\Gamma_m\}$ *of* $\{\Gamma_n\}$ *and* $\Gamma \in \mathrm{SL}(2, \mathbb{R})$ *such that* $\Gamma_m \to \Gamma$, *and* $\mathfrak{H}_1/\Gamma$ *is a Riemann surface of topological type of* M.

Alas, the Mumford theorem is appropriate for the case of the Riemann moduli spaces (the group $\mathcal{D} = Diff^+$) and *not* for the case of Teichmüller space $\mathcal{T}_p(M)$ (the group $\mathcal{D}_0 = Diff_0^+$), but this difficulty could be overcome. The Mumford compactness theorem (or its equivalents) are used in proofs of properness of the energy $E(u, g)$ on Teichmüller space.

The reason why Mumford theorem does not apply to $\mathcal{D}_0$ (diffeomorphisms homotopically equivalent to identity) is the following. On a negatively curved surface every geodesic arc (with fixed end points) is globally minimizing. Thus any two geodesic arcs with the same end points cannot be homotopic; otherwise there would exist a non minimal geodesic joining the end points.

The following observation of Schoen and Yau links Mumford theorem with incompressibility condition.

THEOREM (SHOEN–YAU, 1978-9). *Let (M, g) be closed, oriented surface with $g \in Met_{-1}(M)$ and let $u : M \to \mathbb{R}^d$ be a map of class $C^0 \cap W_2^1(M, \mathbb{R})$ such that for all homotopically non trivial closed C^1 loops α on M the length of $u \circ \alpha$ is bounded from below by $l_0 > 0$.*

Then the length $l(\gamma)$ of any closed geodesic on (M, g) is estimated by

$$l(\gamma) \geqslant \min\left\{1, \frac{1}{2} l_0 \left(\pi - 2\arctan\frac{\sqrt{5}}{4}\right) E(u, g)^{-1}\right\}.$$

There is a parallel, different version of solution of Plateau–Douglas problem due to Jürgen Jost which holds even for the considerable generalization of it: $u : S \to N$, where now N is a complete Riemann manifold of dimension d. But, of course, one restricts N somehow, e.g., the absolute value of the sectional curvature of N has an upper bound and the injectivity radius has a positive lower bound. Moreover, one has to introduce a sort of the so called Douglas condition which is complicated and sophisticated, but is natural for the experts. No wonder, for such a great generality one has to pay a price! But in the case $N = \mathbb{R}^d$ considered by Tomi and Tromba these conditions drop out, so we can write down this magnificent result (Corollary 4.7.1. in the Jost monograph.)

THEOREM (JOST, 1991). *Let $c = (c_1, \ldots, c_k)$ be a system of Jordan curves in $\mathbb{R}^d$ and $p \in \mathbb{N}$. If the infimum of area (or energy) of surfaces of genus p spanning $c_1, \ldots, c_k$ is strictly less than the infimum of (area of) surfaces of genus less than p or consisting of more than one component, the sum of the genera of whose does not exceed p, then there exists a minimal surface of genus p spanning c, and this surface is of least area among all such surfaces.*

Let close this chapter with the words of one of the best experts on the topic (Anthony Tromba):

'... The first to study general Plateau problem for minimal surfaces of higher topological type was Jesse Douglas; his work is truly pioneering, and his ideas and insight are as exciting and important nowadays as at the time when they were published, more than half century ago.'

Chapter 6

Rescuing Riemann's Dirichlet Principle. Potential Theory

As we have mentioned already many times, the proofs of Riemann of his magnificent theorems (mapping theorem, Riemann–Roch, ...) based on his Dirichlet principle contain, as Riemann realized well himself, some gaps. As a of Weierstrass criticism, the method of Dirichlet principle was abandoned for many years.

This situation was immensely interesting and heavy with blessed consequences – one want to exclaim after Paul *o felix culpa!* Although the trust and believe in Riemann's method has been lost, but the truth of His theorems, so fascinating indeed, was never doubted. And thus in the second half of nineteenth century a number of methods and theories was created to prove Riemann's theorems in analytical way, and in the case of the Riemann–Roch theorem, in the algebraically-arithmetic way opened by celebrated work of Dedekind and Weber. Here, I will talk about the former ones.

Herman Amadeus Schwarz, although he was a beloved pupil of Weierstrass, was fascinated by the Riemann work. Schwarz lemma, his works on conformal mappings, hypergeometric equation, and his alternating method are deeply rooted in Riemann's works. The famous sweeping out method of Poincaré is an ingenious modification of Schwartz method. At the end of this chapter, I will mention an unification of these two methods proposed by Browder; the latter applies not only to the Dirichlet problem for Laplace equation, but also to the wide scale of elliptic operators.

Rescuing of the Riemann's Dirichlet principle by Hilbert in 1900 has opened a completely new chapter of calculus of variations and was, along with Fredholm theory, one of the major impulses of, young at the time, functional analysis and distribution theory.

Here I would like to say few words about an important, and conceptually simple method developed by Perron and Remak, called the *Perron method*, whose main tool are subharmonic functions; we mentioned them already in relation with plurisubharmonic functions.

Monotone limits of smooth functions are, in general, only semi-continuous. Similarly, limits of monotonously decreasing (point wisely) sequences of harmonic functions is not, in general, harmonic, but subharmonic; it is only upper semicontinuous.

6.1 Subharmonic functions. Riesz decomposition

As we know a harmonic function $u : \mathcal{O} \to \mathbb{R}$, where $\mathcal{O}$ is a domain in $\mathbb{R}^2$ or $\mathbb{R}^n$ is characterized by the fact that for every sphere $S_x^n \subset \mathcal{O}$ its mean value on the sphere is equal $u(x)$, that is, the value at the center of the sphere. (Similarly one can characterize harmonic function by their values at the center of the balls $B_x^n \subset \mathcal{O}$.)

Subharmonic functions are characterized by the inequality

$$\Delta u \geqslant 0.$$

Since the Laplace operator can be invariantly defined on arbitrary Riemann space (M, g) as $\Delta := (dd^* + d^*d)$, one can define the subharmonic function by the inequality $\Delta u \geqslant 0$ understood in the sense of distribution theory. Thus the notion of a subharmonic function can be generalized to Riemann manifolds, and hence to Riemann surfaces. Analogously a superharmonic function satisfies by the inequality

$$\Delta v \leqslant 0.$$

For a subharmonic function on $\mathcal{O} \subset \mathbb{R}^n$ the famous Riesz decomposition holds.

THEOREM (RIESZ DECOMPOSITION). *(u is superharmonic on $\mathcal{O} \subset \mathbb{R}^n$)$\Longleftrightarrow$($u = P^\mu + h$, where h is harmonic and P^μ is a potential of the measure (mass) μ)*

This famous decomposition was generalized in a simple way by Schwartz for linear partial differential operators with constant coefficients; this was one of the most spectacular successes of the young theory of distributions. Let L be a linear differential operator with constant coefficients; then

THEOREM (SCHWARTZ, 1949). *(A distribution u satisfies $Lu \geqslant 0$) $\Longleftrightarrow$(There exists a measure $\mu \geqslant 0$ with support $\subset \mathcal{O}$ and a solution of h of $Lh = 0$ such that $u = P^{\mu} + h$, where $P^{\mu} := E * \mu$).*
E is a fundamental solution of L defined by $LE(x) = \delta(x)$.

Proof. Since Lu is a positive distribution, it is a positive measure $\mu \geqslant 0$. Therefore we have, by virtue of elementary properties of distributions,

$$Lu = A * \mu = A * E * \mu + A * h = \begin{cases} \mu + A * h & \text{on } \mathbb{R}^n \\ \mu & \text{in } \mathcal{O} \end{cases}$$

Thus $h = u - E * \mu$. □

REMARK. Each L with constant coefficients can be represented as a convolution with the distribution A. The 'only' difficulty is the existence of the fundamental solution E of L, which was proven independently by Malgrange and Hörmander in 1956.

For $L = \Delta$, the existence of fundamental solution (logarithmic potential for $n = 2$ and Newtonian potential for $n > 2$) is the classical fact. Thus the spectacular Schwartz proof of Riesz decomposition of subharmonic functions follows as an easy corollary.

6.2 Poisson integral and Harnack theorems

For an r disc $\mathbb{D}_r = \{z \in \mathbb{C} : |z| < r\}$, the so called Poisson integral gives the solution of the Dirichlet problem: For continuous $\varphi : \partial\mathbb{D}_r \to \mathbb{R}$,

$$u(z) := \frac{1}{2\pi}\int_0^{2\pi} \frac{r^2 - |z|^2}{|re^{i\varphi} - z|^2} f(re^{i\varphi})d\varphi, \quad z \in \mathbb{D}_r$$

is harmonic on $\mathbb{D}_r$, continuous on $\overline{\mathbb{D}}_r$, and satisfies the Dirichlet boundary condition $\partial u|_{\partial\mathbb{D}_r} = f$.

There is a corresponding formula for the ball $B(o, r) \subset \mathbb{R}^n$, $n > 2$; $p(z, \zeta) := \frac{1}{2\pi}\frac{|\zeta|^2 - |z|^2}{|\zeta - z|^2}$, $z \neq \zeta$ is called the Poisson kernel. In $\mathbb{R}^n$, $n \geqslant 3$, the Newtonian kernel $k(x, y) := |unit\ sphere|^{-1}|x - y|^{2-n}$.

Now we give the classical (not distributional)

DEFINITION OF SUBHARMONIC FUNCTIONS. A function $U : X \to \mathbb{R} \cup \{\infty\}$ is subharmonic on a Riemann surface X if

1. $\not\equiv -\infty$;
2. u is upper semicontinuous;
3. $u(x) \leqslant$ its spherical average on any spherical cell S_x^2.

EXAMPLES. 1. Any harmonic u is subharmonic; u is harmonic if and only if u and $-u$ are both subharmonic.

2. For any holomorphic f, $\log|f|$ is subharmonic.

3. If u_1, u_2 are subharmonic $\Longrightarrow \sup(u_1, u_2)$ is subharmonic.

4. If $\{u_n\}$ is a monotonously decreasing sequence of subharmonic functions $u_n \downarrow u$ such that all of the functions are bounded from below at some point, then u is subharmonic.

For subharmonic functions we have

MAXIMUM PRINCIPLE (FOR SUBHARMONIC FUNCTIONS). *If u is subharmonic on the Riemann surface X, then $u(x) < \sup u(X)$, or u is a constant function.*

For the theory of harmonic functions the following theorem is of great importance.

THEOREM (HARNACK).

I. *A locally uniformly convergent sequence of harmonic functions converges to harmonic function.*

II. *A monotonous sequence $\{u_n\}$ of harmonic functions which is bounded at a single point x_0 is compactly convergent to harmonic function.*

Proof of II. It follows from the Poissson integral that in the coordinate patch where $u > 0$ the so called *Harnack inequality* holds:

$$(H) \qquad \frac{1-|z|}{1+|z|}u(0) \leqslant u(z) \leqslant \frac{1+|z|}{1-|z|}u(0).$$

If $U_1 \leqslant u_2 \leqslant \cdots \leqslant u_n \leqslant \cdots$ and $(\mathcal{O}, z)$ is a coordinate neighborhood with $x_0 \in \mathcal{O}$, then taking $u := u_m - u_n$, $m > n$ in (H), the compact convergence of $\{u_n\}$ in $\mathcal{O}$ follows. But since we can connect $\mathcal{O}$ with every coordinate cell, we have compact convergence on the whole of X.

Historical remarks. Axel Carl Gustav Harnack (1851–1888) was an outstanding and versatile mathematician. He was a twin brother of the great Lutherian theologian Adolf von Harnack, who was not only a great historian of early Christianity and the founder of 'Liberal theology school', but also, was for several dozen of years a director of the famous Keiser Wilhelm Institute Gesellschft (which later became the Max Planck Institute.) This institution made the pre-Nazi Germany the leader of research in natural sciences and medicine. It is interesting that the director of the institution in which many Nobel laureates worked was a great theologian. Among its institute there was no not only humanistic institute, but even no mathematical one. In this respect the breakthrough was made by the Max Planck Institute for Mathematics (in Bonn) established by Fritz Hirzebruch, and led by him for thirty years. Presently this institute is one of the major centers of world mathematics.

Harnack's theorems and their generalizations are pearls not only of the potential theory. So let us say a few words about

6.3 History of the potential theory

Of course, as the term suggests, the sources of the notion of potential are in physics: Newton's gravity, electrostatics, etc.

Today, we say that the potential U^μ of the positive Radon measure μ in $\mathbb{R}^n$ is given by

$$U^\mu(x) := \int k(x,y)d\mu(y),$$

where $k(\cdot,\cdot)$ is either Newtonian (for $n \geqslant 3$) or logarithmic kernel. Clearly $0 \leqslant U^\mu \leqslant +\infty$. For an arbitrary measure μ, we have $\mu = \mu^+ - \mu^-$, where both $\mu^+, \mu^- \geqslant 0$. To define the corresponding potential for linear elliptic equation, k is replaced by a fundamental solution (or a parametrix.) For simplicity we will only consider Newtonian potentials.

For domains $\mathbb{D}$ more general than balls with smooth boundaries, G. Green introduced in 1828 the Green function, and in 1840 Gauss introduced the famous *Gauss integral*

$$G(\mu) := \int (U^\mu - 2f)d\mu, \quad \text{where } f \in C^0(\partial\mathbb{D}).$$

Gauss considered all masses (measures) which have continuous density and took for granted the existence of a measure μ_0 with continuous density such that $G(\mu_0) = \inf G(\mu)$. It follows from this (not proved?) assumption that $U^{\mu_0}(x) = f(x) + \gamma$ on $\partial\mathbb{D}$, where γ =const.

Riemann considered the Dirichlet integral (energy) $\mathcal{D}(u) = \int_{\mathbb{D}} |\nabla u|^2$, where $\mathbb{D}$ is a bounded domain and u is a 'sufficiently smooth' function such that $u|_{\partial\mathbb{D}} = f$, a given continuous function on the boundary $\partial\mathbb{D}$ of $\mathbb{D}$. He took as evident the existence of u_0 such that

$$\mathcal{D}(u_0) = \inf \mathcal{D}(u).$$

Therefore $\Delta u_0 = 0$ in $\mathcal{D}$ and $\lim_{x\to y} u_0(x) = f(y)$, $y \in \partial\mathbb{D}$. This is the famous Riemann's Dirichlet principle.

In this way Riemann obtained his conformal mapping theorem which we proved, in connection with Douglas–Rado solution of the Plateau problem, at the beginning of this part.

The thesis of Otto Frostman, Lund 1935, opened a new chapter of rapid development of modern potential theory. He constructs the sequence $\{\mu_n\}$ of measures which converges *weakly* to a measure μ_0 such that for the Gauss integral $G(\mu_0) = \inf G(\mu_n)$. Frostman proves also his 'principle of equilibrium' and 'principle of energy.' He was probably the first who introduced the notion of weak convergence of measures (which are now linear functionals on the space of measures) — for us this notion is almost obvious and necessary. But *great* ideas of mathematics become obvious after some time. It is significant that Frostman have his inspiration in the classical research of Gauss which, in turn, has its source in physics.

To Gauss (and Frostman) variational method refers also Henri Cartan in his works of 1943–50 by introducing the scalar product of measures $(\mu|\nu) := \int U^\mu d\nu$, where the measures have finite energy, $(\mu|\nu) < \infty$. Using the method of orthogonal projection of the theory of Hilbert spaces, Henri Cartan measures μ_0 giving the lower bound to the energy (integral) $I(\mu_0) = \inf I(\mu)$; this is the so called principle of energy of potential theory. He also proves the important

THEOREM (H. CARTAN). *Infimum of a family of superharmonic functions is a superharmonic function, with an exception of a set of (outer) capacity zero.*

In 1938 M. Brelot started his investigations of the Perron method, to which we turn now.

6.4 Perron method

gives a simple construction of harmonic functions as an upper envelope $u^* := \sup_{u\in\mathcal{F}} u$ of an ingeniously constructed family of subharmonic functions, this is therefore (also) a variational method.

This construction makes use of a linear mapping $P_{\mathbb{D}}$ imitating the Poisson integral which has the following properties. For every open set $\mathcal{O} \subset X$, let $Reg(\mathcal{O})$ denotes the family of all subdomains $\mathbb{D} \Subset \mathcal{O}$ (that is, the closure $\bar{\mathbb{D}}$ is a compact set $\subset \mathcal{O}$) such that the Dirichlet problem can be solved on $\mathbb{D}$ for arbitrary continuous boundary values $f : \partial\mathbb{D} \to \mathbb{R}$. We saw that $Reg(\mathcal{O})$ is not empty: for every $u \in C(\mathcal{O}, \mathbb{R})$ and $\mathbb{D} \in Reg(\mathcal{O})$, denote by $P_{\mathbb{D}}(u)$ a function which coincides with u on $\mathcal{O} - \mathbb{D}$ and solves the Dirichlet problem on $\bar{\mathbb{D}}$ for boundary values $u|_{\partial\mathbb{D}}$. The map $u \to P_{\mathbb{D}}(u)$ has the following properties

1. $P_{\mathbb{D}}(u+v) = P_{\mathbb{D}}(u) + P_{\mathbb{D}}(v)$,
2. $P_{\mathbb{D}}(\lambda u) = \lambda P_{\mathbb{D}}(u)$,
3. $u \leqslant v \Longrightarrow P_{\mathbb{D}}(u) \leqslant P_{\mathbb{D}}(v)$,

For all $u, v \in C(\mathcal{O}, rz)$, $\lambda \in \mathbb{R}$. Clearly

$(u \in C(\mathcal{O}, \mathbb{R})$ is harmonic$) \Longleftrightarrow (P_{\mathbb{D}}(u) = u$ for $\mathbb{D} \in Reg(\mathcal{O}))$

It is not difficult to prove

LEMMA. *($u \in C(\mathcal{O}, \mathbb{R})$ is subharmonic and $\mathbb{D} \in Reg(\mathcal{O}, \mathbb{R})$) $\Longrightarrow P_{\mathbb{D}}(u)$ is subharmonic.)*

Now we have

Perron principle. *Let $\mathcal{F} \subset C(\mathcal{O}, \mathbb{R})$ be a* non empty *family of subharmonic functions $\mathcal{O}$ such that*

(a) $u, v \in \mathcal{F} \Longrightarrow \sup(u, v) \in \mathcal{F}$;

(b) $u \in \mathcal{F}$, $\mathbb{D} \in Reg(\mathcal{O}) \Longrightarrow P_{\mathbb{D}}(u) \in \mathcal{F}$;

(c) There exists a constant $M \in \mathbb{R}$ such that $u \leqslant M$ for every $u \in \mathcal{F}$.

Then the upper envelope $u^* := \sup u(x)$, $u \in \mathcal{F}$ is harmonic on $\mathcal{O}$.

PROOF follows from Lemma, Harnack theorem, and Maximum principle.

The next step of the Perron method of construction of solution of the Dirichlet problem for a wide class of open subsets $\mathcal{O}$ of the Riemann surface X, that is, the ones having regular boundary $\partial\mathcal{O}$, is the construction, for every continuous and bounded function $f : \partial\mathcal{O} \to \mathbb{R}$, of a sufficiently rich family $\mathcal{P}(f)$ of continuous functions on $\bar{\mathcal{O}}$, so that its upper envelope gives solution of the Dirichlet problem u for the value of the function u on the boundary equal f.

DEFINITION. If $f : \partial\mathcal{O} \to \mathbb{R}$ is a bounded continuous functions, and $M := \sup\{f(x) : x \in \mathcal{O}\}$, then the set of all functions $u \in C(\bar{\mathcal{O}})$ such that

1. $u|\mathcal{O}$ is continuous and subharmonic,
2. $u|\partial\mathcal{O} \leqslant f$, $u \leqslant M$

is called the *Perron class* of f and denoted $= \mathcal{P}(f)$.

By Perron principle, the upper bound u^* of the class $\mathcal{P}(f)$ is harmonic on $\mathcal{O}$; but for arbitrary $\mathcal{O}$ *not* all boundary points x will be 'regular' in the sense that

$$\lim_{y \to x} u^*(y) = f(x).$$

REMARK. Such u^* is sometimes called a *generalized solution* of the Dirichlet problem. The problem of regular points is crucial in modern potential theory. In order to characterize such points, Lebesgue introduced the notion of *barrier at* $x \in \partial\mathcal{O}$. This notion makes it possible to give a practical, simple criteria for regularity.

DEFINITION. A function β_x, $\xi \in \partial\mathcal{O}$ is called a barrier at x if there is an open neighborhood $\mathcal{U}$ of x such that $\beta_x \in C(\bar{\mathcal{O}} \cap \mathcal{U})$

1. $\beta_x|\mathcal{O} \cap \mathcal{U}$ is subharmonic,
2. $\beta_x(x) = 0$ and $\beta_x(y) < 0$ for every $y \in \bar{\mathcal{O}} \cap (\mathcal{U} - \{x\})$.

The point $x \in \mathcal{O}$ is *regular* if there exists a barrier β_x at x.

The following proposition shows that the above definition of regularity is sufficient for 'regularity' in the sense above.

PROPOSITION (PERRON). *Let $\mathcal{O}$ be an open subset of a Riemann surface X and let $f : \partial\mathcal{O} \to \mathbb{R}$ be a bounded continuous function, and $u^* = \sup\{u : u \in \mathcal{P}(f)\}$. Then if there exists the barrier β_x of $x \in \partial\mathcal{O}$, then x is regular, that is,*

$$\lim_{y \to x} u^*(y) = f(x).$$

Finally we have a desired solution of the Dirichlet problem.

THEOREM (PERRON). *If $\mathcal{O}$ is an open subset of a Riemann surface such that all boundary points are regular (each of them possesses a barrier), then for every bounded continuous function $f : \partial\mathcal{O} \to \mathbb{R}$ the Dirichlet problem has a solution.*

REMARK. It is obvious that this Perron solution is unique since the difference of solutions $u_1 - u_2$ is a harmonic function which vanish on $\partial\mathcal{O}$, hence $u_1 - u_2 = 0$.

The following theorem gives a simple sufficient condition for regularity.

THEOREM. *Let $X = \mathbb{C}$, $\mathcal{O} \subset \mathbb{C}$, and $x \in \partial\mathcal{O}$. If there exists a disc $\mathbb{D}_r(m) = \{z \in \mathbb{C} : |z - m| < r\}$, $m \in \mathbb{C}$, $r > 0$ such that $x \in \partial\mathbb{D}_r(m)$ and $\overline{\mathbb{D}_r(m)} = \emptyset$. Then x is regular.*

PROOF. Put $a := (x + r)/2$; then $\beta_x(z) := \log \frac{r}{2} - \log|z - a|$ is a barrier at x. □

6.5 Rado theorem. Theorem of Poincaré–Volterra

The Perron solution makes it possible to give a simple proof of the famous

RADO THEOREM (1925). *Every Riemann surface X has a countable topology.*

PROOF follows from the simple general criterion for countability of topological spaces

THEOREM (POINCARÉ, VOLTERRA).

1. If X is a connected topological space, Y is a Hausdorff space with countable topology, and $f : X \to Y$ is a continuous discrete mapping. Then X has countable topology.

2. If $h : X \to Y$ is continuous, open, and surjective, and X has countable topology, then Y has also countable topology.

REMARKS. 2. is obvious. Ad 1. Poincaré and Volterra formulated the theorem in quite different terms: in 1888 the notion of Hausdorff space was not known, it appeared only 26 years later (in 1914.)

Also the precise notion of Riemann surface was formulated first in 1913 by Weyl. Therefore 1. was formulated and proved for Weierstrass 'Analytisches Gebilde' – a family of holomorphic function elements (germs.) The modern proof of 1. is quite elementary, and is given, for example in Forster's classic.

PROOF OF RADO THEOREM (FOLLOWING FORSTER). Let $\mathcal{U}$ be a coordinate neighborhood on X. Choose two *disjoint* compact discs $K_0, K_1 \subset \mathcal{U}$, and put $\mathcal{O} := X - (K_0 \cup K_1)$. Since the boundary $\partial\mathcal{O} = \partial K_0 \cup \partial K_1$ is regular by the criterion, we have the solution $u : \bar{\mathcal{O}} \to \mathbb{R}$ of the Dirichlet problem with boundary condition $u|\partial K_0 = 0$ and $u|\partial K_1 = 1$.

Therefore $\omega := d''u$ is a nontrivial holomorphic one-form. If $\pi : \tilde{\mathcal{O}} \to \mathcal{O}$ is the universal covering, take any holomorphic primitive f for $\pi^*\omega$ (on $\tilde{\mathcal{O}}$.) Since $f \neq$const, the function $f : \tilde{\mathcal{O}} \to \mathbb{C}$ satisfies assumptions of the Poincaré–Volterra (1.), and thus $\tilde{\mathcal{O}}$ has countable topology. $\mathcal{O}$ has then countable topology as well. Since $X = \mathcal{O} \cup \mathcal{U}$ we are done. □

We see once again the power of the reasoning along the lines of the Riemann's Dirichlet principle. When the solvability of the Dirichlet problem was guaranteed by Perron and Remak, the way is open to fundamental theorems of the type of Weierstrass and Mittag–Lefflerr theorems proved by Behnke for arbitrary noncompact Riemann surfaces, and the Riemann's idea is the simplest method of proving these magnificent theorems. As it turned out, the fundamental and almost miraculous fact is the theorem of Runge type proved by Benke and Stein in 1948.

Let us return to our short history of potential theory. Usually in presentation of Perron theory, a cofounder of it is being forgotten: this is Robert Erich Remak (born in 1888 in Berlin and murdered in 1942 in Auschwitz). He was a pupil and PhD student of Frobenius in Berlin. He wrote papers on algebraic number theory and Remark decomposition appears in group theory. The method of solving the Dirichlet problem with the help of subharmonic functions was developed by Remak independently of Perron in 1923.

The theory of Perron–Remak constructs a generalized solution for *every* domain $\mathcal{O} \subset \mathbb{C}$, but, as observed by Lebesgue and Zaremba, for regions with nasty boundary $\partial\Omega$, a solution of the classical Dirichlet problem *does not* exist (the famous examples of Lebesgue–Zaremba); this is the reason why, in 1912 Lebesgue introduced his barriers.

De la Vallée Poussin, a great Belgian mathematician, known also from his works on analytical number theory, introduced in 1932 measures (masses) which were omitted by Poincaré in his sweeping out method, and he presented a solution of the Dirichlet problem in the form of $u(x) = \int f d\mu_x$, where μ_x is the so called harmonic measure; he also introduced a *new* notion of *capacity* which was introduced before by Wiener in 1924. The notion of capacity of a set is the mathematical (exact) analogue of capacity of a conductor (capacitor) in electrostatics: If K is an isolated conductor in $\mathbb{R}^3$, then the relation between its total charge Q and the potential created by this charge $U(x) := \int \frac{d\mu(y)}{|x-y|}$ is $Q = CU$, where the constant C is called the capacity of (the conductor) K. This notion can be generalized for arbitrary Radon measures μ. As before one defines the potential U^μ of the measure (mass) μ by the formula

$$U^\mu := \int k(x,y) d\mu(y),$$

where $k(x,y)$ is the Newton potential in $\mathbb{R}^n$ $n \geqslant 3$. One defines the first De la Vallée Poussin capacity C_K of the compact subset $K \subset \mathbb{R}^n$ as follows. $C_K := \sup \mu(\mathbb{R}^n) : \mu \geqslant 0$ and $U^\mu(x) \leqslant 1$ for all $x \in \mathbb{R}^n$ and the support of $\mu \subset K$. If $C_K < \infty$, then there exists such measure μ_0 that $C_K = \mu_0(\mathbb{R}^n)$ (Frostman, H. Cartan, Choquet.) One defines in an obvious way the inner capacity C_A^i of every subset $A \subset \mathbb{R}^n$ as $\sup_{K \subset A}(C_K)$ and corresponding exterior C_A^E as $C_A^e := \inf(C_{\mathcal{O}}^i)$ for open $\mathcal{O} \supset A$, $C_A^i \leqslant C_A^e$. If $C_A^i = C_A^e$ then this number is the *capacity of the set A*.

A fundamental Choquet theorem (1954) asserts that for every Borel set $B \subset \mathbb{R}^n$, $C_B^i = C_B^e$.

This theorem is proved in the huge memoir (160 pages long) of Gustave Choquet *Theory of Capacities*, Ann. Inst. Fourier **5** (1954), 131–295.

It was a burning problem as to how large is the set of the boundary $\partial\mathcal{O}$ of points x which are *not* regular, that is, where the generalized solution $u^*(x)$ does not attain the value $f(x)$? The answer is $C_B^e = 0$ for the set B of irregular points in $\partial\mathcal{O}$. Thus something goes wrong on the set of external capacity zero. A physicist would probably say – of course – I have known that for a long time!

When the abstract potential theory matured, time had came for axiomatics and for general topological spaces on whose the Perron–Wiener–Brelot theory can be practiced. An elegant theory of harmonic spaces has been created. In some sense these investigations were completed in the monograph of Heinz Bauer *Harmonische Räume und ihre Potential Theorie*, Lect. Notes in Mathematics, 22 (1966).

A new approach to potential theory is due to Wiener, who discovered an unexpected and shocking – for the first sight – relation between Dirichlet problem for $\mathcal{O} \subset \mathbb{R}^3$ and the theory of Brown motions. In retrospect such relation were to be expected: we know what a great role is played by the heat transport equation in the theory of harmonic forms and general harmonic mappings.

Today perhaps the most convenient approach to the theory of (linear) elliptic operators is the heat kernel method. In this way all great theorems of the type of Hirzenbuch–Riemann–Roch and Atiyah–Singer–Bott index theorem have been obtained (cf. the excellent monographs in the 'Yellow Series' by Berline et. al. and the Princeton Lectures by G. Faltings on his *Algebraic Riemann–Roch Theory.*).

Chapter 7

The Royal Road to Calculus of Variations (Constantin Carathéodory)

To the great master of calculus of variations, Constantin Carathéodory

7.1 Introduction

The classical extremal problems were aimed at finding a minimum or a maximum of functionals, that is, functions whose arguments run through 'infinite dimensional sets', for example families of curves or surfaces constrained by some 'boundary' conditions. Thus we have the *isoperimetric problem*, the famous *problem of Dido*: among all curves of given length on the plane find such that bounds the greatest area. 'Of course' the solution is a circle.

In physics the least action principle (of Euler, Lagrange, and Hamilton) or the principle of the least light path of Fermat turned out not to be principles concerning minima (or maxima), but critical (or stationary) values, that is, for whose the first functional (sometimes called variational) derivative vanishes. In the calculus of variations such objects are called *extremals.* They satisfy the Euler–Lagrange equations: for the Lagrangian L being a functions of $2n+1$ variables $L = L(t, x, p)$, $x = (x_1, \ldots, x_n)$, $p = (p_1, \ldots, p_n)$, they read

$$(E-L) \qquad \frac{\mathrm{d}L_{p_i}}{\mathrm{d}t} = L_{x_i}, \quad i = 1, \ldots, n.$$

These equations were found first with the help of geometrical method by Euler, and then, with the help of general variational method, by nineteen years old Lagrange.

Of course, the (E-L) equations are only necessary conditions for extremum of the functional $I(x) = \int_c L(t, x, \dot{x}) = \int_c L(t, x, p)$. The problem to find *sufficient* conditions for $x \to u(x)$ to be a minimum turned out to be much more complex. In analogy with functions of one or several variables this condition was expected to be of the form of an inequality saying that the sign of the second variation of the functional is positive. Legendre, and 40 years later Jacobi were mistaken thinking that it was sufficient to investigate the second variation. Nevertheless the theory of second variations of Jacobi is a magnificent achievement. These investigations led Weierstrass to his concept of *fields of extremals*, the $\mathfrak{E}$ function, and Weierstrass criterion for strong minimum in the case $n = 1$. But the Weierstrass theory cannot be generalized automatically to the case $n > 1$. And only in the general case, the beautiful theory of geodesic fields, being a special case of extremals, arises.

Hilbert, generalizing the notion of *Beltrami integrals* rooted in the theory of geodesics, introduces in his famous Paris lectures of 1900, as the last 28th problem (a new theory of calculus of variations) his famous invariant integral, called nowadays the *Hilbert integral*, as a new approach to field theory. Surely, these investigations were not simple, and Costantin Carathéodory, a student of Hilbert and the leading expert in calculus of variations, instinctively felt that there must be a simple, "royal' approach to calculus of variations of single integrals. He was obsessed by the question *Why extremals play such distinguished role?* They are being used and they function fabulously, but how one can reason out and show their necessity and simplicity? The wonderful *Methode der geodetischen Äquidistanten* (Acta Mathematica **47** (1925), 199-230) was the result of these investigations.

The Carathéodory approach gives, in an almost miraculous way

1. Invariant Hilbert–Beltrami integral;
2. Natural and very convincing notion of geodesic fields;
3. Weierstrass' $\mathfrak{E}$ function and the sufficient condition for strong minimum;
4. Legendre sufficient condition for weak minimum;
5. Euler–Lagrange equations without de Bois–Reymond lemma;

6. Hamilton–Jacobi equation and its geometrical interpretation;
7. Notion of transversality and Adolf Kneser transversality theorem;
8. Solution of variational problem with free ends;
9. Problems of broken extremals, so important in optics;
10. Relation of Hilbert integral to integral invariants of Poincaré and its geometric interpretation by Vessiot–Prange (and characterization of geodesic fields by Lagrange brackets.) This gives a natural way to symplectic geometry.

    And last but not least

11. Geometrization of variational problems: extremals as geodesics in the corresponding Finsler space.

This last point was foretold already by Riemann in his 'Habilitationvortrag' and this was, perhaps, the main impulse for Carathéodory method.

7.2 Fields

We look for sufficient conditions for the curve $[t_1, t_2] \ni t \to e(t) \in \mathcal{F} \subset \mathbb{R}^{n+1}$ called the extremal of the functional

(2.1)
$$\mathcal{I}(c) = \int_c L(t, x, p) = \int_{t_1}^{t_2} L(t, x, \dot{x})dt, \quad x = (x_1, \ldots, x_n),\ p = (p_1, \ldots, p_n)$$

to be a local (*strong*) minimum in the class (smooth) curves c neighboring e in the sense of C^0 topology. In the case of a *weak* minimum, one takes a C^1 metric. (In fact, the introduction of these two different metrics was the starting point of not only functional analysis, but also of general topology.)

Thus we demand that

$$\mathcal{I}(c) - \mathcal{I}(e) > 0 \quad \text{for } c \neq e. \tag{2.2}$$

To this end, let us consider an n-parameter family of curves given as integral lines of a vector field

$$\dot{x}^i = \varphi^i(t, x), \quad i = 1, \ldots, n,\ (t, x) \in \mathcal{F}, \tag{2.3}$$

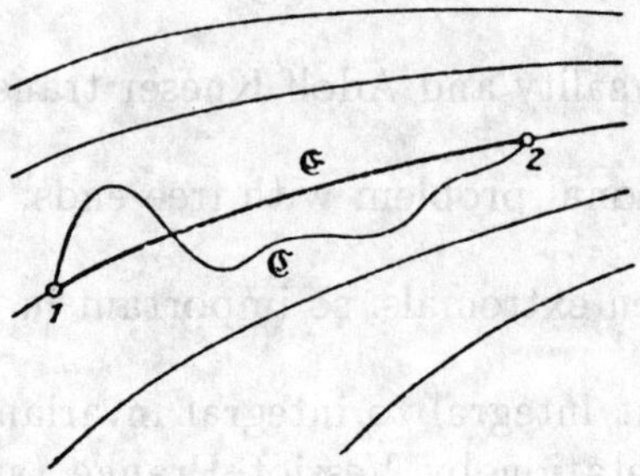

Fig. 1

where $\mathcal{F}$ is a simply connected domain in $\mathbb{R}^{n+1}$. We assume that the Cauchy problem for (2.2) has a unique solution (for example, φ are in $C^1(\mathcal{F})$). If one substitute φ for p in (2.1),

$$\mathcal{F} \ni (t, x) \to L(t, x, \varphi(t, x)) \in \mathbb{R}.$$

If (this, of course, does not happen in general)

$$L(t, x, \varphi) = 0 \quad \text{and} \quad L(t, x, p) > 0 \quad \text{for} \quad p \neq \varphi,$$

then, obviously, the piece of the extremal e of the field $\mathcal{F}$ connecting points 1 and 2 and belonging to $\mathcal{F}$ would be a strong minimum of the integral (2.1), as compared to other curves in $\mathcal{F}$ with the end points 1 and 2.

REMARK. Here we make use of the traditional, suggestive, but not very precise terminology: by a field of curves we understand the n-parametric family of integral curves of the system of ordinary differential equations (2.3) and the support $\mathcal{F} \subset \mathbb{R}^{n+1}$ of the vector field (2.3), which is covered by curves of the field, that is, through every point of $\mathcal{F}$ there pass one and only one curve of the family. Such a notion of the 'field' corresponds to the Faraday's concept and to the classical image of the 'field of force', the field of light rays of geometrical optics, or 'normal congruencies of lines' in algebraic geometry. There is no danger that the reader would mistake a support of vector field with the field itself!

7.3 An equivalent problem

Even though such a wonderful field, as described above, usually does not exist, one can modify the initial functional $\mathcal{I}(c)$ subtracting from it an integral $\mathcal{H}(c)$ which is path-independent, that is, an integral of an exact form $\overset{1}{\omega}= dS$ defined on $\mathcal{F}$. This was the essence of the idea of Hilbert.

$$\mathcal{H}(c) := \int_c dS = \int_c (S_t + S_{x^i} p_i)dt = S_2 - S_1, \tag{3.1}$$

where S_1, S_2 are values of S in the endpoints of c.

It is obvious that (for fixed endpoints) the variational problems for the functionals $\mathcal{I}(c)$ and $\mathcal{K}(c) = \mathcal{I}(c) + \mathcal{H}(c)$ have the same solution. Indeed, since $\mathcal{H}(c) = \mathcal{H}(e)$ (c and e have the same endpoints),

$$\mathcal{K}(c) - \mathcal{K}(e) = \mathcal{I}(c) - \mathcal{H}(c) - (\mathcal{I}(e) - \mathcal{H}(e)) = \mathcal{I}(c) - \mathcal{I}(e).$$

Let us try to check conditions (2.2) for the problem $\mathcal{K}(c)$. For a given field φ we will try to define the function $S(t, x)$ such that the integrand of

$$\mathcal{K}(c) = \mathcal{I}(c) - \mathcal{H}(c) = \int_c (L(t,x,p) - S_t - S_{x^i} p_i)dt \tag{3.2}$$

possesses the required property of the form (2.2):

$$L(t, x, \varphi) - S_t - S_{x^i}\varphi_i = 0 \quad \text{and} \quad L(t, x, p) - S_t - S_{x^i} p_i > 0 \quad \text{for } p \neq \varphi. \tag{3.3}$$

This last function, for fixed (t, x), considered as a function of p, has a minimum at $p = \varphi$. Therefore, taking partial derivative with respect to p, we obtain

$$S_{x^i} = L_{p_i}(t, x, \varphi), \tag{3.4}$$

and if we use (3.3) and (3.2), we obtain

$$S_t = L(t, x, \varphi) - \varphi^i L_{p_i}(t, x, \varphi). \tag{3.5}$$

For existence of a function S satisfying (3.4) and (3.5) it is sufficient that the integrability conditions for the right hand sides of these equations are satisfied.

7.4 Integrability conditions. Geodesic fields. (Independent) Hilbert integral

The integrability conditions are

$$\frac{\partial L_{p_i}}{\partial x^k} = \frac{\partial L_{p_k}}{\partial x^i}, \tag{4.1}$$

$$\frac{\partial L_{p_i}}{\partial t} = \frac{\partial}{\partial x^i}\left(L - \varphi^i L_{p_i}\right), \tag{4.2}$$

where the arguments are, of course, (t, x, φ). If (4.1) is satisfied, then the right hand side of (4.2) is

$$L_{x^j} - \varphi^i \frac{\partial L_{p_i}}{\partial x^j},$$

but the left hand side is

$$\frac{\mathrm{d}L_{p_i}}{\mathrm{d}t} - \frac{\partial L_{p_i}}{\partial x^j}\varphi^j,$$

where $\frac{\mathrm{d}}{\mathrm{d}t}$ denotes the derivative along the field line. Using (4.1) again, we find

$$\frac{\mathrm{d}L_{p_i}}{\mathrm{d}t} = L_{x_i}, \quad i = 1, \ldots, n \tag{4.3}$$

which is nothing but the Euler–Lagrange equations. The field lines are therefore solutions of the Euler–Lagrange equations, the *extremals.* But *not every field of extremals is a geodesic field*; we still have to satisfy condition (4.1).

REMARK. In the simplest case considered already by Weierstrass, $n = 1$ and (3.1) is always satisfied: every field of extremals is a geodesic field; but for $n > 1$ condition (4.1) is very important, and is not an easy problem itself.

The Hilbert–Beltrami 'independent integral' $\mathcal{H}(c)$ is now

$$\mathcal{H}(c) := \int_c \left\{L(t, x, \varphi(t, x)) + (p_j - \varphi_j(t, x))L_{p_j}(t, x, \varphi(t, x))\right\} dt. \tag{4.4}$$

We will investigate this wonderful functional in the next section.

7.5 Weierstrass excess function and condition for strong minimum

Let us now substitute (3.4) and (3.5) into (3.3).

$$L(t,x,p) - (L(t,x,\varphi) - (p_j - \varphi^j)L_{p_j}L(t,x,\varphi) > 0, \quad \text{for } p \neq \varphi. \tag{5.1}$$

Now we define the $\mathfrak{E}$ function (the 'excess function' of Weierstrass)

$$\mathfrak{E}(t,x,p,\bar{p}) := L(t,x,\bar{p}) - L(t,x,p) - (\bar{p}_j - p_j)L_{p_j}(t,x,p), \tag{5.2}$$

we obtain the famous

THEOREM (WEIERSTRASS–HILBERT–CARATHÉODORY) *If a geodesic field φ satisfies the Weierstrass condition*

$$\mathfrak{E}(t,x,\varphi(t,x),p) > 0 \quad \textit{for } p \neq \varphi,$$

then every segment of the field line (e) *with endpoints* 1 *and* 2 *gives the* strong *minimum because*

$$\mathcal{I}(c) - \mathcal{I}(e) = \mathcal{I}(c) - cH(c) = \int_c \mathfrak{E}(t,x,\varphi,p) > 0$$

for all curves c in $\mathcal{F}$ with the endpoints 1 *and* 2.

REMARK. As we will see, the positivity of $\mathfrak{E}$ expresses the convectness of the inditricatrix in a corresponding Finsler space.

But *appetit vient en mangeant*: in a similar way we want to obtain

7.6 Legendre condition for weak minimum

If we expand $L(t,x,\bar{p})$ in a neighborhood of $\bar{p} = p$ in powers of $(\bar{p} - p)$ to quadratic terms, and put linear terms to the left hand side, by virtue of (5.2) we obtain

$$\mathfrak{E}(t,x,p,\bar{p}) = \frac{1}{2}\tilde{L}_{p_ip_k}(\bar{p}_i - p_i)(\bar{p}_k - p_k), \tag{6.1}$$

where the tilde indicates that as the arguments we have to take (t,x) and some point $\tilde{p}$ in the interval $[\bar{p}, p]$. It follows from the continuity of $L_{p_ip_k}$ that

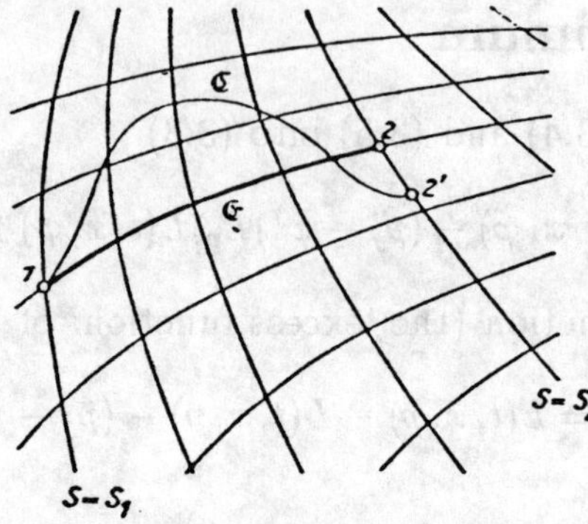

Fig. 2

the quadratic form $(L_{p_i p_k})$ is positive definite for some (t_0, x_0, p_0), and thus the right hand side of (6.1) is positive in some neighborhood of (t_0, x_0, p_0). Therefore we have

THEOREM. *The (generalized) Legendre condition*

$$L_{p_i p_k} \lambda^i \lambda^k > 0 \tag{5.2}$$

for the tangent vector (t, x, p) *of the field extremal* e *joining* 1 *and* 2 *in* $\mathcal{F}$ *and arbitrary* λ *is sufficient for the* weak *minimum of* $\mathcal{I}(c)$ *for all curves* c *with endpoints* 1 *and* 2 *in a neighborhood of* e *in the field* $\mathcal{F}$.

7.7 Complete figure of variational problem

With a geodesic field $\varphi : \mathcal{F} \to \mathbb{R}^{n+1}$ (an n parameter family of field extremals) there is associated a one parameter family of hypersurfaces S =const in $\mathcal{F}$. These hypersurfaces are called *transversal trajectories of the field* φ, and the objects of both families (of curves and transversal hypersurfaces) were called by Carathéodory *complete figure of the variational problem.*

It is intriguing (and almost obvious) that the value of the Hilbert integral $\mathcal{H}(c) = S_1 - S_2$ for any curve c in $\mathcal{F}$ with endpoints on the transversal trajectories $S = S_1$, $S = S_2$, and the value $S_1 - S_2$ is equal $\mathcal{I}(e) = \int_e L$, where the field extremal has endpoints $1 \in (S = S_1)$ and $2 \in (S = S_2$. (Hilbert integral vanishes for any curve on a hypersurface S =const.) Therefore every two transversal trajectories cut out from the geodesic field segments of extremals of the same 'geodesic length' $\mathcal{I}(e)$; hence these transversal hypersurfaces are called 'geodesically equidistant'. This fact is called

TRANSVERSALITY THEOREM OF ADOLF KNESER

For Riemann functional

$$\mathcal{R}(c) = \int_c (g_{ij}\dot{x}^i\dot{x}^j)^{1/2}\, dt, \tag{7.1}$$

(or the energy functional $E(c) = \int_c g_{ij}\dot{x}^i\dot{x}^j\, dt$) the extremals are geodesics and transversal trajectories are orthogonal trajectories of the field, hence the name 'geodesic field'.

In general case, the gradient of S on S =const has components $(L - \varphi^j L_{p_j}, L_{p_i})$, and one says that this hypersurface element is transversal to the line element (t, x, p). More generally, to every line element (t, x, p) there corresponds a transversal hypersurface element $(L - p^j L_{p_j}, L_{p_i})$.

EXAMPLE. FERMAT (–RIEMANN) PRINCIPLE. Let us consider an important case of geometric optics: in a multidimensional inhomogeneous, but isotropic medium (a Riemann manifold (X, g)) travels a light quantum (photon) with the velocity $X \ni x \to v(x) \in \mathbb{R}_+$ (v is understood here in the same way as it was understood by Fermat, namely, as a scalar equal to the norm of velocity.) The optical path of the photon traveling along the curve $c \in X$ with endpoints 1 and 2 is the time needed to travel through the path

$$T(c) = \in_c \frac{ds}{v} = \int_c \frac{(g_{ij}(x)\dot{x}^i\dot{x}^j)^{1/2}}{v(x)} = \int_{t_1}^{t_2} \frac{(g_{ij}(x)\dot{x}^i\dot{x}^j)^{1/2}}{v(x)}\, dt, \tag{7.2}$$

where t is the parameter of the curve c: $[t_1, t_2] \ni t \to c(t) \in X$. According to Fermat principle, light travels along extremals of the functional T, in other words along a geodesic of the Riemann space X equipped with the new metric $h = \frac{g}{v^2}$, that is, $h_{ij} = g_{ij}v^{-2}$. Transversality is then orthogonality in the sense of the metric h, and thus the metric g. Therefore, Fermat principle is a particular case of the theory of geodesics in Riemann space; this is why we called it Fermat–Riemann principle. In order to be able to investigate such geodesics in inhomogeneous and not isotropic optical media, one must leave the framework of Riemann manifolds and use (as Riemann suggested himself) the geometry developed by a pupil of Carathéodory, Paul Finsler.

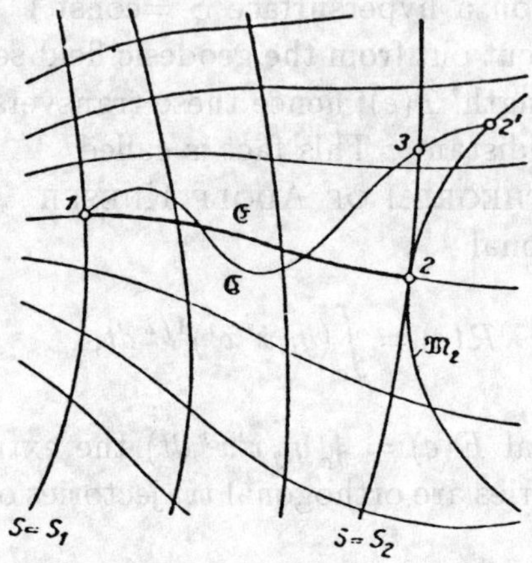

Fig. 3

7.8 Problems with free endpoints. Broken extremals

If the endpoint 2 of c is not fixed, but moves on a surface M, then from the properties of the Hilbert integral $\mathcal{H}$ (it has the same value when the endpoints 1, (resp., 2) of the curve c are on the transversal trajectory S_1 (resp., S_2) and $\mathcal{H}(c) = S_1 - S_2$) there follows a condition for strong and weak minimum which is clear from Fig. 3.

The surface M has to touch the transversal trajectory $S = S_2$ at the endpoint 2 *from outside* (cf. the figure) and the Weierstrass condition ($\mathfrak{E} > 0$) for the strong minimum, and/or the Legendre condition for weak minimum should be satisfied.

Broken extremals. This problem originated from geometric optics: if the medium is not continuous, light passes from one part to another and exhibits refraction. Mathematically, this fact is expressed such that the Lagrange function is discontinuous on some surface D. The *field* φ is *discontinuous* along D but the function S is *continuous* (but not smooth.) The fundamental equation for Hilbert integral $\mathcal{H}(c) = S_1 - S_2$ still holds in this case.

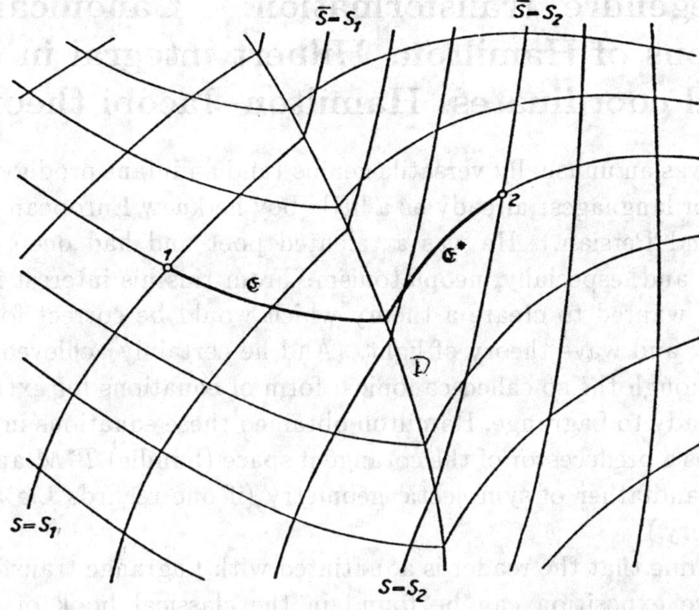

Fig. 4

Since the derivatives of S are given by

$$(8.1) \qquad S_x = L_{p_i}(t,x,\varphi), \quad S_t = L(t,x,\varphi) - \varphi^i L_{p_i}(t,x,\varphi),$$

they are discontinuous on D, but since the values of S on D ant *tangential* derivatives of S on D are well defined, S and *projection* of gradient of S onto D are both *continuous*. Therefore all previous variational problems can be solved for this case as well. We have only to assume the following angle condition of Erdmann

the projection of the vector $(L - \varphi^i L_{p_i}, L_{p_j})$ *onto the surface* D *is continuous*

7.9 Legendre transformation. Canonical equations of Hamilton. Hilbert integral in canonical coordinates. Hamilton–Jacobi theory

Hamilton was an unusually versatile genius (and an infant prodigy with a superb gift for languages: already as a little boy he knew European languages, Turkish, and Persian.) He was a talented poet and had deep interest in philosophy and, especially, neoplatonism. From this his interest in light resulted. He wanted to create a theory which would be correct for both the corpuscular and wave theory of light. (And he certainly achieved that!)

Even though the so called canonical form of equations for extremals was known already to Lagrange, Hamilton obtained these equations in a different way: he was a predecessor of the cotangent space (bundle) T^*M and, in some sense, a grandfather of symplectic geometry (if one regards Lie as a father of this theory.)

We assume that the reader is aquatinted with Lagrange transformations. A very nice exposition can be found in the classical book of Arnold on Mechanics. Here the existence of the Legendre transformation follows from the Legendre condition – and this coincidence is by no means accidental.

Let us denote coordinates on the extended tangent bundle $\mathbb{R} \times TX$ by (t, x^i, p^j), and coordinates on the extended cotangent bundle by (t, x^i, y_j). Let $y = (y_i)$ be the right hand side of

$$L_{p^i} =: y_i, \tag{9.1}$$

and we denote the solution of (9.1) with respect to p^i by ψ^i

$$p^i = \psi^i(t, x, y). \tag{9.2}$$

Since the matrix $(L_{p^i p^j})$ of the Jacobian of (9.1) is positive definite in a neighborhood of $(t, x, \varphi(t, x))$, where φ is a geodesic field, (9.2) is legitimate by virtue of Legendre condition. The hamiltonian function $H : \mathbb{R} \times T^*X \to \mathbb{R}$ of the variational problem is defined to be

$$H(t, x, y) := -L(t, x, \psi(t, x, y)) + \psi^i(t, x, y) y_i. \tag{9.3}$$

This definition is quite natural in our approach (cf. equations (3.4) and (3.5).)

From (9.2) and (9.1) we immediately obtain derivatives of H:

$$H_t = -L_t, \quad H_{x^i} = -L_{x^i}, \quad H_{y_i} = \psi^i. \tag{9.4}$$

For the geodesic field φ, from (3.4) and (3.5), we have

$$S_{x^i} = y_i, \quad S_t = -H(t,x,y), \tag{9.5}$$

and in even more compact form

$$S_t + H(t,x,S_x) = 0 \quad (H-J) \tag{9.6}$$

This is the famous *Hamilton–Jacobi equation.*

Now we see that the function $S(t,x)$ which was the starting point of the whole theory, in the Hamilton theory is the 'action function' of analytical mechanics.

To find a geodesic field, one can start from any solution of equation (9.6): the field φ^i can be then immediately obtained from equation (9.4) and (9.6)

$$\varphi^i(t,x) = \psi^i(t,x,S_x) = H_{y_i}(t,x,S_x).$$

Thus the lines of geodesic field are obtained by integration of an ordinary differential equation

$$\dot{x}^i = H_{y_i}(t,x,S_x). \tag{9.7}$$

If one sets $y_i = S_{x^i}$ and differentiates along so obtained field lines, one gets $S_{tx^i} + H_{x^i} + H_{y_j} S_{x^i x^j} = 0$. Therefore, the field curves satisfy also the equation

$$\dot{y}_i = -H_{x^i}(t,x,S_x). \tag{9.8}$$

Equations (9.7) and (9.8) form the hamiltonian form of Euler–Lagrange equations. Thus we have constructed geodesic field and complete figure.

Instead of starting from the function S, one can start from the curves: to this and, one should replace the Euler–Lagrange equations which form a system of n ordinary differential equations of order 2 with $2n$ ordinary differential equations of order 1

$$\dot{x}^i = H_{y_i}(t,x,y), \quad \dot{y}_i = -H_{x^i}(t,x,y) \tag{9.9}$$

(canonical Hamilton equations) for $2n$ functions $x^i(\cdot)$ and $y_i(\cdot)$. A general solution of the system (9.9) depends on $2n$ parameters. The construction of a geodesic field is a choice of n parameter family of solutions which covers an $(n+1)$ dimensional domain satisfying integrability condition (4.1). These fields have now the following form (for $y_i = \psi_i(t,x)$)

$$\frac{\partial \psi_i}{\partial x^k} = \frac{\partial \psi_k}{\partial x^i}, \tag{9.10}$$

and

$$\frac{\partial y_i(t,x,\psi(t,x))}{\partial t} = -\frac{\partial H(t,x,\psi(t,x))}{\partial x^i}.$$

Finally, the *Hilbert integral in canonical coordinates* takes the form

$$\mathcal{H}(c) = \int_c y_i dx^i - H dt, \tag{9.11}$$

where $H = H(t,x,y) := H(t,x,\psi(t,x))$, $y_i = \psi_i(t,x)$.

REMARK. The differential one form $\overset{1}{\omega}10 := y_i dx^i - H dt$ on the extended phase space $\mathbb{R} \times M^{2n} = \mathbb{R} \times T^*\mathbb{R}^n$ was introduced by Poincaré and is called *Poincaré–Cartan integral invariant.* The Hilbert integral is obtained from this invariant by integration, as in (9.11).

It is natural to ask what is the

7.10 Physical meaning of functions H, S, and L

In mechanics the Lagrangian $L = T - U$, where T and U are, respectively, the kinetic and potential energy of the system. Then

$$\psi^i y_i = p^i L_{p^i} = p^i T_{p^i} = 2T$$

(we assume that the kinetic energy is a quadratic form in (p^i), and the last equality follows from Euler formula for homogeneous functions.) Therefore

$$H = -L + 2T = T + U = \text{total energy}.$$

Thus the Hamilton function H is *the total energy of the system.* Next

$$\frac{\mathrm{d}H}{\mathrm{d}t} = H_t + H_{x^i}\dot{x}^i + H_{y_i}\dot{y}_i = H_t$$

by virtue of (Hamilton) canonical equations. If $L_t = 0$, then $H_t = 0$ and thus both L and H are time independent. Then $\frac{\mathrm{d}H}{\mathrm{d}t} = 0$ and *the total energy is constant on every extremal.*

The (characteristic) function $S(t,x) = \int_{t_1}^{t} L(t,x,p)dt$ is called *action* in mechanics and *eiconal* in optics. This function, as a function of endpoints 1 and 2 is denoted by $W(1,2)$ or $W(P_1,P_2)$.

HISTORICAL REMARK. Historically the 'royal road' proceeds in opposite order: first there were Newton equations, formulated in modern, complete way only by Euler

$$(*) \qquad \frac{\mathrm{d}}{\mathrm{d}t}(m\dot{x}) = -U(x).$$

Since $T = \frac{1}{2}m|\dot{x}|^2$, $L_{\dot{x}^i} = T_{\dot{x}^i} = m\dot{x}^i$, $L_{x^i} = -U_{x^i}$, we obtain

$$(E-L) \qquad \frac{\mathrm{d}L_{\dot{x}^i}}{\mathrm{d}t} = L_{x^i},$$

therefore *the trajectories of the system* $(*)$ *are extremals of the functional* $\mathcal{I}(c) = \int_c L = \int_{t_1}^{t_2} L dt$, where $L = T - U$. This is the classical form of the Hamilton principle of least action.

Here it is worth, and even necessary to present the words of the eighteen years old young man from his epochal work *Theory of Systems of Rays* (Part I)

> *The reasoning I have presented here is independent of any assumptions concerning the nature and velocity of light, but, by analogy, I will keep the name 'Principle of least action'.* In the appendix to Part II, Hamilton writes *In my preceding treatise I proposed to call the result the 'Principle of least action'. In this way it was to be stressed that this principle gives directly the differential equations for the right class of surfaces, which under assumption of wave hypothesis are wave surfaces (wave fronts K.M.), and under assumption of emission hypothesis should be called surfaces of constant action. In the present appendix I propose to call the fundamental formula the 'equation of characteristic function' which is free of any assumption on the nature of light.*

The importance of these few sentences of Hamilton in the history of mathematics and physics can be compared only with Riemann's *'Habilitationsvortrag'*.

This is the reason why Hamilton optics and mechanics survived the major crisis and birth of new theories in physics. It was an impulse leading to Hilbert variational principle, letting him derive Einstein–Hilbert equations, and to the quantum theory of Schrödinger. The Hamilton theory influenced most directly the Lie theory of vector fields which culminated in the modern symplectic geometry, about which we will tell in the following chapter.

But let us return to the integrability conditions for geodesic field in the canonical form which immediately leads us to

7.11 Lagrange bracket and geodesic fields

Let

$$x^i = \xi^i(t, u_1, \ldots, u_n), \quad y_i = \eta_i(t, u_1, \ldots, u_n), \quad i = 1, \ldots, n \tag{11.1}$$

be an n parameter family of solutions of canonical equations.

The Lagrange bracket is defined to be

$$[u_\alpha, u_\beta] := \frac{\partial \xi^i}{\partial u_\alpha}\frac{\partial \eta_i}{\partial u_\beta} - \frac{\partial \eta_i}{\partial u_\alpha}\frac{\partial \xi^i}{\partial u_\beta} \quad \alpha, \beta = 1, \ldots, n. \tag{11.2}$$

The integrability condition $\frac{\partial \psi_i}{\partial x^k} = \frac{\partial \psi_k}{\partial x^i}$ gives $\eta_i(t, u) = y_i(t, \xi)$, $\frac{\partial \eta_i}{\partial u_\beta} = \frac{\partial y_i}{\partial x^j}\frac{\partial \xi^j}{\partial u_\beta}$, and thus

$$[u_\alpha, u_\beta] = \frac{\partial \xi^i}{\partial u_\alpha}\frac{\partial \xi^j}{\partial u_\beta}\left(\frac{\partial y_i}{\partial x^j} - \frac{\partial y_j}{\partial x^i}\right) \tag{11.3}$$

and we proved

PROPOSITION. *(a) A family* (11.1) *of solutions of Hamilton equations is a geodesic field, that is, a solution of the integrability conditions* (9.10) *if and only if all Lagrange brackets vanish*

$$[u_\alpha, u_\beta] = 0 \quad \alpha, \beta = 1, \ldots, n. \tag{11.4}$$

(b) Moreover if an n parameter family of solutions of canonical equations forms a field, then the Lagrange brackets are constant on every trajectory of the field, that is, the Lagrange brackets are first integrals of canonical equations.

PROOF OF (B) follows from the equation

$$\frac{\partial}{\partial t}\left(\frac{\partial \xi^i}{\partial u_\alpha}\frac{\partial \xi^j}{\partial u_\beta}\right) = H_{y_j y_k}\frac{\partial \eta^j}{\partial u_\alpha}\frac{\partial \eta^k}{\partial u_\beta} - H_{x_j x_k}\frac{\partial \xi^j}{\partial u_\alpha}\frac{\partial \xi^k}{\partial u_\beta}$$

and the symmetry of the right hand side with respect to α and β.

REMARK. The role of Lagrange brackets is fully appreciated in symplectic geometry. If Ω is the symplectic two form, then for two hamiltonian vector fields X and Y, one defines their Lagrange bracket to be

$$(X, Y) = \Omega(X, Y). \tag{11.5}$$

7.12 Canonical transformations

Let X be a differentiable manifold, usually the configuration space of a mechanical system, and T^*X its cotangent bundle (in mechanics, the phase space of a system) provided with the canonical symplectic two form $\Omega := dp_i \wedge dq^i$, where (q^i, p_i) are canonical coordinates on T^*X (in the preceding sections we denoted q^i by x^i and p_i by y_i.) The symplectic form is non degenerated and closed, $d\Omega = 0$. In general case, a pair (M^{2n}, Ω), where M^{2n} is an $2n$ dimensional differentiable manifold and Ω is a non degenerated and closed two form is called symplectic manifold. By Darboux theorem any symplectic manifold is locally isomorphic to some cotangent bundle.

Automorphisms of $(T^*M, dp \wedge dq)$ are called *canonical transformations* , and automorphism of symplectic manifolds are called symplectic transformations or symplectomorphisms.

A vector field ξ_H on T^*X, given in local coordinates (q^i, p_i) by

$$\xi_H f = \frac{\partial f}{\partial q^i}\frac{\partial H}{\partial p_i} - \frac{\partial f}{\partial p_i}\frac{\partial H}{\partial q^i},$$

where $H : T^*X \to \mathbb{R}$ has as its trajectories solutions of canonical equations

$$\dot{q}^i = \frac{\partial H}{\partial p_i}, \quad \dot{p}_i = \frac{\partial H}{\partial q^i}.$$

The one parameter group given by ξ_H (a phase flow) is a one parameter group of canonical transformations and will be denoted by $\varphi_H(t)$, $t \in \mathbb{R}$.

Now we can present the invariant definition of Lagrange bracket. Let $\mathcal{U}$ be a coordinate domain in X and $\Phi : \mathcal{U} \to T^*X$ a differentiable map; then $\Phi^*(\Omega) = \Phi^*(dp \wedge dq)$ is called Lagrange form. In coordinates

$$\Phi^*(\Omega) = \sum \frac{\partial(p_i, q^i)}{\partial(u_\alpha, u_\beta)} du_\alpha \wedge du_\beta = \sum [u_\alpha\, u_\beta] du_\alpha \wedge du_\beta,$$

where $[u_\alpha\, u_\beta] = \frac{\partial(p_i,q^i)}{\partial(u_\alpha,u_\beta)}$ is the Lagrange bracket of section 11; they are the u_α, u_β components of the form $\Phi^*(\Omega)$. From the invariance of Ω with respect to canonical transformation the properties of Lagrange brackets follow: they are first integrals of canonical equations.

An interesting geometric characterization of Lagrange brackets we owe Prange and Vessiot.

We know that invariance of the Hilbert integral $\mathcal{H}(c)$ is equivalent that the field $\mathcal{F}$ (of extremals) is geodesic. For an arbitrary field of extremals $\mathcal{F}$

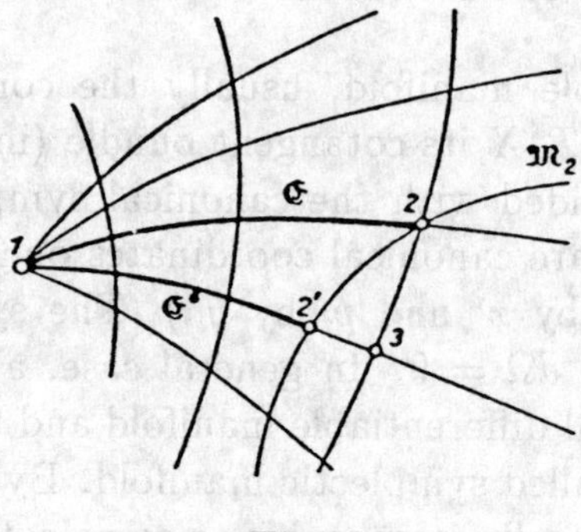

Fig. 5

the integral $\mathcal{H}(c)$ along closed curve does not vanish in general, however it has the same value for any two closed curves c_1 and c_2 embracing a tube $\overset{2}{\sigma}$ of field curves (see Fig. 5.)

This fact follows almost immediately from the Poincaré–Stokes theorem. Indeed, denoting by ω the one form in the Hilbert integral

$$\mathcal{H}(c) = \int_c \omega$$

we find

$$\mathcal{H}(c_1) - \mathcal{H}(c_2) = \int_{\overset{2}{\sigma}} d\omega = 0.$$

Therefore ω is a relative integral invariant since it is constant on closed lines, but the integral on the right hand side vanishes because $d\omega|_{\overset{2}{\sigma}} = 0$: the field lines are characteristic lines of the form $d\omega$. We have

COROLLARY (PRANGE–VESSIOT). (Fig. 6.) *Let $c = \gamma \cup e$, where γ is a line on the tube $\overset{2}{\sigma}$ transversal to the field lines and e is an extremal. Then $\mathcal{H}(\gamma) = \int_e \omega$.*

We have an interesting 'geometric' or 'optical' interpretation of the Lagrange bracket $[u_\alpha, u_\beta]$ (see Fig. 7.) In u_α, u_β plane consider a 1 parameter family of simply connected domains $D(\tau)$ such that $D(\tau) \downarrow (0,0)$ for $\tau \to 0$. Let $\overset{2}{\sigma}\,(\tau)$ be the tube of field extremal emanating from the boundary $c(\tau) := \partial\, \overset{2}{\sigma}$. Then

$$[u_\alpha, u_\beta]_{\tau=0} = \lim_{\tau\to 0} \int_{\overset{2}{\sigma}(\tau)} [u_\alpha, u_\beta] du_\alpha \wedge du_\beta |\, \overset{2}{\sigma}\, |^{-1} =$$

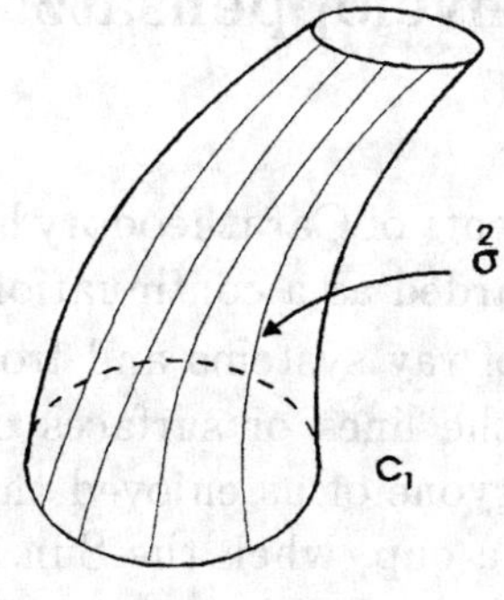

Fig. 6

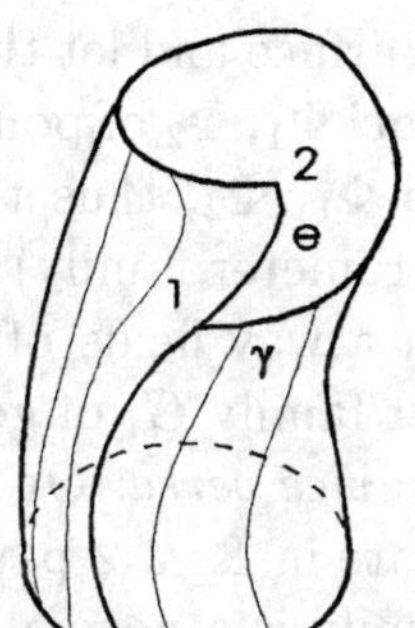

Fig. 7

$$= \lim_{t\to 0} \int_{c(\tau)} dp_i \wedge dq^i | \overset{2}{\sigma} |^{-1}.$$

Therefore we once more proved that $[u_\alpha, u_\beta]$ is constant on each field line.

THEOREM OF MALUS (1808) which is fundamental for classical geometric optics asserts that (in $\mathbb{R}^3$) a field of light rays after reflection and refraction is still a field of rays is therefore a very special case of the theorem saying that Lagrange brackets are constant. It is very interesting that in the same year (1808) Lagrange introduced his bracket and proved that they are constant on extremals.

7.13 Caustics. 'Enveloppensatz' of Carathéodory. Singularities

A number of great achievements of Carathéodory has their roots in geometric optics, and thus can be regarded as a continuation of ideas of Hamilton.

Caustics are envelopes of ray systems and from ancient times they fascinated people – they are the lines or surfaces of 'burning', since light is concentrated on them. Everyone of us enjoyed the metamorphoses of caustics on the inner surface of a cup, when the Sun shines on it. As noted by Arnold, a rainbow in the sky is also due to a caustic of the system of rays that passes through a droplet of water with complete internal reflection.

In the theory of surfaces in $\mathbb{R}^3$ there is a famous theorem of Weingarten (1861) which says, among others, what follows.

Let $S \subset \mathbb{R}^3$ be a smooth surface and let there exist a field of straight lines orthogonal to S. Take both loci Φ_1, Φ_2 of points of principal curvatures. The normals of S are tangent to Φ_1, Φ_2, thus they are caustics (envelopes) of the normal congruence (2 parameter family) of geodesics in $\mathbb{R}^3$. Weingarten proves that obtained in such a way fields of vectors tangent to Φ_i, $i = 1, 2$ generate on Φ_i a 1 parameter family G_i of geodesics, and thus two geodesic fields on Φ_1 and Φ_2. But also *vice versa*: On a surface $\Phi \subset \mathbb{R}^3$ take a field of tangent vectors. They generate in $\mathbb{R}^3$ a 2 parameter family of straight lines, therefore, geodesics. This family is a normal congruence if the resulting one parameter family of curves is a family of geodesics of the surface Φ.

CARATHÉODORY THEOREM of 1923 generalizes the Weingarten theorem into the case of a field of extremals of general variational problem for the functional $\mathcal{I}(c) = \int_c L(t, x^1, \ldots, x^n, p_1, \ldots, p_n)$ as follows

An n parameter family of extremals of $\mathcal{I}(c)$ which are tangent to an n dimensional hypersurface Φ in $\mathbb{R}^{n+1}$ is a geodesic field if and only if the line elements generated by these extremals form a geodesic field of a new variational problem on the space Φ.

In other words

(Extremals of $\mathcal{I}(c)$ for a geodesic field) $\Longleftrightarrow$ *(Extremals of the induced problem $\mathcal{I}(c)|_\Phi$ form a geodesic field on Φ)*

If one regards the Weingarten's theorem as a statement concerning optics in the space $\mathbb{R}^3$ which is homogeneous and isotropic, then the Carathéodory theorem may be regarded as a generalization of the former to the case $n > 2$ and/or optical media that are neither homogeneous nor isotropic.

EXERCISE. Prove this theorem by making use of Hilbert–Beltrami integral (in the canonical form), Carathéodory criterion, and Lagrange brackets.

7.14 Finsler geometry and geometric optics

As we saw, Carathéodory himself proved a number of theorems which are considered now to belong to the theory of Finsler spaces. He encouraged his pupil Paul Finsler (1894–1970) to develop differential geometry of n dimensional manifolds with length of a curve c given by $\int_c L(x, \dot{x})$, where $L(x,)$ is a positively homogeneous function of order one with respect to $\dot{x} = (\dot{x}_1, \ldots, \dot{x}_n)$, symmetric, and convex. (Such a research program was suggested already by Riemann.) In other words, the tangent spaces T^*M are normalized (Minkowski) spaces. We have mentioned such spaces already several times. Thus Riemannian manifolds are Finsler manifolds whose tangent spaces are Euclidean, that is, the indicatrices $F(x, \cdot) \leqslant 1$ are Euclidean balls.

Under the supervision of Carathéodory the famous Ph.D. thesis of Finsler has been written in 1918. We know how fundamental role the Finsler metric plays in the Teichmüller theory.

Finsler was an exceptionally original scientist. In 1922 he got the habilitation degree form the University of Köln. Since 1927 he was a professor of applied mathematics, and in 1944 he got the position of professor at the University of Zürich. He wrote a number of papers on astronomy, discovered 2 comets, worked on probability theory and elementary theory of numbers. He had very original approach to foundations of mathematics. His metaphysical inclinations were exemplified in the work *Über das Leben nach dem Tod.*

In this section we do not intend to present a detailed exposition of Finsler geometry, but to draw the reader's attention to the problem of geometrization of the variational problem of the functional (t independent)

$$\mathcal{I}(c) = \int_c L(x, \dot{x}), \tag{14.1}$$

which, without doubts, concerned already Riemann, I will say a few words about Finsler geometry.

As we know, the tangent spaces T_xM of Finsler space (M, L) are spaces introduced by Minkowski in his monograph *Geometrie des Zahlen.* Minkowski observes that to introduce a metric in $\mathbb{R}^n$ it is sufficient and necessary to define a *convex body* $W \subset \mathbb{R}^n$ defined by the norm $\| \quad \|$, $W = \{x \in \mathbb{R}^n : \| x \| = 1\}$. And *vice versa*, given Minkowski convex body W, the metric $\| \quad \|_W$, called also the Minkowski functional is given by $\| x \|_W = \sup_{t \geqslant 0}\{tx \in W\}$. The set W is clearly symmetric with respect to 0 (since $\| -x \|_W = \| x \|_W$) and convex (since $\| x + y \|_W \leqslant \| x \|_W + \| y \|_W$). In the dual space $(\mathbb{R}^n)^*$ we have a polar set, denoted usually by $W^0 = \{p \in (\mathbb{R}^n)^* : < x, p > \leqslant 1, \quad \text{for all } x \in W\}$. In the calculus of variations, the boundary of W, ∂W is called *indicatrix* and will be denoted by F; it is given by equation $\| x \|_W = 1$. Its dual (polar) hypersurface was called by Carathéodory the *figuratrix* of the variational problem (14.1).

Carathéodory introduced these hypersurfaces in a general case as well (without convexity and symmetry condition for $L(x, \cdot)$), and then the positivity of the Weierstrass $\mathfrak{E}$ function means the convexity of the indicatrix. The investigation of *not* convex indicatrices by Carathéodory was motivated not by the desire of maximal generality: in the optics of birefractional crystals, the wave surface is not convex and has a very interesting geometry (Fresnel wave surface; we will talk more about it while discussing the $K3$ surfaces.) This is an algebraic surface of (real) dimension 4 with singularities; the indicatrix F (wave surface) of such variational problem has singular tangent planes which are tangent to F along a *circle.* Now light is not refracted in a well defined direction, as in the case of doubly tangent plane which contacts F in two points, but along the whole of a cone.

This is the famous *conical refraction* discovered theoretically by Hamilton in 1832. Hamilton's prediction was verified experimentally shortly afterwards by H. Lloyd. This phenomenon was regarded as an important argument in favor of the wave theory of light. The results of investigations of Carathéodory on indicatrices are contained in his Habilitationarbeit written in 1905 in Göttingen (Math. Ann. **62** (1906), 449–503), less than a year after his Ph.D. This work was considered quite sensational because it was thought that refraction of extremals may be only due to non smoothness of the Lagrangians. Carathéodory showed that this is a necessary phenomenon: if one extends an arbitrary strong extremal sufficiently far, a moment comes when it cease to be strong. He proved that at that point, the line element p, together with another line element $\bar{p}$, satisfy the Erdmann condition, and both segments form a discontinuous strong extremal.

In geometric optics, indicatrix is called the *wave surface* or the *ray sur-*

face, and figuratrix, the *normal surface*. (Blaschke gave a simple 'geometric' interpretation of figuratrix.) Now the notion of transversality has a very natural interpretation: vector $\dot{x}$ is transversal to covector p if $p_i\dot{x}^i = 0$.

EXAMPLE. In Riemannian geometry (M, g)

$$L(x, \dot{x}) = (g_{ik}\dot{x}^i\dot{x}^k)^{1/2} \tag{14.2}$$

and transversality means orthogonality. The equation of indicatrix reads

$$g_{ik}q^iq^k = 1, \tag{14.3}$$

and of figuratrix

$$g^{ik}p_ip_k = 1.$$

Without any problem the reader will write down the form of the Hilbert integral. Extremals of the functional (14.2) are geodesics of Finsler geometry. The complete figure of Carathéodory has now a beautiful geometric interpretation: the geodesic field is now a family of geodesics transversal to a hypersurface in M. The 'Enveloppensatz' of Carathéodory is a truly geometric theorem: canonical equations (of Hamilton are equations of geodesics (in dual coordinates) on Finsler manifold.

The reader wonders perhaps if Huygens principle has a natural interpretation in Finsler geometry? The answer is yes.

7.15 General Huygens principle and Finsler geometry

Now we can present the fundamental theorems of calculus of variations in the following optical–Finslerian language.

Let the indicatrix $F(x) \subset T_xM$, $x \in M$ be convex (thus the dual surfaces $H(x) \subset T_x^*M$ are convex as well), and let the corresponding Hamilton function $H(x, p)$ be positively homogeneous of degree 1 with respect to $p = (p_1, \dots, p_n)$. Then the one parameter family of homogeneous canonical transformations $\Phi^H(t)$, $t \in \mathbb{R}$ given by canonical equations $\dot{q} = H_p$, $\dot{p} = -H_q$ generates a complete figure: a geodesic field transversal to the hypersurface $S_0 = \{S(x) = 0\}$, where S is a solution of the Hamilton–Jacobi equation.

We can interpret the family $S_t = \{x \in M : S(x) = t\}$ as a wave emanating from S_0 and S_t as a family of the wave fronts, and the time t as an optical distance between surfaces S_0 and S_t. The $\dot{x}$ convexity of L (respectively p

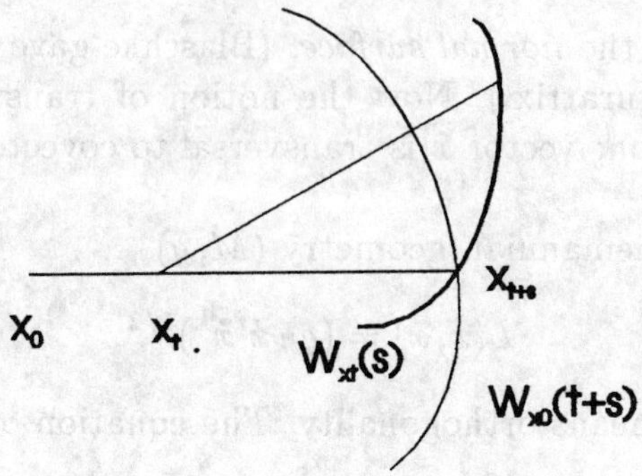

Fig. 8

convexity of H) assures the positivity of the corresponding Weierstrass $\mathfrak{E}$ function, hence the trajectories of our field (light rays) are extremals which provide a strong minimum of the problem with fixed endpoints, and even among all curves joining S_{t_1} and S_{t_2}.

In this way we have almost proved the generalization of the celebrated Huygens principle. Denote by $W_x(t)$ the wave front of the wave emanating from the point x after time t. Consider the wave front $W_{x_0}(t)$; for every $x \in W_{x_0}(t)$ take the wave front $W_x(s)$. Then the wave front $W_{x_0}(t+s)$ is an envelope of the family $\{W_x(s) : x \in W_{x_0}(t)\}$ of wave fronts.

PROOF. Let $x_{t+s} \in W_{x_0}(t+s)$, then there exist a field extremal (ray) with endpoints x_0 and x_{t+s} which is of shortest optical distance. Take the point x_t on this ray with optical distance t from x_0; clearly this segment realizes the shortest way between x_0 and x_t, otherwise the field extremal x_{t+s} would not be shortest. therefore $x_t \in W_{x_0}(t)$. The same reasoning gives $x_{t+s} \in W_{x_t}(s)$. We prove that the wave fronts $W_{x_t}(s)$ and $W_{x_0}(t+s)$ are tangent at x_{t+s}. Indeed if the fronts intersect, then the ray could reach some points of the front $W_{x_0}(t+s)$ from the point x_t in time shorter than s, and thus from the point x_0 in time shorter than $t+s$. This contradicts the definition of the wave front $W_{x_0}(t+s)$. Thus the wave fronts $W_{x_t}(s)$ and $W_{x_0}(t+s)$ are tangent. □

Of course we do not know how Huygens found his principle. There are no recipes for brilliant intuition and inspiration. But we can trace the way of the great Dutch scientist to his theory of envelops. Huygens invented cycloid pendulum: a simple (mathematical) pendulum has constant period only for small amplitudes. In 1673 Huygens discovered that if the thread of the pendulum is a family of cycloid arcs, then the period of the pendulum is constant for any amplitude (cycloid is tautochronic.) As a result of this discovery, Huygens formulated the theory of evolutes, evolvents, and envelopes, and he started the theory of caustics.

Cycloid is a wonderful curve and it played an important role in early times of calculus of variation: it was the line of steepest descent. The famous 'brachystochrone problem' (brachystos=shortest, chronos=time) was a challenge to the mathematical world and was posed in June 1696 by Johannes Bernoulli in the following words:

Given points A and B in a vertical plane, find the path AMB down which a movable point M must, by virtue of weight, proceed from A to B in the shortest possible time.

Johannes Bernoulli published his solution of the brachystochrone problem in the 1697 issue of 'Acta Eruditorum', when he remarked that 'one will be astounded when I say that cycloid, the tautochrone of Huygens, is the sought brachystochrone.'

At almost the same time the solution was found by Leibnitz and Newton. Leibnitz notes that Galileo struggled with the same problem, without success, already in 1638. We see that some problems and their solutions 'are in the air' and 'condense' in minds of, sometimes, several geniuses. This is the reason for the embarrassing discussions concerning priority.

The Huygens ideas however appeared only in his works, and perhaps for that reason were not accepted by his contemporaries; they wait 125 years to be waken up by Fresnel and Hamilton, in the most glorious form. The Huygens principle is of, I think, not only physical and mathematical importance: this is a general philosophical principle; it can be regarded as a principle of life in the Neo-Platonic '*Kosmos noetos*': one wants to realize great ideas: all people, the elements of the spiritual organism – the humanity – are analogs of elementary Huygens waves; a perturbation (for example, moral or social idea etc.) reach somebody, who becomes virtually this elementary wave which may be a source emitting this idea or spiritual impulse farther on. In this way a (global) wave reaching larger and larger circles, an 'envelope' of this elementary waves.

But let us return to the classical Huygens principle. As we have already

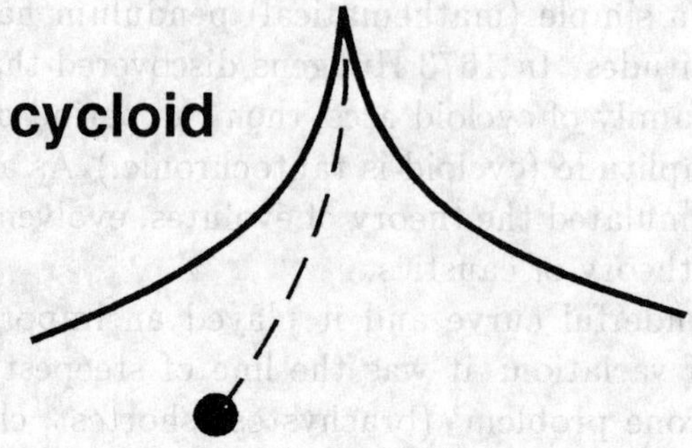

Fig. 9

observed, we must operate with 1. geodesic fields, or, in the language of Finsler geometry, transversal congruencies. Huygens operated with central fields which, as we know, after removal of the center, are geodesic; 2. To be sure that field lines are strong minima of the functional $\mathcal{I}(c) = \int_c L(x, \dot{x})$, positivity of the Weierstrass $\mathfrak{E}$ function must be assumed; this property, in the case of Finsler geometry means convectness of indicatrices. Huygens operated with spheres in $\mathbb{R}^3$, which are obviously convex, thus his reasoning is justified.

7.16 Field theories for calculus of variation for multiple integrals

Perhaps the most original result of all Carathéodory research on calculus of variations was creation of geodesic fields for multiple integrals.

As we have stressed, already in the case of single integrals, and thus extremals being lines, there is an abyss between the situation $n = 2$ and $n \geqslant 3$: not every field of extremals is geodesic for $n \geqslant 3$. But there is a *single* notion of geodesic. As history showed, for $\mu > 1$, there is a number of inequivalent notions of geodesics and corresponding theories. The first such theory was built by Carathéodory: in his approach there exist an analog of the Hilbert integral, it can be applied for investigations of problems with moving boundary, there exists an analog of Legendre transformation, but, alas, there does not exist a natural notion of transversality.

A completely different theory was developed by Belgian mathematician de Donder in a cycle of works published in 1935. Independently, in the same year H. Weyl published an important work *Geodesic fields in calculus of variations for multiple integrals* Ann. of Math. **36** (1935), 607–629, in which he went beyond formal formulas of de Donder: Weyl was able to construct appropriate geodesic fields.

The Belgian mathematician Lepage (whom we owe the term geodesic fields) observed that geodesic fields of Carathéodory and de Donder–Weyl are examples of an infinite family of geodesic fields corresponding to special differential forms, leading to different invariant integrals, which, in the case $\mu = 1$ reduce to the Hilbert–Beltrami integral. It was shown by Boerner that the Carathéodory theory is a single field theory that can be applied to all problems with moving boundaries.

Since the Lepage–Boerner theory makes use of the theory of differential forms (which was not well known at the time), I will formulate, following Boerner, the main results of the Carathéodory theory for single integrals in the language of Pfaff forms, so that the generalization to multiple integral will be straightforward.

Thus again the functional $\mathcal{I}(c)$ has the form

$$\mathcal{I}(c) = \int_c L(t, x, \dot{x})dt, \quad x = (x_1, \ldots, x_n). \tag{16.1}$$

Consider the one form $\omega := L(t, x, p)dt$ in $\mathbb{R}^{2n+1}$ and n Pfaffian forms (one forms)

$$\omega_i = dx_i - p_i dt, \quad i = 1, \ldots, n. \tag{16.2}$$

Instead of considering (16.1), we take the integral $\int \omega$ in $\mathbb{R}^{2n+1}$ taken along the integral curve of the Pfaff system $\omega_i = 0$, $i = 1, \ldots, n$, obtained by taking x_i as functions of t and putting $p_i(t) = \dot{x}_i(t)$. We have indeed

$$\mathcal{I}(c) = \int_{\tilde{c}} \omega, \tag{16.3}$$

where $\tilde{c} \subset \mathbb{R}^{2n+1}$ corresponds to c.

Now the decisive moment comes.

It is clear that we obtain the same (minimum) problem an the same solution of it, if we replace the form ω by a one form Ω such that

$$\Omega \equiv \omega \ (\mathrm{mod} \omega_i), \quad \text{that is,} \ \Omega = \omega + A_i(t, x, p)\omega_i. \tag{16.4}$$

It is useful to take the function A_i such that $d\Omega = 0(\mathrm{mod}\omega_i)$. It can be easily checked that then one should take $A_i = L_{p_i}$ and we obtain

$$\Omega = Ldt + L_{p_i}\omega_i = (L - p_i L_{p_i})dt + L_{p_i}dx_i, \tag{16.5}$$

the well known Hilbert–Poincaré–Cartan form, and

$$d\Omega = (dL_{p_i} - L_{x_i} dt)\omega_i. \tag{16.6}$$

The geodesic field appears in a very natural way: in $\mathbb{R}^{2n+1}$ we take an $(n+1)$ dimensional surface

$$p_i = p_i(t, x).$$

Integrating the system of ordinary differential equations $\dot{x}_i = p_i(t, x_j)$ on the surface we obtain integral curves of the Pfaffian system $\omega_i = 0$. The corresponding curves in $\mathbb{R}^{n+1}$ form a field $\mathcal{F}$.

We adopt the following convention: if in $f(t, x_j, p_j)$ we replace p_j by a given function of x_j and t, the resulting function on $\mathbb{R}^{n+1}$ is denoted by $[f]$.

DEFINITION. A field $\mathcal{F}$ is called *geodesic field* if $d[\Omega] = 0$, that is, if the form $[\Omega]$ is closed.

We check that

$$d[\Omega] = \left(\frac{\mathrm{d}L_{p_i}}{\mathrm{d}t} - L_{x_i}\right) dt \wedge \omega_i + \sum_{i<j} \left(\frac{\partial L_{p_j}}{\partial x_i} - \frac{\partial L_{p_i}}{\partial x_j}\right) \omega_i \wedge \omega_j,$$

and thus we obtain

PROPOSITION. *The geodesic field must satisfy*

$$\frac{\mathrm{d}L_{p_i}}{\mathrm{d}t} - L_{x_i} = 0 \tag{$E-L$}$$

and the integrability condition

$$\frac{\partial L_{p_j}}{\partial x_i} - \frac{\partial L_{p_i}}{\partial x_j}. \tag{16.7}$$

The invariant (Hilbert) integral is given by

$$\mathcal{H}(c) = \int_c [\Omega] = \int_{\bar{c}} [\Omega] = \int_{\tilde{c}} \omega \tag{16.8}$$

for every $\bar{c}$ with the same endpoints as c.

Now we turn to $\mathfrak{C}$ function. In a simply connected domain $\mathcal{F} \subset \mathbb{R}^{n+1}$ it follows from $d[\Omega] = 0$ that there exist a function $S(t, x)$ such that

$$[\Omega] = dS, \tag{16.9}$$

and the Hamilton–Jacobi equations

$$[L_{p_i}] = \frac{\partial S}{\partial x_i}, \quad [L - p_j L_{p_j}] = \frac{\partial S}{\partial t}. \tag{16.10}$$

Thus we obtained the basic elements of the Carathéodory theory, and it is not necessary to repeat definitions of transversality, complete figure, etc.

Instead, we turn to

7.17 Lepage theory of geodesic fields

In what follows we will omit the symbol of exterior multiplication: instead of $\omega_1 \wedge \cdots \wedge \omega_k$ we will simply write $\omega_1 \cdots \omega_k$. Here ω_i denotes a one form (called in classical analysis Pfaff form.)

We do not intend to present here a complete exposition of the theory, our goal is to show the existence of an infinite number of geodesic fields in the case of multiple integrals and to distinguish, among them, two which are particularly simple: they both have their virtues and drawbacks, and are, certainly, not 'given by God.'

First we introduce the notion of *class* and *rank* of a differential form ω. These notions are basic in the theory of the so called Pfaff problem, about which we will tell more in the next section.

The rank of the form ω is the minimal number of one forms needed to represent ω. For a p form $\overset{p}{\omega}$ the rank $\overset{p}{\omega} = p$, since $\overset{p}{\omega} = \omega_1 \cdots \omega_p$.

The class of the form ω is the minimal number of variables (coordinates) needed to represent ω. Of course, in general class, $\omega >$ rank ω, but there is an important

THEOREM. *For a closed form ω its class and rank are equal:* $(d\omega = 0) \Longleftrightarrow$(rank $\omega =$ class ω).

Our integrand is now $L(t^1, \ldots, t^\mu, x_1, \ldots x_n, p_{11} \ldots, p_{n\mu}) \equiv L(t^\alpha, x_i, p_{i\alpha}) \equiv L(t, x, p)$, where $\mu > 1$. We assume $L > 0$. We have to find a μ surface, that is, a μ chin c_μ in $\mathbb{R}^{n+\mu}$, $x_i = x_i(t^1, \ldots, t^\mu) \equiv x_i(t)$ such that

$$\mathcal{I}(c_\mu) = \int_{c_\mu} L(t, x, p) = \int_{c_\mu^*} L\left(t, x, \frac{\partial x_i}{\partial t^\alpha}\right) dt \tag{17.1}$$

is minimal; c_μ^* is the projection of c_μ onto $(t^1, \ldots, t^\mu)$ surface. In the problem with fixed boundary, the boundary ∂c_μ of c_μ is fixed.

Thus in the space $\mathbb{R}^{\mu+n+\mu n}$ of $(t^\alpha, x_i, p_{i\alpha})$ we have a μ form $\overset{\mu}{\omega} \equiv \omega := L(t,x,p)dt^1 \cdots dt^\mu$. As in section 16 we introduce n Pfaffian forms

$$\omega_i := dx_i - p_{i\alpha}dt^\alpha, \quad i = 1, \ldots, n \tag{17.2}$$

and the integral manifold of the Pfaff system $\omega_i = 0$. Since we are going to integrate only on these manifolds, we can replace ω with any $\Omega \equiv \omega(\mathrm{mod}\,\omega_i)$, that is, by

$$\begin{aligned}\Omega := & \, Ldt^1 \cdots dt^\mu + A_{i\alpha}dt^1 \cdots dt^{\alpha-1}\omega_i dt^{\alpha+1} \cdots dt^\mu + \\ & + \sum_{i<j,\,\alpha<\beta} A_{i\alpha,j\beta}dt^1 \cdots dt^{\alpha-1}\omega_i dt^{\alpha+1} \cdots dt^{\beta-1}\omega_j dt^{\beta+1} \cdots dt^\mu + \cdots\end{aligned} \tag{17.3}$$

As in section 16 we define Ω such that $d\Omega \equiv 0(\mathrm{mod}\,\omega_i)$. To this and one must take $A_{i\alpha} = L_{p_{i\alpha}}$ and all other terms can be arbitrary and can be arbitrarily fixed. Therefore, in what follows we will consider the μ forms Ω of the form

$$\begin{aligned}\Omega := & \, Ldt^1 \cdots dt^\mu + L_{p_{i\alpha}}dt^1 \cdots dt^{\alpha-1}\omega_i dt^{\alpha+1} \cdots dt^\mu + \\ & + \sum_{i<j,\,\alpha<\beta} A_{i\alpha,j\beta}dt^1 \cdots dt^{\alpha-1}\omega_i dt^{\alpha+1} \cdots dt^{\beta-1}\omega_j dt^{\beta+1} \cdots dt^\mu + \cdots\end{aligned} \tag{17.4}$$

We will also make use of the [ ] symbol to denote functions and forms in $\mathbb{R}^{\mu+n}$ obtained by substitution of $p_{i\alpha}(t^\beta, x_j) \equiv \varphi_{i\alpha}(t,x)$ for $p_{i\alpha}$.

Definition (Lepage). The field $\mathcal{F}$ (that is, the n parameter μ surfaces given by $\varphi_{i\alpha}$ which covers the $(\mu+n)$ domain of definition of $\varphi_{i\alpha}$ which is simply connected) is Ω geodesic if $d[\Omega] = 0$.

We see from (17.4) that we have in our disposal an infinite number of different geodesic fields, since the terms in the second line can be taken quite arbitrary. The simplest possibility is that all these terms vanish, then

$$\Omega_{WD} := Ldt^1 \cdots dt^\mu + L_{p_{i\alpha}}dt^1 \cdots dt^{\alpha-1}\omega_i dt^{\alpha+1} \cdots dt^\mu \tag{17.5}$$

is the so called *Weyl–De Donder field or divergence method.* Such theory leads to simple formulas which are quite similar to the one dimensional case, but it is of use only in the problems with fixed boundaries.

Let us then consider the general situation.

A μ chain e_μ is embedded into a (geodesic) field if e_μ is an integral surface of the system $\omega_i = 0$. It is sufficient, of course, the these equations are integrable along this surface. If e_μ is embedded into the geodesic field $\mathcal{F}$, and c_μ is a μ chain in $\mathcal{F}$ with the same boundary $\partial c_\mu = \partial e_\mu$, then since $d[\Omega] = 0$, we have from the Poincaré–Stokes theorem

$$\mathcal{H}(e_\mu) = \int_{e_\mu} [\Omega] = \int_{c_\mu} [\Omega] = \mathcal{H}(c_\mu). \tag{17.6}$$

Thus

$$\mathcal{H}(c_\mu) := \int_{c_\mu} [\Omega] \tag{17.7}$$

is an *invariant* integral, a natural generalization of the Hilbert– Beltrami integral. As in the one dimensional case, we obtain the $\mathfrak{E}$ function and sufficient condition for strong minimum

$$\mathcal{I}(c_\mu) - \mathcal{H}(c_\mu) = \int_{c_\mu} \mathfrak{E} dt, \tag{17.8}$$

where

$$\mathfrak{E}(t^\alpha, x_i, p_{i\alpha}, \bar{p}_{i\alpha} := \bar{L} - L - (\bar{p}_{i\alpha} - p_{i\alpha}) L_{p_{i\alpha}} - \\ -\frac{1}{2} A_{i\alpha,\psi\beta}(\bar{p}_{i\alpha} - p_{i\alpha})(\bar{p}_{j\beta} - p_{j\beta}) + \cdots \tag{17.9}$$

In this way we obtained the following sufficient condition

THEOREM (LEPAGE). *For the strong minimum of the extremal e_μ of the problem with fixed boundary it is sufficient that*

1. *It is possible to embed of e_μ into and Ω geodesic field; and*
2. *$\mathfrak{E} > 0$ for $(t^\alpha, x_i, \varphi_{i\alpha})$ and all $\bar{p}_{i\alpha}$.*

REMARK. It is obvious that an Ω geodesic field cannot be Ω' geodesic for different Ω'.

The Weyl–De Donder theory is the simplest one; it is defined by vanishing of all terms starting from $A_{i\alpha,j\beta}$, therefore

$$\mathfrak{E}_{WD}(t, x, p, \bar{p}) = \bar{L} - L - (\bar{p}_{i\alpha} - p_{i\alpha}) L_{p_{i\alpha}}. \tag{17.10}$$

The corresponding invariant integral is obtained as follows.

Let μ functions $S_1(t,x),\dots S_\mu(t,x)$ be given such that the $(\mu-1)$ form $\overset{\mu-1}{S} := \sum_\alpha (-1)^{\alpha-1} S_\alpha dt^1 \cdots dt^{\alpha-1} dt^{\alpha+1} dt^\mu$ satisfies $d\overset{\mu-1}{S} = [\Omega]$. Then

$$[\Omega] = d\overset{\mu-1}{S} = \sum_\alpha \left(\frac{\partial S_\alpha}{\partial t^\alpha}\right) dt^1 \cdots dt^\mu,$$

and we have the invariant integral

$$\mathcal{H}_{WD}(c_\mu) = \int_{c_\mu^*} \sum_\alpha \frac{\partial S_\alpha}{\partial t^\alpha} dt. \tag{17.11}$$

For that reason the Weyl–De Donder method was called by Boerner the divergence method. Now we can write down formulas corresponding to (16.10)

$$[L_{p_{i\alpha}}] = \frac{\partial S_\alpha}{\partial x_i}, \quad [L - p_{j\gamma} L_{p_{j\gamma}}] = \sum_\beta \frac{\partial S_\beta}{\partial t^\beta}. \tag{17.12}$$

Without any difficulty the reader will write down the Legendre transformation and Hamilton–Jacobi equation.

H. Weyl went much further than De Donder: he succeeded in embedding the extremal in the geodesic field, while De Donder was content with a formalism (though applicable in the case of Lagrangians depending on higher derivatives as well.)

As we know in calculus of variations of single integrals, the problems with moving ends, transversality, and complete figure played important roles. The Weyl–De Donder theory fails here, but Carathéodory was able to construct such a theory. This theory leads however to the much more complicated formulas. Here I will only present the invariant Carathéodory integral

$$\mathcal{H}_C(c_\mu) := \int_{c_\mu} \det(S_{\alpha\beta} + S_{\alpha j} p_{j\beta}) dt = \int_{c_\mu^*} \det\left(\frac{\partial S_\alpha}{\partial t^\beta}\right) dt. \tag{17.13}$$

If one assumes that $\det(S_{\alpha\beta} + S_{\alpha j} p_{j\beta}) > 0$, then through

$$S_\alpha(t, x(t)) = \lambda_\alpha, \quad \alpha = 1, \dots, \mu \tag{17.14}$$

the set C_μ^* is mapped onto the $(\lambda_1, \dots, \lambda_\mu)$ domain in a one-to-one way, that is, through any point in the set c_μ there passes exactly one n dimensional surface

$$S_\alpha(t, x) = \lambda_\alpha, \quad \alpha = 1, \dots, \mu. \tag{17.15}$$

Now the Carathéodory integral

$$\mathcal{H}_C(c_\mu) := \int_{h_\mu} d\lambda_1 \wedge \cdots \wedge d\lambda_\mu \tag{17.16}$$

equals the volume of h_μ.

Thus the integral $\mathcal{H}_C(c_\mu)$ does not change value, even if one allows for change of the chain ∂c_μ, if only the boundary of a new chain moves along the passing through ∂c_μ surfaces of the family (17.15), forming in this way a $(n+\mu-1)$ dimensional tube. For the Carathéodory $p_{i\alpha} = \varphi_{i\alpha}(t,x)$, the integral $\mathcal{H}_C(c_\mu) = \mathcal{I}(c_\mu)$.

THEOREM ON TRANSVERSALS (CARATHÉODORY). *An $(n+\mu-1)$ tube of surfaces $S_\alpha = \lambda_\alpha$ cuts out from the field $\mathcal{F}$ extremals of the Carathéodory geodesic field segments of equal volume equal to the value of the integral $\mathcal{I}(e_\mu)$.*

This leads to a natural notion of transversality and complete figure.

The keynote of Carathéodory investigation (as it was stressed by Boerner) was the following

It follows from (17.15) and (17.16) that the Carathéodory form

$$d[\Omega] = dS_1 \cdots dS_\mu. \tag{17.17}$$

It follows from the Carathéodory theorem that the class of $[\Omega]$ equals μ, class$[\Omega] = \mu$. But conversely, if class$[\Omega] = \mu$ then there exist μ functions $S_1, \ldots, S_\mu$ such that $d[\Omega] = dS_1 \cdots dS_\mu$. Since $d[\Omega] = 0$ it follows that rank$[\Omega] = \mu$ as well. Clearly, rank$\Omega \geqslant$ rank$[\Omega] = \mu$.

Thus if one fixes the arbitrary functions in the definition of Ω (17.4) such that rank$\Omega = \mu$, then the Ω geodesic field will possess all properties required in the case of problems with moving boundaries. Such construction is provided by Carathéodory theory.

We complete this chapter, devoted to important concepts of Carathéodory with the achievement that made his name known outside mathematics: his axiomatic approach to thermodynamics. This approach is known to every chemist.

7.18 Carathéodory and thermodynamics (second law). Pfaff problem and Frobenius theorem

A great physicist, a Nobel prize laureate, Max Born recollects that in 1908/1909 in Göttingen he discussed foundations of thermodynamics 'with my mathematical friend Carathéodory.' He writes that they were not satisfied with the formulation of two first principles of thermodynamics which used heat machines. Even though – Born writes – heat machines were older than thermodynamics, but such a generally applicable physical theory should have been formulated on general grounds and use an adequate mathematical language. These conversations have soon led to the famous work of Carathéodory *Untersuchungen über Grundlagen der Thermodynamik* Math. Ann. **67** (1909), 355–386.

As we saw in section 17 Carathéodory was a master of the theory of differential equations, and in particular Pfaff systems. Their relation to calculus of variations, and canonical and contact transformations of Lie theory were described in the famous monograph of Carathéodory *Variationsrechnung und partielle Gleichnagen erster Ordnung.*

PFAFF PROBLEM. Let ω be a one form on a domain $\mathcal{U} \subset \mathbb{R}^n$ (or a differentiable manifold M^n) $n \geqslant 3$. The equation

$$\omega = 0 \tag{18.1}$$

is called the Pfaff equation. A manifold $M^k \subset \mathbb{R}^n$ is an *integral manifold* of (18.1) if $\omega_{M_k} = 0$. Equation (18.1) is *completely integrable* if through every point of the domain $\subseteq \mathcal{U}$ there passes one and only one integral manifold of maximal dimension $n-1$. Therefore if an integral line l passes through x_0 lying on some integral hypersurface M^{n-1} of a completely integrable (18.1), then $l \subset M^{n-1}$. Thus in any neighborhood $\mathcal{O}$ of $x_0 \in \mathcal{U}$ there are points *inaccessible* from x_0 along integral lines of (18.1). The famous inaccessibility theorem of Carathéodory asserts the converse

(inaccessibility) $\Longleftrightarrow$ *(complete integrability)*

The most famous former criterion of integrability is

THEOREM OF FROBENIUS (1877). *(Equation* (18.1) *is completely integrable)* $\Longleftrightarrow$ *(*$d\omega \wedge \omega = 0$*).*

In $\mathbb{R}^3$, $\omega = a_1 dx^1 + a_2 dx^2 + a_3 dx^3$ and
(18.2)

$$0 = d\omega \wedge \omega = a_1 \left(\frac{\partial a_2}{\partial x^3} - \frac{\partial a_3}{\partial x^2} \right) + a_2 \left(\frac{\partial a_3}{\partial x^1} - \frac{\partial a_1}{\partial x^1} \right) + a_3 \left(\frac{\partial a_1}{\partial x^2} - \frac{\partial a_2}{\partial x^1} \right).$$

This condition of complete integrability was known already to Euler (1755). Euler knew also that (18.2) $\Longrightarrow$existence of smooth function S and $\mu \neq 0$ such that

$$\omega = \mu dS. \tag{18.2$'$}$$

Such μ is called *integrating factor* : $\mu^{-1}\omega = dS$. The family $S(x)$ =const gives a foliation of $\mathcal{U} \subset \mathbb{R}^3$ by integrable manifolds, the complete integrability of (18.2). Also for $n \geqslant 3$ we have the implications

(1. existence of integrating factor) $\Longrightarrow$(2. complete integrability) $\Longrightarrow$(3. Frobenius) $\Longrightarrow$(4. inaccessibility)

The famous theorem of Carathéodory asserts the implication 4 $\Longrightarrow$1.

THEOREM (CARATHÉODORY). *(a) Inaccessibility condition (4) assures the existence of the integrating factor μ, $\omega = \mu dS$. Therefore*
(b) All conditions 1, 2, 3, 4 are equivalent.

In thermodynamics the existence of an integrating factor μ is of fundamental importance: there $S = \mu^{-1}\omega$ is the *entropy* and $\mu^{-1} = T$ is the *temperature.* The integral lines of $\omega = 0$ are called adiabatics. Therefore the Carathéodory inaccessibility criterion for existence of an integrating factor is of fundamental relevance; according to the Carathéodory formulation of the II law of thermodynamics, *in any neighborhood of a state x_0 there are states which are inaccessible by adiabatic process.* The postulate of inaccessibility is very 'physical'; it generalizes experience of the mankind.

Like any great discovery Carathéodory formulation raised a lot of enthusiasm on the one hand side, and on the other a wave of critics.

REMARK. The original formulation of Frobenius theorem was, of course, different; the theory of differential forms did not exist at that time. Frobenius and Lie formulated the criterion of complete integrability making use of the notion of involutive system of vector fields. We will present that formulation, which is dual to the one presented here, in the next chapter dedicated to Sophus Lie.

Pfaff problem is a generalization of (18.1) for a Pfaff system

$$\omega_1 = 0, \ldots, \omega_s = 0, \quad s < n. \tag{18.3}$$

Integration of partial differential equations can be reduced to the Pfaff problem; thus, as we know, integration of the partial differential equation $F(x^i, u, \frac{\partial u}{\partial x^i}) = 0$ reduces to Pfaff equation $\omega := du - p_i dx^i = 0$ on the submanifold M of $\mathbb{R}^{2n+1}$ given by $F(x^i, u, p_i) = 0$. Complete integrability of the system (18.3) is defined in the same way as in the case of (18.1): through every point $x \in \mathcal{U}$ there passes one and only one integral manifold of maximal dimension. This dimension is equal $n - r$, where r is the rank of the matrix $(a_{jk}(x))$, $\omega_j(x) = \sum_{k=1}^{n} a_{jk} dx^k$, $j = 1, \ldots, s$. We have

THEOREM (FROBENIUS). *(Pfaffian system $\omega_j = 0$, $j = 1, \ldots, s$ is completely integrable) $\Longleftrightarrow$ ($d\omega_j \wedge \omega_1 \wedge \ldots \wedge \omega_s = 0$, $j = 1, \ldots, s$) $\Longleftrightarrow$ (every $d\omega_j$ belongs to the ideal generated by $\omega_1, \ldots, \omega_s$).*

Pfaff problem is a vast domain. In the analytic case it was only solved by E. Cartan. There is a generalization and algebraization of Cartan's theory in the classic *Ergebnisse* volume by E. Kähler. In this volume the notation $d\omega$ for exterior derivative was introduced to replace the more clumsy notation of Cartan.

A natural question arises: does it exist a 'Complex Frobenius'? The answer is yes, and it is an important theorem, closely related to the famous Newlander–Nirenberg criterion for almost complex structure to be a complex one. We have used it in the chapter on Teichmüller theory, and we still will return to the subject.

7.19 Carathéodory and the beginning of calculus of variations

Constantin Carathéodory was not only one of leading scientists of his time but also the great expert in the history of geometric optics and calculus of variations (he was an editor of Euler's *Opera Omnia*.)

Let us present now some excerpts of his lectures (of August, 1936) *The Beginning of Research in the Calculus of Variations* presented during the tercentenary celebration of Harvard University.

> The name itself, 'Calculus of Variations', was used for the first time by L. Euler in 1756 and again in the third volume of his Integral calculus (1770). It was meant exclusively as a denomination of a new method which J.L. Lagrange had invented shortly before (1755) and from which Euler developed with his usual skill all the possible consequences. Lagrange himself seldom used this combination of words; he speaks simply of the method of variations.

In 1698 John Bernoulli discussed the subject of geodesics with Leibnitz and he remarked that geodesics have always the property of having oscillating planes that cut the surface at right angles, and that this property leads to differential equations for these curves. About 1729

> Euler who was still in his early twenties developed the analytic geometry and solved the problem of geodesics to the satisfaction of his teacher Bernoulli. Nearly in the same year the same discovery was published independently by Clairault in Paris, who was even younger than Euler.
>
> It was a known fact that the whole work of Calculus of Variations during the eighteenth century deals only with *necessary conditions* for the existence of minimum and the most of the method employed during that time did not allow even to separate the cases in which the solution yields a maximum from those in which the minimum is attained. According to general belief Gauss in 1823 was the first to give a method of calculation for the problem of geodesics which was equivalent to the sufficient conditions emphasized fifty years later by Weierstrass for more general cases.
>
> It is therefore important to know that the very first solution which John Bernoulli found for the problem of quickest descent contains the demonstration of the fact that the minimum is really attained for the cycloid and it is more important still to learn from a letter which Bernoulli addressed to Basnage, in 1697, that he himself was thoroughly aware of the advantages of his method. But just as in the case of the problem of geodesics Bernoulli did not publish this most interesting result until 1718.
>
> Thus this method of Bernoulli, in which something of the field theory of Weierstrass appears for the first time did not attract the

attention even of his contemporaries and remained completely ignored for nearly two hundred years.

These two pages of Bernoulli, which I discovered by chance more than thirty years ago, have had a very decisive influence on the work I myself did in the Calculus of Variations. I succeeded gradually in simplifying the exposition of this theory and came finally to a point where I found to my astonishment that the method to which I have directed through long and hard work was contained, at least in principle, in the *Traité de la lumiére* of Christian Huygens.

This little book of some two hundred pages is justly celebrated because it contains in the first chapter the wave theory of light. But the following chapters ought to be read not only by physicists but also by mathematicians, because in their essence they are most splendid, pure geometry. The bending of light rays through the atmosphere, the explanation of double refraction of Iceland spar and the discovery of the aplanatic points of the sphere are apt to elate every true geometer.

The book was printed in 1690 but written twelve years before. Let me tell you the circumstances which have brought Huygens to these ideas.

In 1636, the very year in which Harvard College was founded, Réne Descartes was putting his last hand to his *Discours sur la méthode de bien conduire sa raison* which contained among others his geometry and also his dioptrics. In September 1637 the book came into the hands of Pierre Fermat, who during his lifetime was a higher judge of the Court of Toulouse and after his death became one of the most celebrated mathematicians of all times. Fermat wrote at once to Pére Mersenne, who had presented him with the book, that he objected to the theory of Descartes because by this theory the velocity of light was supposed to increase with the density of the medium it was passing through. There ensured a long and tedious discussion, which lasted many years, but Fermat could not be persuaded, though the experiment showed that the law which Descartes had predicted for refraction of light was accurate to the utmost.

In August 1657 Fermat received from the physician of the king and of mazarin, Cureau de la Chambre, in his time a very reputed man, who was also a physicists of note, a treatise about optics. In the letter in which he acknowledged the receipt of this book, he stated for the first time his idea that the law of refraction might be deduced from the minimum principle, just like Heron of Alexandria had done for the reflection of light. But he was not quite sure himself that such a principle would fit with experiment, and it was not until late in 1661 that he could be persuaded by his friends to put his principle to a mathematical test. In a letter dated January 1, 1662, he announces to Cureau de la Chambre that he had done so and that he found to his amazement that his principle was yielding a new demonstration of the law of Descartes. But the followers of Descartes who were omnipotent in the Parisian scientific world of the time were not satisfied with this discovery of Fermat. As early as in May, 1662, Fermat received from the celebrated Clerselier a letter in which he was told that his principle, which is equivalent to what we cal to-day the principle of least action, was at best a moral principle but not a principle of physics and that his theorem was simply a result of pure geometry.

Huygens, who was at the time living in Paris and who was in constant touch with the learned world of that city, adopted also this view as we informed by a letter to one of his brothers in June, 1662. But it was not until fifteen years later that these very vogue ideas took a definite shape. The result was the admirable book I have just mentioned.

What Huygens did for optics is exactly the same thing which was done two hundred years later by Maxwell for electricity. He replaced the long distance principle of Fermat by a contact principle. I have already mentioned that the book of Huygens was based on the wave theory of light. But the *waves* of Huygens were exactly as hypothetical as the *atoms* of Democritos. Accordingly the exposition he gave of that theory does not correspond to 'Wave mechanics' of de Broglie and Schrödinger but to the classical mechanics of Hamilton and Jacobi. *And this is the reason why the Huygens book can be considered also as a genuine book on the Calculus of Variations.*

The end of this fairy tale is very said indeed. The theory of Huygens was shortly afterwards killed by Newton who showed that it was inconsistent with the propagation of longitudinal waves and the possibility of the existence of transversal waves had not been devised as yet. In consequence the influence of Huygens was delayed for one hundred and twenty-five years if we consider the progress in optics and was lost altogether for the progress of Calculus of Variations.

I must close here this very subjective, very incomplete, and very superficial review of the history of one of the most attractive subjects connected with the development of scientific thought.

I will be glad if I have succeeded in impressing the idea that it is not only pleasant and entertaining to read at times the works of the old mathematical authors, but that this may occasionally be of use for the actual advancement of science.

Besides this there is a great lesson we can derive from the facts which I have just referred to. We have seen that even under conditions which seems most favorable very important results can be discarded for a long time and whirled away from the main stream which is carrying the vessel of science. Sometimes it is of no use even if such results are published in very conspicuous places. It may happen that the work of most celebrated men may be overlooked.

If their ideas are too far in advance of their times, and if the general public is not prepared to accept them, these ideas may sleep for centuries on the shelves of our libraries. Occasionally, as we have tried to do to-day, some of them may be awakened to life. But I can imagine that the greater part of them is still sleeping and is awaiting the arrival of the prince charming who will take them home.

These lines reveal Carathéodory as a true philosopher and a man of nobility!

Chapter 8

Symplectic and Contact Geometries. Conservation Laws

To Sophus Lie, the father of symplectic and contact geometries

8.1 Introduction

Allan Weinstein is a leading expert in symplectic geometry. The discovery of momentum map by him and Marsden led them to a very important theorem, called now the Marsden–Weinstein symplectic reduction theorem proved in 1974. It was a great shock for Weinstein when, while studying the thick three volumes of forgotten monograph of Lie *Theorie der Transformationgruppen*, called earlier shortly 'Lie–Engel', and has discovered that a number of famous theorems proved in nineteen sixties – nineteen seventies was known to Lie already 90 years earlier. 'Lie–Engel' knew also the coadjoint map $\mathrm{Ad}_G^* : \mathfrak{g}^* \to \mathfrak{g}^*$ and the *momentum map* $\mu : \mathbb{R}^{2n} \to \mathfrak{g}^*$ and its Ad^*-equivariance. He knew that the momentum map μ gives the *first* integrals of Hamiltonian action (for a G-invariant Hamiltonian.) In a sense Lie knew the famous Emmy Noether theorem on conservation laws (in hamiltonian formalism) already before her birth. And it is known how impressive and influential this theorem of Noether of 1918 was, and still is.

Emmy Noether (1882–1935) was the daughter of a friend of F. Klein, Max Noether (1844–1921), the founder of modern algebraic geometry. Her, two years younger brother Fritz Noether, introduced a notion of index of an operator. What is today called Fredholm operators theory, was, in fact, invented by Fritz Noether (after his emigration to the Soviet Union in 1933) and not by Fredholm. Therefore in this book I write 'Fredholm–Noether mapping.'

E. Noether wrote her Ph.D. thesis on the theory of invariants under supervision of P. Gordan (a friend of Klein) in 1908. In 1915 Klein and Hilbert invited Noether to Göttingen, when she worked till 1933 (when Hitler came to power), and created a circle of prominent algebraists (among others Van der Waerden.) After short stay in the USSR, she settled in the USA, where she died of cancer 14.04.1935.

As we stressed it many times, Hilbert obtained equations of general relativity from his action integral. It is natural that in the years 1915–1918 in Göttingen people worked intensively on relativity theory, and Klein suggested the young expert on invariants theory to investigate G-invariant variational principles. This resulted in two famous papers of Noether (first for the group G depending on finite, and second for G depending on infinite number of parameters.) It become customary to call the theorems saying that for any one parameter subgroup in G there corresponds a conservation law, the Noether theorem, even though, as we will see, the corresponding theorems for G-invariant Hamiltonian systems were known already to Sophus Lie.

These oversights are understandable today: who would want to read more than 1200 pages of complex mathematics written in difficult language. But, as Van der Waerden says in his delightful *History of Algebra* (Springer,1985)

> *An ... exposition of the whole theory was given in a monumental three-volume work of Sophus Lie and Fritz Engel entitled* Theorie der Transformationgruppen. *This work was* very influential: we have all learnt *the principles of Lie theory* from it ... (emphasized K.M.)

In mathematics it is by no means rare that great works are buried in the libraries, to be woken up by a new generation. The theory is born anew, one might say reincarnates, in a new form, in a new context. Sophus Lie felt lonely, but the fact is that many of his theorems lives in works of other researchers, and even in textbooks without mentioning the name of their Creator.

The history of symplectic geometry is fascinating: its sources can be found in optics of Hamilton, who extended his brilliant concepts from optics to analytical mechanics. As Max Born notes, the formulation of mechanics in the language of Hamilton survived two major revolutions of physics: relativity theory and quantum theory. Perhaps the most important reason why the Hamilton theory (called by physicists 'Hamiltonian formalism') is considered so fundamental is its invariance which is become manifest in the language of

Symplectic geometry. Let us recall that a symplectic manifold is a differentiable manifold (M, ω) equipped with the two-form ω which is closed $(d\omega = 0)$ and non degenerate. The most important

EXAMPLES (which are much more than mere examples indeed!) are

1. $M = V \times V^*$, where V is a (real) n dimensional vector space, and V^* its dual; the canonical form ω on $V \times V^*$, is a bilinear form given by

$$\omega((x,\xi),(x',\xi')) := < x', \xi > - < x, \xi' >,$$

where $< \cdot, \cdot >$ is the duality between V and V^*.

Clearly this definition works also for Hilbert space (and, more generally, for reflexive Banach space) V. The space $\mathbb{R}^n \times (\mathbb{R}^n)^*$ is isomorphic to $T^*(\mathbb{R}^n)$ which leads us to even more important, second example.

2. Let V be a differentiable manifold of dimension n. The cotangent bundle T^*M equipped with two-form ω, given in local coordinate system $\omega = \sum_{i=1}^n dp_i \wedge dq^i$ (for short, $\omega = dp \wedge dq$) has the symplectic structure.

Two symplectic manifolds (M, ω), (M', ω') are isomorphic if there exist a diffeomorphism $\varphi : M \to M'$ such that $\varphi^*\omega' = \omega$. The famous

DARBOUX THEOREM asserts that *every $2n$ dimensional symplectic manifold (M, ω) is locally isomorphic with the tangent bundle T^*V, where V is an n dimensional manifold, and therefore locally isomorphic with $T^*\mathbb{R}^n$*. This theorem shows a fundamental difference between symplectic and Riemannian structures:

The Darboux theorem asserts that *all symplectic manifolds of given dimension are locally isomorphic.*

The Riemann manifolds of the same dimensions are by far not locally isomorphic: the necessary condition is equality of the Riemann– Christoffel tensor R^∇. For example, the two dimensional Euclidean space and the two-sphere are not locally isomorphic! In a complex domain we have a very important relation between

Kähler and symplectic manifolds:

(complex projective manifold) $\subset$ (Kähler manifold) $\subset$ (symplectic manifold)

As we know the Fubini–Study metric on $\mathbb{P}^n(\mathbb{C})$ is symplectic; every submanifold of Kähler manifold is Kählerian; by Chow theorem every projective space is a submanifold of some $\mathbb{P}^n(\mathbb{C})$, therefore it is Kählerian. By definition, a Kähler manifold M is such hermitian manifold (M, h) that $\omega := \operatorname{Im} h$ is symplectic. Therefore Kähler $\Longrightarrow$symplectic.

By virtue of Darboux theorem, in the case of local problems of symplectic geometry, we can work in T^*V bundle with the canonical symplectic structure $\omega = dp \wedge dq$. Lie theory is local and concerns $\mathbb{R}^{2n}$, therefore Lie theorems hold on any symplectic manifold. It was a great achievement of Poincaré to show that the proper domain of analytic dynamics is the cotangent bundle T^*M — this discovery was so fundamental that now it seems to be natural and obvious.

Let us now give a

Short proof of Darboux theorem. We are to show that in some neighborhood of an arbitrary point, any symplectic structure ω can be written, in adequate coordinates, as $\omega = dp \wedge dq$.

Let us recall the homotopy formula of Lie: Let $\mathcal{L}_v$ be the Lie derivative and i_v the contraction defined by $i_v\omega(\xi) := \omega(v, \xi)$, then

$$\mathcal{L}_v = i_v \circ d + d \circ i_v. \tag{1}$$

Now the proof is almost immediate: we proceed by homotopic method. Let ω_t, $t \in [0, 1]$ be a smooth (with respect to t) family of germs of non degenerated two-forms ω_t at the point x. We still need a family of diffeomorphisms $\{g_t\}$ which do not move x such that $g^*\omega_t = \omega_0(x)$. Differentiating this equation with respect to t, we have

$$\mathcal{L}_{v_t}\omega_t = -\gamma_t, \tag{2}$$

where $\gamma_t = \frac{d\omega_t}{dt}$ and v_t is the vector field we are looking for. (1) and (2) give $di_t\omega_t = -\gamma_t$. We take such one form α_t that $\alpha_t(x) = 0$ and $d\alpha_t = i\gamma_t$. The equation for v_t, $i_{v_t}\omega_t = \alpha_t$ has a unique solution, since ω_t is non degenerate. Since the field v_t vanishes at x and generates the one parameter family $\{g_t\}$ of diffeomorphism we are done. □

Sophus Lie regarded the theory of groups of contact transformations (1871) as his greatest discovery and achievement.

If we have an n dimensional smooth manifold $X^n = X$, then a *contact element* of X at the point $x \in X$ is the $(n-1)$ dimensional (hyper)plane tangent to X at x, that is, a $(n-1)$ dimensional subspace of $T_x X$. The family of all contact elements of X^n has a structure of $(2n-1)$ dimensional smooth manifold.

Since a field on hyperplanes is defined locally by differential one-form α which does not vanish, $\alpha|_x \neq 0$ is the equation of hyperplane at the point x. We are led to the following

DEFINITION. A field of hyperplanes on a $(2n+1)$ dimensional manifold is called a *contact structure* if the form $\alpha \wedge (d\alpha)^n$ is non degenerate. Diffeomorphisms of contact manifolds (odd dimensional manifolds equipped with contact structure) which preserve the contact structure are called *contact transformations.*

In Lie's formulation those are 'transformations which map contact elements onto contact elements'. It was a great day for mathematics when Lie discovered his contact transformations and showed that they form a group. The question arises as to if all contact manifolds of the same dimension are locally 'contactomorphic'. The answer is yes, and we have the corresponding

DARBOUX THEOREM FOR CONTACT MANIFOLDS. *Contact manifolds of the same dimension are locally contactomorphic. Moreover, in the neighborhood of every point of a* $2n+1$ *dimensional contact manifold there exist coordinates* $(z, q_1, \ldots, q_n, p_1, \ldots, p_n)$ *in which the contact structure* α *has the form*

$$\alpha = \sum_{i=1}^{n} p_i dq^i - dz.$$

In fact $dz - p_i dq_i$ is a contact structure on $\mathbb{R}^{2n+1}$. This structure is *standard* and the coordinates (z, q, p) are *contact Lie–Darboux coordinates.*

Thermodynamics and contact structures. It is of high philosophical interest that roughly *at the same time* Josiah Willard Gibbs (1839–1903) conceived at Yale his Graphical method in thermodynamical fluids. As remarked by Vladimir Arnold in his witty style at the beginning of his Gibbs Lectures (1989) *Contact geometry: the Geometric Method of Gibbs Thermodynamics*

'... *Every mathematician knows that it is impossible to understand any elementary course in thermodynamics. The reason is that* thermodynamics is based – *as Gibbs has explicitly proclaimed – on rather complicated mathematical theory on* contact geometry' (emphasized K. M.)

Integral submanifolds of dimension n in $(2n+1)$ dimensional contact manifolds are called (by Arnold) *Legendre submanifolds.* A smooth fibration of a contact manifold, all of whose fibers are Lagrangian is called the *Legendre fibration.* We have

COROLLARY. *All Legendre submanifolds are locally contactomorphic. The same holds for Legendre fibrations.*

According to Gibbs, the geometric structure of thermodynamics is described by $2 \cdot 2 + 1 = 5$-contact manifold equipped with the contact form whose zeros define the laws of thermodynamics. In Gibbs notation this form is

$$d\varepsilon = t d\eta - p dv,$$

where ε is the energy; t, the temperature; η, the entropy; p, the pressure; and v, the volume. The Gibbs thesis is that

Chemical substances are Legendre submanifolds of this five dimensional contact manifold.

For different substances, the manifolds are different, but they are always submanifolds of the universal (Gibbs) manifold.

Contact geometry does for geometric optics and the theory of wave propagation what symplectic geometry does for mechanics. But both these geometries and their automorphisms were conceived by the single man – Sophus Lie. The thesis that Lie is the father of both geometries is firmly founded.

The Lie–Darboux one form is our old friend: it appears as an integrand in the Hilbert–Beltrami integral and in the Poincaré integral invariant. Now perhaps we can better understand the aggressive title of the Lie short note *Die Theorie der Integralinvarianten is ein Corollar der Theorie der Differentialinvarianten*, Ber. Sächs. Ges. Leipzig, 1897, 342–357.

8.2 Lie approach to hamiltonian mechanics

The abstract (general) groups theory originated from the theory of groups of transformation of the Euclidean space and general manifolds. At Lie times general topology has not existed yet, and thus his theory has a local character. In the mathematical language of his times, Lie expressed this fact talking about infinitesimal transformations.

Today infinitesimal transformations of a manifold M are called vector fields on M; we will denote them $Vect(M) := C^\infty(TM)$ – the set of smooth sections of the tangent bundle TM of M. Thus Sophus Lie created the theory of vector fields. To denote vector fields we will use small letters $u, v, w, \ldots$. The great discovery of Lie was

THEOREM (LIE). *1. The commutator $[u, v]$ of two vector fields is a vector field $u, v \in Vect(M) \Longrightarrow [u, v] \in Vect(M)$.*

2. Jacobi identity holds.

Let us recall that $[u, v]f := u(v(f)) - v(u(f))$. Thus $Vect(M)$ is a vector space equipped with the structure $[\cdot, \cdot]$ of the Lie bracket. Such space is, deservedly, called Lie algebra.

Let (M, ω) be a symplectic manifold. The automorphism group $\mathrm{Aut}(M, \omega)$ is called the group of canonical, or symplectic, transformations. Let $I : TM \to T^*M$ be an isomorphic bundle map given by the inner product (contraction) with ω,

$$I(v) := i_v\omega.$$

A vector field is locally Hamiltonian, or symplectic if it preserves ω, that is, $\mathcal{L}_v\omega = 0$. v is Hamiltonian if there exists $f \in C^\infty(M)$ such that $I(v) = df$. From these definitions we immediately have

PROPOSITION. *1. (A vector field v is locally Hamiltonian) $\Longleftrightarrow$ ($dI(v) = 0$).*

2. If we denote by $Ham(M)$ Hamiltonian vector fields, and by $Symp(M)$ symplectic, that is, locally Hamiltonian ones then there is an exact sequence of Lie algebras

$$0 \to Ham(M) \to Sympl(M) \to H^1(M, \mathbb{R}) \to 0$$

3. $0 \to \mathbb{R} \to C^\infty(M) \to Ham(M) \to 0.$

PROOF. 1. follows from the homotopy formula

$$0 = \mathcal{L}_v\omega = i_v d\omega + d(i_v\omega) = d(I(v))$$

2. If α is a closed one-form, then $v = I^1(\alpha)$ is clearly symplectic. If $v = I^{-1}(df)$, we write $v := v_f$.

For $f, g \in C^\infty(M)$ one defines the Poisson bracket

$$\{f, g\} := \omega(I^{-1}g, I^{-1}f).$$

THEOREM (LIE). *The Poisson bracket has the property that $\{f, g\} = v \cdot g$, where v is a vector field such that $df = I(v)$. Therefore $\{f, g\} = v_v \cdot g$.*

EXAMPLE. In Darboux coordinates $(M, \omega) = (T^*(X), dp \wedge dq)$, then

$$\{f, g\} = \frac{\partial f}{\partial q^i}\frac{\partial g}{\partial p_i} - \frac{\partial g}{\partial q^i}\frac{\partial f}{\partial p_i}.$$

From the theorem above we obtain

Conservation laws of Hamiltonian mechanics.
The law of conservation of energy. Let $H \in C^\infty(M)$; then H is a first integral of its hamiltonian flow. Indeed $0 = \{H, H\} = v_H \cdot H$ □

THEOREM (LIE–NOETHER). *A Hamiltonian function $F \in C^\infty(M)$ whose flow preserves the Hamiltonian H is a first integral of the flow with Hamiltonian H.*

PROOF.

$$0 = v_F \cdot H = \{F, H\} = -\{H, F\} = -v_H \cdot F = 0.$$

□

REMARK. If (φ_t^H) denotes the Hamiltonian flow generated by the vector field v_H, then

$$(F \text{ is a first integral of } (\varphi_t^H)) \Longleftrightarrow (v_H \cdot F = 0) \Longleftrightarrow (\{F, H\} = 0)$$

If G is a Lie group such that the function $H \in C^\infty M$ is G-invariant: $H(G\cdot x) = H(x)$ for any $x \in M$, then it is invariant (conserved) with respect to any one-parameter subgroup of G. If $F_1, \ldots, F_r \in C^\infty M$ are generators of the Lie algebra of G, then $F_1, \ldots, F_r$ are first integrals of the flow (φ_t^H). Therefore *to every one-parameter subgroup of G there corresponds a conservation law.* Theorems of this kind are usually called Noether theorems (though we see that Lie proved them many years before Noether.) Because these theorems are of great importance in physics (and philosophy as well), we will devote the whole section to them.

8.3 Conservation laws and 'Postulates of impotence'

As a special case of the Lie–Noether theorem of the previous section, take for G the Galileo group and its subgroups:

1. temporal translations;
2. spacial translations; and
3. rotations.

We obtain the conserved quantities

1$'$. energy (of the system)
2$'$. linear momentum; and
3$'$. angular momentum.

To them there correspond three 'impotencies' (Whittaker): we do not know

1$''$. origin of time;
2$''$. origin of space;
3$''$. privileged direction.

This impressive correlation is summarized in the following table

Non observability of absolute entities (impotence)	Symmetry group	Conservation law of
origin of time	time translations	energy
origin of space	space translations	linear momentum
privilegged direction	rotations	angular momentum

E. Whittaker writes: 'A *postulate of impotence* is not the direct result of an experiment, or of any finite number of experiments; it does not mention any measurment, or any numerical relation or analytical equation; it is the assertion of conviction, that all attempts to do a certain thing, however made, are bound to fail' (Whittaker, *From Euclid to Eddington*, Dover 1958.)

In this context let us recall Carathéodory formulation of the second law of thermodynamics, his inaccessibility postulate: *in any neighborhood of a state there are staes which are inaccessible by adiabatic process.*

But the present author cannot share what seems to be Whittaker conviction that the theoretical physics of the future will begin with some postulates of impotency, and then will derive everything else from them by syllogistic reasoning ...

The table above shows impressively the dualities which have their correspondence in quantum mechanics (Heisenberg complementarities), and which were codified in the Hilbert invariant integral. As Hermann Weyl stressed, the objectivity of something means invariance with respect to an automorphism group (symmetry group.) The great philosophical relevance of these theorems can be formulated as follows: *relativity signifies impossibility of measuring of absolute entities* and *the correspondence between relativity (non observability of some absolute entities) symmetry (invariance with respect to some symmetry groups), and conservation laws: conservation of certain quantities and therefore their observability.*

8.4 Momentum map and symplectic reduction. (Reduction of phase space for systems with symmetries)

Sophus Lie introduced another fundamental notion which is a generalization of Hamiltonian for a one-parameter group. This is the concept of *momentum map.*

If the Lie group G acts on (M, ω) preserving the symplectic form ω, then for every vector field $v \in Vect(M)$ generated by $\xi \in \mathfrak{g} = T_eG$, we have (by homotopy formula)

$$0 = \mathcal{L}_v\omega = (i_vd + di_v)\omega = di_v\omega.$$

Thus the one form $i_v\omega$ is closed.

The *momentum map* for the G-action is the map m

(1) $$m : M \to \mathfrak{g}^*$$

such that

(2) $$d(< m, \xi >) = i_{v(\xi)}\omega \quad \text{for all} \quad \xi \in \mathfrak{g}.$$

here $< m, \xi >$ is the function on M obtained by the dual pairing between $\mathfrak{g}$ and $\mathfrak{g}^*$.

The idea of Lie was to associate with every $\xi \in \mathfrak{g}$ in a natural way an observable $F_\xi \in C^\infty(M)$ such that

(3) $$v_{F_\xi} = v(\xi),$$

where $v(\xi)$ is the vector field on M generated by $\exp(t\xi)$. Now, F_ξ are first integrals for every G-invariant Hamiltonian H. Indeed we have the fundamental generalization of Lie–Noether theorem

THEOREM (LIE). *If $H : M \to \mathbb{R}$ is G-invariant, that is, $H(Gx) = H(x)$, $x \in M$, then the momentum map m is the set of first integrals for v_H.*

PROOF. Differentiating the invariance condition $H(g \cdot x) = H(x)$ in $v(\xi)$ direction, we obtain for fixed $x \in M$ $dH(x) \cdot v(\xi)(x) = 0$, thus $0 = \{H, < m, \xi >\} = -d < m.\xi > \cdot v_H$ and $< m, \xi >$ is conserved on v_H trajectories for every $\xi \in \mathfrak{g}$. □

The notion of momentum map m was rediscovered by Bertram Kostant and, independently, by J-M. Souriau in their 'geometric quantization' method and was called 'momentum map' (or 'moment' for short) because if M is a phase space of mechanical system and G is a group of translations or rotations, then the components of momentum map are just linear and angular momenta, respectively.

Construction of equivariant momentum map. Poisson mapping. Let us assume that the Lie group G acts symplectically on the connected symplectic manifold (M, ω). Then to any element $a \in \mathfrak{g}$ there corresponds a symplectic vector field $v(a)$ on M. We will assume that all these vector fields have single valued Hamiltonians H. If we choose such Hamiltonians for a single basis of $\mathfrak{g}$, we obtain the linear mapping

$$\mathfrak{g} \to C^\infty(M, \{\cdot,\cdot\}), \quad a \to H_a.$$

The Poisson bracket $\{H_a, H_b\}$ may differ from $H_{[a,b]}$ by a constant

$$\{H_a, H_b\} = H_{[a,b]} + C(a, b)$$

A symplectic G action on (M, ω) is called *Poisson action* if the basis of Hamiltonians is chosen so that $C(a, b) = 0$ for all $a, b \in \mathfrak{g}$.

REMARK. In general case $C(a, b)$ is bilinear, skewsymmetric $C(a, b) = -C(b, a)$, and satisfies the Jacobi identity

$$C([a, b], c) + C([b, c], a) + C([c, a], b) = 0;$$

therefore C is a two-cocycle on $\mathfrak{g}$. Thus the symplectic action determines a cohomology class $H^2(\mathfrak{g}, \mathbb{R})$, and this action is Poisson if and only if the class in $H^2(\mathfrak{g}, \mathbb{R})$ is zero. We have the following

THEOREM (LIE–MARSDEN–WEINSTEIN). *A Poisson G-action defines the moment map which is equivariant, that is, the following diagram is commutative*

$$\begin{array}{ccc} M & \xrightarrow{\;G\;} & M \\ {\scriptstyle m}\downarrow & & \downarrow{\scriptstyle m} \\ \mathfrak{g}^* & \xrightarrow{\;Ad_G^*\;} & \mathfrak{g}^* \end{array}$$

PROOF. It is sufficient to check that the derivatives of the functions $H_a(\exp s \cdot x)$ and $H_{Ad_g a}(x)$ with respect to s are equal for $s = 0$. $\square$

REMARK. If we take the classical case considered by Lie $M = T^*X$, $\omega = dp \wedge dq$, then we have the

THEOREM (LIE). *The natural G-action on T^*X is Poisson.*

A natural question arises if the coadjoint orbits $\mathrm{Ad}_G^* \cdot f$, where $f \in \mathfrak{g}^*$, have a natural symplectic structure. This question was posed already by Lie, who answered in affirmative.

THEOREM (LIE, 1880, KIRILLOV, 1962–70). *Let N be an orbit of the coadjoint representation, that is, $N := G \cdot f$, where $f \in \mathfrak{g}^*$ and we simply write $g \cdot f := \mathrm{Ad}_g^* \cdot f$. The vector fields v^N are elements of $C^\infty(TN)$, the sections of TN, and span the tangent space at each point $f \in N$; we have*

$v_f^N = -a \cdot f$, *where* $-(a \cdot f)(b) := < f, [a,b] >$, $a, b \in \mathfrak{g}$. *Let* $m(a)$ *be a restriction to* N *of the linear form* $f \to f(a)$ *on* $\mathfrak{g}^*$. *Then*

1. The two-form $\omega(a^N, b^N) := - < f, [a,b] >$ *is symplectic on* N*, the* G*-orbit in* $\mathfrak{g}^*$ *(this is called the* Kirillov form*);*

2. The G*-action on* N *is Poisson; and*

3. The equivariant momentum map of a *is the function* $m(a)$*.*

PROOF. The form ω defined in 1. is clearly non degenerate and G-invariant. As a restriction of a linear function $m(a)$ satisfies

$$dm(b^N)_f = (b^N)_f(a) = - < f, [a,b] >,$$

thus

$$dm(a) = i_{a^N}\omega \tag{1}$$

and it follows that $m(x)$ is the Hamiltonian for the vector field a^N. By virtue of the a^N invariance of ω, $\mathcal{L}_{a^N}\omega = 0$, we obtain from the homotopy formula

$$0 = \mathcal{L}_{a^N}\omega = di_{a^N}\omega + i_{a^N}d\omega,$$

and (1) gives

$$0 \equiv d^2m(a) = di_{a^N}\omega = -i_{a^N}d\omega.$$

Since the vector fields a^N span the tangent space it follows from the last formula that $d\omega = 0$. □

The last theorem is of great importance for Kirillov classification of symplectic G-homogeneous manifolds and his methods of orbits in the representation theory of Lie groups and in geometric quantization.

THEOREM (KIRILLOV). *Every symplectic manifold with a transitive and symplectic action of a (connected) Lie group* G *is locally isomorphic to the orbit of the coadjoint representation of the group* G*, or of its central extension* $\tilde{G}$*.*

REMARK. We recall that the central extension $\tilde{G}$ of a group G is given by the exact sequence of groups

$$G_1 \to \tilde{G} \to G,$$

where $\tilde{G}$ is a group containing G_1 in its center and such that $G = \tilde{G}/G_1$.

All the above constructions are taken, in principle, from the classical works of Jacobi on integrability of Hamiltonian systems: Knowing r first integrals $F_1, \dots, F_r$ of a system having G-symmetry, Sophus Lie gathered them into his momentum map. At the same time, the knowledge of these integrals (which in abstract geometric setting we denote m_j, $m = (m_1, \dots, m_r)$, made it possible for Lie to decrease the dimension of phase space by $2r$. What's more, Lie observed that this reduced phase space $M_c := m^{-1}(c)$ can be identified with the coadjoint orbit $\mathrm{Ad}_G^* \cdot c \subset \mathfrak{g}^*$, where $\mathfrak{g}^*$ is the dual space to the Lie algebra $\mathfrak{g}$ of G. This fact is proved in Chapter 19 of the Lie–Engel treatise, where Lie shows the Ad^* equivariance of his mapping

$$m : \mathbb{R}^{2n} \to \mathfrak{g}^*.$$

Let us return to the 'abstract' theory.

The construction of the symplectic quotient of (M, ω) by the group action. Given an *equivariant* momentum map $m : M \to \mathfrak{g}^*$, we assume for simplicity that the group G acts freely on M. This means that the momentum map has the maximal rank at every point and that the fiber $m^{-1}(0) \subset M$ is a smooth submanifold of M. Since m is equivariant, the set $m^{-1}(0)$ is preserved by G action and the symplectic quotient is defined to be

$$U := m^{-1}(0)/G.$$

Now on U we define a symplectic structure induced from (M, ω). Let $x \in m^{-1}(0)$, that is, $m(x) = 0$. We have linear maps $\mathfrak{g} \xrightarrow{\rho} T_xM \xrightarrow{dm} \mathfrak{g}$, where ρ is the map $\rho : \mathfrak{g} \to Vect(M)$ and $\ker(dm)$ is the anihilator of the image of v under the skew pairing ω_x. Thus ω passes down to $T_{[x]}U$, the tangent space of U at $[x]$:

$$T_{[x]}U = Ker(dm)/\operatorname{Im}(\rho).$$

One checks that in this way we obtain a closed non degenerate form on U — the symplectic reduction of (M, ω).

Thus we have obtained *two* symplectic manifolds:

1. The coadjoint orbit $G \cdot f = \mathrm{Ad}_G^* f =: N \subset \mathfrak{g}$; and
2. The symplectic quotient $m^{-1}(0)/G$.

Sophus Lie identified these two objects

THEOREM (LIE, MARSDEN & WEINSTEIN (1982)). *Both symplectic manifolds 1. and 2. above are symplectically isomorphic.*

We leave the proof of this theorem as an exercise.

There is an important generalization of Hitchin of Lie–Marsden–Weinstein symplectic reduction to hyperkähler manifolds which plays an important role in Yang–Mills theory, gravitational instantons, and in the theory of Klein singularities.

8.5 Hyperkähler quotients

The preceding construction of symplectic quotients clearly applies to Kähler quotients. In the case when the G-space X has a quaterionic structure, the situation is similar but, of course 'three times' more involved. Eugenio Calabi investigated such manifolds and called them hyperkählerian. We recall that Kähler manifold X is a Riemann manifold equipped with the a complex structure $\mathcal{I}$ which is integrable, that is, $\nabla\mathcal{I} = 0$ (∇ is the Levi–Civita connection.)

DEFINITION. A hyperkählerian manifold X is a Riemann manifold of $4n$ dimensions whose holonomy group is a subgroup of the symplectic group $\mathrm{Sp}(n)$. As proved by Calabi this is equivalent to the following.

The manifold X has a Riemann metric g which Kähler for three (integrable) complex structures $\mathcal{I}, \mathcal{J}, \mathcal{K}$, which, in turn, satisfy the quaterionic relation

$$\mathcal{I}\mathcal{J} = -\mathcal{J}\mathcal{I} = \mathcal{K},\ \mathcal{J}\mathcal{K} = -\mathcal{K}\mathcal{J} = \mathcal{I},\ \mathcal{K}\mathcal{I} = -\mathcal{I}\mathcal{K} = c\Psi.$$

The tangent bundle TX is a hyperkählerian manifold is thus a quaterionic vector bundle.

The case of $\dim_{\mathbb{R}} X = 4$ is of great importance in the theory of complex surfaces $\dim_{\mathbb{C}} X = 2$. The important examples of hyperkählerian manifolds are T^4, the four dimensional torus with its flat metric and the surface $K3$.

The construction of hyperkählerian quotient X makes use of the idea of hyperkähler map and proceeds as follows:

Given a hyperkählerian manifold X with action of compact Lie group G preserving *three* Kähler forms ω_1, ω_2, ω_3, to each ω_j we apply the homotopy formula $0 = \mathcal{L}_v\omega_j = d(i_v\omega_j) + i_v d\omega_j$, $j = 1, 2, 3$ for every vector field v

generated by G. Since ω_i is closed ($d\omega_j = 0$), we obtain $d(i_v\omega_j) = 0$. Hence, if X is simply connected there exist functions f_j such that

$$i_v\omega_j = df_j$$

When v runs over vector fields generated by the Lie algebra $\mathfrak{g}$ of G, the functions f_j define a mapping

$$m_j : X \to \mathfrak{g}^*,$$

which, if G invariant, is called the (equivariant) momentum map. Putting three symplectic forms together, we obtain a hyperkählerian momentum map

$$m : X \to \mathfrak{g}^* \otimes \mathbb{R}^3.$$

The important theorem of Hitchin, Karlhede, Lindström, and Rocek asserts the *natural metric* on $m^{-1}(0)/G$ is *hyperkählerian.*

REMARK. In the theory of moduli of anti-self-dual (ASD) Yang–Mills connections, the infinite dimensional space $\mathcal{A}$ of these connections is hyperkählerian in a natural way. On this space acts the group $\mathcal{G}$ of gauge transformations[1] ; there exists a L^2 metric on $\mathcal{A}$ and three operators $\mathcal{I}, \mathcal{J}, \mathcal{K}$. The question arises what is the corresponding momentum map? The answer is

$$m(\nabla) := (F^\nabla \wedge \omega_1, F^\nabla \wedge \omega_2, F^\nabla \wedge \omega_3) \in A^4(M, \mathfrak{g}) \otimes \mathbb{R}^3,$$

and then

$$m^{-1}(0)/G = \{\text{the moduli space of ASD connections}\}.$$

Here M is a compact, oriented, simply connected Riemannian manifold, ∇ the connection on the principal SU(2) bundle. We will discuss these notions in the next chapter.

[1]If $E \to X$ is a vector bundle over manifold X, then automorphisms $u : E \to E$ of E respecting the structure of the fibers and covering the identity map on X form a group $\mathcal{G}$ called the *gauge group* of E

CHAPTER 9

Direct Methods in Calculus of Variations for Manifolds with Isometries. Equivariant Sobolev Theorems. Yamabe Problem and its Relation to General Relativity

The calculus of variations is (and was), through the Euler–Lagrange equations, the main source of equations of physics and differential geometry.

As we know most important linear and nonlinear ordinary and partial differential equations such as Laplace equation, equation of geodesics in Riemann space, the Einstein–Hilbert equations of general relativity, equations of minimal surfaces, equations of harmonic maps, the Yang–Mills and Yang–Mills–Higgs equations, the equation of Yamabe problem, all of them can be cast to the form of a variational problem.

But variational problems are not only the richest source of differential (mostly nonlinear), but also of new methods in already existing theories, and even they prompted creation of new, great chapters of mathematics: functional analysis, topology (Morse theory, and topology of functional spaces, and so on. The previous chapter was intended to show the beauty and richness of methods related to optics and analytical dynamics. However, the proofs of existence and construction of critical (or stationary) points of functionals of variations led to creation of the so called direct methods of calculus of variations, whose roots, as Hilbert stressed, are in the Dirichlet principle of Riemann. We have already talked about these methods in connection with the Plateau problem and the method of Perron of solving the Riemann's Dirichlet problem and Riemann surfaces.

On the way to unification, the longing for a great wholeness, we would like to possess a *unique* methods making it possible, in principle, to solve *all* variational problems. This dream will be, probably, never fulfilled, but as a wise man said 'a man is worth as much as his longings.' It seemed for a long time that such a method is

The method of steepest descent. This ancient method of the largest negative gradient, stemming from physics, expects that a critical point (local extremum) of the function $f : X \to \mathbb{R}$ is reached on the gradient curve which is orthogonal to the level curves $f =$const. of the function f. This method was codified, independently, in nineteen sixties by S. Palais and S. Smale. Their celebrated Palais–Smale condition guaranteed the existence of a minimal sequence converging to the critical point of a functional. The basis of this approach, as it could have been expected, are theorems of Relich and Sobolev on compactness of imbedding of (Hilbert) functional spaces. Alas, the method could have been applied 'only' in the case of existence of geodesic lines on Riemann manifolds. Besides, this problem solved for the first time in the celebrated Hilbert's treatise, which is customary to be considered as the beginning of direct methods.

To see clearly how this method works in the case of a single integral, let us recall

Sobolev imbedding theorem in general relativity. Let (M, g) be a compact n dimensional manifold, and let $E \to M$ be a vector bundle over M with a fixed fiber metric and compatible connection ∇. The space $C^\infty(E)$ of smooth sections of E can be made into Hilbert, or Banach, space $W_k^2(E)$ as follows.

For $k \geqslant 0$ and $p \geqslant 1$ (the most important is the Hilbert case $p = 2$), the Sobolev space $W_k^p(E)$ is the completion of $C^\infty(E)$ with respect to the norm

$$\| u \|_{k,p} := \left(\sum_{l=0}^{k} \int_M |\nabla^l u|^p \right)^{1/p},$$

where $\nabla^l u = \nabla \circ \cdots \circ \nabla u \in C^\infty \left(\overset{l}{\otimes} T^*M \otimes E \right)$ is the lth covariant derivative of $u \in C^\infty(E)$.

The Hölder space $C^{k,\alpha}(E)$ is defined by the norm

$$\| u \|_{k.\alpha} := \sum_{l=0}^{k} \sup |\nabla^l u| + \sup_{x,y} \frac{\nabla^k u(x) - \nabla^k u(y)|}{|dist(x,y)|^\alpha},$$

where the last supremum is taken over all $x \neq y$ contained in a normal coordinate neighborhood of $x \in M$, and $\nabla^l u(y)$ denotes the tensor at y obtained by the parallel transport from x to y. (Here we follow closely the excellent paper of Thomas H. Parker *Equivariant Sobolev Theorems and Yang–Mills–Higgs Fields* in the R. Palais volume *Global Analysis in Modern Mathematics*, 1994, pp. 33–71.) Now we turn to famous

SOBOLEV EMBEDDING THEOREM. *Let* $1 \leqslant p \leqslant q$, $k \geqslant l$, $0 < \alpha < 1$. *Then*

(a) For $k - \frac{n}{p} \geqslant l - \frac{n}{q}$ *the identity map induces a continuous inclusion* $W_k^p(E) \hookrightarrow W_l^q(E)$, *and this inclusion is compact if* $k > l$, *and* $k - \frac{n}{p} > l - \frac{n}{q}$.

(b) For $k - \frac{n}{p} \geqslant l + \alpha$ *the inclusion* $W_k^p(E) \hookrightarrow C^{l,\alpha}(E)$ *is continuous and is compact if* $k - \frac{n}{p} > l + \alpha$.

EXAMPLES. 1. $W_1^2(E) \hookrightarrow L^2(E)$ $(\equiv W_0^2(E))$ is compact;
2. For a 1-manifold M, $W_1^2(E) \hookrightarrow C^0(E)$ is compact;
3. For a 4-manifold M, $W_1^2(E) \hookrightarrow L^4(E)$ is continuous, but *not* compact. This is the so called *borderline case* of the Sobolev theorem.

Let us look closer at inequality (b) guaranteeing compactness of the imbedding $W_k^p(E) \hookrightarrow C^{l,\alpha}(E)$; we see that it is easier to fulfill $k - \frac{n}{p} > l + \alpha$ if the dimension n of M is small. If a compact Lie group G acts isometrically on a compact Riemann manifold $M = M^n$, one can define Sobolev spaces of G-invariant functions, and, more generally, of G-equivariant sections of a vector bundle $E \to M$, then we have the beautiful

EQUIVARIANT SOBOLEV EMBEDDING THEOREM (PARKER, 1994) states that *if all the G-orbits in M have dimension greater or equal d, then the space* $W_{G,k}^p(E)$ *and* $C_G^k(E)$ *of G-equivariant sections of* $E \to M$ *satisfy the Sobolev embedding theorem with n replaced by* $n - d$.

DEFINITION of smooth *equivariant sections* $C_G^\infty(E)$. We say that a compact Lie group G acts isometrically on the vector bundle if every $g \in G$ gives a smooth metric-preserving bundle map $\tilde{g} : E \to E$ covering the isometry

$g : M \to M$ (that is, $\pi\tilde{g} = g\pi$, and the diagram

$$\begin{array}{ccc} E & \xrightarrow{\tilde{g}} & E \\ \pi \downarrow & & \downarrow \pi \\ M & \xrightarrow{g} & M \end{array}$$

is commutative.)

A section $u \in C^\infty(E)$ is *G-equivariant* if $u(gx) = \tilde{g}(u(x))$ for all $x \in M$.

The group G acts on connections on E as well: $g \in G$ takes the connection ∇ to the connection $g \cdot \nabla$ defined by

$$(g \cdot \nabla)_v u := \tilde{g}^{-1}(\nabla_{g_* v}(\tilde{g}u)),$$

where v is a vector field on M, $u \in C^\infty(E)$, and $g_* v$ denotes derivative of g.

Averaging (integrating) over the group G, we obtain the invariant connection ∇^0, fixed once for all. Completing the space $C_G^\infty(E)$ with respect to the norms 1. and 2., and using now the invariant connection ∇^0, one obtains the Sobolev spaces $W_G^p(E)$ and $C_G^k(E)$ in the Parker's theorem.

Remark. We saw in the preceding Chapter that the existence of G-invariant solution for variational problems of mechanics or of corresponding differential equations made it possible to reduce the original problem to a simpler one; this was the idea of Jacobi and Lie which culminated in Lie's momentum map and his reduction of phase space which was a forerunner of the symplectic reduction on Kähler and hyperkählerian manifolds.

Thus the fundamental principle is: look for symmetries of a variational problem and then work with equivariant solutions (or invariant functions.)

The Palais–Smale theory generalizes the classical criteria for existence of minimal sequences for linear variational problems to the *nonlinear* ones. Palais and Smale considered a complete C^2 Riemann–Hilbert manifold $\mathcal{M}$ (without boundary) and a function (functional) $F : \mathcal{M} \to \mathbb{R}$. F satisfies the Palais–Smale condition if every sequence u_k on $\mathcal{M}$ with $F(u_k)$ bounded and $\| (\operatorname{grad} F)_{u_k} \| \to 0$ has a convergent subsequence. The Palais–Smale condition means that the downward gradient flow lines of F converge. Thus this is the sort of a method of *steepest descent.*

Palais and Smale proved their famous theorem by showing that the basic lemmas of Morse theory apply in *infinite dimensional* situation as well.

THEOREM (PALAIS, SMALE). *If $F : \mathcal{M} \to \mathbb{R}$ satisfies the Palais–Smale condition and has only non degenerate critical point, then the Morse theorems hold:*

1. The critical values of F are isolated and there is only a finite number of critical points at any critical level;

2. If there are no critical values of F in $[a,b]$ then $\mathcal{M}_b := F^{-1}((-\infty, b))$ is diffeomorphic to $\mathcal{M}_a$;

3. If $a < c < b$ and c is the only critical value of F in $[a,b]$, and $U_1, \ldots, u_k$ are the critical points at level c, then $\mathcal{M}_b$ deformation retracts to $\mathcal{M}_a \cup H_1 \cup \cdots \cup H_k$, where H_j is a cell, called the 'handle', of dimension equal to the index of F at u_j, attached to $\partial\mathcal{M}_b$ by a homomorphism of the boundary spaces.

This theorem gives beautiful results in the problem of existence of geodesics on Riemann manifolds (this problem was considered in the classic works of M. Morse in nineteen thirties): one considers a complete Riemann manifold (M, g), fixes $x, y \in M$, and considers the energy of the path γ

$$E(\gamma) := \int_0^1 |\dot{\gamma}(t)|^2 dt,$$

where $\gamma : [0,1] \to M$ with $\gamma(0) = x$, $\gamma(1) = y$.

To topologize the path space $\mathcal{P}$ of all W_1^2 paths connecting x and y, one fixes an embedding $M \hookrightarrow \mathbb{R}^N$. This defines an inclusion of $\mathcal{P}$ into the Hilbert space $H := W_1^2([0,1], \mathbb{R}^N)$ and the Sobolev embedding $W_1^2 \hookrightarrow C^{1/2}$ shows that $\mathcal{P}$ is a closed subspace of H with one component for each homotopy class of paths from x to y. One computes the (variational) derivative of E:

$$(dE)_\gamma(v) = \int_0^1 g(v, \nabla_T T)\, dt,$$

where $T = \dot{\gamma}$. The critical points are solutions of the geodesic equation $\nabla_T T = 0$.

The fundamental result of Palais and Smale is

THEOREM (PALAIS, SMALE). *(a) The energy $E : \mathcal{P} \to \mathbb{R}$ satisfies the Palais–Smale condition. Each critical point is a smooth geodesic from x to y.*

(b) On a compact simply connected Riemann manifold (M, g) there are infinitely many geodesics joining x and y.

Another result deals with *closed* geodesics: one considers the space $\bigwedge M$ of W_1^2 maps $S^1 \to M$ and the energy function $E : \bigwedge M \to \mathbb{R}$ to obtain the following classical theorem ('proved' by Poincaré for M being a Riemann surface).

THEOREM. *On a compact Riemann manifold every element of the fundamental group $\pi_1(M)$ is represented by closed geodesic.*

As we noted earlier, the Palais–Smale theory applies to one dimensional variational problems like the geodesic problem. Parker's methods made it possible to extend the Palais–Smale theory to higher dimensional problems in presence of symmetry.

Suppose that a compact Lie group G acts on a Riemann manifold M and that all G-orbits have dimension $d \geqslant 1$. If we are given a functional E defined on a Hilbert or Banach manifold $\mathcal{M}$ associated with M (for example, $\mathcal{M} = W_k^2(E)$ for some vector bundle $E \to M$), we have an induced action of G on $\mathcal{M}$. Then we have the following result.

THEOREM (PARKER). *Let G be a Lie group acting smoothly on a Banach manifold and let $F : \mathcal{M} \to \mathbb{R}$ be smooth G invariant function. If*
(a) G is compact or
(b) G acts isometrically on M,
then the set $\mathcal{M}^G \subset \mathcal{M}$ of fixed points is a smooth closed invariant manifold. If $u \in \mathcal{M}^G$ is a critical point of $F : \mathcal{M}^G \to \mathbb{R}$, then it is a critical point of $F : \mathcal{M} \to \mathbb{R}$.

Let us turn to a very difficult problem of global differential geometry, solved finally by variational methods by R. Schoen after the cumulative work of H. Yamabe, N. Trudinger, and T. Aubin. This is

Yamabe problem. We are given a compact Riemann manifold (M^n, g) of dimension $n \geqslant 3$. Find a metric conformal to g which has *constant scalar curvature.*

The history of this problem is dramatic. Hudihiko Yamabe (to whom (along with Montgomery and Zippin) we owe the solution of the 5th Hilbert problem) posed the problem and 'solved' it in his influential paper of 1960 *On a deformation of Riemann structure on compact manifolds*, Osaka J. Math. **12** (1960), 21–37. In fact the Yamabe's original goal was to solve the three dimensional Poincaré conjecture. The resolution of the Yamabe problem is the first step in a general program for constructing canonical Riemannian metrics on smooth manifolds. R. Schoen writes in 1986: *If M is a smooth compact manifold, then an Einstein metric on M is a metric whose Ricci curvature is proportional to g, that is, $Ric(g) = cg$, for a constant c. If M is three dimensional, then an Einstein metric necessarily has constant curvature. In particular, the existence of and Einstein metric on a simply connected* 3*-manifold implies the Poincaré conjecture.* In 1968 (several years after Yamabe death), Trudinger found that Yamabe original paper contains a severe error. The Yamabe problem was finally solved by R. Schoen only in 1984.

The problem can be cast into variational form as follows. Set $p = 2n/(n-2)$, $a = n - 2/4(n-1)$ and let s_g denote the scalar curvature of the metric g. The critical points $u \in W_1^2$ of the functional

$$L(u) := \frac{1}{2}\int_M |du|^2 + as_g u \tag{1}$$

subject to the constraint $\int_M |du|^p = 1$ are solutions of the nonlinear equation

$$\Delta u + as_g = \lambda u^{p-1} \tag{2}$$

If u is a *minimal* critical point of L then (2) and the maximum principle imply that $|u| > 0$. The metric $g' := |f|^{p-2}g$ has then *constant* scalar curvature.

It is very interesting that the solution of the Yamabe problem relies heavily on the ideas from general relativity. The solution of the energy problem by Schoen and Yau (1979) was an important step towards the final solution of the Yamabe problem. A self-contained and very readable proof of the Yamabe problem is given in the beautiful paper of J.M. Lee and T.H. Parker in Bull. Am. Math. Soc. **17**, 31–91.

As we remarked already, variational methods are a powerful tool in solving nonlinear partial differential equations. As it was proved by Hilbert in 1916, the equation of general relativity, called now the Einstein–Hilbert equations, are Euler–Lagrange equations for a natural variational integral, the Hilbert–Einstein action.

Let $\mathfrak{m}_1$ denote the space of Riemann metrics g on M satisfying the volume constraint $\int_M dvol(g) = 1$, then the Hilbert–Einstein action is a functional $\mathcal{R} : \mathfrak{m}_1 \to \mathbb{R}$ given by

$$\mathcal{R}(g) := \int_m s_g dvol(g),$$

where s_g is the scalar curvature of g. Clearly $\mathcal{R}(\cdot)$ is invariant under action of the diffeomorphism group $Diff(M)$ and hence is defined on $\mathfrak{m}_1/Diff(M)$. $\mathcal{R}$ is minimized with respect to conformal class of g (the metrics $\bar{g} = e^f g$, $f \in C^\infty(M)$), and maximized in directions orthogonal to the conformal class of g. This suggests that one may find an Einstein metric by a two-step procedure (Schoen): fix a metric $g_0 \in \mathfrak{m}_1$, and consider the problem

$$I(M; g_0) := \inf_{g \in [g_0]} \mathcal{R}(g),$$

where $[g_0]$ is the conformal class of g_0. Now maximize over g_0

$$I(M) := \sup_{g_0 \in \mathfrak{m}_1} I(M; g_0).$$

The first step is the Yamabe problem: the existence of a smooth $g \in [g_0]$ with $\mathcal{R}(g) = I(M; g_0)$.

As shown by Parker, in the case of Riemann manifolds with symmetries, solution of the Yamabe is by far easier.

THEOREM (PARKER, 1994). *Let (M, g) be a compact Riemann manifold of dimension $n \geqslant 3$. Suppose that a compact Lie group acts isometrically on M with each orbit having dimension d, $1 \leqslant d \leqslant n$. Then there exists a sequence (u_k) of smooth, G-invariant solutions of the equation*

$$\Delta u + a s_g = \lambda u^{p-1}$$

with $L(u_k) \to \infty$, which, therefore, give infinitely many solutions of the Yamabe problem.

As the last application of the Parker's equivariant Morse theory, we give the theorem on existence of

Yang–Mills–Higgs fields on compact four dimensional Riemann manifolds M. Yang–Mills–Higgs fields are, by definition, the critical points of

$$L(\nabla, u) := \frac{1}{2} \int_M \left(|F^\nabla|^2 + |\nabla u|^2 + \frac{\lambda}{2}(|u|^2 - \mu^2)^2 \right) dv,$$

where ∇ is a connection on a bundle over M with the curvature F^{∇}, u is a section of the associated vector bundle, and $\lambda \neq 0$ and μ are coupling constants.

These fields first appeared in physics. The Yang–Mills connections are the critical points of the functional

$$YM(\nabla) := \frac{1}{2}\int_M |F^{\nabla}|^2 dv.$$

Analytical properties of Yang–Mills–Higgs fields were studied already fifteen years ago; however the existence of Yang–Mills–Higgs fields was entirely unknown on any 4-manifolds until very recently, when existence was established on S^4 with the help of equivariant Morse theory (Parker.) There is the following important

THEOREM (PARKER). *Let P be a principal* SU(2) *bundle on S^4 and let E be the associated vector bundle. Then:*

(a) For $\lambda < 0$ there is infinitely many distinct solutions of the Yang–Mills–Higgs equations on (P, E) which are neither uncoupled, nor reducible.

(b) For $\lambda > 0$ and $\lambda\mu$ sufficiently large there is a solution of the Yang–Mills–Higgs equations on (P, E) which is neither uncoupled, nor reducible.

Part V

Riemann and Complex Geometry

CHAPTER 1

Introduction

This part is dedicated to Hans Graver and his great friend and co-author Reinhold Remmert

Motto. The Riemann–Dirichlet principle is this inner eye which enables us to perceive the divine nature of the idea of Riemann surface culminating in construction of Riemann surfaces.

A variation on Hermann Weyl

In the previous parts of this book we have many times encountered multidimensional complex differential manifolds, but usually these manifolds were compact, for example complex tori, and equipped with Kähler metrics. The culminating point of the analysis on these spaces is the Hodge theory and, in particular, the generalization of the Riemann–Roch theorem by Hirzenbruch and Atiyah–Singer.

Riemann saw his theory as a theory of potential on Riemann surfaces, that is, as a theory of the operator d'' and the Laplace operator $d'd''$ ($= \Delta dz \wedge d\bar{z}$). In the multidimensional case the theory of differential forms $A^{p,q}(X)$ started with research of Wirtinger who introduced the operators $d'' : A^{p,q}(X) \to A^{p,q+1}(X)$, $d' : A^{p,q}(X) \to A^{p+1,q}(X)$ and created the so called Wirtinger calculus.

Along with holomorphic functions (solutions of the equation $d''u = 0$), in Riemann approach an important role is played by harmonic and subharmonic functions, the latter satisfy $\Delta u \geqslant 0$. The theory of these functions made it possible for Perron to secure the Riemann's Dirichlet principle on Riemann surfaces: they satisfy the maximum primciple. In the multidimensional case their role is played by plurisubharmonic functions $\varphi : X \to \overline{\mathbb{R}}$; for them the complex Hessian $d'd''\varphi$ is (strictly) positive definite, therefore they are (holo-) convex and upper semi continuous. We know already the role played by these properties in the calculus of variations.

Riemann knew that every domain $G \subset \mathbb{C}$ is a domain of holomorphy, but Hartog's construction of domains $G \subset \mathbb{C}^n$, $n \geqslant 2$ which are not domains of holomorphy and discovery of E.E. Levis that domains of holomorphy in $\mathbb{C}^2$, characterized by plurisubharmonic functions, have a convexity property opened a new chapter of complex analysis and culminated in the theory of Stein manifolds and the solution of generalized Levi problem due to Grauert:

Domains of holomorphy in $\mathbb{C}^n$ are Stein manifolds.

Stein manifolds can be wonderfully characterized: on them (smooth) plurisubharmonic exhausting functions live.

The Riemann moduli problem was developed to the stage of the magnificent Teichmüller theory which showed without doubts the need for complex spaces with singularities: the moduli space $\mathcal{M}(p)$ of (equivalence classes of) Riemann surfaces of genus $p > 1$ is the quotient space $\mathcal{T}(p)/\Gamma(p)$ of the Teichmüller space $\mathcal{T}(p)$ (which is a Stein manifold) by the Teichmüller modular group $\Gamma(p)$. Since the action of $\Gamma(p)$ has fixed points, the moduli space $cM(p) = \mathcal{T}(p)/\Gamma(p)$ has singularities. Thus the Riemann moduli problem was a mighty impulse of creation of a theory of complex spaces.

But the Riemann moduli problem was only the first problem in plethora of moduli problems: construction of complex spaces parametrizing (equivalent classes of) important mathematical and physical objects (stable vector bundles, instantons, and hermitian Einstein connections on vector bundles $E \to M$ over compact (Kähler) manifolds M); such connections are minima of the 'Dirichlet' integral $\mathcal{I}(h) = \int_M \| K^h \| \, dM$, where K^h is the mean curvature of the connection h on E.

More explicitly, on the compact base M we have a *fixed* Kähler metric g, and on the vector bundle we have a variable hermitian metric h. The Kobayashi functional $\mathcal{I}(h)$ is similar to the Hilbert–Einstein functional of general relativity $\mathcal{H}(g) = \int_M R(g)dM$, where $M = (M, g)$ is a space-time equipped with the pseudo– Riemann metric g (the tangent bundle $TM \to M$ is equipped with g.) here the metric of M (that is, of TM) is variable. The same situation is in the case of the Kähler–Einstein (hermitian) metrics on M. But these interesting metrics are defined as critical points of the energy integrals $\mathcal{H}(h)$ and $\mathcal{I}(h)$.

As we will see the hermitian Einstein connections can be characterized as critical points of yet another very interesting integral: the Donaldson integral. One could 'define' the Riemann's Dirichlet principle as a characterization of the important mathematical and physical objects, the critical points of functionals $\mathcal{F}(h)$ defined by integrals which often have the form of 'energy'.

We saw that geometry of Teichmüller space is a beautiful chapter of Riemann hermitian geometry. And as Nigel Hitchin remarked *We are at the beginning of a huge chapter of complex geometry of moduli spaces – perhaps their origin (as moduli spaces) is not so important.*

Therefore the present part divides naturally into two blocks: the complex analysis of several variables and the differentail geometry of complex vector bundles and their moduli spaces.

We close this introduction with the words of one of the best experts of complex analysis – Hans Grauert:

'It is more difficult to construct holomorphic meromorphic functions of several variables than those of one variable. Up to today they are modular functions or Feynman integrals, or functions with group symmetries if they are defined properly for more than one variable. In general the detailed analysis of such functions is very difficult. It is no wonder that till now essentially only general facts were obtained. Complex analysis (of several complex variables) is rather a special kind of geometry than an analysis of properties of functions.'

CHAPTER 2

On Complex Analysis in Several Variables

In one dimension, every domain $D \subset \mathbb{C}^1$ is a domain of holomorphy, that is, there is a holomorphic function f in D, $f \in \mathcal{O}(D)$ which cannot be holomorphically extended to larger domain because on the boundary ∂D of D f has a dense set of singularities. This fact was known already to Riemann and Weierstrass. For $n > 1$ the situation is quite different: Fritz Hartogs has constructed simple domains $D \subset \mathbb{C}^2$ which are not domains of holomorphy; every $f \in \mathcal{O}(D)$ can be extended to a larger convex domain $\hat{D}$. Soon afterwards E.E. Levi discovered that domains of holomorphy in $\mathbb{C}^2$, characterized by strictly plurisubharmonic exhausting functions φ have a convexity property.

As we know, a complex manifold X on which an exhausting plurisubharmonic function φ lives (that is, every set $X_c = \{x \in X : \varphi(x) < c\}$ is relatively compact) is called Stein manifold. We have already encountered such manifolds in the Riemann moduli problem; the Teichmüller space $\mathcal{T}_p$ of a Riemann surface of genus $p > 1$ is a Stein manifold. Therefore domains of holomorphy in $\mathbb{C}^n$, $n > 1$ are Stein manifolds. They are the most important examples of non compact complex manifolds and can be characterized in many different ways. For these manifold the famous B-theorem of Cartan–Serre holds. Covering by Stein subsets is a very powerful tool in the theory of complex manifolds.

But let us return to Riemann surfaces. Every domain $G \subset \mathbb{C}^1$ is an open Riemann surface. But for any domain $G \subset \mathbb{C}^1$ the famous classic theorem of Mittag–Leffler (on existence of a meromorphic function with prescribed principal part), and Weierstrass theorem (on existence of a meromorphic function with prescribed divisor of zeros and poles) hold.

As noted by Hörmander, the Mittag–Leffler theorem is equivalent to the following theorem concerning the (Cauchy–Riemann) operator d''

THEOREM. *For any domain $G \subset \mathbb{C}^1$ and any $f \in C^\infty(G)$, equation $d''u = f$ has a solution $u \in C^\infty(G)$.*

A natural question arises as to if analogues of Mittag–Leffler and Weierstrass theorems hold for any open Riemann surface? The affirmative answer was given in the seminal paper of H. Behnke and K. Stein; this paper revealed the fundamental importance of Runge theorem and Runge domains in the case open Riemann surfaces X. Behnke and Stein proved that *every open Riemann surface is a (1 dimensional) Stein manifold.*

Pierre Cousin, a student of Poincaré, formulated, for a polycylinder $D_n \subset \mathbb{C}^n$, $n > 1$, an analogue of the Mittag–Leffler problem, called now the additive (or first) Cousin problem; and the multiplicative (or second) Cousin problem and presented solutions of these problems. This was in 1895, thus ten years before the discovery of domains of holomorphy by Hartogs. Nowadays it is evident that one has to formulate the Cousin problems only for domains of holomorphy and, perhaps, on Stein manifolds. The theory of sheaves, developed by Leray, and codified and simplified by Henry Cartan, made it possible for him and his brilliant young student J.P. Serre (born 1926) to prove the famous theorems B and A and to provide solution to the first Cousin problem on any Stein manifold and to give sufficient conditions for solvability of the second Cousin problem. I strongly recommend the reader to get acquainted with the charming talk of Remmert *The Golden Fifties of Complex Analysis* given on the occasion of 60th birthday of Hans Grauert. The great expert on the field presents the atmosphere of this period.

We have seen already the role of the Riemann operator d'' in complex analysis in one dimension. Lars Hörmander has shown that the Riemann approach, that is, the theory of the d'' operator, is the most natural and powerful method of obtaining solutions of the problems of complex analysis on complex manifolds. The role of plurisubharmonic functions in this approach is really illuminating. Below we will present examples of the Hörmander–Riemann method in action. But we start with the general Runge type theorem of Malgrange.

As we have noted already many times, Riemann was 'forced' to considering special complex manifolds of higher dimension (namely certain tori) in his research on abelian functions. The moduli problem posed in the famous work *Theorie der Algebraischer Functionen* of 1857 has led Teichmüller to essentially modern definition of complex manifolds and deformations of complex structures.

The wonderful continuation of Riemannian ideas were works of Hodge on Kähler manifolds which culminates in the main interior characteristic of the (projective) algebraic varieties given by K. Kodaira in 1954 in his famous vanishing theorem for holomorphic functions on compact Kähler manifolds. The Kodaira theorem would be impossible without the method of S. Bochner. In this connection one should recall that, forgotten for many years, Weitzenbröck identities of early 1920s become more and more popular nowadays.

But let us return to Riemann. As I remarked, the Riemann's moduli problem resulted in unbelievable development of complex analysis. As we know Riemann perceived variational methods as a powerful tool to obtain deep existential theorems on Riemann surfaces. This idea of Riemann turned out to be extremely fruitful in multidimensional geometry and analysis. Anthony Tromba and his collaborators showed that this idea leads in the fastest and most natural way to the fundamental theorems of Teichmüller theory. In turn, the Teichmüller theory has led Tromba to deep theorems in the theory of minimal surfaces. The variational Riemann idea flourished in the theory of moduli of yang–Mills connections, the theory of instantons and of moduli of holomorphic stable and simple vector bundles. And all this started from investigations of algebraic functions of one variable.

As we know, in the Riemann approach to the theory, the singular points appear. Completing them with the help of algebraic germs, one obtains a compact Riemann surface (now without singularities.) Riemann was probably the first to stress the importance of singularities. The elementary extension theorem of Riemann for one complex variable has its analogue in the so called Riemann and Levi extension theorems on complex manifolds, which, in one dimension reduce to the theorems of Riemann, and without which the modern theory of complex spaces due to Grauert and Remmert would be impossible.

In the second half of nineteenth century in Italy the classical algebraic geometry was born. The object of investigations are common zeros of polynomials. They are of finite dimensions but may have singularities; outside the singular loci these algebraic sets are complex manifolds.

But even if one considers fibers $f^{-1}(y)$ of holomorphic maps $f : X \to Y$ between complex manifolds (of finite dimension), they are zero sets of a finite number of holomorphic functions, called *analytic sets* in X. Analytic sets are complex manifolds outside their singular loci. This clearly shows that for investigations of complex manifolds which are spaces more general than complex manifolds are necessary. Such spaces were introduced by H. Cartan and J.P. Serre, and are called today *reduced complex spaces.* But even these general objects are not sufficient in practice.

Only in 1960 H. Grauert found sufficiently general topological spaces (called by him *complex spaces*) and for them he proved a number of fundamental theorems, the most magnificent of whose is the direct image theorem for coherent sheaves. As we will see, from this theorem not only famous theorem of Remmert on analytic sets and generalization of the Cartan–Serre finiteness theorem follow, but they also are indispensable tools in the general theory of deformations of complex structures.

In all these marvelous theorems the coherent analytic sheaves play a fundamental role. Coherence is a local property, in a sense it is a *local principle of analytic continuation.* But proofs of coherence of relevant sheaves such like the structure sheaf $\mathcal{O}_X$ of a complex space, or the ideal sheaf $i(A)$ of any analytic set A are very difficult. The support $Supp\,\mathcal{F}$ of a coherent sheaf is an analytic set.

Since the $\mathbb{C}$-ringed space $(\mathbb{C}^n, \mathcal{O}_{\mathbb{C}^n})$, where $\mathcal{O}_{\mathbb{C}^n}$ is the sheaf of germs of local holomorphic functions in $\mathbb{C}^n$ is the departure point for the whole complex analysis, the coherence of this sheaf (proved by K. Oka in 1948) is so fundamental. But all proofs of Oka theorem start with the Weierstrass preparation theorem or division theorem of Stickelberger, therefore we will present the proof of the division theorem of Grauert–Remmert which holds for any field with complex valuation.

Many of these theories could be regarded as extensions of the theory of compact Riemann surfaces to some classes of complex manifolds, usually Kähler manifolds, which are distinguished by the fact that the complex structure J is beautifully related to the (geometric) Riemann structure; for the Levi–Civita connection ∇^{LC}, we have $\nabla^{LC}J = 0$. Another great impulse was provided by

Domains of holomorphy. As Riemann has shown, *every simply connected domain $G \subset \mathbb{C}$ is a domain of holomorphy of a holomorphic function $f \in \mathcal{O}(G)$*, that is on the boundary ∂G of G there is a dense subset of singular points of f; therefore f cannot be holomorphically extended to any larger domain.

PROOF. Riemann made use of the mapping theorem. Since $G \simeq dz$ it is sufficient to prove the theorem for $G = \mathbb{D} = \{z \in \mathbb{C} : |z| < 1\}$. It is sufficient to construct $f_0 \in \mathcal{O}(\mathbb{D})$ with appropriate singularities. Such a function is

$$f_0 = \sum_{k=0}^{\infty} z^{k!}.$$

Indeed f_0 is not bounded for $z = e^{i\varphi}$ with rational $\varphi = p/q$, $p, q \in \mathbb{N}$:

$$|f(re^{i\varphi})| \geqslant \left| \sum_{k=q+1}^{\infty} r^{k!} e^{ik!} \right| - \left| \sum_{k=0}^{q} r^{k!} e^{i\frac{p}{q}k!} \right| =$$

$$\sum_{k=q+1}^{\infty} r^{k!} - \left| \sum_{k=0}^{q} r^{k!} e^{i\frac{p}{q}k!} \right| \to \infty$$

for $r \to 1$. But the set of rational φ is dense in $[0, 2\pi]$; therefore f_0 has singularities on the whole of $\partial\mathbb{D}$. □

This remarkable theorem of Riemann was proved much later by Weierstrass as a corollary to his classic Weierstrass theorem.

Let z_j, $j = 1, 2, \ldots$ be a discrete sequence of different points in an open set $G \subset \mathbb{C}^1$ and let n_j be arbitrary integers. Then there is a meromorphic function f on G such that it is holomorphic and not equal zero except at the points z_j, and $f(z)(z-z_j)^{-n_j}$ is holomorphic and not equal zero in neighborhood of z_j for every j. Thus f has prescribed zeros and poles of given orders.

COROLLARY 1. *Any meromorphic function F in $G \subset \mathbb{C}^1$ can be written as f/g, where $f, g \in \mathcal{O}(G)$ – the space of holomorphic functions on G.*

PROOF. Such F satisfies the Weierstrass theorem; for every j construct a holomorphic function g with zero of order n_j at z_j. Then $Fg = f \in \mathcal{O}(G)$, therefore $F = f/g$. □

COROLLARY 2 (RIEMANN THEOREM). *If D is a disc with the center in G, and g is meromorphic in D and equal f near the center of D then $D \subset G$, then there exists $f \in \mathcal{O}(G)$ which cannot be continued analytically to any larger set, not even as a meromorphic function.*

PROOF (FOLLOWING HÖRMANDER). Label all points with rational coordinates in G so that all these points occur *infinite* number of times, and let r_i be the distance $d(z_j, \mathbb{C} - G) =: r_j$ of z_j to the boundary ∂G of G. Take the sequence K_j of compact subsets exhausting G, and for every j choose $w_j \in \mathbb{C} - K_j$ such that $|w_j - z_j| < r_j$. Since the sequence $\{w_j\}$ is discrete in G, apply Weierstrass theorem: there exist $f \in \mathcal{O}(G)$ with zeros at w_j and no other zeros. This f solves the problem.

Indeed, if $a \in G$ has rational coordinates and $r = d(a, \partial G)$, the disc $D := \{z : |z - a| < r\}$ contains infinitely many points w_j, because $z_j = a$ for infinitely many j. Thus f cannot be extended to a meromorphic function on an open disc containing D, because zeros of non-vanishing meromorphic function are isolated. □

It can be easily imagined how great was the impression caused by the work of Fritz Hartogs of 1906, who presented the following domain $G \subset \mathbb{C}^2$ (the Hartogs figure)

$$G = \{z = (z_1, z_2) \in \mathbb{C}^2 : |z_1| < \frac{1}{2}, |z_2| < 1, \text{ or } |z_1| < 1, \frac{1}{2}|z_2| < 1\}.$$

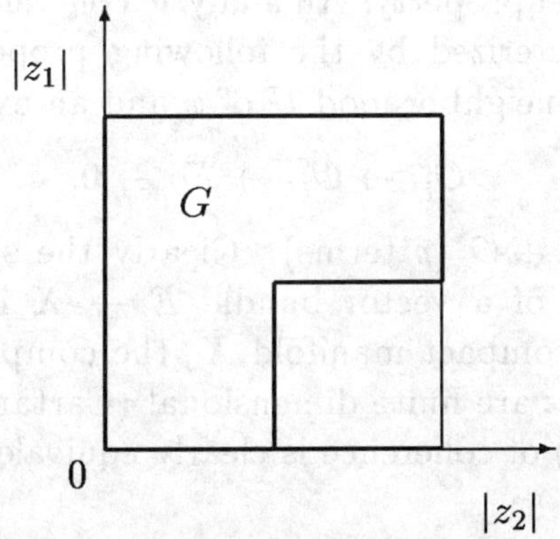

THEOREM (HARTOGS, 1905). *Every function f holomorphic in G can be analytically continued to a holomorphic function F in the polydisc $P = \{z \in \mathbb{C}^2 : |z_1| < 1, |z_2| < 1\}$.*

PROOF. Use the Cauchy integral in one complex variable. In every vertical slice z_1 =const we extend f by the Cauchy formula

$$F(z_1, z_2) := \frac{1}{2\pi i} \int_{w_2=1} \frac{f(z_1, w_2)}{w_2 - z_2} dw_2.$$

F is holomorphic in P and equal to f on an open subset of G. Thus, by the identity theorem F is a holomorphic extension of f. $\square$

Before we turn to the circle of questions related to domains of holomorphy (the Levi problem), solved so brilliantly by Hans Grauert and connected with the name of Karl Stein (the teacher of Grauert and Remmert), we have to bring to the reader attention yet another fundamental difference between analytical function of one and many variables.

REMARK. Of course, relatively early, at the end of nineteenth century people tried to extend the theorems of Mittag–Leffler and Weierstrass to some domains in $\mathbb{C}^n$, $n > 1$ (Poincaré, his pupil Cousin, and others.) It was soon observed that the analogues of these problems, the first and second Cousin problems, should be considered in domains of holomorphy, and, more generally, on Stein manifolds (and spaces.) Soon the language of sheaf theory appeared, in particular of analytic *coherent sheaves* (that is, $\mathcal{O}$ sheaves.) Such is, for example, the sheaf of germs $(X, \mathcal{O}_X)$ of holomorphic functions on a complex manifold, and, more generally, a sheaf of germs of holomorphic sections of a holomorphic vector bundle.

Coherence is a local property. An analytic coherent sheaf $\mathcal{F}$ over complex manifold X is characterized by the following property: Given any point $x \in X$, there exists a neighborhood U of x and an exact sequence

$$(C) \qquad \mathcal{O}_U^p \to \mathcal{O}_U^q \to \mathcal{F}_U \to 0,$$

where $\mathcal{O}^p := \mathcal{O} \oplus \ldots \oplus \mathcal{O}$ (p terms). Clearly the sheaf $\mathcal{O}(E)$ of germs of holomorphic sections of a vector bundle $E \to X$ is coherent. For a coherent sheaf $\mathcal{F}$ over compact manifold X, the complex vector spaces finite $H^p(X, \mathcal{F})$, $p = 0, 1, \ldots$ are finite dimensional (Cartan–Serre, 1953.)

The definition (C) of coherence is clearly equivalent to the following two conditions.

1. An $\mathcal{O}$ sheaf $\mathcal{F}$ is *finite* if every point of X has a neighborhood U such that $\mathcal{F}_U$ is generated by finitely many sections of $\mathcal{F}(U)$, that if $\mathcal{F}_U$ is a quotient sheaf of $\mathcal{O}^p$ for suitable p. Clearly, all sheaves $\mathcal{O}^k$, $k = 1, 2, \ldots$ are finite.

2. An $\mathcal{O}$ sheaf $\mathcal{F}$ is *relationally finite* if every open $U \subset X$ and every $\mathcal{O}_U$ homomorphism $\psi : \mathcal{O}_U^q \to \mathcal{F}_U$, $\ker \psi$ is finite on U, that is, for every $s_1, \ldots, s_q \in \mathcal{F}(U)$ the sheaf of relations $Rel(s_1, \ldots, s_q) := \bigcup_{x \in U} \{(a_{1x}, \ldots, a_{qx}) \in \mathcal{O}_x^q : \sum_1^q a_{ix} s_{ix} = 0\}$ is finite on U.

Thus $\mathcal{O}$ module $\mathcal{F}$ is coherent if $\mathcal{F}$ is finite and relationally finite. Coherence permits to pass from *point* properties to *local* properties. *All the proofs of coherence are very difficult but the starting point of all of them is the Weierstrass preparation theorem and its generalizations.*

At this point we must leave Riemann and say a few words on theorems due to Karl Weierstrass. Alongside with fundamental theorems for functions of one complex variables and investigations of hyper-elliptic integrals, mathematics owes Weierstrass something even more fundamental, namely

Algebraization of complex analysis which is inseparable from the famous Weierstrass preparation theorem. This theorem was proved first by Stickelberger, and during the following 80 years it was being proved anew by many distinguished mathematicians, who did not realize that the proof had existed already. The approach of Weierstrass has led to top achievements of modern complex analysis like Grauert theorem on coherence of direct image of coherent analytic sheaves.

Let us then return to coherent sheaves and look closer at

Coherent sheaves in complex analysis. The notion of the sheaf $\mathcal{O}_X$ of germs of local holomorphic functions in domain $D \subset \mathbb{C}^1$ and in Riemann surface X (covering $\mathbb{P}^1$) is implicitly contained in Riemann–Weierstrass concept of analytic continuation: The sheaves $\mathcal{O}_D$ and $\mathcal{O}_X$ are Hausdorff sheaves in distinction to the sheaf $\mathcal{C}_D$ of germs of (local) continuous functions.

In analysis we have *point properties* (at $x \in X$), *local properties* (in some neighborhood $U(x)$ of the point x), and *global properties* which hold on the whole topological space X.

The notion of sheaf $(\mathcal{F}, \pi, X)$, where $\mathcal{F}$, X are topological spaces, and $\pi : \mathcal{F} \to X$ is a surjection which is locally homeomorphic was conceived by J. Leray in order to pass from local to global properties. The sheaf theory is thus indispensable in global analysis and global differential geometry. Leray developed the cohomology theory: cohomology spaces $H^k(X, \mathcal{F})$ with values in the sheaf $\mathcal{F}$. But in complex analysis one has to have sheaves $\mathcal{F}$ that are coherent: a property of every stalk $\mathcal{F}_x$ at $x \in X$ should be preserved for all stalks $\mathcal{F}_y$ for $y \in U(x)$, in a neighborhood $U(x)$ of x. Thus coherence is, in a vogue sense, a *local principle of analytic continuation* (H. Cartan.) As Grauert–Remmert remark: A typical situation is the following. Let $\mathcal{F}' \xrightarrow{\varphi} \mathcal{F} \xrightarrow{\psi} \mathcal{F}''$ be a sequence of coherent sheaves on a complex space X.

If for a certain point $x \in X$ the sequence of stalks $\mathcal{F}'_x \overset{\varphi_x}{\to} \mathcal{F}_x \overset{\psi_x}{\to} \mathcal{F}''_x$ is exact, then the same holds for all $y \in U(x)$. Thus coherence of sheaves makes it possible to *pass from point properties to local properties* and not only from local to global ones! But, alas, such a deep property cannot be proved easily. The most important sheaf of complex analysis $\mathcal{O}_{\mathbb{C}^n}$ is coherent; this is the famous theorem of Oka. Therefore (since coherence is a local property) the structure sheaf $\mathcal{O}_X$ of a complex space $X = (X, \mathcal{O}_X)$ is coherent; then $\mathcal{O}_X^p$ is coherent, and thus the sheaves of germs of local holomorphic sections of holomorphic vector bundles $E \to X$ (of rank p) are coherent analytic sheaves.

Stickelberger division theorem (1887) was *not* known to Weierstrass (but is often incorrectly called Weierstrass formula); it was proved in the Stickelberger paper *Über einen Satz von Herrn Noether*, Math. Ann. **30** (1887), 401–409.

The beautiful proof of Grauert–Remmert of the division theorem, we are going to reproduce is valid not only for $\mathbb{C}$ but also for any complete field, that is, a field with complete valuation. Since this theorem, as well as the preparation theorem, are so fundamental, we will present complete proof of both of them. We cannot prove another theorems; this would require a voluminous monograph.

NOTATION. $z = (z_1, \ldots, z_{n-1}) \in \mathbb{C}^{n-1}$, $n > 1$, $(z, w) = (z_1, \ldots, z_{n-1}, w) \in \mathbb{C}^n$.

$\mathcal{O}$ is the structure sheaf of $\mathbb{C}^n$, $\mathcal{O}_0 = \mathbb{C}\{z_1, \ldots, z_n\}$;

$\mathcal{O}'$ is the structure sheaf of $\mathbb{C}^{n-1}$, $\mathcal{O}'_0 = \mathbb{C}\{z_1, \ldots, z_{n-1}\}$.

$\mathcal{O}'_0$ is an integral domain by the identity theorem; ($f \in \mathcal{O}_0$ is a unit)$\Longleftrightarrow$($f(0) \neq 0$).

Stalks (at 0) $\mathcal{O}_0$ (respectively $\mathcal{O}'_0$) can be identified with rings of convergent power series in (z, w) (respectively, in $z = (z_1, \ldots, z_{n-1})$) around 0. We can consider $\mathcal{O}'_0$ and the algebra $\mathcal{O}'_0[w]$ of polynomials in w with coefficients in $\mathcal{O}'_0$ as subalgebras of $\mathcal{O}_0$. Degree of a polynomial $p \in \mathcal{O}'_0[w]$ is denoted by $\deg p$.

In elementary algebra there is well known *Euclidean division algorithm* which states

DIVISION THEOREM (EUCLIDEAN). *Let* $g = g_0 + g_1 T + \cdots + g_b T^b \in R[T]$, *where* R *is an integral domain. Then for every polynomial* $f \in R[T]$ *of degree b there exist uniquely determined polynomials* $q, r \in R[T]$ *such that*

$$(*) \qquad f = qg + r, \quad \text{where } \deg r < b \ .$$

REMARK. It was noted by Dedekind that this powerful division algorithm works in every integral domain R equipped with a map $\delta : R - \{0\} \to \mathbb{N}$ so that for $f, g \in R$, $g \neq 0$, there exist $q, r \in R$ with $f = qg + r$, where $\delta(r) < \delta(g)$, or $r = 0$. δ is called degree or norm of *Euclidean ring R*.

EXAMPLES OF EUCLIDEAN RINGS.

1. $\mathbb{Z}$ with $\delta(a) := |a|$;
2. Let $\mathbf{K}$ be a field, then the ring $\mathbf{K}[T]$ of polynomials is Euclidean; now, of course, $\delta(f) = \deg f$;
3. The ring $\mathbb{Z}[i] := \{x + iy : x, y \in \mathbb{Z}\}$ of Gauss integers is Euclidean if we set $\delta(x + iy) := x^2 + y^2 = |x + iy|^2$.

There is another definition of Euclidean rings.

DEFINITION. An integral domain R equipped with a function (called *height*) $h : R - \{0\} \to \mathbb{N}$ with

1. $h(ab) \geqslant h(a)$,
2. for $a, b \in R - \{0\}$ there exists $c \in R$ such that $h(a - bc) < h(b)$ or $a = bc$

is an Euclidean ring. Clearly, $h = \delta$.

By the classic Euclidean procedure one has

DEFINITION. An Euclidean ring R is *factorial* if every element $u \in R$ is a product of irreducible (prime) elements $u_1, \ldots, u_l$; u_i are unique up to multiplication by units.

We recall that $e \in R$ is a *unit* if there exists $v \in R$ such that $ev = 1$. The set of units in R is denoted by R^*.

The famous Gauss lemma asserts that (R is factorial)$\Longrightarrow$($R[T]$ is factorial). Another important property of Euclidean rings is given by

THEOREM. *In an Euclidean ring there exists a great common divisor g of every subset A of R which is of the form*

$$(**) \qquad g = r_1 a_1 + \cdots + r_s a_s \quad \text{with } r_i \in R, a_i \in A.$$

PROOF. We can assume that $A \neq \{0\}$. Consider all elements of the form (**) and take the element g of minimal height.

COROLLARY. *In an Euclidean ring R every ideal $\mathfrak{a}$ is a principal ideal, that is, $\mathfrak{a}$ is generated by one element.*

$p \in R$ is *prime* if $p|xy$ with $x, y \in R \Longrightarrow p|x$ or $p|y$ (that is, p divides x or y.) Therefore in Euclidean ring every principal ideal is prime.

After these algebraic remarks let us formulate and prove Stickelberger and Weierstrass theorems.

We say that $g \in \mathcal{O}_0$ has order $b \in \mathbb{N}$ in w if

$$g = \sum_0^\infty g_i w^i, \ \ g_i \mathcal{O}'_0, \ \ g_0(0) = \cdots = g_{b-1}(0) = 0, \ \ g_b(0) \neq 0.$$

DIVISION THEOREM (STICKELBERGER 1887). *Let $g \in \mathcal{O}_0$ have order b in w. Then for every germ $q \in \mathcal{O}_0$ and polynomial $r \in \mathcal{O}'_0[w]$,*

(i) $f = qg + r$ and degr $< b$;

(ii) the elements q, r are uniquely determined by f.

It is convenient to introduce a notion of *Weierstrass polynomial* $\omega \in \mathcal{O}'_0[w]$ in w:

$$\omega := w^b + a_1 w^{b-1} + \cdots + a_b \in \mathcal{O}'_0[w], \ \ a_1(0) = \cdots = a_b(0) = 0, \ \ b > 1.$$

OBSERVATION. For such polynomials if $q \in \mathcal{O}_0$ and $q\omega \in \mathcal{O}'_0[w]$, then $q \in \mathcal{O}'_0[w]$.

From division theorem one quickly obtains

WEIERSTRASS PREPARATION THEOREM ($\sim$ 1860).

(a) Let $b \in \mathcal{O}_0$ have order $b \geqslant 0$ in w. Then there exists a unique Weierstrass polynomial $\omega \in \mathcal{O}'_0[w]$ and a unit $e \in \mathcal{O}_0$ such that $g = e\omega$.

(b) Moreover, $(g \in \mathcal{O}'_0[w]) \Longrightarrow (e \in \mathcal{O}'_0[w])$.

PROOF. From division theorem we have $w^b = qg + r$ with degr $< b$. Then $g(0, w) = w^b \hat{e}(w)$ with $\hat{e}(0) \neq 0$; therefore $r(0, w) = 0$ and $q(0, w) = hate(w)^{-1}$; hence q is a unit in $\mathcal{O}_0$. Set $e := 1/q$, $\omega := w^b - r$, then $g = ew$. The uniqueness is immediate.

(b) follows from Observation above. □

A very important corollary is

COROLLARY. *If $g \in \mathcal{O}_0$ has finite order and $g = e\omega$, then the injection $\mathcal{O}'_0[w] \to \mathcal{O}_o$ induces a $\mathbb{C}$-algebra isomorphism $\mathcal{O}'_0[w]/\mathcal{O}'_0[w]\omega \tilde{\to} \mathcal{O}_0/\mathcal{O}_0 g$. Moreover*

(ω is prime in $\mathcal{O}'_0[w]$)$\Longleftrightarrow$(ω is prime in $\mathcal{O}_0$).

REMARK. But also (Preparation theorem)$\Longrightarrow$(Division theorem). The reader will prove this fact as an exercise. Thus we have the equivalence

$$\text{(Weierstrass)} \Longleftrightarrow \text{(Stickelberger)}.$$

For formal power series

$$f = \sum a_{\mu_1 \ldots \mu_{n-1}\mu_n} z_1^{m_1} \cdots z_{n-1}^{\mu_{n-1}} w^{\mu_n}$$

with coefficients in $\mathbb{C}$ Grauert and Remmert introduced the following, very clever norm:

Let $\rho := (r_1, \ldots, r_{n-1}, r_n)$ be an n-tuple of positive numbers, then

$$\| f \|_\rho := \sum |a_{\mu_1 \ldots \mu_{n-1}\mu_n}| r_1^{m_1} \cdots r_{n-1}^{\mu_{n-1}} r^{\mu_n},$$

$$B_\rho = \{ f : \| f \|_\rho < \infty \}.$$

Clearly B_ρ is a normal $\mathbb{C}$ algebra which is a subalgebra of holomorphic functions in the polydisc of polyradius ρ around $0 \in \mathbb{C}^n$. For any $f \in \mathcal{O}_0$ there exists ρ such that $\| f \|_\rho < \infty$ and $\lim_{\rho \to 0} \| f \|_\rho = |f(0)|$.

LEMMA (GRAUERT–REMMERT). *B_ρ is complete, that is, it is a Banach algebra.*

PROOF. We use the abbreviated notation $\rho^\mu := r_1^{\mu_1} \cdots r_n^{\mu_n}$, $f = \sum a_\mu z^{\mu'} w^{\mu_n}$, $\mu' \in \mathbb{N}^{n-1}$, $\mu \in \mathbb{N}^n$. For the Cauchy sequence $f_j = \sum a_{j\mu} z^{\mu'} w^{\mu_n}$, $f_j \in B_\rho$, $j = 1, 2, \ldots$, all sequences $a_{1\mu}, a_{2\mu}, \ldots$ are Cauchy sequences in $\mathbb{C}$ because $\rho^m |a_\mu| \leqslant \| f \|_\rho$. Now put $a_\mu := \lim_{j\to\infty} a_{j\mu}$, $f = \sum a_m z^{m'} w^{\mu_n}$. We have to prove that $\| f - f_j \|_\rho \to 0$ for $j \to \infty$. But for given $\epsilon > 0$ and large j, $\| f - f_j \|_\rho = \sum |a_\mu - a_{j\mu}| \rho^m \leqslant 2\epsilon$ if $\sum |a_{j+i,\mu} - a_{j\mu}| \rho^\mu = \| f_{j+i} f_i \|_\rho \leqslant \epsilon$ for all $i \geqslant 0$, $j \geqslant m$. Hence $\lim_{j\to\infty} f_j = f$

with $f \in B_\rho$. □

PROOF OF DIVISION THEOREM. Let $f = \sum_{k=0}^{\infty} f_k w^k$, $\hat{f} := \sum_0^{b-1} f_k w^k$, $\check{f} = \sum_b^{\infty} f_k w^{k-b}$ with $f \in B_\rho$ and b being a given integer. Thus $\check{f}$ is a polynomial of degree $< b$ and $f = \hat{f} + \check{f} b^n$. We have

$$(*) \qquad \| \check{f} \|_\rho \leqslant r_n^{-b} \| f \|_\rho \quad \text{and} \quad \| \hat{f} \|_\rho \leqslant \| f \|_\rho .$$

Let $0 < e < 1$ be a fixed real number. Since $g \in \mathcal{O}_0$ has order b in w, we can choose g such that $g \in B_\rho$; then $\check{g}$ is a unit in $\mathcal{O}_0$. We can assume that $\check{g}^{-1} \in B_\rho$ and $\| w^b - g\check{g}^{-1} \|_\rho \leqslant r_n^b \epsilon$ (since $g = \hat{g} + \check{g} w^b$.) Let now $f \in \mathcal{O}_0$ be given; take ρ sufficiently small that $f \in B^\rho$, Grauert and Remmert define recursively elements $v_j \in B_\rho$ by taking $v_0 = f, \ldots, v_{j+1} := (w^b - g\check{g})\check{v}_j$, ... Now it follows from (*) that $\| \check{v}_j \|_\rho \leqslant r_n^{-b} \| v_j \|_\rho$, therefore $\| v_{j+1} \|_\rho \leqslant \epsilon \| v_j \|_\rho$. Thus $v := \sum_0^\infty v_j \in B_\rho$ is well defined. Let $q := \check{g}^{-1}\check{v} \in B_\rho$ and $r := \check{v} \in B_\rho$ be polynomials of degree $< b$ in w. But

$$f = w_0 = \sum_0^\infty (v_j - v_{j+1} = \sum_0^\infty (\hat{v}_j + \check{v}_j w^b + g\check{g}^{-1} v_j - w^b \check{v}_j) =$$

$$= \sum_0^\infty (g\check{g}\check{v}_j + \hat{v}_j) = g\check{g}^{-1} \sum_0^\infty \check{v}_j + \sum_0^\infty \hat{v}^j = g\check{g}^{-1}\check{v} + \sum_0^\infty \hat{v}^j =$$

$$= qg + \hat{v} = qg + r.$$

For uniqueness of q and r it is sufficient to prove that

$$(qg + r = 0) \Longrightarrow (q = r = 0).$$

We can assume that $g, q, r \in B_\rho$ for suitable ρ. Since $g_b(0) \neq 0$, we can arrange $g_b^{-1} \in B_\rho$ and write $g = g_b(w^b + h)$, where $h(0) = 0$ and $h \in B_\rho$. Since $qg + r = 0$, $qg_b w^b + qg_b h + r = 0$; therefore $qg_b w^b + r = -qg_b h$ and $\deg r < b$. Since $\| w^b \|_\rho < r_n^b$,

$$M := \| qg_b \|_\rho \, r_n^b = \| qg_b w^b \|_\rho \leqslant \| qg_b w^b + r \|_\rho =$$

$$\| qg_b h \|_\rho \leqslant \| qg_b \|_\rho \, r_n^b \epsilon = M\epsilon.$$

Since $0 < \epsilon < 1$, $M = 0$, thus $\| qg_b \|_\rho = 0 \Longrightarrow qg_b = 0 \Longrightarrow q = 0 \Longrightarrow r = 0$. □

From the division theorem and Hensel lemma, there follows generalized division theorem of Grauert–Remmert, but before we turn to this theorem, let us observe the important

COROLLARY (LASKER 1905, RÜCKERT 1933). *The ring $\mathcal{O}_0$ is factorial and Noetherian, moreover it is Henselian, that is, it satisfies the following version of Hensel lemma:*

HENSEL LEMMA (FOR POWER SERIES). *Let $\omega = \omega(z,w) = w^b + a_1 w^{b-1} + \cdots + a_b \in \mathcal{O}'_0[w]$ and let $\omega(0,w) = (w-c_1)^{b_1}\cdots(w-c_t)^{b_t}$ with different $c_1,\ldots,c_t \in \mathbb{C}$. Then there exists unique monic polynomials $\omega_1,\ldots,\omega_t \in \mathcal{O}'_0[w]$ of degree $b_1,\ldots,b_t$ such that $\omega = \omega_1\cdots\omega_t$ and $\omega_j(0,w) = (w-c_j)^{b_j}$, $j = 1,\ldots,t$.*

The geometric meaning is clear: if the zero set N of a monic polynomial in w meets the w axis in t different points $a_1,\ldots,a_t$, then $N = N_1 \cup \cdots \cup N_t$, where $N_1,\ldots,N_t$ are the zero sets of t monic polynomials in w such that $a_j \in N_j$.

PROOF follows from preparation theorem and induction in t. □

In order to clarify the idea of the proof of Oka of the fundamental theorem asserting coherence of $\mathcal{O}_X$, the structural sheaf of complex space X, Grauert and Remmert gave the following version of division theorem

GENERALIZED DIVISION THEOREM. *Let $\omega \in \mathcal{O}_0[w]$ be of degree $b \geqslant 1$, and let $c_1,\ldots,c_t$ be distinct roots of $\omega(0,w)$. Let $x_j := (0,c_j) \in \mathbb{C}^{n+1}$ and $\mathcal{O}_{x_j}$ be the ring of germs of holomorphic functions at x_j. Then for every choice of t germs $f_j \in \mathcal{O}_{x_j}$ there exist t germs $q_j \in \mathcal{O}_{x_j}$ and a polynomial $r \in \mathcal{O}_0[w]$ of degree $< b$ such that $f_j = q_j\omega_{x_j} + r_{x_j}$ for $j = 1,\ldots,t$. The germs q_j and the polynomial r are uniquely determined by f_j and ω.*

To finish this chapter we present the famous Malgrange theorems which play important role in the theory of unfolding of singularities (Thom, Mather, ...).

MALGRANGE PREPARATION THEOREM (1966). *Let f be a C^∞ function of $(t,x) \in \mathbb{R}^{1+n}$ near $(0,0)$ which satisfies*

$$f = \frac{\partial f}{\partial t} = \frac{\partial^{k-1} f}{\partial t^{k-1}} = 0, \quad \frac{\partial^k f}{\partial t^k} \neq 0, \quad at\ (0,0).$$

Then there exists a factorization

$$f(t,x) = c(t,x)\left(t^k + a_{k-1}(x)t^{k-1} + \cdots + a_0(x)\right),$$

where a_i *and* c *are* C^∞ *functions near* 0 *and* (0, 0), *respectively,* $c(0,0) \neq 0$, $a_j(0) = 0$.

MALGRANGE DIVISION THEOREM. *If* f *satisfies the hypothesis of the preceding theorem and* g *is a* C^∞ *function in a neighborhood of* (0, 0), *then*

$$g(t,x) = q(t,x)f(t,x) + \sum_0^{k-1} t^j r_j(x),$$

where q *and* r_j *are* C^∞ *functions in a neighborhood of* (0, 0) *and* 0, *respectively.*

Note that *uniqueness* holds in the case of both theorems. The beautiful proof of these, highly nontrivial theorems can be found in Volume 1 of the Hörmander monumental treatise. The original proof was complex and was based on the fundamental Łojasiewicz inequality.

CHAPTER 3

Ellipticity, Runge Property, and Runge Type Theorems

In this chapter we tell about the general notion of ellipticity codified into the concept of Fredholm complexes and their cohomology. Below, using the method of B. Malrange we present the promised result that open Riemann surfaces are Stein spaces. We are eager to do that because this method is a continuation of Riemann ideas and makes use of investigation of action of the operator d'' on holomorphic vector bundles on a complex manifold X. If X is Kählerian, then $d'd'' = 4\Delta$ and Riemann surfaces are Kählerian. This is, perhaps, the reason why the theory of the Laplace operator Δ is so fundamental for Riemann potential theory on X.

The classic monograph of Lars Hörmander on complex analysis is written in the Riemann spirit: Hörmander consequently works with the Cauchy–Riemann operator d'' and obtains all the main results with astonishing speed and precision. (Of course, for complex spaces with singularities another methods are necessary!)

Also the theorems of Atiyah–Singer–Bott (index theorems, Lefshetz formula, etc.) which work with elliptic complexes are in the true Riemann spirit. A farther extension of these methods is the heat kernel method which also can be traced back to Riemann works.

Fredholm and elliptic complexes. Let E_j be a sequence of Hilbert spaces and d_j a sequence of linear continuous operators, to wit

$$(1) \qquad 0 \xrightarrow{d_{-1}} E_0 \xrightarrow{d_0} E_1 \xrightarrow{d_1} \cdots \xrightarrow{d_{k-2}} E_{k-1} \xrightarrow{d_{k-1}} E_k \xrightarrow{d_k} 0.$$

This sequence is a *complex* if

$$d_{j+1}d_j = 0 \tag{2}$$

for all $j = 0, \ldots, k-2$. Let $Z^j := \ker d_j$, $B^j := \operatorname{im} d_{j-1}$, $H^j = z^j/B^j$. The spaces H^j are *cohomologies* of the complex (1); this complex is called *Fredholm complex* if $\dim H^j < \infty$. For Fredholm complex B^j is a closed subspace of Z^j. One immediately proves that

$$(\text{the complex (1) is Fredholm}) \Longleftrightarrow$$

$$(\Delta_j := d_j^* d_j + d_{j-1} d_{j-1}^* \text{ is Fredholm on } E_j). \tag{3}$$

We recall that an operator $A : E \to F$ is Fredholm if $\dim \ker A < \infty$ and $\dim \operatorname{Coker} A < \infty$ for Banach E, F.

Moreover

$$\dim H^j = \dim \ker \Delta_j. \tag{4}$$

For the Fredholm complex (1), the Euler characteristic equals

$$\chi(E) := \sum_{j=0}^{k} (-1)^j \dim H^j.$$

Clearly, for $k = 1$ the Euler characteristic of the complex E: $0 \to E_0 \overset{d_0}{\to} E_1 \to 0$ is the index of the operator d_0

$$\chi(E) = \operatorname{ind} d_0. \tag{5}$$

EXERCISE. If $\dim E_j < \infty$ for all $j = 0, \ldots, k$, then

$$\chi(E) = \sum_{j=0}^{k} (-1)^k \dim E_j.$$

The preceding facts constitute a natural, abstract formulation of Hodge theorems.

Elliptic complexes. Let $V_j \to X$, $j = 0, \ldots, k$ be smooth vector bundles over compact manifold X. Let $d_j : C^\infty(V_j) \to C^\infty(V_{j+1})$ be elliptic (pseudo) differential operators of the same order m. Denote by $W_1^2(V_j)$ the corresponding Sobolev spaces. Consider the complex V:

$$0 \overset{d_{-1}}{\to} C^\infty(V_0) \overset{d_0}{\to} C^\infty(V_1) \overset{d_1}{\to} \cdots \overset{d_{k-1}}{\to} C^\infty(V_k) \overset{d_k}{\to} 0. \tag{6}$$

The complex (6) is elliptic if for all $(x,\xi) \in T^*X - \{0 \text{ sections}\}$ the corresponding sequence of principal symbols

$$0 \to V_{0,x} \overset{d_0(x,\xi)}{\to} V_{1,x} \overset{d_1(x,\xi)}{\to} \cdots \overset{d_{k-1}(x,\xi)}{\to} V_{k,x} \to 0 \tag{7}$$

is exact. By standard elliptic methods one obtains.

THEOREM (ATIYAH–BOTT).
(i) (The complex (6) *is elliptic)* $\Longleftrightarrow$ *(all operators* $\Delta_j = d_j^* d_j + d_{j-1} d_{j-1}^*$ *are elliptic);*

(ii) If (7) *is an elliptic complex, then for all* $s \in \mathbb{R}$ *the corresponding Sobolev complex*

$$0 \overset{d_{-1}}{\to} W_s^2(V_0) \overset{d_0}{\to} W_{s-m}^2(V_1) \overset{d_1}{\to} \cdots \overset{d_{k-1}}{\to} W_{s-km}^2(V_k) \overset{d_k}{\to} 0. \tag{8}$$

is a Fredholm complex and dimension of its cohomologies is finite and independent of s.

(iii) The cohomology spaces of the complex (7)

$$H^j = \ker d_j / d_{j-1}(C^\infty(V_j)).$$

COROLLARY 1. *The de Rham complex is elliptic, therefore all cohomologies are finite dimensional.*

COROLLARY 2. *Let* $A^{p,q}(X)$ *be a space of smooth* (p,q) *forms and* $d'' : A^{p,q}(X) \to A^{p,q+1}(X)$ *be the Riemann–Dolbeault operator. Then the Dolbeault complex*

$$0 \to \Omega^p \overset{i}{\to} A^{p,0}(X) \overset{d''}{\to} A^{p,1}(X) \to \cdots \to A^{p,n} \to 0, \tag{9}$$

where Ω^p *denotes the space of holomorphic* p *forms is elliptic. Therefore, the cohomology spaces* $H^q(X,\Omega^p)$ *are finite dimensional and we have the isomorphisms*

$$H^q(X,\Omega^p) \simeq \frac{\{\omega \in A^{p,q} : d''\omega = 0\}}{d'' A^{p,q-1}(X)}. \tag{10}$$

In particular,

$$(11) \qquad H^q(X, \mathcal{O}) \simeq \frac{\{\omega \in A^{0,q} : d''\omega = 0\}}{d''A^{0,q-1}(X)}.$$

The main step of the proof of exactness of a sequence of the corresponding sheaves of germs is the

LEMMA (DOLBEAULT–GROTHENDIECK). *For any polycylinder $\mathcal{U} \subset \mathbb{C}^n$ and $q \geqslant 1$, $p \geqslant 0$, $H^q(\mathcal{U}, \Omega^p) = 0$.*

This is a fancy form of the following proposition. *Let $\omega \in A^{p,q}(\mathcal{U})$ with $d''\omega = 0$. Then there exists a $(p, q-1)$ form $\varphi \in A^{p,q-1}(\mathcal{U})$ such that $\omega = d''\varphi$.*

REMARKS. 1. This lemma was probably known already to Fritz Hartogs, and certainly to Erich Kähler.

2. For a polycylinder $\mathcal{U} \subset \mathbb{C}^n$ and, more generally, for any holomorphic domain $\mathcal{U} \subset \mathbb{C}^n$, $n \geqslant 2$ (which is a deep fact), we have a THEOREM B: $H^q(\mathcal{U}, \Omega^p) = 0$ for all $q \geqslant 1$, $p \geqslant 0$.

3. Since all sheaves $A^{p,q}$ are fine, that is, the restrictions $A^{p,q}(X) \to A^{p,q}(Y)$ are surjective for any closed Y (the proof by partition of unity), the resolution of Ω^p

$$0 \to \Omega^p \xrightarrow{i} A^{p,0} \xrightarrow{d''} A^{p,1} \xrightarrow{d''} \cdots$$

is, by the lemma *acyclic*, whence, by H. Cartan abstract de Rham theorem, we have the isomorphisms (10), (11). Corollary 2 is obtained, seemingly, without ellipticity of the operator d''.

Carl Runge (1856–1927) was a versatile, outstanding scientist. He began his studies in München in 1876, where he made friends with Max Planck and under his influence became interested in mathematics and physics. Since 1877 he worked in Berlin with Weierstrass (Ph.D. in differential geometry) and with Kronecker (habilitation in number theory.) During 18 years (1886–1904) he was a professor in Hannover, working on physical problems (spectroscopy and spectral analysis.) He was elected a full professor in Göttingen, and he was the first professor of applied mathematics in Germany. He is famous for his works on numerical analysis.

The work of Runge, we are interested in, *Zur Theorie der eidentigen analytischen Functionen*, Acta Math., **6** (1885), 229–244 gives a full answer to the following question: For which domains $Y \subset \mathbb{C}$ any holomorphic function $u \in \mathcal{O}(Y)$ can be compactly approximated by a sequence of polynomials (u_j)?

THEOREM (RUNGE, 1885 AND 1889). *($\mathbb{C}-Y$ has no compact connected components)$\Longleftrightarrow$(every $u \in \mathcal{O}(Y)$ is a limit of polynomials u_j, $j = 1, 2, \ldots$ converging compactly on Y).*

From this theorem the famous theorems of Weierstrass and Mittag– Leffler follow. It is interesting that the famous Weierstrass theorem on approximation of a function on interval by polynomials appeared in the same year.

Today an open subset Y of a non compact manifold X is called the Runge set if $X - Y$ has no compact connected components. The hull of Y, $h(Y)$ is the sum of Y and all connected components of $X - Y$. Thus Y is a Runge set if $Y = h(Y)$. Let us collect some important properties of the hull operation $h : Y \to h(Y)$.

1. $h(h(Y)) = h(Y)$ and $Y \to h(Y)$ is increasing;
2. $Y \to h(Y)$ preserves compactness and closeness.

The fundamental role of Runge sets in the theory of non compact Riemann surfaces X was recognized by Behnke and Stein in their paper of 1948 *Entwicklungen analytischer Functionen auf Riemannschen Flächen*, Math. Ann. **120**, 430–461, where they proved the following Runge type theorem.

THEOREM (BEHNKE–STEIN, 1948). *For any non compact Riemann surface X and any Runge subset $Y \subset X$ every holomorphic function $u \in \mathcal{O}(Y)$ can be compactly approximated by globally holomorphic $u_j \in \mathcal{O}(X)$, $j = 1, 2 \ldots$.*

The notion of hull $h(Y)$ is very closely related (and identical in the case $\mathbb{C}^1$) to the notion of *holomorphic hull* $\hat{Y} = \hat{Y}_{\mathcal{O}(X)}$ of an open subset of complex manifold X which is very important in complex analysis.

DEFINITION. Let X be a complex manifold, $\mathcal{F} \subset \mathcal{O}(X)$, and $K \Subset X$ (that is, K is relatively compact in X, i.e., $\bar{K}$ is compact), $\hat{K}_{\mathcal{F}} = \{x \in X : |f(x)| \leqslant \sup |f(K)|.$ for $f \in \mathcal{F}\}$. If $\mathcal{F} = \mathcal{O}(X)$, we write $\hat{K} := \hat{K}_{\mathcal{O}(X)}$. The set $\hat{K}_{\mathcal{F}}$ is called the $\mathcal{F}$ hull of K.

PROPOSITION. *($G \subset \mathbb{C}^n$ is domain of holomorphy) $\Longleftrightarrow$ (for every $K \Subset G$, $\hat{K} \Subset G$).*

EXERCISE. Prove that if $G \subset \mathbb{C}^n$ is convex in the geometric sense, then G is a domain of holomorphy.

DEFINITION. A complex manifold (space) X is *holomorphically convex* if for every compact set $K \subset X$ the set $\hat{K} = \{x \in X : |f(x)| \leqslant \sup|f(K)| \quad \text{for all} \quad f \in \mathcal{F}\}$ is compact, that is, the holomorphically convex hull $\hat{K}$ of K is compact.

EXERCISE. Prove that this definition is equivalent to

DEFINITION′. X is holomorphically convex if for any discrete sequence $\{x_i\}$ in X there is a function $f \in \mathcal{O}(X)$ such that the set $\{f(x_i) : i \in \mathbb{N}\}$ is unbounded.

In the seminal paper of Behnke and Stein, the notion of Runge pair was introduced: (Y_1, Y_2) is a Runge pair if $Y_1 \Subset Y_2$ and the restriction $r : \mathcal{O}(Y_2) \to \mathcal{O}(Y_1)$ is dense.

The Behnke–Stein theorem is an important assertion concerning the Cauchy–Riemann operator d'' on an open Riemann surface X. We know that the operator d'' is elliptic. The role of ellipticity is crucial as shown in the following Runge type

PROPOSITION (MALGRANGE–HÖRMANDER). *Let P be an elliptic differential operator on $\mathbb{R}^m$ with constant coefficients of order m, that is, $P = \sum_{|\alpha| \leqslant m} a_\alpha \partial^\alpha$ with $P_m(\xi) := \sum_{|\alpha|=m} a_\alpha \xi^\alpha \neq 0$ for $0 \neq \xi \in \mathbb{R}^m$. Let $Y \subset X \subset \mathbb{R}^m$ be open sets and let Y be a Runge set in X. Then any solution $u \in C^\infty(Y)$ of $Pu = 0$ is a limit in $C^\infty(Y)$ of restrictions to Y of solutions v of $Pv = 0$ in X.*

REMARK. For Runge property of solutions of a partial differential operator P with constant coefficients, ellipticity is also necessary.

Perhaps the most important of Runge type theorems we owe B. Malgrange (1955/6). Such theorem was suggested by Grothendieck in 1952.

THEOREM (MALGRANGE, LAX). *Let X be a real analytic manifold, and let $P : C^\infty(E) \to C^\infty(F)$ be a linear elliptic operator with analytic coefficients acting between bundles $E \to X$ and $F \to X$ of the same rank. Then ($h(Y) = Y$ is an open subset in X)$\Longleftrightarrow$(every $u \in C^\infty(Y, E)$ with $Pu = 0$ can be, on compact subsets of Y approximated uniformly with derivatives of arbitrary high order by global solutions $u \in C^\infty(X, E)$, $Pu = 0$).*

This theorem embraces many former Runge type theorems and gives an elegant proof of the Behnke–Stein theorem.

DEFINITION. A complex manifold X is Stein if

(i) X is $\mathcal{O}$ convex (that is, holomorphically convex);

(ii) $\mathcal{O}$ separates points of X, that is, if $a, b \in X$, $a \neq b$, there is $f \in \mathcal{O}(X)$ such that $f(a) \neq f(b)$.

From the Malgrange–Lax theorem we have

THEOREM (BEHNKE–STEIN). *Every open Riemann surface is a Stein manifold.*

PROOF. (ii) is almost immediate. Let $a, a' \in X$ and $(\mathcal{U}, \varphi)$ and $(\mathcal{U}', \varphi')$ be coordinate maps around a and a' with $\mathcal{U} \cap \mathcal{U}' = \emptyset.$, $\varphi(\mathcal{U}) = \{z \in \mathbb{C} : |z| < 1\} = \varphi'(\mathcal{U}')$, and $\varphi(a) = \varphi'(a') = 0$. If $D = \{x \in \mathcal{U} : |\varphi(x) < r < 1\}$, $D' = \{x \in \mathcal{U}' : |\varphi'(x) < r < 1\}$, then the sets $X - D$, $X - D'$, and $X - D \cup D'$ are connected. then $Y := D \cup D'$ is a Runge set. Take

$$u(x) = \begin{cases} 0, & x \in D \\ 1, & x \in D' \end{cases}$$

Then $d''u = 0$ on Y; thus by Malgrange–Lax theorem u can be uniformly approximated on the (compact) set $\{a\} \cup \{a'\}$ by elements $f \in \mathcal{O}(X)$. In particular, there exists $f \in \mathcal{O}$ with $|f(a)| < \frac{1}{2}$, $|f(a')| > \frac{1}{2}$, hence (ii) is proved.

(i) Since for compact K, $h(K)$ is compact, it is sufficient to prove $\hat{K}_{\mathcal{O}} = h(K)$. It follows from the maximum principle that $h(K) \subset \hat{K}_{\mathcal{O}}$. Now, for $a \notin h(K)$, let $K_1 = \{a\} \cup h(K)$, then $K_1 = h(K_1)$. Let $u(x) = 1$ is x is in a neighborhood of a and let $u(x) = 0$ if x is in a neighborhood of $h(K)$. Then there is $f \in \mathcal{O}$ with $\sup_{x \in K_1} |f(x) - u(x)| < \frac{1}{2}$. Therefore $|f(a)| > \sup |f(K)|$, thus $a \notin \hat{K}_{\mathcal{O}}$, hence $\hat{K}_{\mathcal{O}} \subset \hat{K}$ and $\hat{K}_{\mathcal{O}} = \hat{K}$ is compact.

□

As a corollary one obtains that Cousin I and Cousin II problems are always solvable for an open Riemann surface X; this fact was proved for the first time by Behnke and Stein in their celebrated paper *On Cousin problems on Stein Manifolds* (more on this can be found in the chapter devoted to the Cousin problems below.) The original proof was quite different: the leading idea was the Runge method of translating poles.

REMARK. The condition $Y = h(Y)$ is a topological property, but for analysis another property of subsets $Y \subset X$ of a compact manifold (or even a compact space X) is more interesting. The restriction map

$$\mathcal{O}(Y) \to \mathcal{O}(X)$$

has dense image in the topology of compact convergence. Such a pair (X, Y) is called now the *Runge pair.* The first theorem on a Runge pairs, in the case $X = \mathbb{C}^2$, Y an analytic polyeder (called also the *Oka–Weil domain*) was conceived by André Weil during his two years of stay in India about 1932, and published in his seminal paper *Integral de Cauchy et les functions de plusiners variables*, Math. Ann. **111** (1935), 178–182.

The following theorem of Runge type was obtained by Grauert, Forster, and others, and, in homage to the pioneers, was called Oka–Weil theorem.

THEOREM. *Let X be a Stein manifold and $Y \subset X$ an open subset of X such that for every compact $K \subset Y$*

$$\hat{K}_{X,Y} = \{y \in Y : |f(y)| \leqslant \sup |f(K)| \quad \textit{for all } f \in \mathcal{O}(X)\}$$

is compact; then (X, Y) is a Runge pair, that is, the restriction map $\mathcal{O}(Y) \to \mathcal{O}(X)$ has dense image and every $f \in \mathcal{O}(Y)$ on Y can be compactly approximated by functions in $\mathcal{O}(X)$.

In the next chapter we demonstrate how this marvelous theorem of Grauert could be obtained by the Hörmander method of 'hard analysis' of Riemann–Wistinger operator d''. Grauert's original proof was obtained in a quite different way.

We have already introduced subharmonic and plurisubharmonic functions on complex manifold X as upper semicontinuous functions $p : X \to$

$[-\infty, \infty)$ for whose the Levi form (that is, the $d'd''$ Hessian) is (strictly) positive in the sense of distribution theory of L. Schwartz. An exhausting function $\varphi : X \to [-\infty, \infty)$ has all its sublevel sets $X_c = \{x : \varphi(x) < c\}$ relatively compact in X, that is, $X_c \Subset X$.

For an open Riemann surface X there always exists a subharmonic function p which exhausts X. Such functions can be constructed with the help of exhausting sequences $Y_0 \Subset Y_1 \Subset \cdots$ of relatively compact Runge domains Y_j with $\bigcup Y_j = X$ such that every Y_j has regular boundary with respect to the Dirichlet problem.

But we know that every open Riemann surface is a Stein space and the natural question arises as to if it is possible to characterize Stein manifolds (and general Stein spaces) as manifold (spaces) on which non constant, strictly plurisubharmonic exhausting function live? This would be a truly Riemannian characteristic. The answer is *yes*!

THEOREM (GRAUERT, 1958, NARASIMHAN, 1962). *A connected manifold X (complex space) is a Stain manifold if and only if continuous, strictly plurisubharmonic exhausting function lives on X.*

In the proof of this theorem, the following Runge type theorem plays an important role.

THEOREM (GRAUERT, NARASIMHAN). *Let p be a continuous, strictly plurisubharmonic function on a complex space X. Let $c \in \mathbb{R}$ and $Y = \{x \in X : p(x) < c\}$. Then the restriction map $\mathcal{O}(X) \to \mathcal{O}(Y)$ has a dense image in the topology of compact convergence, that is, Y is a Runge set (with respect to X) or (X, Y) is a Runge pair.*

Stein manifolds were the product of long endeavors aimed to characterize domains of holomorphy. A great idea of E.E. Levi was to use plurisubharmonic functions. Lars Hörmander showed in an impressive way that all outstanding problems of global complex analysis on complex manifolds can be solved with the help of Riemann method, that is, in the framework of the theory of d'' operator.

CHAPTER 4

Hörmander Method in Complex Analysis

Let us start by introducing, following Hörmander, the following notation. Denote by $\mathcal{F}_{(p,q)}$ the linear space of (p,q) forms with coefficients belonging to some space $\mathcal{F}$ of distributions (sections of a holomorphic vector bundle $E \to X$ over manifold X.)

EXAMPLES. 1. $\mathcal{F} = C_0^\infty(X)'$ (distributions on X);

2. $\mathcal{F} = W^s(X)$ the Sobolev space of functions belonging to $L^2(X)$ together with all derivatives of order $\leqslant s$ (in the sense of distribution theory.)

THEOREM 1. (HÖRMANDER, 1965). *Let X be a Stein manifold and $E \to X$ a holomorphic vector bundle over X. Then for every $f \in W^s_{(p,q+1)}(X, E)$ such that $d''f = 0$ the equation $d''u = f$ has a solution $u \in W^{s+1}_{(p,q)}(X, E)$.*

There is another theorem of Hörmander of Runge type.

THEOREM 2. (HÖRMANDER, 1965). *Let X be a Stein manifold and φ a strictly plurisubharmonic function on X which is exhausting, that is, $K_c = \{x \in X : \varphi(x) < c\}$ is compact for every real number c. Let $E \to X$ be a holomorphic vector bundle over X. Then every holomorphic section of E over neighborhood K_0 can be uniformly approximated (in the sense of hermitian metric on E) by holomorphic sections belonging to $\mathcal{O}(X, E)$.*

Let us recall the notion of pseudoconvexity: An open set $X \subset \mathbb{C}^n$ is *pseudoconvex* if there exists an exhausting, continuous, plurisubharmonic function on X such that $X_c = \{x \in X : \varphi(x) < c\} \Subset X$ for every $c \in \mathbb{R}$. On every pseudoconvex open set X there exists a smooth, *strictly* plurisubharmonic, exhausting function φ. The same holds for Stein manifold.

As we have mentioned already, Stein manifolds are characterized by the existence of smooth plurisubharmonic exhausting functions. This is the famous Grauert theorem which is a solution of vastly generalized Levi problem. With the help of his method, Hörmander proves both parts of Grauert theorem.

THEOREM 3. *Let X be a Stein manifold, K a compact subset of X, and $\mathcal{U}$ an open neighborhood of $h(K)$ (the holo- convex hull of K.) Then there exists a function $\varphi \in C^\infty(X)$ such that*

(a) φ is strictly plurisubharmonic;

(b) φ is exhausting;

(c) $\varphi < K$, but $\varphi > 0$ on $X - \mathcal{U}$.

The proof is not difficult.

From both preceding theorems, as a corollary one obtains the following Runge type theorem for Stein manifolds.

THEOREM 4. (GRAUERT). *If X is a Stein manifold and K a compact subset of X such that $K = h(K)$, then every function u which is holomorphic in an neighborhood of K can be uniformly approximated on K by functions from $\mathcal{O}(X)$.*

This theorem makes it possible to obtain a theorem converse to Theorem 3.

THEOREM 5. (GRAUERT). *let X be a complex manifold on which a smooth strictly plurisubharmonic function φ lives. Let for this function $X_c = \{x \in X : \varphi(x) < c\} \Subset X$ (thus φ is exhausting.) Then*

1. X is a Stein manifold;

2. The closure $\bar{X}_c$ of X_c is $\mathcal{O}(X)$ convex and every (X, X_c) is a Runge pair.

Since every point of a Stein manifold is holomorphically convex, we obtain a corollary to Theorem 2.

COROLLARY. *Let $E \to X$ be a holomorphic vector bundle over Stein manifold X. Then for every $x_0 \in X$ and every $e_0 \in E$ there is a global holomorphic section $u \in \mathcal{O}(X, E)$ such that $u(x_0) = e_0$.*

Of course, we cannot present the proofs of these magnificent theorems in which the notion of strict plurisubharmonicity is so fundamental. Instead we will try to demonstrate the role of this notion in the Hörmander method. This theory makes use of the methods of Hilbert spaces, more precisely, of important L^2 estimates that makes it possible to obtain theorems on existence of solutions of linear equations $Au = f$ in appropriately chosen Hilbert spaces. Already in his Ph.D. thesis Hörmander turned out to be a great master of these methods. The apparatus of linear mappings $A : H_1 \to H_2$ of Hilbert spaces is very simple; this is a great charm of this theory. The difficulties lies elsewhere: in construction of appropriate Hilbert spaces (these will turn out to be some Sobolev spaces) and in obtaining interesting inequalities for norms.

Let X be an open set in $\mathbb{C}^n$. If φ is a continuous, real valued function on X, one denotes by $L^2(X, \varphi)$ the space of square integrable functions with respect to the measure $e^{-\varphi} d\lambda$, where $d\lambda$ is the Lebesgue measure. $L^2(X, \varphi)$ is a subspace of the space $L^2(X, loc)$ of functions on X which are locally square integrable with respect to the Lebesgue measure. Clearly, every $f \in L^2(X, loc)$ belongs to $L^2(X, \varphi)$ for some φ. In what follows we will work with strongly plurisubharmonic functions φ and the spaces $L^2_{(p,q)}(X, \varphi)$ and $L^2_{(p,q)}(X, loc)$.

For $f \in A_0^{p,q}(X)$ we introduce a hermitian inner product as follows: $\| f \|^2_\varphi := \int |f|^2 e^{-\varphi} d\lambda$, where for $f = \sum f_{I,J} dz^I \wedge dz^J$, $|f|^2 = \sum |F_{I,J}|^2$. The completion of $A_0^{p,q}(X)$ with respect to the norm $\| \cdot \|_\varphi$ is denoted by $D_{(p,q)}$ and is close in $L^2_{(p,q)}(X, \varphi)$ for every φ.

If φ_1, φ_2 are two continuous functions on X, then the Cauchy–Riemann operator d'' defines a linear, closed, densely defined operator $T : L^2_{(p,q)}(X, \varphi_1) \to L^2_{(p,q+1)}(X, \varphi_2)$ (here d'' is understood in the sense of distribution theory): $u \in L^2_{(p,q)}(X, \varphi_1)$ belongs to the domain D_T of T if $d''u \in L^2_{(p,q+1)}(X, \varphi_2)$, and we set $Tu := d''u$. The operator T is closed[1] since in distribution theory differentiation is a continuous mapping, and the domain of T, D_T, is dense since $D_T \supset D_{(p,q)}(X)$. In the same way we

[1] A map $T : H_1 \to H_2$ between Hilbert (Banach) spaces H_1 and H_2 is *closed* if its graph $G(T) \subset H_1 \times H_2$ is closed (in the norm $\| \cdot \|_{H_1 \times H_2}$.) If T is closed then T^{**} (adjoint to adjoint) equals T.

construct the operator

$$S : L^2_{(p,q+1)}(X, \varphi_2) \to L^2_{(p,q+2)}(X, \varphi_3)$$

defined by $d'' : A^{p,q+1} \to A^{p,q+2}$.

Now Hörmander proves the important inequality

$$(H) \qquad \| f \|_{\varphi_2} \leqslant C\,(\| T^* f \|_{\varphi_1} + \| Sf \|_{\varphi_3}) \quad \text{for } f \in D_{T^*} \cap D_S.$$

If $f \in \ker S =: F$, then the last term drops out, and we have

$$(H') \qquad \| f \|_{\varphi_2} \leqslant C \, \| T^* f \|_{\varphi_1}$$

From the Hahn–Banach theorem we have

PROPOSITION (HÖRMANDER, 1965). *The equation $Tu = f$ has a solution for every $f \in L^2_{(p,q+1)}(X, \varphi_2)$ with $f \in \ker S$ if and only if (H') holds.*

For X being a pseudoconvex set, by carefully choosing φ_1, φ_2, φ_3, the following theorem which is fundamental for the Hörmander approach follows.

THEOREM (HÖRMANDER, 1965). *Let X be a pseudoconvex open set in $\mathbb{C}^n$. Then the equation $d''u = f$ has (in the weak sense, that is, in the sense of distribution theory) a solution $u \in L^2_{(p,q)}(X, loc)$ for every $f \in L^2_{(p,q+1)}(X, loc)$ such that $d''f = 0$.*

PROOF (idea). Let $p \in C^\infty(X)$ be a strictly plurisubharmonic exhausting function, $K_c = \{x \in X : p(x) < c\} \Subset X$ with

$$\sum_{i,k=1}^{n} \frac{\partial^2 p}{\partial z_j \partial \bar{z}_k} w_j \bar{w}_k \geqslant m \sum_{j=1}^{n} |w_j|^2, \quad w = (w_1, \ldots, w_n) \in \mathbb{C}^n,$$

where $0 < m \in C^0(X)$. If $h \in C^\infty(\mathbb{R})$ is a convex, increasing function, and $\varphi := h(p)$, we obtain

$$\sum_{i,k=1}^{n} \frac{\partial^2 \varphi}{\partial z_j \partial \bar{z}_k} w_j \bar{w}_k \geqslant m h'(p) \sum_{j=1}^{n} |w_j|^2.$$

Then we have

$$(H'') \qquad \sum_{i,k=1}^{n} \frac{\partial^2 \varphi}{\partial z_j \partial \bar{z}_k} w_j \bar{w}_k \geqslant 2(|d''\psi|^2 + e^\psi) \sum_{j=1}^{n} |w_j|^2,$$

if $h'(t) \geqslant \sup_{K_t} 2(|d''\psi|^2 + e^{\psi})/m$. Set now $\varphi_1 : \varphi - 2\psi$, $\varphi_2 := \varphi - \psi$, $\varphi_3 := \varphi$. We obtain the inequality (H).

Hörmander obtains now the corresponding theorem for Sobolev spaces $W^s_{(p,q+1)}(X, loc)$ for every $s \geqslant 0$. By Sobolev lemma he gets the promised

COROLLARY (HÖRMANDER). *If X is a pseudoconvex subset of $\mathbb{C}^n$, then the equation $d''u = f$ has a solution $u \in C^\infty_{(p,q)}(X) \equiv A^{p,q}(X)$ for every $f \in A^{p,q+1}(X) = C^\infty_{(p,q+1)}(X)$ such that $d''f = 0$.*

How one can get a solution of the Levi problem from this corollary? The answer is provided by

THEOREM (HÖRMANDER). *Let X be an open set in $\mathbb{C}^n$ such that the equation $d''u = f$ has a solution $u \in C^\infty_{(p,q)}(X) \equiv A^{p,q}(X)$ for every $f \in C^\infty_{(p,q+1)}(X) \equiv A^{p,q+1}(X)$ satisfying $d''f = 0$. Then X is a domain of holomorphy.*

From this last assertion we get the famous solution of the Levi problem obtained in 1954 independently by Oka, Bremermann, and Norguet.

SOLUTION OF LEVI PROBLEM. *An open pseudoconvex subset $X \subset \mathbb{C}^n$ is a domain of holomorphy.*

The missing link – the proof of the last theorem – is obtained by induction in dimension n. For $n = 1$ the assertion is obvious. We assume that it is true for $n-1$. It suffices to prove that for every open convex set $V \subset X$ such that some point $x_0 \in \partial V$ is on the boundary ∂X of X there exists a holomorphic function on X which cannot be analytically continued to a neighborhood of x_0. We assume that the coordinates are chosen such that $x_0 = \{0\}$ and the plane $z_n = 0$ has a non-empty intersection V_0 with V. By convexity of V we see that $\{0\} \in \partial V_0$, and thus $\{0\} \in \partial\mathcal{U}$, where $\mathcal{U} = \partial\{x \in X : z_n = 0\}$, and $\mathcal{U}$ is an open subset of $\mathbb{C}^{n-1}$. Let $j : \mathcal{U} \hookrightarrow X$ and π be projection of $\mathbb{C}^n$ onto $\mathbb{C}^{n-1}$: $\pi(z) = (z_1, \ldots, z_{n-1})$. Now for every $f \in A^{0,q+1}(\mathcal{U})$, $q \geqslant 0$ satisfying $d''f = 0$ we construct $F \in A^{0,q}(X)$ with $d''F = 0$ and $f = j^*F$.

Now we check that the inductive hypothesis is fulfilled if X is replaced by $\mathcal{U}$. Indeed, for given $f \in A^{0,q+1}(\mathcal{U})$ with $d''f = 0$, we have proved the existence of the form $F \in A^{0,q+1}(X)$ with $d''F = 0$ and $j^*F = f$. By hypothesis the euation $d''\Phi = F$ has a solution $\Phi \in A^{0,q}(X)$. Put $\varphi := j^*\Phi$, then $d''\varphi = j^* d''\Phi = j^* F = f$.

It follows from the inductive hypothesis that $\mathcal{U}$ is a domain of holomorphy, whence there exists a holomorphic function f on $\mathcal{U}$ which cannot be analytically extended to a neighborhood of $\bar{V}_0$. If we take F holomorphic on X so that $j^* F = f$, that is, $F = f$ on $\mathcal{U}$, then F cannot be extended analytically to a neighborhood of $\bar{V}$. Therefore X is a domain of holomorphy. □

Exactly the same method works for an arbitrary Stein manifold X and this made it possible for Hörmander to obtain the famous Grauert theorems presented above.

At this point it is worth to look back. About 1900 Hilbert introduced in complex analysis the infinite dimensional spaces $L^2(X)$. In his school in Göttingen the theory of linear operators on Hilbert spaces was soon afterwards created (Erhardt Schmidt, H. Weyl, Friedrichs, Rellich, von Neumann, and many others.) Friedrichs, Rellich, and Sobolev introduce Sobolev spaces. Friedrichs and J. Schauder develop methods of solving (by inequalities) elliptic, linear partial differential equations, which in the hands of Malgrange, J.J. Kohn, and Hörmander reaches the ideal beauty, simplicity and power. For complex manifolds, the Hörmander method (which is a realization of the Riemann program) is most effective. But also in the case of some problems on general complex spaces it gives excellent results (cf. the paper of G. Dethloff (1990), where he proves the 'main theorem' of Grauert–Remmert by using the Hörmander method.)

Hörmander simplified also the famous Bishop–Narashimhan theorem on proper holomorphic embedding of a Stein manifold into some $\mathbb{C}^N$. Also the proof of the classic theorem of Newlander–Nirenberg (1957) asserting that every integrable almost complex structure is defined by a unique complex structure was greatly simplified by Hörmander.

We mentioned this very important theorem in the chapter devoted to Teichmüller theory. Even though I cannot present here even simplified version of the proof, I cannot resist presenting some equivalent definitions of

Integrability conditions of almost complex structure

A C^∞ manifold M is equipped with almost complex structure J, if J is a tensor field of type $(1,1)$ such that $J(J\xi) = -\xi$ for every vector field ξ on X. Briefly, *J is a section of* $\mathrm{End}(TX)$ *satisfying* $J^2 = -1$.

The tangent space of almost complex manifolds splits, upon complexification, into two parts called holomorphic and antiholomorphic tangent spaces

$$TM \otimes_{\mathbb{R}} \mathbb{C} = T^{1,0} \oplus T^{0,1} \quad (T'M \oplus T''M)$$

on whose J acts by i and $-i$. In terms of a local frame X_j, Y_j, $j = 1, \ldots, n$, $JX_j = Y_j$, $JY_j = -X_j$ we see that $T^{1,0}M$ is spanned by $(X_j - iY_j) = W_j$ and $T^{0,1}M$ by $(X_j + iY_j) = \bar{W}_j$. For differential forms we have

$$dA^{0,0} \subset A^{1,0} \oplus A^{0,1}, \quad dA^{1,0} \subset A^{1,1} \oplus A^{2,0} \oplus A^{0,2},$$

$$dA^{0,1} \subset A^{1,1} \oplus A^{2,0} \oplus A^{0,2}.$$

Therefore

$$dA^{p,q} \subset A^{p+1,q} \oplus A^{p,q+1} \oplus A^{p-1,q+2} \oplus A^{p+2,q-1}.$$

For a *complex* structure, we have

$$dA^{p,q} \subset A^{p+1,q} \oplus A^{p,q+1}.$$

Thus for a complex structure

$$(*) \qquad dA^{0,1} \subset A^{1,1} \oplus A^{0,2}$$

and there is no component $A^{2,0}$.

DEFINITION. An almost complex structure is integrable if equation $(*)$ is satisfied.

There is another, equivalent condition of integrability which can be called *complex Frobenius condition.* Denote by $P_k := \sum_{j=1}^{2n} a_{kj}\partial_{x_j}$, $k = 1, \ldots, n$, a complex vector fields on M, and by $\bar{P}_k$ their conjugates.

An almost complex structure J is integrable if $P_1, \ldots, P_n, \bar{P}_1, \ldots, \bar{P}_n$ are linearly independent and the Lie commutators

$$[P_k, P_l] = \text{linear combination of } P_1, \ldots, P_n,$$

that is, if the distribution $\mathcal{H}$ generated by $P_1, \ldots, P_n$ is involutive

$$(**) \qquad [\mathcal{H}, \mathcal{H}] \subset \mathcal{H}.$$

The third equivalent condition of integrability of an almost complex structure J was introduced by A. Nijenhuis, who considered the torsion tensor N of J, called also the Nijenhuis tensor, defined for any two vector fields X and Y by

$$N(X,Y) := -[X,Y] - [JX,JY] - J[JX,Y] - J[X,JY].$$

An almost complex structure J is integrable if

$$(***) \qquad N(X,Y) = 0 \quad \text{for all } X, Y.$$

The equivalence of all these three conditions is almost obvious. Perhaps $(***)$ is the most useful one.

The notion of (integrable) almost complex structure arises in geometry of many complex variables since *in the case of two dimensional manifolds any almost complex structure is integrable.* Since the condition $(***)$ is local, we can take $M = \mathbb{R}^2$ equipped with local Cartesian coordinates (x_1, x_2). For every $x \in M$, the endomorphism J_x of T_xM is given by

$$J_x : a\left(\frac{\partial}{\partial x_1}\right)_x + b\left(\frac{\partial}{\partial x_2}\right)_x \to -b\left(\frac{\partial}{\partial x_1}\right)_x + a\left(\frac{\partial}{\partial x_2}\right)_x, \quad \text{for } a, b \in \mathbb{R}.$$

Clearly $J_x^2 = -1$ and $x \to J_x$ is an almost complex structure on M. We immediately check that $N\left(\frac{\partial}{\partial x_1}, \frac{\partial}{\partial x_1}\right) = 0$. □

In the case $2n > 2$, in general, an almost complex structure is not integrable; moreover *not every C^∞ $2n$ dimensional manifold can be equipped with an almost complex structure.*

G **structures**. If M is a C^∞ manifold of $\dim_{\mathbb{R}} M = m$, then its tangent bundle TM has the structure group $\mathrm{GL}(m, \mathbb{R})$, that is, the transition maps $a_{jk} : \mathcal{U}_j \cap \mathcal{U}_k \to \mathrm{GL}(m, \mathbb{R})$ are smooth. Consider now a subgroup G of $\mathrm{GL}(m, \mathbb{R})$. We say that the structure group of TM reduces to G if all $a_{jk}(x) \in G$. Such a reduction will be called the *G structure.*

EXAMPLE 1. An $O(m)$ structure (where $O(m)$ is the orthogonal group) is a Riemann structure.

EXAMPLE 2. For $m = 2n$ consider $G = \mathrm{GL}(n, \mathbb{C})$ as a subgroup of $\mathrm{GL}(2n, \mathbb{R})$, then the $\mathrm{GL}(n, \mathbb{C})$ is an almost complex structure.

The reader may consider $U(m)$, $SU(m)$, $SO(m)$ structures, the almost hermitian structure, etc. The theory of characteristic classes gives some *necessary* conditions for existence of some G structures.

CHAPTER 5

Wirtinger Theorems. Metric Theory of Analytic Sets

Wilhelm Wirtinger was a versatile mathematician. We owe him a number of achievements and approaches to geometry and (complex) analysis, for example, introduction of the forms of the type (p, q), operators $d' : A^{p,q} \to A^{p+1,q}$, $d'' : A^{p,q} \to A^{p,q+1}$ and investigation of their properties. This apparatus was called the Wirtinger calculus. For that reason we call d'' the Riemann–Wirtinger operator.

In this chapter we present a number of fundamental theorems due to Wirtinger on hermitian geometry which show a fundamental difference of this geometry as compared to the Riemannian one. Such unexpected phenomena always used to fascinate scientists, and in the case of physics, they led to a passionate opposition, for example, in the case of quantum mechanics and the theory of relativity. Man is reluctant to change his habits and feels deeply offended when his philosophy of life turns out to be too narrow and even wrong. Wirtinger discoveries are also interesting from the psychological point of view: they were made by a man in his seventies, during the period of life when mathematicians are usually not creative.

Wilhelm Wirtinger was born on 19 July 1865 in a little town Ybbs, and died there on 16 January 1945. Already as a young man he was under the spell of Riemann surfaces. A long stay in Göttingen and friendship with Felix Klein resulted in flourishing of Wirtinger genius. He became a great master of the theory of abelian functions (and varieties) and of Θ functions. His classic, gigantic monograph published in 1895 *Untersuchungen über Thetafunktionen* is still a standard monograph on Θ functions. In this work a theory of *Wirtinger varieties* (being a multidimensional generalization of

Kummer surfaces) was presented. Felix Klein invited Wirtinger to write review articles and to co-edit the monumental *Enzyklopedie Mathematischer Wissenschaften.*

The theorems of Wirtinger appeared in a short paper *Eine Determinantmidentität und ihre Anwendung auf analytische Gebilde in euclidischer Massbestimmung* published in 1936 in the journal Monatshafte für Mathematic und Physik, a main Austrian mathematical journal, of which Wirtinger was an editor for 40 years.

WIRTINGER THEOREM. *Let (X, h) be an n dimensional hermitian manifold equipped with hermitian metric*

$$ds^2 = \sum_{i,j=1}^{n} h_{ij} dz_i d\bar{z}_j.$$

Denote by ω the corresponding fundamental $(1,1)$ form, that is,

$$\omega = \frac{\sqrt{-1}}{2} h_{ij} dz_i \wedge d\bar{z}_j \;(= -\frac{1}{2}\operatorname{Im} ds^2) \in A^{1,1}(X).$$

Let $M \subset X$ be a complex submanifold of dimension k. Then the restriction $\frac{1}{k!}\omega^k|_M$ is the volume form on M, that is,

$$(W) \qquad vol_{2k}(M) = \frac{1}{k!}\int_M \omega^k, \quad \omega^k = \omega \wedge \cdots \wedge \omega \quad k \text{ times},$$

where $vol_{2k}(M)$ is the Riemannian volume of the manifold M of real dimension $2k$.

Before sketching the proof o the formula (W) let us stress how unexpected this equality is. The volume of the *complex* manifold $M \subset X$ is expressed as an integral over M of a *globally* defined differential (k, k) form on X. In the *real* C^∞ case the situation is quite different.

EXAMPLE. Let $t \to (\xi_1(t), x_2(t)) \in \mathbb{R}^2$ be an C^∞ arc in $\mathbb{R}^2$; the element of the arc length is given by $\sqrt{(\xi_1'(t))^2 + (\xi_2(t))^2}\, dt$ which is not, in general, a pullback of any differential form in $\mathbb{R}^2$.

PROOF OF THE WIRTINGER EQUALITY. If $M \subset X$ is a complex submanifold, then for every $z \in M$ we have the natural inclusion $T_z'(M) \subset$

$T'_z(X)$. More generally, if $f : Y \to X$ is any *holomorphic* map such that $df : T'_z(Y) \to T'_{f(z)}(X)$ is *injective* for all $z \in Y$, the metric on X induces the metric on Y, as follows

$$\left(\frac{\partial}{\partial v_\alpha}, \frac{\partial}{\partial v_\beta}\right)_z := \left(d\frac{\partial}{\partial v_\alpha}, d\frac{\partial}{\partial v_\beta}\right)_{f(z)}.$$

For small $\mathcal{U} \subset Y$ we can find a coframe $(\varphi_1, \ldots, \varphi_n)$ on $f(\mathcal{U}) \subset X$ such that $(\varphi_{k+1}, \ldots, \varphi_n) \in \ker f^* : T'^*_{f(z)}(X) \to T'^*_z(Y)$; then $(f^*\varphi_1, \ldots, f^*\varphi_n)$ is a coframe on $\mathcal{U}$ for the induced metric on Y. The associated fundamental $(1,1)$ form ω_Y on Y is therefore given by

$$\omega_Y = \frac{\sqrt{-1}}{2}\sum_{i=1}^{k}(f^*\varphi_i \wedge f^*\bar{\varphi}_j) = f^*\left(\frac{\sqrt{-1}}{2}\sum_{i=1}^{k}\varphi_i \wedge \bar{\varphi}_j\right) =$$

$$f^*\left(\frac{\sqrt{-1}}{2}\sum_{i=1}^{n}\varphi_i \wedge \bar{\varphi}_j\right) = f^*\omega_X. \tag{1}$$

Thus the fundamental $(1,1)$ form ω_Y of the induced hermitian metric h_Y on Y is the pullback of the fundamental $(1,1)$ form ω_X associated with the hermitian metric h_X on X.

If we write $\varphi_i = a_i + \sqrt{-1}b_i$, then the Riemann metric associated with $\mathrm{Re}(ds^2) = \sum(a_i \otimes a_i + b_i \otimes b_i)$, and the corresponding volume form $vol = a_1 \wedge b_1 \wedge \cdots \wedge a_n \wedge b_n$. But $\omega = \sum a_i \wedge b_i$, therefore $\omega^n = n!\, a_1 \wedge b_1 \wedge \cdots \wedge a_n \wedge b_n = n!\, vol$. But as we proved in (1), the $(1,1)$ form ω_M induced on M by ds^2 is the restriction $\omega|_M$ and (W) is proved. □

Perhaps even more interesting is another Wirtinger result called the *Wirtinger inequality* which gives the *minimalness of complex varieties.*

The restriction $(k!)^{-1}\omega^k|_M$ to an arbitrary $2k$ submanifold in X is *not* a volume element.

EXAMPLE. In $\mathbb{C}^2$ the restriction of ω to the plane $z_2 = \bar{z}_2$ is zero.

But there exists an important inequality.

THEOREM (WIRTINGER INEQUALITY, 1936) *Let M be any oriented, real $2k$ dimensional submanifold imbedded in the C^1 way in n dimensional hermitian manifold (X, h) with the fundamental form ω. Then*

$$\frac{1}{k!}\int_M \omega^k \leqslant vol_{2k}(M). \tag{W_1}$$

Moreover, in the case of finite volume of M, the equality in (W_1) *holds if and only if M is a complex submanifold of X with canonical orientation.*

Analytic sets as minimal varieties. From the Wirtinger inequality and elementary properties of de Rham currents one obtains an important property of analytic sets: they have the minimal volume in the family of relatively compact submanifolds with given boundaries.

In 1936 de Rham introduced his currents as linear continuous functionals on the space $C^\infty_{(p,q),0}(X) \equiv A_0^{p,q}(X)$ of differential forms with compact supports. This was several years before Laurent Schwartz's theory of distributions! We can regard currents of (p,q) type as generalized differential forms, that is, as (p,q) forms $\sum c_{IJ} dz^I \wedge d\bar{z}^J$, where c_{IJ} are Schwartz distributions, that is, $c_{IJ} \in C_0^\infty(X)'$. Since we can differentiate distributions, there is no problem to extend the operators d, d', d'', d_c onto currents. Since for smooth maps $f : Y \to X$, we have the notion of pullback f^* of distributions, the pullback f^* of currents is well defined. We know that a positive distribution T ($T(\varphi) \geqslant 0$ for all $\varphi \geqslant 0$) is represented by Radon (positive) measure $\mu = \mu_T$: $T(\varphi) = \int \varphi d\mu$. There are fundamental applications of currents due to P. Lelong and Ph. Griffiths and their schools (cf. the monographs of Lelong and Griffiths–Harris.) Here we present only the simplest ones.

EXAMPLE 1. A smooth $(1,1)$ form $\omega = \frac{\sqrt{-1}}{2} \sum h_{ij}(z) dz_i \wedge d\bar{z}_j$ is *real* if $\bar{h}_{ij} = h_{ij}$, and is *strictly positive* if (h_{ij}) is positive definite. A real, locally integrable function $p \in L^1(X, loc)$ is called plurisubharmonic if $\sqrt{-1} d'd''p$ is a positive $(1,1)$ current. There is an important

THEOREM ($d'd''$ POINCARÉ LEMMA OF LELONG). *Let T be a closed, positive* $(1,1)$ *current. Then locally*

$$T = \sqrt{-1} d'd''p$$

for a plurisubharmonic function p called the potential *of T which is unique up to addition of real part of a holomorphic function.*

EXAMPLE 2. Let $M \subset X$ be an analytic subvariety of codimension k, and denote by $M^* : M - singM$, the set of its smooth points. The map $\varphi \to \int_{M^*} \varphi$, $\varphi \in A_0^{n-k,n-k}(X)$, $n = \dim_{\mathbb{C}} X$ defines a closed positive current T_M. The cohomology class $[M]$ defined by T_M together with the (de Rham) isomorphism $H^\bullet_{dR}(X) \simeq H^\bullet(\mathcal{T}(X), d)$ is the fundamental class of M.

EXAMPLE 3. If $T = [M]$, where M is orientable, has codimension k and boundary ∂M in X, from Poincaré–Stokes we have $d[M] = (-1)^{k+1}[\partial M]$. We see that in the language of currents the boundary operator ∂ is 'the same' as the differential d. This shows the great unifying power of de Rham theory.

After this remarks let us turn to the third Wirtinger theorem.

THEOREM (WIRTINGER, 1936). *Let $(A, \partial A)$ be a k dimensional analytic set with boundary in $\mathbb{C}^n$ such that $A \cap \partial A$ is compact. Let M be a $2k$ dimensional, C^1 oriented manifold in $\mathbb{C}^n$ such that the closure $\bar{M}$ is compact and $\partial A \subset \bar{M} - M$ and $[\partial A] = -[M]$ as currents (that is, for the pair $(M, \partial A)$ there is a meaningful Poincaré–Stokes formula.) Then*

$$vol_{2k}A \leqslant vol_{2k}M$$

PROOF. Since in $\mathbb{C}^n$ the Euclidean volume form

$$vol = dx_1 \wedge dy_1 \wedge \cdots \wedge dx_n \wedge dy_n,$$

in complex coordinates ($dz_k = dx_k + \sqrt{-1}dy_k$) we have

$$vol = \frac{\sqrt{-1}}{2} dz_1 \wedge d\bar{z}_1 \wedge \cdots \wedge \frac{\sqrt{-1}}{2} dz_n \wedge d\bar{z}_n = \frac{1}{n!}\omega^n,$$

where $\omega = \frac{\sqrt{-1}}{2} d'd'' \parallel z \parallel^2 = \frac{1}{4} dd^c \parallel z \parallel^2$ ($d^c := \sqrt{-1}(d'' - d')$) is the Euclidean fundamental form.

Now, according to the Wirtinger theorem (W), we have

$$vol_{2k}A = \frac{1}{k!}\int_A \omega^k = \frac{1}{4k!}\int_A d(d^c \parallel z \parallel^2 \wedge \omega^{k-1}) = \text{(by Stokes)} =$$

$$= \frac{1}{4k!}\int_{\partial A} d^c \parallel z \parallel^2 \wedge \omega^{k-1} = \frac{1}{k!}\int_M \omega^k \leqslant vol_{2k}M.$$

□

REMARK 1. The same proof holds in general case: replace $\mathbb{C}^n$ by any hermitian n manifold (X, h, ω) with the exact fundamental form ω (that is, $\omega = d\gamma$; it is sufficient that ω^k is d exact.)

REMARK 2. In his thesis (1954) (Math. Ann. **131** (1956), 38– 75) Hans Grauert proved the following important theorems.

THEOREM (GRAUERT). *1. A hermitian metric h in domain $\mathcal{U} \subset \mathbb{C}^n$ with smooth boundary is Kähler if and only if with respect to h all analytic sets in $\mathcal{U}$ are minimal manifolds.*

2. $\mathcal{U}$ above is a convex domain of holomorphy if and only if $\mathcal{U}$ can be equipped with a complete Kähler metric.

Theorem 2. was improved 1983 by Mok and Yau.

THEOREM (MOK–YAU). *A bounded domain in $\mathbb{C}^n$ is a Stein manifold if it admits a complete Kähler–Einstein metric.*

The second Cousin problem once more. The notion of divisor is so important that we will present a geometric interpretation of the group of divisors $\mathrm{Div}(X)$. As we know any divisor D on manifold X is locally represented by f/g with holomorphic $f \neq 0$, $g \neq 0$ modulo non vanishing holomorphic functions, $\mathcal{O}^*(X)$. The functions f, g have zero hypersurfaces of positive order. If we count the order of g with the minus sign, we can consider the abelian group $\mathrm{Div}(X)$ as an additive group of linear combinations $\sum n_i V_i$, $n_i \in \mathbb{Z}$, where $\{V_i\}$ is a locally finite family of irreducible analytic hypersurfaces in X with $V_i \neq V_j$ for $i \neq j$. Clearly we can regard the divisor $D = \sum n_i V_i$ as a current.

THEOREM (POINCARÉ–LELONG FORMULA). *Let φ be a globally meromorphic function on complex manifold X. Then*

$$(\varphi) \equiv D_\varphi = \frac{\sqrt{-1}}{\pi} d'd'' \log \varphi$$

(in the sense of currents), where D_φ is the divisor of φ (denoted also by (φ).)

The Poincaré–Lelong formula can be also regarded as an equation

$$(P - L) \qquad \frac{\sqrt{-1}}{\pi} d'd'' \log \varphi = D_\varphi$$

with a given number on the right hand side; this is exactly the second Cousin problem of construction of a meromorphic function with given divisor. The great advantage of such formulation is clear: the Oka–Cartan theory yields 'only' existence theorem for solutions. The Riemann–Hörmander theory of d'' operator gives solutions as well with estimates which depend on estimates for the right hand side of $(P-L)$.

We have already remarked that the 'input' in the second Cousin problem is precisely a system of transition functions for a holomorphic line bundle on a complex manifold X. The existence of solutions of this problem is equivalent to existence of a solution of the equation $d''u = f$ for sections of the corresponding line bundle. This is the reason why the impressive Hörmander theorems are so precious for complex analysis.

Chapter 6

The Problem of Poincaré and the Cousin Problems

As we have said already the theorems of Weierstrass and Mittag–Leffler proved originally for domains in $\mathbb{C}$ were proved in 1948 by Behnke–Stein for arbitrary open Riemann surfaces. Carathéodory conjectured that on every open Riemann surface there exists a non constant global holomorphic function $f \in \mathcal{O}(X)$. The conjecture that on a surface X every meromorphic function $h \in \mathcal{M}(X)$ is a quotient of two meromorphic functions was proved in the Behnke–Stein paper as well. In this chapter we will reproduce proofs of these theorems, and the fact that *every open Riemann surface is a Stein manifold* using the method due to B. Malgrange based on sheaf theory and the theory of d'' operator.

Already in 1883 Henri Poincaré proved that every meromorphic function on $\mathbb{C}^2$ is a quotient of two holomorphic functions on $\mathbb{C}^2$, that is, that the field $\mathcal{M}(\mathbb{C}^2)$ is a quotient field of the ring $\mathcal{O}(\mathbb{C}^2)$. Naturally the problem arose as to for which domains in $\mathbb{C}^n$, generic manifolds X (or, even more generally, complex spaces X), $\mathcal{M}(x)$ is the quotient ring of $\mathcal{O}(X)$ relatively to the set $\mathcal{O}^*(X)$ of units? This is the so called Poincaré problem. The Carathéodory conjecture is its one dimensional case.

In his Ph.D. thesis (1894), the aim of which was to generalize Poincaré theorem to higher dimensions and more general domains, Pierre Cousin formulated two problems, called Causin I and II problem. In a moment we will formulate these problems in the language of the sheaf theory. Cousin himself solved these problems for product domains $X = B_1 \times \cdots \times B_n \subset \mathbb{C}^n$.

Let X be a complex manifold. Denote by $\mathcal{M}_x := \mathcal{Q}(\mathcal{O}_x)$, the quotient ring defined by $\mathcal{O}_x$, and by $\mathcal{O}_x^*$ (respectively, $\mathcal{M}_x^*$) the group of units in $\mathcal{O}_x$ (resp. $\mathcal{M}_x$). Similarly $\mathcal{M}^* := \bigcup \mathcal{M}_x^* = \mathcal{M} - \{\text{zero section}\}$. Then $\mathcal{M} = \mathcal{M}_X := \bigcup_{x \in X} \mathcal{M}_x$ is the sheaf of germs $\mathcal{M}_x$. The sections $\mathcal{M}(X) = H^0(X, \mathcal{M})$ of $\mathcal{M}$ are called *meromorphic functions* on X. The sheaf $\mathcal{O}^*$ is a subsheaf of $\mathcal{O}$ of (germs of) holomorphic functions which do not vanish in any connected component of X. We form the quotient sheaf $\mathcal{D} \equiv \mathcal{D}_X := \mathcal{M}^*/\mathcal{O}^*$ called the sheaf of germs of sections of $\mathcal{D}$. Sections of $\mathcal{D}$ are called *divisors.* The additive group of divisors is denoted by $\mathcal{D}(X)$ or $Div(X) := H^0(X, \mathcal{D})$.

Thus we have the short exact sequence of sheaves

$$1 \to \mathcal{O}^* \to \mathcal{M}^* \to \mathcal{D} \to 0$$

and we obtain the long exact cohomology sequence

$$1 \to \mathcal{O}^*(X) \to \mathcal{M}^*(X) \xrightarrow{\psi} \mathcal{D}(X) \xrightarrow{\eta} H^1(X, \mathcal{O}^*) \to H^1(X, \mathcal{M}^*) \to \dots$$

Thus

$$(*) \qquad 0 \to \mathcal{D}(X)/\psi\mathcal{M}^*(X) \to H^1(X, \mathcal{O}^*) \to H^1(X, \mathcal{M}^*) \to 0.$$

For any meromorphic function $h \in \mathcal{M}^*(X)$ we have its divisor $\psi(h) \in Div(X)$. Divisors $\psi(h)$ of (nontrivial) meromorphic functions $h \in \mathcal{M}^*(X)$ are called *principal divisors* and are traditionally denoted by (h). We have

$$(gh) = (g) + (h) \quad \text{for all } g, h \in \mathcal{M}^*(X).$$

For every divisor $D \in Div(X)$ there exist a cover $\{\mathcal{U}_i\}$ of X and meromorphic functions $h_i \in \mathcal{M}^*(\mathcal{U}_i)$ with $\psi(h_i) = D|_{\mathcal{U}_i}$. On $\mathcal{U}_i \cap \mathcal{U}_j$ we have $g_{ij} := h_j h_i^{-1} \in \mathcal{O}^*(\mathcal{U}_i \cap \mathcal{U}_j)$, where the family g_{ij} is an alternating 1 cocycle in $Z^1(\{\mathcal{U}_i\}, \mathcal{O}^*)$ which represents the cohomology class $L_D := \eta(D) \in H^1(X, \mathcal{O}^*)$ of holomorphic line bundles on X.

Any family $(\mathcal{U}_i, h_i)$, $h_i \in \mathcal{M}^*(X)$ with $h_j h_i^{-1} \in \mathcal{O}^*(\mathcal{U}_i \cap \mathcal{U}_j)$ determines a divisor $D \in Div(X)$. If all $h_i \in \mathcal{O}(\mathcal{U}_i)$ we say that D is *non negative* and write $D \geqslant 0$.

From the cohomology sequence $(*)$ we see that if $H^1(X, = \mathcal{M}^*) = 0$, then the map η is surjective: $\eta\mathcal{D}(X) = H^1(X, \mathcal{O}^*)$, that is, every line bundle L_D is of the form ηD, $D \in \mathcal{D}(X)$. (Recall that $H^1(X, \mathcal{O}^*)$ is the space of (equivalence classes of) holomorphic line bundles on X.) But for every line bundle ξ on X we have $\Omega(\xi) \equiv \mathcal{O}(X, \xi) \subset \mathcal{M}(X, \xi) = \{$ global meromorphic sections of $\xi\}$, that is, (global) sections of the sheaf $\mathcal{O}(\xi) \otimes_{\mathcal{O}} \mathcal{M}$, because $\mathcal{O}(X,\xi) = \{f \in \mathcal{M}(X,\xi) : (f) \geqslant 0\}$, where by (f) we denoted the divisor of the meromorphic section of the bundle ξ. Clearly, if $f_i \in \mathcal{M}(\mathcal{U}_i)$, $f_i/f_j \in \mathcal{O}^*(\mathcal{U}_i \cap \mathcal{U}_j)$. Therefore on X constant, holomorphic, non zero functions live. Thus if $D \in \mathcal{D}(X)$ is any divisor such that $\xi \in L_D$, that is, if $\xi = \eta(D)$, then there exists a meromorphic section f of ξ with $(f) = D$, and for any meromorphic sections f of ξ, $\xi = [(f)]$. We have therefore obtained the following.

PROPOSITION. *1. (ξ is a line bundle associated with some divisor D on X) $\Longleftrightarrow$(The bundle ξ has a global meromorphic section $\not\equiv 0$.)*

2. (ξ is a line bundle of an effective divisor D ($D \geqslant 0$))$\Longleftrightarrow$(ξ has a non trivial, global holomorphic section).

It follows that the map $H^1(X,\mathcal{O}^*) \to H^1(X,\mathcal{M}^*)$ is surjective; thus $(H^1(X,\mathcal{M}^*) = 0) \Longleftrightarrow (H^0(X,\mathcal{M}(\xi) \neq 0$ for any holomorphic line bundle $\xi)$. Thus we have important

THEOREM. *The following assertions on the manifold X are equivalent:*

1. $H^1(X,\mathcal{M}^*) = 0$;

2. Every holomorphic line bundle $\xi \to X$ has a nontrivial meromorphic section;

3. Every holomorphic line bundle is a bundle of divisors.

REMARK. If X is a compact Riemann surface then the assertions of the theorem holds, and, moreover, assertion 2. holds even for a holomorphic *vector* bundle ξ. Thus $H^1(X,\mathcal{M}^*) = 0$ *for any compact Riemann surface.* This fact follows immediately from the Riemann–Roch theorem. Gunning calls this assertion the *fundamental theorem for Riemann surfaces.*

In the theory of functions in one dimension there are two well known important problems, the Mittag–Leffler and the Weierstrass problems. In a moment we will formulate them in the language required in the multidimensional case, but first let us recall the classical formulation.

Mittag–Leffler problem. Let X be an open subset of $\mathbb{C}$, and let $\{x_j\}$ be a sequence of different points in X. denote by

$$(*) \qquad f_j(z) = \sum_{1}^{n_j} a_{jk}(z - x_j)^{-k}.$$

Thus f_j is meromorphic in a neighborhood $\mathcal{U}_j$ of x_j, that is, $f \in \mathcal{M}(\mathcal{U}_j)$. Find a meromorphic f such that for all j $f|_{\mathcal{U}_j} - f_j \in \mathcal{O}(\mathcal{U}_j)$. A function of the form $(*)$ is called the *principal part* of the meromorphic function f, and the Mittag–Leffler theorem asserts the existence of a meromorphic function f with an (arbitrary) given principal part.

It is not difficult to see that the problem can be phrased in a more elegant way.

THEOREM (MITTAG–LEFFLER). *Let X have an open covering $\{\mathcal{U}_j\}$. Let $f_j \in \mathcal{M}(\mathcal{U}_j)$ and $f_j - f_k \in \mathcal{O}(\mathcal{U}_j \cap \mathcal{U}_k)$ for all j, k. (Such family $\{\mathcal{U}_j, f_j\}$ is called the M–L distribution.) Then there exists $f \in \mathcal{M}(X)$ such that $f|_{\mathcal{U}_j} - f_j \in \mathcal{O}(\mathcal{U}_j)$ for every j.*

The same formulation can be made for any complex manifold (space).

The Cousin I problem Let $\{\mathcal{U}_j\}$ be an open covering of the complex space (manifold) X and let for j, k, $f_j \in \mathcal{M}(\mathcal{U}_j)$ be such that $f_j - f_k \in \mathcal{O}(\mathcal{U}_j \cap \mathcal{U}_k)$. Find $f \in \mathcal{M}(X)$ such that $f|_{\mathcal{U}_j} - f_j \in \mathcal{O}(\mathcal{U}_j)$ for every j.

Since $\mathcal{O} = \mathcal{O}_X$ is a subsheaf of $\mathcal{M}$, one can form the quotient sheaf $\mathcal{M}/\mathcal{O}$ called the *sheaf of germs of principal parts*. Clearly, we have the exact sequence of sheaves

$$0 \to \mathcal{O} \to \mathcal{M} \xrightarrow{\varphi} \mathcal{M}/\mathcal{O} \to 0$$

and the corresponding long, exact cohomology sequence

$$0 \to \mathcal{O}(X) \to \mathcal{M}(X) \to H^0(X, \mathcal{M}/\mathcal{O}) \xrightarrow{\delta} H^1(X, \mathcal{O}) \to H^1(X, \mathcal{M}) \to \dots$$

The first Causin distribution $\{f_j\}$ determines an element $s \in H^0(X, \mathcal{M}/\mathcal{O})$ and the point in solving the first Cousin problem is to find $f \in H^0(X, \mathcal{M})$ with $\varphi(f) = 1$. Such f exists if and only if $\delta(s) = 0$. If $H^1(X, \mathcal{O}) = 0$, then $\delta(s) = 0$. Therefore we have

THEOREM (CARTAN). *The additive first Cousin problem can be solved for all spaces X with $H^1(X, \mathcal{O}) = 0$. Therefore, by Theorem B, it can be always solved for Stein spaces.*

COROLLARY (BEHNKE–STEIN, 1948) *Since connected Riemann surfaces are Stein, the Mittag–Leffler problem can be solved for all first Cousin distributions.*

The second Cousin problem (called also the multiplicative Cousin problem) is much more difficult than the first Cousin problem.

The family $(\mathcal{U}_i, h_i)$ is called a divisor D representing the *second Cousin distribution.* A meromorphic function $h \in \mathcal{M}^*(X)$ has divisor D, that is, $D = \psi(h)$ if and only if $h_i h^{-1} \in \mathcal{O}^*(\mathcal{U}_i)$ for all i. We see the reason for the term *multiplicative* distribution. A divisor D is *positive* and denoted by $D \geqslant 0$ if there exists a representative of the second Cousin distribution $(\mathcal{U}_i, h_i)$, where for all i, $h_i \in \mathcal{O}(\mathcal{U}_i)$. From these definitions it follows that

$$(h \in \mathcal{M}^*(X) \text{ is holomorphic}) \Longleftrightarrow ((h) \geqslant 0).$$

In the **second Cousin problem** (or multiplicative Cousin problem) one asks for characterization of principal divisors in $Div(X)$. The exact cohomology sequence above provides as with the answer.

THEOREM 1. (H. CARTAN–SERRE). *Divisor $D \in Div(X) \equiv \mathcal{D}(X)$ is a divisor of a meromorphic function $h \in \mathcal{M}^*(X)$ if and only if*

$$H^1(X, \mathcal{O}^*) \ni \eta(D) = 0.$$

DEFINITION. The second Cousin problem is universally solvable for a complex manifold X whenever the map ψ in $(*)$ is surjective.

Therefore we have immediately.

THEOREM 2. (H. CARTAN). *The multiplicative Cousin problem is universally solvable for a manifold X (space X) if and only if the natural homomorphism*

$$H^1(X, \mathcal{O}^*) \to H^1(X, \mathcal{M}^*)$$

is injective. In particular, the second Cousin problem is universally solvable for manifolds (spaces) X with $H^1(X, \mathcal{O}^*) = 0$.

The famous Theorem B of Cartan and Serre asserts that for every Stein manifold X (more generally, Stein space) and every coherent sheaf $\mathcal{F}$ on X all cohomology groups $H^k(X, \mathcal{F}) = 0$ for $k \geqslant 1$. Since $\mathcal{O}^*$ is coherent, we have important

COROLLARY (CARTAN–SERRE). *On any Stein manifold (space) the second Cousin problem is universally solvable.*

Since non compact Riemann surfaces are Stein manifolds, as a corollary to this Corollary we have the famous

THEOREM (BEHNKE–STEIN, 1948). *For any non compact Riemann surface $H^1(X, \mathcal{O}^*) = 0$, therefore the Weierstrass problem (the one dimensional version of the second Cousin problem) is universally solvable. Moreover,*

(i) any line bundle on an open Riemann surface X is holomorphically trivial.

More generally

THEOREM (RÖHRL–GRAUERT, 1956). *On any non compact Riemann surface X*

$$H^1(X, \mathrm{GL}(k, \mathcal{O})) = 0. \tag{ii}$$

Since the sheaf of germs of holomorphic sections of any holomorphic vector bundle $E \to X$ of rank k is coherent.

This is equivalent to the statement that

Any holomorphic vector bundle on a non compact Riemann surface X is holomorphically trivial, and therefore it is (holomorphically) parallelizable:

$$H^1(X, \mathcal{O}) = 0. \tag{iii}$$

REMARK 1. We should not have any illusions: in mathematics there are no 'miracles', even though great theorems are wonderful events: Theorem B for Stein spaces is a deep theorem and its proof is very difficult. It is also difficult to prove the theorem that open Riemann surfaces are Stein spaces, that is, that they satisfy one of the definitions of Stein space, for example, that for such surfaces there exists a strongly plurisubharmonic exhausting function, or that holomorphic functions $\mathcal{O}(X)$ separate points in X, and that X is holomorphically convex.

REMARK 2. In his *Habilitationarbeit* Grauert proved much more

GRAUERT–OKA PRINCIPLE *On Stein manifolds (spaces)* topological *fiber bundles are always* holomorphic *fiber bundles.*

More precisely, one considers first a holomorphic fiber bundle $E \to X$ over a (reduced) complex space X (for example, X is a complex manifold)

whose fiber is a complex Lie group G. One forms the sheaf $\mathcal{E}^a$ (resp. $\mathcal{E}^c$) of germs of holomorphic (resp. continuous) sections of $E \to X$ end the associated cohomology sets $H^1(X, \mathcal{E}^a)$ (resp. $H^1(X, \mathcal{E}^c)$.) Then the natural injection $\mathcal{E}^a \hookrightarrow \mathcal{E}^c$ induces the mapping

$$H^1(X, \mathcal{E}^a) \to H^1(X, \mathcal{E}^c).$$

The famous Grauert theorem (1956) asserts that

(X is a Stein space)$\Longrightarrow$($H^1(X, \mathcal{E}^a) \stackrel{\sim}{\to} H^1(X, \mathcal{E}^c)$ is bijective)

The cohomology classes from $\mathcal{E}^a$ (resp. $\mathcal{E}^c$) are represented by holomorphic (resp. continuous) E-principal bundles.

A very important step in the proof of this magnificent theorem is the following

LEMMA (GRAUERT). *In a holomorphic E-principal bundle over (reduced) Stein space every continuous section is homotopic to holomorphic section.*

The following corollaries of Grauert–Oka principle are important and fascinating.

COROLLARY 1. *If X is a Stein manifold and $E \to X$ has a continuous section, then it has holomorphic sections.*

COROLLARY 2. *Take now $E = TX$ where X is Stein. Then every topologically parallelizable X is holomorphically parallelizable, that is, if* $\dim X = n$, *then there exist vector fields $\xi_1, \ldots, \xi_n$ on X which at every $x \in X$ are complex linearly independent; for every $\xi(x) \in T_xX$ it is possible to define at arbitrary $y \in X$ a parallel vector $v(y) \in T_yX$.*

COROLLARY 3. *For any n on $\mathbb{C}^n$, domains of holomorphy $H \subset \mathbb{C}^n$, open Riemann surfaces, and, more generally, Stein manifolds, all holomorphic vector bundles are holomorphically trivial.*

Let us return to the second Cousin problem. We have seen that the cohomology group $H^1(X, \mathcal{O}^*)$ is crucial for solvability of this problem. But this group is much more complicated than the group $H^1(X, \mathcal{O})$. So let us first prove the important

THEOREM. *The sequence*

$$0 \to \mathbb{Z} \to \mathcal{O} \to \mathcal{O}^* \to 1,$$

where $\mathbb{Z}$ is a constant sheaf over X of integers $\mathbb{Z}$, is exact.

The main tool will be the exponential homomorphism $\exp : \mathcal{O} \to \mathcal{O}^*$ of the sheaves over an arbitrary complex space X which is a natural generalization of the classic exponential function $e : \mathbb{C} \to \mathbb{C}^\times$.

For every open $\mathcal{U} \subset X$, $\mathcal{O}(\mathcal{U})$ is Fréchet algebra. therefore, every series $\sum_0^\infty \frac{f^n}{n!}$, $f \in \mathcal{O}(\mathcal{U})$ converges to an element $\exp f \in \mathcal{O}(\mathcal{U})$. But $\exp(f + g) = \exp f \cdot \exp g$, thus $\exp f \in \mathcal{O}^*(\mathcal{U})$. Therefore $e_{\mathcal{U}} : \mathcal{O}(\mathcal{U}) \to \mathcal{O}^*(\mathcal{U})$, $f \to \exp(2\pi i f)$ is a homomorphism, whence the family $\{e_{\mathcal{U}}\}$ defines a sheaf homomorphism $e : \mathcal{O} \to \mathcal{O}^*$ (of sheaves over the space X.) Before proving the Theorem we rewrite the short sequence above

$$(exp) \qquad 0 \to \mathbb{Z} \to \mathcal{O} \xrightarrow{e} \mathcal{O}^* \to 1.$$

PROOF OF THE THEOREM. Clearly $\ker e \supset \mathbb{Z}$, but we have to prove that $\ker e = \mathbb{Z}$. Let $\mathfrak{m}_x$ be the maximal ideal in $\mathcal{O}_x$; to prove that $\ker e = \mathbb{Z}$ it is sufficient to show that that any germ $f \in \mathfrak{m}_x$ with $\exp f = 1$ is the zero germ. But if $\exp f = 1$, then

$$f = -\sum_{k=2}^{\infty} \frac{f^k}{k!} = f^2 g, \quad \text{where } g := \frac{1}{2!} + \frac{f}{3!} + \cdots \in \mathcal{O}_x.$$

Now

$$(f = gf^2 = g^2 f^3 = \cdots = g^n f^{n+1} = \cdots) \Longrightarrow f \in \bigcap_{k=1}^{\infty} \mathfrak{m}_x^k \Longrightarrow f = 0,$$

where we made use of

KRULL INTERSECTION THEOREM. *For every submodule of a finitely generated module M over Noetherian local ring S*

$$\bigcap_{n=1}^{\infty} (N + \mathfrak{m}^k M) = N,$$

where $\mathfrak{m}$ is a (unique) maximal ideal of M. (If we take $N = 0$, the intersection $D := \bigcap_{n=1}^{\infty} \mathfrak{m}^k M = 0$.)

Let us return to the proof. We have proved that $\ker e = \mathbb{Z}$. Every unit $a := 1 + f$, $f \in \mathfrak{m}_x$ has a logarithm $h := \log(1+f) := \sum_{n=1}^{\infty} \frac{(-1)^{n+1}}{n} f^n$ with $\exp h = a$. Therefore e is a surjection. □

The short exact sequence (exp) is associated with the long cohomology sequence

$$\cdots \to H^1(X,\mathbb{Z}) \to H^1(X,\mathcal{O}) \to H^1(X,\mathcal{O}^*) \xrightarrow{c} H^2(X,\mathbb{Z}) \to H^2(X,\mathcal{O}) \to \cdots$$

where c is the Chern homomorphism (see below.)

THEOREM. *(i) If $H^k(X,\mathcal{O}) = 0$ for $k \geqslant 1$, then $H^k(X,\mathcal{O}^*) \simeq H^{k+1}(X,\mathbb{Z})$. For X being Stein, from Theorem B we have that all $H^k(X,\mathcal{O}) = 0$, $k \geqslant 1$, and in particular*

(ii) If X is Stein then $H^k(X,\mathcal{O}^*) \simeq H^{k+1}(X,\mathbb{Z})$.

COROLLARY. *If X is a connected non compact Riemann surface, then $H^2(X,\mathbb{Z} = 0$.*

PROOF. By Behnke–Stein theorem X is Stein, therefore $0 = H^1(X,\mathcal{O}^*) \simeq H^2(X,\mathbb{Z})$. □

We have the important

THEOREM (CARTAN–SERRE). *Let X be a complex space with $H^1(X,\mathcal{O}) = 0 = H^2(X,\mathbb{Z})$. Then the second Cousin problem is universally solvable. In particular, the second Cousin problem is universally solvable for any open Riemann surface (Behnke–Stein theorem).*

Let us now compose the *Chern homomorphism* $c : H^1(X,\mathcal{O}^*) \to H^2(X,\mathbb{Z})$ with the homomorphism $\eta : Div(X) \to H^1(X,\mathcal{O}^*)$. We have

$$c_1 : Div(X) \xrightarrow{\eta} H^1(X,\mathcal{O}^*) \xrightarrow{c} H^2(X,\mathbb{Z}), \quad c_1 = c \circ \eta.$$

For any divisor D we have the *Chern class $c_1(D)$ of the divisor D* as an element of the group $H^2(X,\mathbb{Z})$. For a holomorphic line bundle $L \in H^1(X,\mathcal{O}^*)$, the class $c(L) \in H^2(X,\mathbb{Z})$ is the *Chern class of the line bundle L*. We immediately obtain the following generalization of the Cartan– Serre theorem above.

THEOREM. *(i) For a principal divisor D on X, $c_1(D) = 0$.*

(ii) If $H^1(X, \mathcal{O}) = 0$ and $c_1(D) = 0$ then D is a principal divisor, that is, the exists $f \in \mathcal{M}$ such that $D = (f)$.

PROOF. (i) (D is principal)$\Longleftrightarrow$($\eta(D) = 0$). Therefore $c_1(D) = c(\eta(D)) = \delta(0) = 0$ for all principal divisors.

(ii) ($H^1(X, \mathcal{O}) = 0$)$\Longrightarrow$(c is injective). Therefore ($c_1(D) = 0$)$\Longleftrightarrow$($\eta(D) = 0$), that is, D is principal. □

Thus we have

COROLLARY 1. *On Stein manifold every divisor is principal.*

COROLLARY 2. *Any divisor D on an open Riemann surface is principal.*

COROLLARY 3. *Let X be an open Riemann surface and let a_r, $r \in \mathbb{N}$ be a discrete sequence of points on X. Then for any given $c_r \in \mathbb{C}$ there exists a holomorphic function $f \in \mathcal{O}(X)$ such that $f(a_r) = c_r$ for all $r \in \mathbb{N}$.*

PROOF. It follows from Corollary 2 that there exists $h \in \mathcal{O}(X)$ which has zeros of order 1 at a_r and $h(x) \neq 0$ for $x \neq a_r$. Let $\mathcal{U}_i := X - \bigcup_{r \neq i}\{a_r\}$, then $\{\mathcal{U}_i\}$ is an open cover of X. We define $g_i \in \mathcal{M}(\mathcal{U}_i)$ by c_i/h. For $i \neq j$ we have $\mathcal{U}_i \cap \mathcal{U}_j = X - \{a_r : r \in \mathbb{N}\}$. Thus $1/h \in \mathcal{O}(\mathcal{U}_i \cap \mathcal{U}_j)$, and $\{g_i\}$ is a second Cousin distribution on X which has a solution $g \in \mathcal{M}(X)$. Put $f := gh$. On $\mathcal{U}_i$ we have $f = gh = g_ih + (g - g_i)h = c_i + (g - g_i)h$. But $(g - g_i)$ is holomorphic on $\mathcal{U}_i$ and $h(a_i) = 0$; therefore $f \in \mathcal{O}(X)$ and $f(a_i) = c_i$ for all $i \in \mathbb{N}$. □

The starting point of Cousin investigations was the Poincaré problem whose solution let to impressive results on Cousin problem on Stein spaces. It was shown by Cartan and Serre that every meromorphic function on Stein manifold is a quotient of two holomorphic functions.

THEOREM (CARTAN–SERRE). *On a (connected) Stein manifold X the Poincaré problem has solution: For every $h \in \mathcal{M}^*(X)$ there exist $f, g \in \mathcal{O}(X)$ such that $h = f/g$.*

PROOF. Let X be connected and let $f \in \mathcal{M}^*(X)$. We know that the sheaves $\mathcal{O}$, $\mathcal{O}h$ and $\mathcal{O} + \mathcal{O}h$ are coherent subsheaves of $\mathcal{M}$. Therefore $\mathcal{O} \cap \mathcal{O}h$ is a coherent $\mathcal{O}$ sheaf. But (by definition) every stalk $\mathcal{M}_x : \mathcal{Q}(\mathcal{O}_x)$ is a quotient field of $\mathcal{O}_x$, therefore $(\mathcal{O} \cap \mathcal{O}h)_x \neq 0$ for all $x \in X$. the epimorphism $\varphi : \mathcal{O} \to \mathcal{O}h$, $f_x \to f_x h_x$ defines a coherent ideal $\mathcal{I} := \varphi^{-1}(\mathcal{O} \cap \mathcal{O}h)$ with stalks $\mathcal{I}(x) \neq 0$ for all $x \in X$. By Theorem A there exist global sections $0 \neq g \in \mathcal{I}(X)$. For such a section we have $g_x \neq 0$, $x \in X$ and $f := gh \in \mathcal{O}(X)$; thus $h = f/g$. □

We cannot present here a proof of Theorems A and B; instead we prove the following important

THEOREM. *On Stein manifold X (Theorem B)$\Longrightarrow$(Theorem A), that is, the global sections of coherent $\mathcal{F}$ generate every stalk $\mathcal{F}_x$ as $\mathcal{O}_{X,x}$-module.*

PROOF. Fix $x \in X$ and let $\mathfrak{m}_x \subset \mathcal{O}_{X,x}$ be the maximal ideal. We can identify $\mathfrak{m}_x$ with the ideal sheaf of $\{x\}$ in X. Then

$$0 \to \mathfrak{m}_x \to \mathcal{O}_X \to \mathcal{O}_X/\mathfrak{m}_x \to 0$$

is exact. Here $\mathcal{O}_X/\mathfrak{m}_x$ is clearly the sheaf $\mathcal{O}_{\{x\}}$ of holomorphic functions on the reduced space $\{x\}$; hence $\mathcal{O}_X/\mathfrak{m}_x = \mathbb{C}$ on x and 0 outside x. Let $\mathcal{F}$ be a coherent sheaf on X and let $\mathfrak{m}_x\mathcal{F}$ be the image of the natural map $\mathfrak{m}_x \otimes \mathcal{F} \to \mathcal{F}$. We have therefore the exact sequence

$$0 \to \mathfrak{m}_x\mathcal{F} \to \mathcal{F} \to \mathcal{F} \otimes \mathcal{O}_X/\mathfrak{m}_x \to 0.$$

We have to prove that the global sections of $\mathcal{F}$ generate each stalk $\mathcal{F}_x$ as $\mathcal{O}_{X,x}$-module. But, by Nakayama lemma, this is equivalent to surjectivity of

$$H^0(X, \mathcal{F}) \overset{\nu}{\to} H^0(X, \mathcal{F} \otimes \mathcal{O}_X/\mathfrak{m}_x) \simeq \mathcal{F} \otimes \mathcal{O}_X/\mathfrak{m}_x,$$

where $\nu : s \to s(x)$ for every section $s \in H^0(X, \mathcal{F})$ of $\mathcal{F}$. But in the long cohomology sequence

$$\cdots \to H^0(X, \mathcal{F}) \overset{\nu}{\to} H^0(X, \mathcal{F} \otimes \mathcal{O}_X/\mathfrak{m}_x) \to H^1(X, \mathfrak{m}_x\mathcal{F}) \to \cdots$$

the map ν is surjective if $H^1(X, \mathfrak{m}_x\mathcal{F}) = 0$ which is assured by Theorem B because X is Stein manifold and $\mathfrak{m}_x\mathcal{F}$ is coherent. □

CHAPTER 7

Ringed Spaces and General Complex Spaces

The pair $(X, \mathcal{O}_X)$, where X is a topological space and $\mathcal{O} = \mathcal{O}_X$ is called the *ringed space with the structure sheaf* $\mathcal{O}_X$ By definition the stalks are (commutative) rings and even local $\mathbb{C}$ algebras. Classification of complex spaces is (in principle) classification of these local rings. We owe this classification Richard Dedekind, a student, friend ad co-editor of Riemann's *Collected Works.* Dedekind created his theory to meet the needs of arithmetic's and algebraic geometry, and also the theory of Riemann surfaces.

To help the reader, let us recall a number of fundamental notions and facts of Dedekind theory. A ring R is *reduced* if it has no non zero nilpotent elements. If $\mathfrak{a}$ is an ideal in R, then the *radical* $\mathfrak{r}(\mathfrak{a}) = \{r \in R : r^n \in \mathfrak{n}$ for some $n \geqslant 1\}$. The set $\mathfrak{n}(R)$ of all nilpotent elements of R is called the *nil-radical* of R. The quotient ring *red* $R := R/\mathfrak{n}(R)$ is reduced and is called the *reduction* of R. Let $R_1 \supset R$ be a ring containing the ring R; then $s \in R_1$ is the *integral* over R if there exist elements $r_1, \ldots, r_b \in R$ with

$$s^b + r_1 s^{b-1} + \cdots + r_{b-1} s + r_b = 0.$$

The set $\hat{R}$ of all elements of R_1 which are integral over R is called the *integral closure* of R. If $R_1 = \mathcal{Q}(R)$ is the quotient ring of R, that is, $\mathcal{Q}(R) := \{a/b : a'b \in R$ and b is a non zero divisor in $R\}$, then $\hat{R}$ is an integral closure of R in $\mathcal{Q}(R)$. R is *integral closed ring* if $\hat{R} = R$.

If R is reduced, then $\hat{R}$ is the *normalization* of R $R \subset \hat{R} \subset \mathcal{Q}(R)$. Therefore

(R is normal)$\Longleftrightarrow$(R is reduced and integral closed)

For any integral domain R and any R-module M, the set $T(M) = \{m \in M :$ there exists $R \ni r \neq 0$ such that $rm = 0\}$ is called the *torsion module* of M.

The space $(X, \mathcal{O}_X)$ is reduced if all stalks $\mathcal{O}_x$ are reduced, that is, have no nilpotent elements. $\mathcal{M}_x = \mathcal{Q}(\mathcal{O}_x)$ and $\mathcal{O}_x \subset \hat{\mathcal{O}}_x \subset \mathcal{M}_x$. Moreover $\hat{\mathcal{O}}_x$ is the maximal ring extension of $\mathcal{O}_x$ in $\mathcal{M}_x$, the ring of meromorphic germs. It is customary to call $\hat{\mathcal{O}}_x$ the *normalization* of $\mathcal{O}_x$ in $\mathcal{M}_x$. The *normalization of the sheaf* $\mathcal{O}_X$ in $\mathcal{M}_X$ is by definition $\hat{\mathcal{O}}_X := \bigcup_{x\in X} \hat{\mathcal{O}}_x \subset \mathcal{M}_X$ and one proves that $\hat{\mathcal{O}}_X$ is an analytic subsheaf containing $\mathcal{O}_X$.

Let us now recall the fundamental notion of *analytic set.*

Let $\mathcal{U}$ be an open subset of $\mathbb{C}^n$ A subset $A \subset \mathcal{U}$ is an analytic set if

(1) A is closed in $\mathcal{U}$;

(2) For every $a \in A$ there is an open neighborhood V $V \subset \mathcal{U}$ and holomorphic functions $f_1, \dots, f_k \in= \mathcal{O}(V)$ such that

$$A \cap V = \{z \in V : f_1(z) = \cdots = f_k(z) = 0\}.$$

A subset $A \subset \mathcal{U}$ is *locally analytic* if only 2. holds.

REMARK 1. Conditions 1. and 2. are equivalent to the following

2′. For every $a \in \mathcal{U}$ there exists an open neighborhood $V \subset \mathcal{U}$ and holomorphic functions $f_1, \dots, f_k \in= \mathcal{O}(V)$ such that

$$A \cap V = \{z \in V : f_1(z) = \cdots = f_k(z) = 0\}.$$

REMARK 2. If $A \subset \mathcal{U}$ has $\operatorname{codim} A \geqslant 1$, then $\mathcal{U} - A$ is dense in $\mathcal{U}$.

REMARK 3. If $f = (f_1, \dots, f_k) \equiv 0$, then $A = \mathcal{U}$.

THEOREM. *If D is a domain (that is, it is open and connected) in $\mathbb{C}^n$ and $A \subset D$ is an analytic subset with an inner point, then $A = G$ - domain of holomorphy.*

COROLLARY. *If $D \subset \mathbb{C}^n$ is a domain, $A \subset D$ with $A \neq G$, then* $\operatorname{codim} A \geqslant 1$.

REMARK. An analytic subset of a domain $D \subset \mathbb{C}^1$ is either the whole of D or it consists of isolated points only.

The classic (and elementary)

RIEMANN EXTENSION THEOREM in 1 dimension asserts that
If x is a point in domain $D \subset \mathbb{C}$, then every bounded holomorphic function f in $D - \{x\}$ has a unique holomorphic extension to D.

The same holds true for domains D in $\mathbb{C}^n$
I. if $\{x\}$ is replaced by a nowhere dense analytic set in D;
II. if A has dimension $\leqslant n-2$ everywhere it is *not* necessary to assume that f is bounded in $D - A$.

These two statements are know as the I and II Riemann extension theorems.

Riemann extension theorems on complex manifolds are immediate consequences of preceding assertions since they are of local character. In order to have a more general setting, Grauert and Remmert introduced the following notion: A closed set in a complex space X is *thin* (of order $k \geqslant 1$) in X if every point x has an open neighborhood $\mathcal{U}$ such that $A \cap \mathcal{U}$ is contained in a nowhere dense analytic subset $\tilde{A}$ of $\mathcal{U}$. A is *thin of order* $k \geqslant 1$ if $\mathcal{U}$ and $\tilde{A}$ can be chosen such that $\dim_x \tilde{A} \leqslant \dim_x X - k$. Both Riemann theorems have now the following unified formulation.

RIEMANN EXTENSION THEOREM ON COMPLEX MANIFOLDS. *Let X be a complex manifold and A be a thin subset of X, $f \in \mathcal{O}(X - A)$.*
I. If f is bounded near A; or
II. If A is thin in X of order 2,
then f has a unique holomorphic extension $\hat{f}$ to X.

COROLLARY. *If X is a* connected *complex manifold, then for every thin set A in X the manifold $X - A$ is connected.*

PROOF. If $X - A = X_0 \cup X_1$ is a disjoint union of open sets, the function $f(z) :\equiv n$, $n = 0, 1$ is holomorphic on $X - A$ and bounded near A, thus f has holomorphic extension $\hat{f}$ on X. Since X is connected and $\hat{f}$ takes only values 0 and 1, it is constant. Hence either $X_0 = \emptyset$, or $X_1 = \emptyset$, thus $X - A$

is connected. □

A natural question arises in which complex spaces an analogue of the extension theorem holds. As it can be expected, such spaces must be analogues of 'abstract Riemann surfaces'. Behnke and Stein, generalizing in a natural way the classical Riemann theory, introduced, in 1951/52, finite analytically ramified coverings of domains in $\mathbb{C}^n$ as 'local models' or 'prototypes'. These spaces are now called *normal complex spaces.* About the same time, H. Cartan in his famous seminar at ENS introduced, in an abstract way, another class of complex spaces which are called now *reduced complex spaces*: its structural sheaf $\mathcal{O}_X$ has stalks $\mathcal{O}_x \subset \mathcal{C}_x$ and every ring $\mathcal{O}_x$ is without any nilpotent elements. For normal spaces $(X, \mathcal{O}_X)$, every ring $\mathcal{O}_x$ is normal, is integral closed. In 1958 Grauert and Remmert proved that the Behnke–Stein spaces are identical with normal spaces in Cartan's sense. The original, very difficult proof has been recently essentially simplified by Dethloff who made use of Hörmander method.

A deep and fundamental theorem of Cartan–Oka asserts that *For every reduced space* $(X, \mathcal{O}_X)$ *the normalization sheaf* $\hat{\mathcal{O}}(X)$ *is coherent.*

Analytic coverings of reduced spaces. The idea of analytic covering is perhaps the greatest and most seminal idea of Riemann. This idea plays important role in many branches of mathematics and physics.

At the early stages of general theory of complex spaces (Behnke, H. Cartan, Stein) it was a quite natural endeavor to extend Riemann ideas to analytic continuation of germs coverings, branched coverings etc.

As it was discovered by Grauert and Remmert about 1958, *one-sheeted* analytic coverings are fundamental for the construction of normalization spaces. We recall that an 'abstract' Riemann surface is normal and is obtained by normalization procedure (by attaching branched germs.)

DEFINITION. A finite, surjective, holomorphic map $\pi : X \to Y$ between reduced spaces is an *analytic covering* of Y if there exists a thin subset T of Y with the following properties

(i) $\pi^{-1}(T)$ is thin in X;

(ii) the induced map $X - \pi^{-1}(T) \to Y - T$ is locally biholomorphic.

The set T is called the *critical locus* of the covering π. The set $B \subset X$ of all points where π is *not* locally biholomorphic is the *branch locus* of π. A covering is *unbranched* if B is empty.

For a given (analytic) covering $\pi : X \to Y$, denote by $b(y)$ the finite number of different points in the fiber $\pi^{-1}(y)$, $y \in Y$. It is not difficult to see that if $T \subset Y$ is a critical locus of the covering, than the function $y \to b(y)$ is locally constant in $Y - T$. In the case when $b(y) = b$=const for $y \in Y - T$ the covering $\pi : X \to Y$ is a *b-sheeted analytic covering.* This is clearly the case if $Y - T$ is connected. We have the following.

PROPOSITION. *Every analytic covering $\pi : X \to Y$ of a connected complex manifold is a b-sheeted analytic covering; $b = b(y)$, $y \in Y - T$.*

We have to extend – in some sense – the notion of analytic covering for locally compact spaces X. We recall that the map $f : X \to Y$ between topological spaces is *finite* if it is continuous, closed (that is, $f(V)$ is closed for every closed V), and if every fiber $f^{-1}(y)$, $y \in Y$ is a finite set, that is, it consists only of a finite number of points.

A closed and nowhere dense subset A of X does not *separates X locally* if for every $a \in A$ and every connected open neighborhood $\mathcal{U}$ of a in X there exists a neighborhood V of a contained in $\mathcal{U}$ such that $V - A$ is connected. We can give important

DEFINITION. Let X be a locally compact space, D a domain in $\mathbb{C}^n$, and $\pi : X \to D$ a finite, surjective map. If there exists an *analytic* subset $A \neq D$ of D such that

(1) $\pi^{-1}(A)$ does not separate X locally;

(2) $\pi : X - \pi^{-1}(A) \to D - A$ is locally topological,

then $\pi : X \to D$ is called the *analytically branched covering with critical locus A.*

A point $x \in X$ is *of order k* if it has a basis of neighborhoods $\{\mathcal{U}_j\}$ such that every $\mathcal{U}_j$ has an analytically branched covering with k sheets. One denotes k by $o(x)$; if $o(x) = 1$, x is the *schlicht point*; otherwise x is called the branching point (of the covering π.) All these notions are well known in the theory of Riemann surfaces, and indeed, this theory was the main impulse of topology. Let us make a number of simple, but important remarks.

REMARK 1. For every $\alpha \in D$ there exists a neighborhood $\mathcal{U}$ of α such that $\pi^{-1}(\mathcal{U}) = V_1 \cup \cdots \cup V_r$ such that every V_i contains exactly one point $q_i := \pi^{-1}(\alpha)$. The restriction $\pi|_{V_i} : V_i \to \mathcal{U}$ has $o(q_i)$ sheets and $\pi|_{V_i}$ is topological if and only if q_i is schlicht. We obtain the formula

$$\sum_{i=1}^{r} o(q_i) = b \quad \text{if } \pi : X \to D \quad \text{has } b \text{ sheets.}$$

REMARK 2. The critical locus A is *not* uniquely determined: if A' is any nowhere dense subset of D, then $A \cup A'$ is a critical locus.

REMARK 3. The projection $\pi(B)$ of the set B of *all* branching points of $\pi : X \to D$ gives the *minimal* critical locus which can be empty, or a purely 1-codimensional analytic set in D.

Now we have to define *holomorphic functions* on X as continuous functions $f : X \to \mathbb{C}$ which are holomorphic in schlicht points in the sense of domains over $\mathbb{C}^n$. More explicitly: Let $\mathcal{V}$ be an open subset of X; a continuous function $f : \mathcal{V} \to \mathbb{C}$ is *holomorphic if for every schlicht point* $x \in \mathcal{V}$ *there exists an open neighborhood* $\mathcal{U}(x) \subset \mathcal{V}$ *such that* $\pi : \mathcal{U}(x) \to \pi(\mathcal{U}(x))$ *is topological and the function* $f \circ (\pi|_{\mathcal{U}})^{-1})$ *is holomorphic in* $\pi(\mathcal{U}(x))$. *The set of such functions is denoted by* $\mathcal{O}'(\mathcal{U})$ *and the sheaf given by the presheaf* $\mathcal{U} \to \mathcal{O}'(\mathcal{U})$ *will be denoted by* $\mathcal{O}'_X$.

PROPOSITION. *(A continuous function* $f : X \to \mathbb{C}$ *is holomorphic)* $\Longleftrightarrow$*(There exist holomorphic functions* $a_1, \ldots, a_r : D \to \mathbb{C}$ *such that* $f(x)^r + \sum_{i=1}^r a_i(\pi(x)) f(x)^{r-i} \equiv 0$ *on* X*).*

It is clear that the notions above are faithful extensions of the fundamental concepts of the theory of Riemann surfaces. This was the main idea of Behnke and Stein to extend the notion of complex manifolds by taking as a model the analytic branched coverings. Independently, and basically at the same time, the theory of normal spaces $(X, \mathcal{O}_X)$ of H. Cartan was created. This theory makes use of the completely different language. In their fundamental paper Grauert and Remmert characterized analytic branched coverings in

MAIN THEOREM ON ANALYTIC BRANCHED COVERINGS (GRAUERT-REMMERT, 1958).

1. Every analytic branched coverings is an analytic covering over a connected open domain in $\mathbb{C}^n$, *the covering of whose is a normal complex space. Equivalently, the covering space* $(X, \mathcal{O}'_X)$ *is a normal space.*

2. The converse holds as well.
3. Every normal space yields (locally) an analytic branched coverings.

Therefore, defining the *Behnke-Stein space* (the α-space of Grauert-Remmert) as a Hausdorff space for whose there exists an open covering $\{R_j\}$ with the following properties:

(i) For every j there exists an analytic branched coverings $\pi_j : Z_j \to D_j$ and a topological map $\varphi_j : R_j \to Z_j$.

(ii) If $R_{j_1} \cap R_{j_2} \neq \emptyset$, the map $\varphi_{j_1} \circ \varphi_{j_2}(R_{j_1} \cap R_{j_2}) \to \varphi_{j_2}(R_{j_1} \cap R_{j_2})$ is biholomorphic.

COROLLARY (GRAUERT-REMMERT, 1958). *The Behnke-Stein spaces are normal spaces.*

The original proof of the Main theorem is very difficult, but it was greatly simplified in 1990 by G. Dethloff who applied the method of Hörmander to prove the sharpened version of another Grauert- Remmert theorem which was crucial in their proof of the Main theorem.

THEOREM. *Let $\pi : X \to D$ be an analytic branched covering with critical locus A. Assume that D is bounded and pseudoconvex. Let $z_0 \in D - A$ and $\pi^{-1}(z_0) = \{x_1, \ldots, x_b\}$. Then there exists a holomorphic function $f \in \mathcal{O}'_X$ with pairwise different $f(x_i)$, $i = 1, \ldots, b$.*

It is hard not to feel satisfied since 1. Such apparently distant notions turn out to be identical; 2. The Hörmander-Riemann method of the L^2 theory for the Cauchy-Riemann operator d'' simply leads to the desired result. But the original Grauert-Remmert proof made use of another powerful methods.

In the theory of deformations of complex structures, the development proceeded in the opposite direction: the first famous theorems by Kodaira-Spencer, Kuranishi, and Kodaira-Nirenberg-Spencer have been obtained with the help of the theory of elliptic partial differential equations. Then Grauert obtained much more general and (partially) sharpened results by his Direct Image Theorem for Coherent Sheaves.

Normalization of reduced spaces. In the theory of compact Riemann surfaces, one obtains the smooth 1-dimensional manifold, the 'abstract Riemann surface' $\hat{X}$ as an 1-sheeted analytic covering π of a 'pointed manifold' X which is a reduced space. The same procedure (but, of course, much more difficult) is possible in general multidimensional case.

DEFINITION. An 1-sheeted analytic covering $\pi : \hat{X} \to X$ of a reduced space $X = (X, \mathcal{O}_X)$ is a *normalization* of X if $\hat{X}$ is a normal space.

We know that (by definition) a point $a \in X$ is normal if $\hat{\mathcal{O}}_a = \mathcal{O}_a$, and all points of normal space are normal. Thus the complex structure of a normal space Y is completely determined by the underlying topological structure $|Y|$ and the complex structure of the regular part $Y - S(Y)$ ($S(Y)$ denotes the singular part of Y).

Reinhold Remmert writes *Already in the days of Weierstrass it was intuitively clear that – locally – all hypersurfaces in $\mathbb{C}^n$ admit normalizations ... Oka makes, in his paper of 1951, large strives towards an existence proof ... Oka's ideas and results were put in a rigorous setting by Cartan in his seminar (1953/54).*

The normalization theorem was proved in 1961 by N. Kuhlman.

NORMALIZATION THEOREM (KUHLMAN, 1961). *Every reduced complex space $(X, \mathcal{O}_X)$ admits (up to an isomorphism) exactly one normalization $\xi : \hat{X} \to X$ such that $\hat{\mathcal{O}}_X = \xi_*(\mathcal{O}_X)$.*

REMARK 1. If X is locally irreducible, then the map $\xi : \hat{X} \to X$ is homeomorphic.

REMARK 2. Normalization of a direct product is the direct product of normalizations; in particular ($X \times Y$ is normal) $\Longleftrightarrow$(X and Y are normal).

REMARK 3. If $\xi : \hat{X} \to X$ is a normalization of X, then the lifting map $\mathcal{M}(X) \to \mathcal{M}(\hat{X})$ is a $\mathbb{C}$-algebra isomorphism.

The notion of normalization makes it possible to visualize normal spaces: Every normal complex space of dimension k is, locally, a normalization of a complex hypersurface in $\mathbb{C}^{k+1}$.

Examples of normalizations. Denote by $V(Y; f_1, \ldots, f_m)$ a subvariety determined by (zeros of) $f_1, \ldots, f_m \in \mathcal{O}(Y)$.

EXAMPLE 1. Whitney umbrella: $Y := V(\mathbb{C}^3; x^k - y^k z)$, $k \geqslant 2$; then the surface Y in $\mathbb{C}^3$ has the line $x = 0 = y$ as a singular locus $S(Y)$. The map $\xi : \mathbb{C}^2 \to Y$ given by $(u, v) \to (uv, v, u^k)$ is a 1-sheeted covering with the critical locus $S(Y)$. The branch locus is the line in $\mathbb{C}^2$ given by $v = 0$. The fiber $\xi^{-1}(0)$ has *one* point; all fibers $\xi^{-1}(y)$, $y \in S(Y) - \{0\}$ have k points. The space Y is irreducible at 0 and has k prime components at all other points.

EXAMPLE 2. $X := V(\mathbb{C}^3; z_1z_2z_3 - z_1^2 - z_2^3)$. The normalization is $\xi : \mathbb{C}^2 \to X, (u, v) \to (u^2v, uv, u + v)$.

EXAMPLE 3. Neil parabola $X := V(\mathbb{C}^2; z_1^2 - z_2^3)$. $S(X) =$ the origin.

EXAMPLE 4 (trivial). If $f = (f_1, \ldots, f_k) \equiv 0$, then $A_f = D \subset \mathbb{C}^n$.

The reduced spaces are, however, not general enough for all purposes. The simplest example is the 'double point': the function $w = z^2$ determines a 2- fold covering of $\mathbb{C}$ by itself with the origin 0 being the double point. It is quite natural to associate with 0 the 2-dimensional $\mathbb{C}$-algebra $\mathcal{O}_0/\mathcal{O}_0z^2$ which has nilpotent, non-zero elements. Hence, holomorphic functions may be nilpotent and, as Grauert-Remmert say, 'they are invisible for the geometric eye.'

In algebraic geometry the necessity of more general spaces was stressed by S. Zariski, and the road to this most general notions of complex analysis was paved in algebraic geometry by A. Grothendieck. This notion, in the case of complex analysis and geometry, was introduced by H. Grauert. Grauert spaces are now called

Complex spaces. We recall that a (commutative) ring $\mathcal{R}$ is called *local ring* if it has a unique maximal ideal $\mathfrak{m}$. This maximal ideal of $\mathcal{R}$ will be denoted by $\mathfrak{m}(\mathcal{R})$. In the complex analysis the most interesting are local $\mathbb{C}$-algebras $\mathcal{R}$. The reason is that since $\mathcal{R}/\mathfrak{m}(\mathcal{R}) \simeq \mathbb{C}$, we can identify $\mathcal{R}$ with $\mathbb{C} \oplus \mathfrak{m}(\mathcal{R})$: $\mathcal{R} \simeq \mathbb{C} \oplus \mathfrak{m}(\mathcal{R})$. Let $\mathcal{H} = (\mathcal{H}, \pi, X)$ be a sheaf of local $\mathbb{C}$-algebras over topological space X, that is, $\pi : \mathcal{H} \to X$ is locally topological and all stalks $\mathcal{H}_x$, $x \in X$ are local $\mathbb{C}$-algebras: $\mathcal{H}_x \simeq \mathbb{C} \oplus \mathfrak{m}(\mathcal{H}_x)$.

Functions and sections. For every sheaf of $\mathbb{C}$-algebras over X, with any section $s \in \mathcal{H}(Y), Y \subset X$ we can associate a $\mathbb{C}$ valued function $[s] : Y \to \mathbb{C}$ as follows: Every germ $s_y \in \mathcal{H}_y = \mathbb{C} \oplus \mathfrak{m}(\mathcal{H}_y)$ can be uniquely written as $s_y = c_y + r_y$, where $c_y \in \mathbb{C}$, $r_y \in \mathfrak{m}(\mathcal{H}_y)$, and $[s](y) := c_y \in \mathbb{C}$; c_y is called the *complex value* of s at y. Clearly, $([s](y) = 0) \Longleftrightarrow (s_y \in \mathfrak{m}(\mathcal{H}_y))$. We can call $[s]$ the *function induced by the section s*.

By far the most important example is the sheaf $\mathcal{C}_X$ of germs of continuous local functions on X. This is a sheaf of local $\mathbb{C}$ algebras; the maximal ideal $\mathfrak{m}(\mathcal{C}_X)$ is the set of all germs $f_x \in \mathbb{C}_X$ which are represented around x by continuous functions f that vanish at x. Therefore one can identify every section f in $\mathcal{C}(Y)$ with a continuous complex function $f : Y \to \mathbb{C}$. But if stalks $\mathcal{H}_x$ contain nilpotent elements one has to be more careful!

$\mathbb{C}$-ringed spaces and their isomorphisms. A $\mathbb{C}$-ringed space is a pair $(X, \mathcal{H})$ where X is a Hausdorff space and $\pi : \mathcal{H} \to X$ is a sheaf of $\mathbb{C}$ local algebras over X. It is natural to call two such $\mathbb{C}$ ringed spaces $(X_1, \mathcal{H}_1)$ and $(X_2, \mathcal{H}_2)$ isomorphic if there is a pair $(f, \tilde{f})$ with the following properties:

1. $f : X_1 \to X_2$ is a homeomorphism (topological);
2. $\tilde{f} : \mathcal{H}_1 \to \mathcal{H}_2$ is a homomorphism;
3. $\tilde{f}$ is stalk preserving with respect to f, that is, the following diagram is commutative

$$\begin{array}{ccc} \mathcal{H}_1 & \xrightarrow{\quad \tilde{f} \quad} & \mathcal{H}_2 \\ \Big\downarrow {\scriptstyle \pi_1} & & \Big\downarrow {\scriptstyle \pi_2} \\ X_1 & \xrightarrow{\qquad\quad} & X_2 \end{array}$$

4. For every $x \in X_1$ the restriction $\tilde{f}|_{(\mathcal{H}_1)_x} : (\mathcal{H}_1)_x \to (\mathcal{H}_2)_{f(x)}$ is an isomorphism of $\mathbb{C}$-algebras.

REMARK. If one replaces the word 'topological' by 'continuous' in 1. and 2., and 'isomorphism' by 'homomorphism' in 4., one obtains a (homo)*morphism* of $\mathbb{C}$-ringed spaces. Clearly, the $\mathbb{C}$-ringed spaces form a category.

The great idea of Riemann was to introduce spaces which have as local models (prototypes) some well known spaces, for example, the Euclidean space $\mathbb{R}^n$; in such a way he obtained Riemann surfaces, and in multidimensional case, one obtains complex spaces. In his theory of algebraic functions Riemann considered as local models algebraic curves. In complex analysis in several dimensions, one considers more general

Model complex spaces. Let D be a domain in $\mathbb{C}^n$ and $A = A_f$ be an analytic set $A \subset D$ being a zero set of k holomorphic functions $f_1, \ldots, f_k$, that is, $A_f = \{z \in D : f_1(z) = \cdots = f_k(z) = 0\}$; $f = (f_1, \ldots, f_k)$. We consider the ideal sheaf $\mathcal{I}_f \subset \mathcal{O}_D$ which is coherent in D, that is, for every

$x \in D$ there exists a neighborhood $\mathcal{U} \subset D$ such that the sheaf $\mathcal{I}_f$ is generated by $f = (f_1, \ldots, f_k)$: $\mathcal{I}_{\mathcal{U}} := \mathcal{O}_{\mathcal{U}} f_1 + \cdots + \mathcal{O}_{\mathcal{U}} f_k$. The quotient sheaf $\mathcal{O}_D/\mathcal{I}_f$ is a sheaf of $\mathbb{C}$-algebras on D. Consider the support $Y := \mathrm{Supp}(\mathcal{O}_D/\mathcal{I}_f)$; thus

$$Y = \{z \in D : (\mathcal{O}_D/\mathcal{I}_f)_z \neq 0, \text{ that is, } \mathcal{I}_{f,z} \neq \mathcal{O}_z\}.$$

Therefore, *locally* $Y = A_f$. The restriction

$$\mathcal{O}_Y := (\mathcal{O}_D/\mathcal{I}_f)|_Y$$

is a sheaf of local $\mathbb{C}$-algebras and the $\mathbb{C}$ ringed space $(Y, \mathcal{O}_Y)$ is called the *model complex space* defined by the coherent ideal $\mathcal{I}_f \subset \mathcal{O}_D$. Now we can give the fundamental

DEFINITION. A *complex space* is a $\mathbb{C}$ ringed space $(X, \mathcal{O}_X)$ which is locally isomorphic to a complex model space, that is, every point $x \in X$ has an open neighborhood $\mathcal{U}$ such that the open $\mathbb{C}$ ringed subspace $(\mathcal{U}, \mathcal{O}_{\mathcal{U}})$ of $(X, \mathcal{O}_X)$ is isomorphic to the complex model space. Thus complex space can be locally realized as a $\mathbb{C}$ ringed space by the zero set of finitely many holomorphic functions in some domain $D \subset \mathbb{C}^n$.

Thus every stalk $\mathcal{O}_x$ is isomorphic to a quotient algebra of convergent power series by an ideal. One proves that all stalks are Noetherian and Henselian. A (homo)morphism between complex spaces are called *holomorphic maps*, and isomorphism are called *biholomorphic maps.* We have a simple observation.

PROPOSITION. *Let* $(X, \mathcal{O}_X)$ *be a complex space. Then every section* $s \in c\mathcal{O}(Y)$ *induces a* continuous *function* $[s] \in \mathcal{O}(Y)$. *The map* $\tau_{\mathcal{U}} : \mathcal{O}_X(\mathcal{U}) \to \mathcal{C}_X(\mathcal{U})$, $s \to [s]$ *define a* $\mathbb{C}$*-algebra homomorphism* $\tau : \mathcal{O}_X \to \mathcal{C}_X$ *(but in general this map is* not *injective.)*

REMARK (important!) The famous *Rückert Nullstellensatz* asserts that the kernel of the map $\mathcal{O}_X \to \mathcal{C}_X$ is a *nilradical* of $\mathcal{O}_X$. But reduced complex spaces $(X, \mathcal{O}_X)$ are defined as spaces with empty nilradical of the structure sheaf $\mathcal{O}_X$. Therefore, *for reduced spaces* $(X, \mathcal{O}_X)$ the homomorphism $\mathcal{O}_X \to \mathcal{C}_X$ can be regarded as a subsheaf of $\mathcal{C}_X$, the sheaf of germs of continuous functions. This was always assumed till 1960 when Grauert introduced his general complex spaces, on whose live sections which may be invisible for the geometric eye.

Below we will need the notion of complex subspaces of Banach manifolds. The extension of the notion of complex spaces is immediate.

Let X be an analytic Banach manifold (real or complex); denote by $\mathcal{O}_X$ the sheaf of germs of analytic functions on X. Let E, F be Banach spaces, $\mathcal{U} \subset E$ an open subset, and $f : \mathcal{U} \to F$ an analytic map. Then $A_f := \ker f = \{x \in \mathcal{U} : f(x) = 0\}$ can be endowed with the structure of a $\mathbb{C}$ ringed space:

For $x \in \mathcal{U}$, let $\mathcal{I}_{f,x}$ be an ideal in $\mathcal{O}_{\mathcal{U},x}$ consisting of germs at x of all functions $< a, f >$, where $a : \mathcal{V}(x) \to F^*$ (the dual space of F) is an analytic function defined in an neighborhood $\mathcal{V}(x)$ of x. As in the finite dimensional case above, we obtain a sheaf $\mathcal{I}_f$ of ideals in $\mathcal{O}_{\mathcal{U}}$. By Hahn–Banach theorem, the quotient $\mathcal{O}_{\mathcal{U}}/\mathcal{I}_f$ has support A_f. In this way we obtained the model complex space $(A_f, \mathcal{O}_{\mathcal{U}}/\mathcal{I}_f)$. By definition, analytic spaces are $\mathbb{C}$ (or $\mathbb{R}$) ringed spaces locally isomorphic to the model complex space $(A_f, \mathcal{O}_{\mathcal{U}}/\mathcal{I}_f)$.

Let $(X, \mathcal{O}_X)$ be an analytic Banach manifold. Let $\mathcal{I}$ be an ideal sheaf in $\mathcal{O}_X$ such that for a sufficiently small open subset in the covering of X there exists a Banach space F and an analytic map $f : \mathcal{U} \to F$ with $\mathcal{I}|_{\mathcal{U}} = \mathcal{I}_f$. Then the pair $(\mathrm{Supp}(\mathcal{O}_X/\mathcal{I}), \mathcal{O}_X/\mathcal{I})$ is an analytic space called the *analytic subspace* of X.

Let us look back at the long and dramatic way we passed: the real Platonic ascent from the 'cave', a domain D of a number space $\mathbb{C}^n$, through complex manifolds (like Behnke–Stein spaces), reduced spaces, to the *idea of Grauert complex spaces.* But we have also descent from the *cosmos noethos* of the idea of complex spaces by the following reductions: 1. if the morphism $\tau : \mathcal{O}_X \to \mathcal{C}_X$ is injective, then the space $(X, \mathcal{O}_X)$ is reduced; 2. if every stalk $\mathcal{O}_x$ is integrally closed in its quotient ring $\mathcal{Q}(\mathcal{O}_x)$, then the reduced space $(X, \mathcal{O}_X)$ is normal (Behnke–Stein); 3. if the normal complex space $(X, \mathcal{O}_X)$ is locally isomorphic to a domain $D \subset \mathbb{C}^n$, then it is an n dimensional complex manifold.

Stein spaces are now defined by Grauert as complex spaces $(X, \mathcal{O}_X)$ having the following two properties:

(i) For every infinite, discrete set $M \subset X$ there is a holomorphic function h in X such that $[h]|_M$ is not bounded. This means that X is *holomorphically convex.*

(ii) For every point $x_0 \in X$ there are finitely many (global) holomorphic functions $h_1, \ldots, h_k$ on X such that x_0 is an isolated point of the set $\{[h_1] = \cdots = [h_k] = 0\}$. One says that X is holomorphically spreadable.

REMARK 1. Condition (i) for X being a domain in $\mathbb{C}^n$ characterizes X as a domain of holomorphy; this is the famous theorem of Peter Thullen and Henri Cartan (1932).

REMARK 2. The axiom (ii) implies (by the maximum principle) (ii′) *finiteness of compact analytic sets in* X.

REMARK 3. It follows from (ii) that Stein spaces are *never* compact.

REMARK 4. Grauert proved that every space satisfying (i) is paracompact; this is a deep theorem and can be regarded as a vast generalization of the famous theorem of Tibor Rado: *Every Riemann surface has a countable topology (and therefore is paracompact).*

REMARK 5. For complex manifolds X the axiom (i) is equivalent to the following two conditions

(a) X is holomorphically separable: for $x_1 \neq x_2 \in X$ there exists $h \in \mathcal{O}(X)$ such that $h(x_1) \neq h(x_2)$;

(b) *Uniformization*: For every $x_0 \in X$ there are functions $h_1, \ldots, h_n \in \mathcal{O}(X)$ ($n = \dim X$) such that in local coordinates $\det\left(\frac{\partial h_\nu}{\partial z_\mu}\right)(x_0) \neq 0$.

REMARK 6. As we know, in the original definition Karl Stein assumed more axioms; Grauert reduced them to the now classic set.

REMARK 7. For non schlicht domains over $\mathbb{C}^n$, axiom (a) is not satisfied, and for locally uniformizable branching points there do not exists coordinate systems with globally holomorphic functions. But in such situations axiom (ii) holds.

Let us return to the general complex spaces. A fruitful new notion must satisfy two conditions: it must be sufficiently general to contain important situations which happen in the life of mathematics and are *not* encompassed yet, and second, it cannot be too general, and should imply a number of beautiful and useful theorems.

The remarks above show that the idea of Grauert space satisfied this first condition. As shown by Grauert in his celebrated works, this idea has the second property as well. He showed perhaps the most important theorem of complex analysis of several variables, namely the famous Grauert direct image theorem (1960). Before formulating this theorem, we introduce an important notion of sheaf theory

DEFINITION. Let $f : X \to Y$ be a continuous map of topological spaces and let $\mathcal{F}$ be a sheaf of abelian groups over X. We denote by $R^q f_*(\mathcal{F})$ the sheaf (over Y) of (abelian) groups associated with the presheaf of $\mathcal{U} \to H^q(f^{-1}(\mathcal{U}), \mathcal{F})$, the cohomology group with values in $\mathcal{F}$. For $q = 0$ one writes $f_*(\mathcal{F})$ for $R^0 f_*(\mathcal{F})$.

GRAUERT THEOREM. *Let $F : X \to Y$ be a proper and holomorphic map of complex spaces, and let $\mathcal{F}$ be a coherent sheaf on X. Then the sheaf $R^q f_*(\mathcal{F})$ is coherent for every $q > 0$.*

The theorem is deep: it has many interesting and important consequences; its proof is very difficult. Even simplified proofs of Grauert theorem are long and not simple. Such proofs were given by Malgrange, Forster-Knorr, Kiel-Verdier, and others.

An important ingredient of the proof is the fact that one can introduce a (unique) Fréchet topology in the space of Čech cocycles. One introduces first the natural Fréchet topology in the space $\mathcal{F}(X)$ of sections of the sheaf $\mathcal{F}$.

We recall that a locally convex linear space E is a Fréchet space if its topology is defined by enumerable many seminorms and the space is complete in this topology. Clearly, every Fréchet space is metrizable.

A closed linear subspace of Fréchet space is a Fréchet space. If $\{E_n\}$ is an enumerable family of Fréchet spaces, the product space $\mathsf{X}_i E_i$ and the projective limit $\lim\limits_{\leftarrow} E_i$ are Fréchet spaces. In the proof of Grauert theorem the following classic theorems are indispensable.

BANACH THEOREM. *Given Fréchet spaces E and F and $f : E \to F$, a continuous linear surjection; then f is open.*

For the finiteness theorem of cohomology groups the following theorem of Laurent Schwartz is crucial.

THEOREM (L. SCHWARTZ, 1952). *Given Fréchet spaces E, F and $f, g : E \to F$ linear continuous maps with f surjective and g compact; then the map $f - g : E \to F$ has a finite codimension in F.*

Both theorems are of classic simplicity. One wonders why the Schwartz theorem was not known to the Banach school. Perhaps they were not interested in cohomology theory?

Grauert develops first the Stein theory. He obtains theorem B (and A) of Cartan-Serre for general Stein spaces, and works with Stein coverings of his general complex spaces. The tremendous power of Grauert direct map theorem is demonstrated by the observation that one obtains as immediate corollaries the following famous theorems.

FINITENESS THEOREM (CARTAN-SERRE, GRAUERT). *Let X be a compact Grauert complex space and let $\mathcal{F}$ be a coherent sheaf on X. Then all $\mathbb{C}$ vector spaces $H^q(X, \mathcal{F})$ are of finite dimension.*

PROOF. Take for Y a point $Y = \{y_0\}$; since X is compact, then every holomorphic map $f : X \to y_0$ is proper. From the coherence of $\mathcal{F}$ it follows the finite dimension of $H^q(X, \mathcal{F}) = R^q f_* \mathcal{F}$.

REMARK 1. If X is not compact, the spaces $H^q(X, \mathcal{F})$ may be of infinite dimension.

REMARK 2. If f is not proper, $R^q f_* \mathcal{F}$ may be not coherent.

REMARK 3. For compact complex manifolds the finite dimension $H^q(X, \mathcal{F})$ for a holomorphic vector bundle $\mathcal{F}$ was first proved with the help of elliptic theory (of Laplace-Beltrami- Weitzenbröck operator.)

Another simple corollary is the famous

REMMERT THEOREM ON ANALYTIC SETS (1956). *Let $f : X \to Y$ be a holomorphic proper map of complex spaces, and let $A \subset X$ be a closed analytic subset. Then the set $f(A)$ is analytic in Y.*

PROOF. We recall that the support of every $\mathcal{O}_Y$-coherent sheaf is an analytic set in Y. By Oka theorem the structure sheaf of any complex space is coherent. Equip now A with reduced structure. Then $\mathcal{O}_A$ is a coherent $\mathcal{O}_X$ module (since the ideal sheaf $\mathcal{I}_A$ is coherent) (Oka-Cartan theorem.) Therefore, by Grauert theorem $f_*(\mathcal{O}_A)$ is a coherent $\mathcal{O}_Y$ module. But $\mathrm{Supp}(f_*(\mathcal{O}_A)) = f(A)$, thus $f(A)$ is analytic.

Remmert theorem may be regarded as a far reaching generalization of the classical *elimination theory*. Remmert called his theorem the 'proper mapping theorem'.

A bit of history. The direct mapping theorem is an endpoint of a long heroic history which started with

RIEMANN OBSERVATION. We recall that a meromorphic function f on a Riemann surface is a holomorphic map $f : X \to \mathbb{P}^1$.

1. If X, Y are Riemann surfaces with X compact, and $f : X \to Y$ a non constant holomorphic map, then Y is compact and f is surjective.

It follows immediately from 1. that on a compact Riemann surface every holomorphic function is compact, and

2. Every meromorphic function on $\mathbb{P}^1$ is rational, that is, it is a quotient of two polynomials.

The next step was

THEOREM OF HURWITZ–WEIERSTRASS (1879). *Meromorphic functions on $\mathbb{P}^n$ are rational.*

PROOF. For a lifting $g \in \mathcal{M}(\mathbb{P}^{n+1} - 0)$ of f we have $g(\lambda z) = g(z)$. Extend g meromorphically to $0 \in \mathbb{C}^{n+1}$ and write $g = \frac{u}{v}$, $u = \sum u_\nu$, $v = \sum v_\nu$, where u_ν, v_ν are homogeneous polynomials of degree ν. Then $u(\lambda z) \cdot v(z) = u(z) \cdot v(\lambda z)$ yields $\sum v u_\nu \lambda^\nu = \sum u v_\nu \lambda^\nu$; thus $g = u_k/v_k$. $\square$

Remmert nostaligically recalls his first steps in the world of mathematics in his full of humor style:

'My first problem – given to me by Karl Stein - was to understand the Chow theorem that every analytic set A in $\mathbb{P}^n(\mathbb{C})$ is algebraic (1949). This is - in today's language a GAGA-theorem' (that is, that analytic objects in $\mathbb{P}^n$ are algebraic; the 'GAGA' stems from the first letters of the classical paper of J.P. Serre *Géométrie algébraique et géométrie analitique*, Ann. Inst. Fourier **6**, (1955-6), 1-42. K.M.)

'Can we generalize the preceding arguments (that is, of Riemann and Hurwitz–Weierstrass K.M.) to get Chow's theorem?' asks Remmert. He answers in affirmative.

'Pass from $A \subset \mathbb{P}^n$ to its analytic cone $B := C(A) \subset \mathbb{C}^{n+1} - \{0\}$ of dimension greater than 0. Pass from the topological closure $\bar{B} = B \cup \{0\}$, be courageous, and believe that $\bar{B}$ is analytic at 0. Then in a ball $\mathcal{U}$ around

0 we have $f_1, \ldots, f_k \in \mathcal{O}(\mathcal{U})$ such that

$$\bar{B} \cap \mathcal{U} = \{w \in \mathcal{U} : f_1(w) = \cdots = f_k(w) = 0\}.$$

Now $\lambda w \in \bar{B} \cap \mathcal{U}$ for all $w \in \bar{B} \cap \mathcal{U}$ and all λ with $|\lambda| < 1$ (cone property.) Then it follows that if $\sum p_{kj}$ is the extension of f_k into homogeneous polynomials (of degree j, K.M.) $\bar{B} \cap \mathcal{U} = \{w \in \mathcal{U} : p_{kj} = 0\}$. Therefore

$$\bar{B} = \{w \in \mathbb{C}^{n+1} : p_{kj} = 0\}.$$

Since the polynomial ring is Noetherian, we are done. □

The preceding argument depends crucially on the fact that $\bar{B}$ is analytic at the origin.'

This is a special case of the following general and rather powerful

Theorem on removable singularities (Remmert–Stein, 1952/3). *Let S be an analytic set of dimension $< k$ in complex space X, and let A be an analytic set in $X - S$ of dimension $\geqslant k$ everywhere. Then the topological closure $\bar{A}$ of A in X is an analytic set in X.*

From this theorem the original proof of Remmert proper mapping theorem follows. In 1954 Remmert gets an insight into the structure of the fields of meromorphic functions on complex spaces.

Theorem (Remmert, 1954). *Let X be a compact, connected,* normal *complex space. Then analytically dependent meromorphic functions on X are always algebraically dependent. The field $\mathcal{M}(X)$ is a finite extension of the field $\mathbb{C}(f_1, \ldots, f_n)$, where $f_k \in \mathcal{M}(X)$ and $n \leqslant \dim X$. Thus* $\operatorname{tr}\deg_{\mathbb{C}} \leqslant \dim X$.

This theorem has a long and dramatic history. The theorem for X being a complex torus was *stated* by Weierstrass in 1869 but with gaps in the proof. The same holds for a Poincaré paper. The first complete proof for X being a *manifold* was obtained in 1939 independently by W. Thimm in his Dissertation and, for compact connected manifold X, by C.L. Siegel (Math. Ann. **116** (1939), 617–657.) Siegel gave yet another proof in his mimeographed Princeton lectures (1948-49) *Analytic Functions of Several Complex Variables*, and the last, quite elementary, short proof (using Schwarz lemma) in Nachr. Akad. Wiss. Göttingen, Math. Phys. Ia, 1955, 71–77. The Remmert's generalization of Siegel–Thimm theorem appeared in Math. Ann. **132** (1956), 277–288 and makes use of Chow theorem.

The number $a(X) = \operatorname{tr}\deg_{\mathbb{C}} \mathcal{M}(X)$ is called the *algebraic dimension* of X and thus Remmert–Siegel–Thimm theorem asserts that the algebraic dimension of X, $a(X) \leqslant \dim X$. Boris Moisheson, a student of Shafarevich, in his famous gigantic paper of 1966 (the English translation *On n dimensional compact varieties with n algebraically independent meromorphic functions*, Am. Math. Soc. Trans. II Ser. **63** (1967), 51–177) investigates compact connected varieties X for which $a(X) = \dim X$. Such spaces were called by Michael Artin *Moisheson spaces.* We know that Kähler compact manifold is algebraic if and only if it is Hodge: the fundamental form $\operatorname{Im} h \equiv \omega$, $[\omega] \in H^2(X, \mathbb{R})$ – this is the famous Kodaira theorem. Grauert extended this theorem to general normal Hodge spaces.

THEOREM (GRAUERT, 1960). *Every normal Hodge space is projectively algebraic.*

We owe Moisheson another characterization.

THEOREM (MOISHESON, 1966). *A compact Kähler manifold X is algebraic if and only if $a(X) = \dim X$, that is, if it is a Moisheson space.*

CHAPTER 8

Construction of Complex Spaces by Gluing and by Taking Quotient

The appearance of complex spaces (with singularities) was necessary: we know that if on a complex manifold X acts a (complex) Lie group G with fixed points, then the quotient space, the space of G orbits X/G has singular points (corresponding to the fixed points of G action.) The spaces playing a role in physics and geometry are very often spaces of orbits. This fact has a very deep philosophical meaning: In his famous remark Hermann Weyl says 'objective is what is invariant with respect to some automorphism group.'

And thus in the everyday life, as well as in philosophy, events are considered 'objective', that is, 'true' if they are invariant with respect to permutations of some collection of persons, for example, representing some culture or civilization. The rare virtue of tolerance is based on this deep observation. For this reason there is a truth in the statement of Oswald Spengler (from his classic book *Untergang des Abendlandes*) 'every civilization has its own music, arts, mathematics, and physics.' And that for Plato and Aristotle our physics would be a curiosity without deeper meaning.

Theory of Relativity and mathematical theory of invariants grew out from the deep longing and the quest for such invariant, objective events, for example, invariant with respect to time translations, thus being eternal, everlasting.

On the other hand, we know that the majority of mathematical notions and objects is obtained 'by abstraction', that is, by decomposition into equivalence classes. In the world of X 'symmetric' means the invariance property with respect to some subgroup of the automorphism group $\mathrm{Aut}(X)$. In turn, the very notion of a group appeared 'by abstraction' of the space of automorphisms.

Quotients X/R. The first general investigations of the quotient spaces X/R (or $X/\sim$), where R (or $\sim$) is an equivalence relation on a reduced complex space X, for example, a complex manifold, we owe Henri Cartan (introductory lecture at his Bombay International Colloquium, January 1960, *Quotient and Complex Spaces*.) Let X be a reduced space, R and equivalence relation on X, X/R the quotient space relative to this relation, $\pi : X \to X/R$ a natural projection, and let X/R be endowed with natural quotient topology. On X/R we consider the presheaf obtained by associating with every open set $\mathcal{U} \subset X/R$ the algebra $\mathcal{F}_{\mathcal{U}}$ of all continuous functions $\varphi : \mathcal{U} \to \mathbb{C}$ with $\varphi \circ \pi$ holomorphic on $\pi^{-1}(\mathcal{U}) \subset X$. The corresponding sheaf of $\mathbb{C}$ algebras $\mathcal{O}_X/R$ turns X/R into a $\mathbb{C}$ ringed space $(X/R, \mathcal{O}_X/R)$. The question is when this space is a complex space? In general, the preceding natural construction does not lead to a complex space.

EXAMPLE. $X = \mathbb{C}^2$. The relation R is given by the map $f : X \to \mathbb{C}^2$, $f(z_1, z_2) := (z_1, z_1 z_2)$. Clearly $\mathbb{C}^2/R$ (denoted also by $\mathbb{C}^2/f$) is not locally compact at $\pi(0)$ and we cannot endow $\mathbb{C}^2/f$ with complex structure.

We have therefore to work with a more restrictive class of equivalence relations; these were introduced by Bourbaki (clearly, H. Cartan!)

DEFINITION. An equivalence relation R on the space X is *proper* if X/R is locally compact and $\pi : X \to X/R$ is proper.

We have the following, fundamental

THEOREM (H. CARTAN, 1957). *Let X be a reduced complex space (for example, a complex manifold.) The $\mathbb{C}$ ringed space $(X/R, \mathcal{O}_X/R)$ is complex if and only if every point in X/R has a neighborhood $\mathcal{U}$ separated by functions φ in the algebra $\mathcal{F}_{|}cU$, that is, for $y_1 \neq y_2 \in \mathcal{U}$ there is such φ that $\varphi(y_1) \neq \varphi(y_2)$.*

We have

COROLLARY 1 (H. CARTAN, 1957). *Let G be a subgroup of* Aut(X) *acting properly discontinuously on the reduced space X, that is, for every compact $K \subset X$ there exist only finitely many $g \in G$ such that*

$g(K) \cap K \neq \emptyset$. Then $(X/R, \mathcal{O}_X/R)$ is a reduced complex space and the projection $\pi : X \to X/G$ is holomorphic.

We see therefore that the *moduli space of Riemann surfaces of genus $p > 1$ is a complex space* (of dimension $3p - 3$). This follows from the fact that the moduli space $\mathcal{M}_p = \mathcal{T}_p/\Gamma_p$ is the quotient of the Teichmüller $\mathcal{T}_p$ (which is a Stein manifold of dimension $3p - 3$) by the Teichmüller modular group Γ_p. We recall that Γ_p (for $p > 2$) can be regarded as a group of isometries of $\mathcal{T}_p$ (equipped with a Kähler metric – the Petersson–Weil metric.) For $p > 2$, the moduli space is even a normal complex space and the branch locus of $\mathcal{T}_p \to \mathcal{M}_p$ is the set of singularities of $\mathcal{M}_p$. Locally the moduli space $\mathcal{M}_p$, for $p > 2$, can be identified with an open subset of $\mathbb{C}^{3p-3}$ divided by the *finite* group of automorphisms $\mathcal{G}_p$. (We will reconsider the Teichmüller theory from the general point of view later.)

COROLLARY 2. *Let $f : X \to Y$ be a holomorphic proper map into* reduced *space Y and let R be an equivalence relation defined by (fibers of) f. Then X/f is a complex space and by Remmert theorem, $f(X)$ is an analytic set in Y. The natural bijective map $X/f \to f(X)$ is holomorphic, and if $f(X)$ is normal, then it is an isomorphism.*

EXERCISE 1. Consider in this context the Kummer surface (and the Wirtinger surface.)

EXERCISE 2. From this general perspective reconsider the Klein singularities and binary groups of platonic bodies.

8.1 Construction of complex spaces by gluing

Riemann's dissertation (1851) was also a source of the fundamental method of construction of important mathematical objects from the local ones: this was the gluing device which we exercised several times, for example in construction of fiber bundles. The general procedure is as follows:

DEFINITION. Let M be a topological ('base') space and let an open covering $\{\mathcal{V}_i\}$, $i \in I$ of X be given. A family $\{X_i, f_i\}$ labeled by I of topological spaces with continuous maps $f_i : X_i \to \mathcal{V}_i$ (an 'atlas') is said to be *topologically glued* over $\{\mathcal{V}_i\}$ if for every pair $i, j \in I$ there is a topological map g_{ij} from $X_{ij} := f_j^{-1}(\mathcal{V}_i \cap \mathcal{V}_j) \subset X_j$ onto from $X_{ji} := f_i^{-1}(\mathcal{V}_i \cap \mathcal{V}_j) \subset X_i$ satisfying the *gluing condition*

$$(*) \qquad f_j = f_i \circ g_{ij} \quad \text{on } X_{ij}, \quad g_{ik} = g_{ij} \circ g_{jk} \quad \text{for all } i, j, k \in I.$$

The following theorem is almost obvious

THEOREM. *There exists a unique (up to isomorphism) topological space X, a continuous map $F : X \to M$, and a homomorphism $g_i : f^{-1}(\mathcal{V}_i) \to X_i$ such that for all $i, j \in I$, $f = f_i \circ g_i$ on $f^{-1}(\mathcal{V}_i)$ and $g_{ij} = g_i \circ g_j^{-1}$ on $f^{-1}(\mathcal{V}_i \cap \mathcal{V}_j)$. X is a quotient of the disjoint union $\coprod X_i$ by the equivalence relation $\sim$ given by the definition: $x \in X_i$ and $y \in X_j$ are $\sim$ equivalent if $x = g_{ij}(y)$.*

The proof is identical to the construction of a fiber bundle from transition maps.

REMARK 1. If X_i are sheaves over $\mathcal{V}_i$ and g_{ij} is a sheaf isomorphism, then X is a sheaf over M.

REMARK 2. If the objects are differentiable (complex, holomorphic, ...), we obtain differentiable (complex, holomorphic, ...) gluing.

Here we are mostly interested in the complex case. Therefore we have the following.

PROPOSITION. *Let X_i, M be complex spaces, all f_i be holomorphic, and all g_{ij} biholomorphic. Then for the holomorphically glued family (X_i, f_i, g_{ij}) there exists (up to isomorphism) a unique complex space X, a holomorphic map $X \to M$, $g_i : f^{-1}(\mathcal{V}_i) \to X_i$ such that $f = f_i \circ g_i$ on $f^{-1}(\mathcal{V}_i)$ and $g_{ij} = g_i \circ g_j^{-1}$ on $f^{-1}(\mathcal{V}_i \cap \mathcal{V}_j)$.*

8.2 On deformations of regular families of complex structures (Grauert theory)

As we know, the Riemann problem of moduli (for compact Riemann surfaces) was a mighty impulse for development of such vast theories as Teichmüller theory and moduli theory of stable fiber bundles. In turn, the Teichmüller theory influences the development of tremendous theory of deformations of complex structures initiated by Kodaira and Spencer. One of the most impressive achievements of this theory is the very difficult theorem of Kuranishi on the existence of (versal) deformations for any compact complex manifold, obtained with the help of a very difficult theory of elliptic systems of partial differential equations.

Soon it was realized that in the theory of deformations of complex structures the Grauert's direct image theorem was indispensable. On the other hand, Grauert has himself became interested in the problems of deformations of complex structures already in 1958, when, during his stay at the Institute of Advanced Studies in Princeton he was, along with Andreotti, Nirenberg, Weil, and others an active participant of the seminar led by Kodaira and Spencer. At the time the gigantic papers of Kodaira and Spencer were being written. It seems that the coherence of the direct image coherent sheaf was proved for the need of moduli theory of complex structures. It was observed already at that time that it is necessary to operate with complex spaces with singularities and nilpotent elements.

In the years 1962-64 the theory of deformations of complex structures was enriched by famous Kuranishi theorems. These theorems were based on the methods of potential theory, that is, on the theory of elliptic partial differential equations. But, as we noted above, the necessity to operate with general complex spaces has led to another methods which were developed by Grauert. Eventually, in 1972-74 Grauert was able to show not only the far reaching (and necessary!) generalization of the Kuranishi theorem, but also to enrich the theory in a substantial way. These results were published in the fundamental and, understandably, very difficult paper *Der Satz von Kuranishi für kompakte komplexe Räume*, Inven. Math. **25** (1974), 107-142. Similar results were obtained (by completely another methods) by Duady, Forster–Knorr, Palamadov, and others.

A fundamental notion is the notion of

Regular families, which was, in principle, introduced already by Teichmüller in his fundamental papers on moduli theory for Riemann surfaces. Let us formulate this notion in the case of complex manifolds.

DEFINITION. Let X, Y be connected, n and m dimensional manifolds and let $f : X \to Y$ be a proper, holomorphic, and regular map of X onto Y, that is, f has rank $m = \dim Y$ everywhere. Such f is called a *regular family.*

Since f is proper, the fibers $X_y : f^{-1}(y)$, $y \in Y$ are compact, complex submanifolds of X of dimension $n-m$; thus $\{X_y, y \in Y\}$ is an induced family of compact submanifolds of X of the same dimension $d := \dim X - \dim Y$. If Y is homeomorphic to a ball, the fiber space $f : X \to Y$ is differentiably trivial: there exists a compact $2d$ dimensional *real* analytic manifold F and a fiber preserving diffeomorphism $X \overset{\sim}{\to} F \times Y$. If we identify X_y with $F \times y$, all X_y are diffeomorphic, but the complex structure on X_y depends on $y \in Y$. Therefore, *the regular family $f : X \to Y$ is a deformation of complex structure on F.*

Let now $E \to X$ be holomorphic vector bundles of rank r on X. The restrictions $E_y := E|X_y$ are then holomorphic vector bundles of rank r on X_y, $y \in Y$. Denote by $\underline{E}_y$ the sheaf of local holomorphic sections of E_y. We have very natural

PROBLEM. How the cohomology groups $H^i(X_y, \underline{E}_y)$ depend on $y \in Y$?

Recall now the notion of (higher) direct image $R^i f_*(\mathcal{F})$ which we will denote for short $f_{(i)}(\mathcal{F})$ (thus $f_{(0)}(\mathcal{F}) = f_*(\mathcal{F})$), defined by the presheaf $\mathcal{U} \to H^i(f^{-1}(\mathcal{U}), \mathcal{F})$. There is a very important sub-vector bundle $\Theta \subset TX$ (of the tangent bundle TX) introduced by Kodaira and Spencer: Θ is formed by vectors tangent to X pointing at the direction of the fibers X_y.

Since TX has rank n, $\Theta \to X$ has rank $d = n - m$. $T_Y = T/\Theta$ ($T \equiv TX$), and we have an exact sequence of the corresponding sheaves (of local holomorphic sections)

$$0 \to \Theta \to T \to T/\Theta \to 0$$

which gives a long exact sequence of direct images ($\underline{T/\Theta} = \underline{T}/\underline{\Theta} = \underline{T}_Y \simeq$ tangent sheaf of Y)

$$(*) \qquad 0 \to f_*(\underline{\Theta}) \to f_*(\underline{T}) \overset{\rho}{\to} \underline{T}_Y \to f_{(1)}(\Theta) \to \cdots$$

The map ρ above is called the Kodaira–Spencer map (we have encountered the sheaf $\underline{\Theta}$ and the map ρ already while discussing the Teichmüller theory.) All holomorphic tangent fields over (small) neighborhoods $\mathcal{U}(y)$, $y \in Y$ have inverse images in $I(f^{-1}(\mathcal{U})$ if and only if $\rho = 0$. We have thus the following

PROPOSITION (KODAIRA–SPENCER, 1958). *The holomorphic fiber space $X \to Y$ is locally trivial if and only if $\rho = 0$.*

If the family $f : X \to Y$ is locally trivial, then all fibers X_y, $Y \in Y$ are biholomorphically equivalent. Karl Stein suggested, and Grauert knew already in 1960 that the converse is true:

THEOREM (GRAUERT–W. FISHER, 1964-5). *If all fibers X_y, $y \in Y$ of the regular family $f : X \to Y$ are biholomorphic, then the family (X, f, Y) is locally trivial.*

We recall that the topology of Y whose all closed sets are analytic sets is called the *analytic Zariski topology.* The ordinary topology is obviously much finer. Therefore the following semicontinuity theorem of Grauert is much stronger than the corresponding Kodaira–Spencer result.

GRAUERT SEMICONTINUITY THEOREM. *Let $f : X \to Y$ be a regular family and let $E \to X$ be a (holomorphic) vector bundle on X. Then the functions $Y \ni y \to \dim_{\mathbb{C}} H^i(X_y, \underline{E}_y) \to \mathbb{R}$ are upper semicontinuous with respect to the analytic Zariski topology, that is, for every $y_0 \in Y$ there exists such neighborhood $\mathcal{U}(y_0)$ that $\dim_{\mathbb{C}} H^i(X_y, \underline{E}_y) \leqslant \dim_{\mathbb{C}} H^i(X_{y_0}, \underline{E}_{y_0})$ for $y \in \mathcal{U}(y_0)$.*

REMARK. Oskar Zariski introduced his topology in *algebraic geometry*: closed sets were *algebraic* subsets.

Another important result of Grauert is the following

THEOREM. *If $y \to \dim_{\mathbb{C}} H^i(X_y, \underline{E}_y)$ is independent of y, then all sheaves $f_{(i)}(\underline{E})$ are locally free (that is, they correspond to vector bundles), and all maps*

$$f_{y,i} : f_{(i)}(\underline{E}/\mathfrak{m}_y f_{(i)}(\underline{E})) \to \dim_{\mathbb{C}} H^i(X_y, \underline{E}_y)$$

are isomorphisms. Here $\mathfrak{m}_y \subset \mathcal{O}_y$ is the coherent sheaf of germs of holomorphic functions vanishing at y.

REMARK. Both preceding theorems are much stronger than the corresponding Kodaira–Spencer results. Grauert was able to prove them for general complex spaces X and Y. But, since on spaces with singular points the regularity of $f : X \to Y$ does not make sense, one has to replace this notion by more general one which was introduced in algebraic context by Grothendieck:

Flatness (in complex analysis). We denote by R a commutative ring with unit element. An R module M is *flat over* R if for any injective homomorphism $N_1 \to N_2$ of R modules the induced homomorphism $N_1 \otimes_R M \to N_2 \otimes_R M$ is injective.

PROPOSITION. *(An R module M is flat) $\Longleftrightarrow$ (For any finitely generated ideal $I \subset R$ the map $I \otimes M \to M$ is injective).*

DEFINITION. Let X, Y be complex spaces. A holomorphic map $f : X \to Y$ is *flat at* x if $\mathcal{O}_{X,x}$ is a flat $\mathcal{O}_{Y,f(x)}$-module. The map f is flat if f is flat everywhere (on X).

Let $\mathcal{F}$ be a coherent sheaf on X; $\mathcal{F}$ is f flat at $x \in X$ if $\mathcal{F}_x$ is a flat $\mathcal{O}_{Y,f(x)}$-module. $\mathcal{F}$ is f-flat if it is f-flat everywhere.

A coherent $\mathcal{O}_X$ module $\mathcal{F}$ is *flat over* X if it is id-flat.

PROPOSITION (KAUP–KERNER). *Let X, Y be complex manifolds, then holomorphic, regular $f : X \to Y$ is flat. If f is open, then f is flat.*

PROPOSITION (DUADY). *Let $f : X \to Y$ be a finite holomorphic map of complex space X and Y. Then*

(f is flat) $\Longleftrightarrow$ ($f_*(\mathcal{O}_X)$ is locally free).

Let us pause for a moment to remark that there are three possibilities how to introduce the notion of dimension at $x \in X$, which is denoted by $\dim_x X$:

1. *topological* (Urysohn–Menger), as the dimension m of the topological space X at x;

2. *algebraical* (Krull), as the smallest number d of germs $f_{1x}, \ldots, f_{dx} \in \mathfrak{m}_x$ such that the ring $\mathcal{O}_x | f_{1x}, \ldots, f_{dx}) \mathcal{Q}_x$ is a finite dimensional $\mathbb{C}$ vector space;

3. *analytical* (Chevalley), as the smallest integer n such that n holomorphic have x as an isolated zero.

It is satisfactory that all three these definitions coincide: $m = d = n$. Fundamental – in this context – is the following

THEOREM (KRULL). *The dimension of $\mathcal{O}_x$ is the largest integer d such that there exists a chain $\mathfrak{p}_0 \subsetneqq \mathfrak{p}_1 \subsetneqq \cdots \subsetneqq \mathfrak{p}_d$ of prime ideals in $\mathcal{O}_x$. The dimension is invariant with respect to finite, holomorphic, and open maps*

$f : X \to Y$.

After this remark let us return to flat maps.

COROLLARY. *If $f : X \to Y$ is flat, then*
1. $\dim_x X = \dim_{f(x)} Y + \dim_x X_{f(x)}$, *where* $X_{f(x)} := f^{-1}(f(x))$;
2. f is open.

Now we can formulate the general semicontinuity theorem of Grauert. This theorem is a natural extension of the preceding theorem. We have to replace the sheaf E with a general coherent sheaf on X, and regularity of the map $f : X \to Y$ with flatness:

THEOREM (GRAUERT, 1960). *Let $f : X \to Y$ be a surjective, holomorphic, proper, and flat map. Let $\mathcal{F}$ be a f-flat sheaf on Y. Then*

(a) For any $i \in \mathbb{N}$ the function $y \to H^i(X_y, \mathcal{F}|f^{-1}(y))$ is upper semicontinuous;

(b) If the canonical restriction $f_{(i)}(\mathcal{F})_y \to f_{(i)}(\mathcal{F}/\mathfrak{m}_y, \mathcal{F})_y$, then the function $y \to H^i(X_y, \mathcal{F}|X_y)$ is locally constant. The converse holds if Y is reduced.

(c) If Y is reduced, then the Euler characteristic

$$\chi(y) := \sum_i (-1)^i \dim_{\mathbb{C}} H^i(X_y, \mathcal{F}_y)$$

is locally constant and independent of y.

We saw that every regular family $f : X \to Y$ of complex spaces defines a deformation of a complex structure (of X_{y_0}). Two very important problems arise:

1. Existence of a regular family containing a given compact complex space X_0. Such a deformation should be, in some sense, complete and unique.

2. Such a deformation would be effective if the derivative dg at distinguished points could be uniquely determined.

This program is very ambitious but inevitable if the theory of deformations of complex structures is to satisfy our expectations. The program was partially analyzed by Kuranishi, and definitely completed by Grauert in 1974. But before describing their results we must first define some notions and fix the notation.

8.3 Grauert solution of main problems of deformation theory of complex structures

Let us start with notation. Let X be a (fixed) compact complex space and let (Y, y_0) be a complex space with distinguished point y_0. A (holomorphic) deformation of X over (Y, y_0) consists of a regular family

$$f : \hat{X} \to Y$$

(f is, therefore, proper and flat) along with an isomorphism $X \to \hat{X}_{y_0}$. A holomorphic map $g : (Y', y_0') \to (Y, y_0)$ which is defined by the family $f' : \hat{X}' := \hat{X} \times_Y Y' \to Y'$ together with the induced isomorphism between X and the fiber $\hat{X}'_{y_0}$. Such g is called *base change.* An isomorphism of deformations is the isomorphism of regular families which induces the identity on X.

DEFINITION. A holomorphic deformation of manifold X is *complete* if it generates all deformations by means of the base change g up to an isomorphism after replacing the base spaces (Y, y_0), (Y', y_0') by neighborhoods of the distinguished points y_0, y_0', if necessary.

A complete deformation is *versal* if the differential $dg(y_0')$ at the distinguished point y_0' is uniquely determined. Therefore, *versal deformations of X are, up to an isomorphism, uniquely determined.*

THE MAIN THEOREM OF DEFORMATION THEORY (GRAUERT, 1974). *Every compact complex space X has a holomorphic deformation $(\hat{X}, f, Y, y_0)$ which is versal at y_0 and complete at all points y of the base Y.*

This generalization of the Kuranishi theorem together with the semicontinuity theorem of Grauert solves all the pivotal problems of deformation theory.

REMARK. If $f : \hat{X} \to Y$ and $g : Y' \to Y$, then

$$\hat{X}' := \hat{X} \times_Y Y' := \{(x, y') \in \hat{X} \times Y' : f(x) = g(y')\}$$

$$f' : \hat{X}' \to Y', \quad \hat{g} : \hat{X}' \to \hat{X}$$
$$(x, y') \to y', \quad (x, y') \to x$$

Therefore $f' : \hat{X}' \to Y'$ is a family obtained by the base change. The fiber $f'^{-1}(y')$ is identified with $f^{-1}(g(y'))$.

8.4 On differential calculus on complex spaces

In analysis and geometry of differentiable manifolds the decisive role is played by differential forms (or vector fields) being sections of the cotangent bundle T^*X (resp. tangent bundle TX.) In complex analysis differential forms (Wirtinger calculus) play perhaps more important role than vector fields. In Part II we showed that the cotangent space T^*X_a at a point $a \in X$ can be defined algebraically as $\mathfrak{m}_a/\mathfrak{m}_a^2$, where $\mathfrak{m}_a$ is a maximal ideal in the local algebra C_a^∞ of function germs vanishing at a, and $\mathfrak{m}_a^2 \subset \mathfrak{m}_a$ are function germs which vanish at a up to second order.

This algebraic definition of the cotangent space is called *Zariski cotangent space* and can be applied in the case of complex spaces with singularities.

PHILOSOPHICAL REMARK. It is very interesting to note that in mathematics the *co-notions* play a more fundamental role than the notions themselves: they are more flexible than the 'notions' and have better functional properties. For example, cohomology groups $H^i(X, \mathcal{F})$ are more important than the homology groups; differential forms behave nicely with respect to the maps $f : X \to Y$ (the pullback f^*.)

In complex analysis the sheaf Ω^1_X of germs of holomorphic one forms is most important. Ω^1_X can be defined for every complex space $(X, \mathcal{O}_X)$. Let V be a model space in domain $D \subset \mathbb{C}^n$ with the ideal $\mathcal{I}_V = \mathcal{I}$. The map $\mathcal{I} \to \Omega^1_D$, $f \to df$ sends $\mathcal{I}^2$ into $\mathcal{I}\Omega^1_D$, and passing to residue classes we obtain a (homo) morphism $\alpha : \mathcal{I}/\mathcal{I}^2 \to \Omega^1_D/\mathcal{I}\Omega^1_D$. Thus we obtain a coherent sheaf $\Omega^1_V := \operatorname{coker} \alpha$ on V. The general case is obtained by gluing: let $\{\mathcal{U}_i\}$ be an open covering of X such that there is a biholomorphic map $f_i : \mathcal{U}_i \xrightarrow{\sim} V_i$ onto the model space V_i. Therefore the sheaves $\Omega^1_{V_i}$ and $\Omega^1_i := \Omega^1_{\mathcal{U}_i}$ are well defined. The isomorphisms $f_i^{-1} \circ f_j : \mathcal{U}_i \cap \mathcal{U}_j \to \mathcal{U}_i \cap \mathcal{U}_j$ define by differentiation and passing to quotients the transition maps $g_{ij} : \Omega^1_i | \mathcal{U}_i \cap \mathcal{U}_j \to \Omega^1_j | \mathcal{U}_i \cap \mathcal{U}_j$ with $g_{ik} g_{kj} = g_{ij}$. Thus we have an $\mathcal{O}_X$-sheaf atlas, and therefore, an $\mathcal{O}_X$-sheaf on X such that $\Omega^1_X \simeq \Omega_{\mathcal{U}_i}$.

DEFINITION. The sheaf Ω^1_X on X is the *cotangent sheaf* of X or the *sheaf of holomorphic* 1 *forms* on X.

The tangent sheaf $\mathcal{T}_X$ of a complex space is defined as a dual sheaf of Ω^1_X:

$$\mathcal{T}_X := \operatorname*{Hom}_{\mathcal{O}_X}(\Omega^1_X, \mathcal{O}_X).$$

The sections $s \in \Gamma(\mathcal{T}_X)$ of the sheaf $\mathcal{T}_X$ are called *holomorphic vector fields* on X. The $\mathbb{C}$-vector space $\mathcal{T}_{X,x}/\mathfrak{m}_x\mathcal{T}_{X,x} := T_xX$ is the tangent space of X. Here $\mathfrak{m}_x$ is the ideal corresponding to the embedding $\{x\} \to X$ of the point $x \in X$, and we have the exact sequence of sheaves:

$$\mathfrak{m}_x/\mathfrak{m}_x^2 \xrightarrow{\alpha} \Omega^1_{X,x}/\mathfrak{m}_x\Omega^1_{X,x} \to \Omega^1_{\{x\}} \to 0.$$

LEMMA. *The map α is bijective.*

PROOF.
(α is bijective) $\Longleftrightarrow$($\alpha^* : \mathrm{Hom}_{\mathbb{C}}(\Omega^1_{X,x}/\mathfrak{m}_x\Omega^1_{X,x}) \to \mathrm{Hom}_{\mathbb{C}}(\mathfrak{m}_x/\mathfrak{m}_x^2, \mathbb{C})$. We see that at every *smooth point* $x \in X$ the sheaf Ω^1_X is locally free and $\mathrm{rank}_x\Omega^1_X := \dim_{\mathbb{C}} \mathfrak{m}_x/\mathfrak{m}_x^2 = \dim_x X$. Thus

$$(x \text{ is a singular point}) \Longleftrightarrow (\mathrm{rank}_x\Omega^1_X > \dim_x X).$$

Let $f : X \to Y$ be a holomorphic map of complex spaces. In the regular case, if X and Y are manifolds, the differential $df : f^*\Omega^1_Y \to \Omega^1_X$ defines $\Omega^1_{X/Y} := \mathrm{coker}\, df$ – the sheaf of relative one forms with respect to f. Now, in the singular case, since the notion is a local one, assume that X and Y are model spaces $W \subset \mathbb{C}^m$, $V \subset \mathbb{C}^n$ and that f can be lifted to a holomorphic map $\hat{f} : W \to \mathbb{C}^n$. Then for a local holomorphic one form ω, $df([\omega]) := [d\hat{f}(\omega)]$, where [ ] denotes the residue class (cf. the definition of the map $\alpha : \mathcal{I}/\mathcal{I}^2 \to \Omega^1_D/\mathcal{I}\Omega_D$). Thus we have the following

DEFINITION. A holomorphic map $f : X \to Y$ defines the $\mathcal{O}_X$-sheaf

$$\Omega^1_{X/Y} := \mathrm{coker}(df : f^*\Omega^1_Y \to \Omega^1_X) \tag{1}$$

the *sheaf of germs of relative* 1 *forms with respect to* f (or forms along the fibers of f.)

The dual sheaf

$$\mathcal{T}_{X/Y} := \mathrm{Hom}_{\mathcal{O}_X}(\Omega^1_{X/Y}, \mathcal{O}_X) \tag{2}$$

is called the *relative tangent sheaf of* f.

From these definitions we have the exact sequence

$$f^*\Omega^1_Y \xrightarrow{df} \Omega^1_X \to \Omega^1_{X/Y} \to 0. \tag{3}$$

If Y is a closed complex subspace of X with the defining ideal $\mathcal{I} \subset \mathcal{O}_X$, then the coherent $\mathcal{O}_Y$-sheaves $\mathcal{N}^*_{X/Y} := (\mathcal{I}/\mathcal{I}^2)|Y$ and $\mathcal{N}_{X/Y} := \mathrm{Hom}_{\mathcal{O}_Y}(\mathcal{N}^*_{X/Y}, \mathcal{O}_Y)$ are called the conormal (resp. normal) sheaves of Y in X.

PROPOSITION. *Let $\Omega^1_X|Y := \Omega^1_X/\mathcal{I}\Omega^1_X|Y$, then we have the exact sequence of $\mathcal{O}_Y$ sheaves:*

$$\mathcal{N}^*_{X/Y} \stackrel{\alpha}{\to} \Omega^1_X|Y \to \Omega^1_X \to 0. \tag{4}$$

If Y is a submanifold of manifold X, we have the dual version of (4): there is an exact sequence

$$0 \to \mathcal{O}_Y \to \mathcal{T}_X|Y \to \mathcal{N}_{X/Y} \to 0, \tag{5}$$

that is, the tangent spaces of X are spanned by tangent vectors of Y and the normal vectors of Y in X.

DEFINITION. A projective variety $Y \subset \mathbb{P}$ of codimension n is a complete intersection if it is an intersection of n hypersurfaces $E_1, \ldots, E_n$, $Y = E_1 \cap \cdots \cap E_n$ that meet transversally at any point of intersection. More generally, if the map α is injective then the closed complex subspace Y of complex manifold X is *locally a complete intersection* if the ideal $\mathcal{I}$ can be generated, locally, by $\operatorname{codim}(Y, X)$ holomorphic functions.

Then $\mathcal{I}/\mathcal{I}^2$ is locally of rank $= \operatorname{codim}(Y, X)$. If, in addition, the space Y is reduced (for example, if Y is a manifold), then the sequence

$$0 \to \mathcal{I}/\mathcal{I}^2 \stackrel{\alpha}{\to} \Omega^1_X|Y \to \mathcal{O}^1_Y \to 0$$

is exact. Indeed $\ker\alpha \equiv 0$ everywhere since $\mathcal{I}/\mathcal{I}^2$ is locally free.

PROPOSITION. *The singular locus $S(X)$ of X is the singular locus $S(\Omega^1_X)$ of the coherent sheaf Ω^1_X, and therefore, $S(X)$ is an analytic set in X. Similarly, if $f : X \to Y$ is holomorphic and flat, then the subsets $S(f) = \{x \in X :$ the fiber $X_{f(x)}$ is not a manifold at $x\}$, $N(f) = \{x \in X :$ the fiber $X_{f(x)}$ is not normal at $x\}$, $R(f) = \{x \in X :$ the fiber $X_{f(x)}$ is not reduced at $x\}$ are all analytic.*

REMARK. Let $f : X \to Y$ be holomorphic; then the exact sequence of sheaves

$$(KS) \qquad 0 \to \mathcal{T}_{X/Y} \to \mathcal{T}_X \to N_{X/Y}$$

is called the *Kodaira–Spencer sequence of f*.

As we know, the moduli problems, that is, the construction and investigation of complex spaces being the isomorphic classes of important geometric or analytic objects are very difficult, but of extreme interest. Could it have been expected that short remark of Riemann would result in such an extraordinary development of geometric complex analysis? In the next chapter we will deal with an important example of moduli spaces for hermitian Einstein connections.

8.5 From Riemann period relations to theorems of Kodaira and Grauert

This long and dramatic story, in which great mathematicians were involved for more then hundred years[1] has began in 1857 with the famous paper of Riemann *Theorie der Abelschen Functionen*, Crelle Journ. **54** (1857); *Riemann Werke* pp. 82-142. In this seminal work Riemann introduces the jacobian torus Jac $(M) = \mathbb{C}^n/\Lambda$ of a surface M of genus p and derives for it his famous bilinear period relations. These relations are formulated for any complex torus V/Λ of p dimensions, where $\Lambda = \Lambda_{2p}$ is a lattice in $V \simeq \mathbb{C}^p$. Riemann relations can be formulated in a geometrical way as follows: On $V \times V$ there exists a positive definite hermitian form $h : V \times V \to \mathbb{C}$ such that its imaginary part Im h restricted to the lattice Λ is integer-valued: $\operatorname{Im} h(\lambda_1, \lambda_2) \in \mathbb{Z}$, $\lambda_1, \lambda_2 \in \Lambda$. Such hermitian form h is called *Riemann form.*

A fundamental problem arises: when complex torus is a projective algebraic variety? Such objects are called *algebraic varieties.* The answer is given by

THEOREM (POINCARÉ, WIRTINGER). *(The torus V/Λ is an abelian variety)* $\Longleftrightarrow$ *(It can be equipped with the Riemann form h).*

[1]Among others, Hermite, Weierstrass, Picard, Frobenius, Poincaré, Wirtinger, Lefschetz, Kähler, Bergmann, Bochner, Weitzenbröck, Kodaira.

Lefschetz solved the above problem in 1921 with the help of Θ functions - the method that originated in works of Riemann. 'But' - as Remmert writes – 'in their core the period relations were not understood until 1954; at that time, Kodaira, using the ideas of Hodge, provided his great theorem that Hodge manifolds are always projectively algebraic'.

The great paper of Grauert *Über Modificationen und exzeptionelle analytische Mengen*, Math. Ann. **146** (1960), 331-362 extends Kodaira theorems to normal spaces and gives even deeper insight into the realm of Riemann ideas.

Before presenting these magnificent theorems, let us recall the fundamental notion of

Positivity (negativity) of hermitian line bundles. Let (F, h) be a hermitian line bundle over complex manifold X. Writing $\hat{h}(\xi)$ for $h(\xi, \xi)$, we regard $\hat{h}$ as a functional on F with the properties

$$\hat{h}(\xi) > 0 \quad \text{for } 0 \neq \xi \in F;$$

$$\hat{h}(c\xi) = |c|^2 \hat{h}(\xi) \quad \text{for } \xi \in F,\ c \in \mathbb{C}. \tag{1}$$

We know that $\hat{h}$ satisfying (1) defines, conversely, a hermitian structure in $F \to X$. From the Chern-Weil theory we know that the first Chern class $c_1(F)$ can be represented by a $(1,1)$ form

$$\omega := \frac{-1}{2\pi i} d'd'' \log H = \frac{-1}{2\pi i} \sum R_{\alpha\bar{\beta}} dz^\alpha \wedge d\bar{z}^\beta, \tag{2}$$

where H is given by $h(s, s)$ for a non vanishing section s of F.

DEFINITION. The bundle (F, h) and its class $c_1(F)$ is *negative* (resp. positive) if the curvature $R_{\alpha\bar{\beta}}$ is negative (resp. positive) definite.

The role of positivity is stressed in the first

KODAIRA THEOREM. *(The bundle $F \to X$ is ample) $\Longleftrightarrow$ ($c_1(F)$ is negative).*

Thus if on the compact complex manifold X there exists a positive line bundle (F, h), then X is a projective algebraic variety.

Let now $N \subset X$ be an analytic set of singular points of the complex (reduced) space X. Since N is nowhere dense in X, we can extend the notions of Kähler and Hodge metrics into complex spaces: The hermitian metric $ds^2 = \sum g_{\alpha\bar{\beta}} dz^\alpha d\bar{z}^\beta$ (with continuous coefficients) in $X - N$ is called Kähler if for every $x \in X$ there exists a neighborhood $\mathcal{U} = \mathcal{U}(x)$ and a strictly plurisubharmonic function p such that $d'd''p = \varphi$ on $\mathcal{U} - N$, where φ is the $(1,1)$ form associated with g, $\varphi = i \sum g_{\alpha\bar{\beta}} dz^\alpha \wedge d\bar{z}^\beta$. This means that there exists the potential p of φ. By de Rham theory φ represents a cohomology class $[\varphi] \in H^2(X, \mathbb{R})$. If X is compact and if the image of $[\varphi] \in H^2(X, \mathbb{Z})$, ds^2 is called Hodge metric, and a space which can be equipped with Hodge metric is called the Hodge space. Thus the *class of periods* of ds^2 is in the image of $H^2(X, \mathbb{Z})$.

The famous Kodaira theorem asserts that (compact) Hodge manifold can be equipped with a positive hermitian line bundle and therefore *every Hodge manifold is a projective algebraic variety.*

Grauert generalizes both Kodaira theorems into arbitrary compact normal complex spaces. To achieve this, he brilliantly extends the notion of positivity (negativity) of a line bundle $F \to X$.

PROPOSITION (GRAUERT). *Let (F, h) be a hermitian line bundle over complex manifold. Then*

((F,h) is negative) $\Longleftrightarrow$ (the function $\hat{h}$ on F is strictly plurisubharmonic outside the zero section $0 = \{0_x\}$ of F).

This means that $c_1(F)$ is negative if and only if the zero section 0 of F has a strongly pseudoconvex neighborhood in F. Such vector bundle over X is called *weakly negative* since this notion is valid for any complex space.

In turn, as discovered by Grauert, this notion is in a very close relation with the notion of *exceptional analytic set* which is fundamental in transcendental algebraic geometry.

DEFINITION. Let X be an analytic space and A a compact subvariety of X. If there exists an analytic space Y and *holomorphic* map $f : X \to Y$ such that $f(A) = \{y_0\}$ and $f : X - A \simeq Y - \{y_0\}$, we say that A is an exceptional subvariety of X and that A is *blown down* (collapsed to y_0).

THEOREM (GRAUERT). *A nowhere discrete compact analytic set A in a complex space X can be blown down if and only if there is a strictly pseudoconvex neighborhood $\mathcal{U}(A) \subset X$ such that A is a maximal compact analytic* *set in $\mathcal{U}(A)$.*

We see now that the following criterion is a generalization of the Kodaira criterion for projectivity:

PROJECTIVITY CRITERION OF GRAUERT. (A reduced complex space X is a projective algebraic variety) $\Longleftrightarrow$(There is a weakly negative vector bundle F over X) $\Longleftrightarrow$(There exists a vector bundle $F \to X$ with zero section which is an exceptional analytic set).

Grauert proves that every normal Hodge space can be equipped with a weakly negative line bundle, and as generalization of the fundamental Kodaira theorem he obtains

THEOREM (GRAUERT). *Every normal Hodge space is projectively algebraic.*

We see that Grauert's characterization of Stein spaces (as spaces equipped with strictly plurisubharmonic functions) is crucial for all this magnificent constructions. Of course, we cannot prove these theorems here. Let us just present the proof of Grauert proposition (after Kobayashi.)

We will prove that the complex Hessian $d'd''\hat{h}$ is positive definite. We fix a point x_0 in X and choose a local section of F such that $H = 1$ and $dH = 0$ at x_0. Then at x_0 $d'd'' \log H = d''d'H$. Let $\mathcal{U}$ be a neighborhood of x, where s is defined and let $F|\mathcal{U} \simeq \mathcal{U} \times \mathbb{C}$ be a trivialization given by s. Hence $(x, c) \in \mathcal{U} \times \mathbb{C}$ corresponds to $c \cdot s(x) \in F_x$ and $\hat{h}(c \cdot s(x)) = H(x)c\bar{c}$. Thus the complex Hessian of $\hat{h}$ at $\xi_0 = (x_0, c_0) \in \mathcal{U} \times \mathbb{C}$ is given by

$$(d'd''\hat{h})_{\xi_0} = (d'd''H)_{x_0} c_0 \bar{c}_0 + H(x_0)(dc \wedge d\bar{c})_{c_0}.$$

Therefore the complex Hessian of $\hat{h}$ is positive definite at ξ_0 if and only if the complex Hessian of H is positive definite at ξ_0. But it follows from (2) that this happens if an only if the curvature $R_{\alpha\bar{\beta}}$ is negative definite at x_0. □

REMARK. The proofs of Grauert theorems are reproduced in the last section of the excellent textbook of Cunning and Rossi, but the classical paper is very lucid and now available in Volume I of 'Selected Papers' of Hans Grauert, pp. 271-308.

8.6 Concluding remarks

There is an interesting relation between negative vector bundles $E \to X$ and Finsler structure investigated by Kobayashi in 1975 (*Negative vector bundles and Finsler structures*, Nagoya Math. J. **57**, 153-166.) Let $\pi : \mathbb{P}(E) \to X$ be the projectivization of the vector bundle $E \to X$. The *tautological line bundle* $L(E)$ over $\mathbb{P}(E)$ is a subbundle of the pullback bundle π^*E. The natural map $\tilde{\pi} : \pi^*E \to E$ restricted to $L(E)$ gives the map $\tilde{\pi} : L(E) \to E$ which is biholomorphic outside the zero sections of $L(E)$ and E and blows down the zero sections of $L(E)$ (identified with $\mathbb{P}(E)$ to the zero section of E (identified with X.)

The vector bundle $E \to X$ is *negative* if the line bundle $L(E)$ is negative. One immediately obtains

PROPOSITION. *(The vector bundle $E \to X$ is negative)* $\Longleftrightarrow$*(The zero section of E has a strongly pseudo convex neighborhood $\mathcal{U}$, that is, there exists a strongly plurisubharmonic function on $\mathcal{U}$); in other words* $\Longleftrightarrow$*(E is weakly negative).*

Following Kobayashi we have now

DEFINITION. A *Finsler structure* f in a vector bundle E is a smooth positive function defined outside of the zero section of E such that

$$f(c\xi) = f(\xi)c\bar{c} \quad \text{for } 0 \neq c \in \mathbb{C}, \ \xi in E.$$

If we put $\hat{h} := f \circ \tilde{\pi}$, then $\hat{h}$ is a function on $L(E)$ satisfying $\hat{h}(\xi) > 0$ for $0 \neq \xi \in E$ and $\hat{h}(c\xi) = |c|^2\hat{h}(\xi)$, $\xi \in E$, $c \in \mathbb{C}$. But conversely, every $\hat{h}$ satisfying the above conditions originates from a Finsler structure in E.

COROLLARY. *There is a natural bijection between hermitian structures in $L(E)$ and Finsler structures in E.*

We obtain the following interesting

PROPOSITION (KOBAYASHI,1975). *A holomorphic vector bundle E is negative if and only if it admits a Finsler structure f which is strictly plurisubharmonic on E outside the zero section.*

REMARK. This proposition is in the truly Riemannian spirit: we know that Riemann introduced in his habilitation thesis the notion of Finsler space (which was, 50 years later investigated by Finsler) and the (plurisub) harmonicity was the property so dear to him.

CHAPTER 9

Differential Geometry of Holomorphic Vector Bundles over Compact Riemann Surfaces and Kähler manifolds. Stable Vector Bundles, Hermite-Einstein Connections, and their Moduli Spaces

To Soshishi Kobayashi, the great master of hermitian geometry

As we know from the Riemann-Roch theorem the topological classification of complex vector bundles $E \to M$ over compact Riemann surface is 'very simple': they are classified by one integer $c_1(E)$, the first Chern class of $E \to M$. The situation is quite different in the case of classification of *holomorphic category.* This could have been expected from our experience with the Teichmüller theory, the main chapter of the Riemann moduli problem.

Reach structures - in the case in hands, holomorphic and hermitian - lead to the new, beautiful, and deep events and relations. To investigate them, the powerful methods of algebra and nonlinear analysis are necessary. Already in the simplest case, the theory of geodesics, we deal with a system of nonlinear ordinary differential equations. We have seen that the theory of *harmonic* maps between Riemann and Kähler manifolds requires nonlinear *partial* differentail equations. All these equations have their sources in variational problems and are solved usually with the help of 'direct methods of calculus of variations' - perhaps the greatest idea of Riemann. These equations are *elliptic*, but as we saw in the Hodge theory of harmonic forms, the powerful method of solving these problems is the method of parabolic equations: the prototype was the heat kernel equation: for the time parameter $t \to \infty$ the temperature u_t has a limit u_∞ being a harmonic map (that is, a hermitian form.)

But let us return to holomorphic vector bundles and their moduli. In order to obtain reasonable moduli spaces, one has to restrict the class of holomorphic objects. As R. Bott characterizes this situation in his lovely talk (Surveys in Differential Geometry **1** (1991), 1-18) ' The primary difficulty of extending the classical results (rank$E = 1$, i.e., line bundles on M) to higher rank bundles $E \to M$ is that the problem immediately becomes infinite dimensional. This comes about because bundles E of rank > 1 will in general have many nontrivial automorphisms, and to properly deal with this situation the Mumford notion of *stability is indispensable.*'

This notion was discovered also by F. Tokamato in 1972, and its simplified definition is

DEFINITION. The holomorphic vector bundle E is *stable* if for any subbundle $F \subset E$

$$\frac{c_1(F)}{\operatorname{rank} F} < \frac{c_1(E)}{\operatorname{rank} E}.$$

The following simple observation illustrates the power of the stability condition.

OBSERVATION. *Let E be a holomorphic vector bundle of rank 2 on compact Riemann surface. If E is stable, then E cannot split (holomorphically).*

PROOF. Let $L \subset E$ be a line subbundle of E. There always exists a line bundle Q such that the sequence

$$0 \to L \to E \to Q \to 0$$

is exact. But then $c_1(E) = c_1(L) + c_1(Q)$. By stability of E, $c_1(L) < \frac{1}{2}c_1(E)$, thus $c_1(Q) > c_1(E)$. Hence Q is not a subbundle of E. □

PROPOSITION. *A stable bundle $E \to M$ with* $\text{rank}E = 2$ *has no nontrivial automorphisms.*

Prove it! Hint: Let $\varphi : E \to E$ be an automorphism. The characteristic polynomial of φ is constant because M is compact. If eigenvalues of φ were different, then φ would split E into two line bundles which is impossible by Observation. Hence eigenvalues of φ are equal $\lambda_1 = \lambda_2 = \lambda$. Therefore $\psi := \varphi - \lambda 1$ is a nilpotent endomorphism $\psi^2 = 0$. If ψ is not identically zero and has constant rank, we have the exact sequence

$$0 \to \ker \psi \to E \to \psi E \to 0.$$

By stability of E $c_1(\ker \psi) < \frac{1}{2}c_1(E)$; therefore this alternative is impossible either. Hence $\psi \equiv 0$, thus $\varphi = \lambda 1$. (The non constant rank is ruled out similarly.)

There is a beautiful characterization of stable bundles due to M.S. Narasimhan (brother of Ragavan Narasimhan) and C.S. Seshardi.

THEOREM (NARASIMHAN-SESHARDI). *A holomorphic vector bundle $E \to M$ over compact Riemann surface is stable if and only if the associated projective bundle $\mathbb{P}(E)$ comes from en* irreducible *representation of the fundamental group $\pi_1(M)$ of M on the projective unitary group $\mathbb{P}U(r)$, where r is the rank of E. In other words*

($E \to M$ is stable) $\Longleftrightarrow$(E admits a flat hermitian structure).

For an adequate notion of stability in the case of Kähler manifolds M of dimension $n > 1$, the notion of analytic coherent sheaf is indispensable (Takemoto, 1972). In 1982 Shoshishi Kobayashi introduced in his seminal paper *Curvature and Stability of Vector Bundles*, Proc. Jap. Acad. **58** (1982), 158-162 a fundamental notion, called by him the *Einstein condition* which is a natural generalization of the famous Kähler-Einstein condition. The latter is the condition on the curvature of the tangent bundle $Ric(g) = ag$, a =const, and g is a Riemann metric on $M = (M, g)$. In the case of Riemann surface the Kähler-Einstein condition means that the surface has constant curvature.

The condition of Kobayashi is now called Hermite–Einstein condition (H-E condition) and is formulated in the following way: Let (M, g) be a Kähler manifold and $E \to M$ a holomorphic vector bundle over M equipped with hermitian structure h. There is a unique holomorphic covariant derivative $\nabla = \nabla^E$ on E.

If R is the curvature of ∇, that is, $R = \nabla \circ \nabla$, then the curvature form $\Omega = (\Omega^i_j)$ with respect to the holomorphic frame field $s \in \mathcal{O}(E)$, $s = (s_1, \ldots, s_r)$, $r = \text{rank}(E)$ is given by

$$R(s_j) = \sum_{i=1}^{r} \Omega^i_j s_i, \quad j = 1, \ldots, r,$$

and

$$\Omega^i_j = \sum_{\alpha\beta} R^i_{j\alpha\beta} dz^\alpha \wedge d\bar{z}^\beta.$$

If g is a hermitian metric on M, we write $g = g_{\alpha\bar{\beta}} dz^\alpha d\bar{z}^\beta$, and if we denote the metric inverse to $g_{\alpha\bar{\beta}}$ by $g^{\alpha\bar{\beta}}$, then $K^i_j := g^{\alpha\bar{\beta}} R^i_{j\alpha\bar{\beta}}$. Therefore $K = (K^i_j)$ defines an endomorphism of the bundle E. We say that the bundle $E = (E, h, M, g)$ is an H-E bundle , or satisfies the H-E condition if

$$(H - E) \qquad K = c\, Id_E, \quad \text{that is,} \quad K^i_j = c\delta^i_j.$$

If we introduce $K_{j\bar{k}} := h_{i\bar{k}} K^i_j$, then we find that hermitian holomorphic vector bundle (E, h, M, g) satisfies $(H - E)$ if $K_{i\bar{j}} = c h_{i\bar{j}}$.

The aim of Kobayashi was to give a differential geometry condition for a holomorphic vector bundle over Kähler manifold to be stable or semi-stable, and the H-E condition is indeed the required one. The fundamental result reads

THEOREM (KOBAYASHI, 1982). *Let (E, h) be a hermitian vector bundle over compact Kähler manifold (M, g). If (E, h, M, g) satisfies the condition $(H - E)$, then E is ω-semi-stable and (E, h) is isomorphic to a direct sum $(E_1, h_1) \oplus \cdots \oplus (E_k, h_k)$ of ω-stable hermitian vector bundles (E_i, h_i) $i = 1, \ldots, k$. ω is a Kähler $(1,1)$ form on M.*

REMARK. The ω-stability was introduces by Takemoto (1972) as follows: Let $\mathcal{F}$ be a coherent sheaf over compact Kähler manifold (M, g) of dimension n. Let $c_1(\mathcal{F})$ be the first Chern class of $\mathcal{F}$ – it is represented by a real, closed $(1,1)$ form on M. If ω is a Kähler form of M, then

$$\deg_\omega(\mathcal{F}) \equiv \deg(\mathcal{F}) := \int_M c_1(\mathcal{F}) \wedge \omega^{n-1} = \int_M c_1(\mathcal{F}) \wedge *\omega.$$

The *degree/rank ratio* $\mu(\mathcal{F})$ (called *slope*) is defined by

$$\mu(\mathcal{F}) := \frac{\deg(\mathcal{F})}{rank(\mathcal{F})}.$$

A sheaf $\mathcal{E}$ is ω-stable if for every coherent subsheaf $\mathcal{F} \subset \mathcal{E}$, with $rank(\mathcal{F}) > 0$, we have

$$\mu(\mathcal{F}) \equiv \frac{\deg(\mathcal{F})}{rank(\mathcal{F})} < \frac{\deg(\mathcal{E})}{rank(\mathcal{E})} \equiv \mu(\mathcal{E})$$

(ω-semistable, if $\mu(\mathcal{F}) \leqslant \mu(\mathcal{E})$.)

If $E \to M$ is a holomorphic vector bundle, then the sheaf $\mathcal{E} := \mathcal{Q}(E)$ of germs of holomorphic sections of E is coherent we talk about ω-stable (semistable) vector bundle.

If $H \to M$ is an ample vector bundle, so that M is a projective algebraic variety and if ω represents the Chern $c_1(H)$, then we say that it is H- stable (or H-semistable) instead of ω-stable (ω-semistable). Since $c_1(H)$ and $c_1(\mathcal{E})$ are integral classes, in this case, the degree or H- degree of $\mathcal{E}$ is an integer.

In the case of Riemann surface M, $\dim_{\mathbb{C}} M = 1$ and $\deg(E) = c_1(E)$ is the Chern number of E, and we have the original definition of Mumford stability.

To appreciate the importance of Narasimhan-Seshardi theorem, we must present some simple facts regarding flat bundles.

9.1 Flat bundles and flat connections

We recall that connection (covariant derivative) ∇ in the C^∞ complex vector bundle $E \to M$ over real manifold M is flat if its curvature $\nabla \circ \nabla = 0$, that is, $R = R^\nabla = 0$. A *flat bundle* $E \to M$ is given by transition maps $\{g_{uv}\}$ which are constant matrices in $\mathrm{GL}(r, \mathbb{C})$, for r being the rank of the bundle E. Clearly, trivial bundles are flat.

If we have a vector bundle $E \to M$ with flat connection ∇, we can construct a flat vector bundle as follows. Let $\pi_1 = \pi_1(M, x_0)$ be the fundamental group of M with the reference point x_0. Since the connection ∇ is flat, the parallel displacement along closed curve c starting at x_0 depends only on the homotopy class $[c]$ of c. Thus one obtains a representation (the ∇ holonomy representation) of $\pi_1(M, x_0)$:

$$\text{(hol)} \qquad \rho : \pi_1(M, x_0) \to \mathrm{GL}(n, \mathbb{C});$$

$\rho(\pi_1)$ is called the *holonomy group of* ∇.

Conversely, given a holonomy representation ρ one can construct a flat vector bundle $E_\rho \to M$ as follows. Let $\tilde{M}$ be a universal covering of M; we have a natural flat structure on the trivial bundle $\tilde{M} \times \mathbb{C}^r$. Denote by $E \equiv E_\rho = \tilde{M} \times_\rho \mathbb{C}^r := \tilde{M} \times \mathbb{C}^r / \{\rho \text{ action of } \pi_1\}$, where the action is given by $\gamma : \tilde{M} \times \mathbb{C}^r \ni (x,v) \to (\gamma(x), \rho(\gamma)v) \in \tilde{M} \times \mathbb{C}^r$, $\gamma \in \pi_1(M, x_0)$, where we regard π_1 as a covering (deck) transformation group acting on $\tilde{M}$ (that is, $M \simeq \tilde{M}/\pi_1$). The flat structure on $E = E_\rho$ descends from the flat structure on $\tilde{M} \times \mathbb{C}^r$.

Thus we have the following

THEOREM. *For a complex vector bundle of rank r over M the following conditions are equivalent:*

1. $E \to M$ is a flat bundle;

2. $E \to M$ admits a flat connection ∇;

3. $E \to M$ is defined by the representation $\rho : \pi_1(X, x_0) \to \mathrm{GL}(r, \mathbb{C})$.

PROOF. 1. $\Longrightarrow$2. is trivial; 2. $\Longleftrightarrow$3. was already proved. 2. $\Longrightarrow$1. If s and s' are two local ∇ parallel frame fields, that is, if $\nabla s = \nabla s' = 0$, then $s = s'a$ for some constant matrix $a \in \mathrm{GL}(r, \mathbb{C})$, since the connection forms of ∇ with respect to s and s' vanish. Hence if $\nabla s_U = \nabla s_V = 0$, the transition map g_{UV} is a constant matrix. Thus the flat connection ∇ gives rise to a flat structure $\{\mathcal{U}, s_{\mathcal{U}}\}$. □

We have a corresponding theorem for hermitian vector bundles: *A flat hermitian structure* in C^∞ complex vector bundle $E \to M$ is given by an open cover $\{\mathcal{U}\}$ of M and a system of local frame fields $\{s_{\mathcal{U}}\}$ such that the transition maps $\{g_{UV}\}$ are constant unitary matrices in $U(r)$, where r is the rank of E.

THEOREM. *For a C^∞ complex vector bundle E of rank r over M the following conditions are equivalent:*

1. $E \to M$ admits a flat hermitian structure;

2. $E \to M$ admits a hermitian structure h and a flat h-connection ∇;

3. $E \to M$ is defined by the representation $\rho : \pi_1(X, x_0) \to U(r)$.

As we know it is often more convenient to work with principal G bundles P. The corresponding transition maps $\{g_{UV}\}$ defined by $\{\mathcal{U}, s_{\mathcal{U}}\}$ are C^∞ maps $\mathcal{U} \cap \mathcal{V} \to G$. A flat structure on P is given by constant g_{UV}. We can regard the universal covering space $\tilde{M} \to M$ as a principal $\pi_1(M, x_0)$ bundle, where π_1 acts on $\tilde{M}$ as a deck transformation. Given a homomorphism $\rho : \pi_1(M) \to G$, we obtain a principal G bundle $P \equiv P_\rho = \tilde{M} \times_\rho G$ in the same way as in the case of E_ρ. Then P inherits the flat structure from the natural flat structure of the product bundle $\tilde{M} \times_\rho G \to \tilde{M}$. As an obvious generalization of the theorem above we get

THEOREM. *For a principal G bundle P over M the following conditions are equivalent:*
1. P admits a flat structure;
2. P admits a flat connection;
3. $P = P_\rho$, that is, P is defined by the representation $\rho : \pi_1(X) \to G$.

We will consider the case $G = \mathbb{P}\,\mathrm{GL}(r, \mathbb{C}) := \mathrm{GL}(r, \mathbb{C})/\mathbb{C}^\times 1_r$, the projective linear group; $\mathbb{C}^\times$ is the multiplicative group $\mathbb{C} - \{0\}$, thus $\mathbb{C}^\times 1_r$ is the center of $\mathrm{GL}(r, \mathbb{C})$. Given a vector bundle $E \to M$, let P be an associated principal $\mathrm{GL}(r, \mathbb{C})$ bundle; then $\hat{P} := P/\mathbb{C}^\times 1_r$ is a principal $\mathbb{P}\,\mathrm{GL}(r, \mathbb{C})$ bundle. E is *projectively flat* if $\hat{P}$ is equipped with flat structure. A connection ∇ in E (that is, a connection in P) is *projectively flat* if the induced connection in $\hat{P}$ is flat.

We have the following

COROLLARY. *For a complex vector bundle E of rank r over M with the associated principal* $\mathrm{GL}(r, \mathbb{C})$ *bundle P, the following conditions are equivalent.*
1. $E \to M$ is projectively flat.
2. E admits a projectively flat connection ∇.
3. The $\mathbb{P}\,\mathrm{GL}(r, \mathbb{C})$ *bundle $\hat{P}/\mathbb{C}^\times 1_r$ is defined by the representation $\rho : \pi_1(M) \to \mathbb{P}\,\mathrm{GL}(r, \mathbb{C})$.*
4. The curvature R^∇ takes values in scalar multiples of the identity endomorphism 1_E of E, that is, there exists a complex 2 form $\alpha \in A^2(M)$ such that $R^\nabla = \alpha 1_E$.

REMARK. We must only comment 4.: Let $u : \mathrm{GL}(r, \mathbb{C}) \to \mathbb{P}\,\mathrm{GL}(r, \mathbb{C})$ be the natural homomorphism and u' its differential $u' : \mathfrak{gl}(r, \mathbb{C}) \to \mathfrak{pl}(r, \mathbb{C})$ (of Lie algebras). If R^∇ is the curvature of the connection ∇ in E, then the curvature of the connection induced on $\hat{P}$ is given by $u'(R^\nabla)$.

Now the statement of the Narasimhan-Seshardi theorem

$$(E \to M \text{ is stable}) \Longleftrightarrow (E \to M \text{ admits a flat hermitian structure})$$

is clear. Of course, this is a deep theorem, and all its proofs (there are several of them) are difficult.

A bit of history. The sources of the theory of stable vector bundles can be found in the treatise of André Weil of 1938. Weil had a presentiment that hermitian flat vector bundle should play a decisive role in the theory moduli of vector bundles over compact Riemann surfaces. As we saw, flat vector bundles are defined by unitary representations $\rho : \pi_1(M) \to U(r)$, they are of the form $E_\rho := \tilde{M} \times_\rho \mathbb{C}^r \to M$. It was clear for Weil that particularly interesting must be the bundles E_ρ associated with irreducible representations of $\rho : \pi_1(M) \to U(r)$. Among other results, Weil proves the interesting

THEOREM. *A vector bundle E over M arises from the representation $\rho : \pi_1(M) \to U(r)$, that is, $E = E_\rho$ if and only if each of its indecomposable components is of degree zero:* $E = E^1 \oplus \cdots \oplus E^k$, $\deg E^i = c_1(E^i) = 0$, E^i *is irreducible.*

The crucial step in the progress of the moduli problem for vector bundles over compact Riemann surfaces was the introduction by David Mumford of the notion of a stable vector bundle. This concept was motivated by the geometric invariant theory (Mumford, 1962). The notion of stability was presented by Mumford in his famous talk *Projective invariants of projective structures and applications* at the Stockholm Congress in 1962 (pp. 526-30). Mumford writes 'in other words a vector bundle is stable if all its subbundles are 'less ample' that itself'. To illustrate the stability he gives the following examples: 1. a line bundle $L \to M$ is always stable; 2. If L is a line bundle, then E is stable if and only if $E \otimes L$ is stable; 3. (E is stable) $\Longleftrightarrow$($E^\times$ is stable). He also proves

THEOREM (MUMFORD, 1962). *The set of isomorphism classes (that is, the moduli space) of stable vector bundles on M of rank r (and degree d) has a natural structure of a smooth quasi-projective variety $r^2(p-1)+1$, where p is the genus of M.*

The next fundamental steps were the classic papers of M.S. Narasimhan and C.S. Seshardi (1965) and the Yang-Mills theory on Riemann surfaces by Atiyah and Bott culminating in their large paper (1982) *The Yang-Mills equations over Riemann surfaces* Phil. Trans. Royal Soc. of London, Ser. A **308** (1982), 523-615. In this paper, which is almost a textbook, the authors showed their in own way all existing theorems of the theory, and added their original results. They hinted at surprising new relations between number theory, gauge theories, and conformal field theory. For us particularly important is the following beautiful theorem.

THEOREM. *Every stable vector bundle $E \to M$ on Riemann surface can be equipped with unique Hermite-Einstein metric.*

About the same time Kobayashi introduced his H-E condition on complex holomorphic vector bundle and formulated his famous

CONJECTURE (KOBAYASHI-HITCHIN) *On every ω stable vector bundle $E \to M$ over compact Kähler manifold (M, ω) there exists a unique H-E structure, and therefore, a unique Hermite-Yang-Mills connection.*

Nowadays this conjecture takes the form of the famous theorem of Karin Uhlenbeck and S.T. Yau (1986-87) published as *On the existence of hermitian Yang -Mills connections on stable bundles over compact Kähler manifolds* C.P.A.M. **39** (1986). K. Uhlenbeck published in 1982 her famous 'bubble theorem' and the 'theorem on removable singularities' that are at the heart of the vital compactness property of instanton moduli spaces, in particular, their compactifications.

The very important particular cases where M is a projective surface and an algebraic manifold were analyzed by Donaldson in 1985 and 1986-7, respectively. There is a beautiful monograph about these themes written by Atiyah and Kronheimer *The Geometry of Four-Manifolds*, Clerendon Press, Oxford, 1990. The reader may also consult the excellent Bourbaki report by Christophe Margegin *Fibrés stables at métrique d'Hermite-Einstein* in 152-3 Astérisque (1987), Exposé 683.

REMARK. Lie-Yau proved in 1987 the existence of H-E connections on every $E \to M$ with compact M. Here we mention only the Kobayashi approach following his excellent monograph *Differential Geometry of Complex Vector Bundles* (1987). In what follows we will use the notation of Kobayashi which is differs slightly from the most common one. Our main goal will be to show the way leading to the construction of

9.2 Moduli spaces of H-E structures

As we know on an (almost) complex manifold M, under complexification the tangent bundle TM, $T^cM = TM \otimes_{\mathbb{R}} \mathbb{C}$ splits into two parts, called holomorphic and antiholomorphic tangent spaces and denoted by $T^{1,0}M$ (or $T'M$) and by $T^{0,1}M$ (or $T''M$):

$$TM \otimes_{\mathbb{R}} \mathbb{C} = T'M \oplus T''M = T^{1,0}M \oplus T^{0,1}M.$$

If $E \to M$ is a complex vector bundle over M with covariant derivative ∇, then ∇ splits into two pieces $\nabla = \nabla^{1,0} \oplus \nabla^{0,1} = \nabla' \oplus \nabla''$, which satisfy the following Leibnitz formulas: if $\alpha \in A(M)$, $v \in A(M, E)$ (that is, they are a form and an E valued form, respectively), then

$$\nabla'(\alpha \wedge v) = d'\alpha \wedge v + (-1)^{|\alpha|}\alpha\nabla''v,$$

$$\nabla''(\alpha \wedge v) = d''\alpha \wedge v + (-1)^{|\alpha|}\alpha\nabla''v,$$

where the exterior differential d splits into two pieces $d = d' + d''$; d' increases the degree of a (p,q) form by $(1,0)$, and d'' by $(0,1)$. Clearly, $d'^2 = d'd'' + d''d' = d''^2 = 0$. Very often d' is denoted by ∂ and d'' by $\bar{\partial}$. We recall

PROPOSITION. *1. If $E \to M$ is a holomorphic vector bundle on a complex manifold M then there is a unique operator d''_E denoted for short $d'' : A^{p,q}(M,E) \to A^{p,q+1}(M,E)$ which in the local holomorphic coordinates equals*

$$\sum_{k=1}^{n} d\bar{z}^k \wedge \frac{\partial}{\partial \bar{z}^k}, \quad n = \dim_{\mathbb{C}} M.$$

The covariant derivative ∇ on a holomorphic vector bundle $E \to M$ is holomorphic if $\nabla'' = d''_E$.

2. If $E \to M$ is a C^∞ complex vector bundle and if ∇ is such that $\nabla^{0,1} \circ \nabla^{1,0} = 0$, then there is a unique *holomorphic structure on E such that $\nabla'' = d''_E$.*

3. If $E \to M$ is a holomorphic vector bundle with hermitian metric $h(s,t)$, then there exists a unique holomorphic *covariant derivative d''_E on $E \to M$ preserving the metric h:*

$$d''h(s,t) = h(d''_E s, t) + h(s, d''_E t).$$

Thus d''_E defines a parallel transport which is an isometry of corresponding fibers.

We will use the following shorthand notation: E for $E \to M$, $A^{p,q}(E)$ for $A^{p,q}(E \to M)$, $A^{p,q}$ for $A^{p,q}(M)$. Denote by $\mathrm{GL}(E)$ the group of C^∞ bundle automorphisms of E which induce identity on the base M (this group is also denoted $\mathrm{Aut}(E)$).

DEFINITION. A *gauge transformation* is a section of $\mathrm{GL}(E)$. The group $\mathcal{G}$ of gauge transformations is called the *gauge group* of the (metric, hermitian, etc.) bundle E.

The space of $C^\infty(\mathrm{End}(E))$ of smooth sections of the endomorphism bundle $\mathrm{End}(E) = E \otimes E^*$ is regarded as $\mathfrak{gl}(E)$, the Lie algebra of $\mathrm{GL}(E)$. The set $D''(E)$ of $\mathbb{C}$ linear maps $\nabla'' : A^0(E) \to A^{0,1}(E)$ is an affine space (of infinite dimension.)

Let $\mathcal{H}''(E) = \{\nabla'' : \nabla'' \circ \nabla'' = 0\}$; then, it follows from 2. in the Proposition that we can identify $\mathcal{H}''(E)$ with the set of holomorphic bundle structures in E.

The group $\mathrm{GL}(E)$ acts on $D''(E)$ by

$$(*) \qquad \nabla'' \to \nabla''f := f^{-1} \circ \nabla'' \circ f = \nabla'' + f^{-1}d''f,$$

for $f \in \mathrm{GL}(E)$, $\nabla'' \in D''(E)$. Therefore $\mathrm{GL}(E) : \mathcal{H}''(E) \to \mathcal{H}''(E)$.

DEFINITION. The set $\mathcal{H}''(E)/\mathrm{GL}(E)$ of $\mathrm{GL}(E)$ orbits in $\mathcal{H}''(E)$ is the *moduli space of holomorphic structures* in $E \to M$. This space, equipped with natural topology is not, in general, Hausdorff.

Denote by $U(E,h)$ the subgroup of $\mathrm{GL}(E)$ consisting of unitary (preserving h) automorphisms of (E,h). Its Lie algebra $\mathfrak{u}(E,h)$ consists of skew-hermitian endomorphisms of (E,h). Let us introduce another sets: $D(E,h) = \{\text{connections } \nabla \text{ in } (E,h) \text{ preserving } h\}$, and $\mathcal{H}(E,h) = \{\nabla \in D(E,h) : \nabla'' \in \mathcal{H}''(E)$, that is, $\nabla'' \circ \nabla'' = 0\}$. Clearly the group $U(E,h)$ acts on $D(E,h)$ and on $\mathcal{H}(E,h)$, by $(*)$. We have the important

THEOREM. *1. The subgroup of $U(E,h)$ fixing given $\nabla_0 \in D(E,h)$ consists of automorphisms which are parallel with respect to ∇_0. It is therefore isomorphic with the centralizer in the unitary group $U(r)$ of the holonomy group of the connection ∇_0; it is therefore* compact. *($r = \mathrm{rank} E$).*

2. The action of $U(E,h)$ *on* $D(E)$ *and on* $\mathcal{H}(E,h)$ *is proper. Therefore:*

3. The moduli spaces $D(E,h)/U(E,h)$ *and* $\mathcal{H}(E,h)/U(E,h)$ *are Hausdorff spaces.*

PROOF of 1. is obvious. Since 3. immediately follows from 2., we have to prove 2. only.

Set $U = U(E,h)$, $D = D(E,h)$. We have to prove that the map $D \times U \ni (\nabla, f) \to (\nabla, \nabla f) \in D \times D$ is proper. The sequences $\nabla_i \to \nabla_\infty$ and $n_i^{f_i} \to \nabla_0$ ($f_i \in U$) for $i \to \infty$. One has to show that the subsequence f_i converges to $f_\infty \in U$ and that $\nabla_\infty^{f_\infty} = \nabla^0$. Let P be a $U(r)$ principal bundle associated with (E,h). Fix $x_0 \in M$ and $v_0 \in P$ over x_0. For a subsequence $\{f_j\}$ we have $f_j(v_0) \to v_0^0 \in P$, $j \to \infty$. But f_j commute with the right action of the structure group $U(r)$ on P; therefore $f_j(v_0 g) = (f_j(v_0))g \to v_0^0$ for all $g \in U(r)$. Let $x \in M$ and c be a curve in M connecting x_0 and x; let $\tilde{c}_j \in P$ be a horizontal lift of c with respect to connection ∇_j such that $\tilde{c}(0) = v_0$. Then $f_j(\tilde{c}_j)$ is the horizontal lift of c with respect to connection $\nabla_j^{f_j}$ such that $f_j(\tilde{c}_j(0))f_j(v_0)$. Since ∇_j converges to ∇_∞, $\tilde{c}_j$ converges to $\tilde{c}_\infty$, where $\tilde{c}_\infty$ is the horizontal lift of c with respect to ∇_∞. Let now $\tilde{c}^0$ be the horizontal lift with respect to ∇^0 starting from v_0^0. But $(\nabla_j^{f_j} \to \nabla^0$ and $f_j(v_0) \to v_0^0) \Longleftrightarrow (f_j(\tilde{c}_\infty) \to \tilde{c}^0)$. We define f_∞ in such a way that $f_\infty(\tilde{c}_\infty) = \tilde{c}^0$ which can be done because f_∞ commutes with the right action of $U(r)$, and thus f_∞ is uniquely determined. We recall that if γ is a closed curve in M and if the horizontal lift $\tilde{\gamma}$ of γ gives an element $g \in U(r)$ of the holonomy group, then for every $f \in U$ $(= Aut(P))$ the horizontal lift $f(\tilde{\gamma})$ gives the same element $g \in U(r)$. □

Let $K^\nabla \equiv K = (K_j^i)$ be the mean curvature transformation form and $\hat{K}$ the mean curvature form of connection ∇ in the bundle $E = E(h, M, g)$, where M is a compact Kähler manifold (M, g). We recall that $\hat{K}(\xi, \eta) = K_{j\bar{k}}\xi^j\bar{\eta}^k$. Denote by $\mathcal{E}(E,h)$ the set of Hermite-Einstein connections ∇ in the bundle (E,h), that is,

$$\mathcal{E}(E,h) = \{\nabla \in \mathcal{H}(E,h) : K^\nabla = c1_E\}.$$

Clearly the action of the group $U(E,h)$ on $D(E,h)$ leaves $\mathcal{E}(E,h)$ invariant. The quotient space $\mathcal{E}(E,h)/U(E,h)$ is called the moduli space of H-E structures (or connections) in E. Thus we have

COROLLARY. *The moduli space $\mathcal{E}(E,h)/U(E,h)$ of Hermite–Einstein structures in E is Hausdorff and injects into the space $\mathcal{H}''(E,h)/\operatorname{GL}(E)$ of holomorphic structures in E.*

9.3 Hermite-Einstein metrics (structures) as critical points of Donaldson functional (variational theory of H-E connections)

We remember that it was a fundamental idea of Riemann to characterize solutions of important problems of geometry and analysis (for example, conformal transformations, geodesics on minimal surfaces, etc.) with the help of the appropriate variational principles. This idea culminated in the theory of harmonic maps introduced by S. Bochner. We saw that this Riemann's idea, promoted by Tromba, Wolf, and Jost, gave rise to the most natural approach to Teichmüller theory.

The most important step in such 'Riemannian approach' is to find an appropriate functional, given by the integral of a 'Lagrangian'. In the theory of harmonic maps, this role was played by the 'energy' of a map $M \to N$ between two Riemann manifolds. Harmonic maps are critical points of the energy.

In the same spirit, Donaldson constructed a remarkable functional $B(h,k)$ in the space $\operatorname{Herm}^+(E)$ of (positive) hermitian structures in the bundle $E \to M$ such that the H-E connections are precisely the critical points of the functional $\operatorname{Herm}^+(E) \ni h \to B(h,k_0) \in \mathbb{R}$ for fixed $k_0 \in \operatorname{Herm}^+(E)$.

Since the Donaldson construction was inspired by the notion of Yang-Mills functional, we first present this latter construction.

Given connection ∇ in a holomorphic vector bundle $E \to M$ of rank r over compact Kähler manifold M of dimension n we consider the mean curvature $K^\nabla \equiv K$ of (E,h). The scalar curvature σ of h is defined by $\sigma := K^i_i = g^{\alpha\bar\beta} R^i_{i\alpha\bar\beta} = g^{\alpha\bar\beta} R_{\alpha\bar\beta}$ $(R \equiv R^\nabla)$. Let $\| K \|^2 := K^i_j K^j_i = \operatorname{tr}(K \circ K)$, $\| R \|^2 := \sum |R_{ji\alpha\bar\beta}|^2$. We consider the functional

$$\mathcal{J}(\nabla) \equiv \mathcal{J}(h) := \frac{1}{2} \int_M \| K \|^2 \, \omega^n,$$

where ω is the Kähler form on M. Let c be a constant defined by

$$rc \int_M \omega^n = \int_M \sigma \omega^n, \quad r = \operatorname{rank} E.$$

Then

$$(*) \qquad 0 \leqslant \| K - c1_E \|^2 = \| K \|^2 + rc^2 - 2c\sigma.$$

From the definition of the first Chern class $c_1(E)$ of E and the relation $K\omega^n = \sqrt{-1}R \wedge \omega^{n-1}$, by taking the trace and integrating, we obtain

$$\int_M \sigma\omega^n = 2\pi n \int_M c_1(E) \wedge \omega^{n-1}.$$

Integrating $(*)$ we obtain the fundamental inequality

$$\mathcal{J}(\nabla) = \mathcal{J}(h) := \frac{1}{2}\int_M \| K \|^2 \omega^n \geqslant rc^2 \int_M \omega^n = \frac{\left(\int_M \sigma\omega^n\right)^2}{r\int_M \omega^n} \geqslant$$

$$\geqslant 2\frac{\left(\pi n \int_M \chi_1(E) \wedge \omega^{n-1}\right)^2}{r\int_M \omega^n} =: c.$$

We have proved the following

THEOREM (KOBAYASHI). *The functional $\mathcal{J}(\nabla) = \mathcal{J}(h)$ is bounded by a constant c which depends only on the cohomology class of the Kähler form ω and $c_1(E)$. The lower bound c is attained by $\mathcal{J}(h)$ if and only if h (that is, ∇) satisfies the Einstein condition with constant factor c: $K = c1_E$.*

REMARK 1. We see now the mysterious role of $\deg_\omega E := \int_M c_1(E) \wedge \omega^{n-1}$ in the definition of (semi) stability of a vector bundle.

REMARK 2. One often consider another functional, the Yang-Mills functional defined by

$$YM(\nabla) = YM(h) := \frac{1}{2}\int_M \| R^\nabla \|^2 \omega^n.$$

But the latter integral differs from $\mathcal{J}(h)$ only by a constant

$$YM(h) - \mathcal{J}(h) = 2\pi^2 n(n-1)\int_M (2c_2(E) - c_1(E)^2) \wedge \omega^{n-2}.$$

This follows from the formula

$$(\| R \|^2 - \| K \|^2)\omega^n = n(n-1)\sum_{i,j=1}^{n} \Omega^i_j \wedge \Omega^j_i \wedge \omega^{n-1},$$

where (Ω_i^j) is the curvature form of connection ∇, and $c_1(E)$, $c_2(E)$ are Chern forms of E.

REMARK. In this context we formulate the important *Lübke inequality theorem*

THEOREM (LÜBKE, 1983). *Let (E,h) be hermitian vector bundle of rank r over compact Kähler manifold (M,g) of dimension n. If (E,h) satisfies the weak Einstein condition, then*

$$\int_M ((r-1)c_1(E,h) - 2rc_2(E,h)) \wedge \omega^{n-1} \leqslant 0.$$

The above integral vanishes if and only if (E,h) is projectively flat.

COROLLARY. *If moreover $c_1(E) = 0$ in $H^2(M,\mathbb{R})$, then*

$$\int_M c_2(E) \wedge \omega^{n-1} \geqslant 0,$$

and (E,h) is flat if and only if $\int_M c_2(E) \wedge \omega^{n-1} = 0$.

If $E \to M$ is the tangent bundle of M, we have

COROLLARY. *If a compact Kähler–Einstein manifold M satisfies $c_1(M) = c_2(M) = 0$, then M is flat.*

From this corollary follows the classical Bieberbach theorem stating that compact flat Riemann manifold is covered by an Euclidean torus, and the famous Yau theorem that every compact Kähler manifold with $c_1(M) = 0$ admits a Ricci flat Kähler metric. From these, we get

COROLLARY (YAU). *A compact Kähler manifold M with $c_1(M) = c_2(M) = 0$ is covered by a complex Euclidean torus.*

We saw that H-E metric (or structure) h was a lower bound of the functional $\mathcal{J}$. A natural question arises as to how one can characterize *all* critical points of $\mathcal{J}$? To this end, we must investigate the first variation of $\mathcal{J}(h_t)$, that is, $\left.\frac{d\mathcal{J}(h_t)}{dt}\right|_{t=0}$, where $t \to h_t$ is a 1 parameter family of hermitian structures in E such that $h_0 = h$. If we set $h_t = h + tv$, by routine calculations we find

$$(var) \qquad \partial_t \mathcal{J}(h_t)|_{t=0} = \int_M g^{\alpha\bar{\beta}} v^i_{j\alpha} K^j_{i\bar{\beta}} \omega^n = < \nabla' v | \nabla' K > .$$

Since v is arbitrary, we obtain

THEOREM (KOBAYASHI, 1982). *For a fixed holomorphic vector bundle E over compact Kähler manifold (M, g) a hermitian structure h is a critical point of $\mathcal{J}$ if and only if the mean curvature K is parallel with respect to the hermitian connection ∇ defined by h, that is, if $\nabla'' K = 0$.*

PROOF. It follows from variational formula (var) that if $\nabla' K = 0$ then h is a critical point of $\mathcal{J}$. But also conversely, if h is a critical point of $\mathcal{J}$, taking $h_{ti\bar{j}} := h_{i\bar{j}} + tK_{i\bar{j}}$, we obtain $v_{i\bar{j}} = K_{i\bar{j}}$. Therefore, (var) $\Longrightarrow(< \nabla' K | \nabla' K >= 0) \Longrightarrow(\nabla' K = 0) \Longrightarrow(\nabla'' K = 0)$ since K is hermitian. □

Let now $\Phi(x)$ be the holonomy group of ∇, that is, of (E, h). If h is a critical point, then K is parallel and we can decompose (E, h) with respect to the eigenspaces (and eigenvalues) of K. The decomposition of all parallel hermitian forms defines a $\Phi(x)$ invariant hermitian form on the fiber E_x. If $E_x = E^0_x \oplus \cdots \oplus E^k_x$ is such orthogonal decomposition of E_x such that $\Phi(x)$ is trivial on E^0_x and irreducible on $E^1_x, \ldots, E^k_x$. By parallel displacement we obtain the orthogonal and holomorphic decomposition

$$E = E^0 \oplus E^1 \oplus \cdots \oplus E^k$$

and we have proved

THEOREM (KOBAYASHI). *Let E be a holomorphic vector bundle over compact Kähler manifold (M, g). If h is a critical point of $\mathcal{J}$, if*

$$E = E^0 \oplus E^1 \oplus \cdots \oplus E^k$$

is the corresponding holomorphic and orthogonal decomposition of (E, h), and if $h_0, h_1, \ldots, h_k$ are the corresponding restrictions of h, then $(E^0, h_0), (E^1, h_1), \ldots, (E^k, h_k)$ are Hermite-Einstein vector bundles with constant factors $c_0, c_1, \ldots, c_k$.

Now there is time to formulate the fundamental theorems of the present chapter.

THEOREM (KOBAYASHI, 1982; LÜBKE, 1983). *Let (E, h) be a Hermite-Einstein vector bundle over compact Kähler manifold (M, g). Then E is ω-semistable and (E, h) is a direct sum*

$$(E, h) = (E^0, h_0) \oplus (E^1, h_1) \oplus \cdots \oplus (E^k, h_k)$$

of ω-stable H-E vector bundles (E_j, h_j) with the same factor c as (E, h). Thus if (E, h) is an irreducible *H-E vector bundle, then E is ω-stable.*

PROOF (by Lübke which is simpler than the original proof of Kobayashi). We have to introduce first some definitions and notions. Let $\mathcal{F}$ be a coherent analytic subsheaf of $\mathcal{O}(E)$. We take $\mathcal{F}^* := \mathrm{Hom}(\mathcal{F}, \mathcal{O})$. $\mathcal{F}$ is *reflexive* if $\mathcal{F}^{**} = (\mathcal{F}^*)^* = \mathcal{F}$. A reflexive sheaf of rank 1 is a holomorphic line bundle. We recall that a locally free sheaf is a holomorphic vector bundle. A reflexive sheaf is locally free outside a subvariety of codimension $\geqslant 2$ (that is, of dimension $\leqslant n-2$.)

Since $\det \mathcal{F}^{**}$ is a line bundle, one can define $c_1(\mathcal{F}) := c_1(\det \mathcal{F}^{**})$, and degree of $\mathcal{F}$ by $\deg_\omega(\mathcal{F}) := \int_M c_1(\mathcal{F}) \wedge \omega^{n-1}$, $\mu(\mathcal{F}) := \deg(\mathcal{F})/rank(\mathcal{F})$. E is semistable if for every coherent subsheaf $\mathcal{F}$ of lower rank $\mu(\mathcal{F}) \leqslant \mu(E)$. From the analytic point of view the sheaf $\mathcal{F}$ can be interpreted as a meromorphic section of the Grassmann bundle $Gr_M(k, r)$, the bundle over M whose fibers are Grassmann manifolds $Gr(k, r)$ of k dimensional subspaces of $\mathbb{C}^r$, $0 < k = rank\, \mathcal{F} = rank\, E$.

PROOF OF SEMISTABILITY. Let $\mathcal{F} \subset \mathcal{E} := \mathcal{O}(E)$ be of rank p. The inclusion map $j : \mathcal{F} \to \mathcal{E}$ induces the homomorphism $\det j := (\Lambda^p)^{**} : \det \mathcal{F} = (\Lambda^p \mathcal{F})^{**} \to= (\Lambda^p \mathcal{E})^{**} = \Lambda^p \mathcal{E}$ which is injective because it is injective outside the singularity set of $\mathcal{F}$. Thus we obtain a nontrivial homomorphism $f : \mathcal{O}_M \to \Lambda^p \mathcal{E} \otimes (\det \mathcal{F})^*$ which, in turn, can be regarded as a holomorphic section of the vector bundle $\Lambda^p E \otimes (\det \mathcal{F})^*$. We know that the factor c in the definition of Einstein condition for (E, h) is given by

$$c = \frac{2\pi n \mu(E)}{n! Vol(M)},$$

where $vol(M) := \frac{1}{n!} \int_M \omega^n$. But since every line bundle admits an H-E structure, we can choose an H-E structure in the line bundle $\det \mathcal{F}$ with the factor c',

$$c' = \frac{2\pi n\mu(\det \mathcal{F})}{n!Vol(M)} = \frac{2\pi np\mu(\mathcal{F})}{n!Vol(M)}.$$

Thus the vector bundle $\Lambda^p E \otimes (\det \mathcal{F})^*$ is H-E with the factor $pc - c'$. Since this bundle admits a nontrivial section f, the vanishing theorem from the preceding chapter implies $pc - c' \geqslant 0$; thus $\mu(\mathcal{F}) \leqslant \mu(\mathcal{E})$. □

The magnificent theorem of Uhlenbeck and Yau asserts that every ω-stable bundle admits a unique H-E structure, therefore

$$(\omega \text{ stability of } E) \Longleftrightarrow (\text{H-E property of the irreducible bundle } E).$$

In order to present this result in its full glory we have to consider another variational principle.

The Donaldson functional. Before we turn to the construction let us recall some facts regarding hermitian symmetric spaces.

Denote by $\mathrm{Herm}(r)$ the space of $r \times r$ hermitian matrices, and let $\mathrm{Herm}^+(r)$ be the subset (which turns out to be a convex domain) of positive definite ones. The group $\mathrm{GL}(r, \mathbb{C})$ acting on $\mathrm{Herm}(r)$ by $h \to {}^t\bar{a}ha$, $a \in \mathrm{GL}(r, \mathbb{C})$ is transitive on $\mathrm{Herm}^+(r)$. The isotropy group of 1_r is the unitary group $U(r)$ which is the maximal compact subgroup of $\mathrm{GL}(r, \mathbb{C})$. The domain $\mathrm{Herm}^+(r) \simeq \mathrm{GL}(r, \mathbb{C})/U(r)$ ant this space is symmetric: the symmetry at the identity 1_r is given by $h \to h^{-1}$. The Cartan decomposition of the Lie algebra $\mathfrak{gl}(r, \mathbb{C})$ or $\mathrm{GL}(r, \mathbb{C})$ is

$$\mathfrak{gl}(r, \mathbb{C}) = \breve{\mathfrak{u}}(r) + \mathrm{Herm}(r).$$

We can identify the tangent space $T_h(\mathrm{Herm}^+(r))$ with $\mathrm{Herm}(r)$ and on $\mathrm{Herm}^+(r)$ we have the Riemann metric $(\breve{u}|v) := \mathrm{tr}(h^{-1}\breve{u} \cdot h^{-1}v)$ at $h \in \mathrm{Herm}^+(r)$. This defines a $\mathrm{GL}(r, \mathbb{C})$ invariant Riemann metric on $\mathrm{Herm}^+(r)$ which coincides with the scalar product at 1_r. We can identify $\mathbb{C}^r$ with a complex vector space V of dimension r and hermitian matrices with the corresponding hermitian bilinear forms.

PROPOSITION. *(A curve $t \to h(t)$ in* $\mathrm{Herm}^+(r)$ *is a geodesic)* $\Longleftrightarrow$*($\frac{\mathrm{d}}{\mathrm{d}t}(h^{-1}\partial_t h) = 0$, that is, $h^{-1}\partial_t h$ is independent of* $\mathfrak{gl}(V)$.*)*

Let us elevate these facts to the **space of hermitian structures in the vector bundle $E \to M$.** Denote by $\mathrm{Herm}(E)$ the set of C^∞ hermitian structures v in E (this is an infinite dimensional vector space.) Let $\mathrm{Herm}^+(E) = \{h \in \mathrm{Herm}(E) : h_x \in \mathrm{Herm}^+(r) \text{ for all } x \in M\}$. Clearly, $\mathrm{Herm}^+(E)$ is a convex domain in $\mathrm{Herm}(E)$ and we can regard $\mathrm{Herm}(E)$ as the tangent space $T_h(\mathrm{Herm}^+(E))$ at every point $h \in \mathrm{Herm}^+(E)$.

Thus we have the infinite dimensional analogues of finite dimensional notions introduced above: $\mathrm{GL}(E)$ – the group of complex automorphisms of E inducing the identity transformations on the base M. If we fix $k \in \mathrm{Herm}^+(E)$, then $U(E) = \{a \in \mathrm{GL}(E) : ak = k\}$ and we can identify $\mathrm{Herm}^+(E)$ with $\mathrm{GL}(E)/U(E)$ and regard it as a symmetric space with the involution at k given by $h \to {}^t\bar{a}ka \to a^{-1}k^t\bar{a}^{-1}$; this is correct since $\mathrm{GL}(E)$ acts transitively on $\mathrm{Herm}^+(E)$: indeed for $k, h \in \mathrm{Herm}^+(E)$ there exists a unique $a \in \mathrm{GL}(E)$ such that $h = {}^t\bar{a}ka$.

Now we assume (as always in this chapter) that $M = (M, g)$ is a compact Kähler manifold of dimension n with canonical form ω. Then on $\mathrm{Herm}^+(E)$ we obtain the Riemann, $\mathrm{GL}(E)$ invariant metric defined for each tangent space $T_h\mathrm{Herm}^+(E)$ by

$$(v|w) := \int_M \mathrm{tr}(h^{-1}v \cdot h^{-1}w)\omega^n = \int_M < v_x|w_x >_x \omega(x)^n.$$

We have again

PROPOSITION. *(A curve $t \to h(t)$ in $\mathrm{Herm}^+(E)$ is a geodesic)* $\Longleftrightarrow$*($\frac{\mathrm{d}}{\mathrm{d}t}(h^{-1}\partial_t h) = 0$, that is, $h^{-1}\partial_t h$ is independent of $\mathrm{Herm}(E)$.)*

Every two hermitian structures $h, k \in \mathrm{Herm}^+(E)$ can be connected by a curve h_t, $t \in [0,1]$ such that $h_0 = k$, $h_1 = h$. Every h_t induces the curvature $R_t = R(h_t) \in A^{1,1}(\mathrm{End}\, E)$. For any $t \in [0,1]$, $v_t := h^{-1}\partial_t h_t \in A^0(\mathrm{End}\, E)$ and (h_t is a geodesic in $\mathrm{Herm}^+(E)$) $\Longleftrightarrow$($\partial_t v_t = 0$). Now we can introduce the Donaldson functional.

DEFINITION. Let c be a constant defined by

$$c := \frac{2\pi n \int_M c_1(E) \wedge \omega^{n-1}}{r \int_M \omega^n},$$

$Q_1(h,k) := \log(\det(k^{-1}h))$, $Q_2 := \sqrt{-1}\int_0^1 \mathrm{tr}(v_t \cdot R_t)dt$, then the Donaldson integral is defined to be

$$B(h,k) := \frac{1}{(n-1)!}\int_M (Q_2(h,k) - \frac{c}{n}Q_1(h,k)\omega)\wedge\omega_{n-1}$$

for $h, k \in \mathrm{Herm}^+(E)$. This definition is correct since $B(h,k)$ does not depend on the curve c joining k and k. Indeed, for a closed curve h_t, $t \in [a,b]$, $h_a = h_b$, $\int_a^b \mathrm{tr}(v_t \cdot R_t)dt \in d'A^{0,1}(M) \oplus d''A^{1,0}(m)$.

$B(\cdot,\cdot)$ has several important and interesting properties which are collected in the fundamental

THEOREM (DONALDSON, 1985).
1. For any $h, k, l \in \mathrm{Herm}^+(E)$, $B(h,k) + B(k,l) + B(l,h) = 0$.
2. $B(h, \lambda h) = 0$ for any positive constant λ.
3. For fixed $k_0 \in \mathrm{Herm}^+(E)$, $B(h) := B(h, k_0)$ has the following properties:

(a) (h is a critical point of $B(\cdot)$) $\Longleftrightarrow$ (h is H-E, that is, $K(h) = c1_E$).

(b) If h_0 is a critical point of $B(\cdot)$, then $B(\cdot)$ attains at h_0 an absolute minimum.

(c) If h_1 and h_2 are critical points of $B(\cdot)$, then h_1 and h_2 define the same hermitian connection ∇.

(d) If E is a simple bundle (that is, $\mathrm{Aut}\, E = \mathbb{C}1_E$) and h_1, h_2 are critical points of $B(\cdot)$, then $h_1 = \lambda h_2$ for some $\lambda \in \mathbb{R}$.

(e) The gradient of $B(\cdot)$, that is, the vector field in $\mathrm{Herm}^+(E)$ dual to the 1 form $dB(\cdot)$ with respect to the Riemann metric $(\cdot|\cdot)$ on $\mathrm{Herm}^+(E)$ is given by $\mathrm{grad}\, B(\cdot) = \hat{K} - ch$.

PROOF is not quite easy and we will only make some comments.

(e) For every fixed t consider $\partial_t h_t \in \mathrm{Herm}(E)$ as a tangent vector in $T_{h_t}(\mathrm{Herm}^+(E))$. The differential $dB(\cdot)$ at $\partial_t h_t$ is given by

$$dB(\partial_t h_t) = \frac{\mathrm{d}B(h_t)}{\mathrm{d}t} = (\hat{K}_t - ch_t|\partial_t h_t).$$

We consider therefore the 'negative gradient flow'

$$(Ev) \qquad \frac{\mathrm{d}h_t}{\mathrm{d}t} = -\,\mathrm{grad}\, B(h_t) \equiv (\hat{K}(h) - ch)$$

for all $t \in [0, \infty)$.

This is the classical method of steepest descent: $h_\infty := \lim_{t \to \infty} h_t$ (if exists) is a critical point of the functional $B(\cdot)$. It can be proved that the evolution equation (Ev) has, for any initial hermitian structure h_0 a unique smooth solution defined for all times $t \geqslant 0$. This method was first developed by Eells and Sampson (1964) in their classic paper on harmonic maps.

9.4 Kähler structures on moduli space $\mathcal{M}^{H-E}(E)$

Our dream is to introduce the Kähler structure on the nonsingular part of the moduli space $\mathcal{M}^{H-E}(E)$, the program, we successfully applied in the case of Teichmüller space. In the latter case, the role of the Kähler metric was played by the Petersson–Weil metric. One of the evidences that our hope is not vain is provided by

THEOREM (KOBAYASHI, 1986). *On the nonsingular part of the moduli space $\mathcal{M}^{H-E}(E)$ of irreducible Hermite–Einstein connections in (E, h) there exists a Kähler metric ω. This metric can be obtained by the method of moment map and symplectic reduction of Lie, Marsden, and Weinstein.*

The idea of the proof is the following infinite dimensional analogue of the classic idea of Sophus Lie. Let (V, ω) be a Banach symplectic manifold, that is, $\omega : TV \times TV \to \mathbb{R}$ is a 2 form such that

(a) For every $x \in V$, $\omega_x : T_xV \times T_xV \to \mathbb{R}$ is continuous;

(b) For every $x \in V$, ω_x is non degenerate;

(c) $x \to \omega_x$ is smooth;

(d) ω is closed, that is, $d\omega = 0$.

REMARK. In what follows V will be a Hilbert–Sobolev manifold, but this general construction for Banach manifolds and Banach Lie groups is quite illuminating.

Let G be a Banach Lie group acting on V by symplectic transformations. We recall that a vector field a on V ($a \in C^\infty(TV)$) is an infinitesimal symplectic transformation if $\mathcal{L}_a\omega = 0$, that is, if the form $i_a\omega$ is closed. Here $\mathcal{L}_a$ is the Lie derivative along a, $\mathcal{L}_a\omega = (d \circ i_a + i_a \circ d)\omega$.

Let $\mathfrak{g} = T_eG$ be the Lie algebra of G and $\mathfrak{g}^*$ its Banach dual.

DEFINITION. A momentum map for the action of G on V is the map $\mu \to \mathfrak{g}^*$ such that

(i) $< a, d\mu_x(v) >= \omega(a_x, v)$ for $a \in \mathfrak{g}$, $v \in T_xV$, $x \in V$, and $< \cdot, \cdot >$ denotes the dual pairing between $\mathfrak{g}$ and $\mathfrak{g}^*$.

(ii) μ is equivariant with respect to the coadjoint action of G $\mu(g \cdot x) = \mathrm{Ad}^*_g(\mu(x))$, $g \in G$ and $x \in V$; that is, we have the commutative diagram

$$\begin{array}{ccc} V & \xrightarrow{\quad g \quad} & V \\ \Big\downarrow \mu & & \Big\downarrow \mu \\ \mathfrak{g}^* & \xrightarrow{\quad \mathrm{Ad}^*_g \quad} & \mathfrak{g}^* \end{array}$$

Therefore the subset $\mu^{-1}(0) \subset V$ is G invariant, whence the quotient space $W := \mu^{-1}(0)/G$ is well defined and is called in symplectic geometry the *reduced space.*

In order to obtain a symplectic (Kähler) structure on W we make natural assumptions:

(iii) $0 \in \mathfrak{g}^*$ is a regular value of μ, that is, $\mu^{-1}(0)$ is a submanifold of V. This condition is satisfied if for every $x \in \mu^{-1}(0)$, $T_x\mu^{-1}(0) = \ker(d\mu_x)$ and $d\mu_x : T_xV \to \mathfrak{g}^*$ is surjective for all $x \in \mu^{-1}(0)$.

(iv) We assume that the action of G on $\mu^{-1}(0)$ is free and that for every $x \in \mu^{-1}(0)$ there is a slice $\mathcal{S}_x \subset \mu^{-1}(0)$ for this action, that is, $\mathcal{S}_x$ is a submanifold of $\mu^{-1}(0)$ through x which is transversal to the orbit $G \cdot x$:

$$T_x\mu^{-1}(0) = T_x\mathcal{S}_x + T_x(G \cdot x).$$

If $\mathcal{S}_x$ is sufficiently small, then the projection

$$\pi : \mu^{-1}(0) \to \mu^{-1}(0)/G =: W$$

defines a homeomorphism of $\mathcal{S}_x$ onto the open subset $\pi(\mathcal{S}_x)$ of W. In this way we obtain a local coordinate system in W which makes W into a manifold – perhaps not Hausdorff. But if the action of G on $\mu^{-1}(0)$ is proper, then this construction leads to a Hausdorff manifold.

REMARK 1. In holomorphic case there are obvious modifications of (i)–(iv).

REMARK 2. The construction of slices $\mathcal{S}_x$ is the standard method of introducing an atlas on homogeneous manifolds – spaces of G orbits.

Thus we have the following situation induced by the G equivariant momentum map $\mu : V \to \mathfrak{g}^*$ (π is a surjection!)

$$\begin{array}{ccc} \mu^{-1}(0) & \stackrel{j}{\hookrightarrow} & V \\ \downarrow \pi & & \\ W := \mu^{-1}(0)/G & & \end{array}$$

The above assumption make it possible to prove the infinite dimensional symplectic reduction theorem, of which we present also the holomorphic Kähler version.

THEOREM (SYMPLECTIC REDUCTION) *Let (V, Ω_V) be a complex Banach symplectic manifold on which a Banach Lie groups act preserving the holomorphic symplectic form Ω_V. If there exists a holomorphic momentum map $\mu : V \to \mathfrak{g}^*$ satisfying (i)–(iv) above, then on the reduced space $W := \mu^{-1}(0)$ there exists a unique holomorphic symplectic form Ω_W such that*

$$\pi^*\Omega_W = j^*\Omega_V.$$

The form Ω_W is defined by $\Omega_W(\pi u, \pi v) = \Omega_V(u, v)$, $u, v \in T_x\mu^{-1}(0)$.

This theorem was probably first proved by Marsden and Ratiu, but was 'known' to many people. A simple and elegant proof is presented in the book of Kobayashi, pp. 271–274. The key role in the proof is played by

LEMMA. *Let X be a Banach space and $\omega : X \times X \to \mathbb{R}$ a continuous, non degenerate skew form. For any closed subspace $Y \subset X$ set $Y^\omega = \{x \in X : \omega(y, x) = 0$ for all $y \in Y\}$. Then $(Y^\omega)^\omega = Y$.*

The theorem on symplectic reduction allowed Kobayashi to introduce

Kähler structures in moduli space $\mathcal{M}^{H-E}(E)$ on the nonsingular part V of the Sobolev space $W_k^2(A^{0,1}(\operatorname{End} E))$ (where $E \to M$ is a hermitian holomorphic vector bundle of rank r over compact Kähler manifold of dimension n) of holomorphic structures on $E \to M$ (where we identify a holomorphic connection ∇'' with the corresponding holomorphic structure.)

Now let the Hilbert manifold V be a singular part of $W_k^2(A^{0,1}(\operatorname{End} E))$ and we regard it as the union of all $W_{k+1}^2(\operatorname{GL}(E)) \cdot \nabla''$-orbits, where $\nabla'' \in \mathcal{H}''(E)$ is such that the cohomology spaces $H^0(M, \operatorname{End}^0(E^{\nabla''})) = 0$ and $H^2(M, \operatorname{End}^0(E^{\nabla''})) = 0$; $E^{\nabla''}$ is a complex vector bundle with the holomorphic structure given by ∇''. The set (union) of these $W_{k+1}^2(\operatorname{GL}(E))$ orbits is an open subset of $\mathcal{H}''(E)$ which lie over the nonsingular parts of the moduli space $\mathcal{M}^s(E)$ of simple connections (that is, structures) in E ($\mathcal{M}^s(E) = \bar{S}(E)/\operatorname{GL}(E)$).

To be able to apply the abstract theorem on symplectic reduction presented above we must: 1. Define on V, being a set of connections, a holomorphic atlas; this we have done already using the method of Kuranishi; 2. Thus we can regard V as a complex submanifold of $W_k^2(D(E))$; 3. Now in $W_k^2(D(E))$ we have to construct a hermitian structure $\hat{h}$ and the corresponding Kähler form $\Omega = 2\operatorname{Im}\hat{h}$ induced by the fixed hermitian h on E and Kählerian ω on M; 4. Let now $G := W_k^2(U(E,h))/U(1)$ and the corresponding Lie algebra $\mathfrak{g} = W_k^2(\operatorname{End}(E,h))/\mathfrak{u}(1)$. Kobayashi defines a momentum map $\mu : V \to \mathfrak{g}$ by

$$< a, \mu_\nabla >:= \int_M \sqrt{-1}\operatorname{tr}(a \circ (K_\nabla - c1))\omega^n, \quad a \in \mathfrak{g},$$

where K_∇ is, of course, the mean curvature of ∇ and $K_\nabla - c1 = 0$ is the Einstein condition; 5. Finally one realizes that

$$W := \mu_\nabla^{-1}(0)/G$$

is the nonsingular part of the moduli space $\mathcal{M}^{H-E}(E)$ of the irreducible Hermite–Einstein connections in the bundle (E,h).

Of course, one has to check that all the conditions (i)–(iv) of the abstract (infinite dimensional) symplectic reduction theorem are satisfied, which is not obvious. In this way Kobayashi obtains the following marvelous theorem.

THEOREM(KOBAYASHI). *On the Sobolev space $W_k^2(A^{0,1}(\operatorname{End}(E,h))$, $k > n$ there exists a hermitian inner product $\hat{h}$ which induces a Kähler* *metric on the nonsingular part of the moduli space $\mathcal{M}^{H-E}(E)$ of the irreducible Hermite–Einstein connections in the bundle (E,h).*

PROOF (IDEA). Construction of $\hat{h}$. Denote by $X = W_k^2(A^1(\mathrm{End}(E,h))$. This space of 1 forms over M with values in $\mathrm{End}(E,h)$ decomposes into $Z = W_k^2(A^{0,1}(\mathrm{End}(E,h))$ and $\bar{Z} = W_k^2(A^{1,0}(\mathrm{End}(E,h))$. Thus, every $\xi \in X$ decomposes $\xi = \xi' + \xi''$. Since ξ is skew hermitian (${}^t\bar{\xi} = -\xi$), $\xi' = -{}^t\bar{\xi}''$. The complex structure $\mathcal{J}$ on X is given by $\mathcal{J}\xi := i\xi' + i\xi''$, where $i = \sqrt{-1}$. The isomorphism $X \tilde{\to} Z$ is given by $\xi \to \xi''$. The hermitian product $\hat{h}$ on Z is given by

$$\hat{h}(\alpha,\beta) := \int_M \frac{n}{i}\,\mathrm{tr}(\alpha \wedge {}^t\bar{\beta}) \wedge \omega^{n-1}$$

for $\alpha, \beta \in W_k^2(A^{0,1}(\mathrm{End}(E,h))$. Thus the corresponding Kähler form $\hat{\omega}$ on X is given by

$$\hat{\omega}(\xi,\eta) := \int_M n\,\mathrm{tr}(\xi \wedge \eta) \wedge \omega^{n-1}$$

for $\xi, \eta \in X = W_k^2(A(\mathrm{End}(E,h))$.

We have to check that the definition of the momentum $\mu_\nabla : V \to \mathfrak{g}^*$ is correct. Let $a_\nabla \in T_\nabla V$ be a tangent vector of V at ∇ induced by the infinitesimal action of $a \in \mathfrak{g}$: $a_n := \partial_t(e^{-at} \circ \nabla \circ e^{at})|_{t=0} = \nabla a$. Thus

$$\hat{\omega}(a_\nabla,\xi) = \int_M n\,\mathrm{tr}(\nabla a \wedge \xi) \wedge \omega^{n-1} = -\int_M n\,\mathrm{tr}(a \wedge \nabla \xi) \wedge \omega^{n-1} =$$

$$= \int_M i\,\mathrm{tr}(a \circ \nabla^* \xi)\omega^{n-1}.$$

But

$$< a, d\mu_\nabla(\xi) >= \int_M i\partial_t(\mathrm{tr}\, a \circ (K(\nabla + t\xi) - s1))|_{t=0} =$$

$$\int_M i\,\mathrm{tr}(a \circ \nabla^* \xi)\omega^{n-1} = \hat{\omega}(a_\nabla, \xi).$$

μ_∇ is Ad_G^* equivariant: Put $\mu_\nabla \equiv \mu(\nabla)$; we have for $f \in G$, $K(\nabla^f) = f^{-1} \circ K(\nabla) \circ f$, where $\nabla^f := f^{-1} \circ \nabla \circ f$ defines a right G-action $\mathrm{GL}(E)$ on $D(E)$. Thus the short calculation gives

$$< a, \mu(\nabla^f) >= \int_M i\,\mathrm{tr}(a \circ f^{-1}(K(\nabla) - c1)f)\omega^n =$$

$$= \int_M i\,\mathrm{tr}(faf^{-1} \circ (K(\nabla) - c1))\omega^n =< faf^{-1}, \mu(\nabla) >,$$

therefore $\mu(\nabla^f) = Ad^*_{\bar{f}}\mu(\nabla)$. □

HISTORICAL REMARKS. 1. If M is a Riemann surface, that is, if $n = 1$, the corresponding theorem was proved in 1982 in the seminal paper of Atiyah and Bott. At the same time Kobayashi introduced his notion of Hermite–Einstein vector bundle over Kähler manifold of arbitrary dimension. It was a great idea to apply the Yang–Mills connections for investigations of stable bundles. We recommend the reader to read at least the introduction of the Atiyah–Bott paper (and, of course, the whole paper as well. This paper and recollections of Raul Bott on creation of this theory have unrepeatable charm.

2. As we know the first moduli problem was posed by Riemann himself. The reader sees perhaps the analogy between the space of moduli $\mathcal{R}(p)$ of holomorphic structures on a given Riemann surface of genus p and the moduli spaces $\mathcal{M}^s(E)$ and $\mathcal{M}^{H-E}(E)$ investigated in this chapter. This relation is particularly clear in the approach due to C.J. Eerle and J. Eells presented in *A fiber bundle description of Teichmüller theory* (1969). A lovely report on this approach can be found in *Deformations of Riemann Surfaces*, Springer Lecture Notes in Mathematics **103**, 1969.

Let M be a Riemann surface of genus $p \geqslant 2$. The totality of holomorphic structures on M with C^∞ topology, viewed as the space of tensor fields is a Fréchet manifold $\mathcal{M}(p)$. Let $Diff^+(M)$ be the group of orientation preserving C^∞ diffeomorphisms on M and let $Diff_0^+(M)$ be its subgroup consisting of diffeomorphisms which are homotopic to the identity (this is a closed normal subgroup). We have a natural action of $Diff^+(M)$ on $\mathcal{M}(p)$:

$$\mathcal{M}(p) \times Diff^+(M) \to \mathcal{M}(p),$$

$$(\mathcal{J}, f) \to \mathcal{J} \cdot f,$$

where $\mathcal{J}$ is a complex structure of M.

THEOREM (EERLE–EELLS, 1967-9). *(a) $Diff^+$ acts effectively and properly on $\mathcal{M}(p)$;*

(b) The Riemann moduli space $\mathcal{R}(M) = \mathcal{M}(p)/Diff^+(M)$ is a normal complex space;

(c) The Teichmüller space $\mathcal{T}(p) := \mathcal{M}(p)/Diff_0^+$ is a simply connected complex manifold (even a Stein manifold) of dimension $3(p-1)$ (for $p = 1$, $\dim \mathcal{T}(1) = 1$; $\dim \mathcal{T}(0) = 0$ since $\mathcal{T}(0)$ is a point);

(d) If we set $\Gamma(p) := Diff^+(p)/Diff_0^+(p)$ *(the Teichmüller modular group for genus p), then* $\Gamma(p)$ *is a discrete topological group,*

$$\mathcal{R}(p) = \mathcal{T}(p)/\Gamma(p) = \mathcal{M}(p)/(Diff^+(M)/Diff_0^+(M)).$$

Thus the Teichmüller space $\mathcal{T}(p)$ *is the universal covering manifold for the Riemann moduli space* $\mathcal{R}(p)$.

REMARK. For $p > 2$ the set of singularities of $\mathcal{R}$ is the branch locus of the map $\mathcal{T}(p) \to \mathcal{R}(p)$. The point (b) is analogous to the Kobayashi–Lübke–Okonek theorem: $\mathcal{M}^s(E) = S(E)/G$ is a complex space of finite dimension.

The construction of moduli spaces of *simple* vector bundle has been achieved in the important paper of M. Lübke and C. Okonek (1987). Let $\mathcal{M}^s(E)$ be the set of isomorphism classes of simple holomorphic bundles *differentiably* equivalent to a differentiable vector bundle $E \to M$ over compact complex manifold M. The moduli space $\mathcal{M}^s(E)$ can be equipped with a locally Hausdorff topology and is a complex space (in general, non reduced) of finite dimension. This space is obtained as a closed subspace of the space $\bar{\mathcal{M}}^s(E)$ of all simple connections $\nabla^{0,1}$ in E. The space $\bar{\mathcal{M}}^s(E)$ is a (non separated) complex Hilbert manifold obtained in the in the following steps:

1. Fix hermitian metric in M and E, then the spaces $A^{p,q}(\mathrm{End}(E))$ are Fréchet in a usual manner, and the curvature $R \in A^{0,2}(\mathrm{End}(E))$ is Fréchet differentiable. Since dR is $\mathbb{C}$ linear, we can regard R as a holomorphic map $R : D(E) \to A^{0,2}(\mathrm{End}(E))$, where $D(E) := \{\nabla'' + A^{0,1}(\mathrm{End}(E))\}$ is an affine space. Therefore $\mathcal{H}(E) := R^{-1}(0) \subset D(E)$ is a complex subvariety. The group $\mathrm{GL}(E)$ of differentiable automorphisms of E acts on $D(E)$ by

$$D(E) \times \mathrm{GL}(E) \ni (\nabla'', g) \to \nabla'' \cdot g = g^{-1} \circ \nabla'' \circ g = \nabla'' + g^{-1} \circ \nabla''(g) \in D(E).$$

Since the subgroup $\mathbb{C}^\times$ of constant automorphisms acts trivially, we regard this action as an action of $\mathcal{G} := \mathrm{GL}(E)/\mathbb{C}^\times$. We denote by $\bar{S}(E)$ the set of all simple connections in E, that is, the connections associated with a simple holomorphic bundle (E is simple if every holomorphic endomorphism of E is constant, $H^0(\mathrm{End}(E)) \simeq \mathbb{C}$.)

2. The subset $\bar{S}(E) \subset D(E)$ is open and $\mathcal{G}$ invariant; $\mathcal{G}$ acts freely on $\bar{S}(E)$.

Since the proof of openness is interesting, we reproduce it here. Fix a hermitian metric in E; then the map $\nabla'' : A^0(\mathrm{End}(E)) \to A^{0,1}(\mathrm{End}(E))$ has an adjoint $(\nabla'')^* : A^{0,1}(\mathrm{End}(E)) \to A^0(\mathrm{End}(E))$. Let $\Box := \nabla''^* \circ \nabla''$. This operator has the symbol

$$\sigma_\Box(x,v)(1 \otimes e) = |v|^2 \otimes e, \quad \text{for } 0 \neq v \in T_x^{\mathbb{R}}M,\, e \in E_x.$$

Since this symbol is invertible, the operator $\Box$ is elliptic and the openness follows from $\ker \nabla'' = \ker \Box$ and semicontinuity of elliptic operators.

3. To endow $\bar{\mathcal{M}}^s \equiv \bar{S}(E)/\mathcal{G}$ with a complex analytic structure one uses the inverse mapping theorem and Kuranishi maps. A fixed hermitian metric in E makes it possible to introduce in the spaces $A^{p,q}(\mathrm{End}(E))$ the W_k^2 Sobolev scalar product and the corresponding W_k^2 completion for sufficiently high order k (of derivatives.) In particular, in this way we obtain the Hilbert–Lie groups $\mathcal{G}_k$ with the Lie algebra $A(\mathrm{End}(E))^0_{k+1}$, where $A^0(\mathrm{End}(E))^0_k = \{g \in A^0(\mathrm{End}(E)) : \int_M \mathrm{tr}(g)\omega^n = 0\}$ acting on $D(E)_k := \nabla'' + A^{0,1}(\mathrm{End}(E))_k$.

A slice through ∇'' for this action restricted to $S(E)_k$ is

$$V_\epsilon = \{\nabla'' + \alpha : \alpha \in A^{0,1}(\mathrm{End}(E))_k, \| \alpha \|_k < \epsilon, D_k''^*(\alpha) = 0\}.$$

For small $\epsilon > 0$ the map $s : V_\epsilon \times G \to A^{0,1}(\mathrm{End}(E))_k$, $(D'' + \alpha, g) \to (\nabla'' + \alpha) \cdot g - \nabla''$ is differentiable. Since ∇'' is simple $\left(\ker \left(\nabla''|_{A^0(\mathrm{End}(E))^0}\right) = \{0\}\right)$, and since $\mathrm{im}\, \nabla''_{k+1}$ and $\ker \nabla''_k$ are orthogonal, it follows that $ds_{(\nabla'',1_E)}$ is injective. One proves its surjectivity by

LEMMA. $A^{0,1}(\mathrm{End}(E))_k = \mathrm{im}\, \nabla''_{k+1} \oplus \ker \nabla''_k$.

The proof of this lemma proceeds as the proof of Hodge theorem. Thus we have proved that the map s is *invertible* locally around $(\nabla'', 1_E)$ and it is locally biholomorphic with respect to the Hilbert analytic structures. If $\pi : \bar{S}(E)_k \to \bar{\mathcal{M}}_k^s(E)$ is the projection to $\bar{\mathcal{M}}_k^s(E) := D(E)_k/\mathcal{G}_{k+1}$, then $\pi|_{V_\epsilon}$ is injective for small $\epsilon > 0$.

4. We equip $U := \pi(V_\epsilon)$ with the analytic structure via π and in this way we obtain the biholomorphic map $\varphi : U \times \mathcal{G}_{k+1} \to \pi^{-1}(U)$, $\varphi(\pi(\nabla'' + \alpha), g) = (\nabla'' + \alpha) \cdot g$. If we show that the local charts $\psi : U \to V_\epsilon$, $\psi := (\pi|_{V_\epsilon})^{-1}$ glue together biholomorphically, we have

THEOREM. *$\bar{\mathcal{M}}_k^s(E)$ is a (non separated) Hilbert analytic manifold. The projection $\pi : \bar{S}(E)_k \to \bar{\mathcal{M}}_k^s(E)$ is a principal $\mathcal{G}_k$ bundle.*

Till now we considered only C^∞ forms. But in order to discuss forms valued in holomorphic vector bundles $V \to M$ over compact M and the corresponding Hodge decompositions and Green operators, it is convenient to work with the corresponding Sobolev spaces of sections of vector bundles $W_k^2(V)$. We recall that the Sobolev space $W_k^2(V)$ is the completion of $C^\infty(V)$ with respect to the norm

$$\| \varphi \|_k \equiv \| \varphi \|_{k,2} := \left(\sum_{l=0}^{k} \int_M |\nabla^l \varphi|^2 \right)^{1/2},$$

where $\nabla^l \varphi := \nabla \circ \nabla \circ \cdots \circ \nabla \in C^\infty(\overset{l}{\otimes} T^*M \otimes E)$ is the lth covariant derivative of the section φ.

In the sequel V is a vector bundle $E \otimes T^*M$, or $E \times A^{p,q}(M) := A^{p,q}(E)$, or $\mathrm{End}(E) A^{p,q}(M)$. The spaces $W_k^2(V)$ are Hilbert spaces and we have natural inclusions

$$C^\infty(V) \subset \cdots \subset W_{k-1}^2(V) \subset W_k^2(V) \subset \cdots \subset W_0^2(V),$$

where each inclusion is compact. Consider two vector bundles V and W over M and a linear differential operator $A : C^\infty(V) \to C^\infty(W)$ of order a; then it extends to a continuous linear operator $A : W_k^2(V) \to W_{k-a}^2(W)$. Its adjoint, $A^* : C^\infty(W) \to C^\infty(V)$ is defined by $(Av|w) = (v|A^*w)$ and is a linear differential operator of order a as well. The elliptic regularity theorem asserts the following:

If A is elliptic of order a (for example, the Laplacian $\Box : A^{p,q}(E) \to A^{p,q}(E)$ has order 2), then $\ker A$ *is a finite dimensional subspaces of* $C^\infty(V)$ *and the kernel* $\ker A$ *agrees with the kernel of the extended operator $A : W_{k+a}^2(V) \to W_k^2(W)$.*

An element of $\ker A$ is called A harmonic (harmonic in the case $A = \Box$), and $\ker A$ will be denoted by $\mathcal{H}(V)$. We have

THEOREM (HODGE). $C^\infty(V) = \mathcal{H}(V) \oplus A(C^\infty(V)) = \ker A \oplus \operatorname{im} A$.

If H is an orthogonal projection $H : C^\infty \to \mathcal{H}(V)$ then there exists a unique continuous, linear operator G called the *Green operator* for A $G : C^\infty(V, \| \cdot \|_k) \to C^\infty(V, \| \cdot \|_{k+a})$ such that $1 = H + AG$, $AG = GA$, $\ker G = \ker A$. G extends to a continuous linear operator called also the Green operator,

$$G : W_k^2(V) \to W_{k+a}^2(V).$$

For an elliptic complex (V^j, D) of vector bundles $j = 0, \ldots, m$ with differential operator $D : C^\infty(V^j) \to C^\infty(V^{j+1})$ of order a such that $D \circ D = 0$ and for every $\xi \in T_x^*M, \xi \neq 0$ the symbol sequence

$$0 \to V_x^0 \overset{\sigma(x)}{\to} V_x^1 \overset{\sigma(x)}{\to} \cdots \overset{\sigma(x)}{\to} V_x^m \to 0$$

is exact. Then $A = DD^* + D^*D : C^\infty(V^j) \to C^\infty(V^j)$, $j = 0, \ldots, m$ is elliptic of order $2a$. If H^j is the jth of the elliptic complex $\{V^j, D\}$

$$H^j := \ker\left(D : C^\infty(V^j) \to C^\infty(V^{j+1}\right) / \operatorname{im}\left(C^\infty(V^{j-1}) \to C^\infty(V^j)\right),$$

then there is the Hodge isomorphisms

$$H^j \simeq \mathcal{H}^j := \mathcal{H}(V^j), \quad j = 1, \ldots, m.$$

Now we can continue the construction of moduli spaces for holomorphic structures of our vector bundle $E \to M$.

Let $S(E)_k := \mathcal{H}(E)_k \cap \bar{S}(E)_k$, $\mathcal{M}(E) := S(E)_k / \mathcal{G}_{k+1}$. The space $\mathcal{M}_k^s(E)$ is closed in $\bar{\mathcal{M}}_k^s(E)$. Then we have the commutative diagram

$$\begin{array}{ccc}
S(E)_k & \hookrightarrow & \bar{S}(E)_k \\
\downarrow \pi' & & \downarrow \pi \\
\mathcal{M}_k^s := S(E)_k/\mathcal{G}_{k+1} & \hookrightarrow & S(E)_k/\mathcal{G}_{k+1} =: \mathcal{M}_k^s
\end{array}$$

Now we have the following important

THEOREM (LÜBKE–OKONEK, 1987). *$\mathcal{M}_k^s(E)$ is a finite dimensional, locally Hausdorff complex space and*

$$\pi' : S(E)_k \to \mathcal{M}_k^s(E) := S(E)_k / \mathcal{G}_{k+1}$$

is a principal $\mathcal{G}_{k+1}$ bundle.

REMARK. For sufficiently large k it follows from Sobolev theorem that all objects above are smooth and in the sequel we can drop Sobolev indices.

The proof of finite dimensionality goes as follows:

If $\nabla'' \in S(E)$ defines a simple holomorphic structure and $\alpha \in A^{0,1}(\mathrm{End}(E))$, then

$$(\nabla'' + \alpha \text{ is holomorphic}) \Longleftrightarrow (\nabla''(\alpha) = -\alpha \wedge \alpha).$$

Therefore V_ϵ is a slice through ∇'' for the $\mathcal{G}$ action on $S(E)$ and is isomorphic with $V_\epsilon(\nabla'') = \{\alpha \in A^{0,1}(\mathrm{End}(E)) : \| \alpha \| < \epsilon, \nabla''^* = 0, \nabla''(\alpha) = -\alpha \wedge \alpha\}$ for sufficiently small $\epsilon > 0$. But ∇'' is holomorphic, therefore we have an elliptic Dolbeault complex $A^{0,\cdot}(\mathrm{End}(E), \nabla'')$ with Laplacian $\Box$ and Green operators G, and harmonic spaces

$$H^{0,i} := \ker(\Box : A^{0,i}(\mathrm{End}(E)) \to A^{0,i}(\mathrm{End}(E)) \simeq$$

$$\simeq H^i(\mathrm{End}(E))$$

which are (by ellipticity of $\Box$) finite dimensional.

The *Kuranishi map* was introduced by Kuranishi in his famous paper *New proof for existence of locally complete families of complex structures*, Proceedings of Conference on Complex Analysis, Minneapolis 1964, pp. 142–152. Kuranishi proves there

THEOREM (KURANISHI, 1964). *For any complex analytic compact manifold M there exists a universal family of deformations.*

As we saw Hans Grauert proved later (with quite different methods) a generalization of Kuranishi theorem for every compact complex space M.

DEFINITION. The Kuranishi map is defined as follows

$$k_{\nabla''} : A^{0,1}(\mathrm{End}(E)) \to A^{0,1}(\mathrm{End}(E))$$

$$\alpha \to \alpha + \nabla'' \circ G(\alpha w \omega).$$

The map $k_{\nabla''}$ extends to Sobolev completion, is holomorphic, and invertible at 0 because $dk_{\nabla''}(0) = \mathrm{id}$. Moreover $k_{\nabla''}$ maps the slice $V_{\nabla'',\epsilon}$ into $H^{0,1}$ (which is not obvious!) and therefore $\mathcal{M}^s(E)$ is finite dimensional. $\Box$

If ∇'' is smooth and holomorphic, then all elements of the slice $V_{\nabla'',\epsilon}$ are smooth by virtue of elliptic regularity. Therefore

THEOREM (LÜBKE–OKONEK, 1987). *The space $\mathcal{M}^s(E)$ is finite dimensional, locally Hausdorff complex space.*

Since $\mathcal{M}^s(E)$ has singularities, the question arises as to when $\pi'(\nabla'')$ is a smooth point of the space $\mathcal{M}^s(E)$? A partial answer is provided by

PROPOSITION. *Let $\mathfrak{sl}(E)$ be the bundle of traceless holomorphic endomorphisms of $E = E(\nabla'')$. Then*
$(H^2(\mathfrak{sl}(E) = 0) \Longrightarrow (\pi'\nabla''$ is a smooth point of $\mathcal{M}^s)$.

PROOF. For $\beta \in A^1(\mathrm{End}(E))$, $\mathrm{tr}(\beta \wedge \beta) = 0$. □

Let us return to

Hermite–Einstein bundles. We recall that ∇ is a H-E connection if $\mathrm{tr}_\omega(R^\nabla) = c1_E$, where $\mathrm{tr}\, A^{1,1}(\mathrm{End}(E)) \to A^0(\mathrm{End}(E))$ denotes contraction by the Kähler form ω.

If the H-E connection ∇ is irreducible, that is, there is no splitting of E parallel with respect to ∇, then E is co-stable by Kobayashi–Lübke theorem. Whence the connection ∇'' is simple . We have the map $\varphi = \varphi_h :=$ $\mathcal{M}(E,h) \to \mathcal{M}^s(E)$, where $\mathcal{M}(E,h) := A^{0,1}(E,h)/U$, $U := U(E,h)/U(1)$ and $A^{0,1}$ is the $(0,1)$ part of h-Einstein connections. Thus $\mathcal{M}(E,h)$ is the moduli space of irreducible H-E connections. But the important theorem of H.J. Kim asserts

THEOREM (KIM, 1985). *The image of the map φ_h, $\mathcal{M}^{H-E}(E) :=$ $\varphi_h(\mathcal{M}(E,h)) \subset \mathcal{M}^s(E)$ is independent of h. Moreover $\varphi : \mathcal{M}^{H-E}(E) \to$ $\mathcal{M}^s(E)$ is an open embedding.*

As a corollary we obtain the important

THEOREM (LÜBKE–OKONEK, 1987). *The moduli space $\mathcal{M}^{H-E}(E)$ of irreducible H-E connections is a (globally) Hausdorff complex space.*

This theorem together with the Uhlenbeck–Yau theorem asserting that every ω stable bundle over arbitrary compact Kähler manifold is an irreducible H-E bundle gives

COROLLARY (LÜBKE–OKONEK, KOBAYASHI). *The moduli space $\mathcal{M}^\omega$ of* *isomorphism classes of ω stable vector bundles is a Hausdorff complex space (of finite dimension).*

EXERCISE. We know that the moduli space $\mathcal{M}^s(E)$ is nonsingular at $[\nabla]$ if $H^2(M, \mathrm{End}^0(E, \nabla'')) = 0$. Prove that such $[\nabla]$, $T_{[\nabla]}\mathcal{M}^s(E) \simeq H^{0,1} = H^1(M, \mathrm{End}(E^{\nabla''}))$.

There are interesting results for $n = 2$, that is, for Kähler surfaces, concerning the dimension of moduli spaces $\mathcal{M}^{H-E}(E)$:

THEOREM (KOBAYASHI, KIM). *Let* $\dim_{\mathbb{C}} M = 2$ *and* $c_1(M) > 0$ *or* $c_1(M) = 0$. *Then if* $\int_M c_1(M) \wedge \omega \geqslant 0$, *the moduli space* $\mathcal{M}^{H-E}(E)$ *is a nonsingular Kähler manifold of dimension* ($r = \mathit{rank}(E)$, $h^{0,1} = \dim H^{n-1}(M, \mathcal{H}_M)$)

$$(a) \qquad \dim \mathcal{M}^{H-E}(E) = 2rc_2(E) - (r-1)c_1(E)^2 + r^2 h^{01}(M) + 1 - r^2$$

if the canonical line bundle of M, $\mathcal{H}_M$ *is nontrivial;*

If $\mathcal{H}_M$ *is trivial, that is if* M *is a torus or a* $K3$ *surface, then*

$$(b) \qquad \dim \mathcal{M}^{H-E}(E) = 2rc_2(E) - (r-1)c_1(E)^2 + r^2 h^{01}(M) + 2 - 2r^2$$

If $c_1(M) > 0$, *then*

$$(c) \qquad \dim \mathcal{M}^{H-E}(E) = 2rc_2(E) - (r-1)c_1(E)^2 + 1 - r^2$$

REMARK. For $K3$ surface $h^{0,1} = 0$. Thus if $E = TM$, M is $K3$, $c_1(TM) = 0$, $r = 2$, $c_2(TM) = 24$, and $\dim \mathcal{M}^{H-E}(TM) = 4 \cdot 24 - 4 = 92$.

Part VI

Riemann and Number Theory

Chapter 1

Introduction

1.1 Introduction

There is a common (but incorrect) view that the role of Riemann in number theory is limited to introduction of the ζ function and his famous hypothesis on zeros of ζ function. This alone, of course, would be a great impulse: this is the starting point of a huge part of mathematics, the analytic theory of numbers.

But the theory of Riemann surfaces has given mathematics a new language and has opened new worlds and new perspectives. Today one cannot even think about number theory without this language.

Erich Hecke theory of modular functions is indispensable for the (algebraic) number theory. The famous solution of the Fermat problem by Andrew Wiles (1994) would be impossible without Hecke theory.

The role of modular functions is surprisingly important in solution of yet another famous problem: the problem of construction of all sporadic groups and the solution of the fascinating 'Monster Moonshine' by Borcherds (1992). Along with the theory of modular curves he makes use of the 'no ghost theorem' of quantum string theory.

One of the greatest ideas of Riemann was his general theory of ϑ functions which is connected with Riemann period relations and his construction of the Jacobi tori $\mathbb{C}^p/\Lambda$ for Riemann surfaces of genus $p \geqslant 1$, the wonderful Θ divisor with its geometry, and Riemann solution of the Jacobi inversion problem. The theory of ϑ functions is a vast, difficult domain, but as shown by Pierre Cartier in his seminal paper *Theta Functions and Quantum Commutativity relation* (1964) it becomes lucid and clear thanks to Bargmann representation of the Heisenberg-Weyl group G.

Riemann road to his ζ function was paved by Euler and Dirichlet: The famous Quadratic Reciprocity

$$\left(\frac{l}{p}\right)\left(\frac{p}{l}\right) = (-1)^{\frac{l-1}{2}\frac{p-1}{2}}$$

was discovered or conjectured by Euler and *almost* rpoved by Legendre (almost because he assumed without proof the existence of *Infinite primes in arithmetic progression.*) This famous conjecture was proved much later by Dirichlet. Euler constructed his wonderful ζ function for *real* x:

$$\zeta(x) := \sum_{n=1}^{\infty} \frac{1}{n^x} = \prod_{p-primes} (1 - p^{-x})^{-1}$$

and found the values $\zeta(2k)$ – he was truly happy and proud of it.

Dirichlet extended Euler ζ to his L series

$$L(s, \chi) := \sum_{n=1}^{\infty} \chi(n) n^{-s} = \prod_{p-primes} \frac{1}{1 - \chi(p)/p^s},$$

with complex s and χ being the Dirichlet characters modulo m: $\chi(\cdot)$ is a multiplicative function on the multiplicative commutative group $G(m) := (\mathbb{Z}/m\mathbb{Z})^{\times}$, $\chi(a) \in \mathbb{C}^{\times}$, $\chi(ab) = \chi(a)\chi(b)$. $\chi(a) = 0$ if a is not prime to m. $L(s, \chi)$ is holomorphic for $\mathrm{Re}\, s > 0$. For $\chi = 1$, $L(s, 1) = F(s)\zeta(s)$, where $F(s) := \prod_{p|m}(1 - p^{-s})$.

Clearly $\lim_{s \to 1} L(s, \chi)$ exists but an important question is if

$$(D) \qquad \lim_{s \to 1} L(s, \chi) \neq 0?$$

We know that $\lim_{s \to 1} \zeta(s) = \sum n^{-1} = \infty$. The great idea of Dirichlet was that assuming (D), $L(1, \chi) = \sum_p \chi(p) p^{-1}$ converges. But $\sum n^{-1}$ diverges. Therefore for a and m relatively prime there exist infinitely many primes p with $p = a \bmod m$.

Riemann extends ζ analytically to meromorphic function on the whole complex plane $\mathbb{C}$, investigates its residua and proves the famous functional equation

$$\xi(s) = \xi(1 - s)$$

with $\xi(s) := \pi^{-s/2}\Gamma(s/2)\zeta(s)$, where Γ is the Euler gamma function. He makes five conjectures concerning zeta function; four of them has been proved, but the fifth, the famous hypothesis that all nontrivial zeros of ζ

lie on the line $\operatorname{Re} z = \frac{1}{2}$ still resists all attempts of proving. Today nobody doubts that this conjecture is correct, but mathematics seems still to be helpless, though 'every three years a new (false) proof appears.' Erich Hecke was meditating on functional equations of Riemann type all his life.

1.2 Automorphic forms, modular functions

Let B be a bounded symmetric domain in $\mathbb{C}^n$, e.g., the Siegel half space $\mathfrak{H}_n$ or the unit disc $\mathbb{D}$, and let Γ be a discrete (discontinuous) subgroup of the automorphism group $\operatorname{Aut} B$ of B acting on B. We consider two canonically isomorphic functional spaces:

1. $H^0(B, \Gamma, k)$ of Γ-automorphic forms of weight k: which equals the set of holomorphic functions $f : B \to \mathbb{C}$ such that for $\gamma \in \Gamma$, $j_\gamma(z)^k f(\gamma \cdot z) = f(z)$, where $j_\gamma(z)$ is the derivative (Jacobi determinant) of the map $\gamma : B \to B$ at $z \in B$.

2. Assume that B/Γ is compact. $H^0(B/\Gamma, k)$ is the space of holomorphic differential forms of weight k locally defined by $a(z_1, \ldots, z_n)[d\bar{z}^1 \wedge \cdots \wedge d\bar{z}^n]^k$, with $a(\cdot)$ holomorphic.

Clearly these spaces are canonically isomorphic $H^0(B, \Gamma, k) \simeq H^0(B/\Gamma, k)$.

It is a very important problem to determine the dimension of these spaces. For the modular group $\Gamma = \Gamma(1) = \mathrm{PSL}(2, \mathbb{Z})$ acting on $\mathfrak{H} = \mathfrak{H}_1$,

$$j_\gamma(z) = \left(\frac{az+b}{cz+d}\right)' = \frac{1}{|cz+d|^2}$$

and, by definition, the $\Gamma(1)$ automorphic forms of weight k are

$$f\left(\frac{az+b}{cz+d}\right) = |cz+d|^{2k} f(z)$$

and $f(z)(d\bar{z})^k \in H^0(\mathfrak{H}/\Gamma(1), k)$. The space $H^0(\mathfrak{H}, \Gamma(1), k)$ is denoted by $S_k(\mathfrak{H})$ and we have $S_2(\mathfrak{H}) \simeq \Omega^{0,1}(\mathfrak{H}/\Gamma)$.

The group $\mathrm{SL}(2, \mathbb{Z})$ is generated by two elements $T = \begin{pmatrix} 1 & 1 \\ 0 & 1 \end{pmatrix}$ and $S = \begin{pmatrix} 0 & 1 \\ -1 & 0 \end{pmatrix}$ satisfying the relations $S^2 = (TS)^3 = 1$. Here are some very important modular forms:

Eisenstein series Let L be a lattice in $\mathbb{C}$ generated by $(1, z)$, $\operatorname{Im} z > 0$, $L = \{z \in \mathbb{C} : z = n + mz, n, m \in \mathbb{Z}, \operatorname{Im} z > 0\}$; then

$$E_k(z) := \sum_{0 \neq m, n \in \mathbb{Z}} (mz + n)^{-2k}, \quad z \in \mathfrak{H}$$

is called the Eisenstein series of weight $k \geqslant 0$. If we denote by $g_2 := 60e_4$, $g_3 := 140e_6$, $\Delta := g_2^3 - 27g_3^2$, $j = 12^3 \frac{g_2^3}{\Delta}$, then these modular forms have weights 4, 6, 12, 0.

The modular function j was discovered by Dedekind in 1872 and is called the *modular invariant*. It is of fundamental importance for the theory of elliptic functions and is one of the most famous functions in mathematics.

If L is a lattice in $\mathbb{C}$, then the L periodic meromorphic functions are called elliptic functions. They can be regarded as meromorphic functions on the torus $\mathbb{C}/L$.

Weierstrass $\wp$ function is defined by

$$\wp(z) \equiv \wp_L(z) := \frac{1}{z^2} + \sum_{0 \neq \omega \in \Gamma} \left(\frac{1}{(z - \omega)^2} - \frac{1}{\omega^2} \right).$$

Since $\wp'(z) = -2 \sum_{\omega \in L} (z - \omega)^3$, the elliptic curve $y^2 = 4x^3 - g_2 x - g_3$ can be parametrized by $\wp$ and $\wp'$, to wit

$$\wp'(z)^2 = 4\wp(z)^3 - g_2 \wp(z) - g_3.$$

Therefore the field $\mathcal{M}(\mathbb{C}/L)$ of meromorphic functions on the torus $\mathbb{C}/L$, that is, the L-elliptic functions is generated by $\wp$ and $\wp'$:

$$\mathcal{M}(\mathbb{C}/L) = \mathbb{C}(\wp, \wp').$$

The modular invariant is so important because it generates the field of modular functions on $X(1) := \overline{\mathfrak{H}/\Gamma}$, $\mathcal{M}(X(1)) = \mathbb{C}(j)$. We will denote by $X_0(N)$ the compactification of the Riemann surface $\mathfrak{H}/\Gamma_0(N)$, where $\Gamma_0(N)$ denotes the discrete subgroup $\mathrm{PSL}(2, \mathbb{R})$ of matrices $\begin{pmatrix} a & b \\ c & d \end{pmatrix} = \begin{pmatrix} * & * \\ c & * \end{pmatrix}$ with $c = 0 \bmod N$, that is $c|N$, $N \in \mathbb{N}^\times$.

Erich Hecke developed the wonderful theory of modular forms of weight k and level N $S_k(\Gamma_0(N))$. The group $\Gamma_0(N)$ is so important because it is a common stabilizer of the lattices $L_1 :=< e_1, e_2 >$ and $L_N :=< Ne_1, e_2 >$. For the number theory the spaces $S_2(\Gamma_0(N))$ are of paramount importance.

CHAPTER 2

The Riemann ζ function

The infinite series $\zeta(m) := \sum_{n=1}^{\infty} \frac{1}{n^m}$ for $m \in \mathbb{N}$ fascinated mathematicians long before Euler. It was early known that the 'harmonic series' $1+\frac{1}{2}+\frac{1}{3}+\cdots$ is divergent. In 1737 Euler introduced, for real $s > 1$, his ζ function

$$\zeta(s) = \sum_{n=1}^{\infty} n^{-s} \tag{1}$$

and investigated its properties in depth. He showed that $\zeta(s)$ can be represented as an infinite product

$$\zeta(s) = \prod_{p} \frac{1}{1-p^{-s}} \tag{2}$$

where p runs over all *prime numbers*. In 1725 came his sensational discovery of the formula

$$\zeta(2) = \frac{\pi^2}{6}. \tag{3}$$

Euler was fascinated and proud of it. Shortly afterwards Euler obtained the values of $\zeta(s)$ for positive and even integer s

$$\zeta(2m) = \frac{2^{2m-1}\pi^{2m} B_{2m}}{(2m)!}, \tag{4}$$

where B_{2m} are Bernoulli numbers, defined by the power expansion

$$\frac{x}{e^x - 1} = 1 - \frac{x}{2} + \sum_{k=1}^{\infty} (-1)^{k+1} B_k \frac{x^{2k}}{(2k)!}. \tag{5}$$

Thus $B_1 = \frac{1}{6}$, $B_2 = \frac{1}{30}$, $B_3 = \frac{1}{42}$, $B_4 = \frac{1}{30}$, ..., $B_{14} = \frac{79349362903}{870}$ and all $B_k \in \mathbb{Q}_+$. Therefore, for example,

$$\zeta(14) = \frac{2\pi^{14}}{3^6 \cdot 5^2 \cdot 7 \cdot 13}. \tag{6}$$

No wonder that mathematical community was fascinated. We owe Euler many important and beautiful functions; among them the Γ function is the best known

$$\Gamma(s) := \int_0^\infty e^{-t} t^s \frac{dt}{t} = \mathbb{M}(e^{-t})(s), \tag{7}$$

where

$$\mathbb{M}(F)(s) := \int_0^\infty F(t) t^s \frac{dt}{t} \tag{8}$$

is called the *Mellin transform* .

REMARK. The Mellin transform is the 'Fourier transform' $\hat{F}(s)$ on the (commutative, locally compact) multiplicative group $\mathbb{R}_+^\times = \{t \in \mathbb{R} : t > 0\}$. The Haar measure μ is given by $\mu(\varphi) := \int_0^\infty \varphi(t) \frac{dt}{t}$; the characters are $\chi_s(t) := t^s$. This transform is of paramount importance in the theory of Dirichlet L functions and in the Hecke theory of modular cusp forms.

As we know a great breakthrough in the theory of ζ function was achieved by Riemann in 1859: he was so incredibly courageous, as to consider the ζ function on the whole complex plane, defining it as

$$\zeta(s) := \sum_{n=1}^\infty \frac{1}{n^s}.$$

This function is holomorphic in the half plane $\operatorname{Re} z > 1$. Riemann extended this function meromorphically onto the whole $\mathbb{C}$ proving his famous *fundamental equation for ζ function*

THEOREM (RIEMANN, 1859). *Denote by*

$$\xi(s) := \pi^{-s/2} \Gamma\left(\frac{s}{2}\right) \zeta(s) \tag{9}$$

then

$$\xi(s) = \xi(1-s). \tag{10}$$

The residuum of ζ at $s = 1$ is 1. Around $s = 1$

$$\zeta(s) = \frac{1}{s-1} + C + O(|s-1|),$$

where C is the Euler constant.

Modular forms of weight $k \in \mathbb{Z}$ are (by definition) holomorphic functions f on the upper half plane $\mathfrak{H} := \{z \in \mathbb{C} : \operatorname{Im} z > 0\}$ which satisfy

$$f(\gamma z) = (cz + d)^k f(z), \quad \text{for all } \gamma = \begin{pmatrix} a & b \\ c & d \end{pmatrix} \in \mathrm{SL}(2, \mathbb{Z}). \tag{11}$$

Modular forms are periodic ($f(z+1) = f(z)$) and have therefore the Fourier expansion

$$f(z) = \sum_{n=-\infty}^{\infty} c_n q(z)^n, \quad q(z) := e^{2\pi i z} =: e(z), \tag{12}$$

called q expansion.

2.1 L functions of cusp forms

The modular form is holomorphic at ∞ if its q expansion (12) has $c_n = 0$ for all $n < 0$. If $c_0 = 0$ as well, f is called *cusp form*. The space of cusp forms of weight k will be denoted by S_k.

Let $f \in S_k$ be a cusp form and let $\sum_{n=0}^{\infty} c_n q(z)^n$ be its q expansion. The *L function of f* is the Dirichlet series

$$L(z, f) := \sum_{n=1}^{\infty} \frac{c_n}{n^z} \tag{13}$$

Thus $L(\cdot, f) = \mathbb{M}(f)$ is the Mellin transform of f.

PROOF. One can prove that $|c_n| \leqslant A n^{k/2}$; then the series (13) converges absolutely for $\operatorname{Re} z > k/2 + 1$ and $L(\cdot, f)$ is holomorphic there. Now write $z = x + iy$ and compute the Mellin transform of $f(iy)$:

$$g(s) := \int_0^{\infty} f(iy) y^s \frac{dy}{y} = \int_0^{\infty} \sum_{n=1}^{\infty} c_n e^{-2\pi n y} y^s \frac{dy}{y} =$$

$$\sum_{n=1}^{\infty} c_n \int_0^{\infty} e^{-t}(2\pi n)^{-s} t^s \frac{dt}{t} = (2\pi)^{-s}\Gamma(s) \sum_{n=1}^{\infty} \frac{c_n}{n^s} =$$
$$= (2\pi)^{-s}\Gamma(s) L(s,f).$$

□

Now let us turn to an important result of Hecke: the functional equation for $L(s,f)$ analogous to the Riemann functional equation for ζ function.

THEOREM (HECKE). *If $f \in S_k$ is a cusp form for* $\mathrm{SL}(2,\mathbb{Z})$, *then the L function $L(s,f)$ initially defined for* $\mathrm{Re}\, s > k/2+1$, *extends meromorphically to the whole complex plane. Moreover, the function*

$$\Lambda(s,f) := (2\pi)^{-s}\Gamma(s)L(s,f) \tag{14}$$

satisfies the functional equation

$$\Lambda(s,f) = (-1)^{k/2}\Lambda(k-s,f). \tag{15}$$

PROOF. From (13) we have $\Lambda(s,f) = \int_0^\infty f(iy)y^{s-1}dy$. But $f(i/y) = i^k y^k f(iy)$, whence

$$\Lambda(s,f) = \int_0^1 f(iy)y^{s-1}dy + \int_1^\infty f(iy)y^{s-1}dy =$$

$$= i^k \int_1^\infty f(iy)y^{k-s-1}dy + \int_1^\infty f(iy)y^{s-1}dy \tag{16}$$

(we have replaced y by y^{-1}.)

The first term is of the same shape as the second one and extends therefore to the holomorphic function on the whole $\mathbb{C}$. Replace s by $k-s$ in (16) and multiply by i^k, then

$$i^k\Lambda(k-s,f) = (-1)^k \int_1^\infty f(iy)y^{s-1}dy + \int_1^\infty f(iy)y^{k-s-1}dy \tag{17}$$

Comparing (16) with (17) and using (14), we obtain the functional equation (15). □

Chapter 3

Hecke Theory

3.1 Petersson Scalar Product

In the theory of Teichmüller manifold $\mathcal{T}_p$ of the moduli space of Riemann surfaces of genus $p > 1$ of great importance is the Petersson-Weil $(1,1)$ form P-W (introduced by André Weil.) We saw that the P-W form was Kählerian. Now we will see what was the reason for introducing this form.

Let $f, g \in S_k$ with $k > 0$. One proves immediately that the measure

$$\mu(f,g) := f(z)\overline{g(z)}y^{2k}\frac{dxdy}{y^2}, \quad y := \operatorname{Im} z$$

is $\mathrm{SL}(2,\mathbb{Z})$ invariant (because on $\mathfrak{H}$ $\frac{dxdy}{y^2}$ is invariant under $\mathrm{SL}(2,\mathbb{R})$ – the isometry (automorphy) group of $\mathfrak{H}$.) Denote by $\mathcal{F}$ the fundamental domain of the modular group $\Gamma := \mathrm{SL}(2,\mathbb{Z})$, then we have

$$(f|g)_P := \int_{\mathfrak{H}/\Gamma} \mu(f,g) = \int_{\mathcal{F}} f(z)\overline{g(z)}y^{2k-2}dxdy.$$

We have proved

Proposition (Petersson). *On S_k the hermitian inner product $(\cdot|\cdot)_P$ – called the Petersson product is independent of the fundamental domain for* $\mathrm{SL}(2,\mathbb{R})$. □

Petersson was a student of Erich Hecke, then professor in Strasbourg, and afterwards in Münster. His scalar product is of great importance in the Hecke theory. On the space of $\mathrm{SL}(2,\mathbb{Z})$ cusp forms (and, more generally, for Hecke congruence groups $\Gamma_0(N)$) Hecke introduced his famous operators $T(n)$, $n \geqslant 1$. These operators commute for n, m coprime, that is for $(n, m) = 1$, and are *hermitian* with respect to the Petersson scalar product $(\cdot|\cdot)_P$; thus their common eigenvalues form and orthogonal basis of S_k and the eigenvalues are real.

Let us briefly describe these important notions.

3.2 Hecke operators

As we have seen elliptic functions $\wp(z, L)$, Eisenstein series $G_k(z, L)$ are functions of two arguments $z \in \mathbb{C}$ and the lattice $L \subset \mathbb{C}$. Let $\mathcal{L}$ be a set, for example, the set of lattices L in $\mathbb{C}$. Denote by $X_\mathcal{L}$ the free abelian group generated by $\mathcal{L}$.

A *correspondence* on $\mathcal{L}$ – with integer coefficients - is a homomorphism $T : X_\mathcal{L} \to X_\mathcal{L}$. It can be given by $T(x)$, $x \in \mathcal{L}$:

(1) $$T(x) = \sum_{y \in \mathcal{L}} n_y(x) y, \quad n_y(x) \in \mathbb{Z},$$

($n_y(x) = 0$ for almost all $y \in \mathcal{L}$.)

If F is a $\mathbb{Z}$ valued function on $\mathcal{L}$, we extend it by $\mathbb{Z}$ linearity onto $X_\mathcal{L}$. We define TF by

$$TF(x) := F(T(x)) := \sum_{y \in \mathcal{L}} n_y(x) F(y).$$

Hecke operators (correspondences) $T(n)$. Let n be an integer $\geqslant 1$. $T(n)$ is the correspondence on $\mathcal{L}$ (the set of lattices) which transforms a lattice L to the sum in $X_\mathcal{L}$ of its sublattices L' of index $n := [L : L']$, that is

(3) $$T(n)L := n^{k-1} \sum_{[L:L']} L', \quad \text{for } L \in \mathcal{L}.$$

Clearly the sum is finite: the lattices L' contains nL and their number is the number of subgroups of order n of L/nL. In n is prime, this number is $n + 1$. Denote by R_λ the homothety $R_\lambda L := \lambda L$, $\lambda \in \mathbb{C}^\times$, $L \in \mathcal{L}$; R_λ and $T(n)$ are endomorphisms of the abelian group $X_\mathcal{L}$ and they satisfy important identities.

THEOREM (HECKE). *The operators (correspondences)* $T(n)$, R_λ *satisfy*

(4) $$R_\lambda R_\mu = R_{\lambda\mu}, \quad \lambda, \mu \in \mathbb{C}^\times;$$

(5) $$R_\lambda T(n) = T(n) R_\lambda, \quad n \geqslant 1, \ \lambda \in \mathbb{C}^\times;$$

(6) $$T(m)T(n) = T(n)T(m), \quad \textit{for coprime } n, m\textit{: } (n,m) = 1 \ ;$$

(7) $$T(p^n)T(p) = T(p^{n+1}) + pT(p^{n-1})R_\lambda, \quad \textit{for } p \textit{ prime, } n \geqslant 1 \ .$$

PROOF. Points (4), (5) are obvious. The commutativity of $T(m)$ and $T(n)$ for the relatively prime n and m (formula (6)), which is the most important assertion, follows from the following observation:

If L'' is a sublattice of L of index $mn = [L : L'']$, then there exists a unique sublattice L' of L containing L'' such that $[L' : L''] = m$ and $[L : L'] = n$. □

DEFINITION. The *Hecke algebra* $\mathbb{T}$ is defined to be generated by $T(p)$ with prime p and R_λ.

PROPOSITION (HECKE). *The Hecke algebra* $\mathbb{T}$ *is commutative and it contains all* $T(n)$.

PROOF. It follows from the observation above that $T(p^n)$, $n > 1$ are polynomials in $T(p)$ and R_p (by (7) and induction in n). □

We can regard $T(n)$ as an operator on functions F on $\mathcal{L}$ of weight k. By definition of F we have $R_\lambda F = \lambda^{-k} F$, $\lambda \in \mathbb{C}^\times$. Therefore by (5)

$$R_\lambda(T(n)F) = T(n)(R_\lambda F) = \lambda^{-k} T(n) F;$$

thus $T(n)F$ is of weight k as well and we immediately have

COROLLARY.

(6′) $$T(m)T(n)F = T(mn)F \quad \textit{if } (m,n) = 1$$

(7′) $$T(p)T(p^n)F = T(p^{n+1})F + p^{k-1}T(p^{n-1})F, \quad p \textit{ prime, } n \geqslant 1$$

Hecke can now regard his $T(n)$ as operators on linear spaces of modular functions M_k and cusp forms S_k of weight k:

THEOREM .

1. (Hecke) For every $n > 1$, $T(n) : M_k \to M_k$, $T(n) : S_k \to S_k$.

2. (Petersson) With respect to $(\cdot|\cdot)_P$ we have the following orthogonal decomposition

$$M_k = S_k \oplus \mathcal{E}_k, \tag{8}$$

where $\mathcal{E}_k$ is $< G_k >$ – the subspace generated by Eisenstein series G_k.

Let us pause for a moment in presenting Hecke theory to formulate one of the most famous conjectures in arithmetic.

The Petersson–Ramanujan conjecture. Let $f(z) = \sum a(n)e(nz)$ be the Fourier series of the cusp form $f \in S_k$. Then the coefficients $a(n)$ satisfy the estimate

$$a(n) = O\left(n^{\frac{k-1}{2}+\epsilon}\right),$$

which is equivalent to

$$|a_p| \leqslant 2p^{\frac{k-1}{2}}, \quad p \text{ prime} \ .$$

REMARK. For the discriminant $\Delta(z)$ which is a cusp form of weight 1/2, Ramanujan conjectured in 1916 that $|a_p| \leqslant 2p^{\frac{11}{2}}$. In this case the function $a(n)$ is denoted by $\tau(n)$ and is called the *Ramanujan tau function.* In 1938 Petersson extended the Ramanujan conjecture to arbitrary $f \in S_k$.

Let us write down explicit formula for the Hecke operator on M_k. We have an isomorphism between functions F of lattices $L \subset \mathbb{C}$ which are homogeneous of weight k and the functions f on the upper half plane $\mathfrak{H}$ transforming as modular forms of weight k; this isomorphism is given by

$$F(L) \to f(\tau) = F(\mathbb{Z}\tau + \mathbb{Z})$$

$$f(\tau) \to F(L) = \omega_2^{-k} f\left(\frac{\omega_1}{\omega_2}\right), \quad L = \mathbb{Z}\omega_1 + \mathbb{Z}\omega_2, \quad \operatorname{Im}\left(\frac{\omega_1}{\omega_2}\right) > 0.$$

Now from (3) we obtain explicit formula for $T(n)f(\tau)$, to wit

$$T(n)f(\tau) = n^{k-1} \sum_{\begin{pmatrix} a & b \\ c & d \end{pmatrix} \in \Gamma(1)/\mathfrak{M}_n} (c\tau + d)^{-k} f\left(\frac{a\tau + b}{c\tau + d}\right), \quad f \in M_n. \tag{3'}$$

The sum extends over all left orbits of $\mathfrak{M}_n$ (the set of all 2×2 integer matrices of determinant n) of the left multiplication by elements of $\Gamma(1) = PSL(2, \mathbb{Z})$. One checks (using (3′)) that $T(n)$ are self adjoint on the Hilbert space $(S_k, (\cdot|\cdot)_P)$:

$$(T(n)f|g)_P = (f|T(n)g)_P.$$

Since $T(n)$ and $T(m)$ are commuting for $(n,m) = 0$, we can consider common eigenvectors f of all $T(n)$

$$T(n)f = \lambda(n)f, \quad \text{for all } n \tag{9}$$

DEFINITION. The form $f(\tau) = \sum_{m=0}^{\infty} a(m)q_m$ satisfying (9) and the normalization condition

$$a(1) = 1 \tag{10}$$

is called the *Hecke form.*

It follows from this definition that Hecke forms have the following property

$$a(n) = \lambda(n). \tag{11}$$

Thus we have the principal discovery of Hecke

THEOREM (HECKE). *1. The Hecke forms in M_k form a basis for every k, and various Hecke forms in S_k are mutually orthogonal.*

2. The Fourier coefficients $\{a(n)\}$ of a Hecke form satisfy

$$a(m)a(n) = a(mn), \quad \textit{if } (m,n) = 1; \tag{12}$$

$$a(p)a(p^n) = a(p^{n+1}) + p^{k-1}a(p^{n-1}), \tag{13}$$

for p prime, $n \geqslant 1$; this is a recursion formula.

PROOF. (12) and (13) follow from 1., (11), (6′), and (7′) which read now

$$T(n)T(m) = T(mn) \quad \text{if } (m,n) = 1 \tag{6''}$$

$$T(p)T(p^n) = T(p^{n+1}) + p^{k-1}T(p^{n-1}), \quad \text{for prime } p,\ n \geqslant 1\ . \tag{7''}$$

1. follows from the observation that the Eisenstein series G_k is an eigenform of all $T(n)$ and that, conversely, any $f \in M_k$ with $a(0) \neq 0$ is a multiple of G_k since $T(0)f(\tau) = a(0)G_k(\tau)$.

Moreover, since $M_k =< G_k > \oplus S_k$, it suffices to show that Hecke forms span S_k and are linearly independent. Let $g \in S_k$ be another Hecke form with $g = \sum b(n)q^n$, then

$$a(n)(f|g)_P = (T(n)f|g)_P = (f|T(n)g)_P = \overline{b(n)}(f|g)_P = b(n)(f|g)_P,$$

since the eigenvalues $a(n)$, $b(n)$ of hermitian operator $T(n)$ are real. Thus if $f \neq g$, then $a(n) = b(n)$ and $(f|g)_P = 0$. □

3.3 Hecke L series

Since the function $n \to a(n)$ is multiplicative, for each modular form $f = \sum a(n)q^n \in M_k$ one can form the Dirichlet series $\sum a(n)n^{-s}$. This is the famous *Hecke L series of f*:

$$(H) \qquad L(f,s) := \sum_{m=1}^{\infty} \frac{a(m)}{m^s}.$$

We get the Euler product for $L(f,s)$

$$(14) \qquad L(f,s) = \prod_{p \text{ prime}} (1 - a(p)p^{-s} + p^{k-1-2s})^{-1},$$

where $f \in M_k$ is a Hecke form.

Example (very important). Take $f = \Delta$, the discriminant, then $\Delta \in S_{12}$ and it is a Hecke form. The Fourier expansion reads

$$(R) \qquad \Delta(s) = \sum_{n=1}^{\infty} \tau(n)q^n,$$

where $n \to \tau(n)$ is the famous Ramanujan tau function. For Δ formula (14) reads

$$(15) \qquad L(\Delta,s) = \prod_{p \text{ prime}} (1 - \tau(p)p^{-s} + p^{k-1-2s})^{-1},$$

and from Hecke theorem we get the following famous

Corollary (Ramanujan (1916), Mordell). *The Ramanujan tau function has the following properties*

$$\tau(m)\tau(n) = \tau(mn), \quad \text{if } (m,n) = 1; \tag{12'}$$

$$\tau(p)\tau(p^n) = \tau(p^{n+1}) + p^{11}\tau(p^{n-1}), \tag{13'}$$

for p prime, $n \geqslant 1$.

Remark. (12′) and (13′) were conjectured by Ramanujan in 1916 and proved by Mordell a year later.

That Hecke L series are holomorphic in a half plane, have meromorphic extension on the whole $\mathbb{C}$, and satisfy a functional equation of the Riemann ζ function type is asserted in another important theorem of Hecke.

Theorem (Hecke). *1. The Fourier coefficients* $a(n)$ *of* $f \in M_k$ *satisfy the estimates*

$$a(n) = O(n^{k-1}) \quad \text{for } f \in M_k \tag{16}$$

and

$$a(n) = O(n^{k/2}) \quad \text{for } f \in S_k \tag{16'}$$

Therefore the Hecke L series converges absolutely and compactly in the half plane $\operatorname{Re} s > k$. *Thus (14) is satisfied.*

2. $L(f,s)$ *has a meromorphic extension to the whole* $\mathbb{C}$. *It is everywhere holomorphic for the cusp form* $f \in S_k$. *For modular form* $f \in M_k$ $L(f,s)$ *has exactly one singularity: a single pole at* $s = k$ *with the residue*

$$\operatorname{res} f(k) = \frac{2\pi i}{(k-1)!}a(0). \tag{17}$$

3. The meromorphically extended $L(f,s)$ *satisfies the* Hecke functional equation

$$\Lambda(f,s) := (2\pi)^{-s}\Gamma(s)L(f,s) = (-1)^{k/2}(2\pi)^{s-k}\Gamma(k-s)L(f,k-s), \tag{18}$$

that is

$$\Lambda(f,s) = (-1)^{k/2}\Lambda(f,k-s).$$

PROOF of (16′). By definition of a cups form $f \in S_k$ we have $|f(\tau)| < c_0 y^{-k/2}$ for some constant $c_0 > 0$ and all $\tau = x + iy \in \mathfrak{H}$. The Fourier coefficients of f are by definition

$$a(n) = \int_0^1 f(x+iy)e^{-2\pi in(x+iy)}\,dx.$$

Thus $|a(n)| < c_0 y^{-k/2} e^{2\pi ny}$. Put $y = n^{-1}$ and we have (16′).

3. Is obtained by the Mellin transformation of

$$\varphi(y) := f(iy) - a(0) = \sum_{n=1}^{\infty} a(n)e^{-2\pi ny}, \quad y > 0. \tag{19}$$

Its Mellin transform $\int_0^\infty \varphi(y)y^{s-1}dy = (2\pi)^{-s}\Gamma(s)L(f,s)$. But for any function $\varphi(t)$, $t > 0$ which is small at $t \to \infty$ and satisfies the functional equation

$$\psi\left(\frac{1}{t}\right) = \sum_{j=1}^{J} A_j t^{\lambda_j} + t^h \psi(t), \quad t > 0 \tag{$*$}$$

for some $h, A_j, \lambda_j \in \mathbb{C}$, the following holds.

DON ZAGIER FUNCTIONAL EQUATION PRINCIPLE. *If ψ satisfies $(*)$ and is small at infinity, then its Mellin transform $\mathbb{M}\psi(s)$ has a meromorphic extension to all $s \in \mathbb{C}$ and is holomorphic everywhere with exception of simple poles at λ_j of residue A_j, $j = 1, \ldots, J$; moreover $\mathbb{M}\psi(h-s) = \mathbb{M}\psi(s)$.*

END OF THE PROOF. In the case in hands $\psi := \varphi$ is exponentially small at $t \to \infty$ and satisfies equation $(*)$ with $\varphi\left(\frac{1}{y}\right) = f\left(-\frac{1}{iy}\right) - a(0) = (iy)^k f(iy) - a(0) = (-1)^{k/2}y^k\varphi(y) + (-1)^{k/2}a(0)y^k - a(0)$. □

PROOF OF ZAGIER PRINCIPLE. Decompose the integral $\mathbb{M}\psi(s)$ into $\int_0^1 + \int_1^\infty$. Replace t with t^{-1} in the first term; then for large $\operatorname{Re} s$ we obtain

$$\mathbb{M}\psi(s) = \int_1^\infty \left(\sum_{j=1}^{J} A_j t^{\lambda_j} + t^h\psi(t)\right) t^{-s-1}dt + \int_1^\infty \psi(t)t^{s-1}dt =$$

$$= \sum_{j=1}^{J} \frac{A_j}{s-\lambda_j} + \int_1^\infty \psi(t)(t^s + t^{h-s})\frac{dt}{t}.$$

The integral is convergent for all s and is invariant with respect to $s \to h - s$. But the first term is also invariant: indeed, apply $(*)$ twice and show that for every j there exists j' with $\lambda_{j'} = h - \lambda_j$, $A_{j'} = -A_j$. $\square$

REMARK. PROOF OF RIEMANN FUNCTIONAL EQUATION FOR ζ FUNCTION. Following Riemann take for $\psi(t)$ the ϑ function, $\psi(t) = \sum_{n=1}^{\infty} e^{-\pi n^2 t}$; thus $\mathbb{M}\psi(s) = \pi^{-s}\Gamma(s)\zeta(2s)$ and $(*)$ is satisfied with $J = 2$, $h = \frac{1}{2}$, $\lambda_1 = 0$, $\lambda_2 = \frac{1}{2}$, $A_2 = -A_1 = \frac{1}{2}$.

REMARKS (VERY IMPORTANT). 1. Formulae (H) and (14) imply (12′), (13′).

2. Hecke proved that conversely: Every Dirichlet series $L(f, s)$ which satisfies the functional equation (18) and we get estimates of the type (16′) for modular forms of weight k. Moreover

(f is a Hecke form) $\Longleftrightarrow$ *($L(f, s)$ = right hand side of* (14)*).*

Thus the great discovery of Hecke is that *L series and modular forms are two aspects of the same thing!*

3.4 Ramanujan-Petersson conjecture and Deligne theorem

For every cusp form of weight $k = 12$ we have estimates (16′) $|a(p)| = O(p^{12/2})$, therefore for the Ramanujan τ function defined by (R) we have $|\tau(p)| = O(12/2)$. Ramanujan claimed the much stronger estimate. The famous Ramanujan conjecture is that $|\tau(p)| \leqslant 2p^{11/2}$ for all prime p. Why such claim? In (15) take $X = p^{-s}$, then we have

$$\sum \tau(p^n)X^n = (1 - \tau(p)X + p^{11}X^2)^{-1}.$$

Ramanujan conjecture is therefore equivalent to the assertion that all zeros of the quadratic polynomial

$$1 - \tau(p)X + p^{11}X$$

are not real.

In 1938 Petersson conjectured that for any *cusp* form of weight k

$$|a(p)| \leqslant 2p^{(k-1)}/2.$$

Both these conjectures follow from the general André Weil conjectures on algebraic varieties over *finite* fields (1949). Pierre Deligne proved in 1974 all these conjectures: they are known now as Deligne theorems. For this magnificent work Deligne was awarded the Fields medal. As N. Katz estimated, the complete proof of Deligne theorems, if written from the scratch affords about 2000 pages. And Y. Manin in his lovely booklet *Mathematics and Physics* writes that this is probably the record for the ration

$$\frac{\text{length of the proof}}{\text{length of the statement}}.$$

But perhaps the gigantic Wiles proof of the Taniyama conjecture that elliptic curves over $\mathbb{Q}$ are modular sets a new record?

3.5 Hecke theory for congruence subgroups

This theory aims at the extension of the marvelous Hecke theorem which holds for $\Gamma(1) := \mathrm{PSL}(2, \mathbb{Z})$ to some subgroups $\Gamma \subset \Gamma(1)$ of finite order $\mu = [\Gamma(1) : \Gamma]$. Hecke himself did not manage to complete his theory, he died of cancer in 1946.

Particularly important is the series of subgroups $\Gamma_0(N)$ (N prime and > 1) of matrices $\begin{pmatrix} a & b \\ c & d \end{pmatrix}$ with $N|c$ (that is, $c \equiv 0 \bmod N$); $\mathrm{SL}(2, \mathbb{Z}) = \Gamma_0(1) = \Gamma(1)$.

Congruence subgroups $\Gamma \subset \Gamma(1)$ were defined and investigated in 1878 by F. Klein and these investigations and that of their modular curves $X(\Gamma) := \mathfrak{H}/\Gamma$ were continued by Robert Fricke (his son-in-law.)

Hecke extends the notion of his operators $T(n)$ to the groups $\Gamma_0(N)$, but in this case with $N > 1$ a new phenomenon arises: there turn out to be the so called *old* and new (modular) forms discovered by Atkin and Lehner. Partial understanding of these phenomena appeared already in the papers of Fricke (1922, 1926), but only introduction of the involution W_N by Atkin and Lehner which commutes with the Hecke operators on $\Gamma_0(N)$ cusp forms $S_k(\Gamma_0(N))$ of level N made it possible to them to obtain a beautiful extension of the Hecke theory for congruence subgroups $\Gamma_0(N)$.

THEOREM (ATKIN–LEHNER, 1970). *1. If $f \in S_k(\Gamma_0(N))$ is a newform , then its equivalence class is one dimensional, that is, it consists of multiplets of f.*

2. The space $S_k^{new}(\Gamma_0(N))$ of newforms is an orthogonal sum of one dimensional equivalence classes of 1.

If f is such an eigenform, then f can be normalized so that its Fourier q expansion $f(\tau) = \sum_{n=1}^{\infty} c_n q^n$ has $c_1 = 1$. In such a case the eigenform of f is a common eigenvector of Hecke operators $T_k(n)$ for all n; and the eigenvalues are as follows

(a) $$T_k(n) f = c_n f \quad \textit{for all } n \ ;$$

(b) *if Q is a prime dividing N,* $w_Q f = \lambda(Q)$ *with* $\lambda(Q) = \pm 1$;

(c) *if $p|N$ and Q corresponds to p,* $w_N f = \prod_{p|N} \lambda(Q) f$.

3. The L series $L(s, f)$ has the Euler product expansion

$$L(s, f) =$$

$$= \prod_{\substack{p \text{ prime} \\ p|N,\ p^2 \nmid N}} \left(\frac{1}{1 + \lambda(p) p^{k/2-1-s}} \right) \prod_{\substack{p \text{ prime} \\ p \nmid N}} \left(\frac{1}{1 - c_p p^{-s} + p^{k-1-2s}} \right).$$

4. $L(s, f)$ satisfies the functional equation

$$\Lambda(s, f) = \epsilon(-1)^{k/2} \Lambda(k - s, f),$$

where $\epsilon = \prod_{p \nmid N} \lambda(Q)$ and

$$\Lambda(f, s) := N^{s/2} (2\pi^{-s}) \Gamma(s) L(s, f).$$

Let us take a closer look at

3.6 Congruence subgroups $\Gamma \subset \Gamma(1)$, their modular curves $X(\Gamma)$, and Fricke subgroups $\Gamma_0(N)$

.

We know that the automorphism group of the upper half plane $\mathfrak{H}$ is $\mathrm{PSL}(2, \mathbb{R}) =: \Gamma(1)$. This group is also a group of isometries of $(\mathfrak{H}, ds^2)$ with respect to the Poincaré metric ds^2 on $\mathfrak{H}$.

The corresponding volume elements reads

$$(1)\qquad dV = \frac{dx \wedge dy}{2\pi y^2} = i\frac{dz \wedge d\bar{z}}{4\pi y^2}, \quad z = x + iy.$$

If $\gamma = \begin{pmatrix} a & b \\ c & d \end{pmatrix} \in \Gamma(1)$, then $d(\gamma z) = \frac{dz}{(cz+d)^2}$.

The geodesics of $(\mathfrak{H}, ds^2)$ are upper halves of circles or lines meeting orthogonally the boundary $\partial\mathfrak{H} = \mathrm{Re}\,\mathbb{C}$ – the real axis. If $\mathcal{D}$ is a geodesic polygon in $\mathfrak{H}$ whose boundary $\partial\mathcal{D}$ consists of k geodesic arcs forming the interior angles $\alpha_1, \ldots, \alpha_k$, then the Gauss–Bonnet formula gives the formula for $|\mathcal{D}|$, the volume of $\mathcal{D}$:

$$(2)\qquad |\mathcal{D}| = \int_{\partial\mathcal{D}} \frac{dx}{2\pi y} = \frac{k-2}{2} - \sum_{i=1}^{k} \frac{\alpha_i}{2\pi}.$$

Some of the vertices of $\mathcal{D}$ may be *cusp points* either on the real axis or at $\infty = +i\infty$.

We know the beautiful *modular figure* of the modular group $\Gamma(1)$:

$$(3)\qquad \mathcal{F} = \left\{ z \in \mathfrak{H} : -\frac{1}{2} \leqslant x < \frac{1}{2}, |z| \geqslant 1 \ \text{ if } x > 0 \ \right\}.$$

The volume $|\mathcal{F}| = \frac{1}{6}$.

Now, more generally, let Γ be a discrete subgroup of $\mathrm{PSL}(2, \mathbb{R})$ such that the volume of the orbit space $\mathfrak{H}/\Gamma$ is finite.

The quotient space $X(\Gamma) := \mathfrak{H}/\Gamma$ has a natural complex structure of a Riemann surface such that *meromorphic functions* on $X(\Gamma)$ are 'the same' as Γ automorphic functions on $\mathfrak{H}$. We have the covering $\mathfrak{H} \to X(\Gamma)$. The stabilizer (stability group) Γ_z of z equals $\{\gamma \in \Gamma : \gamma z = z\}$ is a finite cyclic group of order $e = e_z > 0$ and the local parameter at such *elliptic fixed points* z vanishes at z to order e_z, regardless of the point of $\mathfrak{H}$. The inequivalent elliptic fixed points $z_1, \ldots, z_s$ of orders $e_i = e_{z_i}$, $i = 1, \ldots, s$ correspond to the corners of $\mathcal{F}_\Gamma$ with interior angles $\alpha_i = 2\pi/e_i$. We compactify the Riemann (punctured) surface $\mathfrak{H}/\Gamma$ by adding *cusps*, that is *parabolic fixed points* $z_{s+1}, \ldots, z_t$ corresponding to the corners of $\mathcal{F}_\Gamma$ on the boundary of gH. We denote such a surface $X(\Gamma) \equiv \mathfrak{H}^*/\Gamma$. If, for instance, $\infty = i\infty$ is a cusp and if a is the least positive number such that the map $z \to z + a$ is in Γ, then $e^{2\pi i z/a}$ is the local coordinate at ∞. Now we have important

Genus formula for $X(\Gamma)$. Denote by $p = p_{X(\Gamma)}$ the genus of $X(\Gamma)$. Then

$$|\mathfrak{H}/\Gamma| = |\mathcal{F}_\Gamma| = (2p-2) + \sum_{i=1}^{t}(1 - e_{z_i}^{-1}). \tag{4}$$

If Γ is a subgroup of the modular group $\Gamma(1)$ of index $\mu = \mu(\Gamma) = [\Gamma(1) : \Gamma]$ then from (4) follows the genus formula since now the fundamental domain $\mathcal{F}_\Gamma$ is a union of μ copies of $\mathcal{F} = \mathcal{F}_{X(1)}$ and thus $|\mathcal{F}_\Gamma| = \mu \cdot |\mathcal{F}| = \mu \cdot \frac{1}{6}$. Since we have to do only with elliptic fixed points of order 2 and 3, let n_2 and n_3 be the numbers inequivalent fixed points of order 2 and 3, and $_\infty$ the number of cusps. From (4) we find the desired *genus formula:*

$$p_{X(\Gamma)} = 1 + \frac{\mu(\Gamma)}{12} - \frac{n_2}{4} - \frac{n_3}{3} - \frac{n_2}{2}. \tag{5}$$

COROLLARY.

$g(X_0(N)) = 1$ *for* $N = 11, 14, 15, 17, 19, 20, 21, 24, 27, 32, 36, 49.$

REMARK. If we have a pair $\Gamma' \subset \Gamma$ of groups with finite volumes $|\mathfrak{H}/\Gamma|$, $|\mathfrak{H}/\Gamma'|$ which are called *Fuchs groups of first kind* and $f : X(\Gamma') \to X(\Gamma)$ is the corresponding holomorphic map, then

$$\deg f = [\bar{\Gamma} : \bar{\Gamma}'] \quad \text{and} \quad e_z(f) = [\bar{\Gamma}_z : \bar{\Gamma}'_z]. \tag{7}$$

The image of a subgroup $G \subset \mathrm{SL}(2, \mathbb{R})$ under $\mathrm{PSL}(2, \mathbb{R})$ is usually denoted by $\bar{G}$.

Knowing that

(8)

$$\mu(N) = \mu = [\overline{\Gamma(1)} : \overline{\Gamma(N)}] = |\,\mathrm{PSL}(2, \mathbb{Z}/N\mathbb{Z})| = N^3 \prod_{\substack{p\ prime \\ p|N}} \left(1 - \frac{1}{p^2}\right),$$

in the following table we collect the genera p of some modular curves $X(N)$

N	2	3	4	5	6	7	8	9	10	11
$p_{X(N)}$	0	0	0	0	1	3	5	10	13	26
$\mu(N)$	6	12	24	69	72	168	192	324	360	660

Thus, for $N = 0, \ldots, 5$ we have curves of genus 0, and for $N = 3, 4, 5$ our old acquaintances, the groups of Platonic solids: of tetrahedron G_{12}, octahedron G_{24}, and icosahedron G_{60}. For $N = 7$ we obtain the famous Klein curve $X(7)$ of genus 3 and with automorphism group $\mathrm{Aut}(X(7))$ of order 168.

Felix Klein was truly proud of his discovery of $X(7)$. The fundamental domain of $\Gamma(7)$ consists of $2 \cdot 168 = 336$ double triangles with angles $\pi/2$, $\pi/3$, and $\pi/7$. The surface $X(7)$ has 168 sheets over $\mathbb{P}^1(\mathbb{C})$. Let us note an extremal property of all groups $\Gamma(N)$, $N \geqslant 7$.

For a Riemann surface X of genus $p \geqslant 2$, its automorphism group $\mathrm{Aut}(X)$ is finite (Schwartz). Moreover, A. Hurwitz gave the precise estimate for $|\,\mathrm{Aut}(X)|$:

THEOREM (HURWITZ). *For X of genus $p \geqslant 2$,*

$$|\,\mathrm{Aut}(X)| \leqslant 84(p-1). \tag{9}$$

But for $X(7)$ we have $p = 3$, and the order of $\mathrm{Aut}(X(7))$ is precisely $84 \cdot 2 = 168$.

REMARK. The equality $|\,\mathrm{Aut}(X(7))| = 168$ was known already to Schwarz. But we have the following observation:

For any modular curve $X(N)$ of genus $p \geqslant 2$ the congruence group $\Gamma(N)$ is a normal subgroup of $\mathrm{SL}(2,\mathbb{Z})$ *and it acts on $\mathfrak{H}/\Gamma(N)$.* Indeed, if $\gamma \in \mathrm{SL}(2,\mathbb{Z})$ and $\gamma_1 \in \Gamma(N)$, then $\gamma(\gamma_1 z) = (\gamma\gamma_1\gamma^{-1}) \cdot \gamma z$. Therefore $|\,\mathrm{Aut}(X(N))| = \mu(N) = |\,\mathrm{PSL}(2, \mathbb{Z}/N\mathbb{Z})|$ and since

$$p - 1 = \frac{N-6}{12N}\mu(N)$$

we have

$$|\,\mathrm{Aut}(X(N))| = \frac{12N}{N-6}(p-1) \leqslant 84(p-1). \tag{12}$$

3.7 Modular functions and simple (finite) sporadic groups. The Monstrous Moonshine. Borcherds theorem

From the elementary theory of Riemann surfaces we know that if the compact surface $\overline{\mathfrak{H}/\Gamma}$ is of genus 0, then the theory of Γ automorphic functions

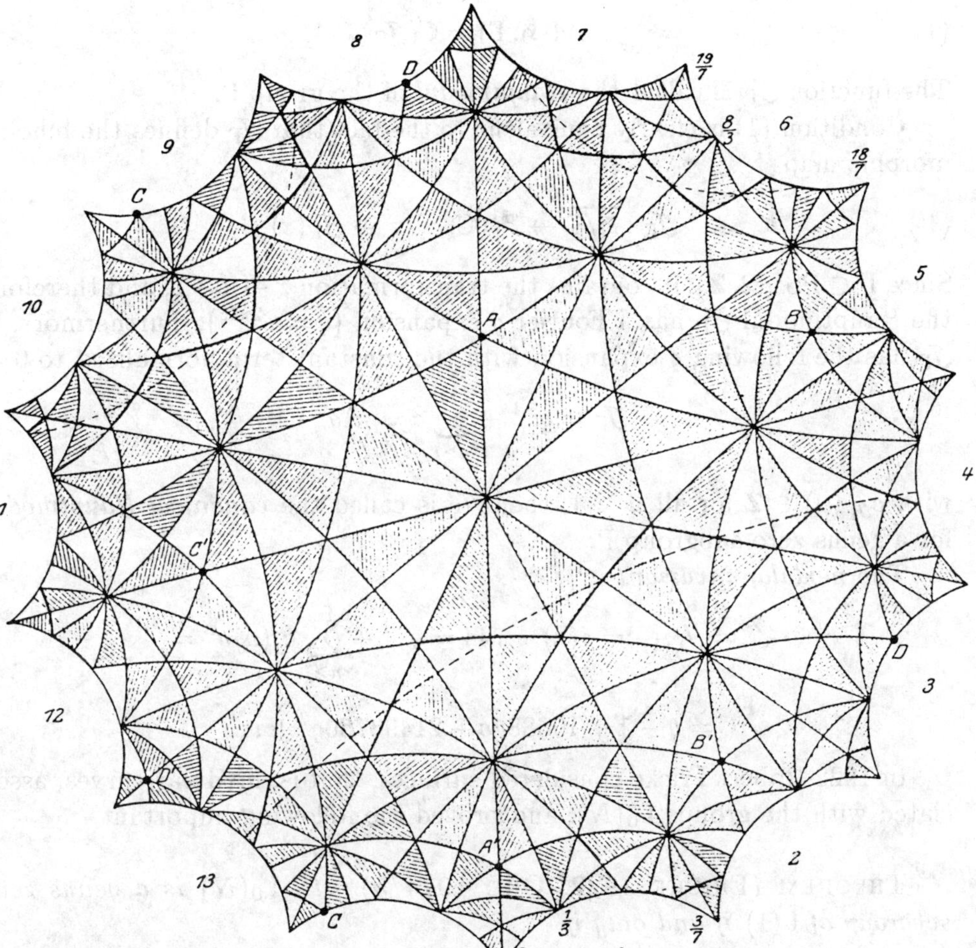

Klein's *Hauptfigur*: A fundamental domain for Γ

on $\mathfrak{H}$ is particularly simple: The field of Γ automorphic functions is generated by only one function denoted by $\mathcal{J}_\Gamma$, determined up to the rational transformations

$$\mathcal{M}(\mathfrak{H},\Gamma) = \mathbb{C}(\mathcal{J}_\Gamma). \tag{1}$$

The function $\mathcal{J}_\Gamma$ is called the *Hauptmodul* of the group Γ.

Condition (1) is clearly equivalent to the fact that $\mathcal{J}_\Gamma$ defines the biholomorphic map

$$\tilde{\mathcal{J}}_\Gamma : \overline{\mathfrak{H}/\Gamma} \to \mathbb{P}^1(\mathbb{C}), \quad z \to \tilde{\mathcal{J}}_\Gamma(z). \tag{1'}$$

Since $\Gamma \subset \mathrm{PSL}(2,\mathbb{Z})$, it contains the transformation $z \to z+1$, and therefore the Hauptmodul $\mathcal{J}_\Gamma$ has a Fourier q expansion ($q = e^{2\pi i z}$). Furthermore, if $\mathcal{J}_\Gamma$ has the following q expansion with the constant term normalized to 0

$$\mathcal{J}_\Gamma = q^{-1} + \sum_{n\geqslant 1} c_{\mathcal{J}_\Gamma}(n) q^n$$

with $c_{\mathcal{J}_\Gamma}(n) \in \mathbb{Z}$ for all $n \geqslant 1$, then $\mathcal{J}_\Gamma$ is called the *canonical Hauptmodul* for a genus zero subgroup Γ.

The *modular invariant*

$$\mathcal{J} = \mathcal{J}_{\Gamma(1)} := j(z) - 744 = q^{-1} + \sum_{n\geqslant 1} c_j(n) q^n =$$

$$= q - 1 + 196884q + 21493760q^2 + \ldots$$

In 1922 Robert Fricke considered surfaces, that is, modular curves, associated with the groups $\Gamma_0(N)$, and proved the following important

THEOREM (FRICKE 1922, OGG 1974/5). *1.* $\Gamma_0(N)$ *is a genus zero subgroup of* $\Gamma(1)$ *if and only if*

$$2 \leqslant N \leqslant 10, \quad \textit{and } N = 12, 13, 16, 18, 25. \tag{3}$$

2. If N is prime, denote by $\Gamma_0(N)+$ the normalizer of $\Gamma_0(N)$ in $\mathrm{PSL}(2,\mathbb{R})$. *Then $\Gamma_0(N)+$ is of genus* 0 *for exactly* 15 *values of N:*

$$N = 2, 3, 5, 7, 11, 13, 17, 19, 23, 29, 31, 41, 47, 59, 71. \tag{4}$$

REMARK 1. We call the groups $\Gamma_0(N)+$ *Fricke subgroups.* One obtains $\Gamma_0(N)+$ by adjoining to the group $\Gamma_0(N)$ the *Fricke involution* $w_N =$

$\begin{pmatrix} 0 & 1 \\ -N & 0 \end{pmatrix}$ that sends $z \in \mathfrak{H}$ to $-1/Nz$.

REMARK 2. We recall that if we have a subgroup B of A, then its *normalizer* N_B is the largest subgroup of A containing B in which B is still normal: $N_B = \{a \in A : aBa^{-1} = B\}$. The normalizers of $\Gamma_0(N)$ provide further examples of genus zero subgroups.

REMARK 3. The cusps of $\Gamma_0(N)$ consist of the finite set $\mathbb{P}^1(\mathbb{Q})/\Gamma_0(N)$ and they are all defined over $\mathbb{Q}$. For prime N the surface $X_0(N) = X(\Gamma_0(N))$ is a double covering of the Fricke surface $X_0(N)+$.

Sporadic groups can be characterized by the following theorem, which is one of the greatest achievements of modern mathematics.

THEOREM. *Every finite simple group is cyclic of prime order, an alternating group, a finite simple group of Lie type, or one of the twenty-six sporadic finite simple groups.*

The *Monster* (denoted by M) is the largest *sporadic* simple group discovered, or better, predicted, in 1973 (independently) by Robert Griess and Bernd Fischer. It was first called the Fischer–Griess group, and later (by Griess) the *Friendly Giant.* Afterwards M was christened by J.H Conway the *Monster.* The order of this gigantic group is

$$(5) \quad |M| = 2^{46} \cdot 3^{20} \cdot 5^9 \cdot 7^6 \cdot 11^2 \cdot 13^3 \cdot 17 \cdot 19 \cdot 23 \cdot 29 \cdot 31 \cdot 41 \cdot 47 \cdot 59 \cdot 71.$$

Thus $|M| \sim 8 \cdot 10^{53}$.

In 1974 A. Ogg, a great expert in modular functions, noticed that primes dividing the order of the Monster group *are exactly the* 15 *primes of the Fricke list* (4).

By virtue of the Hermann Weyl principle it is important to realize the group M as a symmetry group of some canonical object V ('which can come from algebra, geometry, physics, or anything else'.) This object turns out to be a graded vector algebra

$$V = V_{-1} \bigoplus_{n \geqslant 1} V_n, \quad \dim V_{-1}, \dim V_n < \infty$$

and was constructed by Frenkel, Lepovsky, and Meurman in 1988, and thoroughly described in their impressive (500 pages long) monograph *Vertex*

Operator Algebra and the Monster, Academic Press, 1988.

HISTORICAL REMARKS. The term 'sporadic' was introduced by Emil Mathieu in his work of 1861, Crell Journal **6** (1861), 241–322 in which Mathieu found the first five sporadic groups: one of them, the Mathieu group, is denoted by M_{24} and is of order

$$|M_{24}| = 2^{10} \cdot 3^3 \cdot 5 \cdot 7 \cdot 11 \cdot 23.$$

This group was found by Mathieu only in 1871. Conway calls M_{24} 'the most intriguing of all sporadic groups. For one hundred years the five Mathieu M_{11}, M_{12}, M_{22}, M_{23}, and M_{24} constituted the entire list of known sporadic groups. In 1964 Zvanimir Janko provided the first of the list of surprises: he announced the discovery of a new simple group of order 175560 which, at the time, seemed to be quite a large number.' This Janko group is denoted by J_2 and $|J_2| = 2^7 \cdot 3^3 \cdot 5^2 \cdot 7$. The inspection of the impressive table of all 26 sporadic groups shows that order of all but two of them has as prime divisors the numbers from the Fricke list (4). These two exceptions are the group of Lyons–Sims L_y and the group J_4 of Janko–Norton, Parker, Benson, Convey, Thakray (JNPBCT) which have the prime divisor 37.

The important breakthrough in investigations of sporadic groups took place in 1974, where Ogg observed the 'numerology' phenomenon which suggested a close, but mysterious relation between the Monster and modular functions. Further numerological observations by J. McKay, J. Thompson, S. Norton, and J.H. Conway have led Norton and Conway to the famous conjecture, called by them *Monstrous Moonshine* .

In the Introduction to the Frenkel, Lepovsky, and Meurman monograph the authors write 'What is certain, however, is that the following areas of mathematics and physics play basic roles

I. Modular functions

II. Finite groups

III. Lie algebras

IV. String theory.

We are still far from complete understanding.'

These words were written in 1988, and already in 1992 R.E. Borcherds proved the Monstrous Moonshine conjecture in the impressive paper *Monstrous Moonshine and monstrous Lie superalgebras*, Invent. Math. **109** (1992), 405–444. Borcherds makes use of the generalized Kac–Moody algebras and the 'no-ghost theorem' of string theory.

group	order	investigators
M_{11}	$2^4 \cdot 3^2 \cdot 5 \cdot 11$	Mathieu
M_{12}	$2^6 \cdot 3^3 \cdot 5 \cdot 11$	Mathieu
M_{22}	$2^7 \cdot 3^2 \cdot 5 \cdot 7 \cdot 11$	Mathieu
M_{23}	$2^7 \cdot 3^2 \cdot 5 \cdot 7 \cdot 11 \cdot 23$	Mathieu
M_{24}	$2^{10} \cdot 3^3 \cdot 5 \cdot 7 \cdot 11 \cdot 23$	Mathieu
J_2	$2^7 \cdot 3^3 \cdot 5^2 \cdot 7$	Hall, Janko
Sz	$2^{13} \cdot 3^7 \cdot 5^2 \cdot 7 \cdot 11 \cdot 13$	Suzuki
H-S	$2^9 \cdot 3^2 \cdot 5^3 \cdot 7 \cdot 11$	Higman, Sims
McL	$2^7 \cdot 3^6 \cdot 5^3 \cdot 7 \cdot 11$	McLaughlin
Co_3	$2^{10} \cdot 3^7 \cdot 5^3 \cdot 7 \cdot 11 \cdot 23$	Conway
Co_2	$2^{18} \cdot 3^6 \cdot 5^3 \cdot 7 \cdot 11 \cdot 23$	Conway
Co_1	$2^{21} \cdot 3^9 \cdot 5^4 \cdot 7^2 \cdot 11 \cdot 23$	Conway
He	$2^{10} \cdot 3^3 \cdot 5^2 \cdot 7^3 \cdot 17$	Held/Higman, Mc Kay
Fi_{22}	$2^{17} \cdot 3^9 \cdot 5^2 \cdot 7 \cdot 11 \cdot 13$	Fischer
Fi_{23}	$2^{18} \cdot 3^{13} \cdot 5^2 \cdot 7 \cdot 11 \cdot 13 \cdot 17 \cdot 23$	Fischer
Fi_{24}	$2^{21} \cdot 3^{16} \cdot 5^2 \cdot 7^3 \cdot 11 \cdot 13 \cdot 17 \cdot 23 \cdot 29$	Fischer
H-N	$2^{14} \cdot 3^6 \cdot 5^6 \cdot 7 \cdot 11 \cdot 19$	Harada, Norton/Smith
T	$2^{15} \cdot 3^{10} \cdot 5^3 \cdot 7^2 \cdot 13 \cdot 19 \cdot 31$	Thompson/Smith
B	$2^{41} \cdot 3^{13} \cdot 5^6 \cdot 7^2 \cdot 11 \cdot 13 \cdot 17 \cdot 19 \cdot 23 \cdot$ $\cdot 31 \cdot 47$	Fischer/Sims, Leon
M	$2^{46} \cdot 3^{20} \cdot 5^9 \cdot 7^6 \cdot 11^2 \cdot 13^3 \cdot 17 \cdot 19 \cdot 23 \cdot$ $\cdot 29 \cdot 31 \cdot 41 \cdot 47 \cdot 59 \cdot 71$	Fischer/Griess
J_1	$2^3 \cdot 3 \cdot 5 \cdot 7 \cdot 11 \cdot 19$	Janko
$O'Nan$	$2^9 \cdot 3^4 \cdot 5 \cdot 7^3 \cdot 11 \cdot 19 \cdot 31$	O'Nan/Sims
J_3	$2^7 \cdot 3^5 \cdot 5 \cdot 17 \cdot 19$	Janko
Ly	$2^8 \cdot 3^7 \cdot 5^6 \cdot 7 \cdot 11 \cdot 31 \cdot 37 \cdot 67$	Lyons/Sims
Rv	$2^{14} \cdot 3^3 \cdot 5^3 \cdot 7 \cdot 13 \cdot 29$	Rudvalis/Conway, Wales
J_4	$2^{21} \cdot 3^3 \cdot 5 \cdot 7 \cdot 11^3 \cdot 23 \cdot 29 \cdot 31 \cdot 37 \cdot 43$	JNPBCT

The 26 sporadic groups

THEOREM (R.E. BORCHERDS, 1992). *Let* $V = V_{-1} \bigoplus_{n \geqslant 1} V_n$ *be the Frenkel, Lepovsky, and Meurman infinite dimensional graded vector space, where* $\dim V_{-1} = 1$, *and for each* n, V_n *is an* M*-module of dimension equal to* $c_j(n)$, *the coefficient of* q^n *in the* q *expansion of the modular invariant* $J(z) = j(z) - 744$. *For any element* $g \in M$, *let* $\mathrm{tr}(g|V_n)$ *denote the trace of the representation* $\mathcal{U}$ *of the monster* g *acting on* V_n. *Let* $T_g(z)$ *be the* Thompson series *defined by*

$$T_g(z) := q^{-1} + \sum_{n \geqslant 1} \mathrm{tr}(g|V_n) q^n$$

with $q = e^{2\pi i z}$, $z \in \mathfrak{H}$.

Then we have the following assertions:

(a) $\mathrm{tr}(g|V_{-1}) = 1$ *and* $\mathrm{tr}(g|V_n) \in \mathbb{Z}$ *for any* $n \geqslant 1$.

(b) T_g *is a canonical Hauptmodul for a certain genus zero subgroup* Γ *having level* N *divisible by the order of* $g \in M$. *This subgroup* Γ *lies between* $\Gamma_0(N)$ *and its normalizer* $\Gamma_0(N)+$ *in* $\mathrm{PSL}(2, \mathbb{R})$.

REMARK 1. If we have a representation $\mathcal{U} : G \to \mathrm{Aut}\, H$ of a group G in a vector space H, then the function $G \ni g \to \mathrm{tr}\,\mathcal{U}(g) \in \mathbb{C}$ is called the character of G and is a class function on G. $\mathrm{tr}(g|V_n)$ is thus restriction of the trace of $\mathcal{U}(g)$ to V_n.

REMARK 2. Clearly, if $g = 1_M$ is the identity of the Monster M, then $\mathrm{tr}(1_M|V_n) = \dim V_n = c_j(n)$; we have therefore $T_1(z) = J(z) = j(z) - 744$, the canonical Hauptmodul for the modular group $\Gamma_0(1) = \mathrm{PSL}(2, \mathbb{Z})$; J is the famous Dedekind–Klein elliptic modular function, called now the modular invariant.

CHAPTER 4

Dedekind ζ_K function for number field K and Selberg ζ function

In order to define the ζ function of a number field K which in the case $K = \mathbb{Q}$ is the Riemann ζ function, we have to introduce some fundamental notions, all of which have been conceived by Dedekind and codified in his famous XI Supplement to Dirichlet Collected Papers which is one of the most influential piece of work in the whole history of mathematics.

An algebraic number is an *algebraic integer* if it is a root of a polynomial equation $x^n + a_{n-1}x^{n-1} + \cdots + a_0$ with $a_i \in \mathbb{Z}$, $n \geqslant 1$. Given an algebraic number y, there exists an integer $c \in \mathbb{Z}$, $c \neq 0$ such that cy is an algebraic integer: take for c the leading coefficient in the irreducible equation for y over $\mathbb{Z}$. The set of algebraic integers in a number field K is a subring denoted by $\mathcal{O}_K$ and called the *ring of algebraic integers* (of K). Clearly $\mathcal{O}_{\mathbb{Q}} = \mathbb{Z}$.

Dedekind shows that $\mathcal{O}_K$ is a free module over $\mathbb{Z}$ of rank $d = [K : \mathbb{Q}]$, and that there exists a basis $\{\alpha_1, \ldots, \alpha_d\}$ of $\mathcal{O}_K$ over $\mathbb{Z}$. Let $\sigma_1, \ldots, \sigma_d$ be distinct imbeddings $K \hookrightarrow \mathbb{C}$ of K into $\mathbb{C}$. The *discriminant* of K, denoted by D_K is

$$D_K := (\det(\sigma_i \alpha_j))^2$$

An *ideal* of K is a finite generated $\mathcal{O}_K$ module. Ideals in K form the group $I(K)$; this group has as a subgroup the set $K^\times$ of principal ideals. The quotient group $Cl(K) := I(K)/K^\times$ is called the *ideal class group* (of K); its order is finite and is called the *class number* and denoted by $h(K)$ or h_K.

The kernel of the map $i : K^\times \to I(K), \alpha \to (\alpha)$ is $\ker i = \mathcal{O}_K^*$, the group of units in the ring $\mathcal{O}_K$. Thus we have the exact sequence of groups:

$$\{1\} \to \mathcal{O}_K^* \to K^\times \to I(K) \to Cl(K) \to \{1\}.$$

EXAMPLES. 1. Let $\mathbb{Z}$ be the ring of integers. The group $\mathbb{Z}^*$ of units is $\{-1, 1\}$. Prime ideals of $\mathbb{Z}$ are p and $-p$ with prime p.

2. Let $v := K[x]$ be the polynomial ring in x. The units of v are constants $\neq 0$; primes are irreducible polynomials. Both rings 1. and 2. are factorial.

3. Ring of *Gauss integers* $v := \mathbb{Z}[\sqrt{-1}] = \{x + iy : x, y \in \mathbb{Z}\}$. $\mathbb{Q}[\sqrt{-1}] = \{x + iy \in \mathbb{C} : x, y \in \mathbb{Q}\}$ is a quotient field of v. The $\mathbb{Q}$ vector space $\mathbb{Q}(\sqrt{-1})/\mathbb{Q}$ has dimension 2. If $\epsilon = x + iy$ is a unit in $\mathbb{Z}[\sqrt{-1}]$ and if we set $N(\alpha) := x^2 + y^2$ for $\alpha = x + iy$, then clearly $N(\alpha\beta) = N(\alpha)N(\beta)$ and $N(\epsilon) = 1$. Thus $\mathbb{Z}[\sqrt{-1}] = \{\pm 1, \pm i\}$.

It was a great day of mathematics when Gauss proved his famous 'Theoria residuorum biquadraticorum':

THEOREM (GAUSS, 1832). *The ring* $\mathbb{Z}[\sqrt{-1}]$ *is factorial.*

Let $\mathfrak{a}$ be an ideal of $\mathcal{O}_K$; the (absolute) *norm* of $\mathfrak{a}$, $N(\mathfrak{a}) := [\mathcal{O}_K : \mathfrak{a}]$ equals the number of elements in the residue class field $\mathcal{O}_K/\mathfrak{a}$. Clearly $N(\mathfrak{a}\mathfrak{b}) = N(\mathfrak{a})N(\mathfrak{b})$.

FUNDAMENTAL THEOREM OF IDEAL THEORY (DEDEKIND). *Every ideal* $\{0\} \neq \mathfrak{a} \subset \mathcal{O}_K$ *is a product of prime ideals and this decomposition is unique (up to the ordering). Thus* $N(\mathfrak{a}) = \prod_p N(\mathfrak{p})^{m_\mathfrak{p}}$, *with* $\mathfrak{p}$ *being prime ideals.*

Another important result of Dedekind is his

LATTICE THEOREM. *A free* $\mathbb{Z}$ *module in* K *has rank* $\leqslant [K : \mathbb{Q}]$. *If the rank equals* $[K : \mathbb{Q}]$ *then it is called the* lattice *in* K.

THEOREM (DEDEKIND). $\mathcal{O}_K$ *is a lattice in* K.

If we call a finite $\mathcal{O}_K$ module the *ideal* of K (relative to $\mathcal{O}_K$) the we have

COROLLARY (DEDEKIND). *Every ideal of* K *is a lattice in* K.

For a field K Dedekind defines a positive number $\mathcal{R}(K)$ called the *regulator* of K which is the volume of the fundamental domain of a lattice in some $\mathbb{R}^r$ defined by the group $\mathcal{O}_K^*$ of units of $\mathcal{O}_K$.

Now we can define the Dedekind ζ function of a number field K denoted by ζ_K:

$$(D) \qquad \zeta_K(z) := \sum_{\mathfrak{a}} \frac{1}{N(\mathfrak{a})^z} = \prod_{p} (1 - N(p)^{-z})^{-1}, \quad \operatorname{Re} z > 1,$$

where the sum runs over non zero ideals $\mathfrak{a}$ of $\mathcal{O}_K$ and the product runs over all (non zero) prime ideals p. Since for $K = \mathbb{Q}$, $\mathcal{O}_{\mathbb{Q}} = \mathbb{Z}$, the ideals of $\mathbb{Z}$ are integers and prime ideals of $\mathbb{Z}$ are integers, we have $\zeta_{\mathbb{Q}}(z) = \zeta(z)$, the Riemann zeta function. Dedekind extended his $\zeta_{\mathbb{Q}}$ to the whole complex plane as a meromorphic function with the single pole at $z = 1$:

THEOREM (DEDEKIND, 1877). *Let K be a number field. The Dedekind zeta function ζ_K is a meromorphic function on $\mathbb{C}$ having a simple pole at $z = 1$ with the residue*

$$h_K 2^{r_1+r_2} \pi^{r_2} \frac{\mathcal{R}(K)}{w|D(K)|^{1/2}},$$

where h_K is the class number of K, $\mathcal{R}(K)$ is the regulator of K and $D(K)$ is the discriminant of K; r_1 (resp. $2r_2$) is the number of imbeddings (isomorphisms) of K into $\mathbb{R}$ (resp. $\mathbb{C}$) and w is the number of roots of unity of K.

In spite of this marvelous theorem, Dedekind did not succeed to find a functional equation of Riemann type for ζ_K. This was done by Hecke 40 years later!

THEOREM (HECKE, 1917). *Put*

$$(*) \qquad \xi_K(z) := \left(\frac{|D(K)|}{2^{r_2}\pi^{d/2}}\right)^z \Gamma\left(\frac{z}{2}\right)^{r_1} \Gamma(z)^{r_1} \zeta_K(z);$$

then

$$(**) \qquad \xi_K(z) = \xi_K(z-1).$$

REMARK 1. When K is a totally real number field, that is, if $r_2 = 0$ and $r_1 = d := [K : \mathbb{Q}]$, then $(*)$ simplifies to

$$(*') \qquad \xi_K(z) := \left(\frac{D(K)}{\pi^d}\right)^{z/2} \Gamma\left(\frac{z}{2}\right)^d \zeta_K(z).$$

REMARK 2. In the Riemann case $K = \mathbb{Q}$ and both $(*)$ and $(*')$ reduce to

$$\xi_{\mathbb{Q}}(z) = \xi(z) := \pi^{-z/2}\Gamma\left(\frac{z}{2}\right)\zeta(z).$$

REMARK 3. We have $r_1+2r_2 = [K:\mathbb{Q}] = d$. The group of units $U := \mathcal{O}_K^*$ of the ring $\mathcal{O}_K$ can be embedded homomorphically into $\mathbb{R}^{r_1+r_2}$ and the image is contained in the r_1+r_2-1 dimensional Euclidean space as a lattice. The regulator $\mathcal{R}(K)$ of K is the Euclidean volume of the fundamental domain of this lattice.

4.1 Algebraic curves (Riemann surfaces) over Q

Let K be an arbitrary field, $K^n = K \times \cdots \times K$, the n dimensional vector space over K, $\mathbb{P}^n(K) \equiv \mathbb{P}(K^{n+1})$ the projective space of lines in K^{n+1}.

An algebraic variety X over K is defined to be a non zero singular locus of common zeros of a family of polynomials with coefficients in K. By $X(K)$ one denotes the set of points with coordinates in K; these points are called *K-rational points* of X. One dimensional algebraic varieties are *projective algebraic curves* over K. From the point of view of arithmetic, the three particular fields K are of special interest:

1. K is a number field, that is, a finite extension of $\mathbb{Q}$;
2. K is a finite field $\mathbb{F}_q$;
3. K is a function field.

In this section we are interested only in the case 1.

DEFINITION. An algebraic variety (over K) of genus $p = 1$ is called *elliptic curve*; in what follows it will be denoted by E.

There is an important property of elliptic curves:

If the characteristic of $K \neq 2,3$, then $X(K) \neq \emptyset$ if and only if X is isomorphic to a curve defined by the equation (called *Weierstrass form*)

$$y^2 = 4x^3 - g_2x - g_3, \quad \text{with } g_2, g_3 \in K,\ g_2^3 - 27g_3^2 \neq 0.$$

$\Delta := g_2^3 - 27g_3^2$ is the discriminant of the polynomial $4x^3 - g_2x - g_3 = 4(x-\alpha_1)(x-\alpha_2)(x-\alpha_3)$, where α_i are its roots. It is well known that $\Delta = c((\alpha_2-\alpha_1)(\alpha_3-\alpha_2)(\alpha_3-\alpha_2))^2$ and $(\Delta \neq 0) \Longleftrightarrow$(the roots $\alpha_1, \alpha_2, \alpha_3$ are distinct).

The *modular invariant* $j = j_E$ of the elliptic curve E is

$$j := 2^6 3^3 \frac{g_2^3}{g_2^3 - 27 g_3^2} = 1728 \frac{g_2^3}{\Delta}.$$

It was a great sensation when in 1983 a young German mathematician Gerd Faltings proved the famous old *Mordell conjecture* (1922) (which has its roots in a paper of Poincaré):

THEOREM (FALTINGS, 1983). *If X is a curve of genus $p \geqslant 2$, then the set $X(K)$ of K rational points is finite.*

The idea of the proof was presented by Faltings in the excellent, lucid article *Die Vermutungen von Tate and Mordell*, Jahr. Ber. d. Dt. Math.-Vereinigung **86** (1984), 1–13. This paper contains not only extremely interesting historical remarks but also general remarks concerning teaching mathematics at the university level. André Weil has said that the article of Faltings opened a new chapter in number theory. In 1986, in recognition of his work, Faltings was awarded the Fields medal.

4.2 Algebraic curves $X(\Gamma)$ over Q

It is a fundamental problem characterize such classical compact Riemann surfaces X (over $\mathbb{C}$) that can be defined over a number field $\bar{\mathbb{Q}}$, that is, such that X is associated with an irreducible polynomial in $\bar{\mathbb{Q}}(x, y)$. It seems unbelievable, but it was only in 1978, when G. Belyi gave the following beautiful and simple characterization of such arithmetic curves.

THEOREM (BELYI, 1978). *For any compact (connected) Riemann surface X the following conditions are equivalent:*

1. X may be defined over $\bar{\mathbb{Q}}$;

2. There exists a holomorphic (non constant) map $\pi : X \to \mathbb{P}^1(\mathbb{C})$ which is unramified outside $\pi^{-1}(\{0, 1, \infty\})$;

2′. X is isomorphic to $\bar{Y}$, where Y is a finite unramified covering of $\mathbb{P}^1(\mathbb{C}) - \{0, 1, \infty\}$;

3. X is isomorphic to $\overline{\mathfrak{H}/\Gamma}$, where Γ is a subgroup of finite index of the modular group $\mathrm{PSL}(2, \mathbb{Z})$.

REMARK. The map π in 2. is called the *Belyi map*; the implication 1. $\Longrightarrow$2. is called the *Belyi theorem.*

PROOF. The equivalence 2. $\Longleftrightarrow$2$'$. is obvious.

2$'$. $\Longrightarrow$3. follows from the isomorphism (Exercise 1 below)

$$\mathfrak{H}/\Gamma(2) \simeq \mathbb{P}^1(\mathbb{C}) - \{0,1,\infty\} \tag{1}$$

and from the fact that any finite unramified covering of $\mathfrak{H}/\Gamma(2)$ is isomorphic to the covering $\mathfrak{H}/\Gamma \to \mathfrak{H}/\Gamma(2)$, where Γ is a subgroup of finite index of $\Gamma(2)$.

3. $\Longrightarrow$2. is made into Exercise 3.

But, of course, the most exciting is 1. $\Longrightarrow$2.: By definition of Riemann surface (algebraic curve) defined over $\bar{\mathbb{Q}}$, there exists a holomorphic map $\pi : X \to \mathbb{P}(\mathbb{C})$ ramified only over points in $\bar{\mathbb{Q}} \cup \{\infty\}$. By composing π with polynomials with coefficients in $\mathbb{Q}$, it is possible to obtain a holomorphic map $f : X \to \mathbb{P}^1(\mathbb{C})$ over $S \subset \bar{\mathbb{Q}} \cup \{\infty\}$. Superposing this with a homography of the sphere $\mathbb{P}^1(\mathbb{C})$, we can assume that $\{0,1,\infty\} \subset S$. Therefore we have the unramified covering

$$\varphi \circ f : X - (\varphi \circ f)^{-1}\{0,1,\infty\} \to \mathbb{P}^1(\mathbb{C}) - \{0,1,\infty\}$$

which is a quotient of the universal covering $\mathfrak{H} \to \mathbb{P}^1(\mathbb{C}) - \{0,1,\infty\}$ be a subgroup $\Gamma \subset \pi_1(\mathbb{P}^1(\mathbb{C}) - \{0,1,\infty\})$. By attaching points, we recover the whole X. □

EXERCISE 1. There exists an isomorphism defined by the $\Gamma(2)$-modular function $\lambda : \mathfrak{H} \to \mathbb{C} - \{0,1\}$

$$\lambda(s) := \frac{\vartheta(s/2,s)^4}{\vartheta(0,s)^4},$$

where the theta function $\vartheta(z,s) : \mathbb{C} \times \mathfrak{H} \to \mathbb{C}$ is a modular form defined by the series

$$\vartheta(z,s) := \sum_{n\in\mathbb{Z}} \exp(\pi i n^2 s + 2\pi i n z).$$

EXERCISE 2. Prove that the modular invariant equals

$$j = 256\frac{(\lambda^2-\lambda+1)^3}{\lambda^2(\lambda-1)^2}.$$

EXERCISE 3. Prove the implication 3. $\Longrightarrow$2. as follows. As we know the map $\frac{1}{1728}j : \mathfrak{H} \to \mathbb{C}$ is ramified only over $\{0, 1\}$. Consider the diagram

$$\begin{array}{ccc} \mathfrak{H}/\Gamma & \xrightarrow{\quad u \quad} & \mathfrak{H}/\Gamma(1) \\ & \underset{h}{\searrow} & \downarrow \frac{1}{1728}j \\ & & \mathbb{C} \end{array}$$

Since $\Gamma \subset \Gamma(1)$ is of finite index in $\Gamma(1)$, Γ acts properly on $\mathfrak{H}$. Therefore $h = \frac{1}{1728}j \circ u$ is a proper map which is unramified on $h^{-1}(\mathbb{C} - \{0,1\})$ and h extends to the Belyi map $\overline{\mathfrak{H}/\Gamma} \to \mathbb{P}^1(\mathbb{C})$. □

We have yet another formulation of the Belyi theorem.

THEOREM (BELYI). *A compact Riemann surface X may be defined over $\bar{\mathbb{Q}}$ if there exists an unramified holomorphic surjection*

$$f_\Gamma : \mathfrak{H} - \{\textit{finite union of } \Gamma \textit{ orbits}\} \to X - \{\textit{finite set of points}\},$$

where Γ is a subgroup of finite index in $\Gamma(1)$ (that is $[\Gamma(1) : \Gamma] < \infty$), and f_Γ is Γ periodic, that is $f_\Gamma(\gamma z) = f(z)$ for $\gamma \in \Gamma$.

COROLLARY. *Every elliptic curve defined over $\bar{\mathbb{Q}}$ admits f_Γ uniformization with $\Gamma \subset \Gamma(1)$ of finite index in $\Gamma(1)$.*

We know that congruence subgroups, for example, $\Gamma_0(N)$ are of finite index in $\Gamma(1)$, but they are *very exceptional* among subgroups in $\Gamma(1)$ of finite index. Therefore the famous Taniyama–Shimura conjecture that *Every semistable elliptic curve over $\bar{\mathbb{Q}}$ admits a $f_{\Gamma_0(N)}$ uniformization* is very strong indeed and the Belyi theorem is merely a simple particular case of it.

The Taniyama–Shimura conjecture was proved (before Wiles) for some subfamilies of elliptic curves over $\bar{\mathbb{Q}}$, for example, having *complex multiplication*. But this is a very strong assumption: there are only few elliptic curves with complex multiplication. Taniyama–Shimura conjecture brings about very important consequences:

1. The proof of Fermat Last Theorem; and
2. The proof of Hasse–Weil conjectures.

Therefore it was a great sensation when in 1993 Andrew Wiles announced his proof of Taniyama–Shimura conjecture. A gap was found in the huge preprint of Wiles, but in 1994 it was finally filled (by Wiles and R. Taylor) and the monumental work of Wiles is now perfect (Wiles, A. *Modular elliptic curves and Fermat's last theorem*, Annals of Mathematics **168** (1995), 443–572.

In order to gather some more experience in elliptic curves over $\mathbb{Q}$, we have to report an important continuation of Hecke theory by Martin Eichler (1954) and Goro Shimura (1955). We recall that Taniyama put forward his conjecture at 1955 Kyoto Conference, but it was properly modified by his friend Goro Shimura in 1963.

4.3 Eichler–Shimura theory

In this section E denotes an elliptic curve over $\mathbb{Q}$ and $\mathbb{Z}_p$ the ring of p-adic integers: the $|\cdot|$ completion of $\mathbb{Z}$.

Already in 1936 Helmut Hasse proved a very important theorem.

THEOREM (HASSE, 1936). *Let elliptic curve E have integer coefficients, and for every prime p not dividing the discriminant Δ of E, let E_p be a reduction of E modulo p. Then*

$$p + 1 - |E_p(\mathbb{Z}_p)| < 2\sqrt{p}. \tag{1}$$

One defines the L series for E by the Euler product

$$L(s, E) = \prod_{p|\Delta} \left(\frac{1}{1 - a_p p^{-s}} \right) \prod_{p\nmid\Delta} \left(\frac{1}{1 - a_p p^{-s} + p^{1-2s}} \right), \tag{2}$$

where

$$a_p := p + 1 - |E_p(\mathbb{Z}_p)|. \tag{3}$$

Note that $L(s, E)$ has the same form as the $L(s, f)$ function for the Hecke eigenform $f \in S_k(\Gamma_0(N))$. A Corollary to Hecke theorem asserts that $L(s, E)$ converges for $\operatorname{Re} s > 3/2$. But if we do not want to use the difficult Hecke theorem, we immediately have a weaker

PROPOSITION. *The Euler product* (2) *defining* $L(s, E)$ *converges for* $\operatorname{Re} s > 2$ *and in this region is given by an absolutely convergent Dirichlet series.*

One would like $L(s, E)$ to have an analytic continuation to the whole $\mathbb{C}$ and to satisfy a functional equation of Riemann type. The aim of the Eichler–Shimura theory is to construct elliptic curves E (over $\mathbb{Q}$) from modular cusp forms, the newforms $\in S_k(\Gamma_0(N))$ such that $L(s, E) = L(s, f)$. *The Taniyama–Shimura conjecture assures that all semistable elliptic curves* E *are obtained in this way.* Thus the Eichler–Shimura theory together with Belyi theorem make Taniyama–Shimura conjecture more likely. The main result of the Eichler–Shimura theory is the following

THEOREM (EICHLER–SHIMURA) . *Let* $f(s) := \sum_{n=1}^{\infty} c_n e(ns)$ *be a normalized newform in* $S_2(\Gamma_0(N))$ *with* $c_1 = 1$ *and assume that all* a_n *are integers. Then*

1. *There exists an elliptic curve* E *over* $\bar{\mathbb{Q}}$*;*
2. *If*

$$\Lambda(f) := \{\Phi_f(\gamma) := \int_{s_0}^{\gamma(s_0)} f(z)dz : \gamma \in \Gamma_0(N)\}, \tag{4}$$

then $\Lambda(f)$ *is a lattice in* $\mathbb{C}$ *and* $E \simeq \mathbb{C}/\Lambda(f)$ *over* $\mathbb{C}$*;*

3. $L(s, f)$ *and* $L(s, E)$ *coincide as Euler products, except, possibly, at finite many primes.*

REMARK. The integral in (4) is path independent since f is holomorphic. $\Phi_f(\gamma_1\gamma_2) = \Phi_f(\gamma_1)\Phi_f(\gamma_2)$.

In 1986 H. Carayol extended the preceding theorem in a substantial way.

THEOREM (CARAYOL, 1986). *Let* f *be a newform in* $S_2(\Gamma_0(N))$ *and let* E *be an elliptic curve over* $\mathbb{Q}$ *associated with* f *by virtue of the construction in the Eichler–Shimura theory. Then* $L(s, E) = L(s, f)$ *and* N *is the conductor of the curve* E.

REMARK. The conductor N_E of E is a positive integer that can be defined as follows: if $E : y^2 = 4x^3 - g_2 x - g_3$ with $g_2, g_3 \in \mathbb{Z}$ and $\Delta_E = g_2^3 - 27g_3^2$ is its discriminant, we can assume that the equation for E has minimal $|\Delta_E|$.

Then the conductor of E is a natural number such that the prime divisors are equal to the prime divisors of $2\Delta_E$. We give some examples.

E	Δ_E	N_E
$y^2 = x^3 - y$	-27	27
$y^2 = x^3 - xy + y$	-28	14
$y^2 = x^3 - xy - x$	-63	21
$y^2 = x^3 - x$	64	32
$y^2 = x^3 + x$	-64	64

DEFINITION. E is *semistable* if its conductor is quadrat free.

The Hasse–Weil conjecture. The function $L(s, E)$ extends holomorphic to the whole $\mathbb{C}$, and for a certain positive integers the function

$$\xi(s, E) := N^{s/2}(2\pi)^{-s}\Gamma(s)L(s, E)$$

satisfies the functional equation

$$\xi(s, E) = -\epsilon(\xi(2 - s, E),$$

with suitable sign ϵ.

It can be shown that

(Taniyama–Shimura conjecture) $\Longleftrightarrow$(Hasse–Weil conjecture)

The implication $\Longrightarrow$can be easily proved. Indeed, if E is given, choose N to be the smallest positive integer such that we have $\Gamma_0(N)$ parametrization of E. One proves that E is associated (by Eichler–Shimura) to a normalized newform $f \in S_k(\Gamma_0(N))$. The Carayol theorem asserts that $L(s, E) = L(s, f)$, but this is equivalent to the Hasse–Weil conjecture.

4.4 Wiles proof of Last Fermat Theorem

This proof follows from the Wiles solution of the Taniyama–Shimura conjecture and the beautiful

THEOREM (FREY–SERRE–RIBET).
(Taniyama–Shimura conjecture) $\Longrightarrow$*(Last Fermat Theorem).*

This theorem follows from the magnificent idea of Gerard Frey: Suppose that the Last Fermat Theorem is false, that is, the equation $a^p + b^p + c^p = 0$ has for $p > 5$ a solution with $a, b, c \in \mathbb{Z}$. Put $A := a^p$, $B := b^p$, $C := c^p$. Frey constructs a semistable elliptic curve, the Frey curve $E : y^2 = x(x - A)(x + B)$, where $A + B + C = 0$, $ABC \neq 0$. Following to the Taniyama–Shimura conjecture, Serre and Ribel constructed the Galois representation

$$\rho : Gal(\bar{\mathbb{Q}}/\mathbb{Q}) \to \mathrm{Aut}(E_p) \simeq \mathrm{GL}(2, \mathbb{F}_p)$$

on the group of points of order p of the Frey curve E (we have written E_p for $E(\mathbb{Z}_p)$). E is associated with the cusp form $f \in S_2(\Gamma_0(2))$. But we know that $\dim S_2(\Gamma_0(2)) = g(X_0(2)) = 0$, whence such modular cusp form cannot exist, and thus such Frey curve does not exist and Fermat theorem follows!

4.5 ζ functions of elliptic operators on compact Riemann manifolds. The Selberg ζ function

We saw how important role was played by ζ and L functions of an elliptic curve E. As we know the modified classic Riemann ζ function ξ is a Mellin transform of the ϑ function

$$\text{(1)} \qquad \vartheta(t) := \sum_{n=1}^{\infty} e^{-\pi n^2 t}, \quad (\mathbb{M}\vartheta) = \pi^{-s}\Gamma(s)\zeta(2s) =: \xi(2s).$$

For the Riemann ξ function we have the famous relation

$$(RF) \qquad \xi(s) = \xi(1 - s).$$

As we saw all ζ functions are constructed after the fashion of the Riemann ζ function and the aim of the constructors was to prove the analytic continuation and the functional relation similar to (RF). We observed also that the Zagier Functional Equation Principle turns out to be very useful in this context.

In the present section we look at the ζ function from the spectral theory point of view: as a function defined on the spectrum of the simplest differential operator d^2/dt^2, on the simplest manifold M, the circle $\mathbb{C}/\Lambda$. This will make it possible to define, in an analogous way, the function ζ_A of an elliptic operator (for example, the Laplace–Beltrami operator) on arbitrary compact Riemann surface X. In the case of genus of X, $p \geqslant 2$ one obtains the famous Selberg ζ function.

As we know every (positive) self-adjoint elliptic operator A on a compact Riemann manifold has discrete spectrum $0 < \lambda_1 < \lambda_2 < \ldots$. In this book we have already considered the spectral geometry of manifold X and one of the main problems was to describe the asymptotic of the spectrum $\operatorname{Spec} A$ of the operator A: this has led us to the Hermann Weyl asymptotic formula and its generalizations.

Since the complicated objects like $\operatorname{Spec} A$ cannot be analyzed directly, one investigates some functions on $\operatorname{Spec} A$, for example, the function ζ_A:

$$\zeta_A(s) := \sum_{n=1}^{\infty} \lambda_n^{-s}$$

(one follows here the general principle 'if you want to learn something about an object, investigate functions growing on it'.) Riemann made use of this principle in his investigations on ζ function, and he obtained the so called

Explicit formulas for distribution of non trivial zeros of $\zeta(s)$ in the critical strip $0 \leqslant \operatorname{Re} s \leqslant \frac{1}{2}$.

DEFINITION. Let $E \to X$ be a vector bundle over Riemann manifold and let A be a second order differential operator on E such that its principal symbol is

$$\sigma_2(A)(x, \xi) = |\xi|^2 \tag{3}$$

or equivalently

$$A = -\sum g^{ij}(x)\partial_i\partial_j + \text{first order part.} \tag{4}$$

Such operators will be called *generalized Laplacians on $E \to X$*; they are elliptic, and if symmetric and X compact, then they have discrete spectrum $\operatorname{Spec}(A)$.

Let λ be a positive number and X a compact Riemann manifold of dimension n; then the *zeta function* of the self-adjoint generalized Laplacian A on $E \to X$ is given by

$$\zeta_A(s, \lambda) := \Gamma(s)^{-1}\mathbb{M}(\operatorname{tr}(P(\lambda,\infty)e^{-tA})) = \Gamma(s)^{-1} \int_{0\infty} \operatorname{tr}(P(\lambda,\infty)e^{-tA})t^{s-1}\,dt,$$

where $P(\lambda, \infty)$ is the spectral projection of A onto the eigenvalues lying in (λ, ∞). Thus

$$\zeta_A(\lambda, \infty) = \sum_{\lambda_k > \lambda} \lambda_k^{-s} \tag{5}$$

where $\lambda_k \in \mathrm{Spec}(A)$, $k = 1, 2, \ldots$ and the series (5) converges for $s > \frac{1}{2}\dim X$.

PROPOSITION. *The zeta function $\zeta_A(\lambda, \infty)$ has a meromorphic extension to the whole $\mathbb{C}$ and is holomorphic at $s = 0$.*

EXAMPLE. Let $A = -\frac{d}{dx^2}$ on the circle $\mathbb{R}/2\pi\mathbb{Z}$; then the spectrum of $A = \{0, 1, 1, 4, 4, \ldots, n^2, n^2, \ldots\}$. For $0 < \lambda < 1$ and $s > \frac{1}{2}$, we have

$$\zeta_A(s, \lambda) = \frac{2}{\Gamma(s)} \sum_{n=1}^{\infty} \int_0^{\infty} e^{-tn^2} t^{s-1} dt = 2 \sum_{n=1}^{\infty} n^{-2s} = 2\zeta(2s). \tag{6}$$

In many domains of mathematics and physics an important role is played by the

The ζ function regularized determinant of A (Ray–Singer)
In a *finite dimensional* hermitian vector space V, for any positive operator $A : V \to V$ we have $\zeta_A(s) = \operatorname{tr} A^{-s}$. Then $\zeta_A'(0) = -\log\det A$, therefore $\det A = e^{-\zeta_A'(0)}$, and we define

Zeta function of the generalized Laplacian ζ_A (in the Hilbert space $L^2(X, \Omega_{\frac{1}{2}})$ of $\frac{1}{2}$ densities) is defined by

$$\det A := \exp(\zeta_A'(0, 0)). \tag{7}$$

This definition of zeta function of an elliptic operator A as a *renormalized determinant* of A is well adapted to families of operators, for example, Dirac operators parametrized by family B: we have a family of manifolds $X \xrightarrow{\pi} B$ and on every manifold X_b, $b \in B$ of the family we have an operator A_b for which we have the ζ function ζ_{A_b} and the corresponding regularized determinant $\det A_b$. The interest in these families has its roots in quantum field theory.

4.6 Determinant line bundle associated with family of Dirac operators and its Quillen metric

Consider a (smooth) fiber bundle $\pi : M \to B$ with compact fibers $M_b := \pi^{-1}(b)$, $b \in B$ being Riemann manifolds. Thus we have a smooth family M_b o compact Riemann manifolds. Denote by $T(M/B)$ the vertical tangent bundle which is a subbundle of the tangent bundle $T(M)$ consisting of vertical vectors. Given two vector bundles E, F equipped with hermitian metrics, we have L^2 scalar products on the spaces of smooth sections $C^\infty(M_b, E)$ and $C^\infty(M_b, F)$. Finaly, we have a family $D = \{D_b, b \in B\}$ of elliptic operators

$$D_b : C^\infty(M_b, E) \to: C^\infty(M_b, F).$$

By virtue of ellipticity of D_b, $\ker D_b$ and $\ker D_b^*$ are of finite dimension $d(b)$ and $d'(b)$, respectively. We can therefore define

$$DET\, D_b := (\Lambda^{d(b)} \ker D_b)^* \otimes (\Lambda^{d'(b)} \ker D_b^*).$$

In this way we obtained a natural line bundle on B

$$B \ni b \to DET\, D_b \in \mathbb{C}$$

denoted by $DET(D)$ and called the *determinant bundle of the family D*.

We would like to have a C^∞ metric on the bundle $DET(D)$, but the natural condition $\|\cdot\|_{L^2}$ is of no use: since $b \to \dim\ker D_b$ jumps, the $\|\cdot\|_{L^2}$ metric is even not continuous. Now Dan Quillen enters the stage with his brilliant idea: he renormalizes the L^2 metric on $DET(D)$ as follows

$$(Q) \qquad |\cdot|_Q := \det(D_b^* D_b)^{1/2}\, \|\cdot\|_{L^2},$$

where $\det(D_b^* D_b)$ is the ζ renormalized determinant of the self-adjoint operator $(D_b^* D_b)$ on M_b:

$$\det(D_b^* D_b) := \exp(\zeta'(0, (D_b^* D_b)).$$

We have the fundamental

THEOREM (QUILLEN, 1985). *The Quillen metric defined on $DET\, D_b$ by (Q) is a C^∞ metric on the fiber bundle $DET(D)$.*

On $DET(D)$ Quillen defines a natural connection compatible with the Quillen metric:

THEOREM (QUILLEN, BISMUT–FREED, 1986). *Let $D = \{D_b, \in B\}$ be a family of Dirac operators D_b of index zero on a vector bundle $\mathcal{E}$ over the family of manifolds $M \to B$. There is a canonical smooth section* $\det(D^+)$

of the determinant line bundle $DET(D)$ which vanishes precisely when D is not invertible; this section satisfies the formula

$$|\det(D^+)|_Q = \det(D^-D^+)^{1/2}.$$

2. There exists a natural connection ∇_Q on the determinant line bundle $DET(D)$ which is compatible with the Quillen metric.

Now we can apply the Chern–Weil theory of characteristic classes for complex manifolds to the case of hermitian line bundles. Let $\nabla''_{\mathcal{E}}$ be the corresponding covariant derivative; its curvature form is a $(1,1)$ form defined by $R = \nabla''_{\mathcal{E}}\nabla \log |s|^2$, where s denotes a local holomorphic section of $\mathcal{E}$. We have the Chern classes

$$c_1(\mathcal{E}, |\cdot|) = -\frac{1}{2\pi i}R, \quad ch(\mathcal{E}, |\cdot|) = \sum_{k\geqslant 0} \frac{1}{k!}c_1(\mathcal{E}, |\cdot|)^k.$$

For a family $M \to B$ of compact Riemann surfaces and $\mathcal{E}$ being a holomorphic vector bundle over M one obtains beautiful formula

$$c_1(DET\nabla''_{\mathcal{E}}; |\cdot|_Q) = -\int_{M/B} (ch(\mathcal{E}; |\cdot|_Q)Td(T(M/B, |\cdot|_Q),$$

where $\int_{M/B}$ denotes the integral over the fibers of $M \to B$, and $Td(T(M/B, |\cdot|_Q)$ is the Todd class.

Formulas of this type are important in the Teichmüller theory and, generally, in the theory of moduli spaces of compact Kähler manifolds.

4.7 Selberg ζ function and trace formula. The length spectrum

The great Norvegian mathematician Atle Selberg is a great expert in number theory, automorphic functions, discrete groups, and spectral theory. He was awarded the Fields Medal in 1950 (together with L. Schwartz) for his important achievements in number theory: It was a great sensation when he managed to prove the 'Prime number theorem' in elementary way, that is, without making use of the theory of Riemann ζ function.

Selberg investigated the spectrum of the Laplace operator Δ_Γ on the Riemann surface $\mathfrak{H}/\Gamma$ for discrete subgroups Γ of $SL(2,\mathbb{R})$ with *finite* volume of $\mathfrak{H}/\Gamma$. If $\mathfrak{H}/\Gamma$ is not compact, then the Laplacian Δ_Γ has also continuous spectrum with corresponding Eisenstein series. Here we will only consider the case of compact Riemann surface $X = \mathfrak{H}/\Gamma$ of genus $p \geqslant 2$; thus Δ_Γ has only discrete spectrum, and we can form ζ_{Δ_Γ} (ζ_Γ for short) function and the ζ regularized $\det \Delta_\Gamma$ as above.

DEFINITION. The Selberg ζ function is the ζ function of the Laplace operator Δ_Γ. This function is denoted by $Z_\Gamma(s)$.

The Selberg ζ function has a beautiful spectral-geometric expression by means of the Euler product:

$$(S) \qquad Z_M(s) \equiv Z_\Gamma(s) := \prod_{\gamma \in \mathcal{P}} \prod_{k=0}^{\infty} \left(e^{-l(\gamma)(s+k)}\right),$$

where $\mathcal{P}$ is the set of primitive (also called prime) geodesics on the Riemann surface $M := \mathfrak{H}/\Gamma$, and for every $\gamma \in \mathcal{P}$, its length is denoted by $l(\gamma)$.

The Selberg ζ function is $Z_M(s)$ is holomorphic in the half plane $\operatorname{Re} s > 1$, has meromorphic extension to the whole $\mathbb{C}$, and it satisfies a functional equation of Riemann type. Selberg proved a very interesting *trace formula* which provides powerful tools in number theory, spectral geometry, Teichmüller theory. Here is a beautiful trace formula of Cartier and Voros.

THEOREM (CARTIER–VOROS, 1990). *Let $p = p(M)$ be the genus of M. Then*

$$(1) \qquad Z_M(s) = \det\left(\Delta_M - \frac{1}{4} + s^2\right)\left(e^{s^2} \det(\sqrt{\Delta_{S^2} + \frac{1}{4}} + 2)\right)^{2p-2},$$

where Δ_{S^2} is the Laplace operator on the unit sphere S^2 and $\det(A)$ denotes the regularized determinant of the operator A.

The formula (1) is very interesting: on the left hand side we have the length spectrum of closed geodesics on M, and the right hand side is the spectrum of the operator Δ_M.

Originally Selberg defined his ζ function in a more 'algebraic' vein, so that the length spectrum was hidden. In order to stress the geometric interpretation, we change the notation. Denote now by $\gamma \in \Gamma$ elements of the subgroup $\Gamma \subset \mathrm{SL}(2, \mathbb{R})$ operating on the upper half plane $\mathfrak{H}$. In our case, the group Γ has only *hyperbolic* elements γ (no elliptic and parabolic (cusps) ones.) Thus two eigenvalues of $\gamma \in \Gamma$ are distinct real numbers ξ_1, ξ_2 with $\xi_1 \cdot \xi_2 = 1$, $\xi_1 < \xi_2$. Then Selberg calls the number ξ_2^2 the *norm* of γ and denotes it by $N(\gamma) := \xi_2^2$. Clearly Selberg follows the ideas of Dedekind and his ζ_K function for a number field K:

$$\zeta_K := \sum_{\mathfrak{a}} N(\mathfrak{a})^{-s} = \prod_{p} \frac{1}{1 - N(p)^{-s}}, \tag{2}$$

where p runs over prime ideals of the field K (and $\mathfrak{a}$ over integral ideals of K.)

When γ is hyperbolic, γ^n is hyperbolic as well for $n = 1, 2, \ldots$. When $\pm\gamma$ is not a positive power of another hyperbolic element, γ is called *primitive hyperbolic element*. Clearly, elements conjugate to primitive hyperbolic elements are also primitive hyperbolic elements, and have the same norm. Denote by $\mathcal{P}_1, \mathcal{P}_2, \ldots$ the conjugacy classes of primitive hyperbolic elements; Selberg defines his ζ function:

$$Z_\Gamma(s) := \prod_i \prod_{n=0}^{\infty} (1 - N(\gamma_i)^{s-n}) .$$

But the norm $N(\gamma)$ has a simple geometric meaning. Let L be a geodesic on the hyperbolic half plane $(\mathfrak{H}, \rho)$ equipped the Poincaré metric ρ, and $\gamma_i \in \mathcal{P}_i$. Then $\gamma_i a \in L$ is on the same geodesic L and the Poincaré distance (the length of the segment $(a, \gamma_i(a))$ of the geodesic L) is precisely $\log N(\gamma_i)$. If we descend to the surface $\mathfrak{H}/\Gamma$, we obtain a *closed geodesic* in the free homotopy class corresponding to $\mathcal{P}_i$ and we have the geometric interpretation (S) of Z_Γ given by (2).

We have the following important

THEOREM (SELBERG). *1. The Selberg ζ function satisfies the functional equation*

$$Z_\Gamma(1-s) = Z_\Gamma(s) \cdot \exp\left(1 - |\mathfrak{H}/\Gamma| \int_0^{s-\frac{1}{2}} u \tan(\pi u) du\right).$$

2. The function Z_Γ has zeros of order $(2n+1)(2p-1)$ at $-n$, $n = 0, 1, 2, \ldots$. All other zeros lie on the line $\operatorname{Re} s = \frac{1}{2}$, *except for a* finite *number of zeros that lie in the interval* $(0, 1)$.

PRIME NUMBER THEOREM (SELBERG). *Denote by $\pi(x)$ the number of primitive hyperbolic elements γ with $N(\gamma) < x$. Then $\pi(x) \sim \frac{x}{\log x}$ for*

$x \to \infty$.

The classical prime number theorem was proved independently in 1896 by Hadamard and De la Valle Poussin with the help of Riemann ζ function. Gauss conjectured this celebrated theorem 100 years earlier (about 1792). H. Huber proved a more precise result in a series of papers in 1959.

CONCLUDING REMARK. There are beautiful works on length spectra of closed geodesics on arbitrary Riemann manifolds by Gangoli, Chasarin, Colin de Verdiere, Duistermaat, Guillemin that make use of the advanced spectral theory of pseudodifferential operators of first order. These theories can be regarded, in some sense, as quantization of Riemann manifolds.

Concluding Remarks

> Das ganze logische Gespinnst nicht das ist worauf is im Grunde ankommt: es ist nur das *Netz* mit dem wir *die eigentliche Idee*, die in ihrem Wesen nach einfach und gross und göttlich ist aus dem topos atopos, wie Plato sagt – gleich einer Perle aus dem Meere – an die Oberfläche unserer Verstandeswelt heraufholen. Den Kern aber, den dieses Knüpferk von feinen und peinlichen Begriffen umhült zu erfassen – das, was das Leben, den wahren Gehalt, den inneren Wert der Theorie ausmacht – dazu kann ein Buch (und kann selbst ein Lehrer) nur durftüge Fingerzeige geben; hier muss jeder einzelne von neuem für sich um das Verständnis ringen.
>
> Hermann Weyl (from the Foreword to his 'Idee der Riemannschen Fläche' (1913))

Not all important ideas of Riemann were considered in this book: we have omitted two very important seminal papers:

1. *Über die Fortflanzung ebener Luftwellen von endlicher Schwingungsweite*
The great expert Peter D. Lax writes:

'In this paper Riemann lays the foundations of the theory of propagation of non- linear and linear waves governed by hyperbolic equations. The concept introduced there – Riemann invariants, the Riemann initial value problem, jumps conditions for non-linear equations –are *still basic building blocks of the theory today*'.

'Ernst Hölder (1981) has given a detailed up-to-date review of Riemann ideas in the theory of hyperbolic equations and the theory of compressible flows ...'.

2. *Ein Beitrag zu den Untersuchungen über die Bewegung eines flüssingen gleichartigen Ellipsoids*

The great astrophysicist, S. Chandrasekhar writes

'...after a lapse of some one hundred years Riemann's investigations *occupy a central place of theoretical astrophysics today* ...' This is unbelievable, but true!

Our endeavor was to show that Riemann ideas in mathematics and physics are living germs which developed into huge trees –mathematical and physical theories pervading and fecundating one another. These ideas seem to be an illustration of the great parables from XIII chapter of the Gospel of St. Matthew: on the kingdom of heaven 'which is like a grain of mustard seed ...and becomes a tree ...'. And: 'this kingdom of heaven is like unto leaven which a woman took, and hid in the three measures of meal, till the whole was leavened'.

It is not so strange as it might first appear: the kosmos noethos is the work of Logos, the Logos in action, and mathematics is one of the most beautiful works of the Logos who lives, acts, and works in mankind, through mankind.

Suggestions for Further Reading

Many references were included in the main text, here we recall only the most important ones.

A. Indispensable is the new beautiful edition of Bernhard Riemann *Gesamelte Mathematische Werke, Wissenschaftlicher Nachliess und Nachträge*. Springer, 1990.

This is already the third edition of 'Riemann's Collected Papers', very carefully edited by Ragavan Narasimhan and preceded by his wonderful, twenty pages long 'Editors preface (Together with a mathematical commentary of some Riemann's works)'. R. Narasimhan is an expert in complex analysis and student of great Hans Grauert.

To understand some impulses behind Riemann's work one must read his *Fragmente philosophischen Inhalts.* Particularly significant are seven pages 546–552, which give an insight into Riemann's reading of works of Fechner, most of all, his famous *Zend – Avesta.* Riemann confesses that he is 'Herbartian', which, besides, was stressed by some historians. The influence of Fechner is, however, (shamefacedly) concealed. It seems that the end of twentieth century is, in this respect, much more open minded.

B. Concerning textbooks and monographs on the theory of Riemann surfaces, there is a lot of excellent works:

1. Forster's little book *Lectures on Riemann surfaces*, Springer, 1981 is already a classic. Clarity and precision of this work is unrivaled. Forster himself is an expert in 'several complex variables'. It can be seen how the modern 'complex analysis' changed our understanding of, apparently, completely solved problems.

2. Another excellent little book *Quelque aspects de surfaces de Riemann*, Birkhäuser, 1989 written in somehow different style by E. Reyssat par-

tially extends beyond the material covered by the Forster, for example, the draft of the proof of Torelli theorem is presented and modular functions are discussed.

3. Farkas and Kra, *Riemann Surfaces*, Springer, 1980 presents again a different style (Dirichlet principle, Theta functions.)

4. All books by R. Narasimhan.

C. Several complex variables (complex geometry). Written in Riemannian spirit, but using modern tools is the classic work

1. L. Hörmander, *An Introduction to Complex Analysis in Several Variables*, Van Nostrand, 1966. This work of the greatest investigator of partial differential equations (cf. his monumental four volume monograph) shows that his approach to complex analysis understood as the theory of the operator d'' (and thus in truly Riemannian spirit) provides the fastest way to central theorems of complex analysis.

2. Of course all the monographs of the tandem Grauert–Remmert and the two volume selection of Grauert's papers (Springer, 1994).

3. C.L. Siegel *Topics on Complex Function Theory*, Wiley, 1971. These lectures of the great master of analysis, number theory, and mechanics give deep insight into a history of the problems as well.

4. S. Kobayashi *Differential Geometry of Complex Vector Bundles*, Princeton, 1987.

5. M. Lübke and A. Teleman *The KobayashiHitchin Correspondence*, World Scientific, 1996.

D. Teichmüller theory and calculus of variations.

1. A. Tromba *Zürich Lectures on Teichmüller Theory* ant the last chapter of the monumental Springer Monograph of S. Hildebrandt and his students on *Minimal Surfaces*, Springer, 1992.

2. All books by Jürgen Jost (a prominent student of S. Hildebrandt), especially his *Two-dimensional Geometric Variational Problems*, Wiley, 1991.

3. Carathéodory's *Collected Papers*, Teubner.

4. Y. Inayoshi and M. Tariguchi *An Introduction to Teichmüller Spaces*, Springer, 1992.

E. Number theory.

1. The Springer textbook *From Number Theory to Physics* (1992) contains fourteen expository contributions by know experts like P. Cartier (on ζ functions); J.B. Bott (an excellent 150 pages long 'Compact Riemann Surfaces, Jacobians, and Abelian Varieties' –with many historic 'appearances' written with great culture and love); H. Cohen ('Elliptic Curves'); D. Zagier's beautiful ('Introduction to Modular Forms'); H.M. Stark ('Galois Theory, Algebraic Number Theory, and Zeta Functions'); and many other, very readable contributions.

2. Concerning A. Wiles solution of Taniyama–Shimura conjecture and Fermat Last Theorem, an exposition by Darmon, Diamong, and R. Taylor (Current Developments in Mathematics, 1995, 1–154, Harvard–MIT) is available now.

F. For a history of the Riemannian Ideas the lectures of Felix Klein are still indispensable:

1. F. Klein *Vorlesungen über die Entwicklung der Mathematik im 19. Jahrhundert.*

2. All books of great I. Shafarevich, the father and grand-father of Russian mathematical school.

G. Obligatory! Raul Bott, Collected Papers, Vol. 4. In the moments of fatigue and discouragement, which everybody experiences, the reading of Raul Bott will restore enthusiasm for mathematics.

Index

Other *Mathematics and Its Applications* titles of interest:

W. Kuyk: *Complementarity in Mathematics. A First Introduction to the Foundations of Mathematics and Its History*. 1977, 192 pp. ISBN 90-277-0814-2

A.N. Philippou, G.E. Bergum and A.F. Horadam (eds.): *Fibonacci Numbers and their Applications*. 1986, 328 pp. ISBN 90-277-2234-X

C.W. Kilmister (ed.): *Disequilibrium and Self-Organisation*. 1986, 320 pp. ISBN 90-277-2300-1

V.A. Vasiliev, Yu.M. Romanovskii, D.S. Chernavskii and V.G. Yakhno: *Autowave Processes in Kinetic Systems. Spatial and Temporal Self-Organisation in Physics, Chemistry, Biology and Medicine*. 1987, 264 pp. ISBN 90-277-2379-6

R. Feistel and W. Ebeling: *Evolution of Complex Systems. Selforganization, Entropy and Development*. 1989, 248 pp. ISBN 90-277-2666-3

A. Hulanicki, P. Wojtaszcyk and W. Zelasko: *Selected Papers of Antoni Zygmund (3 vols.)*. 1989, 1456 pp. ISBN 0-7923-0474-8 (set)

Wang Hao: *Computation, Logic, Philosophy. A Collection of Essays*. 1990, 372 pp. ISBN 0-7923-0218-4

I.H. Koch: *Introduction to Classical Mathematics*. 1991, 454 pp. ISBN 0-7923-1231-7(hb), ISBN 0-7923-1238-4 (pb)

V.M. Tikomirov (ed.): *Selected Works of A.N. Kolmogorov, Vol. 1: Mathematics and Mechanics*. 1991, 552 pp. ISBN 90-277-2796-1

A.N. Shiryayev (ed.): *Selected Works of A.N. Kolmogorov, Vol. 2: Probability Theory and Mathematical Statistics*. 1992, 598 pp. ISBN 90-277-2795-X

A.N. Shiryayev (ed.): *Selected Works of A.N. Kolmogorov, Vol. 3. Information Theory and the Theory of Algorithms*. 1992, 276 pp. ISBN 90-277-2798-8

K. Maurin: *The Riemann Legacy*. Riemannian Ideas in Mathematics and Physics. 1997, 730 pp. ISBN 0-7923-4636-X